Brain Network Dysfunction in Neuropsychiatric Illness

Vaibhav A. Diwadkar • Simon B. Eickhoff
Editors

Brain Network Dysfunction in Neuropsychiatric Illness

Methods, Applications, and Implications

Springer

Editors
Vaibhav A. Diwadkar
Department of Psychiatry &
Behavioral Neurosciences
Wayne State U School of Medicine
Detroit, MI, USA

Simon B. Eickhoff
Heinrich-Heine Univ Düsseldorf
Institut für Klinische Neurowissenschaft
Düsseldorf, Nordrhein-Westfalen, Germany

Institute of Neuroscience and Medicine
Brain & Behaviour (INM-7)
Research Centre Jülich
Jülich, North Rhine - Westphalia, Germany

ISBN 978-3-030-59796-2 ISBN 978-3-030-59797-9 (eBook)
https://doi.org/10.1007/978-3-030-59797-9

This Springer imprint is published by the registered company Springer Nature Switzerland AG
The registered company address is: Gewerbestrasse 11, 6330 Cham, Switzerland

Preface

What role do edited books play in an increasingly digitized world? Indeed, now that "content" can be accessed through multiple pathways (and is frequently *gratis*), a compendium of curated chapters in any area of work may seem like an oddity. But such a collection is precisely what is missing from the increasingly overlapping fields of neuropsychiatry *and* neuroimaging. In the age of "big data" and "open science," students and researchers have increasingly ready access to remarkably large publicly available *in vivo* functional MRI (fMRI) data from human participants. Thus, this compendium with its inclusive overview of extant methods for fMRI analyses, the applications of these methods to the study of neuropsychiatric disorders, and some implications of what can be learned from these applications becomes an invaluable resource for practitioners and students.

The advent of fMRI in 1990 fundamentally altered the landscape of neuroscience. Collecting data *in vivo* from the human brain, which was extremely challenging (as with PET) to impossible, became remarkably commonplace. Over the years, there has been an exponential increase in published studies and publicly available data. All of this has afforded practitioners in the field with unusual opportunities, but also many conceptual challenges. fMRI is at once *both* well understood *and* misunderstood. Enough of its neurophysiological basis is known for researchers to have confidence in the signal's integrity. fMRI signals hold credible information about fluctuations in the brain's functional states. Yet, not enough is known to understand precisely how these fluctuations arise from the multitude of neurophysiological processes that lie "underneath." Conversely, many researchers (and the general public) frequently attribute qualities and credence to fMRI data that are unwarranted. The resultant effect can be confusion, a profusion of studies that produce "effects" with little recourse to an understanding of mechanisms, and eventually a backlash at what has been termed "mindless neuroscience." It is unsurprising that until recently, practitioners in other fields of neuroscience viewed fMRI results with suspicion and/or disdain.

Yet, like any complex data used in science, the fMRI signal, when acquired under the appropriately constrained conditions, and when interrogated with well characterized methods, provides a valuable source for scientific discovery. In science, whether what is discovered is *meaningful* frequently depends on what the question is and how data are analyzed. To quote Thomas Kuhn, "The answers you get, depend on the questions you ask." Given our extant understanding of the spatial and temporal characteristics of fMRI, an appropriate question might thus be, "How do macroscopic brain networks work in the heathy brain, and how are these networks dysfunctional in neuropsychiatric disorders?" Indeed, fMRI is reasonably well-placed to study some aspects of brain networks at some spatio-temporal scales. And because these network interactions are proximately related to behavior, this question is immediately applicable to the study of neuropsychiatric disorders, for these disorders *are* disorders of behavior.

As the use of fMRI and the sophistication of the analytic methods advances, it is entirely conceivable that the field of neuropsychiatry may abandon a previous commitment to animal models as the primary pathway for identifying *mechanisms*. In part, animal models are holdovers from an era when *in vivo* data from the human brain was unavailable or prohibitively expensive to acquire. Yet, neuropsychiatric disorders are *inherently* human disorders, and the decades of carefully constructed work on mechanistic pathways in animals have done little to alter this fundamentally accepted perception. Thus, in a conceivable future, neuropsychiatry may increasingly rely on network models of the brain to better understand mechanisms.

Brain network neuroscience is a discipline like no other, because of its inherent reliance on inter-disciplinary concepts. Such reliance is demanded because solutions to one of the hardest of the hard problems in all of the natural sciences can only be arrived at from multiple directions. Unsurprisingly, our esteemed colleagues who have graciously provided their expertise to this volume are situated in an array of departments and institutes around the world. They are themselves established leaders or emerging leaders in the field of neuroimaging, analytics, and neuropsychiatry. In addition to their expertise, we value the role played by Simina Calin (neuroscience editor at Springer Nature) in initially promoting this compendium; the extensive editorial support at Springer Nature from Sanjana Sundaram, Vishnu Prakash, Jayashree Dhakshnamoorthy, Henry Rodgers, and Mallaigh Nolan, and additional support from Claudia Voigt (Heinrich Heine University).

Any book, rather than being a finality, represents a point in the evolution of a field of inquiry. This volume on *Brain Network Dysfunction in Neuropsychiatric Illness: Methods, Applications and Implications* highlights the increasing symbiosis between fMRI analyses methods and their clinical applications in neuropsychiatry. The coming years will hopefully see positive expansion of this symbiotic relationship.

Detroit, MI, USA — Vaibhav A. Diwadkar
Jülich, Germany — Simon B. Eickhoff

Contents

Part II Applications

Part III Implications

Contributors

Sora An INSERM, INS, Inst Neurosci Syst, Aix Marseille University, Marseille, France

Peter A. Bandettini Section on Functional Imaging Methods, National Institute of Mental Health, Bethesda, MD, USA

Bharat B. Biswal Biomedical Engineering, NJIT, Newark, NJ, USA

Stefan Borgwardt Department of Psychiatry (UPK), University of Basel, Basel, Switzerland

Department of Psychiatry, Psychosomatics and Psychotherapy, University of Lübeck, Lübeck, Germany

Katherine L. Bottenhorn Department of Psychology, Florida International University, Miami, FL, USA

Paolo Brambilla Department of Pathophysiology and Transplantation, University of Milan, Milan, Italy

Department of Neurosciences and Mental Health, Fondazione IRCCS Ca' Granda, Ospedale, Maggiore Policlinico, Milan, Italy

Steven L. Bressler Center for Complex Systems and Brain Sciences, Florida Atlantic University, Boca Raton, FL, USA

Danilo Bzdok Department of Psychiatry, Psychotherapy and Psychosomatics, RWTH Aachen University, Aachen, Germany

JARA, Translational Brain Medicine, Aachen, Germany

Parietal Team, INRIA, Neurospin, Gif-sur-Yvette, France

Vince D. Calhoun The Mind Research Network, Albuquerque, NM, USA

Department of Electrical and Computer Engineering, University of New Mexico, Albuquerque, NM, USA

Department of Psychiatry, Yale University, New Haven, CT, USA

Julia A. Camilleri Institute of Systems Neuroscience, Heinrich Heine University Düsseldorf, Düsseldorf, Germany

Institute of Neuroscience and Medicine (INM-7), Research Centre Jülich, Jülich, Germany

Henry W. Chase Department of Psychiatry, University of Pittsburgh, Pittsburgh, PA, USA

Samuele Cortese Center for Innovation in Mental Health, School of Psychology and Clinical and Experimental Sciences (CNS and Psychiatry), Faculty of Medicine, University of Southampton, Southampton, UK

Solent NHS Trust, Southampton, UK

Hassenfeld Children's Hospital at NYU Langone, New York, New York, NY, USA

Division of Psychiatry and Applied Psychology, School of Medicine, University of Nottingham, Nottingham, UK

Giuseppe Delvecchio Department of Pathophysiology and Transplantation, University of Milan, Milan, Italy

Vaibhav A. Diwadkar Department of Psychiatry & Behavioral Neurosciences, Wayne State University School of Medicine, Detroit, MI, USA

Simon B. Eickhoff Heinrich-Heine Univ Düsseldorf, Institut für Klinische Neurowissenschaft, Düsseldorf, Nordrhein-Westfalen, Germany

Institute of Neuroscience and Medicine, Brain & Behaviour (INM-7), Research Centre Jülich, Jülich, North Rhine - Westphalia, Germany

Alan C. Evans McConnell Brain Imaging Center, Montreal Neurological Institute, Montreal, QC, Canada

Ludmer Centre for NeuroInformatics and Mental Health, Montreal, QC, Canada

H. New Fei Research Division, Institute of Mental Health, Singapore, Singapore

Sophia Frangou Department of Psychiatry, Icahn School of Medicine at Mount Sinai, New York, NY, USA

Yasser Iturria-Medina McConnell Brain Imaging Center, Montreal Neurological Institute, Montreal, QC, Canada

Ludmer Centre for NeuroInformatics and Mental Health, Montreal, QC, Canada

Viktor Jirsa INSERM, INS, Inst Neurosci Syst, Aix Marseille University, Marseille, France

Teresa M. Karrer Department of Psychiatry, Psychotherapy and Psychosomatics, RWTH Aachen University, Aachen, Germany

Keerthana Karunakaran Biomedical Engineering, NJIT, Newark, NJ, USA

Annegret Krause-Utz Institute of Clinical Psychology, Leiden University, Leiden, The Netherlands

Leiden Institute for Brain and Cognition (LIBC), Leiden, The Netherlands

Angela R. Laird Department of Physics, Florida International University, Miami, FL, USA

Robert Langner Institute of Systems Neuroscience, Heinrich Heine University Düsseldorf, Düsseldorf, Germany

Institute of Neuroscience and Medicine (INM-7), Research Centre Jülich, Jülich, Germany

Annika Linke Brain Development Imaging Laboratories, Department of Psychology, San Diego State University, San Diego, CA, USA

Eleonora Maggioni Department of Neurosciences and Mental Health, Fondazione IRCCS Ca' Granda, Ospedale Maggiore Policlinico, University of Milan, Milan, Italy

Ralph-Axel Müller Brain Development Imaging Laboratories, Department of Psychology, San Diego State University, San Diego, CA, USA

Thomas Nickl-Jockschat Department of Psychiatry, Iowa Neuroscience Institute, Carver College of Medicine, University of Iowa, Iowa City, IA, USA

Georg Northoff University of Ottawa Institute of Mental Health Research, Ottawa, ON, Canada

Center for Cognition and Brain Disorders, Hangzhou Normal University, Hangzhou, China

Center for Brain and Consciousness, Taipeh Medical University (TMU), Taipeh, Taiwan

College for Humanities and Medicine, Taipeh Medical University (TMU), Taipeh, Taiwan

Seventh Hospital for Mental Health, Zhejiang University, Hangzhou, China

Camille Piguet Faculty of Medicine, University of Geneva, Geneva, Switzerland

Andrea B. Protzner Department of Psychology, University of Calgary, Calgary, AB, Canada

Hotchkiss Brain Institute, Calgary, AB, Canada

Gwladys Rey Department of Basic Neurosciences, University of Geneva, Geneva, Switzerland

Alard Roebroeck Department of Cognitive neuroscience, Faculty of Psychology and Neuroscience, Maastricht University, Maastricht, the Netherlands

Christian Schmahl Department of Psychosomatic Medicine and Psychotherapy, Central Institute of Mental Health, Mannheim, Germany

Medical Faculty, University of Heidelberg, Mannheim, Germany

André Schmidt Department of Psychiatry (UPK), University of Basel, Basel, Switzerland

Kang Sim Research Division, Institute of Mental Health, Singapore, Singapore

Letizia Squarcina Department of Pathophysiology and Transplantation, University of Milan, Milan, Italy

June Su Tan Research Division, Institute of Mental Health, Singapore, Singapore

Jordon X. J. Tng Research Division, Institute of Mental Health, Singapore, Singapore

Patrik Vuilleumier Department of Basic Neurosciences, University of Geneva, Geneva, Switzerland

Tom Wassink Department of Psychiatry, Iowa Neuroscience Institute, Carver College of Medicine, University of Iowa, Iowa City, IA, USA

Marie Wolfer Biomedical Engineering, NJIT, Newark, NJ, USA

Qingbao Yu The Mind Research Network, Albuquerque, NM, USA

About the Editors

Vaibhav A. Diwadkar is Professor of Psychiatry & Behavioral Neurosciences at Wayne State University School of Medicine. He received his B.A. in psychology and computer science from Coe College, and his Ph.D. in psychology and cognitive science from Vanderbilt University. Following neuroimaging-related fellowships at Carnegie Mellon University and the University of Pittsburgh, he has served on the faculty at the University of Pittsburgh, and at Wayne State University where he has been since 2005. Dr. Diwadkar uses *in vivo* neuroimaging to understand mechanisms of brain network function (underlying psychological and physiological processes), publishing in the areas of learning, memory, cognition and sensorimotor function, and thermoregulation. His clinical neuroimaging interests lie in understanding brain network dysfunction in psychiatry, leading to publications on schizophrenia, mood disorders, obsessive compulsive disorder, and borderline personality disorder. His research is supported by the National Institutes of Mental Health, the Children's Research Center of Michigan, the Children's Hospital of Michigan Foundation, the Ethel and James Flinn Foundation, the DMC Foundation, the Cohen Neuroscience Endowment, the Prechter World Bipolar Foundation, the Jack Dorsey Endowment, and the National Alliance for Research on Schizophrenia and Depression (now the Brain and Behavior Research Foundation).

Simon B. Eickhoff is a full professor and chair of the Institute for Systems Neuroscience at Heinrich-Heine University in Düsseldorf and the director of the Institute of Neuroscience and Medicine (INM-7, Brain and Behavior) at the Forschungszentrum Jülich. He is furthermore a visiting professor at the Chinese Academy of Science Institute of Automation. Working at the interface between neuroanatomy, data science, and brain medicine, he aims to obtain a more detailed characterization of the organization of the human brain and its inter-individual variability in order to better understand its changes in advanced age as well as neurological and psychiatric disorders. This goal is pursued by the development and application of novel analysis tools and approaches for large-scale, multi-modal analysis of brain structure, function, and connectivity as well as by machine learning for single-subject prediction of cognitive and socio-affective traits and ultimately precision medicine.

http://www.fz-juelich.de/inm/inm-7/EN/Home/home_node.html

Baajour SJ, Chowdury A, Thomas P, Rajan U, Khatib D, Zajac-Benitez C. .. Diwadkar VA (2020) Disordered directional brain network interactions during learning dynamics in schizophrenia revealed by multivariate autoregressive models. Hum Brain Mapp, 41(13), 3594–3607, https://doi.org/10.1002/hbm.25032

Chen J, …, Eickhoff SB. Neurobiological divergence of the positive and negative schizophrenia subtypes identified upon a new factor-structure of psychopathology using non-negative factorization: An international machine-learning study, *Biological Psychiatry 87(3):282–293 (2020)*

Eickhoff SB, Langner R. Neuroimaging-based prediction of mental traits: Road to utopia or Orwell? *PLoS biology*, *17*(11) (2019)

Eickhoff SB, Yeo BT, Genon S. Imaging-based parcellations of the human brain. *Nature Reviews Neuroscience, 19(11):672–686 (2018)*

Eickhoff SB, Langner R. Neuroimaging-based prediction of mental traits: Road to utopia or Orwell? *PLoS biology*, *17*(11) (2019)

Falco D, Chowdury A, Rosenberg DR, Bressler SL, Diwadkar VA (2020) ALE meta-analysis, its role in node identification and the effects on estimates of local network organization. Brain Struct Funct, 225(3), 1089–1102, https://doi.org/10.1007/s00429-020-02061-2.

Genon S, Reid AR, Langner R, Amunts K, Eickhoff SB. How to characterize the function of a brain region. *Trends in Cognitive Science, 22(4):350–364 (2018)*

Müller VI, Cieslik EC, Serbanescu I, Laird AR, Fox PT, Eickhoff SB. ALE meta-analyses on altered brain activity in unipolar depression revisited: A problem of replication. *JAMA Psychiatry, 74(1):47–55 (2017)*

Muzik O, Baajour S, Bressler S, Diwadkar VA (2020) Directional Interactions Between Constituents of the Human Large-Scale Thermoregulatory Network. Brain Topogr, 33(4), 489-503, https://doi.org/10.1007/s10548-020-00778-1.

Eickhoff SB, Etkin A. Going beyond finding the "lesion": a path for maturation of neuroimaging. *American Journal of Psychiatry, 173(3):302–3 (2016)*
Robison AJ, Thakkar KN, Diwadkar VA (2020) Cognition and Reward Circuits in Schizophrenia: Synergistic, Not Separate. Biol Psychiatry, 87(3), 204–214, https://doi.org/10.1016/j.biopsych.2019.09.021
Szczepaniak M, Chowdury A, Soloff PH, Diwadkar VA (2021) Stimulus valence, episodic memory, and the priming of brain activation profiles in borderline personality disorder. Psychol Med, 1–11, https://doi.org/10.1017/S0033291721001136

Brain Network Dysconnection in Neuropsychiatric Disorders: The Practice of "Normal Science"

Vaibhav A. Diwadkar and Simon B. Eickhoff

Introduction

Neuropsychiatric conditions are frequently called "epigenetic puzzles" (Diwadkar et al. 2014; Petronis et al. 1999; Sweatt 2013). The term reflects enduring uncertainties that make scientific psychiatry a perilous exercise. Trained psychiatrists and clinical psychologists generally "know" what these conditions are when they see a patient, and groups of clinicians can rely on reliable nosology (APA 2013; WHO 2016) to reach consensus diagnosis. Nevertheless, symptom heterogeneity and the overlap of "causal" pathways *across* diagnostic categories remain as problems, and criticisms regarding the fundamental value of category-based psychiatric nosology are widespread (Allsopp et al. 2019). Moreover, and as has been repeatedly asserted, psychiatric nosology does not map to, and therefore cannot inform our understanding of, neurobiology (Conway et al. 2019; Cuthbert and Insel 2010; Insel et al. 2010). A resultant effect of decades of research is that we seem to *know* much but *understand* little. Even now the fundamental question remains: "What kinds of things are psychiatric disorders (Kendler et al. 2011)?"

In posing such a question, Kendler and colleagues provided a lucid differentiation between psychiatric conditions as (a) essentialist kinds, resulting from uniquely identifiable origins (e.g., single genes), (b) socio-cultural constructs of the kind proposed (and criticized) by Foucault (Dreyfus 1987), (c) practical kinds, where

V. A. Diwadkar (✉)
Department of Psychiatry & Behavioral Neurosciences, Wayne State University School of Medicine, Detroit, MI, USA
e-mail: vdiwadka@med.wayne.edu

S. B. Eickhoff
Heinrich-Heine Univ Düsseldorf, Institut für Klinische Neurowissenschaft, Düsseldorf, Nordrhein-Westfalen, Germany

Institute of Neuroscience and Medicine, Brain & Behaviour (INM-7), Research Centre Jülich, Jülich, North Rhine - Westphalia, Germany

V. A. Diwadkar, S. B. Eickhoff (eds.), *Brain Network Dysfunction in Neuropsychiatric Illness*, https://doi.org/10.1007/978-3-030-59797-9_1

diagnostic labels provide a shared language for referring to and providing treatment plans for conditions, and d) mechanistic property clusters, a term co-opted from biology (Wilson et al. 2007). Mechanistic property clusters are defined as "… complex, mutually reinforcing networks of causal mechanisms ..", or ".. fuzzy sets defined by mechanisms at multiple levels that act and interact to produce the key features of the kind." Thus, neuropsychiatric disorders are, ".. objectively grounded features of the causal structure of the mind/brain."

This last version now is *the* dominant model of scientific psychiatry: Psychiatric disorders are disorders of behavior, and behavior is inarguably driven by brain network interactions (Genon et al. 2018). It follows (by *modus tollens*) that brain network dysfunction creates disordered behavior. Therefore, neuropsychiatric conditions can *in principle* be understood as such (Cowan et al. 2000; Fornito et al. 2017; Insel and Cuthbert 2015; Walter 2013). If in the modern scientific framework, the etiology of neuropsychiatric conditions lies in (some as yet poorly understood contribution of) genetics, a mechanistic understanding lies in understanding brain networks (Stephan et al. 2016). Indeed, an assertion is that a lack of focus on mechanistic understanding undermines attempts to identifying biologically (as opposed to symptomatically) homogenous subtypes, and has preempted biological psychiatry from transitioning from being descriptive (i.e., simply characterizing phenomenon) to being prescriptive (i.e., informing clinical practice with reliable biological tests) (Kapur et al. 2012). Genetic studies increasingly point to the shared genetic bases of major psychiatric conditions like schizophrenia, mood disorder, autism spectrum disorder, and attention-deficit hyperactivity disorder, notwithstanding the fact that these lie on different axes in psychiatric nosology (Cross-Disorder Group of the Psychiatric Genomics et al. 2013). Even disorders like schizophrenia which have a relatively narrow diagnostic envelope are seen as polygenic in origin (Goh et al. 2007; Schizophrenia Working Group of the Psychiatric Genomics 2014), where multiple genes (that themselves exert pleiotropic effects) confer vulnerability, but wherein the frank symptoms of the disorder themselves emerge from a complex (and plausibly in-deterministic) genetic-development-environmental interplay (Diwadkar et al. 2014; Howes and Murray 2013).

A lack of clarity of the origins of psychiatric conditions is not particularly problematic from a clinical perspective. As is evident in other fields of medicine (notably cardiology and oncology), successful medical treatment and management does not necessarily depend on identifying deterministic and mechanistic pathways toward disease. Rather, a clear characterization of the pathophysiologic *mechanisms* underlying the disease is the key precursor for generating meaningful treatment. Such a grasp of basic biological mechanisms will drive the development of therapies while providing objective biomarkers that are needed to inform and constrain treatment (Goff et al. 2016; Javitt et al. 2008; Kapur et al. 2012). In the realm of neuropsychiatry, it appears to increasingly be the case that characterizing brain network function will provide *an* answer.

Brains and Brain Networks

In the beginning, all that was interesting about brain function was deemed "local" (Fuster 2006), because all that could be assessed was loss of function following focal brain injury (Broca 1861). Thus, lesion-based models were characterized by rudimentary system's-based approaches toward brain–behavior relationships, the Geschwind–Wernicke model of speech comprehension and production (Kertesz 2006) serving as a good example. However, the assumption that brain regions existed in specific one-to-one relationships with overt behavior is obviously untenable (Kosslyn and Intriligator 1992; Price and Friston 2005). Wernicke and Luria created a hybrid model distinguishing between elementary functions expressed by individual brain regions and complex functions that are properties of distributed systems of brain areas (Bressler and Tognoli 2006), that is the principle of the functional integration of information across brain regions. Early phrenological assumptions underlying neurology were coopted by psychiatry, but only after the publication of thousands of papers is it now accepted that the "lesion model" provides no mechanism for psychiatry (Carota and Bogousslavsky 2019). However, lesion studies were the precursors for identifying at least one central defining feature of brain function, the principle of relative specialization of function (Friston 2005), that is the idea that focal brain regions are generally more associated with some discrete set of psychological domains than others.

Notably, hyper-localization approaches also permeated early modern neuroscience, with the "neuron theory" (Lopez-Munoz et al. 2006) emerging in the late nineteenth century, experimentally developing the idea of neurons as functional units, the extracellular outputs of which reflect basic properties of behaviors (both simple and relatively complex). The neuron theory remains a central driving assumption in cellular neuroscience, but theories of structure–function relationships based on single-unit recordings are subject to the same ontological limitations as phrenology (Fuster 2006). Moreover, connectivity is a fundamental property of neurons (which connect through axons and synapses), suggesting that connectivity is a *necessary* condition subserving the emergence of complexity. The application of complex systems theories, and the operationalization of these analytic frameworks for understanding the brain constituted an early step in the direction of addressing system complexity. Ludwig von Bertalanffy's work on general systems theory (Von Bertalanffy 1968) now underpins modern neuroscience's attempt to explain processes not only in terms of components of a system, but also in terms of interactions between such components (Erdi 2008). Thus, the second major organizing principle of brain function is the functional integration of signals across brain regions (Friston 2005; Singer 1994). If complex behavior arises from brain network interactions, and neuropsychiatric conditions are behavioral in their bases, then as noted these disorders result from impaired functional integration of signals across brain regions, i.e., brain network dysfunction.

The role of fMRI, but equally notably, statistical and analytical approaches to unraveling system states hidden in these fMRI signals has been central. This is not surprising because in neuroscience, theories need methods, and methods drive theories (Nicolelis and Lebedev 2009; Singh 2012). fMRI relies on hemodynamic spatio-temporal signals (Logothetis 2002; Logothetis and Wandell 2004) that themselves lie at the apex of a series of complex neuronal (and neurochemical) processes. The fact that these processes exist in uncertain relationships with the observed fMRI signal (Logothetis 2008; Singh 2012) is a limitation, but does not preempt the use of fMRI signals in understanding system states. As the literature on complex systems notes, the goal of any analytic approach is to discover or recover hidden system states that are encoded in any class of signals derived from the system (Granger 1980; Sugihara et al. 2012). Accordingly, recovering brain network function and dysfunction from fMRI time series data is a process of "reverse engineering" or "network discovery"; what must be engineered or discovered, are the "hidden" states of brain function that gave rise to the observed fMRI signals (Friston et al. 2012; Silverstein et al. 2016; Stephan 2004).

What of brain network dysfunction can be gleaned from fMRI data? Brain network interactions at the macroscopic scale, that is where constituent brain regions are defined as relatively large (Eickhoff et al. 2018), can best explain brain–behavior relationships (Passingham et al. 2013). Much has been written about fMRI, its neurophysiological bases, the relative advantages of the signal, and its limitations for assessing brain function (Harel 2012; Logothetis 2008; Lowe et al. 2016; Niessing et al. 2005). Because fMRI provides credible signals, when placed under appropriate analytic constraints these signals can be highly revealing. Thus, more than any other competing in vivo imaging method (like fNIRS, MEG, or PET), fMRI data are well positioned to capture network effects, given its millimeter-based spatial resolution and its seconds-based temporal resolution (Logothetis 2008; Meunier et al. 2009).

Connection and Dysconnection

Connection and dysconnection are generally understood as the normal (or abnormal) integration of signals across the brain (Stephan et al. 2009). Terms like dysconnection can appear colloquial and thus confusing, but a consideration of how the meaning of the term has evolved specifically in the context of schizophrenia is revealing. In the schizophrenia literature, the term dysconnection has two somewhat distinct meanings. In one metaphorical sense, the term represents how Blueler and Kraepelin thought of schizophrenia as a "splitting of the mind" (Bleuler 1924; Kraepelin 1919–1971). This sense was primarily descriptive in origin though it became the precursor for a more literal use of the term. The second, more literal sense of dysconnection (or "disconnection" as originally proposed), is explicitly neurobiological: In this view, schizophrenia emerges from profound alternations in the connectivity of the brain's anatomic and functional pathways (Andreasen 1999; Friston 1996, 1998). Unsurprisingly, the literal sense of the term gained credence only with the advent of in vivo neuroimaging techniques (not restricted only to fMRI). Indeed, the terms schizophrenia and dysconnection are now intertwined to the degree that the issue is not *if* schizophrenia is a "dysconnection syndrome," but *how* it is so (Friston et al. 2016). More specifically, recent focus in schizophrenia has been on functional interactions within the brain, rather than anatomical dysconnectivity. Just as structural connectivity loosely constrains the brain's functional architecture (Behrens and Sporns 2012; Park and Friston 2013), impaired anatomical connectivity loosely constrains the schizophrenic brain's ability to integrate functional signals across regions in both non-task and task-active states (Banyai et al. 2011; Bassett et al. 2008). For example, a cortico-cortical phase synchronization mechanism operates in conjunction with reentrant connections to provide context for local cortical computations (Bressler and Tognoli 2006), may be impaired in schizophrenia, and may be expressed in alterations in acquired fMRI signals (Baajour et al. 2020).

What can be inferred from models for inferring network function and dysfunction?

Distinct analytical techniques are applied to fMRI time series data (Behrens and Sporns 2012; Eickhoff et al. 2011; Stanley et al. 2013; van den Heuvel and Hulshoff Pol 2010), though they are generally subsumed under the aegis of functional connectivity (FC) or effective connectivity (EC) (Friston 2011; Silverstein et al. 2016). These respectively (a) characterize statistical relationships between time series or (b) attempt to capture the relatively precise timing relationships between *neuronal* populations (Aersten and Preisl 1991). The latter class seeks to model "causative" relations between brain regions and depends on generative network architectures of the brain (Daunizeau et al. 2011). The difference between seeking to retrieve causal influences between brain regions against simply co-activations or co-variations is frequently deemed to be significant, though the former class of models often rely on questionable assumptions regarding information in fMRI time series data. For example, an extensively employed method like Wiener–Granger Causality (Bressler and Seth 2011) is based on the idea that causality can be deciphered from asymmetric patterns of statistical predictability of one time series on another. These methods can be applied to fMRI time series to recover functionally induced hierarchically ordered functional relationships in brain structures (Muzik et al. 2020) and ordered interactions between motor circuits (Diwadkar et al. 2017). However, WGC also assumes "separability" between signals. Separability is the notion that information about any causative factor within a source variable that impacts the response of a target variable is independently unique to the source and not contained in the target (Ye et al. 2015). Thus, the causative factors that drive the time series of the dlPFC in a working memory experiment must not be contained in the time series for the parietal cortex from the same experiment. Given that fMRI signals are heavily driven by neuromodulation (Logothetis 2007), and given that in this example, dlPFC neurons synapse on the parietal cortex (Chafee and Goldman-Rakic 2000), separability is likely to be violated. More generally, separability may always be violated given massive anatomical interconnectivity (ascending and re-entrant descending connections) in the human brain, and given that functional and effective connectivity are themselves contextually driven (Mesulam 1998; Park and Friston 2013).

Dynamic causal modeling (DCM) is a more recent alternative for discovering causality in brain networks from fMRI signals (Friston 2009). DCM represents a framework derived from control system's theory and is predicated on specifying and comparing competing generative model architectures of functional brain networks. These are fit to the measured brain responses, with posterior likelihoods attached to each in a Bayesian framework. The resultant metrics (weighted by Bayesian posteriors) represent estimates (across the space of fitted models) of neurophysiologically interpretable measures of connectivity (or modulation of that connectivity) between nodes. DCM has notable advantages over techniques like WGC (Friston et al. 2013; Stephan and Roebroeck 2012), including the fact that the parameters are physiological interpretable (as neuronal activity is transformed through a hemodynamic model with region-specific parameters to model measured responses). However, foundational limitations hinder the recovery of DCM-like causality in macroscopic brain networks. In principle, causality in the brain appears to be probabilistic, not deterministic (Mannino and Bressler 2015). Mutual causality between brain network constituents pervades brain function. This is the phenomenon wherein one neural element influences another even as that target influences the source (Frankel 1986), and such patterns of mutually causal effects are unlikely to fit in strictly deterministic frameworks. In addition, the brain is characterized by the convergence of projections from multiple to individual units (many to one mappings), suggesting that any causal influences of any one of the multiple sources on the target would be probabilistic. Finally, the causal influences of one neuronal population on another may have no observable effect because post-synaptic effects are frequently sub-threshold in nature and therefore not typically observable (Singh 2012). These constraints on interpretation do not apply only to DCM, but they highlight the challenges of drawing mechanistic interpretations.

These limitations do not nullify prospects for meaningful discovery in health and neuropsychiatric disease. After all, multi-factorial and non-linear complexity makes long-term forecasting challenging, but viable probabilistic models make such forecasting feasible (Nicholls 1980). Rather, these models when applied to brain networks in neuropsychiatric conditions can provide viable markers and establish meaningful ranges for ordered and disordered network function. A variety of inductive techniques can be subsequently used in classification based on metrics of network function (Chen et al. 2020; Heinrichs and Eickhoff 2020). A further advantage of some of these techniques (particularly like DCM) is that they profess to provide neurobiologically meaningful parameter estimates, that can render estimation of brain network dysfunction closer to source mechanisms (Brodersen et al. 2014).

The use of previously noted directional connectivity models like WGC and DCM are somewhat secondary to what is a core engine for summarizing fMRI time series data. Static bivariate correlations applied to fMRI time series remain the most widely employed model for summarizing system states in the brain. This summarizing of temporal relationships between regions of interest is "weak" in the sense that correlational models are agnostic about the relationship between hemodynamics and neuronal responses (Stephan et al. 2007). From the perspective of statistical inference, they provide weak statistical tests, generally only testing against the null model, i.e., for an *absence* of significant correlation. However, treated as measures of effect sizes, rather than statistics, they constitute an invaluable mathematical engine for characterizing large-scale functional network structure. By summarizing large swathes of time series data from multiple brain parcels, they provide information about network topology (Hermundstad et al. 2013; Kim and Min 2020; Petersen and Sporns 2015). Characterization of such topology can reveal altered functional characteristics in conditions like schizophrenia (Alexander-Bloch et al. 2011), mood disorders (Wang et al. 2020), and as a measure of characterizing the effects of treatment response (Hadley et al. 2016).

The "weakness" of correlational models also confers some advantages. There is greater statistical reliability with shorter time series than measures that make stronger assumptions, or that require more precise models (as is the case with WGC). Moreover, weak assumptions also mean that these FC analyses are relatively robust to contributions of filtering or slice-time correction applied to fMRI data that can artificially disrupt the fine-scale temporal structure of the signal. Correlational analyses are best seen as providing summaries of large-scale networks that must be submitted to additional analyses, before meaningful inferences regarding brain network function or dysfunction can be drawn. Methodological advances such as meta-analytic connectivity modeling (MACM) use temporal correlations between brain regions as an initial basis from which to estimate probabilities of co-occurrences of activations across brain regions. Then, treating studies as units of analyses, it is possible to discover similarities in activations across studies or tasks, or between task-induced activity and resting state network connectivity (Fox et al. 2014; Heckner et al. 2020). Recent approaches acknowledge that brain network states are unstable or multi-stable, and that sliding windows are needed to model changes in connectivity states over time (within an experimental scan). These approaches quantify dynamics of network activity as changes in the correlational structure of networks over time (Heitmann and Breakspear 2018; Ioannides 2007). Changes in stability are informative of genetic inheritance (Yamin et al. 2019), and are a core characteristic of differences between psychiatric diagnosis including schizophrenia, bipolar disorder, and ADHD (Zhu et al. 2020).

These approaches used in understanding network function and dysfunction are broadly characterized as directed (in which WGC and DCM are included) and undirected methods (of which correlational models are typical examples) (Friston 2011; Silverstein et al. 2016). Understanding directional relations between network constituents can be of particular value in attempting to approximate mechanisms. While various brain areas have reciprocal structural connections, it is unlikely that the directions will have equivalent *functional* connotations. In terms of functional network interactions, the direction of information flow in the brain is of critical importance for organizing cognitive functions and consciousness (Lee et al. 2013; Moon et al. 2015). Moreover, directional effects are also important in the context of brain network hierarchies (Asemi et al. 2015; Fuster 2001; Mesulam 1998; Muzik et al. 2020), where brain regions engaged in *control* enjoy higher hierarchical status within the overall system (Asemi et al. 2015; Paus 2001; Vogt 2016). Capturing these asymmetries will prove highly valuable in neuropsychiatric conditions that are frequently seen as being disorders of control or regulation (Diwadkar et al. 2012; Frangou et al. 2008; Lukito et al. 2020; Phillips et al. 2003).

The study of brain network function and dysfunction as "normal science"

As foreshadowed earlier, the concept of *deterministic* causality is a fundamental assumption in the practice of study brain network function and dysfunction. Unfortunately, brain structure (to the extent that it is fully knowable) provides modest constraints on emergent functional interactions. Furthermore, there is a fundamental indeterminacy with respect to how overt behavior arises from the neural substrate. The "degeneracy" in the brain's structure–function relationships (Price and Friston 2005) renders the process of discovery difficult. For instance, should some metric of connectivity between two regions be established via a statistical model, the hypothetical existence of a third, undetected but mediating member of the triumvirate of regions will undermine our understanding of interactions between system constituents. Time delays may characterize the differences between the activity of regions and can present as difficult to untangle phase-lags. Finally, "neural activity" is itself a highly idealized term (Singh 2012). There are literally dozens of types of "neural activities" that collectively subserve communication within the brain at multiple temporal and spatial scales (Buzsaki et al. 2013), but that also impede a clearer picture of the neuronal bases of the fMRI response (Goense et al. 2016).

Methodological limitations do not inhibit the process of research, but merely constrain the interpretation of the data (Silverstein et al. 2016). In successful instances, these limitations form the bases for successive and better-developed iterations of a research model. In 1962, Thomas Kuhn, in his defining tome on the nature of scientific practice, defined "normal science" as ".... research firmly based on one or more past scientific achievements, achievements that for a time some scientific community acknowledges as supplying the foundation for its further practice .." (p. 10) (Kuhn 1962). Moreover, normal science generates dominant paradigms, that for a time provide problems and solutions for a community of practitioners, attract adherents away from competing modes of scientific activity, and are simultaneously, sufficiently open-ended to leave all sorts of problems for the redefined group of practitioners to resolve. The value of Kuhn's characterization of science is itself a source of discussion, but his broad characterization remains widely accepted. On these terms, the study of brain network function and dysfunction is an example of "normal science" at work. Such a focus has sharpened our models of neuropsychiatric conditions (Baker et al. 2019; Buch and Liston 2020; Roalf and Gur 2017), has led to the explicit neurobiological formulation of what previously were merely metaphorical ideas (Friston et al. 2016), and has provided a largely accepted general framework for neurobiological research (Bressler and Kelso 2016; Fornito and Bullmore 2015; Telesford et al. 2011). Inferring dysconnection in neuropsychiatric conditions is a special case of inferring brain network function in the healthy brain. In that sense, the challenge is perhaps no more or less tractable than understanding how macroscopic brain network interactions can inform brain–behavior relationships. If analyses of fMRI signals have told us anything of significant value, it is that macroscopic brain network interactions captured at the seconds' scale *do* contribute to a viable understanding of how the brain "works." Theories and techniques that illuminate our understanding of the normal brain will unquestionably inform our understanding of diseases of the brain. Thus, the study of brain network dysfunction offers the promise of significant mechanistic understanding of the bases of neuropsychiatric disorders.

References

Aersten, A., Preisl, H., 1991. Dynamics of activity and connectivity in physiological neuronal networks. In: Schuster, H.G. (Ed.), Non-linear dynamics and neuronal networks. VCH Publishing, New York, pp. 281-302.

Alexander-Bloch, A.F., Gogtay, N., Meunier, D., Birn, R., Clasen, L., Lalonde, F., Lenroot, R., Giedd, J., Bullmore, E.T., 2011. Disrupted modularity and local connectivity of brain functional networks in childhood-onset schizophrenia. Front Syst Neurosci 4, 147.

Allsopp, K., Read, J., Corcoran, R., Kinderman, P., 2019. Heterogeneity in psychiatric diagnostic classification. Psychiatry Res 279, 15-22.

Andreasen, N.C., 1999. A unitary model of schizophrenia: Bleuler's "fragmented phrene" as schizencephaly. Arch Gen Psychiatry 56, 781-787.

APA, 2013. Diagnostic Statistical Manual of Mental Disorders, 5th ed. American Psychiatric Association, Arlington, VA.

Asemi, A., Ramaseshan, K., Burgess, A., Diwadkar, V.A., Bressler, S.L., 2015. Dorsal anterior cingulate cortex modulates supplementary motor area in coordinated unimanual motor behavior. Front Hum Neurosci 9, 309.

Baajour, S.J., Chowdury, A., Thomas, P., Rajan, U., Khatib, D., Zajac-Benitez, C., Falco, D., Haddad, L., Amirsadri, A., Bressler, S., Stanley, J.A., Diwadkar, V.A., 2020. Disordered directional brain network interactions during learning dynamics in schizophrenia revealed by multivariate autoregressive models. Hum Brain Mapp 41, 3594-3607.

Baker, J.T., Dillon, D.G., Patrick, L.M., Roffman, J.L., Brady, R.O., Jr., Pizzagalli, D.A., Ongur, D., Holmes, A.J., 2019. Functional connectomics of affective and psychotic pathology. Proc Natl Acad Sci U S A 116, 9050-9059.

Banyai, M., Diwadkar, V.A., Erdi, P., 2011. Model-based dynamical analysis of functional disconnection in schizophrenia. Neuroimage 58, 870-877.

Bassett, D.S., Bullmore, E., Verchinski, B.A., Mattay, V.S., Weinberger, D.R., Meyer-Lindenberg, A., 2008. Hierarchical organization of human cortical networks in health and schizophrenia. J Neurosci 28, 9239-9248.

Behrens, T.E., Sporns, O., 2012. Human connectomics. Curr Opin Neurobiol 22, 144-153.

Bleuler, E., 1924. Textbook of Psychiatry. MacMillan, New York.
Bressler, S.L., Kelso, J.A., 2016. Coordination Dynamics in Cognitive Neuroscience. Front Neurosci 10, 397.
Bressler, S.L., Seth, A.K., 2011. Wiener-Granger causality: a well established methodology. Neuroimage 58, 323-329.
Bressler, S.L., Tognoli, E., 2006. Operational principles of neurocognitive networks. Int J Psychophysiol 60, 139-148.
Broca, P., 1861. Remarks on the Seat of the Faculty of Articulated Language, Following an Observation of Aphemia (Loss of Speech). Bulletin de la Société Anatomique 6, 330-357.
Brodersen, K.H., Deserno, L., Schlagenhauf, F., Lin, Z., Penny, W.D., Buhmann, J.M., Stephan, K.E., 2014. Dissecting psychiatric spectrum disorders by generative embedding. Neuroimage Clin 4, 98-111.
Buch, A.M., Liston, C., 2020. Dissecting diagnostic heterogeneity in depression by integrating neuroimaging and genetics. Neuropsychopharmacology.
Buzsaki, G., Logothetis, N., Singer, W., 2013. Scaling brain size, keeping timing: evolutionary preservation of brain rhythms. Neuron 80, 751-764.
Carota, A., Bogousslavsky, J., 2019. Neurology versus Psychiatry? Hallucinations, Delusions, and Confabulations. Front Neurol Neurosci 44, 127-140.
Chafee, M.V., Goldman-Rakic, P.S., 2000. Inactivation of parietal and prefrontal cortex reveals interdependence of neural activity during memory-guided saccades. J Neurophysiol 83, 1550-1566.
Chen, J., Patil, K.R., Weis, S., Sim, K., Nickl-Jockschat, T., Zhou, J., Aleman, A., Sommer, I.E., Liemburg, E.J., Hoffstaedter, F., Habel, U., Derntl, B., Liu, X., Fischer, J.M., Kogler, L., Regenbogen, C., Diwadkar, V.A., Stanley, J.A., Riedl, V., Jardri, R., Gruber, O., Sotiras, A., Davatzikos, C., Eickhoff, S.B., Pharmacotherapy, M., Outcome Survey, I., 2020. Neurobiological Divergence of the Positive and Negative Schizophrenia Subtypes Identified on a New Factor Structure of Psychopathology Using Non-negative Factorization: An International Machine Learning Study. Biol Psychiatry 87, 282-293.
Conway, C.C., Forbes, M.K., Forbush, K.T., Fried, E.I., Hallquist, M.N., Kotov, R., Mullins-Sweatt, S.N., Shackman, A.J., Skodol, A.E., South, S.C., Sunderland, M., Waszczuk, M.A., Zald, D.H., Afzali, M.H., Bornovalova, M.A., Carragher, N., Docherty, A.R., Jonas, K.G., Krueger, R.F., Patalay, P., Pincus, A.L., Tackett, J.L., Reininghaus, U., Waldman, I.D., Wright, A.G.C., Zimmermann, J., Bach, B., Bagby, R.M., Chmielewski, M., Cicero, D.C., Clark, L.A., Dalgleish, T., DeYoung, C.G., Hopwood, C.J., Ivanova, M.Y., Latzman, R.D., Patrick, C.J., Ruggero, C.J., Samuel, D.B., Watson, D., Eaton, N.R., 2019. A Hierarchical Taxonomy of Psychopathology Can Transform Mental Health Research. Perspect Psychol Sci 14, 419-436.
Cowan, W.M., Harter, D.H., Kandel, E.R., 2000. The emergence of modern neuroscience: some implications for neurology and psychiatry. Annu Rev Neurosci 23, 343-391.
Cross-Disorder Group of the Psychiatric Genomics, C., Lee, S.H., Ripke, S., Neale, B.M., Faraone, S.V., Purcell, S.M., Perlis, R.H., Mowry, B.J., Thapar, A., Goddard, M.E., Witte, J.S., Absher, D., Agartz, I., Akil, H., Amin, F., Andreassen, O.A., Anjorin, A., Anney, R., Anttila, V., Arking, D.E., Asherson, P., Azevedo, M.H., Backlund, L., Badner, J.A., Bailey, A.J., Banaschewski, T., Barchas, J.D., Barnes, M.R., Barrett, T.B., Bass, N., Battaglia, A., Bauer, M., Bayes, M., Bellivier, F., Bergen, S.E., Berrettini, W., Betancur, C., Bettecken, T., Biederman, J., Binder, E.B., Black, D.W., Blackwood, D.H., Bloss, C.S., Boehnke, M., Boomsma, D.I., Breen, G., Breuer, R., Bruggeman, R., Cormican, P., Buccola, N.G., Buitelaar, J.K., Bunney, W.E., Buxbaum, J.D., Byerley, W.F., Byrne, E.M., Caesar, S., Cahn, W., Cantor, R.M., Casas, M., Chakravarti, A., Chambert, K., Choudhury, K., Cichon, S., Cloninger, C.R., Collier, D.A., Cook, E.H., Coon, H., Cormand, B., Corvin, A., Coryell, W.H., Craig, D.W., Craig, I.W., Crosbie, J., Cuccaro, M.L., Curtis, D., Czamara, D., Datta, S., Dawson, G., Day, R., De Geus, E.J., Degenhardt, F., Djurovic, S., Donohoe, G.J., Doyle, A.E., Duan, J., Dudbridge, F., Duketis, E., Ebstein, R.P., Edenberg, H.J., Elia, J., Ennis, S., Etain, B., Fanous, A., Farmer, A.E., Ferrier, I.N., Flickinger, M., Fombonne, E., Foroud, T., Frank, J., Franke, B., Fraser,

C., Freedman, R., Freimer, N.B., Freitag, C.M., Friedl, M., Frisen, L., Gallagher, L., Gejman, P.V., Georgieva, L., Gershon, E.S., Geschwind, D.H., Giegling, I., Gill, M., Gordon, S.D., Gordon-Smith, K., Green, E.K., Greenwood, T.A., Grice, D.E., Gross, M., Grozeva, D., Guan, W., Gurling, H., De Haan, L., Haines, J.L., Hakonarson, H., Hallmayer, J., Hamilton, S.P., Hamshere, M.L., Hansen, T.F., Hartmann, A.M., Hautzinger, M., Heath, A.C., Henders, A.K., Herms, S., Hickie, I.B., Hipolito, M., Hoefels, S., Holmans, P.A., Holsboer, F., Hoogendijk, W.J., Hottenga, J.J., Hultman, C.M., Hus, V., Ingason, A., Ising, M., Jamain, S., Jones, E.G., Jones, I., Jones, L., Tzeng, J.Y., Kahler, A.K., Kahn, R.S., Kandaswamy, R., Keller, M.C., Kennedy, J.L., Kenny, E., Kent, L., Kim, Y., Kirov, G.K., Klauck, S.M., Klei, L., Knowles, J.A., Kohli, M.A., Koller, D.L., Konte, B., Korszun, A., Krabbendam, L., Krasucki, R., Kuntsi, J., Kwan, P., Landen, M., Langstrom, N., Lathrop, M., Lawrence, J., Lawson, W.B., Leboyer, M., Ledbetter, D.H., Lee, P.H., Lencz, T., Lesch, K.P., Levinson, D.F., Lewis, C.M., Li, J., Lichtenstein, P., Lieberman, J.A., Lin, D.Y., Linszen, D.H., Liu, C., Lohoff, F.W., Loo, S.K., Lord, C., Lowe, J.K., Lucae, S., MacIntyre, D.J., Madden, P.A., Maestrini, E., Magnusson, P.K., Mahon, P.B., Maier, W., Malhotra, A.K., Mane, S.M., Martin, C.L., Martin, N.G., Mattheisen, M., Matthews, K., Mattingsdal, M., McCarroll, S.A., McGhee, K.A., McGough, J.J., McGrath, P.J., McGuffin, P., McInnis, M.G., McIntosh, A., McKinney, R., McLean, A.W., McMahon, F.J., McMahon, W.M., McQuillin, A., Medeiros, H., Medland, S.E., Meier, S., Melle, I., Meng, F., Meyer, J., Middeldorp, C.M., Middleton, L., Milanova, V., Miranda, A., Monaco, A.P., Montgomery, G.W., Moran, J.L., Moreno-De-Luca, D., Morken, G., Morris, D.W., Morrow, E.M., Moskvina, V., Muglia, P., Muhleisen, T.W., Muir, W.J., Muller-Myhsok, B., Murtha, M., Myers, R.M., Myin-Germeys, I., Neale, M.C., Nelson, S.F., Nievergelt, C.M., Nikolov, I., Nimgaonkar, V., Nolen, W.A., Nothen, M.M., Nurnberger, J.I., Nwulia, E.A., Nyholt, D.R., O'Dushlaine, C., Oades, R.D., Olincy, A., Oliveira, G., Olsen, L., Ophoff, R.A., Osby, U., Owen, M.J., Palotie, A., Parr, J.R., Paterson, A.D., Pato, C.N., Pato, M.T., Penninx, B.W., Pergadia, M.L., Pericak-Vance, M.A., Pickard, B.S., Pimm, J., Piven, J., Posthuma, D., Potash, J.B., Poustka, F., Propping, P., Puri, V., Quested, D.J., Quinn, E.M., Ramos-Quiroga, J.A., Rasmussen, H.B., Raychaudhuri, S., Rehnstrom, K., Reif, A., Ribases, M., Rice, J.P., Rietschel, M., Roeder, K., Roeyers, H., Rossin, L., Rothenberger, A., Rouleau, G., Ruderfer, D., Rujescu, D., Sanders, A.R., Sanders, S.J., Santangelo, S.L., Sergeant, J.A., Schachar, R., Schalling, M., Schatzberg, A.F., Scheftner, W.A., Schellenberg, G.D., Scherer, S.W., Schork, N.J., Schulze, T.G., Schumacher, J., Schwarz, M., Scolnick, E., Scott, L.J., Shi, J., Shilling, P.D., Shyn, S.I., Silverman, J.M., Slager, S.L., Smalley, S.L., Smit, J.H., Smith, E.N., Sonuga-Barke, E.J., St Clair, D., State, M., Steffens, M., Steinhausen, H.C., Strauss, J.S., Strohmaier, J., Stroup, T.S., Sutcliffe, J.S., Szatmari, P., Szelinger, S., Thirumalai, S., Thompson, R.C., Todorov, A.A., Tozzi, F., Treutlein, J., Uhr, M., van den Oord, E.J., Van Grootheest, G., Van Os, J., Vicente, A.M., Vieland, V.J., Vincent, J.B., Visscher, P.M., Walsh, C.A., Wassink, T.H., Watson, S.J., Weissman, M.M., Werge, T., Wienker, T.F., Wijsman, E.M., Willemsen, G., Williams, N., Willsey, A.J., Witt, S.H., Xu, W., Young, A.H., Yu, T.W., Zammit, S., Zandi, P.P., Zhang, P., Zitman, F.G., Zollner, S., International Inflammatory Bowel Disease Genetics, C., Devlin, B., Kelsoe, J.R., Sklar, P., Daly, M.J., O'Donovan, M.C., Craddock, N., Sullivan, P.F., Smoller, J.W., Kendler, K.S., Wray, N.R., 2013. Genetic relationship between five psychiatric disorders estimated from genome-wide SNPs. Nat Genet 45, 984-994.

Cuthbert, B.N., Insel, T.R., 2010. Toward new approaches to psychotic disorders: the NIMH Research Domain Criteria project. Schizophr Bull 36, 1061-1062.

Daunizeau, J., David, O., Stephan, K.E., 2011. Dynamic causal modelling: A critical review of the biophysical and statistical foundations. Neuroimage 58, 312-322.

Diwadkar, V.A., Asemi, A., Burgess, A., Chowdury, A., Bressler, S.L., 2017. Potentiation of motor sub-networks for motor control but not working memory: Interaction of dACC and SMA revealed by resting-state directed functional connectivity. PLoS ONE 12, e0172531.

Diwadkar, V.A., Bustamante, A., Rai, H., Uddin, M., 2014. Epigenetics, stress, and their potential impact on brain network function: a focus on the schizophrenia diatheses. Front Psychiatry 5, 71.

Diwadkar, V.A., Wadehra, S., Pruitt, P., Keshavan, M.S., Rajan, U., Zajac-Benitez, C., Eickhoff, S.B., 2012. Disordered cortico-limbic interactions during affective processing in children and adolescents at risk for schizophrenia revealed by fMRI and Dynamic Causal Modeling. Arch Gen Psychiatry 69, 231-242.
Dreyfus, H.L., 1987. Foucault's critique of psychiatric medicine. J Med Philos 12, 311-333.
Eickhoff, S.B., Bzdok, D., Laird, A.R., Roski, C., Caspers, S., Zilles, K., Fox, P.T., 2011. Co-activation patterns distinguish cortical modules, their connectivity and functional differentiation. Neuroimage 57, 938-949.
Eickhoff, S.B., Yeo, B.T.T., Genon, S., 2018. Imaging-based parcellations of the human brain. Nat Rev Neurosci 19, 672-686.
Erdi, P., 2008. Complexity Explained. Springer, New York.
Fornito, A., Bullmore, E.T., 2015. Connectomics: a new paradigm for understanding brain disease. Eur Neuropsychopharmacol 25, 733-748.
Fornito, A., Bullmore, E.T., Zalesky, A., 2017. Opportunities and Challenges for Psychiatry in the Connectomic Era. Biol Psychiatry Cogn Neurosci Neuroimaging 2, 9-19.
Fox, P.T., Lancaster, J.L., Laird, A.R., Eickhoff, S.B., 2014. Meta-analysis in human neuroimaging: computational modeling of large-scale databases. Annu Rev Neurosci 37, 409-434.
Frangou, S., Kington, J., Raymont, V., Shergill, S.S., 2008. Examining ventral and dorsal prefrontal function in bipolar disorder: a functional magnetic resonance imaging study. Eur Psychiatry 23, 300-308.
Frankel, L., 1986. Mutual causation, simultaneity and event description. Philos Stud 49, 361-372.
Friston, K., 2009. Causal modelling and brain connectivity in functional magnetic resonance imaging. PLoS Biol 7, e33.
Friston, K., Moran, R., Seth, A.K., 2013. Analysing connectivity with Granger causality and dynamic causal modelling. Curr Opin Neurobiol 23, 172-178.
Friston, K.J., 1996. Theoretical neurobiology and schizophrenia. Br Med Bull 52, 644-655.
Friston, K.J., 1998. The disconnection hypothesis. Schizophr Res 30, 115-125.
Friston, K.J., 2005. Models of brain function in neuroimaging. Annu Rev Psychol 56, 57-87.
Friston, K.J., 2011. Functional and effective connectivity: a review. Brain Connect 1, 13-36.
Friston, K.J., Brown, H.R., Siemerkus, J., Stephan, K.E., 2016. The dysconnection hypothesis (2016). Schizophr Res 176, 83-94.
Friston, K.J., Li, B., Daunizeau, J., Stephan, K.E., 2012. Network discovery with DCM. Neuroimage 56, 1202-1221.
Fuster, J., 2006. Stretch that net or your mind may fall into phrenology. Cortex 42, 1048-1052.
Fuster, J.M., 2001. The prefrontal cortex-an update: Time is of the essence. Neuron 30, 319-333.
Genon, S., Reid, A., Langner, R., Amunts, K., Eickhoff, S.B., 2018. How to Characterize the Function of a Brain Region. Trends Cogn Sci 22, 350-364.
Goense, J., Bohraus, Y., Logothetis, N.K., 2016. fMRI at High Spatial Resolution: Implications for BOLD-Models. Front Comput Neurosci 10, 66.
Goff, D.C., Romero, K., Paul, J., Mercedes Perez-Rodriguez, M., Crandall, D., Potkin, S.G., 2016. Biomarkers for drug development in early psychosis: Current issues and promising directions. Eur Neuropsychopharmacol.
Goh, K.I., Cusick, M.E., Valle, D., Childs, B., Vidal, M., Barabasi, A.L., 2007. The human disease network. Proc Natl Acad Sci U S A 104, 8685-8690.
Granger, C.W.J., 1980. Testing for causality: a personal viewpoint. . J. Econ. Dyn. Control 2, 329–352.
Hadley, J.A., Kraguljac, N.V., White, D.M., Ver Hoef, L., Tabora, J., Lahti, A.C., 2016. Change in brain network topology as a function of treatment response in schizophrenia: a longitudinal resting-state fMRI study using graph theory. NPJ Schizophr 2, 16014.
Harel, N., 2012. Ultra high resolution fMRI at ultra-high field. Neuroimage 62, 1024-1028.
Heckner, M.K., Cieslik, E.C., Eickhoff, S.B., Camilleri, J.A., Hoffstaedter, F., Langner, R., 2020. The Aging Brain and Executive Functions Revisited: Implications from Meta-analytic and Functional Connectivity Evidence. J Cogn Neurosci, 1-36.

Heinrichs, B., Eickhoff, S.B., 2020. Your evidence? Machine learning algorithms for medical diagnosis and prediction. Hum Brain Mapp 41, 1435-1444.
Heitmann, S., Breakspear, M., 2018. Putting the "dynamic" back into dynamic functional connectivity. Netw Neurosci 2, 150-174.
Hermundstad, A.M., Bassett, D.S., Brown, K.S., Aminoff, E.M., Clewett, D., Freeman, S., Frithsen, A., Johnson, A., Tipper, C.M., Miller, M.B., Grafton, S.T., Carlson, J.M., 2013. Structural foundations of resting-state and task-based functional connectivity in the human brain. Proc Natl Acad Sci U S A 110, 6169-6174.
Howes, O.D., Murray, R.M., 2013. Schizophrenia: an integrated sociodevelopmental-cognitive model. Lancet.
Insel, T., Cuthbert, B., Garvey, M., Heinssen, R., Pine, D.S., Quinn, K., Sanislow, C., Wang, P., 2010. Research domain criteria (RDoC): toward a new classification framework for research on mental disorders. Am J Psychiatry 167, 748-751.
Insel, T.R., Cuthbert, B.N., 2015. Medicine. Brain disorders? Precisely. Science 348, 499-500.
Ioannides, A.A., 2007. Dynamic functional connectivity. Curr Opin Neurobiol 17, 161-170.
Javitt, D.C., Spencer, K.M., Thaker, G.K., Winterer, G., Hajos, M., 2008. Neurophysiological biomarkers for drug development in schizophrenia. Nat Rev Drug Discov 7, 68-83.
Kapur, S., Phillips, A.G., Insel, T.R., 2012. Why has it taken so long for biological psychiatry to develop clinical tests and what to do about it? Mol Psychiatry 17, 1174-1179.
Kendler, K.S., Zachar, P., Craver, C., 2011. What kinds of things are psychiatric disorders? Psychol Med 41, 1143-1150.
Kertesz, A., 2006. Wernicke–Geschwind Model. Encyclopedia of Cognitive Science.
Kim, D.J., Min, B.K., 2020. Rich-club in the brain's macrostructure: Insights from graph theoretical analysis. Comput Struct Biotechnol J 18, 1761-1773.
Kosslyn, S.M., Intriligator, J.M., 1992. Is cognitive neuropsychology plausible? The perils of sitting on a one-legged stool. J Cogn Neurosci 4, 96-105.
Kraepelin, E., 1919–1971. Dementia Praecox. Churchill Livingston Inc., New York.
Kuhn, T., 1962. The structure of scientific revolutions. University of Chicago Press, Chicago, Illinois.
Lee, U., Ku, S., Noh, G., Baek, S., Choi, B., Mashour, G.A., 2013. Disruption of frontal-parietal communication by ketamine, propofol, and sevoflurane. Anesthesiology 118, 1264-1275.
Logothetis, N.K., 2002. The neural basis of the blood-oxygen-level-dependent functional magnetic resonance imaging signal. Philos Trans R Soc Lond B Biol Sci 357, 1003-1037.
Logothetis, N.K., 2007. The ins and outs of fMRI signals. Nat Neurosci 10, 1230-1232.
Logothetis, N.K., 2008. What we can do and what we cannot do with fMRI. Nature 453, 869-878.
Logothetis, N.K., Wandell, B.A., 2004. Interpreting the BOLD signal. Annu Rev Physiol 66, 735-769.
Lopez-Munoz, F., Boya, J., Alamo, C., 2006. Neuron theory, the cornerstone of neuroscience, on the centenary of the Nobel Prize award to Santiago Ramon y Cajal. Brain Res Bull 70, 391-405.
Lowe, M.J., Sakaie, K.E., Beall, E.B., Calhoun, V.D., Bridwell, D.A., Rubinov, M., Rao, S.M., 2016. Modern Methods for Interrogating the Human Connectome. J Int Neuropsychol Soc 22, 105-119.
Lukito, S., Norman, L., Carlisi, C., Radua, J., Hart, H., Simonoff, E., Rubia, K., 2020. Comparative meta-analyses of brain structural and functional abnormalities during cognitive control in attention-deficit/hyperactivity disorder and autism spectrum disorder. Psychol Med 50, 894-919.
Mannino, M., Bressler, S.L., 2015. Foundational perspectives on causality in large-scale brain networks. Phys Life Rev 15, 107-123.
Mesulam, M.M., 1998. From sensation to cognition. Brain 121, 1013-1052.
Meunier, D., Lambiotte, R., Fornito, A., Ersche, K.D., Bullmore, E.T., 2009. Hierarchical modularity in human brain functional networks. Front Neuroinformatics 3, 37.

Moon, J.-Y., Lee, U., Blain-Moraes, S., Mashour, G.A., 2015. General Relationship of Global Topology, Local Dynamics, and Directionality in Large-Scale Brain Networks. PLoS Comput Biol 11, e1004225.
Muzik, O., Baajour, S., Bressler, S., Diwadkar, V.A., 2020. Directional Interactions Between Constituents of the Human Large-Scale Thermoregulatory Network. Brain Topogr 33, 489-503.
Nicholls, N., 1980. Long-range weather forecasting: Value, status, and prospects. Reviews of Geophysics 18, 771-788.
Nicolelis, M.A., Lebedev, M.A., 2009. Principles of neural ensemble physiology underlying the operation of brain-machine interfaces. Nat Rev Neurosci 10, 530-540.
Niessing, J., Ebisch, B., Schmidt, K.E., Niessing, M., Singer, W., Galuske, R.A., 2005. Hemodynamic signals correlate tightly with synchronized gamma oscillations. Science 309, 948-951.
Park, H.J., Friston, K., 2013. Structural and functional brain networks: from connections to cognition. Science 342, 1238411.
Passingham, R.E., Rowe, J.B., Sakai, K., 2013. Has brain imaging discovered anything new about how the brain works? Neuroimage 66, 142-150.
Paus, T., 2001. Primate anterior cingulate cortex: where motor control, drive and cognition interface. Nat Rev Neurosci 2, 417-424.
Petersen, S.E., Sporns, O., 2015. Brain Networks and Cognitive Architectures. Neuron 88, 207-219.
Petronis, A., Paterson, A.D., Kennedy, J.L., 1999. Schizophrenia: an epigenetic puzzle? Schizophr Bull 25, 639-655.
Phillips, M.L., Drevets, W.C., Rauch, S.L., Lane, R., 2003. Neurobiology of emotion perception II: Implications for major psychiatric disorders. Biol Psychiatry 54, 515-528.
Price, C.J., Friston, K.J., 2005. Functional ontologies for cognition: The systematic definition of structure and function. Cogn Neuropsychol 22, 262-275.
Roalf, D.R., Gur, R.C., 2017. Functional brain imaging in neuropsychology over the past 25 years. Neuropsychology 31, 954-971.
Schizophrenia Working Group of the Psychiatric Genomics, C., 2014. Biological insights from 108 schizophrenia-associated genetic loci. Nature 511, 421-427.
Silverstein, B., Bressler, S., Diwadkar, V.A., 2016. Inferring the dysconnection syndrome in schizophrenia: Interpretational considerations on methods for the network analyses of fMRI data. Front Psychiatry 7, 132.
Singer, W., 1994. Coherence as an organizing principle of cortical functions. Int Rev Neurobiol 37, 153-183; discussion 203-157.
Singh, K.D., 2012. Which "neural activity" do you mean? fMRI, MEG, oscillations and neurotransmitters. Neuroimage 62, 1121-1130.
Stanley, M.L., Moussa, M.N., Paolini, B.M., Lyday, R.G., Burdette, J.H., Laurienti, P.J., 2013. Defining nodes in complex brain networks. Front Comput Neurosci 7, 169.
Stephan, K.E., 2004. On the role of general system theory for functional neuroimaging. J. Anat. 205, 443-470.
Stephan, K.E., Binder, E.B., Breakspear, M., Dayan, P., Johnstone, E.C., Meyer-Lindenberg, A., Schnyder, U., Wang, X.J., Bach, D.R., Fletcher, P.C., Flint, J., Frank, M.J., Heinz, A., Huys, Q.J., Montague, P.R., Owen, M.J., Friston, K.J., 2016. Charting the landscape of priority problems in psychiatry, part 2: pathogenesis and aetiology. Lancet Psychiatry 3, 84-90.
Stephan, K.E., Friston, K.J., Frith, C.D., 2009. Dysconnection in schizophrenia: from abnormal synaptic plasticity to failures of self-monitoring. Schizophr Bull 35, 509-527.
Stephan, K.E., Harrison, L.M., Kiebel, S.J., David, O., Penny, W.D., Friston, K.J., 2007. Dynamic causal models of neural system dynamics: current state and future extensions. J Biosci 32, 129-144.
Stephan, K.E., Roebroeck, A., 2012. A short history of causal modeling of fMRI data. Neuroimage 62, 856-863.
Sugihara, G., May, R., Ye, H., Hsieh, C.H., Deyle, E., Fogarty, M., Munch, S., 2012. Detecting causality in complex ecosystems. Science 338, 496-500.

Sweatt, J.D., 2013. The emerging field of neuroepigenetics. Neuron 80, 624-632.
Telesford, Q.K., Simpson, S.L., Burdette, J.H., Hayasaka, S., Laurienti, P.J., 2011. The brain as a complex system: using network science as a tool for understanding the brain. Brain Connect 1, 295-308.
van den Heuvel, M.P., Hulshoff Pol, H.E., 2010. Exploring the brain network: a review on resting-state fMRI functional connectivity. Eur Neuropsychopharmacol 20, 519-534.
Vogt, B.A., 2016. Midcingulate cortex: Structure, connections, homologies, functions and diseases. J Chem Neuroanat 74, 28-46.
Von Bertalanffy, K., 1968. General systems theory. Brazilier, New York.
Walter, H., 2013. The third wave of biological psychiatry. Front Psychol 4, 582.
Wang, H., Zhu, R., Dai, Z., Tian, S., Shao, J., Wang, X., Sun, Y., Chen, Z., Li, X., Yao, Z., Lu, Q., 2020. Aberrant functional connectivity and graph properties in bipolar II disorder with suicide attempts. J Affect Disord 275, 202-209.
WHO, 2016. International Classification of Diseases and Related Health Problems. WHO Press, Geneva, Switzerland.
Wilson, R.A., Barker, M.J., Brigandt, I., 2007. When traditional essentialism fails: biological natural kinds. Philosophical Topics 35, 189-215.
Yamin, M.A., Dayan, M., Squarcina, L., Brambilla, P., Murino, V., Diwadkar, V., Sona, D., 2019. Investigating the Impact of Genetic Background on Brain Dynamic Functional Connectivity Through Machine Learning: A Twins Study. 2019 Ieee Embs International Conference on Biomedical & Health Informatics (Bhi), pp. 1-4.
Ye, H., Deyle, E.R., Gilarranz, L.J., Sugihara, G., 2015. Distinguishing time-delayed causal interactions using convergent cross mapping. Sci Rep 5, 14750.
Zhu, J., Zhang, S., Cai, H., Wang, C., Yu, Y., 2020. Common and distinct functional stability abnormalities across three major psychiatric disorders. Neuroimage Clin 27, 102352.

Part I
Methods

fMRI: Blood Oxygen Level-Dependent Contrast and Its Value for Understanding Functional Brain Networks

Peter A. Bandettini

Introduction

In the spring through fall of 1991, three labs performed the first successful fMRI experiments. Inspired by the pioneering work of Seiji Ogawa, who coined the term Blood Oxygenation Level-Dependent Contrast in his seminal paper (Ogawa et al. 1990), describing the effect of hemoglobin saturation alteration on the susceptibility-weighted MRI signal, these labs forged ahead to determine if, with brain activation, blood oxygenation locally changed enough to cause an observable MRI signal change. Visual stimulation or finger tapping was performed. The results were better than expected: the time course of sequential susceptibility-weighted (T2*-weighted) MR images showed an increased signal in the appropriate areas, beginning about 2 s after activation and plateauing after about 9 s, then, on activation cessation, falling back to baseline after a slightly longer time. The field of BOLD-based functional MRI was born.

The birth of fMRI came about a year prior to the discovery of activation-induced BOLD changes. Jack Belliveau, from the Massachusetts General Hospital, had demonstrated that if two sequential bolus injections of gadolinium (Gd), a blood volume indicating contrast agent, were performed, one prior to brain activation and one during, localized blood volume changes with brain activation could be seen. This result was published in the November issue of Science (Belliveau et al. 1991) and has become the most iconic cover page in brain mapping history – that of a surface rendered head with the back sliced off at an angle, revealing the visual cortex with the changes in blood volume with visual stimulation color-coded in shades of red. This was indeed inspiring; however, with the inception of BOLD contrast-based fMRI, a noninvasive and generally much easier approach than using

P. A. Bandettini (✉)
Section on Functional Imaging Methods, National Institute of Mental Health, Bethesda, MD, USA
e-mail: bandettini@nih.gov

V. A. Diwadkar, S. B. Eickhoff (eds.), *Brain Network Dysfunction in Neuropsychiatric Illness*, https://doi.org/10.1007/978-3-030-59797-9_2

exogenous contrast, this innovative yet invasive and more cumbersome technique was obsolete – at least as a brain activation mapping tool – before the ink of that Science paper even dried. Dynamic Gd-enhanced contrast has proven to be an extremely useful clinical tool for characterizing vascular patency in the context of stroke and other brain lesions (Rosen et al. 1990).

In the summer of 1992, the first three BOLD contrast papers were published (Kwong et al. 1992; Ogawa et al. 1992; Bandettini et al. 1992), gaining the attention and imagination of neuroscientists, clinicians, and physiologists worldwide, as it was the first brain mapping method that allowed relatively fast, noninvasive, sensitive, accurate, whole-brain mapping of human brain activation. The leading methods at the time – electroencephalography (EEG), magnetoencephalography (MEG), and positron emission tomography (PET) – suffered from either low resolution in the case of MEG, EEG, and PET; or spatial certainty because of the inverse problem in the case of MEG and EEG; as well as tissue conductivity variability in the case of EEG; and invasiveness because of the need to inject radiolabeled tracers in the case of PET. Cognitive neuroscientists, in particular, embraced fMRI as it offered to effectively elucidate the neural correlates at a systems level of a wide range of behaviors and tasks that they had been studying.

Functional MRI had another unique advantage: About a decade earlier, MRI emerged as the most effective technique for assessment of soft tissue pathology, and therefore, MRI gained rapid clinical acceptance, catalyzing a rapid dissemination of MRI scanners worldwide. By about 1996, most MRI scanners were equipped with the rapid imaging pulse sequence, Echo-Planar Imaging (EPI), that was most effective for functional MRI because of the temporal stability that single-shot acquisition afforded (Stehling et al. 1991). After about 1996, fMRI could spread and flourish rapidly as MRI scanners equipped with high-performance gradients and the EPI pulse sequence were already in place in almost every major hospital in the world.

Functional MRI grew in use and sophistication. Imaging hardware and acquisition methods advanced, allowing for improvements in image signal to noise ratio (SNR), resolution, speed of acquisition, quality, and time series stability. Processing methodology and our understanding of the basis of the contrast grew, increasing confidence in the interpretation of fMRI signal changes as being reflective of underlying neuronal activity. The applications moved beyond visual and motor cortex mapping to those of cognition and emotion. Functional MRI was used by more diverse and interdisciplinary groups to address more nuanced and challenging questions about functional brain organization and how it varied in health and disease. Within the first decade after the initial publications, there were over 2000 papers published per year using fMRI. Currently, there are over 4500 papers published per year.

In 1995, a breakthrough in fMRI occurred. Bharat Biswal, a then graduate student at The Medical College of Wisconsin, realized that if he used the "resting state" time series signal as a reference function or regressor from a voxel in the motor cortex of a subject performing no task at all, the rest of the primary motor cortex as well as the supplementary motor cortex and primary motor cortex from the other hemisphere showed significant temporal correlations (Biswal et al. 1995). The time series signal itself appeared – to the eye – to be a noisy and random but the correlations revealed that even during "rest" the brain was not only spontaneously activated but that these activations were correlated between functionally "connected" regions. The predominant frequency range at which these correlations occur is about 0.05–0.15 Hz. This range, determined primarily by the sluggishness of the hemodynamic response, indicates that there is enough underlying spontaneous neuronal fluctuation power at these low frequencies to effectively drive the hemodynamic response. The discovery of functionally specific resting-state correlations in the data was effectively ignored in the literature for almost a decade, likely because, in part, the field was enjoying immense success with task activation, the processing methods were not optimized for this type of approach, and many simply dismissed this observation as an artifact.

Research in resting-state fMRI picked up by about 2005 as it was being further validated. New findings using resting-state correlations were demonstrated and the field accelerated as more took interest. Studies using simultaneous electrophysiologic measures with fMRI demonstrated a direct relationship between the fluctuations and spontaneous neuronal activity (Shmuel and Leopold 2008; Laufs et al. 2003). Other early studies using resting-state and task-based fMRI uncovered the "default mode" network (Greicius et al. 2003; Lustig et al. 2003) which was previously shown in an earlier PET study that showed increased metabolism in this network during rest and decreased metabolism during cognitive tasks (Raichle et al. 2001). Methods for understanding fMRI time series correlation during "resting state" progressed. Seed voxel analysis, which was the first approach that led to the discovery of resting-state fMRI, is useful for hypothesis-driven probing of specific networks (Greicius et al. 2003), but is relatively inefficient for revealing networks throughout the brain. Early use of Independent Component Analysis (ICA) on resting-state data revealed several more networks that appeared to coincide with the known functional organization (Beckmann et al. 2005). Subsequently, larger number of networks were found and defined. So far, a multitude of networks at different spatial scales and delineation granularities have been revealed by ICA and other approaches. At the moment, it is not certain how deeply one can probe – when the delineations become meaningless. Networks were also shown to be disrupted in a wide range of pathologies. Ultimately, resting-state fMRI maps of whole-brain functional organization, including detailed retinotopy (Arcaro et al. 2015), have emerged in the literature. Figure 1 shows an example of a whole-brain resting-state correlation map obtained using the average of seven subjects, multi-echo ICA denoizing (Kundu et al. 2012), and hierarchical clustering. It was exciting to see that these maps coincided closely with known functional architecture obtained years prior by invasive methods and by standard fMRI approaches – lending further validity to the use of resting-state fMRI for exploring whole-brain network organization in health and disease.

As the field of resting-state fMRI grew, a new term "functional connectivity MRI" (fcMRI), was embraced, as time series correlation during resting state implies functional connectivity. While there are many other reasons why time series may show high correlation or correlation changes, this nomenclature caught on.

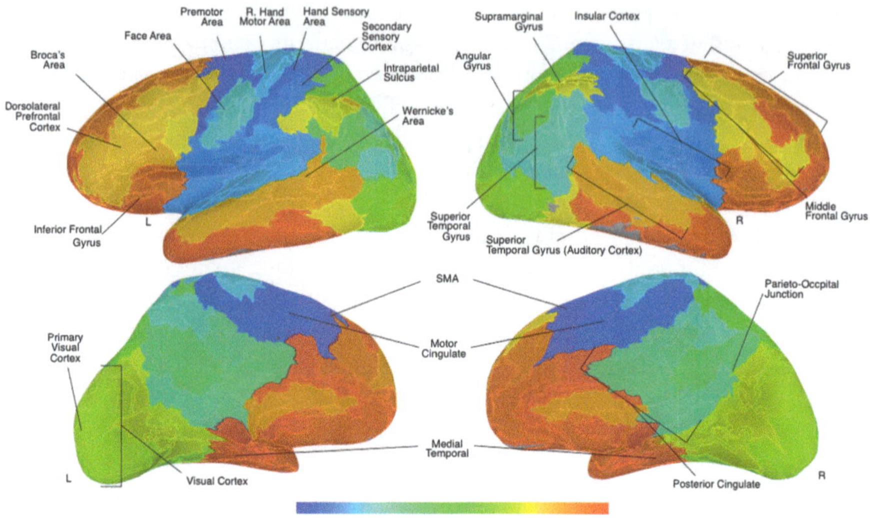

Fig. 1 Whole-brain resting-state connectivity map created from the average of seven subjects. Multi-echo EPI time series were first denoised using ME-ICA. Then hierarchical clustering was performed on the ICA maps produced. The clusters are labeled according to known brain anatomy. (Courtesy Dr. Prantik Kundu)

This chapter first discusses the basics of BOLD contrast with an eye on what the limits of interpretation of fcMRI are. BOLD contrast is fundamentally based on neurovascular coupling to poorly understood "spontaneous" and seemingly random, yet network-specific – temporally correlated neuronal activity. As fcMRI is based on BOLD contrast, it also suffers from all the issues that standard fMRI continues to struggle with – inordinate weighting of the signal to regions of high venous vascularity, motion, breathing, and susceptibility-induced signal dropout, among others. The depth and accuracy by which the interpretation of network structure and activity from BOLD contrast is directly related to the degree to which these potential confounds are appreciated, minimized, and removed.

The development of fcMRI processing methodology will then be briefly discussed. Here, a brief overview of several prominent processing methods as well as applications and findings will be given. Here also, the value of BOLD contrast for understanding networks will be assessed in the context of describing the current advantages and shortcomings relative to other techniques.

Lastly, the chapter includes an overview of some of the most promising current directions that the field is taking as massive datasets are being analyzed, high-resolution and high-field fMRI is growing, and differentiation of individual information or creation of fcMRI-based "biomarkers" are being attempted.

BOLD Contrast

BOLD contrast is based on the tight relationship between neuronal activity and local hemodynamic changes. At a baseline neuronal state, blood oxygenation in capillaries and veins is lower than that of arteries due to the extraction of oxygen from the blood to support neuronal metabolism. Deoxyhemoglobin (deoxy-HB) is paramagnetic relative to the rest of brain tissue and water and oxyhemoglobin (oxy-Hb) has the same susceptibility as brain tissue and water. A compartment (in this case, a deoxy-Hb molecule or a capillary or vein containing deoxy-Hb molecules) that has a different susceptibility than its surrounding medium creates microscopic magnetic field distortions when in the presence of a magnetic field – these distortions are proportional both to the oxygenation of the blood and to the main magnetic field – in this case the scanner field typically ranging from 3T to 7T. Water molecules, the primary signal source in MRI and referred to as "spins" in the context of MRI, precess at a frequency that is directly proportional to the magnetic field that they are experiencing. Within a voxel, if spins are precessing at different frequencies, such as in the vicinity of deoxygenated blood, they rapidly become out of phase. The MRI signal is directly proportional to the coherence of spins. If they are completely out of phase, destructive addition takes place and there is no signal. If they are completely in phase, there is maximal signal. During rest, enough spins are out of phase, due to the many microscopic field distortions around deoxy-Hb compartments in each voxel, to cause the MRI signal to be attenuated. This phenomenon is the basis of susceptibility-weighted imaging (SWI) that highlights venous vasculature (Haacke and Ye 2012; Menon et al. 1993). During activation, blood flow increases locally such that there is an apparent overabundance of oxygenated blood delivered to the active regions. The precise reason for this "overshoot" in oxygenation during activation is still not fully understood; however, it is likely due to the diffusion limitation of oxygen delivery from blood vessels to active neurons furthest away from the blood vessels, requiring an overabundance of oxygen at the source to adequately supply the furthest active neurons. Whether this neurovascular coupling is a feed-forward or feedback mechanism – or both – is still not fully understood. This activation-induced increase in flow and oxygenation is associated with a decrease in the amount of deoxy-Hb and therefore, a decrease in the dephasing effect of the induced microscopic magnetic field distortions, thus increasing the coherence of spins within each voxel, leading to a signal increase of a few percent – which is detectable with MRI (Bandettini and Wong 1995; Ogawa et al. 1993).

While the shape and sign of BOLD signal changes are remarkably consistent across regions and individuals, BOLD has some clear limits. BOLD signal begins to increase approximately 2 s after neuronal activity begins, and plateaus in the "on" state after about 7–10 s. A pre-undershoot is sometimes observed (Hu et al. 1997), and a post-undershoot is more commonly observed (Yacoub et al. 2006). These undershoots are likely due to transient mismatches between flow increases, which decrease the amount of deoxy-Hb and therefore increase the signal, and either blood volume or cerebral metabolic rate ($CMRO_2$) changes, which increase the amount of deoxy-Hb and therefore decrease the signal (Buxton et al. 2004; Zong et al. 2012). The dynamics, location, and magnitude of the BOLD signal change are influenced by the vasculature in each voxel. If voxels happen to capture large vessels, the magnitude of the signal may be large – up to an order of magnitude larger than with capillary effects, the timing a bit more delayed than average – up to 2 s delayed from capillary effects, and the location of the signal somewhat distal – up to a centimeter or more from the true region of activation. While improvements are being made in fMRI methodology such that the effects of this variability are kept to a minimum, the problem of variable vasculature and hemodynamic coupling in fMRI remains at all field strengths and subsequently limits the accuracy with which fMRI can be used as a tool for understanding neuronal activity.

Relative to other fMRI-based functional contrasts such as blood perfusion (Wong et al. 1997) and volume using vascular space occupancy (VASO) (Lu et al. 2003), BOLD contrast is more sensitive, robust, and significantly easier to obtain. It only requires the time series acquisition of T2*-weighted signal and no additional inversion pulses or waiting times as perfusion and volume sensitive techniques do. T2*-weighting is obtained by setting the echo time (TE) approximately equal to the T2* of the tissue in a gradient-echo EPI pulse sequence. At 3 T, the TE is typically 20–40 ms. At 7 T, as T2* becomes shorter, the TE used is about 15–30 ms.

The repetition time (TR) is the rate at which sequential images are obtained. To allow time for the collection of an entire volume of data – usually 15–40 slices, depending on slice thickness and spacing, typical TR values are in the range of 2–3 s. With recent advancements in acquisition approaches such as SENSE (Pruessmann et al. 1999) reconstruction and multi-band excitation (Feinberg et al. 2010), the TR for an entire volume can be practically reduced to about 400 ms. With a shorter TR, the MRI signal, and thus MRI signal to noise ratio (SNR), decreases, as there is less time to allow longitudinal magnetization to recover; however, this decrease in SNR is offset by the increased number of temporal data points, allowing for better characterization and removal of noise components in the signal. Because the hemodynamic response is slow, the increased density of points – beyond about one point per 3–5 s, does not increase the effective degrees of freedom in statistical analysis of the time series. One can argue, however, that when considering the correlation of time series to a regressor, the more accurate description of the hemodynamic waveform shape, afforded by higher sampling rate, allows for increased precision. Figure 2 shows a summary of the BOLD contrast hemodynamic function.

Even though the hemodynamic response is slow and delayed, it can be reliably elicited by very brief events – as short as 20 ms. Additionally, when the brain is driven by a stimulus such as a flashing checkerboard for visual cortex, a hemodynamic response is detectable at an on–off switching rate as high as 0.5 Hz. Detection at this rapid rate brings to focus that the BOLD response is highly nonlinear in many contexts. Very brief stimuli elicit a larger than expected BOLD response (Birn et al. 2001; Birn and Bandettini 2005), as do very weak stimuli (Logothetis et al. 2001). The reasons for this nonlinear coupling are not fully understood.

The spatial resolution of BOLD-based fMRI is determined by the signal to noise available, the vascular weighting, and ultimately by the precision of the

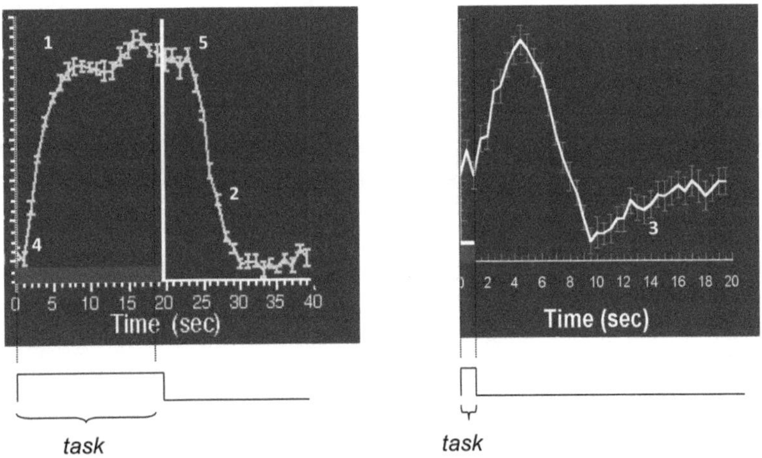

Fig. 2 The hemodynamic response function with blocked design (left) and event-related (right) activation. 1. Time to reach a plateau in the "on" state is about 10 s. 2. Time to fall back to baseline after cessation of activation is about 10–11 s. 3. A post undershoot is commonly observed, lasting up to 40 s following cessation of activation. It has shown approximate proportionality to the magnitude of activation. The hypothesize mechanisms following activation are either a strong decrease in flow, a longer perseveration in venous blood volume, or a longer perseveration in elevated cerebral metabolic rate. 4. Time to start increasing is about 2 s and a pre-undershoot has been seen although is extremely rare. 5. The magnitude of the signal change at TE = T2* of tissue is proportional to field strength and is typically 3–5% for 3T and 4–8% for 7T

hemodynamic response function. Typical resolutions for whole-brain studies range from 2.0 mm to 3.0 mm in plane, with comparable or slightly thicker slices. High-resolution studies, using either single-shot SENSE or multi-band or multi-shot, and 3D sequences at greater than 3T field strength, have obtained in-plane voxel sizes of about 0.5 mm. At higher resolutions than these, the SNR is just too low to obtain functional information in a reasonable time frame. Serendipitously, the practical limits of fMRI acquisition resolution coincide with the size of the smallest neuronal structures (cortical columns and layers) known to elicit a specific hemodynamic response. Venous vascular structures also play a significant role in the functional resolution of BOLD contrast. Larger vessels, up to 1–10 mm distal from the area of activation may show significant signal changes. Novel pulse sequences and processing approaches have been attempted to alleviate this problem. Spin-echo acquisition, flow nulling that removes large vessel intravascular effects, and blood volume sensitive techniques have been used, with varying success – all coming with a cost in sensitivity. Some of the most promising results so far have been with blood volume-sensitive sequences (Huber et al. 2017, 2018) as it appears that the activation-induced blood volume response is much more localized to activity than the oxygenation change, which may propagate "downstream."

BOLD contrast magnitude, shape, and latency, while generally consistent, have shown variation between individuals, regions, and voxels (Handwerker et al. 2004); however, within a voxel, the BOLD response is consistent enough to allow sensitive mapping of neuronal latency modulation between tasks (Bellgowan et al. 2003). The variables that determine the spread in latencies across subjects, regions, and voxels include hematocrit, baseline venous blood volume, vascular patency, blood pressure, and cerebral metabolic rate changes. In spite of these limitations, BOLD contrast has demonstrated, in a wide range of populations, to be exceedingly robust and consistent, allowing it to be used widely as an effective tool for understanding brain physiology and functional organization in normal and clinical populations. Simultaneous invasive and noninvasive electrophysiologic measures have effectively confirmed that BOLD increases as well as decreases follow increases and decreases in neuronal activity, respectively, and more specifically, scale with the local field potential power and, to a lesser degree, spiking rate (Logothetis et al. 2001).

Advancement

The use of BOLD contrast toward understanding functional networks has been driven by advances in imaging hardware and pulse sequences as well as, perhaps most importantly, by advances in processing methods. While the details of the advances are beyond the scope of the chapter, a brief overview of both the technology and processing advances is provided.

Functional MRI has continued its push toward higher field strength because of the higher signal and functional contrast to noise that it offers. BOLD signal changes and SNR both increase with field strength, allowing for high-resolution studies the resolution where physiologic noise and thermal noise are of similar magnitude, and where the SNR at these high resolutions is high enough (above about 40/1) to collect useful data in a reasonable amount of time. If high field imaging is carried out at low resolution, then the intrinsic gains in SNR are lost as physiologic noise sets an upper time series limit of about 150/1, regardless of the image SNR. Currently, 7T has gained FDA clearance thus the market should be offering 7T at lower costs in the near future. Figure 3 shows a map, as of 2018, of all the human 7T and higher field scanners worldwide.

The number and size of RF coil array elements have, respectively, increased and decreased over the years, resulting in increased sensitivity and better ability to perform coil-sensitivity-assisted spatial encoding using SENSE acquisition (Pruessmann et al. 1999) – thus higher resolution. Currently, 64 elements are the state of the art although coils with as many as 96 elements have been used (Wald 2012).

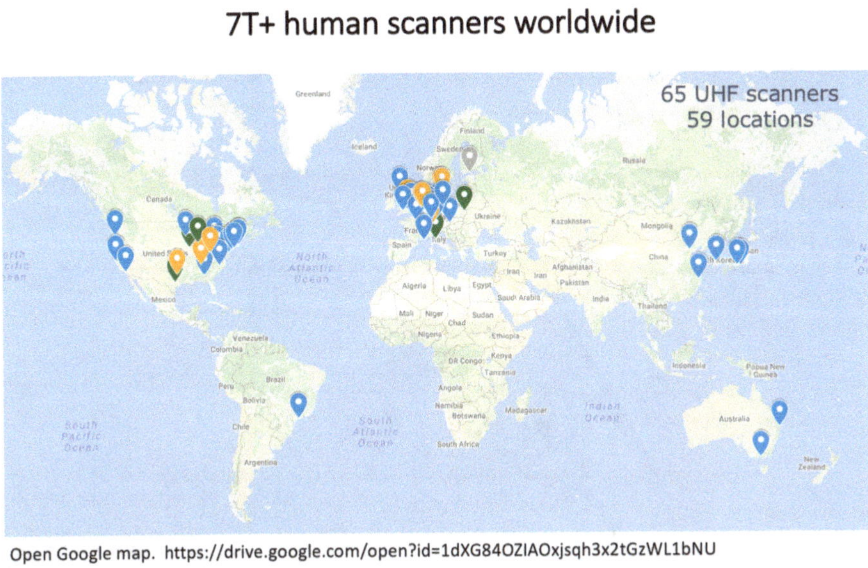

Fig. 3 A map of the 7T and higher field strength human scanners worldwide as of 2018. Blue = Siemens, Orange = Philips, Green = General Electric. (This map is from https://drive.google.com/open?id=1dXG84OZIAOxjsqh3x2tGzWL1bNU and is continually updated)

Processing

Noise and spatial variability of the hemodynamic response remain, after almost 30 years, the central challenge in both task-based and fcMRI. As mentioned, the upper limit on temporal SNR is about 100 to 150 to 1 (Bodurka et al. 2007). This limit is determined primarily by physiologic noise (Triantafyllou et al. 2005). Noise sources include cardiac pulsation, respiration, motion, scanner instabilities, and thermal noise. If physiologic noise could be completely modeled and eliminated from the fMRI time series, then the temporal SNR would be simply a function of intrinsic sensitivity of MRI, determined by field strength and coil sensitivity, allowing up to an order of magnitude higher temporal SNR values. Figure 4 shows a plot of the spatial image signal to noise vs. the temporal signal to noise at 3T from different tissues. The upper limit in temporal signal to noise, due to physiologic fluctuations, is about 100 to 150 to 1. If physiologic noise were removed, the upper temporal signal to noise would only be limited by the image signal to noise.

Regarding cardiac pulsation, with every heartbeat, the base of the brain moves, and blood and cerebral spinal fluid accelerate, causing inflow-related signal enhancement. With respiration, the effects are more complex and varied. With every breath, the chest cavity expands, causing Bo distortion variability that can include the brain – causing breathing-related image distortion and apparent motion where no head motion is present. With variability in breathing rate, the end-tidal CO_2 levels related to the baseline level of CO_2 varies in the blood. As CO_2 is a vasodilator,

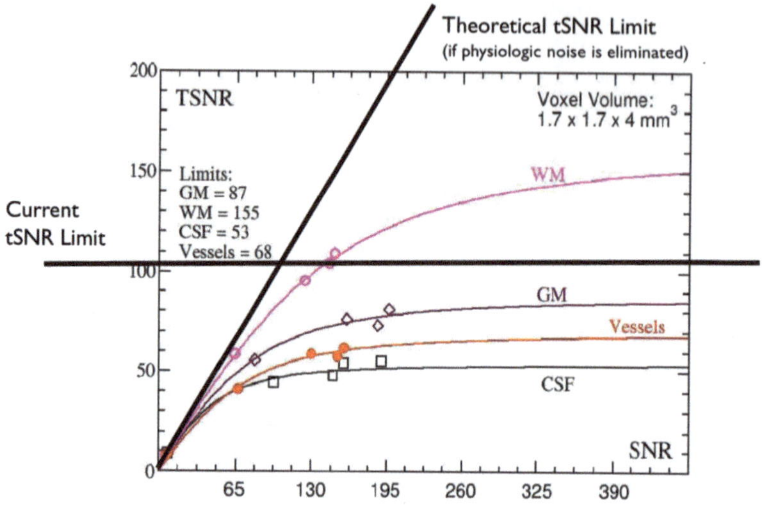

Fig. 4 Plot of temporal vs image (or spatial) signal to noise ratio in white matter, gray matter, and CSF. Because of physiologic noise, the upper limit in temporal signal to noise in gray matter is about 100–150/1. This is depicted by the horizontal line. If physiologic noise could be completely eliminated, the temporal signal to noise ratio would only be limited by the image (or spatial) signal to noise ratio, depicted by the line with slope 1

any variability in CO_2 causes a change in blood flow, leading to BOLD contrast changes that match the timing of breathing depth variations (Birn et al. 2008). Motion causes signal changes particularly in voxels that are adjacent to sharp signal intensity gradients. It also can lead to what is known as "spin-history" effects. If the head moves in the direction that is perpendicular to the imaging plane, then some spins will not experience the next imaging RF pulse, allowing the longitudinal relaxation to continue. If the head then moves back into the plane, the spins that were previously out of plane would have missed an RF pulse and would have had time to experience more T1-relaxation to equilibrium and would resultingly show a higher signal intensity until, several RF pulses later, the magnetization again reaches equilibrium. Regarding scanner instabilities, there are potentially many sources. A goal is not necessarily to eliminate all scanner or image reconstruction imperfections, as this is extremely difficult and not entirely necessary, but to at least keep them constant during the time series. Heating of the gradient amplifiers and other scanner components can cause a change in these instabilities, leading to signal drift. Finally, thermal noise is simply the noise intrinsic to the RF coil and sets the upper limit in SNR of each image.

Over the years much work has gone into eliminating or reducing noise, and specifically, physiologic noise. Most methods generally include attempts to model the noise and then to regress it out in various manners. This approach has met with limited success as these noise sources are challenging to model accurately since they vary over time and across voxels and have nonlinear characteristics. Other methods that are in regular use include image registration, signal intensity normalization, global signal regression (Fox et al. 2005), and multi-echo acquisition to separate pure susceptibility-related fluctuations from nonsusceptibility-related fluctuations (Kundu et al. 2013; Gonzalez-Castillo et al. 2016). Again, all of these approaches can reduce physiologic noise, but they all have drawbacks and trade-offs.

In practical implementation, the typical amount of time used for the collection of resting-state data for fcMRI is about 10–30 min. When pressed for time to fit in multiple different types of scans in a session, one or two 5-min resting-state time series are typically collected. This time period is at the lower limit of what is necessary to create reasonably stable data; however, studies that focus more on resting state during the scan session typically collect up to 30 min, divided into about six 5-min runs. Breaking the runs up into 5-min segments is essential to allow the subject to relax and regroup.

Noise problems aside, processing approaches have been developed toward the goal of extracting meaningful network information from BOLD contrast time series obtained either during rest, continuous tasks, or with blocked or event-related activation paradigms. Mentioned earlier was Independent Component Analysis (ICA) for revealing spatially independent networks in fMRI time series. In the early days of fcMRI, the primary method for determining functional connectivity across regions was to extract regressors from "seed voxels" in specific regions. Any other areas that correlated with these regions were determined to be "connected" as part of a network. While this method is still used today in many contexts, it is relatively inefficient as it allows exploration of only small parts of the brain at any one time.

ICA helped to solve this inefficiency problem by allowing determination of the spatially independent components that had unique temporal signatures. Use of ICA allowed a multitude of "networks" to be revealed in one step (Damoiseaux et al. 2006). There are issues related to this approach, however. The first is in deciding which components are "real" and which are not, as the ordering of the independent components is quasi-random. Interpretation is based on what appears to make sense – or based on spatial template matching. For instance, ICA may reveal a network that consists predominantly of areas that are along sharp spatial intensity gradients (i.e., near clear edges), which would then be discarded as a likely motion artifact. Progress has been made in automating this procedure and expanding it to effective motion mitigation, but there are still issues of how many ICA bins to consider. Another problem is the constraint that the distribution of these components must be spatially independent. Real brain networks are likely not organized in a spatially independent manner. In addition, network node groupings change over time (Hutchison et al. 2013; Gonzalez-Castillo et al. 2014), likely forming multiple layers of subnetworks which are also not spatially independent. Figure 5 illustrates the temporal variability of connections between nodes distributed throughout the brain showing that cross hemisphere correlations are among the most stable and within hemisphere correlations are the least stable. It also shows that network stability plateaus after about 8–10 min; however, if a snapshot is taken at any period during this time, the network configuration can appear to have a spatially different pattern than the average.

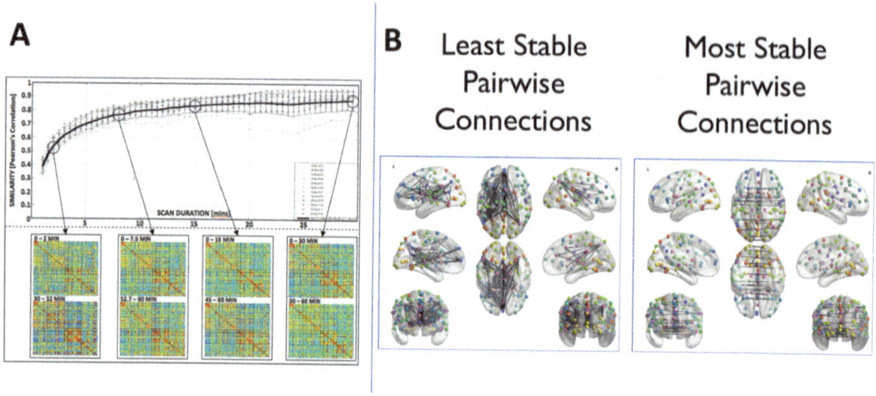

Fig. 5 Depiction of the variability of connectivity patterns over time and space. (**a**) This is a plot of correlation matrix similarity between the first and last 2 min (far left point) to the first and last 30 min (far right point). At about 8 min, the plot stabilizes; however, the matrices at 2 min show different patterns, illustrative of the constant variability in correlation structure over time. (**b**) These figures highlight the connections that are least (left) and most stable (right) over time, revealing that cross-hemisphere connectivity is generally more stable than within hemisphere connectivity

Group ICA was a next step that allowed performing ICA on many resting-state datasets. The problem was that with each dataset – even within the same subject – the ICA components produced were ordered differently and sometimes varied spatially. Group ICA allowed one to extract the most salient independent components from a large multi-session, multi-subject dataset, providing the means to leverage the increased statistical power of group averaging (Calhoun et al. 2009).

Temporal PCA emerged several years ago as a means to work around the spatial independence constraint on networks by performing PCA in the temporal rather domain. As an initial step, the brain was parceled into many functional units, and the temporal independence of each unit was determined (Boubela et al. 2013). This approached perhaps a more biologically feasible compression of the varying cascade of neuronal activity that activated clusters of segments in a varying way over time. The movies produced from this analysis dramatically demonstrated how the various nodes in the brain come online, work together, then change over time.

A commonly used approach for analyzing and displaying BOLD contrast temporal correlations is to first segment the brain into 50–500 parcels, and then to create a two-dimensional matrix showing the correlation coefficient of each parcel's time series with every other time series, as a row or column in that matrix. This method is a powerful way to capture, and subsequently analyze, in one display, whole-brain connectivity patterns. A central challenge related to this method is how to best segment the brain. The field has struggled with creating the optimal parcellation scheme (Eickhoff et al. 2018). In order to properly average or compare whole-brain network connectivity in this manner across subjects, it is critical to ensure that the parcels are capturing the identical regions across brains. The problem is that brains and functional units across brains vary considerably in relative size and shape. Any one parcellation scheme will invariably capture functional units better in some subjects than others, throwing into question any averages or comparisons that are made. Progress has been made toward the establishment of a universal parcellation and parcellation registration scheme (Glasser et al. 2016); however, this work is still in early stages.

The time series correlations between brain parcels – considered here to be nodes for network analysis – can be analyzed in several different manners. Either the entire time series can be used to obtain an average correlation over the entire time period or temporal windows can be considered separately. Such temporal window sampling can be step-wise or continuous and is generally called a "sliding window" approach. This approach captures network dynamics as they evolve over time, revealing a constantly changing network configuration (Hutchison et al. 2013). The significance and potential utility of these transient connectivity states are currently being investigated. One prominent example is a study that explored the "dwell time" of specific networks, observing clear differences between patients with schizophrenia and normal populations (Calhoun et al. 2014).

Evaluating connectivity across regions during a task (Gonzalez-Castillo et al. 2015) rather than simply using task regressor correlation is a powerful way of classifying brain activity that may be missed when using an over-simplified model function – which typically only evaluates when the brain is hypothetically "on" or "off." This approach has been demonstrated to be effective in classifying ongoing brain states during various continuous or ongoing tasks, where the use of a regressor would be ineffective. Figure 6 shows this approach in which a correlation matrix is calculated for each window, and later grouped into four bins. Figure 7 shows that this correlation-based grouping successfully classified ongoing task. Also shown in Fig. 7 is a correspondence of task classification success with subject task engagement measures such as percent correct responses, response time, and percent missing responses. The best performing subjects allowed ongoing classification of nearly 100%. These results suggest that this is potentially a sensitive method to assess ongoing task engagement or even general measures of attention. This result had two other curious findings. The first curiosity was that even with a window width of less than 30 s, classification was above 90% accurate. How could such a short window classify tasks where the fluctuations are at 0.1 Hz? The second curiosity was that the correlated network nodes where significantly more extensive than the regions selectively activated by the tasks – performed in separate standard blocked design runs. Regions that normally would not show significant activation demonstrated task-informative ongoing correlations. How could this be possible? Perhaps we have to rethink what we mean by activation. It may not simply be a shift upwards of the hemodynamic response but a brain-wide concert of unique shapes that never show a net deviation from baseline. Hints of this effect are suggested in another study by Gonzalez-Castillo et al. (2012) in which single subjects performed a simple visual-motor task – with 9 h of averaging (coming back on multiple days of course). From these data, it was apparent that the entire cortex and subcortical regions showed time-locked yet spatially structured fluctuations that did not resemble the canonical

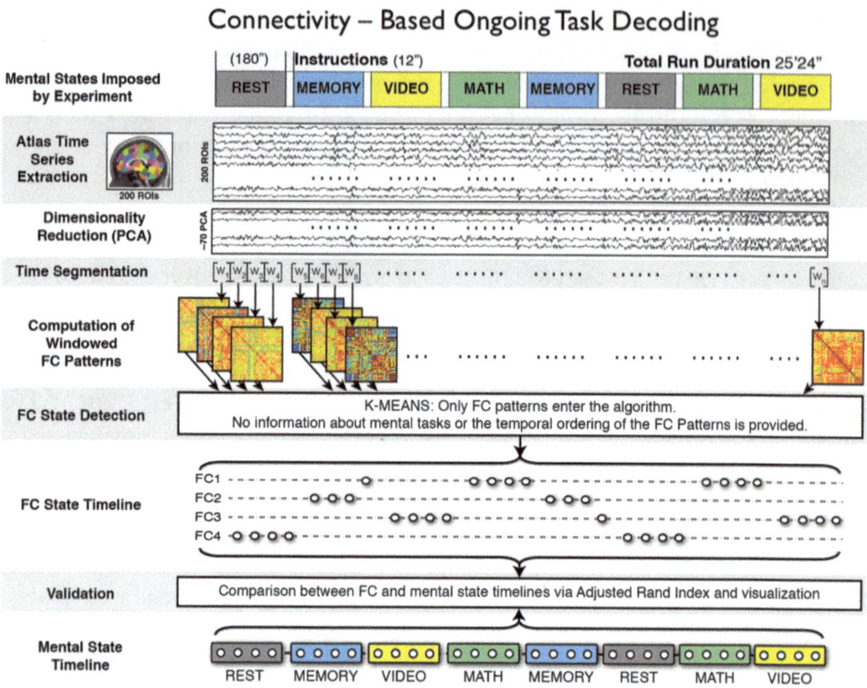

Fig. 6 Processing pipeline for decoding ongoing task from the connectivity matrix pattern across four different continuous task segments. Using this approach, nearly 100% accuracy was obtained with subjects that demonstrated highest engagement in the tasks

and expected waveforms in the primary regions seen for the task (visual and motor systems). Figure 8 shows, on the left, the nodes and connections that most strongly informed the ongoing task classification as discussed above – superimposed on the regions that showed activation with the tasks in a standard functional localizer scan. On the right are the results from the massive averaging study, showing all the regions that demonstrated time-locked activation, with 9 h of averaging, to a simple visual-motor task. These both may be revealing that time-locked fluctuations may be an aspect of brain function that has been missed by typical paradigms but that may appear with ongoing network connectivity assessment.

Other applications have been effective in tracking ongoing attention (Rosenberg et al. 2016) and reading comprehension (Jangraw et al. 2018). A subclass of this approach in which the correlations between pre-specified regions are compared is known as a psychophysiological interaction (PPI) approach (Friston et al. 1997).

Approaches for assessing whole-brain connectivity during rest and during task also include graph-theoretical approaches (Bullmore and Sporns 2009) where parcels or networks are represented as nodes and their connectivities are represented as edges. Connectivity strength is represented by the spatial proximity of the nodes. This approach provides a powerful way to visualize and analyze the data, as regions having close connectivity are clustered in this informational space.

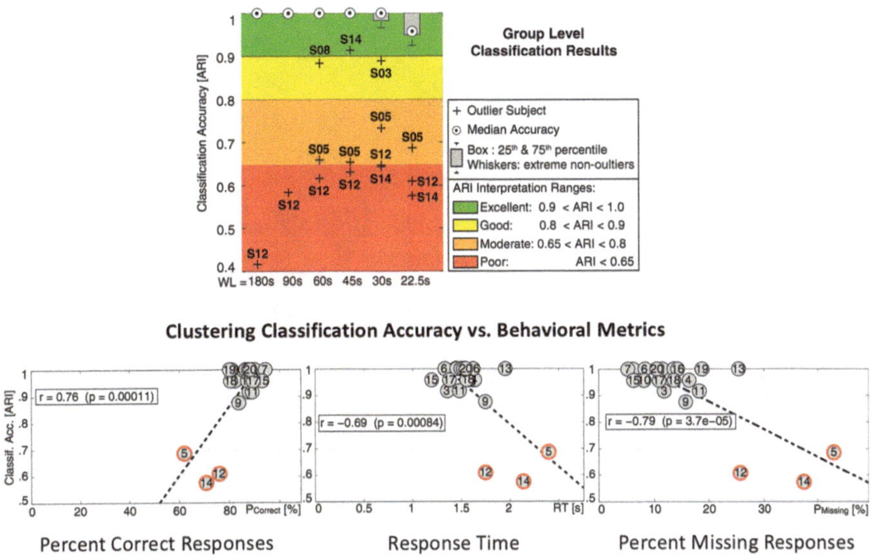

Fig. 7 Task classification results using time series correlation signatures from individual subjects as a function of window width and various measures of task engagement including percent correct responses, response time, and percent missing responses. Surprisingly, even window widths of 22.5 s were able to classify most tasks with above 90% accuracy

A promising recent approach to assessing dynamic connectivity signatures during various task states with a high degree of fidelity is with the use of Topological Data Analysis (Saggar et al. 2018), where such metrics as performance on task are able to be accurately predicted in individual subjects. This new approach maintains high dimensionality and is able to describe individual differences in how brain activity is represented and organized across different cognitive tasks from moment to moment. While such a method is useful mostly for studies attempting to assess what task as well as task engagement, one can imagine using it on resting-state data where state or task changes are tracked, perhaps comparing to a library of paths through this multidimensional space to identify transient thoughts or cognitive states.

Beyond the assessment of state or ongoing task during a time series, an emerging area of research has been of deriving information about an individual's traits from time series data. In a recent study (Finn et al. 2015), individual resting-state "fingerprints" were obtained from whole-brain resting-state data, implying that each subject has a unique resting-state correlation matrix signature. While a concern that these "signatures" might be simply related to differential parcellation of each subject's brain, their study showed that these signatures were relatively insensitive to the parcellation scheme, and also showed a correspondence with an independent measure of fluid intelligence in each subject.

One other paradigm and analysis approach that is growing in popularity is that of using completely naturalistic stimuli such as listening to a story or watching a

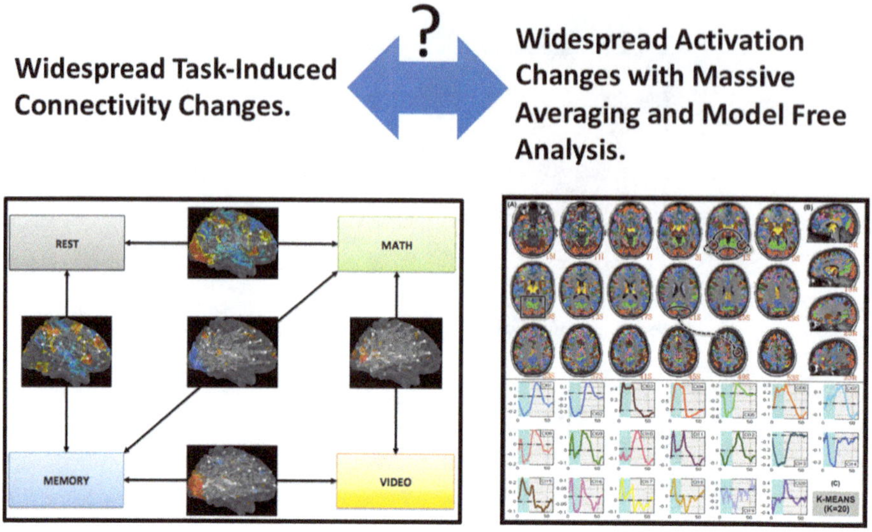

Fig. 8 There is more there with task activation. On the left, more spatially extensive network correlation than is seen with standard task-activation fMRI when assessing ongoing correlated activity during tasks. On the right are the regions demonstrating noncanonical but significant signal changes with a simple visual-motor task. These suggest that during a task, extensive brain regions, while time-locked to the task, do not behave as expected and modeled by a standard convolution of a simple box-car function with a hemodynamic response function

movie. Rather than assessing the network connectivity across regions or parcels within a subject, the degree of connectivity between subject pair responses during the identical naturalistic stimuli timing – either in the entire brain or specific networks – is determined. Using this approach – applied exhaustively across all subject pairs, subtle differences – or cross-subject connectivity variations – are able to be detected, thus allowing individual subject assessment to be carried out (Finn et al. 2018). This approach has been used to reveal varying paranoia levels in individual subjects by assessing the time series correlation in specific networks across subjects as they listened to an ambiguous narrative. The subjects that had a higher degree of paranoid traits showed increases as well as decreases in cross-subject correlations in specific networks.

A growing trend is to perform network connectivity analysis on massive datasets to pull out specific network strengths that relate to very subtle differences in populations. For example, such a study has been carried out on 5 K subjects from the UK Biobank dataset (Smith et al. 2013). Network strengths have been compared against a multitude of lifestyle, physiologic, and genetic characteristics, revealing specific networks that are sensitive to such individual-defining characteristics as recreational drug use and intelligence. Such an approach may soon pave the way for the establishment of connectivity base biomarkers to be used for screening, prediction, and diagnosis of disease.

Several years ago, attempts were made to extract causality information from standard task-based fMRI time courses (Stephan and Roebroeck 2012). Many, but certainly not all, of these approaches relied on subtle differences in fMRI onset latency, that was interpreted as implying that one region or network became active moments before another, thus influenced the later activation. While the analytical approach was sound, the major obstacle was that BOLD latencies vary considerably on a voxel-wise basis in the brain due mostly to hemodynamics. Any underlying influence that neuronal latencies may exert on the spread of latencies seen in any task is likely completely washed out by the intrinsic variability in the hemodynamic response latency (Smith et al. 2012), limiting the certainty with which causal neuronal information can be extracted by comparing relative latencies of regions or networks. Until all latencies can be normalized for their hemodynamic influence, either by completely eliminating downstream draining vein effects or by a normalization procedure in post-processing, extracting causal information from hemodynamic latencies has limited value. A takeaway principle here is that the hemodynamic response shows a considerable spread over space but is consistent in repeated scans within each voxel and can be used to detect task-induced changes in latency rather than absolute latencies corresponding to cascaded neuronal activity in a single task.

Value for Understanding Brain Networks

The BOLD signal or more generally, the hemodynamic response, is of extremely high value for understanding brain networks at a particular spatial and temporal scale. The brain is organized across several spatial scales – ranging from individual neuronal connections (microns) to systems-level networks (centimeters). It also is constantly changing across several time scales – from the order of milliseconds to days and years. The networks and network dynamics that BOLD contrast captures hit upon a spatial and temporal "sweet spot" in characterizing the neuronal correlates of human behavior and differences in behavior over time and between subjects. Because of this serendipitous overlap, BOLD contrast is a valuable and complementary tool to those that delve with more precision into neuronal and molecular underpinnings of brain function.

BOLD contrast-based network analysis has demonstrated clear neuronal correlates of ongoing attention and comprehension of spoken or written text and has been able to reveal differences between many populations as well as individuals with disorders – or even traits that do not fall into disorder classification.

BOLD contrast-based network analysis has been successful in differentiating, based on altered network properties, populations who suffer from depression, Alzheimer's disease, schizophrenia, and autism. It has also helped to quantitate disturbances that occur with trauma or stroke.

A fundamental limitation is that we are observing neuronal activity through a neurovascular coupling filter which varies over space and across subjects and perhaps even populations. The precise limits of this filter remain to be fully clarified;

however, even with this limitation, BOLD contrast has shown to be robust and reliable. In general, the field has enjoyed early success and has a long way to go before realizing its full potential.

Current Trends, Challenges, and Future Prospects

Presently, the field of fcMRI is experiencing a renewed growth not only in methodology sophistication as it draws upon established processing and statistical techniques developed for other domains, but also is branching out in new directions. The use of whole-brain connectivity signatures to assess and differentiate ongoing task-driven states is gaining acceptance. Findings suggest that connectivity measures are more sensitive than magnitude changes in decoding as well as mapping subtle differences within and across tasks.

Dynamic connectivity assessment is well established (Hutchison et al. 2013) and is advancing as it has been found that the brain travels through multiple states during a typical resting-state scan. The sequence, timing, and duration of these states have been shown to vary between normal populations and those with schizophrenia and depression. More study needs to be carried out to more fully understand these dynamic brain states as well as how as well as why they vary.

Extremely high-resolution fMRI at high field has been carried out by a growing number of labs, and the results so far have been extremely promising. BOLD imaging at high field and high resolution has shown the ability to reveal activity in mesoscopic functional structures such as ocular dominance (Menon et al. 1997; Cheng 2012) and orientation columns (Yacoub et al. 2008). Successful results of layer-specific fMRI using in-plane resolutions of about 0.5–1 mm have started to emerge; however, even at high field, large vessel sensitivity has been a challenge – as the venous blood volume heavily weights the magnitude of the BOLD signal. A promising development has been to use blood volume-weighted sequences such as Vascular Space Occupancy (VASO), as it appears that activation-induced blood volume changes are more localized to small veins and capillaries (Huber et al. 2018).

Recently, layer-specific activity and modulation has been demonstrated with the use of VASO fMRI at high field. In the study by Huber et al. (2017), subjects alternated tapping their fingers with simply moving their fingers in the same manner but without tapping, modulating the sensory input to M1 in the upper layers. A reduction in the upper layer activity of M1 with reduced somatosensory input is demonstrated in Fig. 9.

Layer-specific fMRI is potentially an important advance in the ability to not only uncover mesoscale networks, but also in untangling directionality of connections, as the input and output of specific layers are generally understood from a wealth of previous literature on invasively derived brain structure and physiology in animal models. Early results exploring layer activity in primary sensory as well as prefrontal regions during cognitive tasks such as working memory and manipulation of concepts have been promising along these lines.

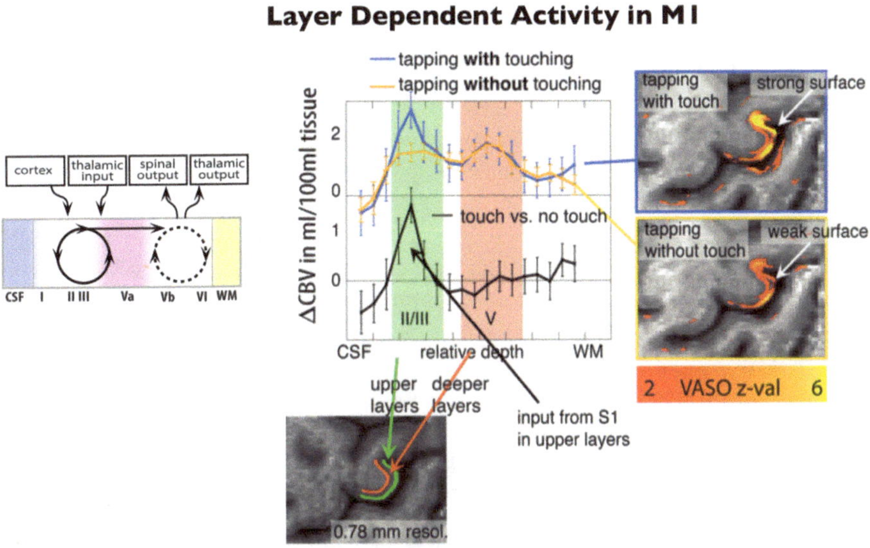

Fig. 9 Layer-specific fMRI using VASO. On the left is a model of input and output connections to and from M1. The two tasks were tapping movement with touching of thumb and index finger, involving higher sensory input to M1; and tapping without touching of the fingers, involving less sensory input to M1. Shown is a modulation of the upper layer activity corresponding to sensory input modulation

In spite of its tremendous success, the field of fcMRI has challenges and unknowns to grapple with in order to continue advancement in shedding light on understanding brain networks. First is the question of what neuronal information is contained in time series fluctuations? It has been shown that correlated fluctuations contain information on intrinsic subject traits, subject states as they vary spontaneously over minutes, and information related to ongoing tasks. To what degree are trait, state, and task information represented in fMRI time series and how might they be cleanly separated? This question remains to be answered.

A second issue that is fundamental and typically bypassed is with regard to all the possible causes of changes in time series correlations that are not in fact related to neuronal connectivity. A trend in the field is to equate time series correlation, and changes in this correlation, directly with connectivity and changes in connectivity. Correlation changes can be influenced by many factors that may have nothing to do with neuronal connectivity. These include the following: (1) Changes in amplitude of one or both of the correlated time series. In the presence of noise, any change in amplitude of the fluctuations will increase the correlation value between two time series. Does this imply increased connectivity? Likely not. (2) Changes in the underlying noise of one or both of the correlated time series. Again, given a constant amplitude of the correlated time series, a change in underlying noise will modulate the correlation value. This also is not likely related to connectivity. (3) A change in latency or shape of one of the correlated time series. A latency shift will certainly change the correlation value between two time series. A latency shift does not necessarily imply a change in connectivity.

What confirmatory measures would allow an interpretation that changes in correlation reflect a clear change in "connectivity?" While the reported connectivities and variations in connectivity support many neuronal connectivity hypotheses and are quite consistent, this central question to the entire field of fcMRI needs to be addressed.

Related to the above issue is the open question of why does whole-brain fcMRI work at all? How is it that unique temporal signatures can be so reliably produced by up to 100 different networks in a 5-min time series (or even in windows as short as 30 s) using a 0.05–0.15 Hz signal? What is the unique temporal signature of each node or network? Is this signature stationary? While modern time series analyses approaches have been used to address this question, no clear finding has emerged. Are the fluctuations in all these unique networks simply random in different ways? This issue also needs to be understood more completely.

The problem of brain parcellation, typically a first and crucial step in the creation and analysis of time series correlation matrices, remains central to fcMRI. Any change in the parcellation scheme may have significant effects on the correlation matrix. The goal is not only to match, with high accuracy, parcels to known functional regions, but to do this on a subject-specific basis such that meaningful group averages or single subject comparisons can be made. Brains vary in shape. Functional regions vary in their relative sizes and shapes. Some parcels may perhaps even emerge or fall away among subjects. The problem is daunting but likely solvable in time.

Related to the above challenge is the question of how to handle multi-subject averaging or comparisons as fMRI resolution increases from the scale of systems to the scale of columns and layers. It is generally thought that brain shows more unique and perhaps random variability the more finely one looks. At some point, the organization becomes arbitrarily variable – as is the case of ocular dominance columns. We all have columns, and there is an organizing principle behind columnar organization; however, any attempt to line up columns across subjects is likely to fail.

Parcellation is essentially the collapsing of data in the spatial dimension to allow for multi-subject comparisons. Brain parcellation on the scale of centimeters tends to work since it is large enough to show regularity across subjects. Once we start regularly collecting data at much higher resolution, we need to find other meaningful dimensions that show minimal variability across subjects in which to perform this data reduction. So far, with layer-specific fMRI, the approach has been to painstakingly isolate the specific anatomic structure for each individual and then to warp the isolated section of anatomy across subjects to fit. This approach works with layers, however is extremely time-consuming and requires manual intervention. How would one automate this process for whole-brain high-resolution comparisons?

The above issue leads to an ongoing challenge of whole-brain acquisition in extremely high-resolution fMRI. Currently, the best method for high-resolution fMRI is with the use of VASO acquisition. This approach involves an additional inversion pulse and an additional second delay before acquisition, limiting it to perhaps 5 slices at a time. Performance of high-resolution VASO on the entire brain would require multiple separate acquisitions, taking an inordinate amount of time. It would be desirable to use a more straightforward and time-efficient acquisition scheme characteristic of BOLD contrast – a simple gradient-echo EPI pulse sequence. The problem is that all the time-efficient methods have large draining vein weighting, reducing functional resolution. So, the search is on to find a time-efficient, sensitive, and capillary or at least small vein-specific functional pulse sequence to allow high resolution, whole-brain fMRI.

The entire field of fMRI, including fcMRI, is increasing focus on deriving meaningful information from *individuals* using functional and structural scans. In the first few years that fMRI existed, success was determined simply by the ability to create reliable maps of neuronal activity corresponding to specific tasks. Then the field moved to performing group averaged comparisons. Typically, normal controls were compared with patient populations. This effort has been successful. However, while group averaged differences may be useful for elucidating disrupted regional or network activity as it lends potential insight into the disorders, these results are relatively useless when it comes to making a statement about any one individual in a group. Put another way, while it is relatively easy to show differences in group distributions, it is a much harder problem to attribute an individual as belonging to one group or another based on their own fMRI data. The reason for this is that the effect size to noise (and subject variability) ratio is much too low to differentiate any one subject. A potential solution is to collect large datasets with many different measures. A hope is that a massively multivariate approach with large pooled datasets will stratify the distributions enough such that clean separations can be made (Smith et al. 2013). Given increasing pressure to push fMRI into the clinical realm, the effort to achieve clinical acceptance along these lines has been recently accelerated – allowing fMRI can be used to help predict, diagnose, and assist treatment on patients with disorders ranging from psychiatric to neuronal as they come into the hospital. Early work along these lines has shown promise; however, the statistical reliability in differentiating different populations is still not yet high enough for clinical efficacy.

Finally, this chapter concludes with consideration of prospects for the future. What is likely in store for fcMRI? Optimistically, the challenges mentioned above are all being addressed. The universal yet subject-specific parcellation problem is solvable. New insight is being reported regularly on the dynamics and the unique temporal signatures inherent to networks. Massive pooled datasets are just starting to be mined for useful multivariate information for the creation of useful biomarkers. With the increased propagation of 7 Tesla scanners and related hardware, as well as with increased distribution of advanced fMRI pulse sequences, progress is being made toward an effective processing pipeline for the routine mapping of layer-specific activity. Once layer-specific mapping takes hold, a new era in mapping mesoscopic networks and directional connectivity will likely emerge. Clinical fMRI protocols, perhaps involving naturalistic stimuli to derive useful biomarkers of disease or physiologic manipulations to derive information on vascular patency, will slowly grow at first, but once fMRI does take hold in the clinical market, vendors will likely start to shift more manpower into developing fMRI, increasing its rate of growth considerably. Realistically, any of these challenges may prove insurmountable, and the fields of clinical and basic neuroscience may shift to more promising methods. The collective ingenuity, skill, and creativity of researchers in fMRI is growing. Methods are growing, and standards are converging. The field is advancing. These challenges will most likely be embraced and mostly solved in the years to come.

Functional MRI has already established itself as an essential tool for understanding the network organization and dynamics of the human brain at a spatial and temporal scale that has previously been unexplored. The reliability and interpretability of fMRI is growing as we develop better models of the signal and the noise, and as we further reveal the neuronal underpinnings of hemodynamic response with multi-modal comparisons in human and animal models. Higher resolution is moving the spatial scale into that of columns and layers – potentially revolutionizing the questions that we may be able to ask with fMRI. Biomarker development is moving fMRI back from group studies to individual assessment, with useful and, importantly, cost-effective, clinical applications likely following.

References

ARCARO, M. J., HONEY, C. J., MRUCZEK, R. E., KASTNER, S. & HASSON, U. 2015. Widespread correlation patterns of fMRI signal across visual cortex reflect eccentricity organization. *Elife,* 4.

BANDETTINI, P. A. & WONG, E. C. 1995. Effects of Biophysical and Physiological-Parameters on Brain Activation-Induced R(2)Asterisk and R(2) Changes – Simulations Using a Deterministic Diffusion-Model. *International Journal of Imaging Systems and Technology,* 6, 133–152.

BANDETTINI, P. A., WONG, E. C., HINKS, R. S., TIKOFSKY, R. S. & HYDE, J. S. 1992. Time course EPI of human brain function during task activation. *Magnetic resonance in medicine: official journal of the Society of Magnetic Resonance in Medicine / Society of Magnetic Resonance in Medicine,* 25, 390–397.

BECKMANN, C. F., DELUCA, M., DEVLIN, J. T. & SMITH, S. M. 2005. Investigations into resting-state connectivity using independent component analysis. *Philos Trans R Soc Lond B Biol Sci,* 360, 1001–13.

BELLGOWAN, P. S., SAAD, Z. S. & BANDETTINI, P. A. 2003. Understanding neural system dynamics through task modulation and measurement of functional MRI amplitude, latency, and width. *Proc Natl Acad Sci U S A,* 100, 1415–9.

BELLIVEAU, J. W., KENNEDY, D. N., MCKINSTRY, R. C., BUCHBINDER, B. R., WEISSKOFF, R. M., COHEN, M. S., VEVEA, J. M., BRADY, T. J. & ROSEN, B. R. 1991. Functional Mapping of the Human Visual-Cortex by Magnetic-Resonance-Imaging. *Science,* 254, 716–719.

BIRN, R., SAAD, Z. & BANDETTINI, P. 2001. Spatial heterogeneity of the nonlinear dynamics in the FMRI BOLD response. *Neuroimage,* 14, 817–826.

BIRN, R. M. & BANDETTINI, P. A. 2005. The effect of stimulus duty cycle and "off" duration on BOLD response linearity. *Neuroimage,* 27, 70–82.

BIRN, R. M., SMITH, M. A., JONES, T. B. & BANDETTINI, P. A. 2008. The respiration response function: the temporal dynamics of fMRI signal fluctuations related to changes in respiration. *Neuroimage,* 40, 644–54.

BISWAL, B., YETKIN, F. Z., HAUGHTON, V. M. & HYDE, J. S. 1995. Functional Connectivity in the Motor Cortex of Resting Human Brain Using Echo-Planar Mri. *Magnetic Resonance in Medicine,* 34, 537–541.

BODURKA, J., YE, F., PETRIDOU, N., MURPHY, K. & BANDETTINI, P. A. 2007. Mapping the MRI voxel volume in which thermal noise matches physiological noise--implications for fMRI. *Neuroimage,* 34, 542–9.

BOUBELA, R. N., KALCHER, K., HUF, W., KRONNERWETTER, C., FILZMOSER, P. & MOSER, E. 2013. Beyond Noise: Using Temporal ICA to Extract Meaningful Information from High-Frequency fMRI Signal Fluctuations during Rest. *Front Hum Neurosci,* 7, 168.

BULLMORE, E. & SPORNS, O. 2009. Complex brain networks: graph theoretical analysis of structural and functional systems. *Nat Rev Neurosci,* 10, 186–98.

BUXTON, R. B., ULUDAG, K., DUBOWITZ, D. J. & LIU, T. T. 2004. Modeling the hemodynamic response to brain activation. *Neuroimage,* 23, S220–S233.

CALHOUN, V. D., LIU, J. & ADALI, T. 2009. A review of group ICA for fMRI data and ICA for joint inference of imaging, genetic, and ERP data. *Neuroimage,* 45, S163–172.

CALHOUN, V. D., MILLER, R., PEARLSON, G. & ADALI, T. 2014. The chronnectome: time-varying connectivity networks as the next frontier in fMRI data discovery. *Neuron,* 84, 262–74.

CHENG, K. 2012. Revealing human ocular dominance columns using high-resolution functional magnetic resonance imaging. *Neuroimage,* 62, 1029–34.

DAMOISEAUX, J. S., ROMBOUTS, S. A. R. B., BARKHOF, F., SCHELTENS, P., STAM, C. J., SMITH, S. M. & BECKMANN, C. F. 2006. Consistent resting-state networks across healthy subjects. *Proceedings of the National Academy of Sciences of the United States of America,* 103, 13848–13853.

EICKHOFF, S. B., CONSTABLE, R. T. & YEO, B. T. T. 2018. Topographic organization of the cerebral cortex and brain cartography. *Neuroimage,* 170, 332–347.

FEINBERG, D. A., MOELLER, S., SMITH, S. M., AUERBACH, E., RAMANNA, S., GUNTHER, M., GLASSER, M. F., MILLER, K. L., UGURBIL, K. & YACOUB, E. 2010. Multiplexed echo planar imaging for sub-second whole brain FMRI and fast diffusion imaging. *PLoS One,* 5, e15710.

FINN, E. S., CORLETT, P. R., CHEN, G., BANDETTINI, P. A. & CONSTABLE, R. T. 2018. Trait paranoia shapes inter-subject synchrony in brain activity during an ambiguous social narrative. *Nat Commun,* 9, 2043.

FINN, E. S., SHEN, X., SCHEINOST, D., ROSENBERG, M. D., HUANG, J., CHUN, M. M., PAPADEMETRIS, X. & CONSTABLE, R. T. 2015. Functional connectome fingerprinting: identifying individuals using patterns of brain connectivity. *Nat Neurosci,* 18, 1664–71.

FOX, M. D., SNYDER, A. Z., VINCENT, J. L., CORBETTA, M., VAN ESSEN, D. C. & RAICHLE, M. E. 2005. The human brain is intrinsically organized into dynamic, anticorrelated functional networks. *Proceedings of the National Academy of Sciences of the United States of America,* 102, 9673–9678.

FRISTON, K. J., BUECHEL, C., FINK, G. R., MORRIS, J., ROLLS, E. & DOLAN, R. J. 1997. Psychophysiological and modulatory interactions in neuroimaging. *Neuroimage,* 6, 218–29.

GLASSER, M. F., COALSON, T. S., ROBINSON, E. C., HACKER, C. D., HARWELL, J., YACOUB, E., UGURBIL, K., ANDERSSON, J., BECKMANN, C. F., JENKINSON, M., SMITH, S. M. & VAN ESSEN, D. C. 2016. A multi-modal parcellation of human cerebral cortex. *Nature,* 536, 171–8.

GONZALEZ-CASTILLO, J., HANDWERKER, D. A., ROBINSON, M. E., HOY, C. W., BUCHANAN, L. C., SAAD, Z. S. & BANDETTINI, P. A. 2014. The spatial structure of resting state connectivity stability on the scale of minutes. *Front Neurosci,* 8, 138.

GONZALEZ-CASTILLO, J., HOY, C. W., HANDWERKER, D. A., ROBINSON, M. E., BUCHANAN, L. C., SAAD, Z. S. & BANDETTINI, P. A. 2015. Tracking ongoing cognition in individuals using brief, whole-brain functional connectivity patterns. *Proc Natl Acad Sci U S A,* 112, 8762–7.

GONZALEZ-CASTILLO, J., PANWAR, P., BUCHANAN, L. C., CABALLERO-GAUDES, C., HANDWERKER, D. A., JANGRAW, D. C., ZACHARIOU, V., INATI, S., ROOPCHANSINGH, V., DERBYSHIRE, J. A. & BANDETTINI, P. A. 2016. Evaluation of multi-echo ICA denoising for task based fMRI studies: Block designs, rapid event-related designs, and cardiac-gated fMRI. *Neuroimage,* 141, 452–468.

GONZALEZ-CASTILLO, J., SAAD, Z. S., HANDWERKER, D. A., INATI, S. J., BRENOWITZ, N. & BANDETTINI, P. A. 2012. Whole-brain, time-locked activation with simple tasks revealed using massive averaging and model-free analysis. *Proc Natl Acad Sci U S A,* 109, 5487–92.

GREICIUS, M. D., KRASNOW, B., REISS, A. L. & MENON, V. 2003. Functional connectivity in the resting brain: a network analysis of the default mode hypothesis. *Proc Natl Acad Sci U S A,* 100, 253–8.

HAACKE, E. M. & YE, Y. 2012. The role of susceptibility weighted imaging in functional MRI. *Neuroimage,* 62, 923–9.

HANDWERKER, D. A., OLLINGER, J. M. & D'ESPOSITO, M. 2004. Variation of BOLD hemodynamic responses across subjects and brain regions and their effects on statistical analyses. *Neuroimage,* 21, 1639–1651.

HU, X., LE, T. H. & UGURBIL, K. 1997. Evaluation of the early response in fMRI in individual subjects using short stimulus duration. *Magn Reson Med,* 37, 877–84.

HUBER, L., HANDWERKER, D. A., JANGRAW, D. C., CHEN, G., HALL, A., STUBER, C., GONZALEZ-CASTILLO, J., IVANOV, D., MARRETT, S., GUIDI, M., GOENSE, J., POSER, B. A. & BANDETTINI, P. A. 2017. High-Resolution CBV-fMRI Allows Mapping of Laminar Activity and Connectivity of Cortical Input and Output in Human M1. *Neuron,* 96, 1253–1263 e7.

HUBER, L., IVANOV, D., HANDWERKER, D. A., MARRETT, S., GUIDI, M., ULUDAG, K., BANDETTINI, P. A. & POSER, B. A. 2018. Techniques for blood volume fMRI with VASO: From low-resolution mapping towards sub-millimeter layer-dependent applications. *Neuroimage,* 164, 131–143.

HUTCHISON, R. M., WOMELSDORF, T., ALLEN, E. A., BANDETTINI, P. A., CALHOUN, V. D., CORBETTA, M., DELLA PENNA, S., DUYN, J. H., GLOVER, G. H., GONZALEZ-CASTILLO, J., HANDWERKER, D. A., KEILHOLZ, S., KIVINIEMI, V., LEOPOLD, D. A., DE PASQUALE, F., SPORNS, O., WALTER, M. & CHANG, C. 2013. Dynamic functional connectivity: promise, issues, and interpretations. *Neuroimage,* 80, 360–78.

JANGRAW, D. C., GONZALEZ-CASTILLO, J., HANDWERKER, D. A., GHANE, M., ROSENBERG, M. D., PANWAR, P. & BANDETTINI, P. A. 2018. A functional connectivity-based neuromarker of sustained attention generalizes to predict recall in a reading task. *Neuroimage,* 166, 99–109.

KUNDU, P., BRENOWITZ, N. D., VOON, V., WORBE, Y., VERTES, P. E., INATI, S. J., SAAD, Z. S., BANDETTINI, P. A. & BULLMORE, E. T. 2013. Integrated strategy for improving functional connectivity mapping using multiecho fMRI. *Proc Natl Acad Sci U S A,* 110, 16187–92.

KUNDU, P., INATI, S. J., EVANS, J. W., LUH, W. M. & BANDETTINI, P. A. 2012. Differentiating BOLD and non-BOLD signals in fMRI time series using multi-echo EPI. *Neuroimage,* 60, 1759–70.

KWONG, K. K., BELLIVEAU, J. W., CHESLER, D. A., GOLDBERG, I. E., WEISSKOFF, R. M., PONCELET, B. P., KENNEDY, D. N., HOPPEL, B. E., COHEN, M. S., TURNER, R., CHENG, H. M., BRADY, T. J. & ROSEN, B. R. 1992. Dynamic Magnetic-Resonance-Imaging of Human Brain Activity During Primary Sensory Stimulation. *Proceedings of the National Academy of Sciences of the United States of America,* 89, 5675–5679.

LAUFS, H., KLEINSCHMIDT, A., BEYERLE, A., EGER, E., SALEK-HADDADI, A., PREIBISCH, C. & KRAKOW, K. 2003. EEG-correlated fMRI of human alpha activity. *Neuroimage,* 19, 1463–1476.

LOGOTHETIS, N. K., PAULS, J., AUGATH, M., TRINATH, T. & OELTERMANN, A. 2001. Neurophysiological investigation of the basis of the fMRI signal. *Nature,* 412, 150–157.

LU, H. Z., GOLAY, X., PEKAR, J. J. & VAN ZIJLL, P. C. M. 2003. Functional magnetic resonance Imaging based on changes in vascular space occupancy. *Magnetic Resonance in Medicine,* 50, 263–274.

LUSTIG, C., SNYDER, A. Z., BHAKTA, M., O'BRIEN, K. C., MCAVOY, M., RAICHLE, M. E., MORRIS, J. C. & BUCKNER, R. L. 2003. Functional deactivations: Change with age and dementia of the Alzheimer type. *Proceedings of the National Academy of Sciences of the United States of America,* 100, 14504–14509.

MENON, R. S., OGAWA, S., STRUPP, J. P. & UGURBIL, K. 1997. Ocular dominance in human V1 demonstrated by functional magnetic resonance imaging. *Journal of Neurophysiology,* 77, 2780–2787.

MENON, R. S., OGAWA, S., TANK, D. W. & UGURBIL, K. 1993. Tesla Gradient Recalled Echo Characteristics of Photic Stimulation-Induced Signal Changes in the Human Primary Visual-Cortex. *Magnetic Resonance in Medicine,* 30, 380–386.

OGAWA, S., LEE, T. M., KAY, A. R. & TANK, D. W. 1990. Brain Magnetic-Resonance-Imaging with Contrast Dependent on Blood Oxygenation. *Proceedings of the National Academy of Sciences of the United States of America,* 87, 9868–9872.

OGAWA, S., MENON, R. S., TANK, D. W., KIM, S. G., MERKLE, H., ELLERMANN, J. M. & UGURBIL, K. 1993. Functional brain mapping by blood oxygenation level-dependent contrast magnetic resonance imaging. A comparison of signal characteristics with a biophysical model. *Biophys J,* 64, 803–12.

OGAWA, S., TANK, D. W., MENON, R., ELLERMANN, J. M., KIM, S. G., MERKLE, H. & UGURBIL, K. 1992. Intrinsic Signal Changes Accompanying Sensory Stimulation – Functional Brain Mapping with Magnetic-Resonance-Imaging. *Proceedings of the National Academy of Sciences of the United States of America,* 89, 5951–5955.

PRUESSMANN, K. P., WEIGER, M., SCHEIDEGGER, M. B. & BOESIGER, P. 1999. SENSE: sensitivity encoding for fast MRI. *Magn Reson Med,* 42, 952–62.

RAICHLE, M. E., MACLEOD, A. M., SNYDER, A. Z., POWERS, W. J., GUSNARD, D. A. & SHULMAN, G. L. 2001. A default mode of brain function. *Proc Natl Acad Sci U S A,* 98, 676–82.

ROSEN, B. R., BELLIVEAU, J. W., VEVEA, J. M. & BRADY, T. J. 1990. Perfusion imaging with NMR contrast agents. *Magn Reson Med,* 14, 249–65.

ROSENBERG, M. D., FINN, E. S., SCHEINOST, D., PAPADEMETRIS, X., SHEN, X., CONSTABLE, R. T. & CHUN, M. M. 2016. A neuromarker of sustained attention from whole-brain functional connectivity. *Nat Neurosci,* 19, 165–71.

SAGGAR, M., SPORNS, O., GONZALEZ-CASTILLO, J., BANDETTINI, P. A., CARLSSON, G., GLOVER, G. & REISS, A. L. 2018. Towards a new approach to reveal dynamical organization of the brain using topological data analysis. *Nat Commun,* 9, 1399.

SHMUEL, A. & LEOPOLD, D. 2008. Neuronal correlates of spontaneous fluctuations in fMRI signals in monkey visual cortex: implications for functional connectivity at rest. *Human Brain Mapping,* current issue.

SMITH, S. M., BANDETTINI, P. A., MILLER, K. L., BEHRENS, T. E., FRISTON, K. J., DAVID, O., LIU, T., WOOLRICH, M. W. & NICHOLS, T. E. 2012. The danger of systematic bias in group-level FMRI-lag-based causality estimation. *Neuroimage,* 59, 1228–9.

SMITH, S. M., VIDAURRE, D., BECKMANN, C. F., GLASSER, M. F., JENKINSON, M., MILLER, K. L., NICHOLS, T. E., ROBINSON, E. C., SALIMI-KHORSHIDI, G., WOOLRICH, M. W., BARCH, D. M., UGURBIL, K. & VAN ESSEN, D. C. 2013. Functional connectomics from resting-state fMRI. *Trends Cogn Sci,* 17, 666–82.

STEHLING, M. K., TURNER, R. & MANSFIELD, P. 1991. Echo-planar imaging: magnetic resonance imaging in a fraction of a second. *Science,* 254, 43–50.

STEPHAN, K. E. & ROEBROECK, A. 2012. A short history of causal modeling of fMRI data. *Neuroimage,* 62, 856–63.

TRIANTAFYLLOU, C., HOGE, R. D., KRUEGER, G., WIGGINS, C. J., POTTHAST, A., WIGGINS, G. C. & WALD, L. L. 2005. Comparison of physiological noise at 1.5 T, 3 T and 7 T and optimization of fMRI acquisition parameters. *Neuroimage,* 26, 243–250.

WALD, L. L. 2012. The future of acquisition speed, coverage, sensitivity, and resolution. *Neuroimage,* 62, 1221–9.

WONG, E. C., BUXTON, R. B. & FRANK, L. R. 1997. Implementation of quantitative perfusion imaging techniques for functional brain mapping using pulsed arterial spin labeling. *Nmr in Biomedicine,* 10, 237–249.

YACOUB, E., HAREL, N. & UĞURBIL, K. 2008. High-field fMRI unveils orientation columns in humans. *Proceedings of the National Academy of Sciences of the United States of America,* 105, 10607–10612.

YACOUB, E., UGURBIL, K. & HAREL, N. 2006. The spatial dependence of the poststimulus undershoot as revealed by high-resolution BOLD- and CBV-weighted fMRI. *Journal of Cerebral Blood Flow and Metabolism,* 26, 634–644.

ZONG, X., KIM, T. & KIM, S. G. 2012. Contributions of dynamic venous blood volume versus oxygenation level changes to BOLD fMRI. *Neuroimage,* 60, 2238–46.

Review of Resting-State Functional Connectivity Methods and Application in Clinical Populations

Keerthana Karunakaran, Marie Wolfer, and Bharat B. Biswal

Introduction

The complex functions of the brain are the outcome of functional networks that go beyond static anatomical pathways. Functional connections integrate multiple systems using numerous permutations and combinations to produce unique networks pertaining to the different cognitive and sensory demands of the body. Detection and quantification of functional connectivity therefore becomes crucial to the investigation of behavioral and clinical abnormalities in brain function. Clinicians and researchers have used fMRI as the gold standard to map brain function in the form of blood oxygen level-dependent (BOLD) signal in humans. The BOLD signal reflects neural activity on the basis that neural firing results in increased local metabolic consumption of glucose, leading to a small, transient increase in local deoxyhemoglobin concentration. This transient increase in deoxyhemoglobin concentration compensated by a surplus increase in regional blood flow creates a spatial contrast in the BOLD signal due to the opposing diamagnetic and paramagnetic properties of oxyhemoglobin and deoxyhemoglobin molecules (Ogawa et al. 1992). A pictorial representation of the neurovascular coupling of fMRI signal resulting in BOLD activation is presented in Fig. 1.

Early fMRI studies classically examined the BOLD signal acquired in response to an exogenous stimulus: known as task fMRI. In task activation fMRI, a stimulus/activation is periodically presented for 20–30 s followed by 20–30 s of rest. It was assumed based on neurovascular coupling that the stimulus/activation results in an increase in FMRI signal intensity (Ogawa et al. 1992). The limitation of this approach is the variability of task performance and participant compliance on brain activation and the need for several trials to cover multiple brain systems. In 1995, we first demonstrated the presence of sensorimotor activity in the absence of a task,

K. Karunakaran · M. Wolfer · B. B. Biswal (✉)
Biomedical Engineering, NJIT, Newark, NJ, USA

V. A. Diwadkar, S. B. Eickhoff (eds.), *Brain Network Dysfunction in Neuropsychiatric Illness*, https://doi.org/10.1007/978-3-030-59797-9_3

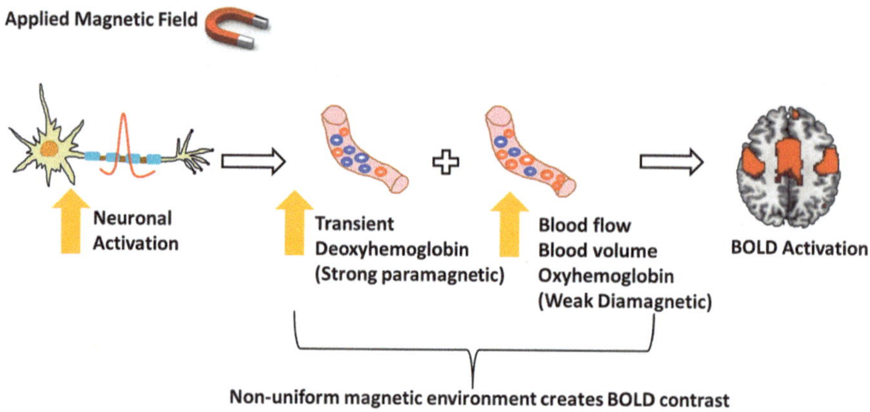

Fig. 1 Neurophysiological principle of fMRI to measure brain function: An increase in neuronal activity results in increased consumption of oxygen (and increase in deoxyhemoglobin) initiating an inflow of cerebral blood flow to replenish oxygen. This difference in oxy and deoxyhemoglobin concentration when inside a large magnetic field (such as the MRI scanner) due to the opposing magnetic properties creates a spatial contrast in the fMRI signal. The spatial contrast in fMRI signal upon reconstruction allows the mapping of brain areas that are active

that is, resting state, where brain function using fMRI could be quantified and used to map spatially independent functional maps in the absence of an explicit stimulus (Biswal et al. 1995). We showed that even during resting state, the brain is temporally correlated between spatially distinct functional regions in the form of synchronized low-frequency fluctuations (<0.1 Hz). Such a state of temporal dependence between anatomically connected regions during the absence of a task was coined as resting-state functional connectivity (RSFC) (Friston et al. 1993; Biswal et al. 1995). Despite the initial skepticism behind the neurophysiological basis of low-frequency fluctuations, subsequent resting-state studies corroborated the presence of highly correlated regions between the hemispheres of the visual cortex, auditory cortex, and amygdala of the brain known as resting-state networks (RSNs) (Lowe et al. 1998; Cordes et al. 2001, 2002; Beckmann et al. 2005).

Numerous studies have reported evidence of negative task activation in nontask regions simultaneous with positive task activation in task-relevant regions, especially in response to attention-demanding stimulus (Shulman et al. 1997; Gusnard et al. 2001; McKiernan et al. 2003). This task-negative network at rest was postulated as the default-mode network (DMN) of the brain, where a set of regions including dorsolateral prefrontal cortex, posterior cingulate cortex/precuneus, bilateral parietal cortex, and bilateral temporal gyrus remains active in the absence of goal-directed behavior but decreases during attention-demanding tasks (Raichle et al. 2001; Greicius et al. 2003). Empirical evidence has demonstrated that DMN actively participates in episodic memory retrieval for self-referential mental activities at rest as well as sleep (Fransson 2006; Horovitz et al. 2009). The different nodes of the DMN have been shown to influence and contribute to the manner in which task-positive regions process external information (Uddin et al. 2009; Sestieri

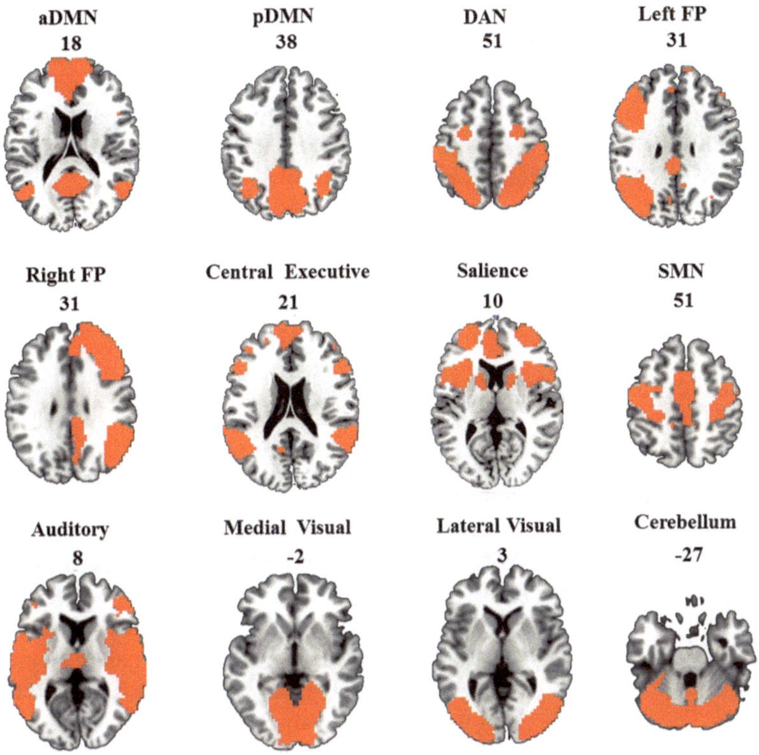

Fig. 2 Resting-state brain networks obtained using group probabilistic independent component analysis. Shown here are 12 standard RSNs frequently identified in fMRI studies, known to be associated with sensory, affective, and cognitive processing. The RSNs include anterior default-mode network (aDMN), posterior default-mode network (pDMN), dorsal attention network (DAN), left frontoparietal (left FP), right frontoparietal (right FP), central executive, salience, sensorimotor network (SMN), auditory, medial visual, lateral visual, and cerebellum networks. Depending on the technique used to derive the networks, they may be further classified as ventral attention network, basal ganglia, and medial temporal networks. The z index of the images is shown in MNI coordinates in neurological orientation

et al. 2011). Multiple studies have since replicated this finding to identify the presence of DMN during rest in addition to several robust RSNs. As shown in Fig. 2, the most standard RSNs include visual, sensorimotor, auditory, left executive, right executive, dorsal attention, DMN, and salient networks (Hampson et al. 2002; Greicius et al. 2003; Fox and Raichle 2007; Toro et al. 2008; Smith et al. 2009; Allen et al. 2011; Yeo et al. 2011). The salient network comprising anterior insular and dorsal anterior cingulate cortices has been indicated to act as a stimulus-dependent switch between the DMN and task-positive states (Sridharan et al. 2008). Consequently, a single resting-state scan covering the whole brain allows the investigation of functional connections both within and between the different RSNs.

Resting-state fMRI has provided reliable and consistent RSNs between cortical and subcortical regions. It has enabled the investigation of the brain as a set of complex integrated networks with functional and structural connections (Damoiseaux et al. 2006; Shehzad et al. 2009; Guo et al. 2012; Li et al. 2012). It has expanded our ability to understand brain function, and researchers are increasingly using RSFC patterns to understand the brain pathophysiology of neurological disorders (Buckner et al. 2008; Greicius 2008; Woodward and Cascio 2015). In this chapter, we discuss the neurophysiological basis of RSFC, give an overview of the most established resting-state analysis techniques, and summarize relevant findings from clinical populations. We also discuss current limitations in the resting-state approach and provide a short outlook for the future.

Neurovascular Origin of RSFC

The neurophysiological origin of resting-state BOLD signal has been a question of debate for several years. Inference of resting-state hemodynamic response is complex due to the number of factors that can contribute to the microvasculature and BOLD contrast during rest. For instance, low-frequency fluctuations from resting-state BOLD signal were initially suspected to be contaminated by aliased cardiac signals (and respiratory signals in certain instances) (Bhattacharyya and Lowe 2004). Respiratory cycles and cardiac pulsations also generate small movements in brain tissue introducing physiological motion in resting-state BOLD signal (Friston et al. 1996; Dagli et al. 1999). As a result, lower repetition time and improved physiological noise removal techniques are recommended to investigate low-frequency fluctuations (Cordes et al. 2001).

The manifestation of sensory RSNs such as visual, auditory, and motor networks comprising regions that are anatomically connected and functionally engaged only helped sustain the neurophysiological argument of low-frequency fluctuations. Using an arterial spina labeling (simultaneous perfusion and BOLD measurement), we demonstrated that neurovascular coupling plays a greater role in generation of RSFC than simply the variations in cerebral blood flow (Biswal et al. 1997). Studies have since substantiated the vascular and metabolic correlates of low-frequency fluctuation measures such as functional connectivity and functional amplitude as secondary to neuronal activity (Biswal et al. 1997; Fukunaga et al. 2008; Wu et al. 2009; Aiello et al. 2015; Nugent et al. 2015). Using a spatial working memory paradigm Lowe and colleagues showed greater RSFC in task-relevant regions during task suggesting that controlled activity results in greater correlation of low-frequency fluctuations (Lowe et al. 2000). As a result, early studies of resting-state fMRI questioned if RSFC could be a result of persistent hemodynamic response to an uncontrolled task or behavioral change. Because RSNs are persistent during different states such as sleep or light anesthesia, RSFC has been considered to reflect intrinsic activity of the brain over uncontrolled behavior (Peltier et al. 2005; Greicius et al. 2008; Larson-Prior et al. 2009).

RSFC deriving from a neurovascular standpoint proves extremely useful to measure baseline functions in healthy and disease populations. However, one of the common questions arising when inferring resting-state phenomenon is what an alteration in functional connectivity translates in terms of local neuronal operations. Early reports indicate that the hemodynamic response obtained from task-induced BOLD response is associated to the local field potential representing the net synchronized input of a population of neurons (Logothetis et al. 2001). A similar study by Shmuel and colleagues involving simultaneous measurement of neurophysiological and fMRI signals report high correspondence of neural spiking rate and local field potential with resting-state BOLD metrics (Shmuel and Leopold 2008). In fact, majority of the studies report gamma power local field potential to be more closely associated to resting-state BOLD response than neural spiking rate as the metabolic cost of neuronal firing is shown to be too small to elicit a BOLD response (Scholvinck et al. 2010; Chen et al. 2017). Alternatively, Mateo and colleagues report vasomotion of arteriole diameter as the missing link between neuronal activation and the hemodynamic response (Mateo et al. 2017). Using two-photon laser scanning microscopy, they show that neuronal activity drives arteriole vasomotor fluctuations which consequently drive hemodynamic response.

With mounting evidence for neuronal origin of resting-state BOLD signal, researchers are using this information to understand the relationship between neurovascular activity and resting-state BOLD signal across different brain states, brain regions, disease conditions, and the life period of an individual. We recommend interested readers to refer to the review on the neurovascular factors of the resting-state BOLD signal given by Liu and colleagues (Liu 2013).

Linking Anatomical Connectivity to Resting-State Functional Connectivity

Greicius and colleagues proposed that functional connectivity is a reflection of underlying structural connectivity (Greicius et al. 2009). Decrease or lack of interhemispheric RSFC in individuals with agenesis of the corpus callosum or callosotomy has also highlighted the essential role anatomical connections play in forming RSFC (Quigley et al. 2003; Johnston et al. 2008). But, the relationship between structure and function of a brain network is not straightforward for two reasons: (1) varying nature of interaction between nodes viz. inhibitory or excitatory, unidirectional, or bidirectional and (2) dynamic functional connections for a fixed anatomical structure. For the same reason, prediction of RSFC using structural connectivity is indicated to be more robust in the presence of direct structural connections which would otherwise be impractical due to the large number of possible functional connections for a single fixed structural framework (Honey et al. 2009).

Structural connections of the brain demonstrate small-world properties and high modularity necessary for efficient information processing of complex brain functions (Bassett et al. 2008). RSNs as a reflection of structural connections are thought to be made of highly dense functional modules comprised of dense short-range connections interconnected with each other through strong long-range connections.

Greicius and colleagues corroborated the presence of long-range (likely cingulum bundles) and short-range fibers from medial prefrontal cortex and middle temporal lobe to the posterior cingulate cortex of the DMN in humans (Greicius et al. 2009). The brain, being the multifaceted system it is, has been shown to be capable of recruiting alternative anatomical pathways. For instance, weak interhemispheric RSFC reported in a case of complete commissurotomy emphasizes the brain's ability to recruit subcortical pathways for the same RSN (Uddin et al. 2008). Hence, spatial patterns of RSNs are thought to be constrained but not limited to underlying anatomical connections as functional connections may exist between separate regions through polysynaptic connections or common anatomical links (Owen et al. 2013). Nonetheless, defining the structural connections can help understand and substantiate the findings of functional brain networks and perhaps could help predict the abnormal dynamics of the functional networks in a neurological disease condition.

Resting-State Analysis Methods

Preprocessing

Depending on the goal of the analysis, a standard set of preprocessing steps precede the computation of RSFC measures. Preprocessing techniques are used to improve the spatial agreement between subjects and minimize confounding effects of physiological noise. A flowchart of a standard preprocessing pipeline is depicted in Fig. 3. There are three major platforms for carrying out different preprocessing and post-processing functions: (a) statistical parametric mapping (SPM, Wellcome Trust Center for Neuroimaging, UK), (b) analysis of functional neuroimages (AFNIs), and (c) FMRIB Software Library (FSL) (Cox 1996; Smith et al. 2004). Other toolboxes with a complete pipeline for processing resting-state fMRI data include RESting-state fMRI data analysis Toolkit (REST), data processing assistant for resting-state fMRI (DPARSF), and connectivity toolbox (CONN) implemented on MATLAB platform (Chao-Gan and Yu-Feng 2010; Song et al. 2011; Whitfield-Gabrieli and Nieto-Castanon 2012).

A general first step of the preprocessing pipeline is the manual alignment of the origin of the functional and anatomical images to the anterior commissure of the brain. By setting the same origin for the individual subject image and standard template, the likelihood of local maxima problem is reduced, resulting in better coregistration between individual subject images and standard brain template. But with improved nonlinear coregistration techniques, anterior commissure alignment is no longer mandatory.

Physiological information, such as auxiliary heart rate or respiratory information if available, is removed from raw BOLD data before any type of spatial or temporal processing. Researchers begin preprocessing with computation of rigid-body

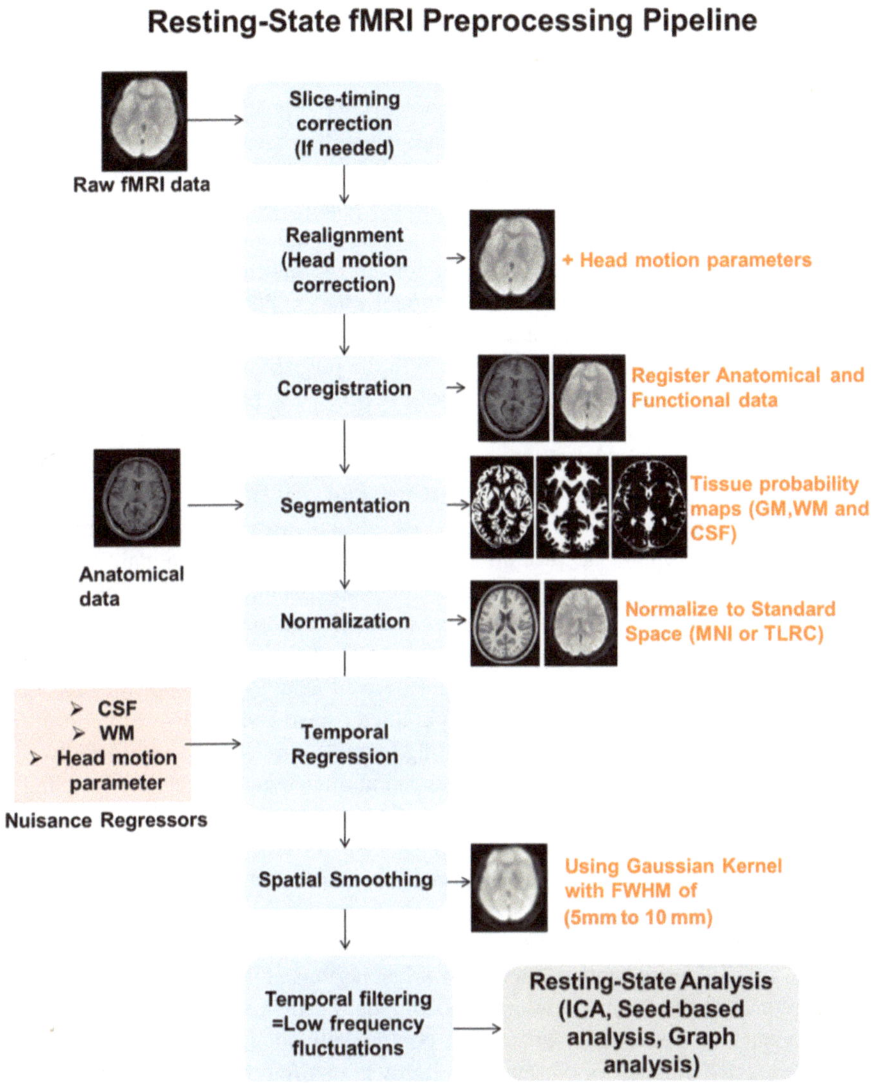

Fig. 3 An overview of the standard preprocessing pipeline used for resting-state fMRI data. The order and steps of the pipeline are often modified depending on the toolbox used as well as the goal of the study. *GM* Gray matter, *WM* white matter, *CSF* cerebrospinal fluid, *MNI* Montreal Neurological Institute, *TLRC* Talairach, *FWHM* full-width at half maximum, *ICA* independent component analysis

parameters (x, y, z, roll, pitch, yaw) of head motion with respect to the first time point or mean image within a session. The resulting head motion parameters of individual subjects are typically regressed from the BOLD time series at a later stage. Other techniques for head motion correction include frame-wise displacement, scrubbing, component-based noise correction method, and ICA-automatic removal of motion artifacts (Behzadi et al. 2007; Power et al. 2012; Pruim et al. 2015). Although slice timing correction is frequently skipped in resting-state data (Wu et al. 2011), it is generally performed before head motion correction to avoid interpolation of BOLD signals from realigned voxels. After head-motion correction, the functional and anatomical images are coregistered to increase anatomical correspondence between the two. Since anatomical images have higher spatial resolution and therefore will more accurately deform to a standard template, an individual's anatomical image is used to normalize their functional images to standard brain template such as MNI_T1_152. Errors in normalization often occur due to inaccurate coregistration of structural and functional images, and therefore it is a good practice to visually inspect the outcome of coregistration before proceeding.

The anatomical data are segmented into tissue probability maps of gray matter, white matter (WM), and cerebrospinal fluid (CSF) using prior tissue maps. These tissue probability maps are used to extract confounding variables of WM and CSF by computing an average of significant WM and CSF voxels or the first five principal components of WM and CSF voxels (Chai et al. 2012). The confounding variables are temporally regressed from functional data using a general linear model to account for effects of head motion correction, WM, and CSF signal. The head motion parameters may include the (1) six rigid body parameters, (2) six parameters plus previous time points of six parameters and their quadratics (Friston 24 model), or (3) six parameters or Friston 24 model plus frame-wise displacement (Friston et al. 1996). Thirty-six parameter estimates derived from 9-parameter model along with previous time point and quadratic parameter have been shown to be slightly superior to 24-parameter model in accounting for motion-related noise (Satterthwaite et al. 2013). Spatial smoothing is recommended after temporal regression using a Gaussian filter of 5–10 mm full-width at half maximum (FWHM). Smoothed regressed signals are then filtered at 0.01–0.08 or 0.01–0.1 Hz to obtain intrinsic low-frequency fluctuations in the functional data (Biswal et al. 2010).

Depending on the goal of the analysis, alternating strategies such as diffeomorphic image registration or other nonlinear registration techniques can be utilized to improve subject normalization, especially when regions of interest include subcortical structures such as amygdala or hippocampus. Further, visual inspection of the independent components for artifactual components such as head motion, large number of small clusters, or clusters in nongray matter regions can help evaluate the efficiency of spatial and temporal preprocessing (Griffanti et al. 2017).

Resting-State Analyses

Several techniques exist in literature to identify and quantify functional segregation and integration of regions in resting-state fMRI. RSFC analysis techniques can be broadly categorized based on (a) functional connectivity, (b) functional activation, (c) dynamic functional connectivity, and (d) effective connectivity.

Functional Connectivity

Functional connectivity was first defined in the context of task activation neuroimaging studies as the "statistical dependency among two or more spatially distinct time series". We defined resting-state functional connectivity as the statistical dependency among two or more spatially distinct time series in the absence of an overt task (Biswal et al. 1995). Functional connectivity reveals spatial patterns in interregional relationship and allows comparison of two or more sample populations. Figure 4 explains the different methods available to compute functional connectivity within the DMN. The most common approach to computing RSFC maps is using a hypothesis-driven or model-driven technique such as seed-based or region-of-interest (ROI)-based correlation analysis (Biswal et al. 1995; Cordes et al. 2000; Fransson 2005; Andrews-Hanna et al. 2007). Seed-based correlation analysis is based on the selection of a reference voxel or seed to cross-correlate with all the other voxels in the brain. The result is a whole-brain RSFC map of a seed region that

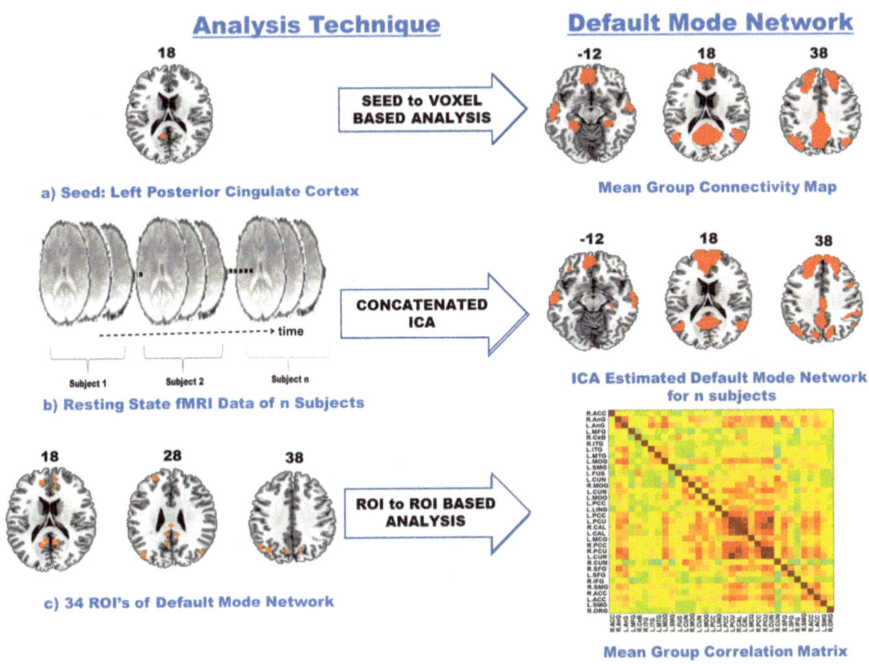

Fig. 4 Different approaches to measuring functional connectivity of the default-mode network at rest. Displayed here are (**a**) a 5-mm spherical seed at left posterior cingulate cortex used as a reference time series to perform whole-brain voxel-wise correlation known as seed-based analysis. Output of seed-based analysis results in whole-brain correlation of left PCC with regions of DMN as well as other regions. (**b**) Probabilistic concatenated independent component analysis (ICA) of n subjects results in DMN as one of the components and (**c**) region of interest (ROI) to ROI-based correlation matrix of DMN comprising 34 seeds defined by Dosenbach atlas (Dosenbach et al. 2010)

can be converted to fisher z-maps for statistical analysis. The advantage of seed-based correlation analysis is its good reliability (Shehzad et al. 2009), making it one of the most frequently used resting-state approaches. The method, however, requires empirical knowledge of the seed region that could bias and limit the scope of the findings. The technique may also be sensitive to structured noise introduced due to head motion, scanner-induced artifacts, or spatially overlapping RSNs.

Alternatively, principal component analysis (PCA) (Friston et al. 1993) and independent component analysis (ICA) (Beckmann et al. 2005) are data-driven techniques that do not require a priori model to compute functional connectivity measures. PCA generates a set of Eigen maps that represents connectivity and explains the maximum amount of variance in the fMRI data. Due to poor performance in noisy data and difficulty in interpretation of Eigen maps, PCA is often used as a precursor to ICA to reduce data dimensionality (Baumgartner et al. 2000). Spatial ICA on the other hand calculates spatial-temporal components of fMRI data that are maximally independent and not necessarily explain the highest variance. ICA using concatenated time series technique can be used to characterize the spatiotemporal patterns common to a group of samples. Further, concatenated ICA technique can be less prone to physiological noise sources such as cardiac and respiratory cycles as well as global signal in the fMRI data. The downside of ICA technique includes run-to-run variability, model order selection, manual classification of components, and neuroanatomical inference of alterations in network maps (Chang and Glover 2010). A comprehensive review by Cole and colleagues and Azeez and colleagues is recommended to readers interested in learning more about the resting-state methods (Cole et al. 2010; Azeez and Biswal 2017).

A popular alternative to voxel-based statistics is the multivariate analysis of different RSNs using graph theory. The brain as a complex biological network is known to topologically exhibit small-world architecture and modular structure for efficient multi-scale integration of information. By defining the brain as a set of nodes (ROIs) and edges (correlation, covariance, mutual information, etc.), one can quantify network metrics of RSNs. Nodes are defined based on parcellation schemes using functional activation patterns or anatomical atlas such as Automated Anatomic Labeling and Harvard-Oxford Atlas. An adjacency matrix consisting of edges from multiple nodes is generated to measure meaningful graph properties, such as local efficiency, global efficiency, and degree. Graph metrics are usually thresholded and compared with a benchmark network such as random network or compared with other subject groups.

Functional Activation

Functional activation or segregation is typically quantified using frequency-dependent analysis or regional homogeneity (ReHo) analysis. Frequency-dependent analysis such as amplitude of low-frequency fluctuation (ALFF) is in simplest terms a measure of the strength of BOLD activity at rest (Zang et al. 2007). It is calculated at the voxel level as the square root of the power of BOLD time series in the low-frequency range. ALFF has been shown to be sensitive to large vessels and ventricles influencing the BOLD time series (Zuo et al. 2010). Hence, a modification of ALFF, known as fractional ALFF (fALFF), is alternatively preferred and computed by measuring ALFF as a fraction of the average power in the entire frequency range (0–2.5 Hz) (Zou et al. 2008). Both ALFF and fALFF have been shown to be fairly reliable and carry useful neuronal information that can complement functional connectivity analysis in neurological populations.

ReHo maps the strength of temporal interdependence between a set of neighboring voxels or clusters (Zang et al. 2004). First introduced by Zang and colleagues, it is under the assumption that spatially adjacent regions of a network have high temporal correlation during a given condition. Therefore, any differences in the functional segregation of a region in a condition or population could be reflected in the extent of local synchronization of a given region. ReHo is computed using Kendall's coefficient of concordance for a fixed set of neighboring voxels (e.g., 7, 19, 27 voxels), where higher ReHo values denote smaller temporal variability within a cluster during a given condition. Group analysis of ReHo simply requires smoothing followed by statistical analysis, although due to the inherently local nature of the technique, the extent of smoothing and the size of the cluster affect the ReHo values (Tian et al. 2012; Zuo et al. 2013). Yet, due to its insensitivity to differences in the hemodynamic response of different regions and runs, ReHo is a robust measure of local homogeneity of regions at resting-state condition.

Dynamic Functional Connectivity

Standard functional connectivity measures are under the assumption that similar to fixed anatomical connections, resting-state inter-regional correlations are stable during the entire duration of the scan. However, exceptions exist as some resting-state studies have demonstrated alterations in within- and between-network connectivity during different behavioral states such as resting eyes closed, resting with visual fixation, and sleep. (Horovitz et al. 2009; Patriat et al. 2013). As resting-state scans can involve undirected attention or "mind-wandering" and other physiological changes, it has led to the hypothesis that RSFC measures also vary within the duration of a standard scan. Consequently, the FC measured during shorter time scales (~seconds) may carry unique spatiotemporal patterns that are characteristic to ongoing cognitive, affective, and autonomic functions at rest. Chang and colleagues confirmed the variability in the strength, phase, and coherence of RSFC between posterior cingulate cortex and other brain regions within the duration of a standard resting-state scan (Chang and Glover 2010). The finding of resting-state dynamic FC (dFC) has led more studies to explore and develop new methods to study and interpret the dynamic features of RSNs in healthy and neurological populations. The most common approach to examine dFC is the sliding window analysis, where the dynamic connectivity of a set of time series representing brain resting-state activity is computed for a predefined temporal window (e.g., 30 s) and shifted by a fixed number of data points (e.g., 2 s for TR = 2 s). This results in a dynamic connectivity metric (often using Pearson's correlation) for the entire duration of the scan calculated over short time windows. Other connectivity metrics utilized within a window include covariance, ICA, ReHo and multiplication of temporal derivatives. Primary measures to characterize dFC include standard deviation, coefficient of variation, and ALFF of dFC time series, while repetitive spatiotemporal patterns known as dFC states are computed using matrix decomposition techniques such as k-mean clustering or PCA. Alternative techniques include time-frequency analysis and point process analysis. For an extensive review on the methodology, interpretation and limitations of dFC refer to publications by Hutchison and colleagues and Preti and colleagues (Hutchison et al. 2013; Preti et al. 2017).

A less explored mechanism of resting-state dFC involves a modulatory interaction of a third region on the functional connectivity of two ROIs. Modulatory interactions between any three ROIs at rest can be modeled using a linear regression framework known as physiophysiological interaction (PPI). First implemented on resting-state data by Di and Biswal (2013), PPI technique is a variant of psychophysiological interaction introduced by Friston et al. (1997). Psychophysiological interaction is used to model the effect of one region on another depending on the psychological process presented to the individual. Similarly, the PPI model at resting state tests whether the relationship between the first target region (dependent variable) and a second region (1st explanatory variable) is modulated based on the interaction (or FC) between the second region and a third region (2nd explanatory variable). The PPI term or beta value that explains the relationship between the connectivity of two regions modulated by a third region can be modeled for specific networks or the whole brain using a general linear model applied using SPM platform. PPI term is computed at the neuronal level by first deconvolving the explanatory time series of the two ROIs by a canonical hemodynamic response function (hrf). Next, the two deconvolved time series are detrended, point multiplied, and convolved with a hrf function once again to generate PPI beta values that explain the variance in the target ROI (Di and Biswal 2013, 2015). The main limitations of PPI technique are the relatively low reliability compared to functional connectivity measures and the need for a priori seed selection to calculate modulatory interactions (Di and Biswal 2017).

Effective Connectivity

Effective connectivity is defined as the influence one neural system has on another (Friston et al. 2003). It provides inference regarding the dependency or causal connectivity between two neural systems or regions. The most standard techniques to measure effective connectivity in resting-state fMRI are granger causality analysis, structural equation modeling, and dynamic causal modeling (DCM). Granger causality analysis is based on the principle of "cause before effect" where an event is caused or influenced by a past event. Using this principle, an autoregressive model is generally used to predict the BOLD neuronal activity in a target region based on the BOLD neuronal activity in one or more seed regions (Kaminski et al. 2001;

Goebel et al. 2003). Structural equation modeling, on the other hand, is a statistical method mainly used as a confirmatory rather than an exploratory technique. Also known as path analysis, it tests the influence of specific nodes or ROIs in a predefined model (James et al. 2009). The drawback is the need for a predefined optimal model and difficulty modeling systems with feedback between a set of nodes. Both of these techniques (granger causality and structural equation modeling) model only the steady-state brain connectivity patterns during resting state; DCM originally introduced by Friston and colleagues for task-fMRI data models the dynamic effective relationship of brain regions at the neuronal level using differential equations (Friston et al. 2003, 2013; Friston 2011). A variation of DCM has been applied on resting-state data using a neuronally plausible model of coupled neuronal states to generate the complex cross spectra among observed hemodynamic responses. A few recent publications on the application of DCM in resting state describe this technique in greater detail (Di and Biswal 2014; Friston et al. 2014).

Resting-State fMRI Changes in Clinical Populations

A vast number of publications is available about resting-state fMRI connectivity and activity in healthy individuals and clinical populations. Here, we will provide a short summary of several interesting findings in normal aging, major depressive disorder, autism, and patients with brain tumors. We chose these conditions to demonstrate the application of resting-state fMRI in the investigation of altered neuronal (autism and major depressive disorder), vascular (brain tumor), and neurovascular factors (aging).

Aging

Due to our demographic change, it is essential to gain substantial knowledge about resting-state changes during normal aging, particularly, to distinguish them from pathological states such as Alzheimer's or Parkinson's diseases. Age has been identified as an important factor in resting-state studies (Biswal et al. 2010). Most reliably, a decrease in anterior and posterior DMN connectivity was found during aging (Bluhm et al. 2008; Damoiseaux et al. 2008; Esposito et al. 2008; Koch et al. 2010; Wu et al. 2011; Toussaint et al. 2014), which may start as early as middle age (Evers et al. 2012) and is associated with reorganization of the network structure (Meunier et al. 2009). In addition, this DMN connectivity decrease has been associated with a decline in cognitive function such as memory performance (Damoiseaux et al. 2008; Mevel et al. 2013). Of note, these studies focused mainly on DMN connectivity and were conducted in relatively small sample sizes. Studies with larger samples, for instance, Tomasi and Volkow (2012) (n = 913) and Salami and colleagues

(Salami et al. 2014) (n = 339), have reproduced DMN connectivity changes but additionally show that the connectivity of the dorsal attention network decreases while the connectivity of somatosensory, motor, hippocampus, and subcortical networks increases with aging. Other studies also demonstrated a decrease in salience network connectivity both within and between networks (salience-auditory; DMN-visual) (Onoda et al. 2012). A decline in time course spectral power and functional connectivity across broad RSNs were also confirmed by Allen et al. (2011) in 603 healthy study participants. In general, long-range functional connectivity density seems to be more strongly altered with age than short-range connections.

Also worth mentioning that several research groups have investigated the effect of age in the same publically available dataset using different methods (Nathan Kline Institute/Rockland Sample, NKI–RS, ages 6–85 years old, (Nooner et al. 2012)), for example, using static functional connectivity (Uddin et al. 2011; Betzel et al. 2014; Tian et al. 2016), graph theory (Cao et al. 2014), ICA (Yang et al. 2014; Vij et al. 2018), Hurst exponent (Dong et al. 2018), and dynamic functional connectivity (Fukushima et al. 2015; Qin et al. 2015; Chen et al. 2017, 2018; Tian et al. 2018). These investigations with multiple brain functional measures can help to better interpret and understand the findings in normal aging. A further interesting aspect is the machine-learning approach for the development of potential biomarkers of aging (Dosenbach et al. 2010; Meier et al. 2012; Smyser et al. 2016) although longitudinal studies with large sample sizes, such as the study by Salami et al. (2016), are also necessary to validate prediction accuracy due to a large number of plausible factors influencing the aging process. Hypotheses about the causes of altered connectivity in aging have already been discussed in comprehensive reviews (Buckner 2004; Ferreira and Busatto 2013).

Major Depressive Disorder (MDD)

Major depressive disorder affects more than 300 million people worldwide (Mathers 2008). It impacts mood, motivation, cognition, vegetative functions, and the socio-economic status of each individual suffering from MDD. Previously found limbic-cortical dysregulation (Mayberg et al. 1997) was confirmed by studies using resting-state fMRI (first reproduced in Anand et al. 2005). For instance, an altered cortical regulation of medial and lateral prefrontal cortices (Yao et al. 2009; Zhu et al. 2012; Zhang et al. 2016) is suggested to disinhibit and increase activation of limbic structures such as amygdala, hippocampus, and subgenual anterior cingulate cortex (Greicius et al. 2007; Cao et al. 2012; Guo et al. 2013). Notably, abnormal connectivity between prefrontal cortex and anterior cingulate cortex, insula, and limbic structures was broadly found in MDD (to mention only few: Cullen et al. 2009; Horn et al. 2010; Kenny et al. 2010; Veer et al. 2010; Lui et al. 2011; Connolly et al. 2017) and also shown via dynamic approach (Kaiser et al. 2016). In addition, connectivity changes of subcortical structures such as thalamus and striatum have been often detected (Anand et al. 2005, 2009; Greicius et al. 2007; Salomons et al.

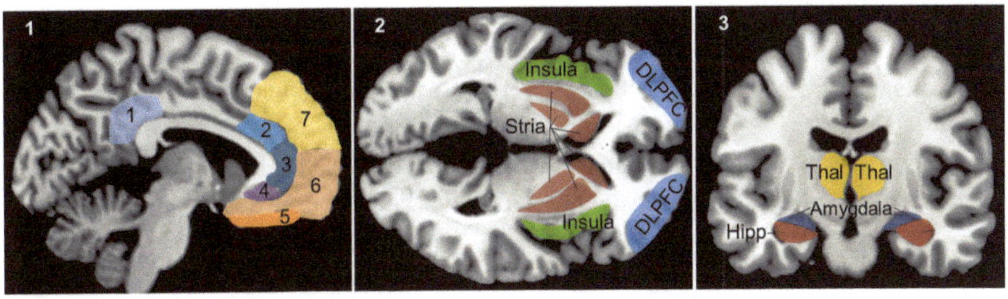

Fig. 5 Frequently reported brain regions with functional changes in major depressive disorder: (1) 1 = posterior cingulate cortex, as key structure of default-mode network, 2 = dorsal anterior cingulate cortex, 3 = perigenual cingulate cortex, 4 = subgenual cingulate cortex, 5 = orbitofrontal cortex, 6 = ventromedial prefrontal cortex, 7 = dorsomedial prefrontal cortex. (2) DLPFC = dorsolateral prefrontal cortex (blue), Insula = insular cortex (green), Stria = corpus striatum (dusky pink). (3) Amygdala (dark blue), Hipp = hippocampus (red), Thal = thalamus (yellow). Figure was created with MRIcron (http://people.cas.sc.edu/rorden/mricron/index.html)

2014). Some studies also reported abnormal connectivity of entire networks in patients suffering from MDD. Briefly, a hyperconnectivity within the DMN and a hypoconnectivity within frontoparietal network as well as an abnormal connectivity between task-positive and task-negative networks were found in depressed patients (Bluhm et al. 2008; Sheline et al. 2010; Kaiser et al. 2015; Zheng et al. 2015). MDD is a highly heterogeneous mental disease that has several subtypes with distinct neurobiological profiles (Woelfer et al. 2018). A promising work from Drysdale and colleagues demonstrated a potential subclassification of MDD into four neurophysiological subtypes based on the dysfunctional resting-state fMRI connectivity of limbic and frontostriatal regions (Drysdale et al. 2017). In addition to providing novel insights into the pathology of MDD, studies seek to develop early resting-state fMRI biomarkers to predict the treatment response before or during the antidepressant therapy (Cheng et al. 2017; Philip et al. 2018; and reviewed in Dichter et al. 2015). A depiction summarizing the regions regularly reported to be altered in MDD is given in Fig. 5.

Autism Spectrum Disorders (ASDs)

Autism spectrum disorders are neurodevelopmental conditions and defined by difficulties in social interactions, communication, and stereotyped behavior (Castillo et al. 2007). The underconnectivity theory (Just et al. 2004, 2012) was initially defined based on several – mostly task – fMRI studies which showed lower anatomical and functional connectivity between the frontal cortex and posterior regions in ASD (Kana et al. 2006; Just et al. 2007; Kleinhans et al. 2008; Koshino et al. 2008). Based on resting-state fMRI studies, a more complex picture of increased and decreased connectivity throughout the whole brain was developed in the following years (Kennedy and Courchesne 2008; Monk et al. 2009; Assaf et al. 2010; Di Martino et al. 2011; Supekar et al. 2013).

Eminent was the release of the Autism Brain Imaging Data Exchange (ABIDE) dataset (Di Martino et al. 2014) which has largely been used by the ASD studies discussed here (only few examples: Cerliani et al. 2015; Abraham et al. 2017; Jung et al. 2017). Di Martino and colleagues demonstrated a comprehensive picture of decreased intrinsic functional connectivity and brain function in ASD patients (n = 360) compared to typically developed individuals (n = 403). However, the authors also show increased connectivity between subcortical and primary sensorimotor areas. The increased functional connectivity in sensorimotor and subcortical (thalamus, basal ganglia) networks was also shown in later studies (Cerliani et al. 2015; Khan et al. 2015). Additionally, boys have been shown to be more likely to develop ASD than girls (Baxter et al. 2015). Therefore, recent resting-state connectivity studies are shifting focus toward understanding sex-dependent and sex-independent changes of ASD (Alaerts et al. 2016; Di and Biswal 2016; Floris et al. 2018). Furthermore, since network organization is expected to differ between developing brains and adolescents' or adults' brains (Dosenbach et al. 2010; Peper et al. 2011), the age of the study participants also requires careful consideration (Uddin et al. 2013; Nomi and Uddin 2015; Neufeld et al. 2018). In parallel to the functional alterations, focal WM reductions in children, adolescents, and adults with autism using meta-analysis (Di et al. 2018) has been observed. An urgent question, however, is how brain network changes are associated with autistic features and can be used as imaging biomarker, particularly, to define subtypes of ASD as well as predict individual outcome (Iidaka 2015; Plitt et al. 2015; Balsters et al. 2016; Chen et al. 2016; Abraham et al. 2017; Hong et al. 2018).

Brain Tumors

Within recent years, resting-state fMRI data from patients with brain tumors have attracted a lot of interest. The location of the tumor is obviously a crucial factor. A small number of studies reported a brain tumor-related neurovascular uncoupling with decreased BOLD signal in the ipsilesional (ipsilateral to tumor side) area when compared to the contralesional area of the sensorimotor cortex (Agarwal et al. 2016). In contrast, a study found significantly altered connectivity in DMN and executive control network with changes mainly observed in contralesional areas of the network (Maesawa et al. 2015). In general, high-grade tumors and brain tumors in the parietal cortex seem to lead to a stronger decrease in DMN connectivity when compared to low-grade tumors or frontal brain tumors (Harris et al. 2014; Mallela et al. 2016). Analyzing RSFC of somatosensory, motor, and language networks not only helps understand the tumor pathology but also proves useful for presurgical planning in patients with brain tumor (Kokkonen et al. 2009; Zhang et al. 2009; Batra et al. 2016; Cochereau et al. 2016; Lee et al. 2016; Lu et al. 2017). Remarkably, intraoperative resting-state fMRI scans of sensorimotor networks have been used to predict postoperative outcomes (Roder et al. 2016).

Outlook and Limitations of Resting-State Approach

As the clinical applicability of resting-state fMRI is rapidly expanding, there is an urgent need to address and overcome the methodological limitations to achieve high test–retest reliability in resting-state metrics. Low intra-individual and high inter-individual variability in resting-state connectivity metrics establishes stability and selectivity necessary for developing clinical tests using neuroimaging biomarkers. Existing resting-state setting and data processing strategies result in fair to almost perfect reliability depending on the different functional connectivity and graph metrics. Local ReHo analysis and intrinsic connectivity analysis using ICA have the highest reliability, while graph measures and frequency-dependent measures such as ALFF and fALFF have a fair to moderate reliability (Han et al. 2011; Zuo et al. 2014). Simple measures such as scanning in a fixed resting-state setting can improve reliability. For instance, studies show that a scan duration of 5 min is sufficient for 50% reliability, whereas increasing the duration to 13 min for intra-session and 9 min for inter-session has been shown to further improve reliability. However, many of the reliability issues arise due to physiological noise and preprocessing procedures (Van Dijk et al. 2010; Birn et al. 2013).

Head motion-induced artifacts are one of the most common concerns of RSFC analyses as between-group neural differences could be driven due to one group moving more than other (e.g., older adults more than young adults, patients more than healthy controls). Head motion results in distortion of BOLD activity by displacing voxels from its initial site of excitation resulting in spurious variance in BOLD time series within nearby voxels. The impact of scanner movement on RSFC measures is well documented in pediatric, adolescent, and aging populations, where in-scanner movement even in the magnitude of micro-movement has been associated with diminishing long-range connections and increasing short-range connections, such as in the homologous primary motor cortex (Power et al. 2012; Satterthwaite et al. 2012; Van Dijk et al. 2012). Subject-level motion correction appears to be inadequate in efficiently removing individual-specific motion effects. Hence, additional steps of correction such as mean frame-wise displacement are recommended at the group-level analysis. Other techniques such as global signal regression and z-standardization, although not explicitly used for motion correction, have also been shown effective in removing motion-related effects in group-level analyses (Power et al. 2014).

Global signal regression is the practice of removing the average time series of the whole brain from the time series of all voxels using linear regression (Murphy and Fox 2017). While the global average time series is expected to account for the variance of global confounds such as head-motion and other physiological noises, there is a major caveat to this. Electrophysiological studies in monkeys suggest a neuronal basis to the average global signal (Scholvinck et al. 2010; Wen and Liu 2016). Therefore, removing the average global signal has been shown to carry the risk of removing true neuronal effects. A lack of consensus on a one-way fix-all approach to remove noise in resting-state fMRI data emphasizes the need for careful design of pre- and postprocessing strategies depending on the specific goals and population studied.

Despite some methodological limitations, resting-state fMRI holds great promise for translation of neuroimaging findings to clinical application. Resting-state fMRI technique inherently possesses the advantage of allowing the investigation of brain's intrinsic activity with greater signal-to-noise ratio while also avoiding confounds associated to task (Fox and Greicius 2010). The advancement of data analysis and noise correction techniques has extended the application of resting-state fMRI in populations that were previously difficult to scan such as neonates or cognitively challenged as well as during sleep and sedation. Further, high temporal resolution and multiband imaging technology have allowed studies to record meaningful spontaneous fluctuations at higher frequencies of up to 0.75 Hz, the functional implication of which is yet to be established (Cordes et al. 2001; Wu et al. 2008; Boubela et al. 2013; Lee et al. 2013; Gohel and Biswal 2015). And of course, combining resting-state data with other modalities could serve to bridge the gap between functional findings and biomarkers for prediction, diagnosis, and prognosis of clinical conditions. However, more work is required to improve the reliability of multi-modal studies before resting-state fMRI can be routinely used in the clinical realm.

References

Abraham, A., M. P. Milham, A. Di Martino, R. C. Craddock, D. Samaras, B. Thirion and G. Varoquaux (2017). "Deriving reproducible biomarkers from multi-site resting-state data: An Autism-based example." Neuroimage **147**: 736–745.

Agarwal, S., H. I. Sair, R. Airan, J. Hua, C. K. Jones, H. Y. Heo, A. Olivi, M. A. Lindquist, J. J. Pekar and J. J. Pillai (2016). "Demonstration of Brain Tumor-Induced Neurovascular Uncoupling in Resting-State fMRI at Ultrahigh Field." Brain Connect **6**(4): 267–272.

Agarwal, S., H. I. Sair, N. Yahyavi-Firouz-Abadi, R. Airan and J. J. Pillai (2016). "Neurovascular uncoupling in resting state fMRI demonstrated in patients with primary brain gliomas." J Magn Reson Imaging **43**(3): 620–626.

Aiello, M., E. Salvatore, A. Cachia, S. Pappatà, C. Cavaliere, A. Prinster, E. Nicolai, M. Salvatore, J.-C. Baron and M. Quarantelli (2015). "Relationship between simultaneously acquired resting-state regional cerebral glucose metabolism and functional MRI: A PET/MR hybrid scanner study." NeuroImage **113**: 111–121.

Alaerts, K., S. P. Swinnen and N. Wenderoth (2016). "Sex differences in autism: a resting-state fMRI investigation of functional brain connectivity in males and females." Soc Cogn Affect Neurosci **11**(6): 1002–1016.

Allen, E. A., E. B. Erhardt, E. Damaraju, W. Gruner, J. M. Segall, R. F. Silva, M. Havlicek, S. Rachakonda, J. Fries, R. Kalyanam, A. M. Michael, A. Caprihan, J. A. Turner, T. Eichele, S. Adelsheim, A. D. Bryan, J. Bustillo, V. P. Clark, S. W. Feldstein Ewing, F. Filbey, C. C. Ford, K. Hutchison, R. E. Jung, K. A. Kiehl, P. Kodituwakku, Y. M. Komesu, A. R. Mayer, G. D. Pearlson, J. P. Phillips, J. R. Sadek, M. Stevens, U. Teuscher, R. J. Thoma and V. D. Calhoun (2011). "A baseline for the multivariate comparison of resting-state networks." Front Syst Neurosci **5**: 2.

Anand, A., Y. Li, Y. Wang, M. J. Lowe and M. Dzemidzic (2009). "Resting state corticolimbic connectivity abnormalities in unmedicated bipolar disorder and unipolar depression." Psychiatry research **171**(3): 189–198.

Anand, A., Y. Li, Y. Wang, J. Wu, S. Gao, L. Bukhari, V. P. Mathews, A. Kalnin and M. J. Lowe (2005). "Activity and connectivity of brain mood regulating circuit in depression: a functional magnetic resonance study." Biol Psychiatry **57**(10): 1079–1088.

Andrews-Hanna, J. R., A. Z. Snyder, J. L. Vincent, C. Lustig, D. Head, M. E. Raichle and R. L. Buckner (2007). "Disruption of large-scale brain systems in advanced aging." Neuron **56**(5): 924–935.

Assaf, M., K. Jagannathan, V. D. Calhoun, L. Miller, M. C. Stevens, R. Sahl, J. G. O'Boyle, R. T. Schultz and G. D. Pearlson (2010). "Abnormal functional connectivity of default mode subnetworks in autism spectrum disorder patients." Neuroimage **53**(1): 247–256.

Azeez, A. K. and B. B. Biswal (2017). "A Review of Resting-State Analysis Methods." Neuroimaging Clin N Am **27**(4): 581–592.

Balsters, J. H., D. Mantini, M. A. J. Apps, S. B. Eickhoff and N. Wenderoth (2016). "Connectivity-based parcellation increases network detection sensitivity in resting state fMRI: An investigation into the cingulate cortex in autism." Neuroimage Clin **11**: 494–507.

Bassett, D. S., E. Bullmore, B. A. Verchinski, V. S. Mattay, D. R. Weinberger and A. Meyer-Lindenberg (2008). "Hierarchical organization of human cortical networks in health and schizophrenia." J Neurosci **28**(37): 9239–9248.

Batra, P., S. K. Bandt and E. C. Leuthardt (2016). "Resting state functional connectivity magnetic resonance imaging integrated with intraoperative neuronavigation for functional mapping after aborted awake craniotomy." Surg Neurol Int **7**: 13.

Baumgartner, R., L. Ryner, W. Richter, R. Summers, M. Jarmasz and R. Somorjai (2000). "Comparison of two exploratory data analysis methods for fMRI: fuzzy clustering vs. principal component analysis." Magn Reson Imaging **18**(1): 89–94.

Baxter, A. J., T. S. Brugha, H. E. Erskine, R. W. Scheurer, T. Vos and J. G. Scott (2015). "The epidemiology and global burden of autism spectrum disorders." Psychol Med **45**(3): 601–613.

Beckmann, C. F., M. DeLuca, J. T. Devlin and S. M. Smith (2005). "Investigations into resting-state connectivity using independent component analysis." Philos Trans R Soc Lond B Biol Sci **360**(1457): 1001–1013.

Behzadi, Y., K. Restom, J. Liau and T. T. Liu (2007). "A component based noise correction method (CompCor) for BOLD and perfusion based fMRI." Neuroimage **37**(1): 90–101.

Betzel, R. F., L. Byrge, Y. He, J. Goni, X. N. Zuo and O. Sporns (2014). "Changes in structural and functional connectivity among resting-state networks across the human lifespan." Neuroimage **102 Pt 2**: 345–357.

Bhattacharyya, P. K. and M. J. Lowe (2004). "Cardiac-induced physiologic noise in tissue is a direct observation of cardiac-induced fluctuations." Magn Reson Imaging **22**(1): 9–13.

Birn, R. M., E. K. Molloy, R. Patriat, T. Parker, T. B. Meier, G. R. Kirk, V. A. Nair, M. E. Meyerand and V. Prabhakaran (2013). "The effect of scan length on the reliability of resting-state fMRI connectivity estimates." Neuroimage **83**: 550–558.

Biswal, B., A. G. Hudetz, F. Z. Yetkin, V. M. Haughton and J. S. Hyde (1997). "Hypercapnia reversibly suppresses low-frequency fluctuations in the human motor cortex during rest using echo-planar MRI." J Cereb Blood Flow Metab **17**(3): 301–308.

Biswal, B., F. Z. Yetkin, V. M. Haughton and J. S. Hyde (1995). "Functional connectivity in the motor cortex of resting human brain using echo-planar MRI." Magn Reson Med **34**(4): 537–541.

Biswal, B. B., M. Mennes, X.-N. Zuo, S. Gohel, C. Kelly, S. M. Smith, C. F. Beckmann, J. S. Adelstein, R. L. Buckner, S. Colcombe, A.-M. Dogonowski, M. Ernst, D. Fair, M. Hampson, M. J. Hoptman, J. S. Hyde, V. J. Kiviniemi, R. Kotter, S.-J. Li, C.-P. Lin, M. J. Lowe, C. Mackay, D. J. Madden, K. H. Madsen, D. S. Margulies, H. S. Mayberg, K. McMahon, C. S. Monk, S. H. Mostofsky, B. J. Nagel, J. J. Pekar, S. J. Peltier, S. E. Petersen, V. Riedl, S. A. R. B. Rombouts, B. Rypma, B. L. Schlaggar, S. Schmidt, R. D. Seidler, G. J. Siegle, C. Sorg, G.-J. Teng, J. Veijola, A. Villringer, M. Walter, L. Wang, X.-C. Weng, S. Whitfield-Gabrieli, P. Williamson, C. Windischberger, Y.-F. Zang, H.-Y. Zhang, F. X. Castellanos and M. P.

Milham (2010). "Toward discovery science of human brain function: Supporting Information." Proceedings of the National Academy of Sciences **107**(10): 4734–4739.

Biswal, B. B., M. Mennes, X. N. Zuo, S. Gohel, C. Kelly, S. M. Smith, C. F. Beckmann, J. S. Adelstein, R. L. Buckner, S. Colcombe, A. M. Dogonowski, M. Ernst, D. Fair, M. Hampson, M. J. Hoptman, J. S. Hyde, V. J. Kiviniemi, R. Kotter, S. J. Li, C. P. Lin, M. J. Lowe, C. Mackay, D. J. Madden, K. H. Madsen, D. S. Margulies, H. S. Mayberg, K. McMahon, C. S. Monk, S. H. Mostofsky, B. J. Nagel, J. J. Pekar, S. J. Peltier, S. E. Petersen, V. Riedl, S. A. Rombouts, B. Rypma, B. L. Schlaggar, S. Schmidt, R. D. Seidler, G. J. Siegle, C. Sorg, G. J. Teng, J. Veijola, A. Villringer, M. Walter, L. Wang, X. C. Weng, S. Whitfield-Gabrieli, P. Williamson, C. Windischberger, Y. F. Zang, H. Y. Zhang, F. X. Castellanos and M. P. Milham (2010). "Toward discovery science of human brain function." Proc Natl Acad Sci U S A **107**(10): 4734–4739.

Biswal, B. B., J. Van Kylen and J. S. Hyde (1997). "Simultaneous assessment of flow and BOLD signals in resting-state functional connectivity maps." NMR Biomed **10**(4–5): 165–170.

Bluhm, R. L., E. A. Osuch, R. A. Lanius, K. Boksman, R. W. Neufeld, J. Theberge and P. Williamson (2008). "Default mode network connectivity: effects of age, sex, and analytic approach." Neuroreport **19**(8): 887–891.

Boubela, R. N., K. Kalcher, W. Huf, C. Kronnerwetter, P. Filzmoser and E. Moser (2013). "Beyond Noise: Using Temporal ICA to Extract Meaningful Information from High-Frequency fMRI Signal Fluctuations during Rest." Front Hum Neurosci **7**: 168.

Buckner, R. L. (2004). "Memory and executive function in aging and AD: multiple factors that cause decline and reserve factors that compensate." Neuron **44**(1): 195–208.

Buckner, R. L., J. R. Andrews-Hanna and D. L. Schacter (2008). "The brain's default network: anatomy, function, and relevance to disease." Ann N Y Acad Sci **1124**: 1–38.

Cao, M., J. H. Wang, Z. J. Dai, X. Y. Cao, L. L. Jiang, F. M. Fan, X. W. Song, M. R. Xia, N. Shu, Q. Dong, M. P. Milham, F. X. Castellanos, X. N. Zuo and Y. He (2014). "Topological organization of the human brain functional connectome across the lifespan." Dev Cogn Neurosci **7**: 76–93.

Cao, X., Z. Liu, C. Xu, J. Li, Q. Gao, N. Sun, Y. Xu, Y. Ren, C. Yang and K. Zhang (2012). "Disrupted resting-state functional connectivity of the hippocampus in medication-naive patients with major depressive disorder." J Affect Disord **141**(2–3): 194–203.

Castillo, R. J., Carlat, D. J., Millon, T., Millon, C. M., Meagher, S., Grossman, S., … & American Psychiatric Association. (2007). "Diagnostic and statistical manual of mental disorders." Washington, DC: American Psychiatric Association Press.

Chicago

Cerliani, L., M. Mennes, R. M. Thomas, A. Di Martino, M. Thioux and C. Keysers (2015). "Increased Functional Connectivity Between Subcortical and Cortical Resting-State Networks in Autism Spectrum Disorder." JAMA Psychiatry **72**(8): 767–777.

Chai, X. J., A. N. Castanon, D. Ongur and S. Whitfield-Gabrieli (2012). "Anticorrelations in resting state networks without global signal regression." Neuroimage **59**(2): 1420–1428.

Chang, C. and G. H. Glover (2010). "Time-frequency dynamics of resting-state brain connectivity measured with fMRI." Neuroimage **50**(1): 81–98.

Chao-Gan, Y. and Z. Yu-Feng (2010). "DPARSF: A MATLAB Toolbox for "Pipeline" Data Analysis of Resting-State fMRI." Front Syst Neurosci **4**: 13.

Chen, H., X. Duan, F. Liu, F. Lu, X. Ma, Y. Zhang, L. Q. Uddin and H. Chen (2016). "Multivariate classification of autism spectrum disorder using frequency-specific resting-state functional connectivity--A multi-center study." Prog Neuropsychopharmacol Biol Psychiatry **64**: 1–9.

Chen, L. M., P. F. Yang, F. Wang, A. Mishra, Z. Shi, R. Wu, T. L. Wu, G. H. Wilson, 3rd, Z. Ding and J. C. Gore (2017). "Biophysical and neural basis of resting state functional connectivity: Evidence from non-human primates." Magn Reson Imaging **39**: 71–81.

Chen, Y., W. Wang, X. Zhao, M. Sha, Y. Liu, X. Zhang, J. Ma, H. Ni and D. Ming (2017). "Age-Related Decline in the Variation of Dynamic Functional Connectivity: A Resting State Analysis." Front Aging Neurosci **9**: 203.

Chen, Y., X. Zhao, X. Zhang, Y. Liu, P. Zhou, H. Ni, J. Ma and D. Ming (2018). "Age-related early/late variations of functional connectivity across the human lifespan." Neuroradiology **60**(4): 403–412.

Cheng, Y., J. Xu, D. Arnone, B. Nie, H. Yu, H. Jiang, Y. Bai, C. Luo, R. A. Campbell, B. Shan, L. Xu and X. Xu (2017). "Resting-state brain alteration after a single dose of SSRI administration predicts 8-week remission of patients with major depressive disorder." Psychol Med **47**(3): 438–450.

Cochereau, J., J. Deverdun, G. Herbet, C. Charroud, A. Boyer, S. Moritz-Gasser, E. Le Bars, F. Molino, A. Bonafe, N. Menjot de Champfleur and H. Duffau (2016). "Comparison between resting state fMRI networks and responsive cortical stimulations in glioma patients." Hum Brain Mapp **37**(11): 3721–3732.

Cole, D. M., S. M. Smith and C. F. Beckmann (2010). "Advances and pitfalls in the analysis and interpretation of resting-state FMRI data." Front Syst Neurosci **4**: 8.

Connolly, C. G., T. C. Ho, E. H. Blom, K. Z. LeWinn, M. D. Sacchet, O. Tymofiyeva, A. N. Simmons and T. T. Yang (2017). "Resting-state functional connectivity of the amygdala and longitudinal changes in depression severity in adolescent depression." J Affect Disord **207**: 86–94.

Cordes, D., V. Haughton, J. D. Carew, K. Arfanakis and K. Maravilla (2002). "Hierarchical clustering to measure connectivity in fMRI resting-state data." Magn Reson Imaging **20**(4): 305–317.

Cordes, D., V. M. Haughton, K. Arfanakis, J. D. Carew, P. A. Turski, C. H. Moritz, M. A. Quigley and M. E. Meyerand (2001). "Frequencies contributing to functional connectivity in the cerebral cortex in "resting-state" data." AJNR Am J Neuroradiol **22**(7): 1326–1333.

Cordes, D. V., M. Haughton, K. Arfanakis, G. J. Wendt, P. A. Turski, C. H. Moritz, M. A. Quigley and M. E. Meyerand (2000). "Mapping functionally related regions of brain with functional connectivity MR imaging." AJNR Am J Neuroradiol **21**(9): 1636–1644.

Cox, R. W. (1996). "AFNI: software for analysis and visualization of functional magnetic resonance neuroimages." Comput Biomed Res **29**(3): 162–173.

Cullen, K. R., D. G. Gee, B. Klimes-Dougan, V. Gabbay, L. Hulvershorn, B. A. Mueller, J. Camchong, C. J. Bell, A. Houri, S. Kumra, K. O. Lim, F. X. Castellanos and M. P. Milham (2009). "A preliminary study of functional connectivity in comorbid adolescent depression." Neurosci Lett **460**(3): 227–231.

Dagli, M. S., J. E. Ingeholm and J. V. Haxby (1999). "Localization of cardiac-induced signal change in fMRI." Neuroimage **9**(4): 407–415.

Damoiseaux, J. S., C. F. Beckmann, E. J. Arigita, F. Barkhof, P. Scheltens, C. J. Stam, S. M. Smith and S. A. Rombouts (2008). "Reduced resting-state brain activity in the "default network" in normal aging." Cereb Cortex **18**(8): 1856–1864.

Damoiseaux, J. S., S. A. Rombouts, F. Barkhof, P. Scheltens, C. J. Stam, S. M. Smith and C. F. Beckmann (2006). "Consistent resting-state networks across healthy subjects." Proc Natl Acad Sci U S A **103**(37): 13848–13853.

Di Martino, A., C. Kelly, R. Grzadzinski, X. N. Zuo, M. Mennes, M. A. Mairena, C. Lord, F. X. Castellanos and M. P. Milham (2011). "Aberrant striatal functional connectivity in children with autism." Biol Psychiatry **69**(9): 847–856.

Di Martino, A., C. G. Yan, Q. Li, E. Denio, F. X. Castellanos, K. Alaerts, J. S. Anderson, M. Assaf, S. Y. Bookheimer, M. Dapretto, B. Deen, S. Delmonte, I. Dinstein, B. Ertl-Wagner, D. A. Fair, L. Gallagher, D. P. Kennedy, C. L. Keown, C. Keysers, J. E. Lainhart, C. Lord, B. Luna, V. Menon, N. J. Minshew, C. S. Monk, S. Mueller, R. A. Muller, M. B. Nebel, J. T. Nigg, K. O'Hearn, K. A. Pelphrey, S. J. Peltier, J. D. Rudie, S. Sunaert, M. Thioux, J. M. Tyszka, L. Q. Uddin, J. S. Verhoeven, N. Wenderoth, J. L. Wiggins, S. H. Mostofsky and M. P. Milham (2014). "The autism brain imaging data exchange: towards a large-scale evaluation of the intrinsic brain architecture in autism." Mol Psychiatry **19**(6): 659–667.

Di, X., & Biswal, B. B. (2016). "Sex-dependent and sex-independent brain resting-state functional connectivity in children with autism spectrum disorder. ." bioRxiv, 038026.
Di, X., A. Azeez, X. Li, E. Haque and B. B. Biswal (2018). "Disrupted focal white matter integrity in autism spectrum disorder: A voxel-based meta-analysis of diffusion tensor imaging studies." Prog Neuropsychopharmacol Biol Psychiatry **82**: 242–248.
Di, X. and B. B. Biswal (2013). "Modulatory interactions of resting-state brain functional connectivity." PLoS One **8**(8): e71163.
Di, X. and B. B. Biswal (2014). "Identifying the default mode network structure using dynamic causal modeling on resting-state functional magnetic resonance imaging." Neuroimage **86**: 53–59.
Di, X. and B. B. Biswal (2015). "Characterizations of resting-state modulatory interactions in the human brain." J Neurophysiol **114**(5): 2785–2796.
Di, X. and B. B. Biswal (2017). "Psychophysiological Interactions in a Visual Checkerboard Task: Reproducibility, Reliability, and the Effects of Deconvolution." Front Neurosci **11**: 573.
Dichter, G. S., D. Gibbs and M. J. Smoski (2015). "A systematic review of relations between resting-state functional-MRI and treatment response in major depressive disorder." Journal of affective disorders **172**: 8–17.
Dong, J., B. Jing, X. Ma, H. Liu, X. Mo and H. Li (2018). "Hurst Exponent Analysis of Resting-State fMRI Signal Complexity across the Adult Lifespan." Front Neurosci **12**: 34.
Dosenbach, N. U., B. Nardos, A. L. Cohen, D. A. Fair, J. D. Power, J. A. Church, S. M. Nelson, G. S. Wig, A. C. Vogel, C. N. Lessov-Schlaggar, K. A. Barnes, J. W. Dubis, E. Feczko, R. S. Coalson, J. R. Pruett, Jr., D. M. Barch, S. E. Petersen and B. L. Schlaggar (2010). "Prediction of individual brain maturity using fMRI." Science **329**(5997): 1358–1361.
Drysdale, A. T., L. Grosenick, J. Downar, K. Dunlop, F. Mansouri, Y. Meng, R. N. Fetcho, B. Zebley, D. J. Oathes, A. Etkin, A. F. Schatzberg, K. Sudheimer, J. Keller, H. S. Mayberg, F. M. Gunning, G. S. Alexopoulos, M. D. Fox, A. Pascual-Leone, H. U. Voss, B. J. Casey, M. J. Dubin and C. Liston (2017). "Resting-state connectivity biomarkers define neurophysiological subtypes of depression." Nature medicine **23**(1): 28–38.
Esposito, F., A. Aragri, I. Pesaresi, S. Cirillo, G. Tedeschi, E. Marciano, R. Goebel and F. Di Salle (2008). "Independent component model of the default-mode brain function: combining individual-level and population-level analyses in resting-state fMRI." Magn Reson Imaging **26**(7): 905–913.
Evers, E. A., E. B. Klaassen, S. A. Rombouts, W. H. Backes and J. Jolles (2012). "The effects of sustained cognitive task performance on subsequent resting state functional connectivity in healthy young and middle-aged male schoolteachers." Brain Connect **2**(2): 102–112.
Ferreira, L. K. and G. F. Busatto (2013). "Resting-state functional connectivity in normal brain aging." Neurosci Biobehav Rev **37**(3): 384–400.
Floris, D. L., M. C. Lai, T. Nath, M. P. Milham and A. Di Martino (2018). "Network-specific sex differentiation of intrinsic brain function in males with autism." Mol Autism **9**: 17.
Fox, M. D. and M. Greicius (2010). "Clinical applications of resting state functional connectivity." Front Syst Neurosci **4**: 19.
Fox, M. D. and M. E. Raichle (2007). "Spontaneous fluctuations in brain activity observed with functional magnetic resonance imaging." Nat Rev Neurosci **8**(9): 700–711.
Fransson, P. (2005). "Spontaneous low-frequency BOLD signal fluctuations: an fMRI investigation of the resting-state default mode of brain function hypothesis." Hum Brain Mapp **26**(1): 15–29.
Fransson, P. (2006). "How default is the default mode of brain function? Further evidence from intrinsic BOLD signal fluctuations." Neuropsychologia **44**(14): 2836–2845.
Friston, K., R. Moran and A. K. Seth (2013). "Analysing connectivity with Granger causality and dynamic causal modelling." Curr Opin Neurobiol **23**(2): 172–178.
Friston, K. J. (2011). "Functional and effective connectivity: a review." Brain Connect **1**(1): 13–36.
Friston, K. J., C. Buechel, G. R. Fink, J. Morris, E. Rolls and R. J. Dolan (1997). "Psychophysiological and modulatory interactions in neuroimaging." Neuroimage **6**(3): 218–229.

Friston, K. J., C. D. Frith, P. F. Liddle and R. S. Frackowiak (1993). "Functional connectivity: the principal-component analysis of large (PET) data sets." J Cereb Blood Flow Metab **13**(1): 5–14.

Friston, K. J., L. Harrison and W. Penny (2003). "Dynamic causal modelling." Neuroimage **19**(4): 1273–1302.

Friston, K. J., J. Kahan, B. Biswal and A. Razi (2014). "A DCM for resting state fMRI." Neuroimage **94**: 396–407.

Friston, K. J., S. Williams, R. Howard, R. S. Frackowiak and R. Turner (1996). "Movement-related effects in fMRI time-series." Magn Reson Med **35**(3): 346–355.

Fukunaga, M., S. G. Horovitz, J. A. de Zwart, P. van Gelderen, T. J. Balkin, A. R. Braun and J. H. Duyn (2008). "Metabolic origin of BOLD signal fluctuations in the absence of stimuli." J Cereb Blood Flow Metab **28**(7): 1377–1387.

Fukushima, M., Betzel, R. F., He, Y., Zuo, X. N., & Sporns, O. (2015). "Characterizing Spatial Patterns and Flow Dynamics in Functional Connectivity States and Their Changes across the Human Lifespan. ." arXiv preprint arXiv:1511.06427.

Goebel, R., A. Roebroeck, D. S. Kim and E. Formisano (2003). "Investigating directed cortical interactions in time-resolved fMRI data using vector autoregressive modeling and Granger causality mapping." Magn Reson Imaging **21**(10): 1251–1261.

Gohel, S. R. and B. B. Biswal (2015). "Functional integration between brain regions at rest occurs in multiple-frequency bands." Brain Connect **5**(1): 23–34.

Greicius, M. (2008). "Resting-state functional connectivity in neuropsychiatric disorders." Curr Opin Neurol **21**(4): 424–430.

Greicius, M. D., B. H. Flores, V. Menon, G. H. Glover, H. B. Solvason, H. Kenna, A. L. Reiss and A. F. Schatzberg (2007). "Resting-State Functional Connectivity in Major Depression: Abnormally Increased Contributions from Subgenual Cingulate Cortex and Thalamus." Biological psychiatry **62**(5): 429–437.

Greicius, M. D., V. Kiviniemi, O. Tervonen, V. Vainionpaa, S. Alahuhta, A. L. Reiss and V. Menon (2008). "Persistent default-mode network connectivity during light sedation." Hum Brain Mapp **29**(7): 839–847.

Greicius, M. D., B. Krasnow, A. L. Reiss and V. Menon (2003). "Functional connectivity in the resting brain: a network analysis of the default mode hypothesis." Proc Natl Acad Sci U S A **100**(1): 253–258.

Greicius, M. D., K. Supekar, V. Menon and R. F. Dougherty (2009). "Resting-state functional connectivity reflects structural connectivity in the default mode network." Cereb Cortex **19**(1): 72–78.

Griffanti, L., G. Douaud, J. Bijsterbosch, S. Evangelisti, F. Alfaro-Almagro, M. F. Glasser, E. P. Duff, S. Fitzgibbon, R. Westphal, D. Carone, C. F. Beckmann and S. M. Smith (2017). "Hand classification of fMRI ICA noise components." Neuroimage **154**: 188–205.

Guo, C. C., F. Kurth, J. Zhou, E. A. Mayer, S. B. Eickhoff, J. H. Kramer and W. W. Seeley (2012). "One-year test-retest reliability of intrinsic connectivity network fMRI in older adults." Neuroimage **61**(4): 1471–1483.

Guo, W., F. Liu, Y. Dai, M. Jiang, J. Zhang, L. Yu, L. Long, H. Chen, Q. Gao and C. Xiao (2013). "Decreased interhemispheric resting-state functional connectivity in first-episode, drug-naive major depressive disorder." Prog Neuropsychopharmacol Biol Psychiatry **41**: 24–29.

Gusnard, D. A., M. E. Raichle and M. E. Raichle (2001). "Searching for a baseline: functional imaging and the resting human brain." Nat Rev Neurosci **2**(10): 685–694.

Hampson, M., B. S. Peterson, P. Skudlarski, J. C. Gatenby and J. C. Gore (2002). "Detection of functional connectivity using temporal correlations in MR images." Hum Brain Mapp **15**(4): 247–262.

Han, Y., J. Wang, Z. Zhao, B. Min, J. Lu, K. Li, Y. He and J. Jia (2011). "Frequency-dependent changes in the amplitude of low-frequency fluctuations in amnestic mild cognitive impairment: a resting-state fMRI study." Neuroimage **55**(1): 287–295.

Harris, R. J., S. Y. Bookheimer, T. F. Cloughesy, H. J. Kim, W. B. Pope, A. Lai, P. L. Nghiemphu, L. M. Liau and B. M. Ellingson (2014). "Altered functional connectivity of the default mode network in diffuse gliomas measured with pseudo-resting state fMRI." J Neurooncol **116**(2): 373–379.

Honey, C. J., O. Sporns, L. Cammoun, X. Gigandet, J. P. Thiran, R. Meuli and P. Hagmann (2009). "Predicting human resting-state functional connectivity from structural connectivity." Proc Natl Acad Sci U S A **106**(6): 2035–2040.

Hong, S. J., S. L. Valk, A. Di Martino, M. P. Milham and B. C. Bernhardt (2018). "Multidimensional Neuroanatomical Subtyping of Autism Spectrum Disorder." Cereb Cortex **28**(10): 3578–3588.

Horn, D. I., C. Yu, J. Steiner, J. Buchmann, J. Kaufmann, A. Osoba, U. Eckert, K. C. Zierhut, K. Schiltz and H. He (2010). "Glutamatergic and resting-state functional connectivity correlates of severity in major depression–the role of pregenual anterior cingulate cortex and anterior insula." Frontiers in systems neuroscience **4**.

Horovitz, S. G., A. R. Braun, W. S. Carr, D. Picchioni, T. J. Balkin, M. Fukunaga and J. H. Duyn (2009). "Decoupling of the brain's default mode network during deep sleep." Proc Natl Acad Sci U S A **106**(27): 11376–11381.

Hutchison, R. M., T. Womelsdorf, E. A. Allen, P. A. Bandettini, V. D. Calhoun, M. Corbetta, S. Della Penna, J. H. Duyn, G. H. Glover, J. Gonzalez-Castillo, D. A. Handwerker, S. Keilholz, V. Kiviniemi, D. A. Leopold, F. de Pasquale, O. Sporns, M. Walter and C. Chang (2013). "Dynamic functional connectivity: promise, issues, and interpretations." Neuroimage **80**: 360–378.

Iidaka, T. (2015). "Resting state functional magnetic resonance imaging and neural network classified autism and control." Cortex **63**: 55–67.

James, G. A., M. E. Kelley, R. C. Craddock, P. E. Holtzheimer, B. W. Dunlop, C. B. Nemeroff, H. S. Mayberg and X. P. Hu (2009). "Exploratory structural equation modeling of resting-state fMRI: applicability of group models to individual subjects." Neuroimage **45**(3): 778–787.

Johnston, J. M., S. N. Vaishnavi, M. D. Smyth, D. Zhang, B. J. He, J. M. Zempel, J. S. Shimony, A. Z. Snyder and M. E. Raichle (2008). "Loss of resting interhemispheric functional connectivity after complete section of the corpus callosum." J Neurosci **28**(25): 6453–6458.

Jung, M., Y. Tu, C. A. Lang, A. Ortiz, J. Park, K. Jorgenson, X. J. Kong and J. Kong (2017). "Decreased structural connectivity and resting-state brain activity in the lateral occipital cortex is associated with social communication deficits in boys with autism spectrum disorder." Neuroimage.

Just, M. A., V. L. Cherkassky, T. A. Keller, R. K. Kana and N. J. Minshew (2007). "Functional and anatomical cortical underconnectivity in autism: evidence from an FMRI study of an executive function task and corpus callosum morphometry." Cereb Cortex **17**(4): 951–961.

Just, M. A., V. L. Cherkassky, T. A. Keller and N. J. Minshew (2004). "Cortical activation and synchronization during sentence comprehension in high-functioning autism: evidence of underconnectivity." Brain **127**(Pt 8): 1811–1821.

Just, M. A., T. A. Keller, V. L. Malave, R. K. Kana and S. Varma (2012). "Autism as a neural systems disorder: a theory of frontal-posterior underconnectivity." Neurosci Biobehav Rev **36**(4): 1292–1313.

Kaiser, R. H., J. R. Andrews-Hanna, T. D. Wager and D. A. Pizzagalli (2015). "Large-scale network dysfunction in major depressive disorder: a meta-analysis of resting-state functional connectivity." JAMA psychiatry **72**(6): 603–611.

Kaiser, R. H., S. Whitfield-Gabrieli, D. G. Dillon, F. Goer, M. Beltzer, J. Minkel, M. Smoski, G. Dichter and D. A. Pizzagalli (2016). "Dynamic Resting-State Functional Connectivity in Major Depression." Neuropsychopharmacology **41**(7): 1822–1830.

Kaminski, M., M. Ding, W. A. Truccolo and S. L. Bressler (2001). "Evaluating causal relations in neural systems: granger causality, directed transfer function and statistical assessment of significance." Biol Cybern **85**(2): 145–157.

Kana, R. K., T. A. Keller, V. L. Cherkassky, N. J. Minshew and M. A. Just (2006). "Sentence comprehension in autism: thinking in pictures with decreased functional connectivity." Brain **129**(Pt 9): 2484–2493.

Kennedy, D. P. and E. Courchesne (2008). "The intrinsic functional organization of the brain is altered in autism." Neuroimage **39**(4): 1877–1885.

Kenny, E. R., J. T. O'Brien, D. A. Cousins, J. Richardson, A. J. Thomas, M. J. Firbank and A. M. Blamire (2010). "Functional connectivity in late-life depression using resting-state functional magnetic resonance imaging." Am J Geriatr Psychiatry **18**(7): 643–651.

Khan, A. J., A. Nair, C. L. Keown, M. C. Datko, A. J. Lincoln and R. A. Muller (2015). "Cerebro-cerebellar Resting-State Functional Connectivity in Children and Adolescents with Autism Spectrum Disorder." Biol Psychiatry **78**(9): 625–634.

Kleinhans, N. M., T. Richards, L. Sterling, K. C. Stegbauer, R. Mahurin, L. C. Johnson, J. Greenson, G. Dawson and E. Aylward (2008). "Abnormal functional connectivity in autism spectrum disorders during face processing." Brain **131**(Pt 4): 1000–1012.

Koch, W., S. Teipel, S. Mueller, K. Buerger, A. L. Bokde, H. Hampel, U. Coates, M. Reiser and T. Meindl (2010). "Effects of aging on default mode network activity in resting state fMRI: does the method of analysis matter?" Neuroimage **51**(1): 280–287.

Kokkonen, S. M., J. Nikkinen, J. Remes, J. Kantola, T. Starck, M. Haapea, J. Tuominen, O. Tervonen and V. Kiviniemi (2009). "Preoperative localization of the sensorimotor area using independent component analysis of resting-state fMRI." Magn Reson Imaging **27**(6): 733–740.

Koshino, H., R. K. Kana, T. A. Keller, V. L. Cherkassky, N. J. Minshew and M. A. Just (2008). "fMRI investigation of working memory for faces in autism: visual coding and underconnectivity with frontal areas." Cereb Cortex **18**(2): 289–300.

Larson-Prior, L. J., J. M. Zempel, T. S. Nolan, F. W. Prior, A. Z. Snyder and M. E. Raichle (2009). "Cortical network functional connectivity in the descent to sleep." Proc Natl Acad Sci U S A **106**(11): 4489–4494.

Lee, H.-L., B. Zahneisen, T. Hugger, P. LeVan and J. Hennig (2013). "Tracking dynamic resting-state networks at higher frequencies using MR-encephalography." NeuroImage **65**: 216–222.

Lee, H. L., B. Zahneisen, T. Hugger, P. LeVan and J. Hennig (2013). "Tracking dynamic resting-state networks at higher frequencies using MR-encephalography." Neuroimage **65**: 216–222.

Lee, M. H., M. M. Miller-Thomas, T. L. Benzinger, D. S. Marcus, C. D. Hacker, E. C. Leuthardt and J. S. Shimony (2016). "Clinical Resting-state fMRI in the Preoperative Setting: Are We Ready for Prime Time?" Top Magn Reson Imaging **25**(1): 11–18.

Li, Z., A. Kadivar, J. Pluta, J. Dunlop and Z. Wang (2012). "Test-retest stability analysis of resting brain activity revealed by blood oxygen level-dependent functional MRI." J Magn Reson Imaging **36**(2): 344–354.

Liu, T. T. (2013). "Neurovascular factors in resting-state functional MRI." Neuroimage **80**: 339–348.

Logothetis, N. K., J. Pauls, M. Augath, T. Trinath and A. Oeltermann (2001). "Neurophysiological investigation of the basis of the fMRI signal." Nature **412**(6843): 150–157.

Lowe, M. J., M. Dzemidzic, J. T. Lurito, V. P. Mathews and M. D. Phillips (2000). "Correlations in low-frequency BOLD fluctuations reflect cortico-cortical connections." Neuroimage **12**(5): 582–587.

Lowe, M. J., B. J. Mock and J. A. Sorenson (1998). "Functional connectivity in single and multislice echoplanar imaging using resting-state fluctuations." NeuroImage **7**(2): 119–132.

Lu, C. M., Y. J. Zhang, B. B. Biswal, Y. F. Zang, D. L. Peng and C. Z. Zhu (2010). "Use of fNIRS to assess resting state functional connectivity." J Neurosci Methods **186**(2): 242–249.

Lu, J., H. Zhang, N. U. F. Hameed, J. Zhang, S. Yuan, T. Qiu, D. Shen and J. Wu (2017). "An automated method for identifying an independent component analysis-based language-related resting-state network in brain tumor subjects for surgical planning." Sci Rep **7**(1): 13769.

Lui, S., Wu, Q., Qiu, L., Yang, X., Kuang, W., Chan, R. C., Gong, Q (2011). "Resting-State Functional Connectivity in Treatment-Resistant Depression." American Journal of Psychiatry: 168(166), 642–648.

Maesawa, S., E. Bagarinao, M. Fujii, M. Futamura, K. Motomura, H. Watanabe, D. Mori, G. Sobue and T. Wakabayashi (2015). "Evaluation of resting state networks in patients with gliomas: connectivity changes in the unaffected side and its relation to cognitive function." PLoS One **10**(2): e0118072.

Mallela, A. N., K. K. Peck, N. M. Petrovich-Brennan, Z. Zhang, W. Lou and A. I. Holodny (2016). "Altered Resting-State Functional Connectivity in the Hand Motor Network in Glioma Patients." Brain Connect.

Mateo, C., P. M. Knutsen, P. S. Tsai, A. Y. Shih and D. Kleinfeld (2017). "Entrainment of Arteriole Vasomotor Fluctuations by Neural Activity Is a Basis of Blood-Oxygenation-Level-Dependent "Resting-State" Connectivity." Neuron **96**(4): 936–948 e933.

Mathers, C. (2008). The global burden of disease: 2004 update, World Health Organization.

Mayberg, H. S., S. K. Brannan, R. K. Mahurin, P. A. Jerabek, J. S. Brickman, J. L. Tekell, J. A. Silva, S. McGinnis, T. G. Glass and C. C. Martin (1997). "Cingulate function in depression: a potential predictor of treatment response." Neuroreport **8**(4): 1057–1061.

McKiernan, K. A., J. N. Kaufman, J. Kucera-Thompson and J. R. Binder (2003). "A parametric manipulation of factors affecting task-induced deactivation in functional neuroimaging." J Cogn Neurosci **15**(3): 394–408.

Meier, T. B., A. S. Desphande, S. Vergun, V. A. Nair, J. Song, B. B. Biswal, M. E. Meyerand, R. M. Birn and V. Prabhakaran (2012). "Support vector machine classification and characterization of age-related reorganization of functional brain networks." Neuroimage **60**(1): 601–613.

Meunier, D., S. Achard, A. Morcom and E. Bullmore (2009). "Age-related changes in modular organization of human brain functional networks." Neuroimage **44**(3): 715–723.

Mevel, K., B. Landeau, M. Fouquet, R. La Joie, N. Villain, F. Mezenge, A. Perrotin, F. Eustache, B. Desgranges and G. Chetelat (2013). "Age effect on the default mode network, inner thoughts, and cognitive abilities." Neurobiol Aging **34**(4): 1292–1301.

Monk, C. S., S. J. Peltier, J. L. Wiggins, S. J. Weng, M. Carrasco, S. Risi and C. Lord (2009). "Abnormalities of intrinsic functional connectivity in autism spectrum disorders." Neuroimage **47**(2): 764–772.

Murphy, K. and M. D. Fox (2017). "Towards a consensus regarding global signal regression for resting state functional connectivity MRI." Neuroimage **154**: 169–173.

Neufeld, J., R. Kuja-Halkola, K. Mevel, E. Cauvet, P. Fransson and S. Bolte (2018). "Alterations in resting state connectivity along the autism trait continuum: a twin study." Mol Psychiatry **23**(7): 1659–1665.

Nomi, J. S. and L. Q. Uddin (2015). "Developmental changes in large-scale network connectivity in autism." Neuroimage Clin **7**: 732–741.

Nooner, K. B., S. J. Colcombe, R. H. Tobe, M. Mennes, M. M. Benedict, A. L. Moreno, L. J. Panek, S. Brown, S. T. Zavitz, Q. Li, S. Sikka, D. Gutman, S. Bangaru, R. T. Schlachter, S. M. Kamiel, A. R. Anwar, C. M. Hinz, M. S. Kaplan, A. B. Rachlin, S. Adelsberg, B. Cheung, R. Khanuja, C. Yan, C. C. Craddock, V. Calhoun, W. Courtney, M. King, D. Wood, C. L. Cox, A. M. Kelly, A. Di Martino, E. Petkova, P. T. Reiss, N. Duan, D. Thomsen, B. Biswal, B. Coffey, M. J. Hoptman, D. C. Javitt, N. Pomara, J. J. Sidtis, H. S. Koplewicz, F. X. Castellanos, B. L. Leventhal and M. P. Milham (2012). "The NKI-Rockland Sample: A Model for Accelerating the Pace of Discovery Science in Psychiatry." Front Neurosci **6**: 152.

Nugent, A. C., A. Martinez, A. D'Alfonso, C. A. Zarate and W. H. Theodore (2015). "The Relationship between Glucose Metabolism, Resting-State fMRI BOLD Signal, and GABAA-Binding Potential: A Preliminary Study in Healthy Subjects and Those with Temporal Lobe Epilepsy." Journal of Cerebral Blood Flow & Metabolism **35**(4): 583–591.

Ogawa, S., D. W. Tank, R. Menon, J. M. Ellermann, S. G. Kim, H. Merkle and K. Ugurbil (1992). "Intrinsic signal changes accompanying sensory stimulation: functional brain mapping with magnetic resonance imaging." Proc Natl Acad Sci U S A **89**(13): 5951–5955.

Onoda, K., M. Ishihara and S. Yamaguchi (2012). "Decreased functional connectivity by aging is associated with cognitive decline." J Cogn Neurosci **24**(11): 2186–2198.

Owen, J. P., Y. O. Li, F. G. Yang, C. Shetty, P. Bukshpun, S. Vora, M. Wakahiro, L. B. Hinkley, S. S. Nagarajan, E. H. Sherr and P. Mukherjee (2013). "Resting-state networks and the functional connectome of the human brain in agenesis of the corpus callosum." Brain Connect **3**(6): 547–562.

Patriat, R., E. K. Molloy, T. B. Meier, G. R. Kirk, V. A. Nair, M. E. Meyerand, V. Prabhakaran and R. M. Birn (2013). "The effect of resting condition on resting-state fMRI reliability and consistency: a comparison between resting with eyes open, closed, and fixated." Neuroimage **78**: 463–473.

Peltier, S. J., C. Kerssens, S. B. Hamann, P. S. Sebel, M. Byas-Smith and X. Hu (2005). "Functional connectivity changes with concentration of sevoflurane anesthesia." Neuroreport **16**(3): 285–288.

Peper, J. S., M. P. van den Heuvel, R. C. Mandl, H. E. Hulshoff Pol and J. van Honk (2011). "Sex steroids and connectivity in the human brain: a review of neuroimaging studies." Psychoneuroendocrinology **36**(8): 1101–1113.

Philip, N. S., J. Barredo, M. van 't Wout-Frank, A. R. Tyrka, L. H. Price and L. L. Carpenter (2018). "Network Mechanisms of Clinical Response to Transcranial Magnetic Stimulation in Posttraumatic Stress Disorder and Major Depressive Disorder." Biol Psychiatry **83**(3): 263–272.

Plitt, M., K. A. Barnes, G. L. Wallace, L. Kenworthy and A. Martin (2015). "Resting-state functional connectivity predicts longitudinal change in autistic traits and adaptive functioning in autism." Proc Natl Acad Sci U S A **112**(48): E6699–6706.

Power, J. D., K. A. Barnes, A. Z. Snyder, B. L. Schlaggar and S. E. Petersen (2012). "Spurious but systematic correlations in functional connectivity MRI networks arise from subject motion." Neuroimage **59**(3): 2142–2154.

Power, J. D., A. Mitra, T. O. Laumann, A. Z. Snyder, B. L. Schlaggar and S. E. Petersen (2014). "Methods to detect, characterize, and remove motion artifact in resting state fMRI." Neuroimage **84**: 320–341.

Preti, M. G., T. A. Bolton and D. Van De Ville (2017). "The dynamic functional connectome: State-of-the-art and perspectives." Neuroimage **160**: 41–54.

Pruim, R. H. R., M. Mennes, D. van Rooij, A. Llera, J. K. Buitelaar and C. F. Beckmann (2015). "ICA-AROMA: A robust ICA-based strategy for removing motion artifacts from fMRI data." Neuroimage **112**: 267–277.

Qin, J., S. G. Chen, D. Hu, L. L. Zeng, Y. M. Fan, X. P. Chen and H. Shen (2015). "Predicting individual brain maturity using dynamic functional connectivity." Front Hum Neurosci **9**: 418.

Quigley, M., D. Cordes, P. Turski, C. Moritz, V. Haughton, R. Seth and M. E. Meyerand (2003). "Role of the corpus callosum in functional connectivity." AJNR Am J Neuroradiol **24**(2): 208–212.

Raichle, M. E., A. M. MacLeod, A. Z. Snyder, W. J. Powers, D. A. Gusnard and G. L. Shulman (2001). "A default mode of brain function." Proc Natl Acad Sci U S A **98**(2): 676–682.

Roder, C., E. Charyasz-Leks, M. Breitkopf, K. Decker, U. Ernemann, U. Klose, M. Tatagiba and S. Bisdas (2016). "Resting-state functional MRI in an intraoperative MRI setting: proof of feasibility and correlation to clinical outcome of patients." J Neurosurg **125**(2): 401–409.

Salami, A., S. Pudas and L. Nyberg (2014). "Elevated hippocampal resting-state connectivity underlies deficient neurocognitive function in aging." Proc Natl Acad Sci U S A **111**(49): 17654–17659.

Salami, A., A. Wahlin, N. Kaboodvand, A. Lundquist and L. Nyberg (2016). "Longitudinal Evidence for Dissociation of Anterior and Posterior MTL Resting-State Connectivity in Aging: Links to Perfusion and Memory." Cereb Cortex **26**(10): 3953–3963.

Salomons, T. V., K. Dunlop, S. H. Kennedy, A. Flint, J. Geraci, P. Giacobbe and J. Downar (2014). "Resting-state cortico-thalamic-striatal connectivity predicts response to dorsomedial prefrontal rTMS in major depressive disorder." Neuropsychopharmacology **39**(2): 488–498.

Satterthwaite, T. D., M. A. Elliott, R. T. Gerraty, K. Ruparel, J. Loughead, M. E. Calkins, S. B. Eickhoff, H. Hakonarson, R. C. Gur, R. E. Gur and D. H. Wolf (2013). "An improved framework for confound regression and filtering for control of motion artifact in the preprocessing of resting-state functional connectivity data." NeuroImage **64**: 240–256.

Satterthwaite, T. D., D. H. Wolf, J. Loughead, K. Ruparel, M. A. Elliott, H. Hakonarson, R. C. Gur and R. E. Gur (2012). "Impact of in-scanner head motion on multiple measures of functional connectivity: relevance for studies of neurodevelopment in youth." Neuroimage **60**(1): 623–632.

Scholvinck, M. L., A. Maier, F. Q. Ye, J. H. Duyn and D. A. Leopold (2010). "Neural basis of global resting-state fMRI activity." Proc Natl Acad Sci U S A **107**(22): 10238–10243.
Sestieri, C., M. Corbetta, G. L. Romani and G. L. Shulman (2011). "Episodic memory retrieval, parietal cortex, and the default mode network: functional and topographic analyses." J Neurosci **31**(12): 4407–4420.
Shehzad, Z., A. M. Kelly, P. T. Reiss, D. G. Gee, K. Gotimer, L. Q. Uddin, S. H. Lee, D. S. Margulies, A. K. Roy, B. B. Biswal, E. Petkova, F. X. Castellanos and M. P. Milham (2009). "The resting brain: unconstrained yet reliable." Cereb Cortex **19**(10): 2209–2229.
Sheline, Y. I., J. L. Price, Z. Yan and M. A. Mintun (2010). "Resting-state functional MRI in depression unmasks increased connectivity between networks via the dorsal nexus." Proceedings of the National Academy of Sciences of the United States of America **107**(24): 11020–11025.
Shmuel, A. and D. A. Leopold (2008). "Neuronal correlates of spontaneous fluctuations in fMRI signals in monkey visual cortex: Implications for functional connectivity at rest." Hum Brain Mapp **29**(7): 751–761.
Shulman, G. L., J. A. Fiez, M. Corbetta, R. L. Buckner, F. M. Miezin, M. E. Raichle and S. E. Petersen (1997). "Common Blood Flow Changes across Visual Tasks: II. Decreases in Cerebral Cortex." J Cogn Neurosci **9**(5): 648–663.
Smith, S. M., P. T. Fox, K. L. Miller, D. C. Glahn, P. M. Fox, C. E. Mackay, N. Filippini, K. E. Watkins, R. Toro, A. R. Laird and C. F. Beckmann (2009). "Correspondence of the brain's functional architecture during activation and rest." Proc Natl Acad Sci U S A **106**(31): 13040–13045.
Smith, S. M., M. Jenkinson, M. W. Woolrich, C. F. Beckmann, T. E. Behrens, H. Johansen-Berg, P. R. Bannister, M. De Luca, I. Drobnjak, D. E. Flitney, R. K. Niazy, J. Saunders, J. Vickers, Y. Zhang, N. De Stefano, J. M. Brady and P. M. Matthews (2004). "Advances in functional and structural MR image analysis and implementation as FSL." Neuroimage **23 Suppl 1**: S208–219.
Smyser, C. D., N. U. Dosenbach, T. A. Smyser, A. Z. Snyder, C. E. Rogers, T. E. Inder, B. L. Schlaggar and J. J. Neil (2016). "Prediction of brain maturity in infants using machine-learning algorithms." Neuroimage **136**: 1–9.
Song, X. W., Z. Y. Dong, X. Y. Long, S. F. Li, X. N. Zuo, C. Z. Zhu, Y. He, C. G. Yan and Y. F. Zang (2011). "REST: a toolkit for resting-state functional magnetic resonance imaging data processing." PLoS One **6**(9): e25031.
Sridharan, D., D. J. Levitin and V. Menon (2008). "A critical role for the right fronto-insular cortex in switching between central-executive and default-mode networks." Proc Natl Acad Sci U S A **105**(34): 12569–12574.
Supekar, K., L. Q. Uddin, A. Khouzam, J. Phillips, W. D. Gaillard, L. E. Kenworthy, B. E. Yerys, C. J. Vaidya and V. Menon (2013). "Brain hyperconnectivity in children with autism and its links to social deficits." Cell Rep **5**(3): 738–747.
Tian, L., Q. Li, C. Wang and J. Yu (2018). "Changes in dynamic functional connections with aging." Neuroimage **172**: 31–39.
Tian, L., L. Ma and L. Wang (2016). "Alterations of functional connectivities from early to middle adulthood: Clues from multivariate pattern analysis of resting-state fMRI data." Neuroimage **129**: 389–400.
Tian, L., J. Ren and Y. Zang (2012). "Regional homogeneity of resting state fMRI signals predicts Stop signal task performance." Neuroimage **60**(1): 539–544.
Tomasi, D. and N. D. Volkow (2012). "Aging and functional brain networks." Mol Psychiatry **17**(5): 471, 549–458.
Toro, R., P. T. Fox and T. Paus (2008). "Functional coactivation map of the human brain." Cereb Cortex **18**(11): 2553–2559.
Toussaint, P. J., S. Maiz, D. Coynel, J. Doyon, A. Messe, L. C. de Souza, M. Sarazin, V. Perlbarg, M. O. Habert and H. Benali (2014). "Characteristics of the default mode functional connectivity in normal ageing and Alzheimer's disease using resting state fMRI with a combined approach of entropy-based and graph theoretical measurements." Neuroimage **101**: 778–786.

Uddin, L. Q., A. M. Kelly, B. B. Biswal, F. X. Castellanos and M. P. Milham (2009). "Functional connectivity of default mode network components: correlation, anticorrelation, and causality." Hum Brain Mapp **30**(2): 625–637.

Uddin, L. Q., E. Mooshagian, E. Zaidel, A. Scheres, D. S. Margulies, A. M. Kelly, Z. Shehzad, J. S. Adelstein, F. X. Castellanos, B. B. Biswal and M. P. Milham (2008). "Residual functional connectivity in the split-brain revealed with resting-state functional MRI." Neuroreport **19**(7): 703–709.

Uddin, L. Q., K. Supekar and V. Menon (2013). "Reconceptualizing functional brain connectivity in autism from a developmental perspective." Front Hum Neurosci **7**: 458.

Uddin, L. Q., K. S. Supekar, S. Ryali and V. Menon (2011). "Dynamic reconfiguration of structural and functional connectivity across core neurocognitive brain networks with development." J Neurosci **31**(50): 18578–18589.

Van Dijk, K. R., T. Hedden, A. Venkataraman, K. C. Evans, S. W. Lazar and R. L. Buckner (2010). "Intrinsic functional connectivity as a tool for human connectomics: theory, properties, and optimization." J Neurophysiol **103**(1): 297–321.

Van Dijk, K. R., M. R. Sabuncu and R. L. Buckner (2012). "The influence of head motion on intrinsic functional connectivity MRI." Neuroimage **59**(1): 431–438.

Veer, I. M., C. F. Beckmann, M.-J. van Tol, L. Ferrarini, J. Milles, D. J. Veltman, A. Aleman, M. A. van Buchem, N. J. van der Wee and S. A. R. B. Rombouts (2010). "Whole brain resting-state analysis reveals decreased functional connectivity in major depression." Frontiers in systems neuroscience **4**.

Vij, S. G., J. S. Nomi, D. R. Dajani and L. Q. Uddin (2018). "Evolution of spatial and temporal features of functional brain networks across the lifespan." Neuroimage **173**: 498–508.

Wen, H. and Z. Liu (2016). "Broadband Electrophysiological Dynamics Contribute to Global Resting-State fMRI Signal." J Neurosci **36**(22): 6030–6040.

Whitfield-Gabrieli, S. and A. Nieto-Castanon (2012). "Conn: a functional connectivity toolbox for correlated and anticorrelated brain networks." Brain Connect **2**(3): 125–141.

Woelfer, M., V. Kasties, S. Kahlfuss and M. Walter (2018). "The Role of Depressive Subtypes within the Neuroinflammation Hypothesis of Major Depressive Disorder." Neuroscience.

Woodward, N. D. and C. J. Cascio (2015). "Resting-State Functional Connectivity in Psychiatric Disorders." JAMA Psychiatry **72**(8): 743–744.

Wu, C. W., C. L. Chen, P. Y. Liu, Y. P. Chao, B. B. Biswal and C. P. Lin (2011). "Empirical evaluations of slice-timing, smoothing, and normalization effects in seed-based, resting-state functional magnetic resonance imaging analyses." Brain Connect **1**(5): 401–410.

Wu, C. W., H. Gu, H. Lu, E. A. Stein, J.-H. Chen and Y. Yang (2008). "Frequency Specificity of Functional Connectivity in Brain Networks." NeuroImage **42**(3): 1047–1055.

Wu, C. W., H. Gu, H. Lu, E. A. Stein, J. H. Chen and Y. Yang (2009). "Mapping functional connectivity based on synchronized CMRO2 fluctuations during the resting state." Neuroimage **45**(3): 694–701.

Wu, J. T., H. Z. Wu, C. G. Yan, W. X. Chen, H. Y. Zhang, Y. He and H. S. Yang (2011). "Aging-related changes in the default mode network and its anti-correlated networks: a resting-state fMRI study." Neurosci Lett **504**(1): 62–67.

Yang, Z., C. Chang, T. Xu, L. Jiang, D. A. Handwerker, F. X. Castellanos, M. P. Milham, P. A. Bandettini and X. N. Zuo (2014). "Connectivity trajectory across lifespan differentiates the precuneus from the default network." Neuroimage **89**: 45–56.

Yao, Z., L. Wang, Q. Lu, H. Liu and G. Teng (2009). "Regional homogeneity in depression and its relationship with separate depressive symptom clusters: a resting-state fMRI study." J Affect Disord **115**(3): 430–438.

Yeo, B. T., F. M. Krienen, J. Sepulcre, M. R. Sabuncu, D. Lashkari, M. Hollinshead, J. L. Roffman, J. W. Smoller, L. Zollei, J. R. Polimeni, B. Fischl, H. Liu and R. L. Buckner (2011). "The organization of the human cerebral cortex estimated by intrinsic functional connectivity." J Neurophysiol **106**(3): 1125–1165.

Zang, Y., T. Jiang, Y. Lu, Y. He and L. Tian (2004). "Regional homogeneity approach to fMRI data analysis." Neuroimage **22**(1): 394–400.

Zang, Y. F., Y. He, C. Z. Zhu, Q. J. Cao, M. Q. Sui, M. Liang, L. X. Tian, T. Z. Jiang and Y. F. Wang (2007). "Altered baseline brain activity in children with ADHD revealed by resting-state functional MRI." Brain Dev **29**(2): 83–91.

Zhang, D., J. M. Johnston, M. D. Fox, E. C. Leuthardt, R. L. Grubb, M. R. Chicoine, M. D. Smyth, A. Z. Snyder, M. E. Raichle and J. S. Shimony (2009). "Preoperative sensorimotor mapping in brain tumor patients using spontaneous fluctuations in neuronal activity imaged with functional magnetic resonance imaging: initial experience." Neurosurgery **65**(6 Suppl): 226–236.

Zhang, X., X. Di, H. Lei, J. Yang, J. Xiao, X. Wang, S. Yao and H. Rao (2016). "Imbalanced spontaneous brain activity in orbitofrontal-insular circuits in individuals with cognitive vulnerability to depression." J Affect Disord **198**: 56–63.

Zheng, H., L. Xu, F. Xie, X. Guo, J. Zhang, L. Yao and X. Wu (2015). "The Altered Triple Networks Interaction in Depression under Resting State Based on Graph Theory." Biomed Res Int **2015**: 386326.

Zhu, X., X. Wang, J. Xiao, J. Liao, M. Zhong, W. Wang and S. Yao (2012). "Evidence of a dissociation pattern in resting-state default mode network connectivity in first-episode, treatment-naive major depression patients." Biol Psychiatry **71**(7): 611–617.

Zou, Q. H., C. Z. Zhu, Y. Yang, X. N. Zuo, X. Y. Long, Q. J. Cao, Y. F. Wang and Y. F. Zang (2008). "An improved approach to detection of amplitude of low-frequency fluctuation (ALFF) for resting-state fMRI: fractional ALFF." J Neurosci Methods **172**(1): 137–141.

Zuo, X. N., J. S. Anderson, P. Bellec, R. M. Birn, B. B. Biswal, J. Blautzik, J. C. Breitner, R. L. Buckner, V. D. Calhoun, F. X. Castellanos, A. Chen, B. Chen, J. Chen, X. Chen, S. J. Colcombe, W. Courtney, R. C. Craddock, A. Di Martino, H. M. Dong, X. Fu, Q. Gong, K. J. Gorgolewski, Y. Han, Y. He, Y. He, E. Ho, A. Holmes, X. H. Hou, J. Huckins, T. Jiang, Y. Jiang, W. Kelley, C. Kelly, M. King, S. M. LaConte, J. E. Lainhart, X. Lei, H. J. Li, K. Li, K. Li, Q. Lin, D. Liu, J. Liu, X. Liu, Y. Liu, G. Lu, J. Lu, B. Luna, J. Luo, D. Lurie, Y. Mao, D. S. Margulies, A. R. Mayer, T. Meindl, M. E. Meyerand, W. Nan, J. A. Nielsen, D. O'Connor, D. Paulsen, V. Prabhakaran, Z. Qi, J. Qiu, C. Shao, Z. Shehzad, W. Tang, A. Villringer, H. Wang, K. Wang, D. Wei, G. X. Wei, X. C. Weng, X. Wu, T. Xu, N. Yang, Z. Yang, Y. F. Zang, L. Zhang, Q. Zhang, Z. Zhang, Z. Zhang, K. Zhao, Z. Zhen, Y. Zhou, X. T. Zhu and M. P. Milham (2014). "An open science resource for establishing reliability and reproducibility in functional connectomics." Sci Data **1**: 140049.

Zuo, X. N., A. Di Martino, C. Kelly, Z. E. Shehzad, D. G. Gee, D. F. Klein, F. X. Castellanos, B. B. Biswal and M. P. Milham (2010). "The oscillating brain: complex and reliable." Neuroimage **49**(2): 1432–1445.

Zuo, X. N., T. Xu, L. Jiang, Z. Yang, X. Y. Cao, Y. He, Y. F. Zang, F. X. Castellanos and M. P. Milham (2013). "Toward reliable characterization of functional homogeneity in the human brain: preprocessing, scan duration, imaging resolution and computational space." Neuroimage **65**: 374–386.

Directed Interregional Brain Interactions

Steven L. Bressler

The Neurocognitive Network Perspective in Neuropsychiatry

A large body of experimental and theoretical work in different fields, including cognitive neurophysiology, cognitive neuropsychology, artificial intelligence, neurolinguistics, biomedical engineering, and neuropsychiatry, indicates that cognitive function emerges from the coordination of brain areas in large-scale networks (Luria 1966; Goldman-Rakic 1988; Mesulam 1990; Fuster 2009; Bressler and Kelso 2016). Large-scale network coordination in the healthy brain is a manifestation of brain areas working together to achieve a functional goal. As a corollary, there is a growing consensus that many kinds of cognitive dysfunction result from the dyscoordination of large-scale brain networks (Afshari and Jalili 2016; Grace 2016; Engels et al. 2015; Rubinov and Bullmore 2013). Abnormal neural synchrony, a direct manifestation of network dyscoordination, has been associated with several different brain disorders, such as schizophrenia, epilepsy, autism, Alzheimer's Disease, and Parkinson's Disease (Uhlhaas and Singer 2006).

The cognitive behavior of brain networks is revealed by the relative timing of coordinated brain areas during cognitive task performance (Bressler and Kelso 2016). Similarly, motor coordination has long been investigated by studying the relative timing of body limb motion during motor task performance (von Holst 1939/1973; Kelso 1995). Brain network coordination has been demonstrated by numerous studies of neural synchrony, based on relative phase relations between oscillatory activity in different brain areas. These studies of brain area coordination have shown function-related synchrony of the EEG, MEG, iEEG, ECoG, or LFP in the 1–200-Hz frequency range where these signals commonly show brain activity. Interregional interaction has also been demonstrated at a slower timescale by coor-

S. L. Bressler (✉)
Center for Complex Systems and Brain Sciences, Florida Atlantic University, Boca Raton, FL, USA
e-mail: bressler@fau.edu

V. A. Diwadkar, S. B. Eickhoff (eds.), *Brain Network Dysfunction in Neuropsychiatric Illness*, https://doi.org/10.1007/978-3-030-59797-9_4

dination studies that use the fMRI BOLD signal (Bressler et al. 2008; Tang et al. 2012; Asemi et al. 2015; Silverstein et al. 2016; Meehan et al. 2017; Diwadkar et al. 2017). From phase synchrony to interregional interaction to large-scale network coordination, an emerging view of brain function sees large-scale systems of distributed and interconnected neuronal populations in the central nervous system as uniquely organized to perform specific cognitive functions. These systems are simply referred to as neurocognitive networks. Thus, a specific cognitive function is performed by a uniquely configured neurocognitive network. Even though neurocognitive network edges are commonly demonstrated by showing a consistent between-area relative phase relation as an undirected network edge (see below), they ultimately depend on directed interactions that transpire between brain regions. In what follows, I will focus on directed interregional interactions in neurocognitive networks as the basis for cognitive function.

Studies of neurocognitive networks since Wernicke distinguish between complex and simple cognitive functions (Wernicke 1973/1874). Simple cognitive functions are low-level sensory and motor neural functions of the brain, whereas complex cognitive functions are high-level neural functions of multiple brain areas coordinated in large-scale neurocognitive networks. The neurocognitive network perspective applies to both simple and complex cognitive functions because they both emerge from interregional brain interactions.

The neurocognitive network perspective has given special importance in modern neuropsychiatry to the methodologies of structural and functional connectivity analysis (Mannino and Bressler 2015). Analysis of the brain's large-scale structural connectivity (tractography) is an ongoing enterprise of great importance for neuropsychiatry. It has traditionally required the use of invasive techniques in experimental animals, but the advent of modern structural neuroimaging techniques has meant that it can also be noninvasively applied to human brain data obtained by magnetic resonance imaging. Techniques such as diffusion tensor imaging (DTI) and diffusion spectrum imaging (DSI) are providing unprecedented access to magnetic resonance imaging tractography of the human brain.

Functional Connectivity

Analysis of the brain's functional connectivity is also an extremely important enterprise in neuropsychiatry. Not all interactions that are structurally allowable actually occur in every instantiation of cognitive function. Only a subset of the total number of potential interactions that are allowed by network structure actually transpire in any given instance. And, since cognitive function is degenerate, a given cognitive function may be implemented by different functional network configurations at different times (Friston and Price 2003). Thus, to characterize the functional interactions underlying a cognitive function requires that all the network configurations of functional interaction existing during execution of that function be described and measured.

By definition, brain networks consist of network nodes and edges (Bressler and Menon 2010). The network nodes are taken to be the brain sources of recorded neural data. Network edges in the brain are measured by functional connectivity analysis of the neural data (Vincent et al. 2008), the goal of which is to characterize network edges as the interrelations between putative network nodes. Brain network edges may be undirected or directed. Two basic approaches exist for determining directed functional network edges in brain networks. First is by perturbation (Paus 2005), using electrical or magnetic stimulation of a brain region and recording of the effect in another region (Moore and Armstrong 2003; Ebitz and Moore 2017; Ruff et al. 2006, 2008, 2009; Bachinger et al. 2017). This approach is intuitively appealing because its reliance on perturbing stimulation fits well with common notions of causality, but it has serious drawbacks. The main drawbacks are (1) it only measures directed influence in one direction (from the stimulated region to the recorded region); and (2) to date, its use has been limited to epochs of time when cognitive task performance is not taking place. Furthermore, it may prove impossible to overcome this second limitation because of technical problems.

Another approach to the study of directed functional connectivity in brain networks is by its computation from simultaneously recorded time series (Weiner 1956). Granger Causality (GC) and transfer entropy are two examples of directed functional connectivity measures computed from simultaneous time series. (These measures are equivalent when the time series are vector auto-regressive processes.) There are at least four major advantages of time series analysis for measuring directed functional connectivity between time series simultaneously recorded from different brain regions. One advantage is that it can be utilized when subjects are performing a cognitive task as well as when they are at rest. A second advantage is that time series analysis can measure directed influence in both directions between pairs of regions. Since time series analysis enables the measurement of directed influence in both directions between recorded time series, when simultaneous time series are recorded from multiple regions, time series analysis of directed influence routinely involves the measurement of directed functional connectivity in both directions between pairs of time series from different brain regions. A third advantage is that time series analysis may be applied in short time windows (Ding et al. 2000), making it possible to chart the time evolution of directed influence either during a cognitive task or in a resting epoch. A fourth advantage of time series analysis is that it opens the door to analysis of spectral directed functional connectivity, which yields results that are resolved at frequencies across the spectrum (Geweke 1982).

Determining directed functional connectivity by time series analysis may usefully be performed on either task-related or resting-state brain data. Whether applied to task-related or resting-state data, the basic requirement for analysis is that the data must be simultaneously recorded from multiple brain locations successively at multiple time steps. The type of neural data that have been analyzed to date includes EEG, MEG, ECoG, LFP, spike trains, and fMRI BOLD signals from either humans or animals.

When applied to electrical or magnetic data, time series analysis may yield frequency-resolved directed influences in the 1–200 Hz frequency range where these data usually show brain activity. Spectral directed functional connectivity analysis of fMRI BOLD data yields frequency-resolved directed influence in the low-frequency range typical of this type of data. Directed influence, either time domain or frequency domain, can be measured between pairs of time series in the recorded data, conditional on other time series in the recorded data set (Ding et al. 2006).

Although fMRI BOLD data are commonly analyzed as static fMRI BOLD images at single time points, or averaged over successive time points, the typical fMRI experiment involves simultaneous recording of neural time series from voxels making up the volume of the brain. Thus, fMRI BOLD data may also be treated as time series recorded with an approximately 2-s sampling interval. Even though this slow sampling keeps fMRI BOLD data from having sufficient temporal resolution to provide frequency-resolved time series in the 1–200 Hz frequency range where brain activity is observed by EEG-like measures, directed interregional brain interactions can still be measured from fMRI BOLD data with low temporal resolution. Because of this low temporal resolution, our research group decided not to measure directed connectivity of fMRI BOLD time series by frequency-resolved analysis. Rather, as discussed below, directed connectivity was measured in the time domain by Granger's method with only one linear predictor in the autoregressive model (order = 1). Essentially then, our time domain directed connectivity analysis of fMRI BOLD time series, following Granger's method with model order 1, was equivalent to a test of the linear prediction of one time series by another.

When prior knowledge is absent about the constituents of the neurocognitive network that underlie a particular cognitive function, each voxel in the brain is treated as a putative network node, and directed influences may be measured between all pairs of voxels, conditional on all other voxels, using sparse regression (Garg et al. 2011). However, prior knowledge about the network in operation in the cognitive function may importantly direct the search for the underlying network nodes (Falco et al. submitted).

Functional connectivity may be undirected or directed, and much current functional connectivity analysis measures the undirected type. Undirected functional connectivity, also called statistical interdependency, measures the common, or shared, activity of two time series. Undirected functional connectivity analysis, which measures the time series similarity of network nodes, but not the directed influences between them, is often performed in the time domain simply by measuring the Pearson product-moment correlation between pairs of putative network-node time series. However, it can also be computed in the spectral domain (Sun et al. 2004), employing spectral measures such as spectral coherence (Salazar et al. 2012), the phase-locking index (Lachaux et al. 1999; Sazonov et al. 2009), and the Hilbert transform (Le Van Quyen et al. 2001).

The measurement of directed functional connectivity by linear autoregressive modeling of time series was introduced by Granger (1969) as a practical method for implementing Wiener's original 1956 concept of causality. An autoregressive model is an equation where the current value of a time series is regressed onto past values of the same and different time series. Granger's method has been widely used in economics since the 1960s. Applications in neuroscience have become popular in the last 10 years. The method compares two autoregressive models, called the restricted and the unrestricted models. The restricted model is an autoregressive model of the current value of one time series simply using past values of the same time series. The model order is the number of past values that go into the model. Each term in the model is the product of a coefficient and the past value of the time series. The unrestricted model is an autoregressive model of the current value of the time series with terms representing the past of the same time series and other terms representing the past of a second time series. Like the restricted model, each term in the unrestricted model is the product of a coefficient and the past value of a time series. The unrestricted and restricted model coefficients are determined using well-known techniques for solving sets of linear equations (Jansen 1991). The measure of directed functional connectivity introduced by Granger is a ratio reflecting comparison between the degree of success of predictability by the restricted and unrestricted models. If the prediction of the current time series value in the unrestricted model is significantly more successful than that in the restricted model, then the second time series contains information that is not found in the first time series, and is said to G-cause the first time series. Since the past of the first time series is always successful at predicting the current value to some degree, G-causality depends on the unrestricted model giving significantly greater success at prediction than the restricted model, not simply greater success. The Wald test determines whether all the coefficients of the unrestricted model are jointly zero, in which case there is no causal influence in the direction tested. Since the model equations are oblivious to which time series is considered first or second, the models can be solved for G-causality in both directions between neural signals from two brain sites.

The Physiological Basis of Directed Functional Connectivity

The analysis of causal influence at synapses between individual or groups of neurons is fundamental to our understanding of brain function (Mannino and Bressler 2015). Since the neuron is its basic functional unit, it is clear that the brain is organized so that causal influences are exerted from one neuron or neuronal group to another. Neurons, with somata in sending brain areas and axons projecting from somata in a sending area and synapsing on other neurons in receiving areas, are the foundation for interregional interactions and large-scale brain networks. It is widely accepted that areas only exert influences on one another by way of synaptic action between neurons (Bressler 1987a, b; Bressler et al. 1993; Bressler and Kelso 2016). Directed influences are carried in a unidirectional manner from the synapse into the dendritic tree to the soma as graded potentials, and from the soma and axon hillock and along the axon trunk into all its branches and terminals as action potential spikes. They continue unidirectionally across the synapses between neurons to the dendrites of postsynaptic neurons. This basic unidirectional flow of activity is not negated by the additional generation of calcium spikes and their reverse-direction propagation into the dendrites. Given that the neuron is the basic unit of communication in the brain, causal influences between neurons must be considered to underlie brain function. That is, brains function is due to the causal influences that are transmitted between neurons, both within and between different brain regions.

In the mammalian neocortex, most connected cortical areas are connected reciprocally, with axons that project from cell bodies in a sending area synapsing in a receiving area, and cell bodies in the receiving area sending axons that synapse in the first. As a result of this arrangement, interregional causal influences in the cortex are commonly exerted bidirectionally between connected cortical areas. Within a neuronal group in a single cortical area, the neurons are also reciprocally connected. This arrangement causes the interactions of neurons within the group to show positive feedback, whereby interacting excitatory neurons within the group enforce greater overall excitation (Freeman 1975). Interactions between excitatory and inhibitory neurons enforce negative feedback loops, the timing of which may produce oscillatory activity, the oscillation frequency depending on the loop transit time in the negative feedback loop.

Physiological Evidence for Directed Functional Connectivity

The first experiment to provide physiological evidence for directed functional connectivity came from Nakamura and Copolla (Bressler et al. 1993), who utilized local field potentials (LFPs) recorded simultaneously from bipolar microelectrodes chronically implanted in multiple areas from V1 to prefrontal cortex (PFC) in one hemisphere of the neocortex of macaque monkeys performing a visual pattern discrimination task with a GO/NO-GO lever-press motor response. Spectral directed functional connectivity analysis of LFPs was first performed in the somatosensory-motor system in the pre-stimulus period when the hand lever was depressed (Brovelli et al. 2004). It revealed directed influences from primary somatosensory cortex (S1) to primary motor cortex (M1) that depended on synchronization in the beta frequency range (Figs. 1 and 2). This result is consistent with the large literature on synchronized beta rhythms in the neocortex during motor behavior (van Wijk et al. 2012), with the operation of a motor-sensory loop in motor maintenance (MacKay and Mendonca 1995; MacKay 1997), in which M1's motor output is influenced by proprioceptive feedback to S1, and with the proposal that neocortical beta rhythms represent maintenance of the current sensorimotor or cognitive state (Engel and Fries 2010) (Fig. 3).

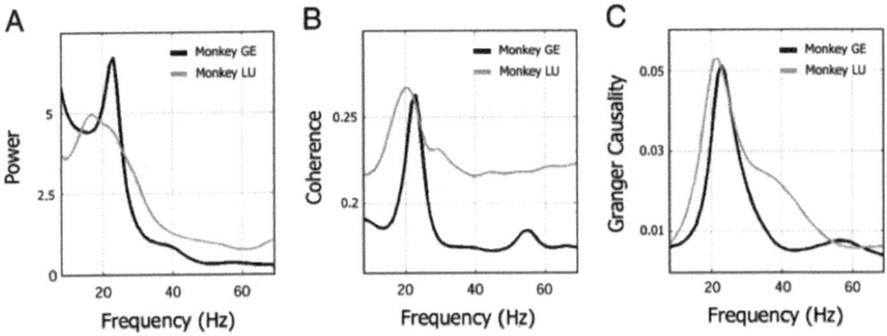

Fig. 1 Power (**a**), Coherence (**b**), and Granger Causality (**c**) spectra of LFPs averaged over sites and site pairs in the somatosensory-motor cortex of two rhesus macaque monkeys (GE and LU) engaged in a visual pattern discrimination task with GO/NO-GO lever-press motor responses. The spectra, all of which show peaks in the beta (~20 Hz) frequency range, were computed from LFPs recorded during a pre-stimulus period in which the monkey had already indicated task engagement by pressing a hand lever, and maintaining pressure on it, but before the stimulus was presented. (Adapted from Brovelli et al. (2004))

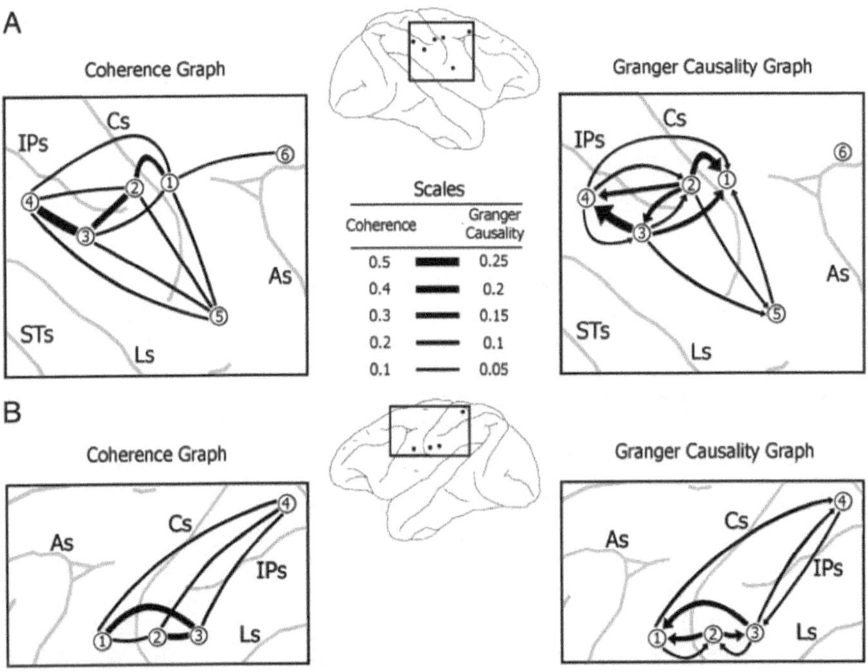

Fig. 2 Coherence and Granger Causality maps of peak activity in the beta frequency range, from the somatosensory-motor cortex, showing the spatial extent of the spectral results shown in Fig. 1. The maps show the specific electrode pairs that are significantly synchronized (with high beta peak coherence) or have significant directed functional connectivity (with high beta peak Granger Causality) in the pre-stimulus period. (Adapted from Brovelli et al. (2004))

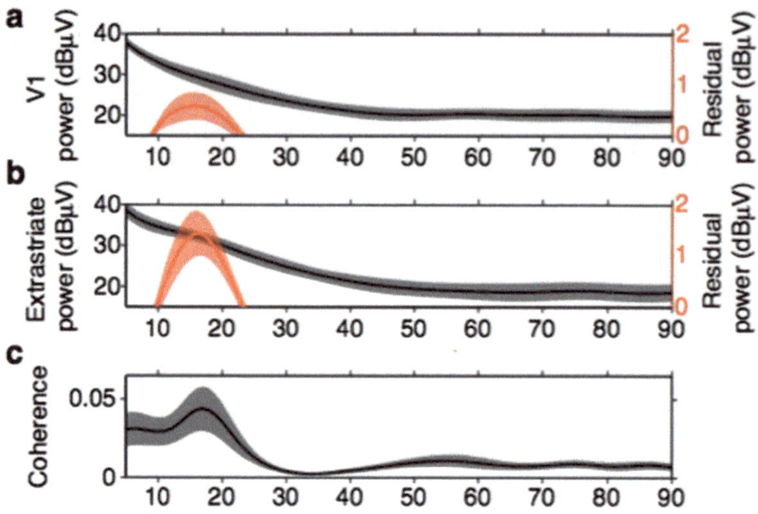

Fig. 3 The same spectral analysis procedures were applied to macaque monkey LFPs from the visual cortex as for the somatosensory-motor cortex. Denoised power spectra from visual striate area V1 (**a**) and extrastriate areas V4 and TEO (**b**), and striate–extrastriate coherence spectra between them (**c**), show beta frequency peaks, as they do in the somatosensory-motor cortex (see Fig. 1). (Adapted from Richter et al. (2016))

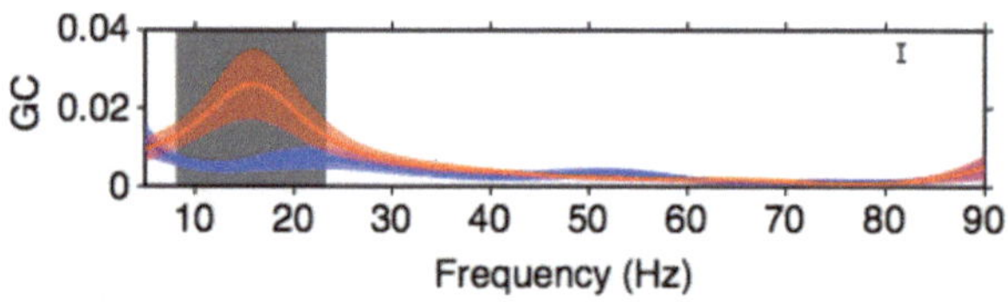

Fig. 4 Granger Causality spectra of LFPs were computed in both top-down (extrastriate➔striate) and bottom-up (striate➔extrastriate) directions between all extrastriate–striate site pairs in the visual cortex by spectral directed functional connectivity analysis in the same monkeys as in Fig. 3. Note that a beta peak in the Granger Causality spectrum only occurs in the top-down (extrastriate➔striate) direction. The similarity of peak frequency of this top-down (extrastriate➔striate) Granger Causality spectrum to that of the coherence spectrum in Fig. 3 suggests that the top-down influence from extrastriate to striate cortex is conveyed by synchronized beta frequency oscillations. (Adapted from Richter et al. (2016))

In the visual system, spectral directed functional connectivity analysis of visual cortical LFPs in the pre-stimulus period revealed top-down influences from V4 and TEO to V1, but not bottom-up influences in the opposite direction (Fig. 4). Like the somatosensory-motor cortex, these visual cortex influences were also specific to, and also depended on synchronization in, the beta frequency range (Bressler et al. 2007). Visual top-down influences were found to convey behavioral context from V4 and TEO to V1: Support Vector Machine pattern classification of visual top-down GC values successfully discriminated between two task condition signifying

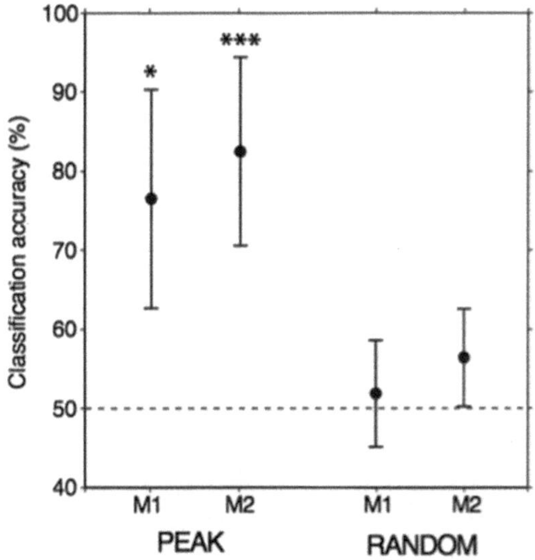

Fig. 5 Pre-stimulus top-down beta peak Granger Causality values were used as features in Support Vector Machine (SVM) pattern classification. The classification accuracies were above 75% for both of the monkeys used in Figs. 3 and 4. By comparison, classification accuracy was near chance (50%) when the same analysis was applied to label-randomized data in both monkeys. (Adapted from Richter et al. (2016))

two different contingencies between the type of visual stimulus pattern (line or diamond) and the type of motor response (GO or NO-GO) (Richter et al. 2016; Fig. 5).

The second experiment aimed at exploring the neural basis of working memory. LFPs and spike trains were simultaneously recorded by the research team of Charles Gray (Salazar et al. 2012) from multiple microelectrodes implanted in both the prefrontal cortex (PFC) and the posterior parietal cortex (PPC) of one hemisphere in macaque monkeys performing an oculomotor delayed match-to-sample (DMTS) task. As in the previous experiment, LFPs from PFC and PPC were found to be synchronized in the beta frequency range (Fig. 6). In fact, fronto-parietal beta synchronization was observed to encode sample object identity during the delay period of the DMTS task. Spectral directed functional connectivity analysis in the delay period revealed roughly balanced beta-band causal influences in both directions between the two regions (i.e., PFC⟶PPC and PPC⟶PFC) (Fig. 7). This result indicated that the PFC and PPC work together as a unified system in support of working memory.

Evidence from Directed Functional Connectivity Analysis of Human fMRI BOLD Time Series

Neurocognitive networks have been found to exist, not only at the faster time scale of electrophysiological time series, but also at the slower time scale of functional magnetic resonance time series. In two series of experiments with human fMRI BOLD time series (Bressler et al. 2008; Tang et al. 2012; Asemi et al. 2015; Silverstein et al. 2016; Meehan et al. 2017; Diwadkar et al. 2017), we used directed

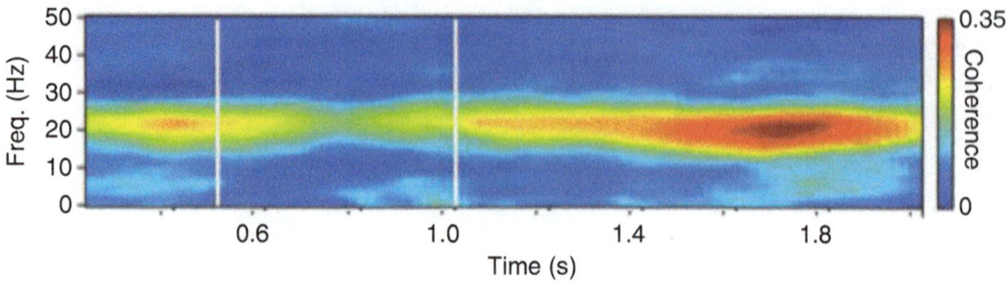

Fig. 6 Time dependence of fronto-parietal coherence demonstrating fronto-parietal synchronization in the beta frequency range. The time–frequency coherence spectrum between event-related LFPs of all correct trials, from lateral prefrontal and posterior parietal cortices of macaque monkeys, is shown, time-locked to the sample presentation, as the monkeys performed a visual delayed match-to-sample task with oculomotor response. During visual fixation, a sample stimulus, consisting of one out of three possible objects positioned at one out of three possible locations, was presented for 500 ms, followed by a random delay of 800–1200 ms. At the end of the delay, a match stimulus was presented, consisting of the previous sample object (target) and a distracter object positioned at two out of three possible locations. A saccadic eye movement to the target was rewarded with juice. The figure is divided into three sections corresponding to the pre-sample, sample, and delay/match periods of the task. Fronto-parietal beta synchronization was observed to encode sample object identity during the delay period of the task. (Adapted from Salazar et al. (2012))

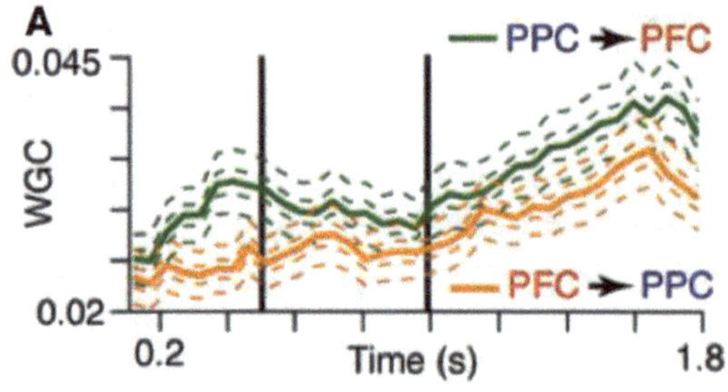

Fig. 7 Beta frequency directed functional connectivity in the fronto-parietal system during the delay period of a delayed match-to-sample visual working memory task. The finding that beta directed influences are exerted in both directions (i.e., both PPC➔PFC and PFC➔PPC) with roughly the same magnitude indicates that prefrontal and posterior parietal cortical areas work together as a unified system in support of working memory. (Adapted from Salazar et al. (2012))

functional connectivity analysis in the time domain (with model order 1) to test for directed interregional interactions in human neurocognitive networks. Measuring directed interregional interactions of the human PFC, all these studies support the concept that the PFC provides top-down executive control in sensory-motor processing (Miller and Cohen 2001). Beyond that, they provide evidence supporting the view that different areas of the PFC act as top-down executive controllers in different cognitive functions, or more simply, that the PFC contains multiple top-down controllers (Dosenbach et al. 2008), each operating in a different cognitive function.

The first of these sets of experiments investigated the interaction between cortical areas in two neurocognitive networks during visuospatial attention. The fronto-parietal dorsal attention network (DAN), the first of these neurocognitive networks, having major nodes in the neocortices of the frontal eye field (FEF) and intraparietal sulcus (IPS) in both hemispheres, is responsible for the voluntary deployment of visuospatial attention (Vossel et al. 2014). We investigated the top-down influences exerted by these DAN areas on areas of the visual occipital cortex (VOC) during performance of a visuospatial attention task. The cue to attend to the spatial location of the visual stimulus was presented in the auditory modality so as to avoid potential interference between the cue-evoked response and attentional processes in the visual modality. Applying directed functional connectivity analysis in the time domain, we discovered a large asymmetry between GC in the top-down (DAN➔VOC) and bottom-up (VOC➔DAN) directions, with GC in the top-down (DAN➔VOC) direction being significantly larger than that in the bottom-up (VOC➔DAN) direction (Bressler et al. 2008; Fig. 8). Both FEF and IPS nodes of the DAN showed this asymmetry. Furthermore, this directional asymmetry was found to be larger during the task than at rest (Meehan et al. 2017; Fig. 9). Overall, these findings are consistent with a functional role for the DAN in exerting top-down directed influence on VOC areas in visuospatial attention. They implicate the FEF, part of the PFC, in top-down visuospatial attentional control, and imply that prefrontal and posterior parietal regions work together in a unified network for top-down control of visuospatial attention.

The second series of experiments with human fMRI BOLD time series investigated the directed functional connectivity of the dorsal anterior cingulate cortex (dACC) in motor behavior. A visuomotor coordination task was employed in which subjects were required to tap the forefinger of their right had as quickly as possible in response to a flashing white stimulus presented in the center of the display panel. Directed functional connectivity was assessed in both directions between the dACC and the supplementary motor area (SMA), and, for comparison, between the dACC and forefinger region of the primary motor area (Brodmann area 6 or M1). The dACC ROI was nonoverlapping with the SMA. Since the dACC is connected anatomically with the motor system in both humans (Paus et al. 1993; Amiez and Petrides 2014) and nonhuman primates (Morecraft & Van Hoesen 1992; Dum & Strick 2002; Morecraft et al. 2012), the experiments tested whether this area indeed has motor function. It was discovered that the directed influence from dACC to SMA, but not from SMA to dACC, was significantly greater in the coordination task than in a rest control condition (Asemi et al. 2015; Fig. 10). Thus, the bottom-up effect from SMA to dACC in the task was no different than that at rest, whereas the top-down effect from dACC to SMA was significantly greater in the task than that at rest, meaning that the top-down effect was significantly greater than the bottom-up effect in the task and so there was a task-specific directional asymmetry in this system. No such task specificity was found between the dACC and M1. These results indicate that the dACC is in fact part of the motor system, and that it exerts a top-down influence on SMA, itself a prominent motor control region, in a task-specific manner during visually coordinated unimanual finger movement.

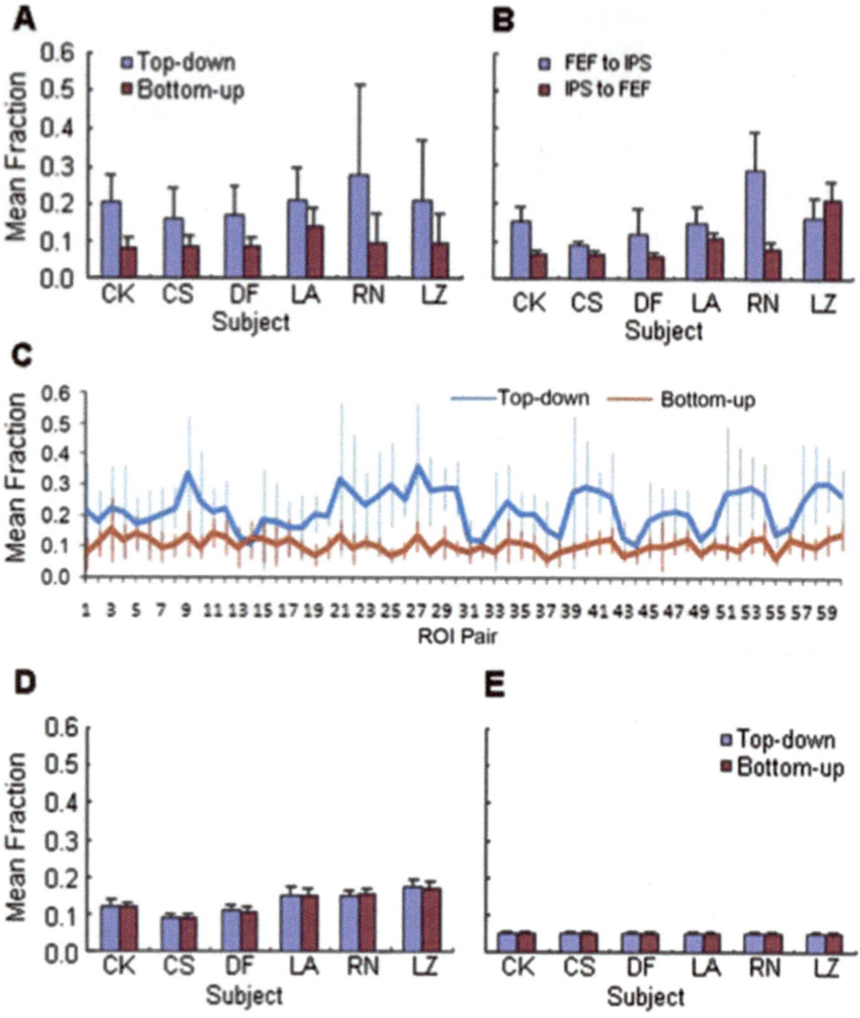

Fig. 8 Top-down vs. bottom-up Granger Causality between the Dorsal Attention Network (DAN) and the Visual Occipital Cortex (VOC) in the delay period following an auditory preparatory cue in a visuospatial attention task. The results indicate that the top-down (DAN→VOC) Granger Causality is consistently greater than the bottom-up (VOC→DAN) Granger Causality both across subjects (**a**, **b**) and ROI pairs (**c**). This asymmetry between top-down and bottom-up Granger Causality is not seen when the data are voxel-randomized (**d**) or trial-randomized (**e**). (Adapted from Bressler et al. (2008))

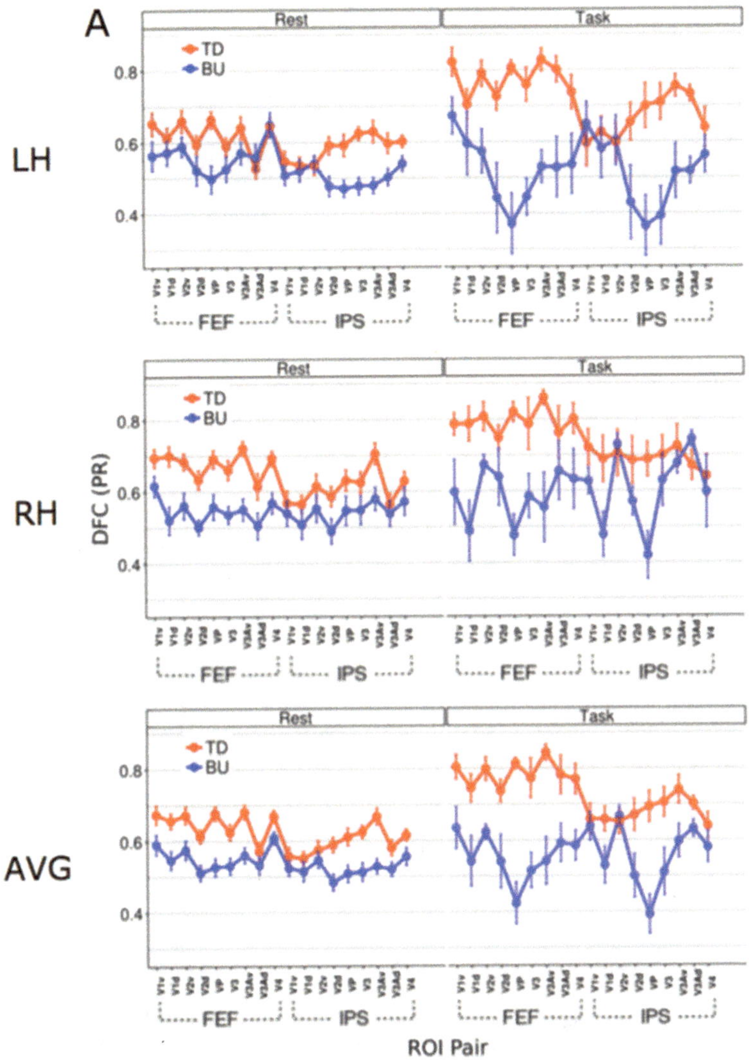

Fig. 9 Directed functional connectivity at rest (left column) and during the cued preparatory delay period of the visual spatial attention task (right column) of DAN–VOC ROI pairs in the left hemisphere (LH) and right hemisphere (RH). Each plot shows a directional asymmetry with top-down (red) significantly greater than bottom-up (blue) Granger Causality. Each row shows that the asymmetry is larger during the task than at rest. (Adapted from Meehan et al. (2017))

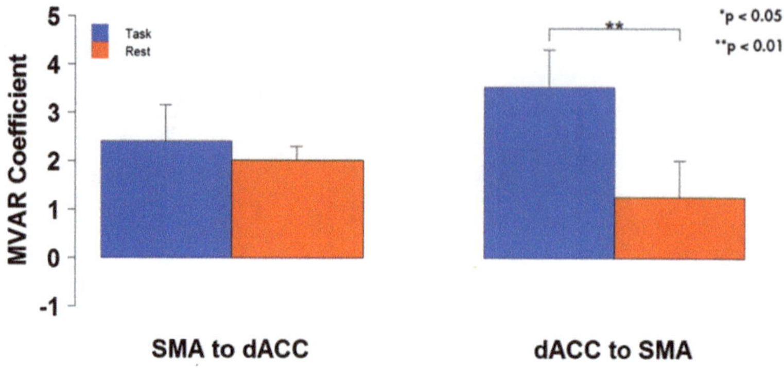

Fig. 10 The group mean and standard error of influence in both directions between dACC and SMA for Task (blue) and Rest (red) conditions, showing that the top-down influence, but not the bottom-up, are significantly larger in the visuomotor task than at rest. (Adapted from Asemi et al. (2015))

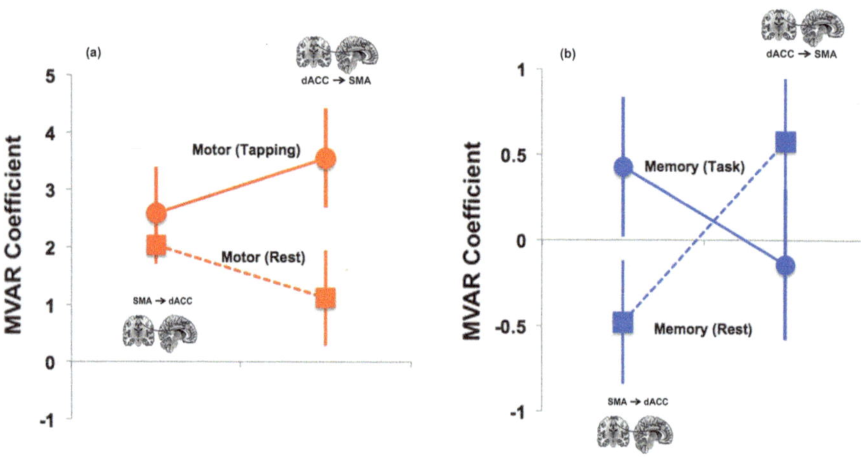

Fig. 11 Directed functional connectivity during task and at rest in motor (visually coordinated unimanual finger movement) and working memory behavior. The dACC➔SMA influence significantly increased when participants transitioned to the motor task from the motor rest state, but not to the memory task from the memory rest state. Moreover, the dACC➔SMA influence was significantly greater in the motor than the working memory task, but the SMA➔dACC effect was not. (Adapted from Diwadkar et al. (2017))

In a follow-up study with a different participant group, the same visuomotor coordination task was compared with a 2-back working memory task (Diwadkar et al. 2017; Fig. 11). It was found that the directed influence from dACC to SMA significantly increased when participants transitioned to the motor task from the motor rest state, but not to the memory task from the memory rest state. Moreover, the dACC to SMA directed influence was significantly larger in the motor task than in the working memory task, whereas the directed influence from SMA to dACC was significantly larger in the rest state of the motor task than in the rest state of the working memory task. These findings suggest that motor sub-networks are potentiated during the motor rest condition to support the dACC exerting motor control of the SMA during active visuomotor coordination.

Conclusions

This chapter argues that cognitive functions are implemented by brain networks, called neurocognitive networks. Brain networks, like other kinds of networks, consist of nodes and edges. The underlying hypothesis here is that the nodes and edges of a neurocognitive network are uniquely configured for the cognitive function that the network implements. Beyond that, it is argued that directed functional connectivity analysis, whether in the frequency or time domain, can capture essential aspects of neurocognitive network function.

The different experiments highlighted here demonstrate that directed influences between brain regions, measured by directed functional connectivity analysis, using different neural signals, and in all cases revealing statistical time series predictability, consistently track the coordination dynamics of neurocognitive networks. Each cognitive function is suggested to be uniquely determined by directed interregional interactions in the neurocognitive networks that underlie that function.

References

Afshari S, Jalili M (2016) Directed functional networks in Alzheimer's Disease: Disruption of global and local connectivity measures. IEEE J Biomed Health Informat. 21:949–955.

Amiez C, Petrides M (2014) Neuroimaging evidence of the anatomo-functional organization of the human cingulate motor areas. Cereb Cortex 24:563–578.

Asemi A, Ramaseshan, Burgess A, Diwadkar VA, Bressler SL (2015) Front Hum Neurosci 9:doi:https://doi.org/10.3389/fnhum.2015.00309.

Bachinger M, Zerbi V, Moisa M, Polania R, Liu Q, Mantini D, Ruff C, Wenderoth N (2017) J Neurosci 37:4766–4777.

Bressler SL (1987a) Relation of olfactory bulb and cortex. I. Spatial variation of bulbocortical interdependence. Brain Res 409:285–293.

Bressler SL (1987b) Relation of olfactory bulb and cortex. II. Model for driving of cortex by bulb. Brain Res 409:294–301.

Bressler SL, Kelso JAS (2016) Coordination dynamics in cognitive neuroscience. Front Neurosci 10:397.

Bressler SL, Menon V (2010) Large-scale brain networks in cognition: emerging methods and principles. Trends Cogn Sci 14:277–290.

Bressler SL, Coppola R, Nakamura R (1993) Episodic multiregional cortical coherence at multiple frequencies during visual task performance. Nature 366:153–156.

Bressler SL, Richter CG, Chen Y, Ding M (2007) Cortical functional network organization from autoregressive modeling of local field potential oscillations. Stat Med 26:3875–3885.

Bressler SL, Tang W, Sylvester CM, Shulman GL, Corbetta M (2008) Top-down control of human visual cortex by frontal and parietal cortex in anticipatory visual spatial attention. J Neurosci 28:10056–10061.

Brovelli A, Ding M, Ledberg A, Chen Y, Nakamura R, Bressler SL (2004) Beta oscillations in a large-scale sensorimotor cortical network: directional influences revealed by Granger causality. Proc Natl Acad Sci USA 101:9849–9854.

Ding M, Bressler SL, Yang W, Liang H (2000) Short-window spectral analysis of cortical event-related potentials by adaptive multivariate autoregressive modeling: data preprocessing, model validation, and variability assessment. Biol Cybern 83:35–45.

Ding M, Chen Y, Bressler SL (2006) Granger causality: basic theory and application to neuroscience 17. In: Schelter B, Winterhalder M, Timmer J (Eds) Handbook of Time Series Analysis: Recent Theoretical Developments and Applications. Wiley-VCH, Verlag GmbH & Co. KGaA.

Diwadkar VA, Asemi A, Burgess A, Chowdury A, Bressler SL (2017) Potentiation of motor sub-networks for motor control but not working memory: interaction of dACC and SMA revealed by resting-state directed functional connectivity. PloS One 12:e0172531.

Dosenbach NU, Fair DA, Cohen AL, Schlaggar BL, Petersen, SE (2008) A dual-networks architecture of top-down control. Trends Cogn Sci, 12(3), 99-105, https://doi.org/10.1016/j.tics.2008.01.001.

Dum RP, Strick PL (2002) Motor areas in the frontal lobe of the primate. Physiol Behav, 77(4-5), 677-682, doi:S0031938402009290 [pii].

Ebitz RB, Moore T (2017) Selective modulation of the pupil light reflex by prefrontal cortex microstimulation. J Neurosci 37:5008–5018.

Engel AK, Fries P (2010) Beta-band oscillations – signaling the status quo? Curr Opin Neurobiol 20:156–165.

Engels MMA, Stam CJ, van der Flier WM, Scheltens P, de Waal H, van Straaten ECW (2015) Declining functional connectivity and changing hub locations in Alzheimer's disease: an EEG study. BMC Neurol 15:145.

Falco D, Chowdury A, DeBusschere S, Rosenberg DR, Diwadkar V, Bressler SL (submitted) From nodes to networks: how defining nodes based on Activation Likelihood Estimates or experimentally derived estimates impacts functional network connectivity analysis.

Freeman WJ (1975) Mass Action in the Nervous System. Academic.

Friston KJ, Price CJ (2003) Degeneracy and redundancy in cognitive anatomy. Trends Cogn Sci 7:151–152.

Fuster JM (2009) Cortex and memory: emergence of a new paradigm. J Cogn Neurosci 21:2047–2072.

Garg R, Cecchi GA, Rao AR (2011) Full-brain auto-regressive modeling (FARM) using fMRI. Neuroimage 58:416–441.

Geweke J (1982) Measurement of linear dependence and feedback between multiple time series. J Amer Stat Assoc 77:304–313.

Goldman-Rakic PS (1988) Topography of cognition: parallel distributed networks in primate association cortex. Ann Rev Neurosci 11:137–156.

Grace AA (2016) Giles de la Tourette's and the disruption of internet-mediated synchrony. Comments on: Hashemiyoon R, Kuhn J, Viser-Vanderwalle V. Brain Topogr 30:1–2.

Granger CWJ (1969) Investigating causal relations by econometric models and cross-spectral methods. Econometrica 37:424–438.

Jansen BH (1991) Time series analysis by means of linear modelling 7. In: Weitkunat R (Ed) Digital Biosignal Processing. Elsevier Science.

Kelso JAS (1995) Dynamic patterns: the self-organization of brain and behavior. MIT.

Lachaux JP, Rodriquez E, Martinerie J, Varela FJ (1999) Measuring phase synchrony in brain signals. Hum Brain Mapp 8:194–208.

Le Van Quyen M, Foucher J, Lachaux J, Rodriquez E, Lutz A, Martinerie J, Varela FJ (2001) Comparison of Hilbert transform and wavelet methods for the analysis of neuronal synchrony. J Neurosci Meth 111:83–98.

Luria A (1966) Higher cortical functions in man. Basic Books.
MacKay WA (1997) Synchronized neuronal oscillations and their role in motor processes. Trends Cogn Sci 176–183.
Macay WA, Mendonca AJ (1995) Field potential oscillatory bursts in parietal cortex before and during reach. Brain Res 704:167–174.
Mannino M, Bressler SL (2015) Foundational perspectives on causality in large-scale brain networks. Phys Life Rev 15:107–123.
Meehan TP, Bressler SL, Tang W, Astafiev SV, Sylvester CM, Shulman GL, Corbetta M (2017) Top-down cortical interactions in visuospatial attention. Brain Struct Funct https://doi.org/10.1007/s00429-017-1390-6.
Mesulam MM (1990) Large-scale neurocognitive networks and distributed processing for attention, language, and memory. Ann Neurol 28:597–613.
Miller EK, Cohen JD (2001) An integrative theory of prefrontal cortex function. Ann Rev Neurosci 24:167–202.
Moore T, Armstrong KM (2003) Selective gating of visual signals by microstimulation of frontal cortex. Nature 421:370–373.
Morecraft RJ, Van Hoesen GW (1992) Cingulate input to the primary and supplementary motor cortices in the rhesus monkey: evidence for somatotopy in areas 24c and 23c. J Comp Neurol, 322(4), 471–489, https://doi.org/10.1002/cne.903220403.
Morecraft RJ, Stilwell-Morecraft KS, Cipolloni PB, Ge J, McNeal DW, Pandya DN (2012) Cytoarchitecture and cortical connections of the anterior cingulate and adjacent somatomotor fields in the rhesus monkey. Brain Res Bull, 87(4–5), 457–497, https://doi.org/10.1016/j.brainresbull.2011.12.005.
Paus T (2005) Inferring causality in brain images: a perturbation approach. Philos Trans Royal Soc B 360:1109–1114.
Paus T, Petrides M, Evans AC, Meyer E (1993) Role of the human anterior cingulate cortex in the control of oculomotor, manual, and speech responses: a positron emission tomography study. J Neurophys 70:453–469.
Richter CG, Coppola R, Bressler SL (2016) Top-down oscillatory signaling conveys behavioral context to primary visual cortex. bioRxiv doi:https://doi.org/10.1101/074609.
Rubinov M, Bullmore E (2013) Fledging pathoconnectomics of psychiatric disorders. Trends Cogn Sci 17:641–647.
Ruff CC, Blankenburg F, Bjoertomt O, Bestmann S, Freeman E, Haynes JD, Rees G, Josephs O, Deichmann R, Driver J (2006) Concurrent TMS-fMRI and psychophysics reveal frontal influences on human retinotopic visual cortex. Curr Biol 16:1479–1488.
Ruff CC, Bestmann S, Blankenburg F, Bjoertomt O, Josephs O, Weiskopf, Deichmann R, Driver J (2008) Distinct causal influences of parietal versus frontal areas on human visual cortex: evidence from concurrent TMS-fMRI. Cereb Cortex 18:817–827.
Ruff CC, Driver J, Bestmann S (2009) Combining TMS and fMRI: from 'virtual lesions' to functional-network accounts of cognition. Cortex 45:1043–1049.
Salazar RF, Dotson NM, Bressler SL, Gray CM (2012) Content-specific fronto-parietal synchronization during visual working memory. Science 338:1097–1100.
Sazonov AV, Ho CK, Bergmans JWM, Arends JBAM, Griep PAM, Verbitskiy EA, Clutmans PJM, Boon PAJM (2009) An investigation of the phase locking index for measuring of interdependency of cortical source signals recorded in the EEG. Biol Cybern 100:129–146.
Silverstein BH, Bressler SL, Diwadkar VA (2016) Inferring the disconnection syndrome in schizophrenia: interpretational considerations on methods for the network analyses of fMRI data. Front Psychiat 7:doi: https://doi.org/10.3389/fpsyt.2016.00132.
Sun FT, Miller LM, D'Esposito M (2004) Measuring interregional functional connectivity using coherence and partial coherence analyses of fMRI data. Neuroimage 21:647–658.
Tang W, Bressler SL, Sylvester CM, Shulman GL, Corbetta M (2012) Measuring Granger Causality between cortical regions from voxelwise fMRI BOLD signals with LASSO. PLoS Comp Biol 8:e1002513.

Uhlhaas PJ, Singer W (2006) Neural synchrony in brain disorders: relevance for cognitive dysfunctions and pathophysiology. Neuron 52:155–168.

van Wijk BCM, Beek PJ, Daffertshofer A (2012) Neural synchrony within the motor system: what have we learned so far? Front Hum Neurosci 6:252.

Vincent JL, Kahn I, Snyder AZ, Raichle ME, Buckner RL (2008) Evidence for a frontoparietal control system revealed by intrinsic functional connectivity. J Neurophysiol, 100(6), 3328-3342, https://doi.org/10.1152/jn.90355.2008.

von Holst E (1939/1973) The behavioral physiology of man and animals. In: Martin R (Ed) The Collected Papers of Erich von Holst. University of Miami.

Vossel S, Geng JJ, Fink GR (2014) Dorsal and ventral attention systems: distinct neural circuits but collaborative roles. Neuroscientist 20:150–159.

Wernicke (1874/1973) The aphasia symptom-complex: a psychological study on an anatomic basis. In: Eggert GH (Ed) Wernicke's Work on Aphasia: A Sourcebook and Review. Mouton de Gruyter.

Weiner N (1956) The theory of prediction. In: Beckenbach EF (Ed) Modern Mathematics for the Engineer. McGraw-Hill.

Meta-Analytic Connectivity Modelling (MACM): A Tool for Assessing Region-Specific Functional Connectivity Patterns in Task-Constrained States

Robert Langner and Julia A. Camilleri

What Is MACM, and What Is It for?

Over the last 30 years, neuroimaging has become an increasingly popular method to further the understanding of the brain, its plasticity, and its disorders through the noninvasive study of functional neuroanatomy and brain–behavior relationships. The surge of neuroimaging use in research has resulted in tens of thousands of individual studies, each reporting findings associated with a specific task (Fox et al. 2014; Müller et al. 2018). This wealth of neuroimaging literature led to the conception of large-scale databases, such as BrainMap (www.brainmap.org; Fox and Lancaster 2002) or Neurosynth (www.neurosynth.org; Yarkoni et al. 2011), that store an increasingly large number of results obtained from published brain activation studies. Such databases usually contain the locations of the activation maxima, as coordinates in standard stereotactic space, reported for each particular experiment together with the associated meta-data describing the experimental design. These databases, in turn, can be exploited by coordinate-based meta-analytic approaches in order to quantitatively synthesize the host of results and test for spatial convergence of activations across experiments. Such unbiased syntheses across many independent results form a potent way of distinguishing between spurious and replicable findings or of gauging whether effects are generalizable or may rather be due to idiosyncrasies of a given study (Fox et al. 2014; Müller et al. 2018).

A popular way of implementing this approach is to derive brain regions consistently found across studies investigating a specific mental function or cognitive process. However, the meta-analytic approach can also be employed to derive brain

R. Langner (✉) · J. A. Camilleri
Institute of Systems Neuroscience, Heinrich Heine University Düsseldorf, Düsseldorf, Germany

Institute of Neuroscience and Medicine (INM-7), Research Centre Jülich, Jülich, Germany
e-mail: robert.langner@uni-duesseldorf.de

V. A. Diwadkar, S. B. Eickhoff (eds.), *Brain Network Dysfunction in Neuropsychiatric Illness*, https://doi.org/10.1007/978-3-030-59797-9_5

regions that are consistently activated together with a given region, across studies investigating different functions (Koski and Paus 2000; Postuma and Dagher 2006; Toro et al. 2008). This latter approach does not focus on a specific mental process but rather on the co-activation of brain regions with a specific a priori defined region of interest across all kinds of different mental processes. Due to this focus on jointly active regions throughout the brain, the approach allows for the identification of a functional network of the given region of interest and, therefore, constitutes a way to investigate this region's *functional connectivity*. Roughly ten years after the pioneering work of Koski and Paus (2000), this approach came to be known as "meta-analytic connectivity modelling" (MACM; Eickhoff et al. 2010, 2011; Laird et al. 2009b; Robinson et al. 2010).

Functional connectivity (FC) has been defined as the temporal correlation of spatially distant neurophysiological events (Friston 1994). This definition is based on the notion that if neural events of the same kind occur together, they should be functionally related to each other. Accordingly, two regions are considered to be functionally connected when there is a statistical relationship between the measures of activity recorded from each of these regions (Eickhoff and Müller 2015). This notion of FC may pertain to any form of neurophysiological event, including the joint activation of spatially distinct brain regions across a multitude of experiments and tasks. The most common strategy for investigating FC, however, is based on the assessment of within-subject correlations between time series of hemodynamic signal fluctuations in different brain regions. This can be done with fMRI data recorded from participants lying in the scanner under task-free, "resting" conditions (i.e., resting-state FC analysis) or performing a specific task (i.e., task-dependent FC). The uncontrolled nature of the resting state, however, raised questions about the generalizability of resting-state FC findings to task-constrained situations, whereas task-dependent FC is at least partly driven by the nature of the task at hand and, therefore, cannot provide information about which regions in the brain are co-activated with one another *in general*. In contrast, MACM rests on co-activation patterns observed across a large number of different tasks and implementations, thus allowing for examining task-based FC in a highly generalizable way (Laird et al. 2009b, 2013; Eickhoff and Grefkes 2011). In other words, MACM answers the following question: If a particular brain region is activated, which other regions are more likely to be consistently co-activated across a diverse set of tasks than would be expected by chance? The answer is a task-based FC map for that particular region that is not specific to one particular task or implementation.

As such, MACM provides an unbiased, data-driven approach to robustly determine the connectional "fingerprint" of a given region of interest (i.e., a region's pattern of FC with the rest of the brain). This makes MACM useful for clinical neuroimaging research in three major ways: First, it can be employed to compare the FC patterns of a given brain region in health and disease, given that the database (or, more generally, the sample of experiments) used contains a sufficiently large and sufficiently diverse set of brain mapping studies done in patients with the clinical condition of interest. Second, MACM allows defining "normative" functional brain networks in healthy participants for the subsequent study of clinical aberrations in their activity, morphology, or connectivity. Third, it enables the post hoc FC-based characterization of brain regions that have shown clinically relevant effects, revealing the functional networks that these regions are embedded in and, thereby, generating testable hypotheses about additional targets of potential clinical change (for nonclinical examples, see Camilleri et al. 2018; Langner et al. 2019).

How Does MACM Work?

MACM is performed by first retrieving all experiments, present in the neuroimaging database of choice, that include at least one focus of activation in the region of interest. This search can be specific to certain kinds of participants or studies, depending on the research question (e.g., healthy adults). The (other) coordinates reported in the individual experiments are then extracted and their spatial convergence (i.e., consistency across experiments) is tested by means of a quantitative coordinate-based meta-analysis. Different algorithms can be used to perform such a meta-analysis (cf. Müller et al. 2018), including activation likelihood estimation (ALE; Eickhoff et al. 2009, 2012; Turkeltaub et al. 2002, 2012), (multilevel) kernel density analysis (Wager et al. 2004, 2007; Pauli et al. 2016), Gaussian-process regression (Salimi-Khorshidi et al. 2011), parametric voxel-based meta-analysis (Costafreda et al. 2009), or signed differential mapping (Radua and Mataix-Cols 2009).

This chapter will focus on the use of ALE for conducting MACM analyses. In this coordinate-based meta-analytic approach, all activation foci included are modeled as centers of three-dimensional Gaussian probability distributions, which capture the spatial uncertainty related to each of the coordinates reported in the literature. In particular, this probability distribution takes into account the limited reliability in peak activity localization that results from functional and neuroanatomical differences between individuals as well as normalization differences between studies. The width of spatial uncertainty depends on the number of participants in the corresponding experiment, with a narrower Gaussian distribution for experiments with larger sample sizes because of their increased power for finding "true" effects (Eickhoff et al. 2009).

The probability distributions of all activation foci in a given experiment are then combined for each voxel, yielding a modeled activation map. Subsequently, the across-experiment voxel-wise union of these modeled activation maps (i.e., activation probabilities of each voxel) is computed, resulting in an ALE map indicating the degree of convergence across experiments at each brain location. This way, an ALE score for each voxel is obtained, representing the co-activation probability of that particular voxel with the given seed region. Finally, the voxels of statistically significant ("true") convergence are determined by comparing the computed ALE map against an empirically derived null-distribution (for details, see Eickhoff et al. 2012). In particular, the significance of an ALE score is given by the proportion of equal or higher values under the null-distribution. Thresholding the ALE map at a given cut-off (usually at $P \geq 0.95$, which corresponds to the conventional p-value of

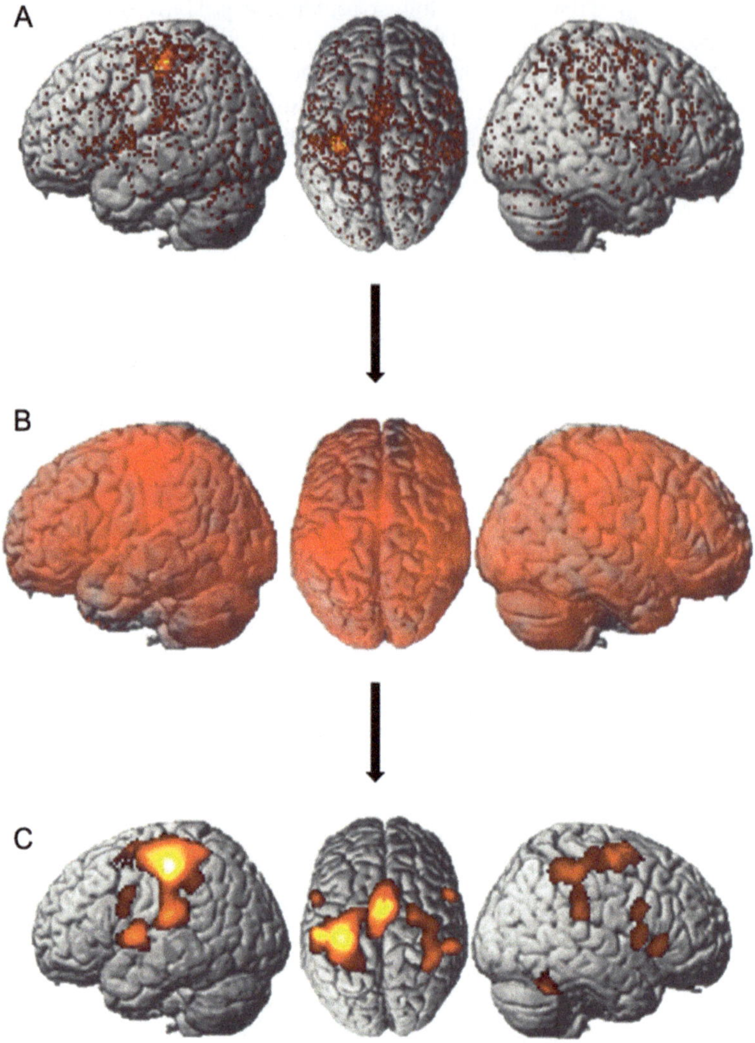

Fig. 1 Schematic depiction of steps performed during an activation likelihood estimation (ALE) meta-analysis
(**a**) All activation foci reported in all experiments activating the seed region (left primary motor area)
(**b**) ALE map reflecting the union of the modeled activation maps across all experiments
(**c**) Thresholded ALE map showing clusters of significant convergence across studies ($p < 0.05$, family-wise error corrected), as assessed by comparison with an empirical null-distribution

0.05) and correcting for multiple comparisons will in turn specify those voxels in which "true" (i.e., above-chance) convergence is observed (for further details, see Eickhoff and Bzdok 2013).

Per definition, all experiments included in such an analysis will feature at least one focus of activation in the region of interest, which, as a result, will always show the highest convergence of activity. Significant convergence of the reported foci outside the region of interest indicates co-activation that is higher than that expected by chance. This meta-analytic co-activation map, in turn, shows the pattern of task-based FC of the a priori selected seed region with the entire brain (Fig. 1).

In the above-described implementation of MACM, significant spatial convergence (i.e., consistent co-activation across experiments) is differentiated from chance findings by comparison against a null-distribution of random spatial associations between experiments. According to this empirical null-distribution, all gray-matter voxels have the same chance of convergence. In the brain, however, the a priori probability of observing increased activity markedly differs between individual locations. For instance, a high-level integrative region such as the anterior insula is likely to be activated more often than a functionally specialized unimodal region such as the secondary somatosensory cortex of the parietal operculum. In addition to this "intrinsic" activation bias, there are also asymmetries in the number of published experiments on particular cognitive domains and, as a result, domain-specific regional activation patterns. These across-study imbalances in the brain locations being reported propagate to databases like BrainMap or Neurosynth, contributing to an unequal distribution of stored activation foci across the brain (Fig. 2). Therefore, using a quasi-rectangular null-distribution in MACM renders the detection of significant convergence more likely in those regions that are more frequently found activated.

To counter this bias, a modified MACM approach was proposed (Langner et al. 2014). In this "specific co-activation likelihood estimation" (SCALE) algorithm, a null-distribution is generated that reflects the base rate of reporting activation in any given voxel and thus equalizes the a priori chance of finding across-study convergence in each voxel of the brain. This modified MACM algorithm is complementary to the standard one described above, providing information on *specific* functional associations between brain regions (see Fig. 3). This heightened specificity may be particularly useful for defining functional core networks or differentiating functional sub-networks.

How to Run a MACM Analysis

Table 1 outlines how MACM can be performed using the BrainMap database as an example, detailing the different analysis steps and corresponding software solutions.

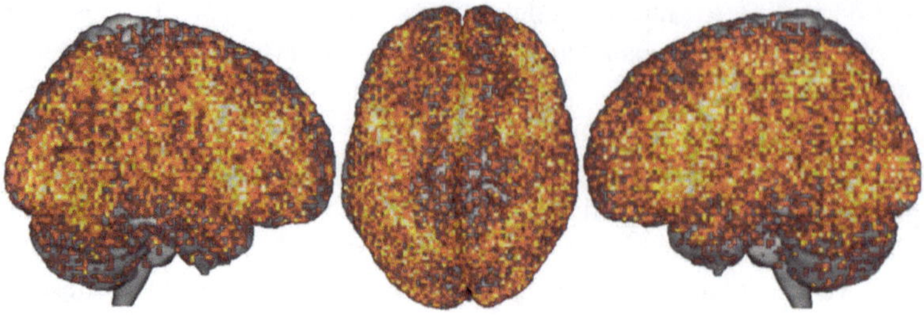

Fig. 2 All activation foci contained in the BrainMap database rendered on a standard brain template. Color codes the frequency of activation of a given voxel in BrainMap, with lighter colors indicating higher frequencies

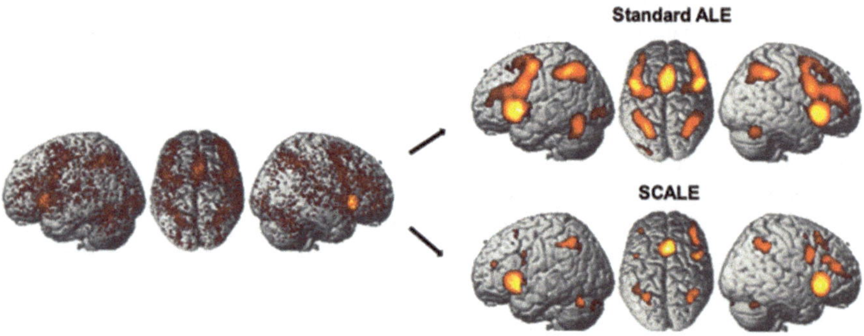

Fig. 3 Schematic comparison of results obtained when using the standard ALE versus SCALE algorithm for a MACM analysis of a seed region in the right anterior insula. Of note, using SCALE yields a more focused pattern of functional connectivity with the seed

Table 1 Work flow overview

Analysis steps	Software[a]
(1) Definition of the voxels of interest ("seed") in 3-D standard space	
(2) Definition of specific inclusion or exclusion criteria for experiments to be eligible (e.g., healthy adults only)	Sleuth
(3) Selection of all eligible experiments activating the seed region	Sleuth
(4) Extraction of all coordinates reported in the identified experiments	Sleuth
(5) Performing a meta-analysis across all coordinates reported in the identified experiments	GingerALE

[a]Available from www.brainmap.org

Meta-Analytic Functional Profiling of Brain Regions

Databases like BrainMap or Neurosynth contain, beside standardized neuroimaging results, information on the context in which the experimental results were obtained. This allows extending the MACM approach to studying brain–behavior relationships at the meta-analytic level. In particular, such databases can be used to objectively characterize a given brain region in terms of its cognitive/behavioral function(s), based on the meta-data stored in the database used (Fox et al. 2005; Laird et al. 2009a, 2011; Yarkoni et al. 2011; see also Genon et al. 2018a, and Poldrack 2011). The key idea behind this approach is to identify all experiments that activate a particular region of interest and then quantitatively analyze the meta-data describing the experimental task employed. This allows statistical inference on the type of tasks and cognitive processes that evoke activation in a particular region. Such a quantitative functional characterization can be achieved using forward and/or reverse inference, as performed in previous studies (Müller et al. 2013; Camilleri et al. 2018; Langner et al. 2018). Forward inference reveals the probability of observing activity in a brain region, given a particular task or cognitive process, whereas reverse inference reveals the probability of a particular task or cognitive process being present, given activation in a particular brain region.

In BrainMap, tasks are coded along two dimensions (Fox et al. 2005): "behavioral domains" (BD) describe the behavioral or mental function probed by an experiment, and "paradigm classes" (PC) label the type of task ("paradigm") used (see www.brainmap.org/scribe for more information on the BrainMap taxonomy). In the BrainMap coding scheme, a BD and PC is recorded for each statistical contrast reported in the studies included to describe the context of the research. The six BD main categories, which are further divided into subcategories, comprise cognition, action, perception, emotion, interoception, and pharmacology (Fig. 4).

In the forward inference approach, the functional profile of a region of interest is determined by identifying those taxonomic labels for which the probability of

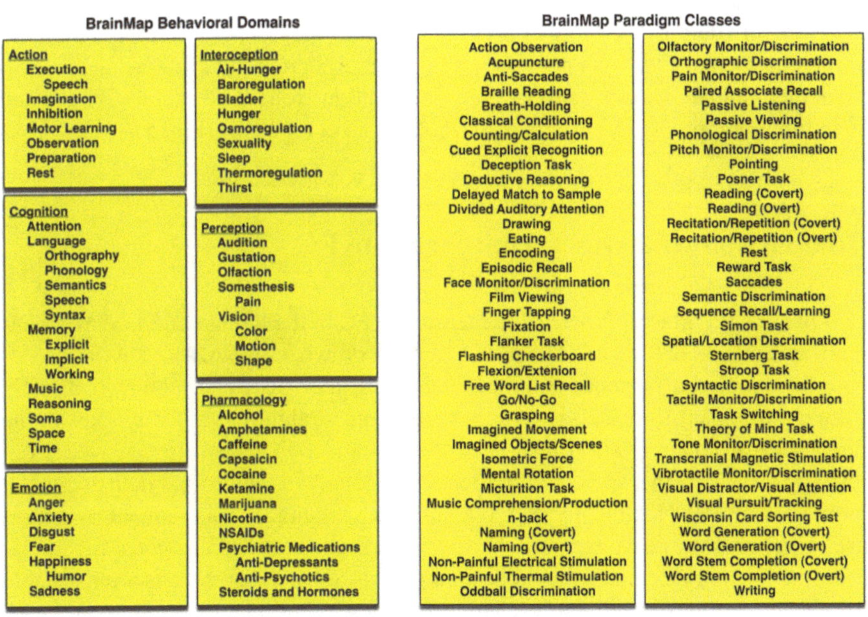

Fig. 4 BrainMap database behavioral domains and paradigm classes (from Laird et al. 2009a)

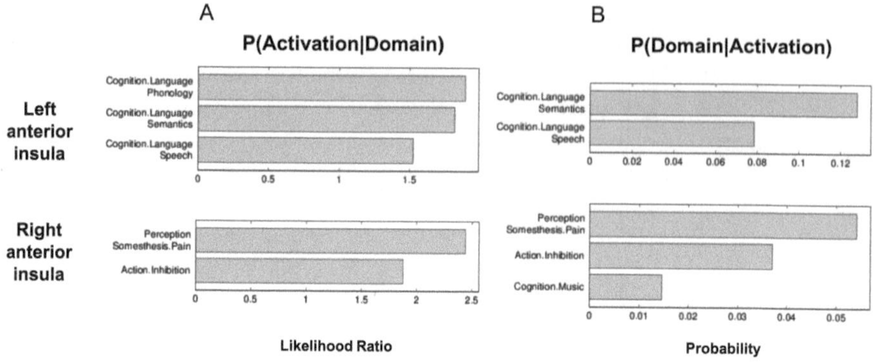

Fig. 5 Example of BrainMap-based functional profiling using (**a**) forward and (**b**) reverse inference: Behavioral domains associated with regions in the left and right anterior insula as defined by a meta-analysis on behavioral and emotional regulation (adapted from Langner et al. 2018)

finding activation in the respective region/set of regions is significantly higher than the overall (a priori) chance across the entire database. Thus, using a binomial test, this approach examines whether the conditional probability of activation given a particular BD or PC label [P(Activation|Task)] is higher than the baseline probability of activating the region(s) in question per se [P(Activation)]. In the reverse inference approach, the functional profile is determined by identifying the most likely taxonomic labels, given activation in a particular region/set of regions. This likelihood P(Task|Activation) can be derived from P(Activation|Task) as well as P(Task) and P(Activation) using Bayes' rule; significance can be assessed via a chi-square test (for more details, see Laird et al. 2013). Of note, these inference approaches aim at relating mental functions/processes to a given brain region of interest in an unbiased, quantitative manner, without claiming a unique role of a brain region for any such function/process (cf. Poldrack 2006; Yarkoni et al. 2011) (Fig. 5).

Comparison with Other Approaches to FC

Compared with other approaches to measuring FC in the human brain, MACM has both advantages and disadvantages. A certain drawback of MACM is its use of coordinates stored in neuroimaging databases rather than data obtained directly from a participant sample. As a result, there is no straightforward way to examine direct FC–behavior relationships. For this purpose, FC measures derived from resting-state fMRI in a sample of participants, along with behavioral data, would be more appropriate. Notably, these two approaches are highly complementary and, if used together, may provide comprehensive information regarding a given functional network. In fact, research using both methods has shown that there is a remarkable degree of overlap between the "layouts" of particular brain networks identified with either approach (e.g., Jakobs et al. 2012; Reid et al. 2016; Smith et al. 2009; Genon et al. 2018b).

Conversely, MACM has several unique advantages when compared to task-free FC measures as obtained from resting-state fMRI. First, MACM yields robust and highly generalizable patterns of coordinated brain activity in response to a broad range of tasks, rather than spontaneous activity observed in a task-free situation with the body (but not the brain) at rest. Second, MACM results arise from the analysis of data obtained across a host of studies, providing a firm basis for normative models of FC for particular brain regions, which can also be statistically compared to test for their specificity as well as similarity or overlap. This, in turn, makes MACM particularly useful for delineating functional modules within a given larger brain region based on commonalities and differences in the FC profiles of the region's constituent voxels (i.e., connectivity-based parcellation; Eickhoff et al. 2011, 2018). Finally, MACM can also be used to identify what kinds of tasks or mental processes yield co-activation between two regions (Müller et al. 2013).

Summary

This chapter reviewed a meta-analytic approach to functional brain connectivity, MACM, which capitalizes on the rich corpus of published neuroimaging studies that have accumulated over the years. Exploiting large databases of neuroimaging results, MACM employs meta-analytic algorithms such as ALE to identify brain regions that are consistently activated together with a given region, across studies investigating different functions. This allows for the identification of a functional network of the given region of interest and, therefore, constitutes a way to investigate this region's FC. The MACM approach can also be extended to objectively characterize a given brain region's cognitive/behavioral function(s), based on the meta-data of the experiments stored in the database used. The key idea of this approach is to identify all experiments that activate a particular region of interest and then quantitatively analyze the meta-data describing the experimental task employed, allowing for statistical inference on the type of tasks and cognitive processes that evoke activation in a particular region. MACM is a useful tool for delineating functional networks in the brain as it yields robust and highly generalizable patterns of coordinated brain activity in response to a broad range of tasks. It is thus complementary to other, task-free approaches to FC, providing a firm basis for normative FC models of particular brain regions.

References

Camilleri JA, Müller VI, Fox P, Laird AR, Hoffstaedter F, Kalenscher T, Eickhoff SB (2018). Definition and characterization of an extended multiple-demand network. NeuroImage, 165:138–147. doi: https://doi.org/10.1016/j.neuroimage.2017.10.020.

Costafreda SG, David AS, Brammer MJ (2009). A parametric approach to voxel-based meta-analysis. NeuroImage, 46(1):115–122.

Eickhoff SB, Bzdok D (2013). Meta-analyses in basic and clinical neuroscience: State of the art and perspective. In: S. Ulmer, O. Jansen (eds.), fMRI – Basics and Clinical Applications (pp. 77–87). Springer: Berlin.

Eickhoff SB, Bzdok D, Laird AR, Kurth F, Fox PT (2012). Activation likelihood estimation meta-analysis revisited. NeuroImage, 59 (3):2349–2361. doi:https://doi.org/10.1016/j.neuroimage.2011.09.017.

Eickhoff SB, Bzdok D, Laird AR, Roski C, Caspers S, Zilles K, Fox PT (2011). Co-activation patterns distinguish cortical modules, their connectivity and functional differentiation. NeuroImage, 57(3):938-949. doi:https://doi.org/10.1016/j.neuroimage.2011.05.021.

Eickhoff SB, Laird AR, Grefkes C, Wang LE, Zilles K, Fox PT (2009). Coordinate-based activation likelihood estimation meta-analysis of neuroimaging data: a random-effects approach based on empirical estimates of spatial uncertainty. Human Brain Mapping, 30:2907–2926.

Eickhoff SB, Grefkes C (2011). Approaches for the integrated analysis of structure, function and connectivity of the human brain. Clinical Neurophysiology, 42(2):107–121.

Eickhoff SB, Jbabdi S, Caspers S, Laird AR, Fox PT, Zilles K, Behrens TE (2010). Anatomical and functional connectivity of cytoarchitectonic areas within the human parietal operculum. Journal of Neuroscience, 30(18):6409–6421.

Eickhoff SB, Müller VI (2015). Functional connectivity. In: A.W. Toga (ed.), Brain Mapping: An Encyclopedic Reference, vol. 2 (pp. 187–201). Academic Press: Elsevier.

Eickhoff SB, Yeo BTT, Genon S (2018). Imaging-based parcellations of the human brain. Nature Reviews Neuroscience, 19(11):672–686. doi: https://doi.org/10.1038/s41583-018-0071-7.

Fox PT, Laird AR, Fox SP, Fox PM, Uecker AM, Crank M, Koenig SF, Lancaster JL (2005). BrainMap taxonomy of experimental design: Description and evaluation. Human Brain Mapping, 25:185–198.

Fox PT, Lancaster JL (2002). Opinion: Mapping context and content: The BrainMap model. Nature Reviews Neuroscience, 3:319–321.

Fox PT, Lancaster JL, Laird AR, Eickhoff SB (2014). Meta-analysis in human neuroimaging: computational modeling of large-scale databases. Annual Review of Neuroscience, 37:409–434.

Friston, KJ (1994). Functional and effective connectivity in neuroimaging: a synthesis. Human Brain Mapping, 2(1-2): 56–78.

Genon S, Reid A, Langner R, Amunts K, Eickhoff SB (2018a). How to characterize the function of a brain region. Trends in Cognitive Sciences, 22(4):350–364. doi:https://doi.org/10.1016/j.tics.2018.01.010.

Genon S, Reid A, Li H, Fan L, Müller VI, Cieslik EC, Hoffstaedter F, Langner R, Grefkes C, Laird AR, Fox PT, Jiang T, Amunts K, Eickhoff SB (2018b). The heterogeneity of the left dorsal premotor cortex evidenced by multimodal connectivity-based parcellation and functional characterization. NeuroImage, 170:400–411. doi: https://doi.org/10.1016/j.neuroimage.2017.02.034.

Jakobs O, Langner R, Caspers S, Roski C, Cieslik EC, Zilles K, Laird AR, Fox PT, Eickhoff SB (2012). Across-study and within-subject functional connectivity of a right temporo-parietal junction subregion involved in stimulus–context integration. NeuroImage, 60(4):2389–2398. doi:https://doi.org/10.1016/j.neuroimage.2012.02.037.

Koski, L, Paus, T (2000). Functional connectivity of the anterior cingulate cortex with the human frontal lobe: a brain-mapping meta-analysis. Experimental Brain Research, 133:55–65.

Laird AR, Eickhoff SB, Kurth F, Fox PM, Uecker AM, Turner JA, Robinson JL, Lancaster JL, Fox PT (2009a). ALE meta-analysis workflows via the Brainmap database: Progress towards a probabilistic functional brain atlas. Frontiers in Neuroinformatics, 3:23. doi:https://doi.org/10.3389/neuro.11.023.2009.

Laird AR, Eickhoff SB, Li K, Robin DA, Glahn DC, Fox PT (2009b). Investigating the functional heterogeneity of the default mode network using coordinate-based meta-analytic modeling. Journal of Neuroscience, 29(46):14496–14505. doi:https://doi.org/10.1523/JNEUROSCI.4004-09.2009.

Laird AR, Eickhoff SB, Rottschy C, Bzdok D, Ray KL, Fox PT (2013). Networks of task co-activations. NeuroImage, 80:505–514. doi:https://doi.org/10.1016/j.neuroimage.2013.04.073.

Laird AR, Fox PM, Eickhoff SB, Turner JA, Ray KL, McKay DR, ... Fox, PT (2011). Behavioral interpretations of intrinsic connectivity networks. Journal of Cognitive Neuroscience, 23(12):4022–4037.

Langner R, Eickhoff SB, Bilalić M (2019). A network view on brain regions involved in experts' object and pattern recognition: Implications for the neural mechanisms of skilled visual perception. Brain and Cognition, 131:74-86. doi:https://doi.org/10.1016/j.bandc.2018.09.007.

Langner R, Leiberg S, Hoffstaedter F, Eickhoff SB (2018). Towards a human self-regulation system: Common and distinct neural signatures of emotional and behavioural control. Neuroscience and Biobehavioral Reviews, 90:400–410. doi:https://doi.org/10.1016/j.neubiorev.2018.04.022.

Langner R, Rottschy C, Laird AR, Fox PT, Eickhoff SB (2014). Meta-analytic connectivity modeling revisited: controlling for activation base rates. NeuroImage, 99:559–570.

Müller VI, Cieslik EC, Laird AR, Fox PT, Eickhoff SB (2013). Dysregulated left inferior parietal activity in schizophrenia and depression: functional connectivity and characterization. Frontiers in Human Neuroscience, 7:268.

Müller VI, Cieslik EC, Laird AR, Fox PT, Radua J, Mataix-Cols D, ...Wager TD (2018). Ten simple rules for neuroimaging meta-analysis. Neuroscience and Biobehavioral Reviews, 84:151–161.

Pauli, WM, O'Reilly, R. C, Yarkoni, T, & Wager, TD. (2016). Regional specialization within the human striatum for diverse psychological functions. Proceedings of the National Academy of Sciences, 113(7):1907–1912.

Poldrack, RA (2006). Can cognitive processes be inferred from neuroimaging data? Trends in Cognitive Sciences, 10:59–63.

Poldrack, RA (2011). Inferring mental states from neuroimaging data: from reverse inference to large-scale decoding. Neuron, 7: 692–697. doi:https://doi.org/10.1016/j.neuron.2011.11.001.

Postuma, RB, Dagher, A (2006). Basal ganglia functional connectivity based on a meta-analysis of 126 positron emission tomography and functional magnetic resonance imaging publications. Cerebral Cortex, 16:1508–1521.

Radua, J, & Mataix-Cols, D (2009). Voxel-wise meta-analysis of grey matter changes in obsessive–compulsive disorder. The British Journal of Psychiatry, 195(5):393–402.

Reid AT, Bzdok D, Langner R, Fox PT, Laird AR, Amunts K, Eickhoff SB, Eickhoff CR (2016). Multimodal connectivity mapping of the human left anterior and posterior lateral prefrontal cortex. Brain Structure & Function, 221(5):2589–2605. doi: https://doi.org/10.1007/s00429-015-1060-5.

Robinson, JL, Laird, AR, Glahn, DC, Lovallo, WR, & Fox, PT (2010). Metaanalytic connectivity modeling: delineating the functional connectivity of the human amygdala. Human Brain Mapping, 31(2):173–184.

Salimi-Khorshidi G, Nichols TE, Smith SM, Woolrich MW (2011). Using Gaussian-process regression for meta-analytic neuroimaging inference based on sparse observations. IEEE Trans Med Imaging, 30(7):1401–1416. https://doi.org/10.1109/TMI.2011.2122341.

Smith, SM, Fox, PT, Miller, KL, Glahn, DC, Fox, PM, Mackay, CE, ... & Beckmann, CF (2009). Correspondence of the brain's functional architecture during activation and rest. Proceedings of the National Academy of Sciences, 106(31):13040–13045.

Toro R, Fox P, Paus T (2008). Functional coactivation map of the human brain. Cerebral Cortex, 18:2553–2559.

Turkeltaub PE, Eden GF, Jones KM, Zeffiro TA (2002). Meta-analysis of the functional neuroanatomy of single-word reading: method and validation. NeuroImage, 16:765–780.

Turkeltaub, PE, Eickhoff, SB, Laird, AR, Fox, M, Wiener, M, & Fox, P (2012). Minimizing within-experiment and within-group effects in activation likelihood estimation meta-analyses. Human Brain Mapping, 33(1), 1–13.

Wager, TD, Jonides, J, & Reading, S (2004). Neuroimaging studies of shifting attention: a meta-analysis. NeuroImage, 22(4):1679–1693.

Wager TD, Lindquist M, Kaplan L (2007). Meta-analysis of functional neuroimaging data: current and future directions. Social Cognitive and Affective Neuroscience, 2:150–158.

Yarkoni, T, Poldrack, RA, Nichols, TE, Van Essen, DC, Wager, TD (2011). Large-scale automated synthesis of human functional neuroimaging data. Nature Methods, 8:665–670.

dMRI: Diffusion Magnetic Resonance Imaging as a Window onto Structural Brain Networks and White Matter Microstructure

Alard Roebroeck

Introduction

Diffusion magnetic resonance imaging (dMRI) can be used to probe the connectivity and microstructure of human brain tissue non-invasively in vivo. The diffusion-weighted MR signal has sensitivity to micrometer-scale tissue properties averaged over the imaging voxel size, for example, intra-cellular and extra-cellular volumes and the 3D orientations of axonal bundles. It derives its contrast from sensitivity to the bulk displacement of water through what is known as thermal motion, Brownian motion, or passive (self-)diffusion. For free diffusion, the displacement probability distribution within an isotropic medium can be modeled using a Gaussian distribution, with its spherical variance and iso-probability contours moving outward with time. However, in the brain, cell membranes, organelles, and myelin sheaths create barriers and form biological compartments that constrain the displacement of water molecules, modifying the statistical behavior of bulk diffusion over time. This review therefore focuses on the use of dMRI for estimates of the local orientations and estimates of microstructure of fiber tracts, as well as on an understanding of dMRI signal mechanisms and appropriate signal processing and modeling for this purpose. Section "Diffusion MRI acquisition and diffusion contrast" discusses basic diffusion MRI acquisition principles and diffusion contrast and the constraints the acquisition places on the modeling of diffusion. Section "Diffusion tensor imaging" sets out the diffusion tensor model, the most used model in dMRI that underlies diffusion tensor imaging, in a way which prepares a discussion of its limitations. The next sections set out advances in dMRI beyond DTI, first focusing on tractography and connectomics, with a need to accurately model spatially complex fiber configurations (in section "Tractography and connectomics") and then on diffusion

A. Roebroeck (✉)
Department of Cognitive Neuroscience, Faculty of Psychology and Neuroscience, Maastricht University, Maastricht, The Netherlands
e-mail: a.roebroeck@maastrichtuniversity.nl

V. A. Diwadkar, S. B. Eickhoff (eds.), *Brain Network Dysfunction in Neuropsychiatric Illness*, https://doi.org/10.1007/978-3-030-59797-9_6

microstructure, with a need to accurately model restricted diffusion and compartmentalization (in section "Diffusion microstructure"). Throughout, the emphasis is on a thorough understanding of basic principles and assumptions underlying techniques, as well as their possibilities and limitations for inference of brain connectivity, with a minimum of technical detail and mathematics. Section "Outlook" ends with an outlook on future developments emanating from current trends.

Diffusion MRI Acquisition and Diffusion Contrast

In the brain, cell membranes, organelles, and myelin sheaths create barriers and form biological compartments that constrain the displacement of water molecules, modifying the statistical behavior of bulk diffusion over time. Water diffusion in biological tissue is typically hindered or restricted by these barriers and often anisotropic in nature. Brain white matter mostly consists of myelinated axons that are often coherently organized in bundles or tracts. This makes diffusion inherently anisotropic, where the dominant orientations of diffusion are equal to the local orientations the dominant fiber tracts or bundles. Furthermore, the diffusion orthogonal to that of fiber tracts gives information about microstructural characteristics such as the axonal diameters and density. Therefore, the aim of diffusion MRI contrast mechanisms is to create a range of measurements, for each imaging voxel, which together have sensitivity for these anisotropic and microstructural characteristics.

Creating diffusion contrast in MRI is typically performed by a pulsed gradient spin echo (PGSE) pulse sequence. A PGSE or Stejskal-Tanner sequence is a spin-echo sequence, with 90° excitation pulse, 180° refocusing pulse and spin-echo readout, with the addition of two identical, large gradient pulses before and after the refocusing pulse (Stejskal and Tanner 1965). These diffusion gradients are characterized by their length δ, amplitude |G|, and spacing Δ, with a higher value for each of these parameters creating greater diffusion weighting and more diffusion contrast in the resulting image (Fig. 1).

In the case of Gaussian diffusion (e.g., as in the case of free diffusion without hindrance), the amount of diffusion weighting is fully specified by the b-value:

$$b = \gamma^2 \delta^2 |G|^2 \left(\Delta - \frac{\delta}{3} \right)$$

where b is *the* b-value measured in s/mm^2 and γ is the gyromagnetic ratio, and δ, |G|, and Δ characterize the diffusion gradients (Fig. 1a). The diffusion contrast is created by a signal attenuation with respect to the baseline signal generated by the spin-echo. Since the baseline is a T_2-weighted signal (acquired without diffusion gradients and often called the b0 signal), it decreases with echo time (TE). Therefore, to avoid excessive T_2 decay, the objective is mostly to minimize TE and thus optimize

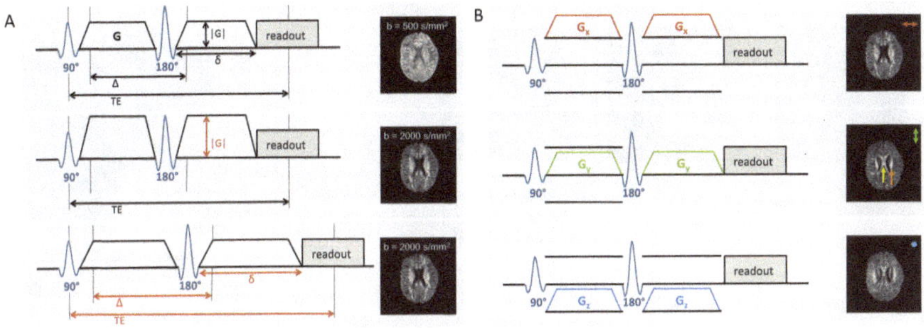

Fig. 1 Principles of diffusion-weighted MRI with the pulsed gradient spin echo (PGSE) pulse sequence. (**a**) Strength of diffusion encoding in a PGSE sequence with 90° excitation pulse, 180° refocusing pulse, and spin-echo readout. Diffusion is encoded by a pair of diffusion gradients G, characterized by their length δ, amplitude |G|, and spacing Δ (top). At low echo time TE, a higher degree of diffusion encoding, as quantified by the b-value, can be achieved by increasing |G| (middle) or increasing δ and Δ (bottom). All depicted images are individually scaled to emphasize contrast, compensating lower signal level for higher b images. (**b**) Diffusion gradients have three orthogonal components G_x, G_y, and G_z that encode diffusion in three independent orientations defined by the scanner coordinate system. Yellow arrow: corpus callosum. Orange arrow: corticospinal tract in the internal capsule

signal-to-noise (SNR) ratio. A higher degree of diffusion encoding, as quantified by the b-value, can be achieved either by increasing |G| (Fig. 1a, middle) or increasing δ and Δ (Fig. 1a, bottom). Since increasing |G| does not carry the penalty of increasing TE and decreasing SNR, this is often preferred. The possibility for low-TE high-SNR diffusion weighting is also the reason why MR systems with a high-performing gradient set, with maximum amplitude in excess of 40mT/m, are desirable for dMRI acquisitions. Furthermore, diffusion encoding is directional (vector valued in 3D space), which makes it sensitive to the orientation of anisotropic diffusion (Fig. 1b).[1] Diffusion gradients set in MR systems have three orthogonal components G_x, G_y, and G_z, which are used to encode diffusion in three independent orientations defined by the scanner coordinate system. Signal for stationary water will first be dephased by the first diffusion gradient and, after the refocusing pulse has been delivered, be rephrased by the second diffusion gradient, with no resulting signal attenuation. In contrast, water molecules that are displaced along the direction of the applied diffusion gradient will not be perfectly rephrased by the second diffusion gradient and, consequently, show signal attenuation. Figure 1b shows how diffusion weighting in the left–right or right–left direction (top) attenuates the signal in the left–right-oriented corpus callosum (yellow arrow), unlike orthogonal diffusion directions (middle and bottom). Conversely, diffusion weighting in the inferior–superior or superior–inferior direction (bottom) attenuates the signal in the

[1] Diffusion gradients have a well-defined direction, with polarity, whereas the symmetry of the diffusion weighting mechanism makes dMRI only sensitive to the orientation, without polarity, of anisotropic structures. This is why this review refers to the direction of diffusion gradients and orientation of anisotropic structures such as white matter fibers.

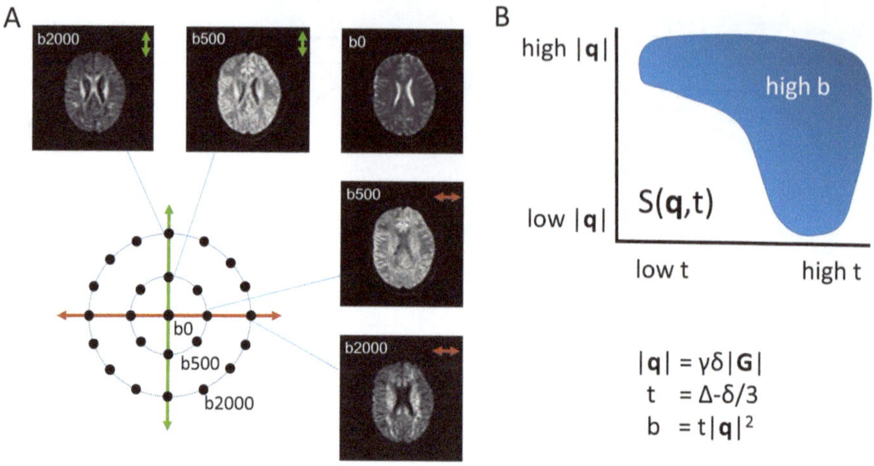

Fig. 2 A more general look at diffusion encoding parameters. (**a**) Regarding diffusion acquisitions as shells vector b-space, here simplified to 2-dimensional (only G_x and G_y are considered). The b0 measurement forms the origin, and measurements with equal b-value but different direction form circular "shells" (spherical in 3D x/y/z b-space). Images are individually scaled to emphasize contrast, compensating lower signal level for higher-b images. (**b**) b-value can be separated into |**q**|, which quantifies diffusion gradient area, and t, which specifies diffusion time (i.e., diffusion gradient spacing). Here, **q** is a vector which inherits its direction from **G**. Measurements S(**q**,t) with high b can result either from high t or from high |**q**|, or a moderate value for both as illustrated by the blue area

inferior–superior-oriented corticospinal tract (orange arrow), unlike orthogonal diffusion directions (middle and top). Diffusion weighting can be achieved in any 3D direction by appropriate simultaneous G_x, G_y, and G_z diffusion gradients with amplitudes proportional to the (x,y,z) components of the desired direction. Therefore, a more general look at diffusion encoding parameters defines **G** and **b** for a set of measurements as vectors in a three-dimensional space (Fig. 2a).

In this way, diffusion acquisitions with equal b-value but different directions are often regarded as a spherical "shell" in **b**-space, with b = |**b**|, the Euclidian length of vector **b** (analogously G = |**G**|). Complete diffusion protocols are often described as single shell (e.g., 30 directions at b1000 + b0 measurements, where b1000 is short for b = 1000s/mm^2) or multi-shell (e.g., 24 directions at b700, 48 directions at b2700 + b0 measurements). Here, it is important that TE and repetition time (TR) for all shells/b-values are kept constant throughout, even if this means that one foregoes a shorter TE, and higher SNR, at lower-b shells. Acquisition of b0 measurements, matched in TE and TR for all TE and TR used in diffusion measurements, is also crucial as baseline signal level for any analysis or post-processing. A more detailed specification of diffusion encoding separates b-value into |**q**|, which quantifies diffusion gradient area (dependent on δ and |**G**|, and often scaled by a constant 2π), and t, which specifies diffusion time (mostly determined by gradient spacing Δ). Here, **q** is a vector which inherits its direction from **G** and defines a **q**-space, analogous to **b**-space but without dependence on t. Measurements S(**q**,t) with high *b* can result either from high t (and low/moderate |**q**|) or from high |**q**| (and low/moderate *t*), or a moderate value for *t* and |**q**| as illustrated by the blue area in Fig. 2b. Although the b-value is sufficient for Gaussian diffusion models, such as in Diffusion Tensor Imaging (DTI) discussed below, its conflation of diffusion time and **q** makes it inadequate for more sophisticated diffusion methods, described below, that consider non-Gaussian diffusion.

Under specific requirements, such as high-t or high-q, alternative pulse sequences for diffusion are sometimes more suitable than PGSE (Fig. 3). Under availability of sufficient gradient amplitude |G|, PGSE is especially applicable for short diffusion time experiments. One alternative to SE-based diffusion preparations are stimulated echoes (STE; Merboldt et al. 1991), in which the SE refocusing pulse is replaced with two 90° pulses (Fig. 3b). In the time interval between these pulses, the magnetization is stored along the longitudinal axis where it decays more slowly according to the longer T_1 relaxation time, rather than T_2. This makes the Pulsed Gradient STimulated Echo (PGSTE) sequence especially suitable for long diffusion time experiments. Unfortunately, STE sequences incur a signal loss overhead of a factor of 2, which corresponds to the amount of magnetization that is not stored longitudinally. In practice, this overhead has limited the attraction of STE for short diffusion time experiments (Lundell et al. 2014). Another alternative is the diffusion-weighted

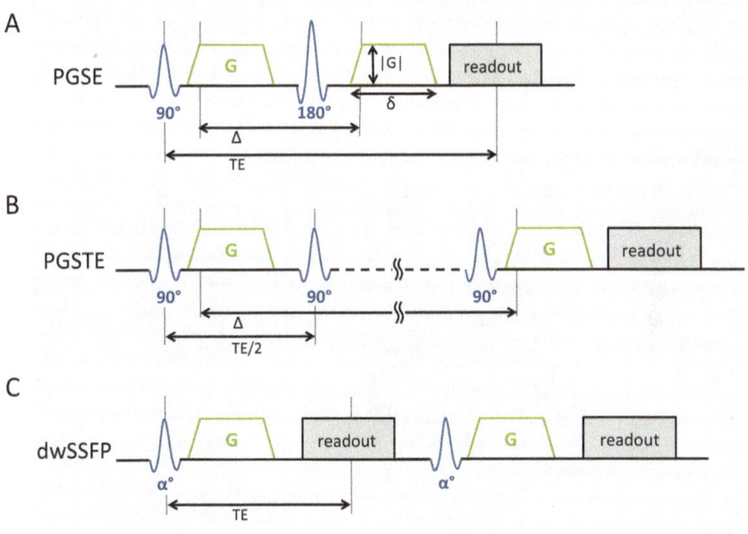

Fig. 3 Diffusion MRI pulse sequence variants. (**a**) Pulsed gradient spin echo (PGSE) especially suitable for short diffusion time experiments (**b**) Pulsed gradient stimulated echo (PGSTE) especially suitable for long diffusion time experiments (**c**) diffusion-weighted steady-state free precession (dwSSFP) for especially suitable high |*q*|, short T_2 experiments without the need for a well-defined diffusion time

steady-state free precession sequence (dwSSFP; Fig. 3c; Mcnab and Miller 2010; Merboldt et al. 1989). SSFP is able to retain signal over multiple TR periods (Buxton 1993) essentially acquiring a plurality of spin and stimulated echoes simultaneously. The result is both strong diffusion weighting and high SNR, even for short T_2 tissues. However, one cannot assign a single b-value or diffusion time in dwSSFP since the signal results from a combination of SE and STE contributions such that the diffusion weighting accumulates over multiple diffusion times. Therefore, dwSSFP is especially suitable for high $|\mathbf{q}|$ experiments on short T_2 tissues without the need for a well-defined diffusion time. One such application is high-quality diffusion imaging in intact human brain samples post mortem (McNab et al. 2009) since T_2 in post-mortem tissue decreases strongly compared to the in vivo case. One final consideration that influences pulse sequence choice is field strength. The considerations above mean that the discussed pulse sequences behave differently under different field strengths, particularly when moving from 3T to ultra-high fields (UHF; 7T and above). T_2 decreases with increasing field strengths, which means that at UHF PGSE only remains effective at high b-values (b > 3000 s/mm^2) at very short TEs. However, excellent high spatial resolution dMRI data can be acquired with PGSE at lower b-values (Heidemann et al. 2010; Vu et al. 2015). For PGSTE and dwSSFP, UHF can be more beneficial because T_2's (which dominate SNR for PGSE) get shorter, but T_1's get longer which is an advantage that is leveraged in SSFP and PGSTE.

Diffusion Tensor Imaging

Most dMRI's possibilities for the elucidation of networks and microstructure in the brain arise from the estimation of quantitative (i.e., relatively measurement independent) parameters embedded in a signal model. Diffusion tensor imaging (DTI; Basser et al. 1994; Basser and Pierpaoli 1996) represents one of the simplest models for dMRI data. DTI was introduced to characterize diffusion in an anisotropic medium, such as brain white matter with a relatively simple parametric model. DTI today still represents a clinical and research standard for analyzing dMRI data, being sensitive to microstructural changes due to, for example, development (Pfefferbaum et al. 2000; Neil et al. 2002; Assaf and Pasternak 2008; Lebel et al. 2010) and pathology (Sotak 2002; Werring et al. 1999; Sundgren et al. 2004; Horsfield and Jones 2002). DTI's success stems from the simplicity of its tensor model and the relatively low requirement on the dMRI data needed to fit it. The DTI model approximates contrast arising from increasing b-values (i.e., radial structure in b-space, Fig. 4a) as resulting from mono-exponential decay (Fig. 4b) and contrast arising from changing diffusion direction (i.e., angular structure in b-space, Fig. 4a) as characterized by a 3D Gaussian (Fig. 4c). In the mono-exponential signal decay model under the narrow-pulse approximation ($\delta \ll \Delta$), the diffusion signal S_i, for the ith diffusion encoding gradient in direction $\mathbf{G}_i$ is modeled as $S_i = S_0 e^{-bADC_i}$, where S_0 is the measured signal without diffusion encoding, that is, the signal for the

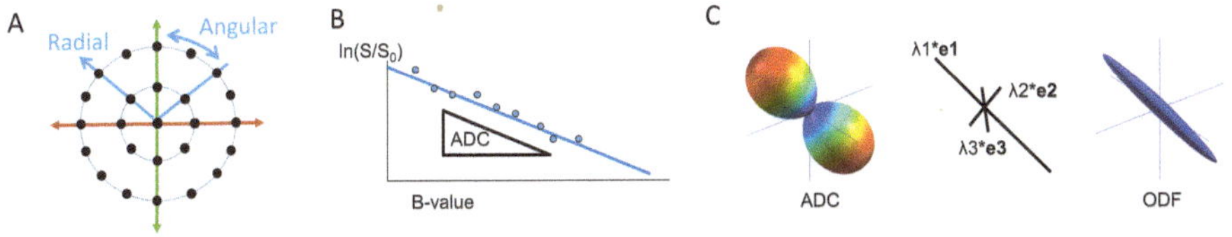

Fig. 4 The Diffusion Tensor model. (**a**) The distinction between radial and angular structure in b-space. (**b**) DTI's characterization of the relationship between data points acquired with different b-values (i.e., radial signal model). (**c**) DTI's characterization of the relationship between data points acquired with different directions (i.e., angular signal model).Thin blue lines: G_x, G_y, and G_z forming the MR scanner's coordinate system. Note that the separation into "radial" model and "angular" model is for clarification purposes only since these are two aspects of the same signal model

b0 measurement. This means the log-normalized signal $ln(S_i/S_0)$ is modeled as a straight line function of b-value, with ADC_i as the slope of the line (Fig. 4b). The apparent diffusion coefficient ADC_i quantifies water diffusion in the direction $\mathbf{G}_i$. ADC_i can then be estimated in every voxel of the image and for every sampled diffusion direction $\mathbf{G}_i$, which leads to a 3D ADC-profile.

The DTI model approximates this ADC-profile (i.e., angular structure) as characterized by the 3D Gaussian of a rank-2 tensor (Fig. 4c). By sampling at least six linearly independent diffusion directions $\mathbf{G}_i$ and a non-diffusion-weighted volume, it is possible to model the diffusion behavior of water molecules in three dimensions as $S_i = S_0 e^{-bG_i D G_i}$, with D the rank-2 symmetric diffusion tensor:

$$D = \begin{bmatrix} D_{xx} & D_{xy} & D_{xz} \\ D_{xy} & D_{yy} & D_{yz} \\ D_{xz} & D_{yz} & D_{zz} \end{bmatrix}$$

This symmetric matrix contains six independent components that summarize the diffusion properties of water molecules within a voxel. The diffusion tensor models water diffusion with a peanut-shaped ADC profile and an ellipsoid as the orientation density function (ODF), that is, the probability function over diffusion directions (Fig. 4c). This means the ODF will be characterized by only one peak and, therefore, with DTI it is not possible to reconstruct more than one independent direction of diffusion. This single main direction can be found by decomposing (i.e., diagonalizing) the tensor matrix into eigenvalues and eigenvectors:

$$D = \begin{bmatrix} e_{1x} & e_{2x} & e_{3x} \\ e_{1y} & e_{2y} & e_{3y} \\ e_{1z} & e_{2z} & e_{3z} \end{bmatrix}^T \cdot \begin{bmatrix} \lambda_1 & 0 & 0 \\ 0 & \lambda_2 & 0 \\ 0 & 0 & \lambda_3 \end{bmatrix} \cdot \begin{bmatrix} e_{1x} & e_{2x} & e_{3x} \\ e_{1y} & e_{2y} & e_{3y} \\ e_{1z} & e_{2z} & e_{3z} \end{bmatrix}$$

where $\mathbf{e}_1$, $\mathbf{e}_2$ and $\mathbf{e}_3$ are the eigenvectors, λ_1, λ_2, and λ_3 are the corresponding eigenvalues and T, the transpose operator. Intuitively, this decomposition reflects the amount of displacement due to diffusion (the eigenvalues) along three orthogonal orientations (the eigenvectors) where $\mathbf{e}_1$ is the orientation of highest diffusion (the primary or major eigenvector), $\mathbf{e}_3$ is the orientation of lowest diffusion (the tertiary or minor eigenvector), and $\mathbf{e}_2$ is the secondary or median eigenvector (Fig. 4c, middle). This means that in a highly isotropic voxel (e.g., one that lies within the ventricles), the eigenvalues λ_1, λ_2, and λ_3 will roughly be the same and have a relatively high value. On the other hand, in anisotropic voxels (e.g., those lying in the corpus callosum), the eigenvalues will be different from each other, with the major eigenvalue λ_1 much larger than the other two. The corresponding major eigenvector e_1 then represents the main direction of diffusion and the estimate of the local direction of axons. The single-peaked ODF with its single main direction of diffusion is the first large limitation of the DTI, which is especially relevant in tractography and connectomics. This is since many white matter voxels contain more than one fiber population, which can only be represented by the multi-peaked ODFs of more sophisticated diffusion models, as discussed below.

Several scalar indices can be estimated from the diffusion tensor. Mean diffusivity (MD) measures the average degree of water diffusion within a voxel and is computed as the average of the three eigenvalues. For low b-values (i.e., <<2000 s/mm^2), MD does not clearly highlight differences between white and gray matter, but it is very useful to discriminate pathological diffusion behavior due to, for example, ischemic stroke. Fractional anisotropy (FA; Basser and Pierpaoli 1996) measures the directional specificity of diffusion as obtained by DTI analysis. FA values, essentially a normalized measure of variance of the eigenvalues, range between 0 (fully isotropic) and 1 (fully anisotropic). Both the MD and FA indices have been widely used to study neurodevelopment (e.g., McKinstry et al. 2002), axonal integrity (e.g., Khayal et al. 2011), clinical conditions (for a review, see: Dong et al. 2004), and plasticity (for a review, see: Zatorre et al. 2012). Although sensitive, DTI indices such as FA are also unspecific, since differences in FA can reflect different axonal properties such as axon density, diameter distribution, and myelination (Santis et al. 2014; Assaf and Pasternak 2008; Assaf et al. 2004; Jones et al. 2013; Beaulieu 2002). This lack of specificity is the second large limitation of DTI, especially relevant in diffusion microstructure studies. Greater specificity requires more sophisticated diffusion methods, described below, that consider non-Gaussian diffusion.

Tractography and Connectomics

Reconstructing the macroscale human connectome, the map of all large-scale (cortico-cortical) structural connections in the human brain (Sporns et al. 2005), represents a challenging and important research topic in neuroimaging. A crucial element to generate dMRI-based macroscale connectomes is tractography, the inference process that takes local directions of diffusion in voxels to long-range virtual tracks that model, as well as possible, the course of large fiber tracts in the brain.[2] Tractography is crucially dependent on accurate local voxel-wise directions, which is why developments in tractography have often been preceded by more sophisticated dMRI signal models, particularly multi-peaked ODF methods or "crossing-fiber" methods.

HARDI and ODFs

One of the main limitations of DTI is its incapability of resolving complex fiber architecture within an imaging voxel such as high curvature, divergence or splaying, or two or more crossing fiber populations. The identification of more than one independent diffusion directions in a voxel enables superior fiber tractography through regions with multiple populations of fibers with different orientations (Wedeen et al. 2008). It has been estimated that at the nominal resolution of current day standard dMRI acquisition (2–3 mm isotropic voxels), more than 90% of voxels contain more than one fiber bundle (Jeurissen et al. 2013). Therefore, more sophisticated models for dMRI data have been proposed that can better represent such complex fiber architecture.

Although DTI can represent the true ADC profile for a single fiber, it cannot represent the true ADC profile for two crossing fibers, which is multi-peaked. In addition, in the case of multiple fiber components with different orientations, the peaks of the ADC profile do not directly correspond to actual fiber orientations, in contrast with the single-fiber case (Fig. 5a). In the case of a 90 degree crossing, for example, there is an offset of 45 degrees in the peaks of the ADC profile with respect to the actual fiber orientations (Alexander et al. 2002; Frank 2002). Furthermore, the ADC is computed under the assumption of mono-exponential signal decay in DTI, which breaks down at higher *b*-values. Therefore, more sophisticated models for complex fiber architecture tend to avoid the ADC formulation. Diffusion spectrum imaging (DSI; Wedeen et al. 2005) reconstructs the diffusion propagator, or displacement probability, in a model-free or nonparametric way by directly computing the Fourier transform of the measured q-space signal as defined above. In order for this approach to work, it is necessary to sample the q-space very densely on a Cartesian grid (i.e., use a high number, often more than 250, of diffusion directions at different q-values). A popular simplification to this acquisition scheme is represented by high angular resolution diffusion imaging (HARDI) acquisitions (Frank 2001). HARDI probes a high number (usually 60 or more) of diffusion encoding directions on only one spherical shell in *q*-space (i.e., using one *b*-value besides b0 measurements). The single-shell HARDI protocols are often acquired at a *b*-value of 2000–3000 s/mm^2 since the higher *b*-value provides more angular contrast for fiber orientation modeling. Using HARDI acquisitions, it is possible to infer

[2]This review uses the conventional terminology of referring to the virtual entities created by tractography (a processing and modeling of dMRI data) as "tracks," and to the actual neuroanatomical bundles of axons or fibers as "tracts."

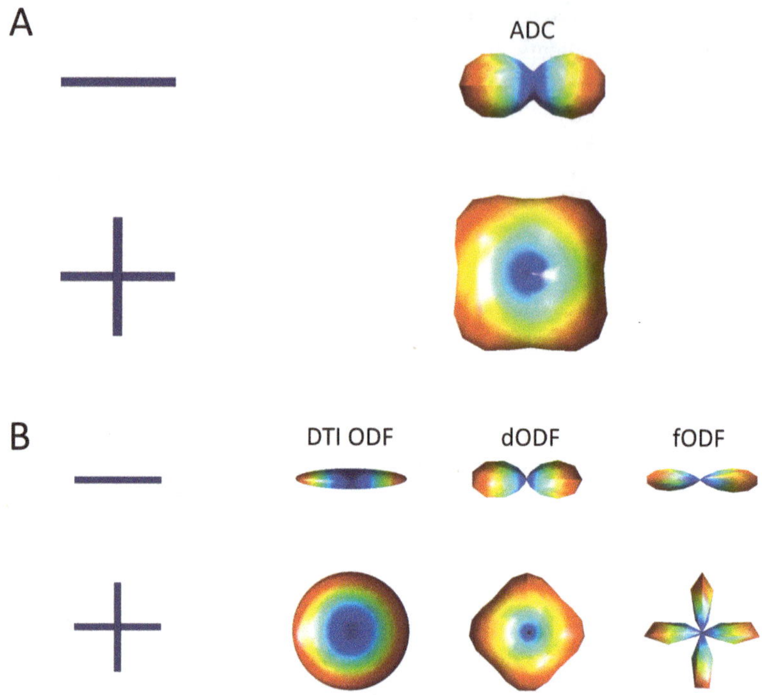

Fig. 5 Crossing fibers and orientation density functions. (**a**) The true ADC profile for a single fiber (top) and two fibers crossing at 90° (bottom). (**b**) The orientation density function (ODF) for a single fiber (top) and two fibers crossing at 90° (bottom) as modeled by DTI, diffusion ODF methods (sODF) and fiber ODF methods (fODF)

non-Gaussian effects in the reconstruction of ADC profiles (Alexander et al. 2002; Frank 2002). Q-ball imaging (QBI; Tuch 2004) was the first attempt to model multiple fiber components in a voxel from HARDI data. Using a model-free approach, QBI reconstructs the spin displacement orientation distribution function (dODF). Further model-free approaches, such as the diffusion orientation transform (Ozarslan et al. 2006) and persistent angular structure MRI (Jansons and Alexander 2003; Parker and Alexander 2005), can be considered as extensions of the QBI *q*-space approach. Like the diffusion tensor ellipsoid, the dODF represents a probability function defined over the sphere that maps the likelihood of water molecules to move along a certain orientation due to diffusion (Fig. 5b, middle). The dODF peaks can then be assumed to represent the orientations of multiple fiber bundles within a specific voxel. By mapping the water displacement, QBI dODFs are not very sharp, as water will still have some directional components orthogonal to the main axon directions. Spherical deconvolution (SD, Tournier et al. 2004) and its subsequent improvement constrained spherical deconvolution (CSD, Tournier et al. 2007) aim at modeling the actual fiber orientation distribution function (fODF, also known as the fiber orientation distribution, FOD), rather than water displacement (for review, see Dell'Acqua and Tournier 2018). The fODF reflects the actual orientations of distinct fiber populations by modelling complex fiber configurations as linear combinations of Dirac's delta functions (Fig. 5b, right).

Moreover, a generalization of single shell (i.e., single b-value) HARDI-based techniques has been proposed. Generalized q-sampling imaging (GQI; Yeh et al. 2010) increases the sensitivity to multiple fiber compartments by sampling q-space using multiple shells. Model-based approaches that are capable of resolving more than one fiber population in a voxel have also been proposed. The multi-tensor model is a generalization of the classical DTI approach, where two or more diffusion tensors are fitted to the data (Tuch et al. 2002). Furthermore, a Bayesian framework has been described by which it is possible to fit an isotropic component and several fiber components ("Ball and Sticks") to the sampled diffusion signal, with an estimate of orientational uncertainty (Behrens et al. 2007).

Some studies have validated local orientations derived from dMRI data against ground truth, particularly in ex vivo studies. Validation of the accuracy of ODFs in human post-mortem tissue was facilitated by automated 2D structure tensor analysis (STA) to detect local image orientation in histological sections (Budde and Annese 2013; Budde and Frank 2012). Budde and Frank (2012) initially applied this method to the rat brain, demonstrated the construction of FODs from STA, and illustrated the effects of multiple fiber bundles and intra-voxel orientation dispersion. Applying STA to human brain sections obtained from a digital public brain bank, Budde and Annese (2013) show the complexity of human white matter fibers at microscopic resolution and the effects of partial volume and small-scale orientation dispersion on ODF estimates at lower resolutions obtainable with dMRI. Seehaus et al. (2015) applied STA to myelin-stained human brain sections aimed at validating DTI orientation estimates in both WM and GM. The agreement was strongest in WM, particularly in unidirectional fiber pathways where high FA values were linked to high myelin density and a sharply tuned histological orientation profile. However, DTI orientations broke down in white matter regions with multiple differently oriented fiber populations, as expected.

Tractography

Any volume of modeled local fiber orientations can be used for tractography or fiber tracking. That is, once the local orientations have been reconstructed in every voxel, it is possible to reconstruct tracts connecting different brain regions by further modeling (Fig. 6a). Several algorithms have been developed to perform tractography. These algorithms can be clustered in different classes from the methodological point of view (Fig. 6b; Bastiani et al. 2012): single-orientation vs multiple-orientation (according to the type of local ODF they use), local vs global (according to whether tracking proceed in discrete steps or a whole set of local orientations is modeled together), and deterministic vs probabilistic (according to whether they take into account uncertainty).

The classic local deterministic approach was based on DTI and uses a deterministic streamline algorithm, which locally integrates the fiber path using a step-wise approach (e.g., Fig. 6c, top). Such a streamline is initiated by the user at a certain seed point and follows the main eigenvector of the diffusion tensor, changing direction according to the orientation in a newly entered local voxel. While the paths of early methods changed direction only at voxel boundaries (Conturo et al. 1999; Mori et al. 1999; Mori and van Zijl 2002), continuous approximation techniques of the orientation field were soon proposed to smooth the results and make them follow more biologically plausible pathways (Basser et al. 2000). Despite their simplicity, local deterministic tractography methods have been successfully used to perform in vivo "virtual dissection" of known fibers analogous to classical post-mortem fiber dissection techniques (Catani et al. 2002). Using prior knowledge, two regions of interest (ROIs) are selected through which the tractography streamlines should run. The problem of false-positive fibers is partially addressed by selecting a third exclusion ROI to remove the known spurious fibers.

However, dMRI data are inherently corrupted by a certain amount noise, which depends on factors such as voxel size and amount of diffusion weighting (i.e., b- or q-value). As a result, the estimation of the diffusion tensor contains a certain amount of uncertainty, which is then reflected in its eigenvalues and eigenvectors (Jones 2003). Probabilistic tractography approaches (Fig. 6d) characterize the variability of tractography results resulting from the uncertainty of the estimated local fiber directions (Behrens et al. 2003b; Parker et al. 2003). This contrasts them with deterministic approaches which give a single deterministic answer for the connection of a given region of interest (ROI) to any part of the brain. In a local probabilistic algorithm, the direction of track propagation will be randomly selected at every step by sampling the local orientation distribution function. Probabilistic local algorithms, therefore, will propagate several hundreds of streamlines from the same seed

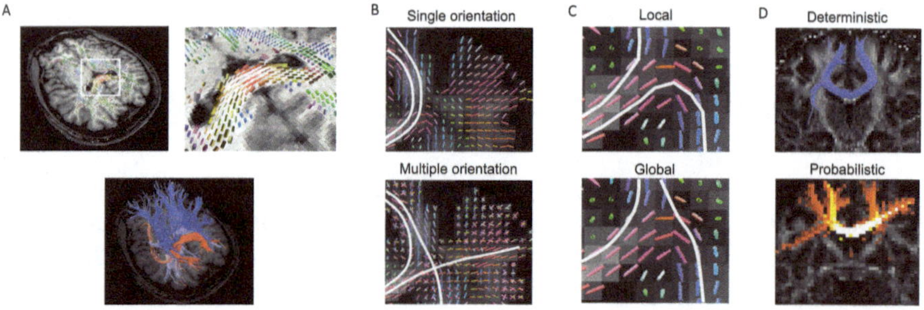

Fig. 6 Tractography methods. (**a**) Local orientations of fibers or diffusion in each voxel are used to infer long-range tracks which model large white matter axon bundles or tracts. (**b**) Multiple-orientation tractography methods can multimodal ODFs and model crossing fibers, whereas single-orientation methods cannot. (**c**) Local tractography methods create streamline tracks which always follow local directions, whereas global methods fit tracks to all data passed through which can ignore some local directions. (**d**) Deterministic tractography methods attach full confidence to each track, whereas probabilistic methods take uncertainty into account and can create maps of the likelihood of track connection to a seedpoint, given the data

point and each iteration will result in a slightly different preferential path. The result of a local probabilistic tractography algorithm for every single well-defined seed point is often represented as a three-dimensional map of visitation counts for fibers through a voxel (e.g., Fig. 6d, bottom). The noisy streamline or diffusing particle principles have been applied both to tensor models and to complex local architecture models (Behrens et al. 2007; Tournier et al. 2010; Parker and Alexander 2005).

Local streamline methods, either deterministic or probabilistic, use only local information to determine the course of tracks. Global approaches move from local step-wise reconstruction of fiber trajectories to a global goodness-of-fit of the entire candidate fiber. Here, the measure of fit quantifies the joint likelihood of the fiber given all voxel data it passes through (Jbabdi et al. 2007; Sherbondy et al. 2008, 2009; Tuch et al. 2002; Zalesky and Fornito 2009). The global fit measure makes tractography less sensitive to modeling errors caused by local noise (Jbabdi et al. 2007). A graph-based global tractography algorithm (Iturria-Medina et al. 2007) and its extension to multiple orientation fiber models (Sotiropoulos et al. 2010) have been proposed. These graph-based algorithms reconceptualize the global tractography problem as a shortest-path search in a graph, in which (in contrast with the connectome graph, described below) nodes are represented by the center of each white-matter voxel. Since graph weights are then defined as the probability of voxel-center connections given the local ODFs, a shortest path from one point to another in this graph constitutes a globally optimized fiber. Since in a shortest-path search all possible nodes are visited and the path lengths recorded, the n% shortest paths then correspond to the n% most likely paths in the probabilistic tractography sense. Thus, graph-based methods – and in fact, global methods in general – are naturally used as probabilistic methods. The only way to force a global tractography approach to be deterministic is to select only its highest percentile results for consideration, which corresponds to looking at only the single most likely path that connects two points. Some global tractography approaches even aim at constructing the entire connectivity pattern between all the voxels in the acquired volume (sometimes called "connectome-global" which consider constraints to all paths over the complete white matter or connectome, in contrast to "fiber-global" or "fascicle-global"; Pestilli et al. 2014). One way toward this that has proven successful is to fit thousands of line segments in the entire sampled volume and optimizing both local fit of the segments to data they pass through and the connection of the segments to each other (Kreher et al. 2008; Reisert et al. 2011). Global approaches are often computationally very expensive, but their decreased dependence on user choices and higher resistance to noise, as well as their fundamentally probabilistic representation of the tractography result, makes them very interesting. An attractive approach to global tractography recently explored is a hybrid approach in which local tractography methods are used to generate a high number of tracks covering white matter (a so-called tractogram), after which the tractogram is filtered to subselect the subset of tracks that together best fit all local voxel ODFs or dMRI data, under appropriate constraints (Pestilli et al. 2014; Smith et al. 2013).

Diffusion MRI-based tractography has made it possible to trace, in vivo and non-invasively, the three principal types of macroscopic white matter tracts that were originally described by Meynert using post-mortem samples. Both the long association tracts and the commissural fibers have been described in great detail in terms of their anatomical location and fiber termination area (Catani and Thiebaut de Schotten 2008; Wakana et al. 2004). Fibers obtained from tractography analysis connecting cortical areas to subcortical structures such as the thalamus have been traced with a level of detail that resembles that of classical post-mortem histological studies (Behrens et al. 2003a). Moreover, advancements in dMRI acquisition techniques and MRI hardware have increased the achievable voxel resolution, allowing short-association tracts (u-fibers) to be identified and segregated in the frontal lobe (Catani et al. 2012).

The model-based inferential nature of tractography makes it prone to pitfalls and interpretation errors (see, for review and discussion, Jeurissen et al. 2017; Jones et al. 2013). This makes validation studies of tractography results against ground truth especially important. Validation studies of tractography results have been performed with different approaches, including use of hardware phantoms (Cote et al. 2013; Fieremans et al. 2008; Pullens et al. 2010), in vivo data (Neher et al. 2015; Bastiani et al. 2012), and histological validation on ex vivo biological tissue. The latter approach, validating tractography results obtained on data from post-mortem dMRI studies, is especially appealing since ground truth can be obtained in a biophysically realistic and neurologically relevant context. In one early experiment, Roebroeck et al. (2008) validated streamline tractography of known human optic chiasm pathways at 9.4T and investigated the effects of spatial resolution of the diffusion data. Despite the complex structure of the fiber paths through the optic chiasm, all known connections could be tracked by a DTI local streamline algorithm, although underrepresented, overrepresented, and erroneous connections were tracked, and errors made by the tractography algorithm at high resolution were shown to increase at lower resolutions closer to those used in vivo. Seehaus et al. (2013) combined DTI at 9.4T with carbocyanine die tracing in the same human temporal lobe tissue block. Since axons are selectively labeled along their length by the die, robust definitions of sensitivity and specificity for DTI tractography could be given and its accuracy could be assessed in a receiver operating characteristic (ROC) curve analysis. DTI streamline tractography was shown to have sensitivity and specificity of greater than 80% over distances of several centimeters with voxel sizes in the 100 s of microns. In these studies, dMRI data were acquired with high spatial resolution but with relatively low angular resolution and b-value, which allowed single-orientation tractography. In post-mortem human HARDI studies, high tractography accuracy has also been reported, for instance, in a comparative study of white matter projections from ventral prefrontal cortex between human and macaque brains (Jbabdi et al. 2013). However, in a study using reference atlas axonal tracer results to assess sensitivity and specificity of several tractography algorithms for long-range projections in the macaque brain, anatomical accuracy was highest for different tractography parameters in different tracts with overall low accuracy (Thomas et al. 2014). In another study of the macaque brain validating tractography accuracy with anterograde tracers and myelin stains, Reveley et al.

(2015) give dense juxta-cortical zones of tangential fibers crossed by radial cortical insertions as a possible reason for this limited accuracy.

Ex vivo tractography and its validation are becoming increasingly important in the context of human neurosurgery. In an ex vivo mesoscopic dMRI study of surgically excised hippocampi in a mesial temporal lobe epilepsy case, Modo et al. (2016) show that individual cell layers can be discerned in dMRI contrasts and that DTI tractography detects extra- and intra-hippocampal connections. The results suggested that there is an aberrant connection between the dentate gyrus and the stratum moleculare of the hippocampus, supporting the hypothesis of mossy fiber sprouting in the dentate gyrus as a reverberant excitatory network in mesial temporal lobe epilepsy. In a validation study of the dentatorubrothalamic tract (DRTT), Mollink et al. (2015) compared post-mortem dMRI tractography with 3D histological reconstruction from serial myelin-stained sections. ROC curve analysis showed a high sensitivity and specificity in general and high spatial accuracy of post-mortem probabilistic tractography of the DRTT when compared to a 3D histological reconstruction, showing prospects for studying structure–function correlations in patients with cerebellar disorders. Plantinga and colleagues (Plantinga et al. 2016) performed an ex vivo tractography study of the connections between the subthalamic nucleus, substantia nigra, and globus pallidus, three nuclei of the human basal ganglia, which play an important role in motor, associative, and limbic processing and have relevance for deep brain stimulation treatments.

Macroscale Connectomics

An individual macroscale connectome is represented by a graph with a set of nodes, representing a parcellation of the brain anatomy, and a set of edges which represent white matter connections of the nodes (Fig. 7). The connectome can be represented by a real-valued connectivity matrix or binary adjacency matrix (Fig. 7b) which can be usefully analyzed using graph theory and network analysis (Rubinov and Sporns 2010).

There are several steps in the post-processing pipeline for dMRI data to create a connectivity matrix that represents the macroscopic structural connectome (Hagmann et al. 2008; Hagmann et al. 2007). Among them are white matter and cortical gray matter segmentation from an anatomical (T1-weighted) MRI, parcellation of the gray matter into nodes of the connectome graph, creation of white matter seedpoints and masks, intra-voxel water diffusion and/or fiber direction modeling, tractography to connect local directions into long-distance connections and computation of edge weights between cortical nodes from tractography results. Several studies have investigated the influence of choices in some of these steps, such as the effects of gray matter parcellation (Zalesky et al. 2010), seedpoint placement strategy for tractography (Li et al. 2011; Cheng et al. 2011), and the reproducibility of network indices (Bassett et al. 2011). For instance, Zalesky et al. (2010) show that the small-worldness graph measures increases with a larger number of

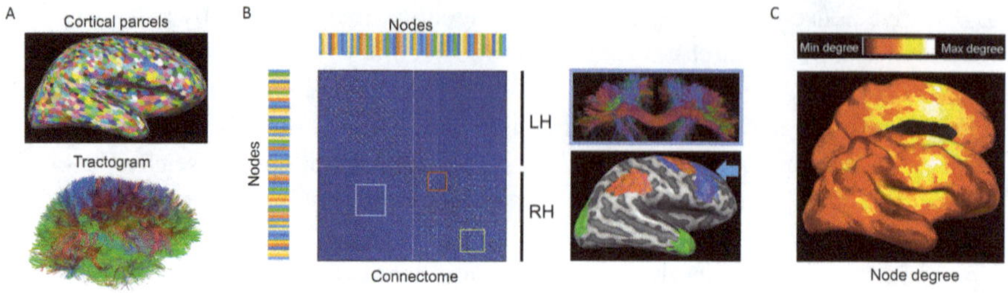

Fig. 7 From cortical parcellation and tractography to Connectome. (**a**) A single-subject connectome is generated from a parcellation of the gray matter or cortical surface (top, on a flattened surface representation) and a tractogram generated by brain-wide tractography (bottom). (**b**) The connectome is represented as a #nodes x #nodes matrix, where each node is a cortical parcel. Each connectome-matrix entry is an edge that represents the connectivity between nodes, derived from the tractogram. With proper node reordering, within-hemisphere edges are organized on block-diagonal submatrices and between-hemisphere edges on off-diagonal blocks. Through node reordering, track-sets corresponding to anatomical tracts are connectome sub-matrices connecting subsets of nodes, for example, the frontal corpus callosum (blue submatrix, left, and blue boxed tracks and node coloring at the arrow, right), the superior longitudinal fasciculus branch II (red submatrix, left, and red node coloring, right), and inferior longitudinal fasciculus (green submatrix, left, and green node coloring, right). LH: left hemisphere, RH: right hemisphere. (**c**) Node degree, the number of edges connecting a node, can be represented for all nodes as a node degree map, highlighting densely connected nodes in the connectome

nodes resulting from a greater degree of parcellation of the gray matter. In Li et al. (2011), it is shown that the identification of network hubs in the brain can depend on the tractography seedpoint placement strategy. Bastiani and colleagues (Bastiani et al. 2012) have focused on the effects of intra-voxel fiber direction modeling and tractography algorithm on connectome graphs. They found the effect of algorithm and parameter choice has two components. First, there is the large effect of tractography algorithm and parameters on global network density, that is, number of edges, and consequently on node degree (Fig. 7c). Second, and more importantly, they found remaining effects on graph indices which range in the tens of percent even when global density is controlled for, including higher density for multi orientation models and global tractography methods, compared to local single orientation tractography. They also investigated the effect in choices in tractography algorithm on sensitivity and specificity of the resulting connections with a connectome dissection quality control approach and find that, generally, multiple orientation models are preferred over single orientation models and global tractography methods are preferred over local tractography methods.

The connectome matrix can be either a weighted real-valued connectivity matrix or binarized adjacency matrix. This means that the edges connecting the different nodes of the connectome can either indicate the strength of a connection between two different cortical areas or simply indicate the existence or lack of a specific neural pathway. The typical approach consists in initially obtaining a weighted connectivity matrix that is then thresholded and binarized. Methods of computing the weights include mean path length between the nodes, averaged FA or MD values along the tracts, or probabilistic and deterministic streamline counts (Li et al. 2012). As discussed in the previous section, estimating "connectivity" between areas using dMRI and tractography is not straightforward. As a consequence, the definition of the weights and the choice of the threshold used to binarize the connectivity matrix can dramatically influence the metrics obtained from connectome analysis (Bastiani et al. 2012). Moreover, the matrix representing the connectome can be symmetric or asymmetric. An asymmetric matrix indicates that connection weights are different when tracking from Node A to Node B in respect to those obtained when tracking in the opposite direction. Since connectivity indices estimated from dMRI-based tractography do not reflect the (afferent or efferent) direction of the connections between two areas, it is common practice to symmetrize the connectivity matrix, for instance, by taking the maximum of the two weights between regions.

The application of network analysis to the connectome graph has allowed neuroscientists to describe different properties of cortical areas based on their connectivity profiles, such as their degree of segregation based on clustering approaches or their centrality in the communication network (Rubinov and Sporns 2010). Using such network analysis, it has been shown that certain cortical regions (e.g., superior parietal and frontal regions) are highly connected "hubs" and form a so-called rich club, in the sense that these hubs are also very strongly connected to each other (van den Heuvel and Sporns 2011). These findings can be used to study several pathological conditions by bringing the analysis to a new level of abstraction (van den Heuvel et al. 2010), in which different brains can be compared based on their connectivity profiles rather than using a standard spatial reference template. This can even be used to compare the macroscale connectomes of two different species and identify similarities and differences in their connectivity profiles (Goulas et al. 2014).

Several issues exist in the pursuit of whole-brain tractography and caution should be exercised when interpreting connectivity results obtained using diffusion MRI tractography (Thomas et al. 2014), particularly in the context of macroscale connectomics (for review, see Sotiropoulos and Zalesky 2017; Van Essen et al. 2014). The selection of tractography algorithm in reconstructing the macroscale connectome has a considerable effect on resulting connectivity estimates (Bastiani et al. 2012; Pestilli et al. 2014). Estimating connectivity indices that are both accurate and specific is not straightforward. Tractography algorithms can reconstruct streamlines that are well spatially correlated to large axonal bundles or tracts, but false positives and false negatives will often exist in node-to-node connections. When using deterministic tractography algorithms, the number of streamlines that successfully reached node B when seeded from node A is usually taken as a measure of connectivity. However, this result is highly dependent on the resolution of the dMRI data and on the initial parameters set by the user, such as the position of the nodes, the tractography step size, angular threshold, and the intra-voxel diffusion model. Moreover, connection probabilities that are estimated when using probabilistic tractography algorithms are not a measure of true anatomical connectivity. The final output of such algorithms will still be a streamline count between two nodes, albeit it being a count of probabilistically propagated streamlines. Moreover, the distance between two nodes influences these total counts. The further apart two nodes are, the more difficult it will be for streamlines to propagate from one to another (Jones 2010). Furthermore, fibers can cross and these configurations can be resolved when using HARDI measurements and sophisticated orientation modeling as discussed above. However, fibers can also kiss (i.e., converge, touch, and diverge again) instead of cross, bend, or fan within a voxel, and these configurations will be undistinguishable when looking at the ODF signal of a single voxel (for review, see Jbabdi and Johansen-Berg 2011). Some studies have tried to solve this issue by fitting helical curves between neighboring voxels to determine whether fibers within them will fan or bend (Savadjiev et al. 2006; Savadjiev et al. 2008). Moreover, it is also possible to determine the polarity of the fanning (i.e., fanning-in or fanning-out), which is very useful when performing tractography (Campbell et al. 2014). Another suggested approach is to model the intra-voxel diffusion fODFs while breaking the assumption of symmetry. Since the recorded dMRI signal is symmetric in a three-dimensional space, most diffusion modeling techniques try to estimate a symmetric PDF. Based on the fact that neuronal fibers are continuous between adjacent voxels (i.e., a fiber that leaves a voxel with a certain orientation should enter the next one with the same orientation), recent work has investigated the benefits of modeling local fODFs as asymmetric functions (Reisert et al. 2012; Reisert and Kiselev 2011; Bastiani et al. 2017).

Diffusion Microstructure

The lack of specificity in indices such as FA is the second large limitation of DTI, especially relevant in diffusion microstructure studies, which aim to quantify different axonal properties such as axonal density, axonal diameter distribution, and axonal orientation dispersion. One of the first diffusion MRI models to parsimoniously account for non-Gaussianity was diffusional kurtosis imaging (DKI; Jensen and Helpern 2010; Jensen et al. 2005), which augments the DTI model with a rank-4 kurtosis tensor and allows mapping of non-Gaussian signal characteristics, such as the mean kurtosis. From the microscopic viewpoint, both white and gray matter show a wide range of different patterns in their microstructural organization. Within white matter, axons themselves are characterized by different diameters and packing densities. In the corpus callosum of both macaque (LaMantia and Rakic 1990) and human (Aboitiz et al. 1992), several clusters of tracts with different density and diameter distributions have been identified. Using light microscopy, Aboitiz et al. (1992) examined the corpus callosum of several individuals and found that thin axons are denser in the genu. Furthermore, their density decreases moving toward the posterior midbody and increases again in the splenium. On the other hand, thicker fibers exhibit the opposite pattern, with a maximum density in the posterior midbody of the corpus callosum. Despite their microstructural heterogeneity, the pattern of callosal fiber diameters is well defined when related to the main fiber termination areas, with dense low diameter axons preferentially connecting occipital visual areas and frontal areas and high diameter axons connecting superior motor areas. The microanatomical differences can be related to the functional specialization of cortical areas, since there is a clear correlation between axonal diameter and conduction velocity of the electrical signals, that is, larger diameter axons will propagate action potentials at greater velocity (Tasaki et al. 1943), whereas small diameter axons allow greater axon density and a higher bandwidth of simultaneous spike-mediated information transfer.

Diffusion MRI has recently proven to be successful in mapping specific aspects of this microstructural architecture. Several new biophysical markers have been developed that show strong correlation to information such as axonal diameters, axonal density, and fiber dispersion both in vivo and post mortem (for review, see Alexander et al. 2017; Assaf et al. 2013). These advances have been achieved by introducing new modelling approaches that aim to estimate and separate the contributions of different biophysical compartments within white matter, most importantly the restricted intra- and hindered extra-axonal compartments (Assaf and Basser 2005; Panagiotaki et al. 2012; Zhang et al. 2012). If modeled correctly, these two separate water pools are characterized by different signal profiles in standard PGSE experiments (Stanisz et al. 1997). This difference in signal properties can be used to fit dMRI microstructural compartment models of white matter to the acquired data. This is illustrated by comparing the Gaussian diffusion assumptions in the DTI model to a biophysical multi-compartment model (Fig. 8). The mono-exponential signal decay with b-value in DTI (equivalently: the linear trend for the log-normalized signal) does not hold beyond b-values of ~1200s/mm^2 (Fig. 8a). One of the reasons for this non-Gaussianity of diffusion in biological tissue is that water diffuses in multiple compartments, prominently the intra-cellular (or intra-axonal) and extra-cellular (or extra-axonal) compartment (Fig. 8b). Although extra-axonal diffusion is hindered diffusion, which is well modeled (for a single fiber population) by the diffusion tensor model, intra-axonal diffusion is restricted diffusion, for which signal differs from that of hindered diffusion at high b-values (Fig. 8c, bottom). The Combined Hindered And Restricted ModEl of Diffusion (CHARMED model) approximates the diffusion signal, as a function of q and t, as a linear combination of intra-axonal restricted diffusion and extra-axonal hindered diffusion terms (Fig. 8c, top). The signal fraction for the restricted term (f_r or FR, Fraction of Restricted) is a scalar parameter, which is more specific to tissue microstructure, specifically the fiber density, than the fractional anisotropy (FA) derived from DTI (Fig. 8d).

Since hindered and restricted diffusion can only be distinguished over a range of b-values, including higher b-values (>3000 s/mm^2), biophysical compartment models such as CHARMED depend on multi-shell acquisitions with 2 or more shells that extend to high b-values. Since the restricted compartment model often depends on $\mathbf{q}$ and t independently (unlike the hindered compartment), biophysical compartment models sometimes depend on knowing $\mathbf{q}$ and t, rather than just the b-value of the measurement (cf. Fig. 2).

Therefore, advances in biophysical multi-compartment modeling have gained popularity because of greater specificity to the underlying cellular microstructure (Assaf et al. 2004, 2008, 2013; De Santis et al. 2014; Santis et al. 2014; Zhang et al.

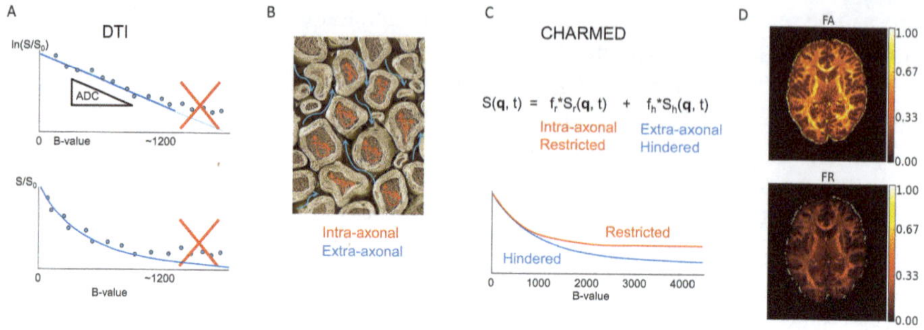

Fig. 8 From DTI to biophysical compartment models. (**a**) Mono-exponential signal decay with b-value in DTI does not hold beyond b-values of ~1200s/mm^2. (**b**) An illustration of diffusion in intra-cellular (or intra-axonal) compartments (in red), and extra-cellular (or extra-axonal) compartments (in blue). (**c**) The Combined Hindered And Restricted ModEl of Diffusion (CHARMED model) approximates the diffusion signal as a linear combination of intra-axonal restricted diffusion and extra-axonal hindered diffusion terms (top) which behave differently at higher b-values (bottom). (**d**) The signal fraction for the restricted term (f_r or FR, Fraction of Restricted) provides a scalar map more specific to tissue microstructure than the fractional anisotropy (FA) derived from DTI

2012; Alexander et al. 2010; Jelescu et al. 2015; Fieremans et al. 2013; Assaf and Basser 2005; Panagiotaki et al. 2012). Although the compartment models are based on simple geometric shapes, and a strong assumption of no inter-compartment exchange is made (cf; Nilsson et al. 2013; Reeves et al. 2016), these models can provide specific measures such as fiber density, orientation dispersion, and axonal diameter distributions, and have been shown to be sensitive to specific white matter alterations due to development (e.g., Jelescu et al. 2015; Kunz et al. 2014) and pathology (Wen et al. 2015; Fieremans et al. 2013; Benitez et al. 2014; Kamagata et al. 2016; Bergers et al. 2002; Timmers et al. 2015). There is a large range of diffusion microstructure models, including popular models such as neurite orientation dispersion and density imaging model (NODDI; Zhang et al. 2012), the CHARMED model (Assaf and Basser 2005), the white matter tract integrity model (WMTI; Fieremans et al. 2013), AxCaliber (Assaf et al. 2008), the minimal model of white matter diffusion (MMWMD in ActiveAx; Alexander et al. 2010; Zhang et al. 2011), including further developments of these models like NODDIDA (Jelescu et al. 2015), Bingham NODDI (Tariq et al. 2016), and diffusion time-dependent CHARMED (De Santis et al. 2016). The general multi-compartment diffusion microstructure model has the form of a weighted sum of single compartments:

$$S(q,t) = S_0 \sum_{n}^{i=0} w_i S_i(q,t)$$

Where S_0 is the signal for the non-diffusion-weighted (or $b0$) acquisitions, w_i the volume fractions (signal weights, signal fractions or water fractions) and S_i is the signal function for the i'th of n total compartments. The CHARMED model is seen to be an instantiation of this general model with 2 compartments and may other diffusion microstructure models differ only in the number (often 2 or 3) and model formulation of compartments (Panagiotaki et al. 2012).

Biophysical multi-compartment models require non-linear parameter optimization to be fit to dMRI data, typically performed using either maximum likelihood estimation (MLE) or the Markov chain Monte Carlo (MCMC) sampling. Whereas MLE provides only a point estimate of the fitted model parameters, MCMC recovers the entire posterior distribution of the model parameters given in the data, providing additional information such as parameter uncertainty and correlations (Harms and Roebroeck 2018). This comes with some challenges. Each of the microstructure models comes with its own, often different, optimization algorithm, noise model, and initialization strategy to estimate its parameter maps. Since data fit, accuracy, and precision are hard to verify, this creates additional challenges to comparability and generalization of results from diffusion microstructure models. In addition, nonlinear optimization is computationally expensive leading to very long run times, which can be prohibitive in large group or population studies. Solutions to these challenges have started to arise. Harms et al. (2017) evaluated whether a single well-performing optimization approach exists that could be applied to many models and would equate both run time and fit aspects. They implemented a general framework, available as an open-source toolbox, containing many models, algorithms, and strategies. The graphics processing unit (GPU) is used to remove run time constraints, achieving whole-brain dataset fits in seconds to minutes, establishing standards by which robust fit and accuracy can be achieved in shorter run times. This is especially relevant for the use of diffusion microstructure modeling in large group or population studies and in combining microstructure parameter maps with tractography results.

Diffusion microstructure models involve, inevitably, a higher complexity of modeling of dMRI data than DTI. That is, models with a higher number of parameters are fit to the data in order to have more specific interpretations of the results. In addition to needing more data (more measurements in ($\mathbf{q}$, t) space), complex parametric models also entail making more assumptions, which leads to new limitations and the need for caution (for review and discussion, see Jones et al. 2018; Novikov et al. 2018a). Limitations of biophysical compartment models are mostly concerned with one of two things. The first is assumption violation: assumptions embedded in the model break down without graceful degradation of robust model behavior. Examples of common assumptions vulnerable to violation include no exchange between compartments, perfectly straight cylindrical axons, the same diffusion behavior at different diffusion times, and the absence or equality of T_1 and T_2 relaxation parameters for compartments. The second is non-identifiability: data acquired on standard MRI systems in realistic time do not support fitting a unique set of model parameters with high precision. Examples of non-identifiability issues are the difficulty of fitting axonal diameter models to data acquired on standard MRI systems, and the degeneracy of models (i.e., a large range of parameters values fits a dataset equally well) unless further strong assumptions on parameter values are made.

Outlook

Developments in diffusion MRI have moved rapidly in the last decade, and have endeavored to move beyond the simple and powerful, yet limited Gaussian description of diffusion in DTI. Although the advances have been on various fronts and multi-pronged, it is useful to think of them having focused on either tractography and connectomics or on diffusion microstructure, as we have done here. In tractography and macroscale connectomics, the focus has been on physical models or signal models that can accurately model multi-peaked ODFs for complex fiber configurations, with an emphasis on HARDI acquisitions. In diffusion microstructure, much focus has been on biophysical models or tissue models that can provide more specific microstructure parameters by modeling restricted diffusion, with an emphasis on multi-shell acquisitions or ($\mathbf{q}$, t)-space acquisitions. Not all developments fit this strict two-legged mold. For instance, spherical deconvolution techniques have arisen which use multi-shell HARDI acquisitions to distinguish tissue types (white matter, gray matter, cerebrospinal fluid) in addition to estimating ODFs (Jeurissen et al. 2014), and diffusion microstructure techniques have seen the light which are based on low-assumption signal models rather than biophysical tissue models (Novikov et al. 2018b). Future developments might continue the idea of mixing tractography with microstructure, with the aim to provide tracks or connectomes with more quantitative interpretability of its connections in terms of true axonal density rather than streamline count (Jones et al. 2013; Sherbondy et al. 2010). Nevertheless, it is likely that tractography developments and microstructure developments will largely run on semi-separate tracks, even if only because of the differing data requirements (high angular resolution for tractography, high ($\mathbf{q}$, t) resolution for microstructure) and diverging focus of model sophistication (subtleties in fiber dispersion or low-angle mixing for tractography, subtleties in compartmental division and exchange for microstructure).

Advances in MRI hardware will probably continue to propel all dMRI advances. Vendors have steadily upgraded gradient system capability which immediately benefits dMRI, with clinical 3T systems moving beyond 40 mT/m to up to 80 mT/m, and bespoke research systems having reached 300 mT/m (Jones et al. 2018; Setsompop et al. 2013), and are already moving beyond. Although ultra-high fields of 7 T and above have mixed blessings for dMRI contrast, as described above, 7T acquisitions have been used to advance spatial resolution to new levels (Heidemann et al. 2010; Vu et al. 2015), mostly for the purpose of ODF modeling and tractography. MRI acquisition acceleration advances that shorten the time taken for a dMRI dataset, such as multiband or simultaneous multi-slice techniques (Setsompop et al. 2013; Ugurbil et al. 2013), will no doubt enlarge their influence in a data-hungry field, and future acceleration techniques will similarly be quickly and thankfully adopted.

A recent trend is the investigation of more complex diffusion gradient waveforms, modifying the simple trapezoid gradient used in PGSE. Sinusoidal oscillating gradients have the potential to probe smaller diffusion times and improve cell size and axon diameter distribution estimations, double diffusion encoding with two probed directions can improve sensitivity to compartmental exchange, and 3D q-space trajectory waveforms can help distinguish microscopic anisotropy with randomly distributed orientations within a voxel.

Finally, multi-modal modeling efforts, integrating diffusion MRI with complementary strengths from other MRI imaging modalities, such as susceptibility MRI (sMRI), quantitative magnetic transfer (qMT) imaging, or myelin water imaging (MWI), are starting to take hold. Diffusion MRI is insensitive to the myelin compartment (due to its very low T_2) but sensitive to relative size of intra- and extracellular compartments. The sensitivity of sMRI, qMT, or MWI is roughly the reverse, with a high sensitivity to myelin. This shows the potentially beneficial complementarity that could be exploited. A case in point is g-ratio imaging (Stikov et al. 2014) in which the g-ratio (the ratio of inner and outer axon diameter including myelin) is estimated in vivo from human dMRI data combined with qMT or MWI. Such multi-modal approaches are likely to grow in number, requiring sophisticated unified modeling of data from different modalities, properly accounting for diverging imaging distortions and artifacts.

References

Aboitiz, F., A. B. Scheibel, R. S. Fisher, and E. Zaidel. 1992. 'Fiber composition of the human corpus callosum', *Brain Res*, 598: 143-53.

Alexander, D. C., G. J. Barker, and S. R. Arridge. 2002. 'Detection and modeling of non-Gaussian apparent diffusion coefficient profiles in human brain data', *Magn Reson Med*, 48: 331-40.

Alexander, D. C., T. B. Dyrby, M. Nilsson, and H. Zhang. 2017. 'Imaging brain microstructure with diffusion MRI: practicality and applications', *Nmr in Biomedicine*.

Alexander, D. C., P. L. Hubbard, M. G. Hall, E. A. Moore, M. Ptito, G. J. Parker, and T. B. Dyrby. 2010. 'Orientationally invariant indices of axon diameter and density from diffusion MRI', *Neuroimage*, 52: 1374-89.

Assaf, Y., D. C. Alexander, D. K. Jones, A. Bizzi, T. E. Behrens, C. A. Clark, Y. Cohen, T. B. Dyrby, P. S. Huppi, T. R. Knoesche, D. Lebihan, G. J. Parker, C. Poupon, Connect consortium, D. Anaby, A. Anwander, L. Bar, D. Barazany, T. Blumenfeld-Katzir, S. De-Santis, D. Duclap, M. Figini, E. Fischi, P. Guevara, P. Hubbard, S. Hofstetter, S. Jbabdi, N. Kunz, F. Lazeyras, A. Lebois, M. G. Liptrot, H. Lundell, J. F. Mangin, D. M. Dominguez, D. Morozov, J. Schreiber, K. Seunarine, S. Nava, C. Poupon, T. Riffert, E. Sasson, B. Schmitt, N. Shemesh, S. N. Sotiropoulos, I. Tavor, H. G. Zhang, and F. L. Zhou. 2013. 'The CONNECT project: Combining macro- and micro-structure', *Neuroimage*, 80: 273-82.

Assaf, Y., and P. J. Basser. 2005. 'Composite hindered and restricted model of diffusion (CHARMED) MR imaging of the human brain', *Neuroimage*, 27: 48-58.

Assaf, Y., T. Blumenfeld-Katzir, Y. Yovel, and P. J. Basser. 2008. 'AxCaliber: a method for measuring axon diameter distribution from diffusion MRI', *Magn Reson Med*, 59: 1347-54.

Assaf, Y., R. Z. Freidlin, G. K. Rohde, and P. J. Basser. 2004. 'New modeling and experimental framework to characterize hindered and restricted water diffusion in brain white matter', *Magn Reson Med*, 52: 965-78.

Assaf, Y., and O. Pasternak. 2008. 'Diffusion tensor imaging (DTI)-based white matter mapping in brain research: a review', *J Mol Neurosci*, 34: 51-61.

Basser, P. J., J. Mattiello, and D. LeBihan. 1994. 'MR diffusion tensor spectroscopy and imaging', *Biophys J*, 66: 259-67.

Basser, P. J., S. Pajevic, C. Pierpaoli, J. Duda, and A. Aldroubi. 2000. 'In vivo fiber tractography using DT-MRI data', *Magn Reson Med*, 44: 625-32.

Basser, P. J., and C. Pierpaoli. 1996. 'Microstructural and physiological features of tissues elucidated by quantitative-diffusion-tensor MRI', *J Magn Reson B*, 111: 209-19.

Bassett, D. S., J. A. Brown, V. Deshpande, J. M. Carlson, and S. T. Grafton. 2011. 'Conserved and variable architecture of human white matter connectivity', *Neuroimage*, 54: 1262-79.

Bastiani, M., M. Cottaar, K. Dikranian, A. Ghosh, H. Zhang, D. C. Alexander, T. E. Behrens, S. Jbabdi, and S. N. Sotiropoulos. 2017. 'Improved tractography using asymmetric fibre orientation distributions', *Neuroimage*, 158: 205-18.

Bastiani, M., N. J. Shah, R. Goebel, and A. Roebroeck. 2012. 'Human cortical connectome reconstruction from diffusion weighted MRI: the effect of tractography algorithm', *Neuroimage*, 62: 1732-49.

Beaulieu, C. 2002. 'The basis of anisotropic water diffusion in the nervous system – a technical review', *Nmr in Biomedicine*, 15: 435-55.

Behrens, T. E., H. J. Berg, S. Jbabdi, M. F. Rushworth, and M. W. Woolrich. 2007. 'Probabilistic diffusion tractography with multiple fibre orientations: What can we gain?', *Neuroimage*, 34: 144-55.

Behrens, T. E., H. Johansen-Berg, M. W. Woolrich, S. M. Smith, C. A. Wheeler-Kingshott, P. A. Boulby, G. J. Barker, E. L. Sillery, K. Sheehan, O. Ciccarelli, A. J. Thompson, J. M. Brady, and P. M. Matthews. 2003a. 'Non-invasive mapping of connections between human thalamus and cortex using diffusion imaging', *Nat Neurosci*, 6: 750-7.

Behrens, T. E., M. W. Woolrich, M. Jenkinson, H. Johansen-Berg, R. G. Nunes, S. Clare, P. M. Matthews, J. M. Brady, and S. M. Smith. 2003b. 'Characterization and propagation of uncertainty in diffusion-weighted MR imaging', *Magn Reson Med*, 50: 1077-88.

Benitez, A., E. Fieremans, J. H. Jensen, M. F. Falangola, A. Tabesh, S. H. Ferris, and J. A. Helpern. 2014. 'White matter tract integrity metrics reflect the vulnerability of late-myelinating tracts in Alzheimer's disease', *Neuroimage Clin*, 4: 64-71.

Bergers, E., J. C. Bot, C. J. De Groot, C. H. Polman, G. J. Lycklama, A Nijeholt, J. A. Castelijns, P. van der Valk, and F. Barkhof. 2002. 'Axonal damage in the spinal cord of MS patients occurs largely independent of T2 MRI lesions', *Neurology*, 59: 1766-71.

Budde, M. D., and J. Annese. 2013. 'Quantification of anisotropy and fiber orientation in human brain histological sections', *Front Integr Neurosci*, 7: 3.

Budde, M. D., and J. A. Frank. 2012. 'Examining brain microstructure using structure tensor analysis of histological sections', *Neuroimage*, 63: 1-10.

Buxton, R. B. 1993. 'The diffusion sensitivity of fast steady-state free precession imaging', *Magn Reson Med*, 29: 235-43.

Campbell, J. S., P. MomayyezSiahkal, P. Savadjiev, I. R. Leppert, K. Siddiqi, and G. B. Pike. 2014. 'Beyond crossing fibers: bootstrap probabilistic tractography using complex subvoxel fiber geometries', *Front Neurol*, 5: 216.

Catani, M., F. Dell'acqua, F. Vergani, F. Malik, H. Hodge, P. Roy, R. Valabregue, and M. Thiebaut de Schotten. 2012. 'Short frontal lobe connections of the human brain', *Cortex*, 48: 273-91.

Catani, M., R. J. Howard, S. Pajevic, and D. K. Jones. 2002. 'Virtual in vivo interactive dissection of white matter fasciculi in the human brain', *Neuroimage*, 17: 77-94.

Catani, M., and M. Thiebaut de Schotten. 2008. 'A diffusion tensor imaging tractography atlas for virtual in vivo dissections', *Cortex*, 44: 1105-32.

Cheng, H., Y. Wang, J. Sheng, O. Sporns, W. G. Kronenberger, V. P. Mathews, T. A. Hummer, and A. J. Saykin. 2011. 'Optimization of seed density in DTI tractography for structural networks', *J Neurosci Methods*.

Conturo, T. E., N. F. Lori, T. S. Cull, E. Akbudak, A. Z. Snyder, J. S. Shimony, R. C. McKinstry, H. Burton, and M. E. Raichle. 1999. 'Tracking neuronal fiber pathways in the living human brain', *Proc Natl Acad Sci U S A*, 96: 10422-7.

Cote, M. A., G. Girard, A. Bore, E. Garyfallidis, J. C. Houde, and M. Descoteaux. 2013. 'Tractometer: towards validation of tractography pipelines', *Med Image Anal*, 17: 844-57.

De Santis, S., M. Drakesmith, S. Bells, Y. Assaf, and D. K. Jones. 2014. 'Why diffusion tensor MRI does well only some of the time: variance and covariance of white matter tissue microstructure attributes in the living human brain', *Neuroimage*, 89: 35-44.

De Santis, S., D. K. Jones, and A. Roebroeck. 2016. 'Including diffusion time dependence in the extra-axonal space improves in vivo estimates of axonal diameter and density in human white matter', *Neuroimage*, 130: 91-103.

Dell'Acqua, F., and J. D. Tournier. 2018. 'Modelling white matter with spherical deconvolution: How and why?', *Nmr in Biomedicine*: e3945.

Dong, Q., R. C. Welsh, T. L. Chenevert, R. C. Carlos, P. Maly-Sundgren, D. M. Gomez-Hassan, and S. K. Mukherji. 2004. 'Clinical applications of diffusion tensor imaging', *J Magn Reson Imaging*, 19: 6-18.

Fieremans, E., A. Benitez, J. H. Jensen, M. F. Falangola, A. Tabesh, R. L. Deardorff, M. V. Spampinato, J. S. Babb, D. S. Novikov, S. H. Ferris, and J. A. Helpern. 2013. 'Novel white matter tract integrity metrics sensitive to Alzheimer disease progression', *AJNR Am J Neuroradiol*, 34: 2105-12.

Fieremans, E., Y. De Deene, S. Delputte, M. S. Ozdemir, Y. D'Asseler, J. Vlassenbroeck, K. Deblaere, E. Achten, and I. Lemahieu. 2008. 'Simulation and experimental verification of the diffusion in an anisotropic fiber phantom', *J Magn Reson*, 190: 189-99.

Frank, L. R. 2001. 'Anisotropy in high angular resolution diffusion-weighted MRI', *Magn Reson Med*, 45: 935-9.

Frank, L. R. 2002. 'Characterization of anisotropy in high angular resolution diffusion-weighted MRI', *Magn Reson Med*, 47: 1083-99.

Goulas, A., M. Bastiani, G. Bezgin, H. B. Uylings, A. Roebroeck, and P. Stiers. 2014. 'Comparative analysis of the macroscale structural connectivity in the macaque and human brain', *PLoS Comput Biol*, 10: e1003529.

Hagmann, P., L. Cammoun, X. Gigandet, R. Meuli, C. J. Honey, V. J. Wedeen, and O. Sporns. 2008. 'Mapping the structural core of human cerebral cortex', *PLoS Biol*, 6: e159.

Hagmann, P., M. Kurant, X. Gigandet, P. Thiran, V. J. Wedeen, R. Meuli, and J. P. Thiran. 2007. 'Mapping human whole-brain structural networks with diffusion MRI', *PLoS One*, 2: e597.

Harms, R. L., F. J. Fritz, A. Tobisch, R. Goebel, and A. Roebroeck. 2017. 'Robust and fast nonlinear optimization of diffusion MRI microstructure models', *Neuroimage*, 155: 82-96.

Harms, R. L., and A. Roebroeck. 2018. 'Robust and Fast Markov Chain Monte Carlo Sampling of Diffusion MRI Microstructure Models', *Front Neuroinform*, 12: 97.

Heidemann, R. M., D. A. Porter, A. Anwander, T. Feiweier, K. Heberlein, T. R. Knosche, and R. Turner. 2010. 'Diffusion imaging in humans at 7T using readout-segmented EPI and GRAPPA', *Magn Reson Med*, 64: 9-14.

Horsfield, M. A., and D. K. Jones. 2002. 'Applications of diffusion-weighted and diffusion tensor MRI to white matter diseases – a review', *Nmr in Biomedicine*, 15: 570-7.

Iturria-Medina, Y., E. J. Canales-Rodriguez, L. Melie-Garcia, P. A. Valdes-Hernandez, E. Martinez-Montes, Y. Aleman-Gomez, and J. M. Sanchez-Bornot. 2007. 'Characterizing brain anatomical connections using diffusion weighted MRI and graph theory', *Neuroimage*, 36: 645-60.

Jansons, K. M., and D. C. Alexander. 2003. 'Persistent Angular Structure: new insights from diffusion MRI data. Dummy version', *Inf Process Med Imaging*, 18: 672-83.

Jbabdi, S., and H. Johansen-Berg. 2011. 'Tractography: where do we go from here?', *Brain Connect*, 1: 169-83.

Jbabdi, S., J. F. Lehman, S. N. Haber, and T. E. Behrens. 2013. 'Human and monkey ventral prefrontal fibers use the same organizational principles to reach their targets: tracing versus tractography', *J Neurosci*, 33: 3190-201.

Jbabdi, S., M. W. Woolrich, J. L. Andersson, and T. E. Behrens. 2007. 'A Bayesian framework for global tractography', *Neuroimage*, 37: 116-29.

Jelescu, I. O., J. Veraart, V. Adisetiyo, S. S. Milla, D. S. Novikov, and E. Fieremans. 2015. 'One diffusion acquisition and different white matter models: How does microstructure change in human early development based on WMTI and NODDI?', *Neuroimage*, 107: 242-56.

Jensen, J. H., and J. A. Helpern. 2010. 'MRI quantification of non-Gaussian water diffusion by kurtosis analysis', *Nmr in Biomedicine*, 23: 698-710.

Jensen, J. H., J. A. Helpern, A. Ramani, H. Lu, and K. Kaczynski. 2005. 'Diffusional kurtosis imaging: the quantification of non-gaussian water diffusion by means of magnetic resonance imaging', *Magn Reson Med*, 53: 1432-40.

Jeurissen, B., M. Descoteaux, S. Mori, and A. Leemans. 2017. 'Diffusion MRI fiber tractography of the brain', *Nmr in Biomedicine*.

Jeurissen, B., A. Leemans, J. D. Tournier, D. K. Jones, and J. Sijbers. 2013. 'Investigating the prevalence of complex fiber configurations in white matter tissue with diffusion magnetic resonance imaging', *Hum Brain Mapp*, 34: 2747-66.

Jeurissen, B., J. D. Tournier, T. Dhollander, A. Connelly, and J. Sijbers. 2014. 'Multi-tissue constrained spherical deconvolution for improved analysis of multi-shell diffusion MRI data', *Neuroimage*, 103: 411-26.

Jones, D. K. 2003. 'Determining and visualizing uncertainty in estimates of fiber orientation from diffusion tensor MRI', *Magn Reson Med*, 49: 7-12.

Jones, D. K., D. C. Alexander, R. Bowtell, M. Cercignani, F. Dell'Acqua, D. J. McHugh, K. L. Miller, M. Palombo, G. J. M. Parker, U. S. Rudrapatna, and C. M. W. Tax. 2018. 'Microstructural imaging of the human brain with a 'super-scanner': 10 key advantages of ultra-strong gradients for diffusion MRI', *Neuroimage*, 182: 8-38.

Jones, D. K., T. R. Knosche, and R. Turner. 2013. 'White matter integrity, fiber count, and other fallacies: the do's and don'ts of diffusion MRI', *Neuroimage*, 73: 239-54.

Jones, Derek K. 2010. 'Challenges and limitations of quantifying brain connectivity in vivo with diffusion MRI', *Imaging in Medicine*, 2: 341-55.

Kamagata, K., T. Hatano, A. Okuzumi, Y. Motoi, O. Abe, K. Shimoji, K. Kamiya, M. Suzuki, M. Hori, K. K. Kumamaru, N. Hattori, and S. Aoki. 2016. 'Neurite orientation dispersion and density imaging in the substantia nigra in idiopathic Parkinson disease', *Eur Radiol*, 26: 2567-77.

Khayal, I. S., S. R. Vandenberg, K. J. Smith, C. P. Cloyd, S. M. Chang, S. Cha, S. J. Nelson, and T. R. McKnight. 2011. 'MRI apparent diffusion coefficient reflects histopathologic subtype, axonal disruption, and tumor fraction in diffuse-type grade II gliomas', *Neuro Oncol*, 13: 1192-201.

Kreher, B. W., I. Mader, and V. G. Kiselev. 2008. 'Gibbs tracking: a novel approach for the reconstruction of neuronal pathways', *Magn Reson Med*, 60: 953-63.

Kunz, N., H. Zhang, L. Vasung, K. R. O'Brien, Y. Assaf, F. Lazeyras, D. C. Alexander, and P. S. Huppi. 2014. 'Assessing white matter microstructure of the newborn with multi-shell diffusion MRI and biophysical compartment models', *Neuroimage*, 96: 288-99.

LaMantia, A. S., and P. Rakic. 1990. 'Cytological and quantitative characteristics of four cerebral commissures in the rhesus monkey', *J Comp Neurol*, 291: 520-37.

Lebel, C., S. Caverhill-Godkewitsch, and C. Beaulieu. 2010. 'Age-related regional variations of the corpus callosum identified by diffusion tensor tractography', *Neuroimage*, 52: 20-31.

Li, L., J. K. Rilling, T. M. Preuss, M. F. Glasser, and X. Hu. 2011. 'The effects of connection reconstruction method on the interregional connectivity of brain networks via diffusion tractography', *Hum Brain Mapp*.

Lundell, H., D. C. Alexander, and T. B. Dyrby. 2014. 'High angular resolution diffusion imaging with stimulated echoes: compensation and correction in experiment design and analysis', *Nmr in Biomedicine*, 27: 918-25.

McKinstry, R. C., A. Mathur, J. H. Miller, A. Ozcan, A. Z. Snyder, G. L. Schefft, C. R. Almli, S. I. Shiran, T. E. Conturo, and J. J. Neil. 2002. 'Radial organization of developing preterm human cerebral cortex revealed by non-invasive water diffusion anisotropy MRI', *Cereb Cortex*, 12: 1237-43.

McNab, J. A., S. Jbabdi, S. C. Deoni, G. Douaud, T. E. Behrens, and K. L. Miller. 2009. 'High resolution diffusion-weighted imaging in fixed human brain using diffusion-weighted steady state free precession', *Neuroimage*, 46: 775-85.

Mcnab, J. A., and K. L. Miller. 2010. 'Steady-state diffusion-weighted imaging: theory, acquisition and analysis', *Nmr in Biomedicine*, 23: 781-93.

Merboldt, K. D., H. Bruhn, J. Frahm, M. L. Gyngell, W. Hanicke, and M. Deimling. 1989. 'MRI of "diffusion" in the human brain: new results using a modified CE-FAST sequence', *Magn Reson Med*, 9: 423-9.

Merboldt, K. D., W. Hanicke, and J. Frahm. 1991. 'Diffusion imaging using stimulated echoes', *Magn Reson Med*, 19: 233-9.

Modo, M., T. K. Hitchens, J. R. Liu, and R. M. Richardson. 2016. 'Detection of Aberrant Hippocampal Mossy Fiber Connections: Ex Vivo Mesoscale Diffusion MRI and Microtractography With Histological Validation in a Patient With Uncontrolled Temporal Lobe Epilepsy', *Hum Brain Mapp*, 37: 780-95.

Mollink, J., K. M. van Baarsen, P. J. Dederen, S. Foxley, K. L. Miller, S. Jbabdi, C. H. Slump, J. A. Grotenhuis, M. Kleinnijenhuis, and A. M. van Cappellen van Walsum. 2015. 'Dentatorubrothalamic tract localization with postmortem MR diffusion tractography compared to histological 3D reconstruction', *Brain Struct Funct*.

Mori, S., B. J. Crain, V. P. Chacko, and P. C. van Zijl. 1999. 'Three-dimensional tracking of axonal projections in the brain by magnetic resonance imaging', *Ann Neurol*, 45: 265-9.

Mori, S., and P. C. van Zijl. 2002. 'Fiber tracking: principles and strategies – a technical review', *Nmr in Biomedicine*, 15: 468-80.

Neher, P. F., M. Descoteaux, J. C. Houde, B. Stieltjes, and K. H. Maier-Hein. 2015. 'Strengths and weaknesses of state of the art fiber tractography pipelines--A comprehensive in-vivo and phantom evaluation study using Tractometer', *Med Image Anal*, 26: 287-305.

Neil, J., J. Miller, P. Mukherjee, and P. S. Huppi. 2002. 'Diffusion tensor imaging of normal and injured developing human brain – a technical review', *Nmr in Biomedicine*, 15: 543-52.

Nilsson, Markus, Danielle van Westen, Freddy Ståhlberg, Pia C. Sundgren, and Jimmy Lätt. 2013. 'The role of tissue microstructure and water exchange in biophysical modelling of diffusion in white matter', *Magnetic Resonance Materials in Physics, Biology and Medicine*, 26: 345-70.

Novikov, D. S., E. Fieremans, S. N. Jespersen, and V. G. Kiselev. 2018a. 'Quantifying brain microstructure with diffusion MRI: Theory and parameter estimation', *Nmr in Biomedicine*: e3998.

Novikov, D. S., J. Veraart, I. O. Jelescu, and E. Fieremans. 2018b. 'Rotationally-invariant mapping of scalar and orientational metrics of neuronal microstructure with diffusion MRI', *Neuroimage*, 174: 518-38.

Ozarslan, E., T. M. Shepherd, B. C. Vemuri, S. J. Blackband, and T. H. Mareci. 2006. 'Resolution of complex tissue microarchitecture using the diffusion orientation transform (DOT)', *Neuroimage*, 31: 1086-103.

Panagiotaki, E., T. Schneider, B. Siow, M. G. Hall, M. F. Lythgoe, and D. C. Alexander. 2012. 'Compartment models of the diffusion MR signal in brain white matter: a taxonomy and comparison', *Neuroimage*, 59: 2241-54.

Parker, G. J., and D. C. Alexander. 2005. 'Probabilistic anatomical connectivity derived from the microscopic persistent angular structure of cerebral tissue', *Philos Trans R Soc Lond B Biol Sci*, 360: 893-902.

Parker, G. J., H. A. Haroon, and C. A. Wheeler-Kingshott. 2003. 'A framework for a streamline-based probabilistic index of connectivity (PICo) using a structural interpretation of MRI diffusion measurements', *J Magn Reson Imaging*, 18: 242-54.

Pestilli, F., J. D. Yeatman, A. Rokem, K. N. Kay, and B. A. Wandell. 2014. 'Evaluation and statistical inference for human connectomes', *Nat Methods*, 11: 1058-63.

Pfefferbaum, A., E. V. Sullivan, M. Hedehus, K. O. Lim, E. Adalsteinsson, and M. Moseley. 2000. 'Age-related decline in brain white matter anisotropy measured with spatially corrected echo-planar diffusion tensor imaging', *Magn Reson Med*, 44: 259-68.

Plantinga, B. R., A. Roebroeck, V. G. Kemper, K. Uludag, M. Melse, J. Mai, M. L. Kuijf, A. Herrler, A. Jahanshahi, B. M. Ter Haar Romeny, and Y. Temel. 2016. 'Ultra-High Field MRI Post Mortem Structural Connectivity of the Human Subthalamic Nucleus, Substantia Nigra, and Globus Pallidus', *Frontiers in Neuroanatomy*, 10: 66.

Pullens, P., A. Roebroeck, and R. Goebel. 2010. 'Ground Truth Hardware Phantoms for Validation of Diffusion-Weighted MRI Applications', *Journal of Magnetic Resonance Imaging*, 32: 482-88.

Reeves, C., M. Tachrount, D. Thomas, Z. Michalak, J. Liu, M. Ellis, B. Diehl, A. Miserocchi, A. W. McEvoy, S. Eriksson, T. Yousry, and M. Thom. 2016. 'Combined Ex Vivo 9.4T MRI and Quantitative Histopathological Study in Normal and Pathological Neocortical Resections in Focal Epilepsy', *Brain Pathol*, 26: 319-33.

Reisert, M., E. Kellner, and V. G. Kiselev. 2012. 'About the geometry of asymmetric fiber orientation distributions', *IEEE Trans Med Imaging*, 31: 1240-9.

Reisert, M., and V. G. Kiselev. 2011. 'Fiber continuity: an anisotropic prior for ODF estimation', *IEEE Trans Med Imaging*, 30: 1274-83.

Reisert, M., I. Mader, C. Anastasopoulos, M. Weigel, S. Schnell, and V. Kiselev. 2011. 'Global fiber reconstruction becomes practical', *Neuroimage*, 54: 955-62.

Reveley, C., A. K. Seth, C. Pierpaoli, A. C. Silva, D. Yu, R. C. Saunders, D. A. Leopold, and F. Q. Ye. 2015. 'Superficial white matter fiber systems impede detection of long-range cortical connections in diffusion MR tractography', *Proc Natl Acad Sci U S A*, 112: E2820-8.

Roebroeck, A., R. Galuske, E. Formisano, O. Chiry, H. Bratzke, I. Ronen, D. S. Kim, and R. Goebel. 2008. 'High-resolution diffusion tensor imaging and tractography of the human optic chiasm at 9.4 T', *Neuroimage*, 39: 157-68.

Rubinov, M., and O. Sporns. 2010. 'Complex network measures of brain connectivity: uses and interpretations', *Neuroimage*, 52: 1059-69.

Santis, S., Y. Assaf, C. J. Evans, and D. K. Jones. 2014. 'Improved precision in CHARMED assessment of white matter through sampling scheme optimization and model parsimony testing', *Magn Reson Med*, 71: 661-71.

Savadjiev, P., J. S. Campbell, M. Descoteaux, R. Deriche, G. B. Pike, and K. Siddiqi. 2008. 'Labeling of ambiguous subvoxel fibre bundle configurations in high angular resolution diffusion MRI', *Neuroimage*, 41: 58-68.

Savadjiev, P., J. S. Campbell, G. B. Pike, and K. Siddiqi. 2006. '3D curve inference for diffusion MRI regularization and fibre tractography', *Med Image Anal*, 10: 799-813.

Seehaus, A. K., A. Roebroeck, O. Chiry, D. S. Kim, I. Ronen, H. Bratzke, R. Goebel, and R. A. Galuske. 2013. 'Histological validation of DW-MRI tractography in human postmortem tissue', *Cereb Cortex*, 23: 442-50.

Seehaus, A., A. Roebroeck, M. Bastiani, L. Fonseca, H. Bratzke, N. Lori, A. Vilanova, R. Goebel, and R. Galuske. 2015. 'Histological validation of high-resolution DTI in human post mortem tissue', *Frontiers in Neuroanatomy*, 9.

Setsompop, K., R. Kimmlingen, E. Eberlein, T. Witzel, J. Cohen-Adad, J. A. McNab, B. Keil, M. D. Tisdall, P. Hoecht, P. Dietz, S. F. Cauley, V. Tountcheva, V. Matschl, V. H. Lenz, K. Heberlein, A. Potthast, H. Thein, J. Van Horn, A. Toga, F. Schmitt, D. Lehne, B. R. Rosen, V. Wedeen, and L. L. Wald. 2013. 'Pushing the limits of in vivo diffusion MRI for the Human Connectome Project', *Neuroimage*, 80: 220-33.

Sherbondy, A. J., R. F. Dougherty, R. Ananthanarayanan, D. S. Modha, and B. A. Wandell. 2009. 'Think global, act local; projectome estimation with BlueMatter', *Med Image Comput Comput Assist Interv*, 12: 861-8.

Sherbondy, A. J., R. F. Dougherty, M. Ben-Shachar, S. Napel, and B. A. Wandell. 2008. 'ConTrack: finding the most likely pathways between brain regions using diffusion tractography', *J Vis*, 8: 15 1-16.

Sherbondy, A. J., M. C. Rowe, and D. C. Alexander. 2010. 'MicroTrack: an algorithm for concurrent projectome and microstructure estimation', *Med Image Comput Comput Assist Interv*, 13: 183-90.

Smith, R. E., J. D. Tournier, F. Calamante, and A. Connelly. 2013. 'SIFT: Spherical-deconvolution informed filtering of tractograms', *Neuroimage*, 67: 298-312.

Sotak, C. H. 2002. 'The role of diffusion tensor imaging in the evaluation of ischemic brain injury – a review', *Nmr in Biomedicine*, 15: 561-9.

Sotiropoulos, S. N., L. Bai, P. S. Morgan, C. S. Constantinescu, and C. R. Tench. 2010. 'Brain tractography using Q-ball imaging and graph theory: Improved connectivities through fibre crossings via a model-based approach', *Neuroimage*, 49: 2444-56.

Sotiropoulos, S. N., and A. Zalesky. 2017. 'Building connectomes using diffusion MRI: why, how and but', *Nmr in Biomedicine*.

Sporns, O., G. Tononi, and R. Kotter. 2005. 'The human connectome: A structural description of the human brain', *PLoS Comput Biol*, 1: e42.

Stanisz, G. J., A. Szafer, G. A. Wright, and R. M. Henkelman. 1997. 'An analytical model of restricted diffusion in bovine optic nerve', *Magn Reson Med*, 37: 103-11.

Stejskal, E.O., and J.E. Tanner. 1965. 'Spin Diffusion Measurements: Spin Echoes in the Presence of a Time-Dependent Field Gradient', *Journal of Chemical Physics*, 42.

Stikov, N., J. Campbell, M. Boudreau, S. Narayanan, T. Stroh, S. Nuara, J. Novek, S. Frey, Ming-Kay Ho, B. Bedell, and B. Pike. 2014. "In vivo histology of the myelin g-ratio." In *Proceedings of the 2014 Organization for Human Brain Mapping annual meeting*, 2249. Hamburg, Germany.

Sundgren, P. C., Q. Dong, D. Gomez-Hassan, S. K. Mukherji, P. Maly, and R. Welsh. 2004. 'Diffusion tensor imaging of the brain: review of clinical applications', *Neuroradiology*, 46: 339-50.

Tariq, M., T. Schneider, D. C. Alexander, C. A. Gandini Wheeler-Kingshott, and H. Zhang. 2016. 'Bingham-NODDI: Mapping anisotropic orientation dispersion of neurites using diffusion MRI', *Neuroimage*, 133: 207-23.

Tasaki, I., F. Ishii, and H. Ito. 1943. 'On the relation between the conduction rate and fiber diameter and the internodal distance of the medullated nerve fiber', *Japanese Journal of Medical Sciences*, 9: 189-99.

Thomas, C., F. Q. Ye, M. O. Irfanoglu, P. Modi, K. S. Saleem, D. A. Leopold, and C. Pierpaoli. 2014. 'Anatomical accuracy of brain connections derived from diffusion MRI tractography is inherently limited', *Proc Natl Acad Sci U S A*, 111: 16574-9.

Timmers, I., H. Zhang, M. Bastiani, B. M. Jansma, A. Roebroeck, and M. E. Rubio-Gozalbo. 2015. 'White matter microstructure pathology in classic galactosemia revealed by neurite orientation dispersion and density imaging', *J Inherit Metab Dis*, 38: 295-304.

Tournier, J. D., F. Calamante, and A. Connelly. 2007. 'Robust determination of the fibre orientation distribution in diffusion MRI: non-negativity constrained super-resolved spherical deconvolution', *Neuroimage*, 35: 1459-72.

Tournier, J. D., F. Calamante, D. G. Gadian, and A. Connelly. 2004. 'Direct estimation of the fiber orientation density function from diffusion-weighted MRI data using spherical deconvolution', *Neuroimage*, 23: 1176-85.

Tuch, D. S. 2004. 'Q-ball imaging', *Magn Reson Med*, 52: 1358-72.

Tuch, D. S., T. G. Reese, M. R. Wiegell, N. Makris, J. W. Belliveau, and V. J. Wedeen. 2002. 'High angular resolution diffusion imaging reveals intravoxel white matter fiber heterogeneity', *Magn Reson Med*, 48: 577-82.

Ugurbil, K., J. Xu, E. J. Auerbach, S. Moeller, A. T. Vu, J. M. Duarte-Carvajalino, C. Lenglet, X. Wu, S. Schmitter, P. F. Van de Moortele, J. Strupp, G. Sapiro, F. De Martino, D. Wang, N. Harel, M. Garwood, L. Chen, D. A. Feinberg, S. M. Smith, K. L. Miller, S. N. Sotiropoulos, S. Jbabdi, J. L. Andersson, T. E. Behrens, M. F. Glasser, D. C. Van Essen, E. Yacoub, and W. U-Minn HCP Consortium. 2013. 'Pushing spatial and temporal resolution for functional and diffusion MRI in the Human Connectome Project', *Neuroimage*, 80: 80-104.

van den Heuvel, M. P., R. C. Mandl, C. J. Stam, R. S. Kahn, and H. E. Hulshoff Pol. 2010. 'Aberrant frontal and temporal complex network structure in schizophrenia: a graph theoretical analysis', *J Neurosci*, 30: 15915-26.

van den Heuvel, M. P., and O. Sporns. 2011. 'Rich-club organization of the human connectome', *J Neurosci*, 31: 15775-86.

Van Essen, D., S. Jbabdi, S. N. Sotiropoulos, C. Chen, K. Dikranian, T. Coalson, J. Harwell, T. E. Behrens, and M. F. Glasser. 2014. 'MAPPING CONNECTIONS IN HUMANS AND NON-HUMAN PRIMATES: ASPIRATIONS AND CHALLENGES FOR DIFFUSION IMAGING.' in, *Diffusion MRI*.

Vu, A. T., E. Auerbach, C. Lenglet, S. Moeller, S. N. Sotiropoulos, S. Jbabdi, J. Andersson, E. Yacoub, and K. Ugurbil. 2015. 'High resolution whole brain diffusion imaging at 7T for the Human Connectome Project', *Neuroimage*, 122: 318-31.

Wakana, S., H. Jiang, L. M. Nagae-Poetscher, P. C. van Zijl, and S. Mori. 2004. 'Fiber tract-based atlas of human white matter anatomy', *Radiology*, 230: 77-87.

Wedeen, V. J., P. Hagmann, W. Y. Tseng, T. G. Reese, and R. M. Weisskoff. 2005. 'Mapping complex tissue architecture with diffusion spectrum magnetic resonance imaging', *Magn Reson Med*, 54: 1377-86.

Wedeen, V. J., R. P. Wang, J. D. Schmahmann, T. Benner, W. Y. Tseng, G. Dai, D. N. Pandya, P. Hagmann, H. D'Arceuil, and A. J. de Crespigny. 2008. 'Diffusion spectrum magnetic resonance imaging (DSI) tractography of crossing fibers', *Neuroimage*, 41: 1267-77.

Wen, Q., D. A. Kelley, S. Banerjee, J. M. Lupo, S. M. Chang, D. Xu, C. P. Hess, and S. J. Nelson. 2015. 'Clinically feasible NODDI characterization of glioma using multiband EPI at 7 T', *Neuroimage Clin*, 9: 291-9.

Werring, D. J., C. A. Clark, G. J. Barker, A. J. Thompson, and D. H. Miller. 1999. 'Diffusion tensor imaging of lesions and normal-appearing white matter in multiple sclerosis', *Neurology*, 52: 1626-32.

Yeh, F. C., V. J. Wedeen, and W. Y. Tseng. 2010. 'Generalized q-sampling imaging', *IEEE Trans Med Imaging*, 29: 1626-35.

Zalesky, A., and A. Fornito. 2009. 'A DTI-derived measure of cortico-cortical connectivity', *IEEE Trans Med Imaging*, 28: 1023-36.

Zalesky, A., A. Fornito, I. H. Harding, L. Cocchi, M. Yucel, C. Pantelis, and E. T. Bullmore. 2010. 'Whole-brain anatomical networks: does the choice of nodes matter?', *Neuroimage*, 50: 970-83.

Zatorre, R. J., R. D. Fields, and H. Johansen-Berg. 2012. 'Plasticity in gray and white: neuroimaging changes in brain structure during learning', *Nat Neurosci*, 15: 528-36.

Zhang, H., P. L. Hubbard, G. J. Parker, and D. C. Alexander. 2011. 'Axon diameter mapping in the presence of orientation dispersion with diffusion MRI', *Neuroimage*, 56: 1301-15.

Zhang, H., T. Schneider, C. A. Wheeler-Kingshott, and D. C. Alexander. 2012. 'NODDI: practical in vivo neurite orientation dispersion and density imaging of the human brain', *Neuroimage*, 61: 1000-16.

Data Mining in the Era of Big Data: The BrainMap Database as a Resource for Characterizing Brain Networks in Psychiatric Illness

Katherine L. Bottenhorn and Angela R. Laird

The BrainMap Database

The volume of published human functional neuroimaging research has more than quintupled in the past 20 years and the sheer amount of data collected doubles every 26 months (Van Horn and Toga 2014). While this means that the scientific community has access to much more information about how the brain works than ever before, it brings with it the challenges of cataloguing and summarizing such a large literature. More than 20 years ago, a group of researchers at the University of Texas Health Science Center San Antonio began addressing this issue with the creation of BrainMap, an online community database of the human neuroimaging literature that presents both context and content in a highly accessible feature space. BrainMap was modeled on existing databases employed by the genome community, initially providing a condensed guide to the literature (Fox and Lancaster 1994). Since its creation as a mechanism by which members of the multidisciplinary, rapidly growing field could easily identify experiments of interest, BrainMap has grown into a robust and comprehensive framework for data mining and quantitative analysis of the published literature (Fox and Lancaster 2002; Laird et al. 2005b). Currently, BrainMap provides not only streamlined access to concise reports of the vast neuroimaging literature, but a set of integrated tools for quantitative meta-analysis of published research.

Conventions of the human neuroimaging literature make it particularly well-suited to such a database, as most results are presented as statistical parametric images with reported local maxima in *x*, *y*, *z* coordinates that refer to locations in a

K. L. Bottenhorn
Department of Psychology, Florida International University, Miami, FL, USA

A. R. Laird (✉)
Department of Physics, Florida International University, Miami, FL, USA
e-mail: alaird@fiu.edu

V. A. Diwadkar, S. B. Eickhoff (eds.), *Brain Network Dysfunction in Neuropsychiatric Illness*, https://doi.org/10.1007/978-3-030-59797-9_7

standardized template space (Fox et al. 2005b). The coordinate data can be extracted directly from publications and are a lightweight representation of brain imaging results that facilitates secondary data analysis (Fox and Lancaster 2002). BrainMap presents these coordinate-based results linked to their experimental meta-data, which provides a condensed, highly structured description of the paper, experiment, and coordinate locations themselves. The meta-data taxonomy provides a hierarchical coding scheme that echoes that found in the literature (Fox et al. 2005a). It provides description at the *Paper* level, including information concerning the citation, population, and conditions of the study. Papers are then broken down into *Experiments* and information is provided concerning meta-data and results (i.e., brain activation foci) of individual statistical contrasts. Information is provided, too, on the statistical value and spatial extent of brain activation at the *Location* level. BrainMap's associated software tools (Laird et al. 2011a) allow users to leverage this coding scheme for straightforward, expeditious data mining and, even further, to uncover underlying properties of brain organization that support cognition (Fox et al. 2014). This is made possible via BrainMap's quantitative meta-analytic approach, termed activation likelihood estimation (ALE), that performs voxelwise assessment of reported brain activations for a given set of coordinates from the database, e.g. from several experiments that all used the same paradigm. Not only does the ALE approach allow users to assess emergent patterns of brain activation across a single experimental paradigm, it provides a mechanism by which researchers can define ROIs for new studies, contrast neural recruitment across related cognitive systems, or assess differences in brain activation across subject groups, i.e., disease effects on cognition (Turkeltaub et al. 2002; Fox et al. 2014).

Activation Likelihood Estimation (ALE) Meta-Analysis

The human brain mapping literature has historically been represented by studies of small sample sizes, inconsistencies across results, and highly variable data analysis strategies, all of which add to observed individual differences in brain structure and function. In the face of high levels of variability across laboratories, experiments, and individuals, quantitative meta-analysis provides an automated, unbiased mechanism by which to obtain statistical maps of consistent findings of the neural basis of mental functions. Conceptually, these meta-analyses present an opportunity to assess consistency in functional neuroimaging findings of a homogenous group of studies, such as from a given task paradigm or collection of related paradigms. In 2002, Turkeltaub and colleagues developed the first activation likelihood estimation (ALE) meta-analysis, comparing a coordinate-based meta-analysis of 11 positron emission tomography (PET) studies of single-word reading with new results of a functional magnetic resonance imaging (fMRI) implementation of that same task (Turkeltaub et al. 2002). In this seminal work, Turkeltaub and colleagues established the following requirements for ALE meta-analyses: statistically powerful brain activation foci reported in standardized, x, y, z coordinate space. The original application of the ALE method proceeded by centering a 15 mm FWHM Gaussian distribution, the average smoothing kernel for PET experiments at the time, at each focus. This essentially converted the focus into the peak of a probability density function to account for uncertainty introduced by differences both between experiments and between individuals. Voxelwise ALE statistics were then calculated from the union of these distributions, representing the probability that one of the 11 studies reported an activation peak in each voxel. To determine significance of ALE scores at each voxel, Turkeltaub et al. computed 1000 null distributions from random foci and used these noise ALE scores to perform an omnibus single threshold (Holmes et al. 1996; Laird et al. 2005a). The resulting ALE map presents voxelwise p values representing regions where foci significantly converge across previous analyses.

Evolution of the ALE Algorithm

Since its introduction in 2002, the ALE approach has undergone a series of modifications, yielding a more statistically robust method. Turkeltaub and colleagues' original permutation testing was updated in 2005 to include corrections for multiple comparisons and statistical testing for significant differences between meta-analyses (Laird et al. 2005a). Following a comparison of the original single threshold test with a false-discovery rate (FDR) correction across four ALE meta-analyses with varying numbers of foci, the ALE approach was extended to calculate voxelwise, FDR-corrected p values. Furthermore, the method was adapted to account for the bias toward higher ALE values for voxels in the middle of the brain, testing ALE values against voxelwise null distributions instead of a single, brain-wide threshold value. In addition to improved statistical thresholding, Laird et al. (2005a) introduced a mechanism to statistically compare ALE maps by determining the significant, voxelwise difference in ALE scores between maps, thresholded at an FDR-corrected p value. Mathematically representing the concept of cognitive subtraction, this method presents a means by which to assess the differences in the neural substrates of two different paradigms or behavioral domains. However, cognitive subtraction is an imperfect strategy for isolating a discrete cognitive process, as in many cases there are elements that do not occur by pure insertion, so this method must be used and its results interpreted with caution (Friston et al. 1996; Price et al. 1997).

Following these improvements, Eickhoff et al. (2009) continued to address the limitations of the ALE method, specifically the modeling of spatial uncertainty, generalizability of ALE results, and lack of anatomical constraints. Initially, Turkeltaub et al. applied a 15 mm FWHM Gaussian kernel to model uncertainty, which was based on smoothing kernels used in neuroimaging analyses. Eickhoff et al. suggested that uncertainty should instead be modeled according to the relative contributions of between-subject and between-template variability and were the first to estimate this variability. Eickhoff et al. quantified between-template and between-subject variability by assessing the Euclidean distances between peaks from 9 near-identical, group-level analyses that differed only in normalization method and from 21 identical single-subject analyses (for each of the 9 normalization methods) respectively. Each of these mean Euclidean distances were then used to estimate the equivalent Gaussian kernel sizes, based in part on the number of subjects included, to be used in ALE meta-analyses. This revised Turkeltaub and colleagues' "one size fits all" 15 mm FWHM Gaussian uncertainty kernel to an empirically based variable-width kernel, scaled to account for the number of subjects in each experiment.

In addition to experiment-specific uncertainty modeling, Eickhoff et al. extended the ALE algorithm to perform random-effects inference, instead of fixed-effects, to improve the generalizability of inferences from ALE meta-analyses (Eickhoff et al. 2009). Formerly, the ALE algorithm merely uncovered above-chance clustering between foci from all experiments included in the meta-analysis. The algorithm did not conserve the distribution of foci within each study, instead pooled them to perform a meta-analysis. Eickhoff et al. modified this procedure so that the algorithm instead computed a 3D volume, or "modeled activation" (MA) map, per experiment (using the experiment-specific Gaussian kernel) and then calculated the union of these MA maps. This new approach also included anatomical constraints, restricting the calculation of voxelwise ALE scores to a liberal gray matter mask (based on the ICBM tissue probability maps, thresholded at >10% probability gray matter), as functional brain activation is largely restricted to gray matter (Evans et al. 1994). Additionally, a null distribution was empirically defined by sampling the gray matter mask of each MA map included in an ALE meta-analysis, to compute an analysis-specific ALE score (per permutation) that represents the value that would be obtained in the absence of neurobiologically convergent activation.

However, this modification to account for groupings of foci within experiments failed to address the influence that a single subject-group could have, if represented in multiple experiments (i.e., inclusion of several experiments from the same paper in a single meta-analysis). In 2012, Turkeltaub and colleagues modified the ALE algorithm to minimize within-experiment and within-group effects that might bias the results of ALE meta-analyses. The group tested these modifications in turn and observed that clusters of foci within a single experiment and multiple experiments from the same subject groups can have small, but noticeable effects on ALE scores. A modified ALE algorithm was presented to address these biases, but the authors noted that within-experiment and within-subject group effects on previous ALE analyses were minimal, and thus did not invalidate any results (Turkeltaub et al. 2012).

In the same year, Eickhoff and colleagues further modified the ALE algorithm to improve how it performs statistical inference, streamlining the procedure for estimating the null distribution and extending the options for multiple comparisons corrections (Eickhoff et al. 2012). Previously, the ALE approach only offered the false discovery rate (FDR) correction, which is flawed in calculations with continuous data, produced by using Gaussian PDFs to model spatial uncertainty in ALE meta-analyses (Chumbley and Friston 2009). This work found that cluster-level thresholding was the most suitable replacement for FDR correction, controlling for familywise error (FWE) rate by setting a spatial extent threshold at the cluster level.

One of the main advantages of ALE is that it allows users to combine previously published experiments, many of which suffer from small sample sizes and similar issues with statistical power, to identify convergence in the neurobiological substrates of mental processes of interest. But how many experiments need to be synthesized to perform statistically robust meta-analyses? Eickhoff and colleagues addressed this in 2016, performing a massive set of simulation analyses to compare sensitivity and selectivity of the various thresholding methods and provide recommendations of how many experiments need to be included for a sufficiently powered ALE meta-analysis (Eickhoff et al. 2016). In correcting for multiple comparisons, this report affirmed that cluster-level FWE correction is the optimal thresholding method and the formerly utilized FDR correction ought to be avoided. Additionally, at least 20 experiments should be included in an ALE meta-analysis to detect a true effect and mitigate undue influence from any one experiment. Also in 2016, a report was published that detailed glaring errors in thresholding procedures used in primary fMRI data analysis, calling into question the validity of thousands of published neuroimaging studies (Eklund et al. 2016). At the same time, errors were discovered in ALE's methods for multiple comparisons corrections (Eickhoff et al. 2017). BrainMap community members noticed implementation errors in the code for both FDR and FWE corrections similar to the algorithmic errors noted by Eklund and colleagues, which resulted in much more liberal thresholding than was reported or intended by researchers. Both errors were corrected in version 2.3.6 of GingerALE, the BrainMap software implementation of the ALE algorithm, which continues to be actively maintained.

Clinical Application of the ALE Approach

In addition to uncovering emergent properties of neural systems, ALE can be applied to provide insight into differences in brain activation between subject groups. The first clinical application of ALE assessed differences in brain activation during the n-back paradigm between healthy controls and patients diagnosed with schizophrenia (Glahn et al. 2005). From 12 functional neuroimaging articles, Glahn et al. performed four ALE meta-analyses: separate meta-analyses for healthy controls and patients with schizophrenia, as well as meta-analyses of healthy subjects greater than schizophrenia patients and patients greater than controls. Prior research had highlighted dysregulation of the dorsolateral prefrontal cortex (dlPFC) in schizophrenia, evidenced by decreased activation of the region during working memory tasks (Spitzer 1993; Goldman-Rakic 1994; Weinberger and Berman 1996). Glahn and colleagues corroborated this characteristic hypofrontality but observed increases in activity in anterior cingulate and frontal polar regions as well. A few years later, this work was extended by Minzenberg and colleagues, who broadened their inclusion criteria to assessed executive function in schizophrenia across a wider range of tasks (Minzenberg et al. 2009). This work found similar results across executive function tasks, suggesting a broad dysfunction of the cognitive control network in schizophrenia. Minzenberg and colleagues noted that the differences in brain activation in schizophrenia go beyond hypofrontality and are better characterized as "displaced functions," as regions of relative hypoactivation were often observed to be adjacent to regions of hyperactivation, which may represent compensatory behavior. Taken together, these meta-analytic results built on primary research and proposed a more comprehensive network model of altered executive function in schizophrenia over the simple notion of hypofrontality.

Since 2005, ALE has been used to identify consensus and synthesize the literature to quantitatively assess the neural correlates of various diseases and disorders. Among these are numerous disorders associated with neurodegeneration, mood and personality, substance use and abuse, eating and sleeping, neurocognition, trauma, and atypical neurodevelopment. Due to the breadth of disorders interrogated, we summarize here several of the most highly-cited ALE meta-analyses of clinical populations. In 2007, Fitzgerald and colleagues performed an ALE meta-analysis of functional neuroimaging studies of patients with major depressive disorder (MDD) in several contexts (Fitzgerald et al. 2008). From meta-analyses of resting state, emotion induction, and treatment (SSRI) studies, this study uncovered complementary networks of consistently altered activation when compared to neurotypical controls. Emotional control circuitry, comprised of regions including the dlPFC, anterior cingulate cortex (ACC), insula, and superior temporal gyrus (STG), exhibited decreased activation in resting state and emotion induction studies in depressed patients, compared to healthy controls, but increased activation in studies of SSRI treatment. Conversely, emotional awareness circuitry, consisting of the medial and inferior prefrontal regions and the basal ganglia, was observed to be overactive at rest and during emotion induction tasks in depressed patients, compared to healthy controls, but showed decreased activation with SSRI treatment. This suggests complicated, reciprocal circuitry aberrations underlying depressive symptoms, which can be partially improved with antidepressant medication.

A common theme across ALE meta-analyses that assess neuropathology of various disorders is the finding that dysfunction occurs in distributed networks, rather than individual regions. Challenging the notion that attention deficit hyperactivity disorder (ADHD) is largely characterized by hypofrontality, Dickstein and colleagues performed an ALE meta-analysis of executive function and working memory across patients diagnosed with ADHD (Dickstein et al. 2006). By performing a subtraction analysis on separate ALE maps of patients with ADHD and healthy controls, Dickstein and colleagues synthesized an incredibly heterogeneous literature and uncovered patterns of altered brain activation across working memory challenges in frontal regions including the dlPFC, ACC, orbitofrontal cortex (OFC), as well as basal ganglia and parietal regions that comprised a distributed network of dysfunction. By separately assessing differences between ADHD patients and healthy controls in response inhibition, specifically, this work also uncovered task-specific hypoactivation of the caudate nucleus, which was absent from the larger, task-general ALE. Similarly, Di Martino and colleagues expanded the formerly held model of autism spectrum disorder (ASD) from a focus on frontal hypoactivation and disruptions in social processing to include network models of complex hypo- and hyperactivation across the brain (Di Martino et al. 2009). Much of the autism neuroimaging literature, prior to 2009, was highly heterogeneous, suffered from small sample sizes, and lacked appropriate group comparisons. In this work, Di Martino et al. performed ALE meta-analyses of ASD patients and neurotypical controls, then compared the two with a subtraction analysis, for all studies of ASD and separately for nonsocial and social paradigms. They found domain-specific patterns of hypoactivation, along with default mode network dysfunction, which had been previously noted in ASD research. Additionally, they uncovered an inappropriate recruitment of face processing and motor regions, in which ASD patients recruited lower level face processing and motor regions (i.e., SMA instead of dACC/preSMA) in social and nonsocial tasks, respectively, than the neurotypical controls.

Because clinical neuroimaging research can often plagued with small sample sizes and a large amount of variation both in clinical presentation of the disorder and in experimental analysis methods, ALE is a particularly useful tool for clinical researchers. By providing an unbiased, quantitative method for uncovering emergent patterns of aberrant brain activation across studies, ALE allows an otherwise fractionated literature to be synthesized to identify the convergent neural correlates of disease.

Independent Component Analysis of the BrainMap Database

Containing results and meta-data from over 3000 published manuscripts, BrainMap serves as a rich and valuable resource for investigating the functional topography of brain networks. Data mining approaches such as independent component analysis (ICA) have allowed researchers to leverage the full power of the BrainMap database to assess large-scale patterns in brain organization. A data mining technique for blind source separation, ICA provides a means for identifying a predefined number of independent patterns, or “components,” that are present in multivariate data. By 2009, ICA had been successfully applied to resting state fMRI data to uncover a handful of so-called resting state networks (RSNs) that were stable across humans and across times (Beckmann et al. 2005). These networks had been observed across individuals and remained stable across a range of mental states (Cordes et al. 2000; Greicius et al. 2004; Filippini et al. 2009), representing functional associations between diffuse brain regions (Fox and Raichle 2007). Smith and colleagues extended these methods to meta-analytic data, comparing ICA of the full complement of the BrainMap database to resting state ICA to investigate patterns of brain activation consistent across big data from both resting state and task-based sources (Smith et al. 2009). At the time, BrainMap was estimated to contain nearly a fifth of all published functional neuroimaging papers, and Smith and colleagues determined that more than half of all identified RSNs matched a component from the BrainMap ICA decomposition. These common networks demonstrated common features in the brain's functional organization across mental states. Two years later, these RSNs were replicated by Biswal and colleagues, who found even more in common between their RSNs and Smith's BrainMap ICA components (Biswal et al. 2010). To further explore how these networks reflect intrinsic organizational properties of the brain, Laird et al. (2011b) mined the BrainMap database for behavioral meta-data related to the activation of each RSN, now termed "intrinsic connectivity networks" (ICNs) due to their persistence across rest and task states. Laird and colleagues interpreted ICNs via data-driven clustering of behavioral domains and paradigm classes, based on brain-behavior groupings throughout BrainMap (Fig. 1). These results were used to better understand the ICNs underlying brain function, and the data-driven groupings of BrainMap meta-data precipitated the development of a cognitive ontology framework. In this way, the work of Laird and colleagues demonstrated that brain imaging data can provide insight to understanding relationships between different mental processes.

Each of the aforementioned studies specified the same ICA model order, which is the number of components returned by ICA, based on the precedent set by Beckmann and colleagues' early investigation of 20 components (2005). Ray et al. (2013) addressed model order selection from a data-driven perspective by mining the BrainMap database to investigate how the data naturally coalesces into components (i.e., networks). By using hierarchical clustering to assess how strongly each network related to each of BrainMap's behavioral domains and paradigm classes, Ray et al. determined the "fit" for each model order. These results represent the dynamic qualities of ICNs across levels of increasing granularity (prescribed by increasing model order), in addition to providing data-driven grouping of different

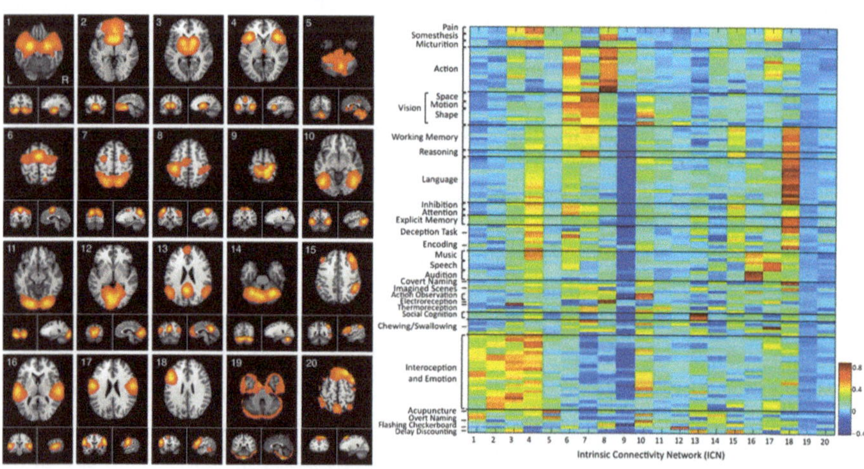

Fig. 1 Decomposition of the BrainMap database into 20 ICNs and their behavioral interpretations Independent component analysis (ICA) of all functional neuroimaging studied included in the BrainMap database yielded 20 ICNs that are stable across task and rest, representing the intrinsic organization of the brain (left) and the interpretation of these networks' functions from an analysis of BrainMap meta-data (right). (*Reprinted with permission from* Laird et al. (2011b))

tasks and behaviors. From this analysis, Ray et al. determined the model orders exhibiting and optimal fit were 70 and 20 (2013). Further probing of these ICNs using meta-analytic connectivity modeling, revealed how each of the 20 ICNs uncovered by Smith et al. (2009) and Laird et al. (2011b) fractionate into 70 ICNs, adding to their functional interpretation (Laird et al. 2017).

Clinical Utility of BrainMap ICA Components

ICNs are rich markers of disease and the works of Smith, Laird, Ray, and others to add unbiased interpretation of the functional roles of each network have contributed greatly to the network-level study of psychiatric conditions, illustrating the utility of large-scale data mining to clinical science. These networks have been used as templates for investigating psychiatric disorders such as depression (Belleau et al. 2015; Wang et al. 2015b), neurodegenerative conditions including Parkinson's (Putcha et al. 2015, 2016), aging (Putcha et al. 2016; Madden et al. 2017), and substance use (Lee and Telzer 2016; Hobkirk et al. 2017). Likewise, researchers have used Laird et al.' (2011b) interpretation of these networks' function in hypothesis generation to study Chron's disease and irritable bowel disorder (IBS; (Thomann et al. 2017), as well as anoxic brain injury (Ishaque et al. 2017).

Beyond informing study design, the functional explanation of these ICNs have helped researchers meaningfully interpret alterations of brain function in clinical populations. Worhunsky et al. (2016, 2017) used the ICN interpretations from Laird et al. (2011b) to understand addictive and substance use disorders. Notably, individuals diagnosed with cocaine use disorder (CUD) and gambling disorder (GD) exhibit increased engagement of executive control networks, citing increased demands on cognitive processing in surrendering loss in CUD and GD and highlighting a the potential of cognitive intervention in treating these disorders (Worhunsky et al. 2017). Similarly, young adults escalating their alcohol consumption showed increased engagement of a frontoparietal network during response inhibition, indicating a greater need for coordination of goal-directed and internal processing in these individuals (Worhunsky et al. 2016).

Understanding the functions related to ICNs have contributed greatly to studies of schizophrenia (Meda et al. 2012; Servaas et al. 2015; Wang et al. 2015a; Poppe et al. 2016; Solé-Padullés et al. 2016) and bipolar disorder (Meda et al. 2012; Khadka et al. 2013; Lois et al. 2014; Solé-Padullés et al. 2016). Researchers have uncovered the neural substrates of aberrant executive function in schizophrenia from medial frontal and parietal hypoactivation (Poppe et al. 2016). Those diagnosed with schizophrenia or bipolar disorder, as well as their first degree relatives, show alterations in frontostriatal networks that connect cognition to emotion (Khadka et al. 2013) and decreased connectivity between networks that facilitate pursuit of top-down executive control (Meda et al. 2012). These findings converge on noted issues of executive functioning and emotional processing, revealing how these symptoms are represented in an underlying dysfunction of neurobiology and demonstrating further commonality between the two disorders. Several authors note that these findings highlight the potential for cognitive therapies to ameliorate not only symptoms of schizophrenia, but the underlying biology (Poppe et al. 2016).

Data mining has shown its value in converging on the functions associated with activity in large-scale brain networks, which previously had been studied from varied perspective, with varied methods, and yielding varied interpretations. Combining the information mined from BrainMap across a large body of literature, much of which was not even targeting these large-scale networks, has allowed researchers to understand the consistent functions of these networks. Here, data-driven study of functional brain networks shows its value to the clinical sciences. Not only have researchers discovered the biological expression of symptomology, they have extended our understanding of inherent similarities between disorders and underlined the promise of cognitive intervention in treatment strategies.

Data Mining via Clustering Approaches

Independent component analysis of the BrainMap database revealed a consistent set of large-scale brain networks and allowed Laird and colleagues to add unbiased interpretations to their functions. However, ICA is only the beginning of what data-driven approaches and big data can tell us about the brain and cognition. Understanding the brain on a macroscopic systems level has greatly added to the study of cognition and psychiatric conditions, but further study has shown that there is even more to learn about specific aspects of mentation and improve knowledge representation of mental processes. By narrowing focus from the entire contents of the BrainMap database to focus on specific paradigms and behavioral domains, researchers have embarked on a more granular study of behavior and the underlying functional neuroanatomy. These studies demonstrate that data mining's contributions to cognitive neuroscience can provide scientific evidence to (i) help revise psychological theory, providing updates to prior models; (ii) support existing theory, contributing information about the supporting neurobiology; and (iii) understand component mental processes and brain networks supporting complex behaviors.

Using the broad paradigm class of "face perception," Laird and colleagues demonstrated a new approach to and application of data mining methods (Laird et al. 2015). By applying a method of data-driven classification to all papers in the BrainMap database that were annotated with the paradigm class "Face Monitor or Discrimination," Laird et al. identified the functional heterogeneity of semantically identical mental processes. Hierarchical clustering provided an unbiased grouping of the activation patterns from all face perception papers *and* showed the relationships between groupings, with respect to their neuroanatomical similarity and commonalities evidenced by BrainMap's rich functional meta-data. Four categories of "face perception" tasks emerged from this analysis, representing visuospatial attention and visuomotor tasks, perception and recognition tasks, social processing and episodic recall tasks, and face naming and lexical retrieval tasks. Not only do these four task groupings highlight the complexity of face perception in the real world, they also proposed an evolution to the well-accepted Gobbini and Haxby model of face perception (Haxby et al. 2000; Gobbini and Haxby 2007). Laird and colleagues demonstrated that face perception consists of a core visual system, which supports the *perception* of faces, in addition to specific systems for processing emotional and personal aspects of the faces and their salient value. These findings illustrated a powerful link between functional neuroimaging research and psychological theory and demonstrated that unbiased data mining and methods are able to contribute greatly to our understanding of these paradigms and the development of a standardized ontology of cognitive processes.

Since then, data mining and clustering methods have been applied to other neuroimaging domains. Similar to Laird et al. (2015), Riedel et al. (2018) performed unbiased clustering of the affective processing domain. By clustering brain activations of all papers annotated with the "Emotion" behavioral domain in BrainMap, Riedel et al. uncovered separate systems underlying affective processing, including systems for sensory input, attentional gating, appraisal and prediction of emotional events, and the production of emotional responses, in emotion induction. This model corresponded with previous work by Ochsner et al. (2012), an example of data mining techniques supporting previously held theory and adding convergent evidence from functional neuroimaging to supplement our understanding of the biological basis of behavior.

Recently, these clustering methods were extended beyond the BrainMap database, to a novel exploration of published neuroimaging results mined from a variety of sources. Bottenhorn et al. (2018) explored the heterogeneous literature associated with naturalistic fMRI paradigms. In contrast to many widely used, static paradigms, *naturalistic* paradigms necessitate the continual integration of dynamic streams of multisensory information to inform behavioral responses in real-time. Naturalistic studies use stimuli including video games, movies, and virtual reality environments to study brain function underlying affective, spatial, and linguistic processes in rich contexts with higher ecological validity than typical experiments, while sacrificing the aspects of experimental control that allow researchers to isolate specific aspects of behavior. However, by applying a similar data-driven clustering method, Bottenhorn et al. uncovered a model similar to Riedel and colleagues' model of emotional processing, which describes that, across the broad naturalistic fMRI literature, there are consistently-activated brain networks that cooperate to facilitate behavior. These networks include sensory systems for visual and auditory input, an attentional system for filtering relevant information, and domain-specific systems for emotional, navigational, and language processing (Fig. 2). While data mined from these papers comes from an array of very different tasks, the results of clustering these data suggest that there are common networks engaged throughout complex behavior. Together, the findings of Laird, Riedel, and Bottenhorn show how researchers can utilize data-driven methods to uncover novel relationships between the brain and behavior.

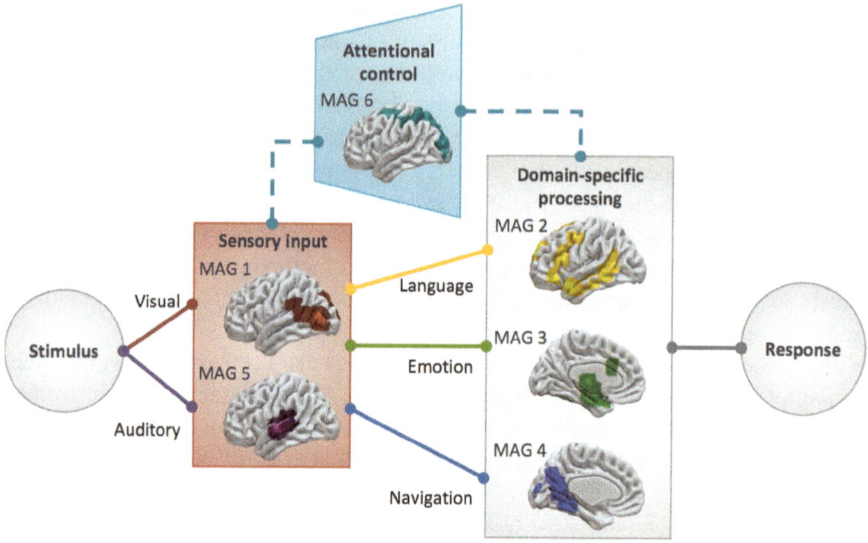

Fig. 2 Cooperating brain networks underlying complex behavior. Five networks that interact to facilitate complex behavior during naturalistic fMRI paradigms, representing sensory input systems, attentional filtering, and domain-specific systems for linguistic, emotional, and navigational processes. (*Reprinted with permission from* Bottenhorn et al. (2018))

Conclusions and Future Work

The BrainMap database and various meta-analytic techniques allow researchers to capitalize on a wealth of previously published functional neuroimaging research to further study the human brain. Activation likelihood estimation (ALE) meta-analysis of a given domain provides information about the consensus reached across decades of research, overcoming the drawbacks of any single study and the differences in perspective of any single paper. With more complex methods, researchers have discovered a set of large-scale brain networks that are consistently engaged across a variety of tasks and mental states *and* used meta-data from every paper in the BrainMap database to functionally understand these networks. The application of these well-defined and described networks to clinical investigations has provided new information about the biological substrates of disorder symptomology, providing motivation for cognitive treatment of these disorders. And finally, unbiased clustering approaches have shown patterns underlying seemingly homogenous behaviors, contributing an improved model of face perception and supporting an existing model of affective processing, and the commonalities between heterogeneous behaviors, providing a model of cooperating brain networks that support complex behavior.

However, there are limitations to the BrainMap method. While it includes a large number of published reports and allows high-powered study of a wide array of behavior, the database relies on manual annotation of papers and, thus, cannot keep up with the accelerating volume of neuroimaging publications. While there exists a fully automated counterpart to BrainMap in Neurosynth (Neurosynth.org; Yarkoni et al. 2011), there remains a need for a comprehensive database that can both keep up with the rate of publication and provide reliable, valuable well-curated annotations about published neuroimaging studies. To address this need, there are several new projects on the horizon. An improved algorithm for automatically extracting relevant information from published manuscripts, the *Automated Text Harvesting and Exploration of Neuroimaging Annotations* (ATHENA) project endeavors to provide a more informative ontology for describing previous research, as well as improved classification of neuroimaging articles within this ontology. Additionally, Brainspell, an open-source project for data curation, aims to add human annotation to the Neurosynth database for improved data synthesis via neuroimaging meta-analysis. In the future, these tools will contribute to the effort to catalogue and synthesize existing literature, thereby providing significant impact to the study of human behavior and disorders.

References

Beckmann CF, DeLuca M, Devlin JT, Smith SM (2005) Investigations into resting-state connectivity using independent component analysis. Philos Trans R Soc B Biol Sci 360:1001–1013 Available at: http://www.ncbi.nlm.nih.gov/pubmed/16087444 [Accessed December 30, 2017].

Belleau EL, Taubitz LE, Larson CL (2015) Imbalance of default mode and regulatory networks during externally focused processing in depression. Soc Cogn Affect Neurosci 10:744–751 Available at: http://www.ncbi.nlm.nih.gov/pubmed/25274576 [Accessed January 30, 2018].

Biswal BB et al. (2010) Toward discovery science of human brain function. Proc Natl Acad Sci 107:4734–4739 Available at: http://www.ncbi.nlm.nih.gov/pubmed/20176931 [Accessed January 8, 2018].

Bottenhorn KL, Flannery JS, Boeving ER, Riedel MC, Eickhoff SB, Sutherland MT, Laird AR (2018) Cooperating yet distinct brain networks engaged during naturalistic paradigms: A meta-analysis of functional MRI results. Netw Neurosci.

Chumbley JR, Friston KJ (2009) False discovery rate revisited: FDR and topological inference using Gaussian random fields. Neuroimage 44:62–70.

Cordes D, Haughton VM, Arfanakis K, Wendt GJ, Turski PA, Moritz CH, Quigley MA, Meyerand ME (2000) Mapping functionally related regions of brain with functional connectivity MR imaging. AJNR Am J Neuroradiol 21:1636–1644 Available at: http://www.ncbi.nlm.nih.gov/pubmed/11039342 [Accessed December 30, 2017].

Di Martino A, Ross K, Uddin LQ, Sklar AB, Castellanos FX, Milham MP (2009) Functional Brain Correlates of Social and Nonsocial Processes in Autism Spectrum Disorders: An Activation Likelihood Estimation Meta-Analysis. Biol Psychiatry 65:63–74 Available at: http://www.ncbi.nlm.nih.gov/pubmed/18996505 [Accessed September 30, 2017].

Dickstein SG, Bannon K, Xavier Castellanos F, Milham MP (2006) The neural correlates of attention deficit hyperactivity disorder: an ALE meta-analysis. J Child Psychol Psychiatry 47:1051–1062 Available at: http://doi.wiley.com/10.1111/j.1469-7610.2006.01671.x [Accessed September 27, 2017].

Eickhoff SB, Bzdok D, Laird AR, Kurth F, Fox PT (2012) Activation likelihood estimation meta-analysis revisited. Neuroimage 59:2349–2361 Available at: http://www.sciencedirect.com/science/article/pii/S1053811911010627.

Eickhoff SB, Laird AR, Fox PM, Lancaster JL, Fox PT (2017) Implementation errors in the GingerALE Software: Description and recommendations. Hum Brain Mapp 38:7–11.
Eickhoff SB, Laird AR, Grefkes C, Wang LE, Zilles K, Fox PT (2009) Coordinate-based activation likelihood estimation meta-analysis of neuroimaging data: A random-effects approach based on empirical estimates of spatial uncertainty. Hum Brain Mapp 30:2907–2926 Available at: http://onlinelibrary.wiley.com.spot.lib.auburn.edu/doi/10.1002/hbm.20718/abstract.
Eickhoff SB, Nichols TE, Laird AR, Hoffstaedter F, Amunts K, Fox PT, Bzdok D, Eickhoff CR (2016) Behavior, sensitivity, and power of activation likelihood estimation characterized by massive empirical simulation. Neuroimage 137:70–85.
Eklund A, Nichols TE, Knutsson H (2016) Cluster failure: Why fMRI inferences for spatial extent have inflated false-positive rates. Proc Natl Acad Sci 113:7900–7905 Available at: http://www.pnas.org/lookup/doi/10.1073/pnas.1602413113.
Evans AC, Kamber M, Collins DL, MacDonald D (1994) An MRI-Based Probabilistic Atlas of Neuroanatomy. In: Magnetic Resonance Scanning and Epilepsy, pp. 263–274. Boston, MA: Springer US. Available at: http://link.springer.com/10.1007/978-1-4615-2546-2_48 [Accessed September 19, 2017].
Filippini N, MacIntosh BJ, Hough MG, Goodwin GM, Frisoni GB, Smith SM, Matthews PM, Beckmann CF, Mackay CE (2009) Distinct patterns of brain activity in young carriers of the APOE-epsilon4 allele. Proc Natl Acad Sci U S A 106:7209–7214 Available at: http://www.ncbi.nlm.nih.gov/pubmed/19357304 [Accessed January 4, 2018].
Fitzgerald PB, Laird AR, Maller J, Daskalakis ZJ (2008) A meta-analytic study of changes in brain activation in depression. Hum Brain Mapp 29:683–695 Available at: http://doi.wiley.com/10.1002/hbm.20426 [Accessed September 27, 2017].
Fox MD, Raichle ME (2007) Spontaneous fluctuations in brain activity observed with functional magnetic resonance imaging. Nat Rev Neurosci 8:700–711 Available at: https://doi.org/10.1038/nrn2201 [Accessed July 9, 2014].
Fox PT, Laird AR, Fox SP, Fox PM, Uecker AM, Crank M, Koenig SF, Lancaster JL (2005a) BrainMap taxonomy of experimental design: description and evaluation. Hum Brain Mapp 25:185–198 Available at: http://www.ncbi.nlm.nih.gov/pubmed/15846810.
Fox PT, Laird AR, Lancaster JL (2005b) Coordinate-based voxel-wise meta-analysis: Dividends of spatial normalization. Report of a virtual workshop. Hum Brain Mapp 25:1–5 Available at: http://doi.wiley.com/10.1002/hbm.20139 [Accessed September 20, 2017].
Fox PT, Lancaster JL (1994) Neuroscience on the net. Science 266:994–996 Available at: http://www.ncbi.nlm.nih.gov/pubmed/7973682 [Accessed March 1, 2018].
Fox PT, Lancaster JL (2002) Opinion: Mapping context and content: the BrainMap model. Nat Rev Neurosci 3:319–321.
Fox PT, Lancaster JL, Laird AR, Eickhoff SB (2014) Meta-Analysis in Human Neuroimaging: Computational Modeling of Large-Scale Databases. Annu Rev Neurosci 37:409–434 Available at: http://www.annualreviews.org/doi/10.1146/annurev-neuro-062012-170320.
Friston KJ, Price CJ, Fletcher P, Moore C, Frackowiak RS, Dolan RJ (1996) The trouble with cognitive subtraction. Neuroimage 4:97–104 Available at: http://www.ncbi.nlm.nih.gov/pubmed/9345501 [Accessed November 10, 2015].
Glahn DC, Ragland JD, Abramoff A, Barrett J, Laird AR, Bearden CE, Velligan DI (2005) Beyond hypofrontality: A quantitative meta-analysis of functional neuroimaging studies of working memory in schizophrenia. In: Human Brain Mapping, pp. 60–69.
Gobbini MI, Haxby J V. (2007) Neural systems for recognition of familiar faces. Neuropsychologia 45:32–41 Available at: https://www.sciencedirect.com/science/article/pii/S0028393206001588?via%3Dihub [Accessed March 1, 2018].
Goldman-Rakic PS (1994) Working memory dysfunction in schizophrenia. J Neuropsychiatry Clin Neurosci 6:348–357.
Greicius MD, Srivastava G, Reiss AL, Menon V (2004) Default-mode network activity distinguishes Alzheimer's disease from healthy aging: evidence from functional MRI. Proc Natl

Acad Sci U S A 101:4637–4642 Available at: http://www.ncbi.nlm.nih.gov/pubmed/15070770 [Accessed January 4, 2018].

Haxby JVJ, Hoffman EEA, Gobbini MIMMII (2000) The distributed human neural system for face perception. Trends Cogn Sci 4:223–233 Available at: http://www.sciencedirect.com/science/article/pii/S1364661300014820%5Cn. http://www.ncbi.nlm.nih.gov/pubmed/10827445.

Hobkirk AL, Nichols TT, Foulds J, Yingst JM, Veldheer S, Hrabovsky S, Richie J, Eissenberg T, Wilson SJ (2017) Changes in resting state functional brain connectivity and withdrawal symptoms are associated with acute electronic cigarette use. Brain Res Bull Available at: https://www.sciencedirect.com/science/article/pii/S0361923017302320?via%3Dihub [Accessed January 31, 2018].

Holmes AP, Blair RC, Watson JDG, Ford I (1996) Nonparametric Analysis of Statistic Images from Functional Mapping Experiments. J Cereb Blood Flow Metab 16:7–22 Available at: http://journals.sagepub.com/doi/10.1097/00004647-199601000-00002.

Ishaque M, Manning JH, Woolsey MD, Franklin CG, Tullis EW, Beckmann CF, Fox PT (2017) Functional integrity in children with anoxic brain injury from drowning. Hum Brain Mapp 38:4813–4831 Available at: http://doi.wiley.com/10.1002/hbm.23745 [Accessed January 31, 2018].

Khadka S, Meda SA, Stevens MC, Glahn DC, Calhoun VD, Sweeney JA, Tamminga CA, Keshavan MS, O'Neil K, Schretlen D, Pearlson GD (2013) Is aberrant functional connectivity a psychosis endophenotype? A resting state functional magnetic resonance imaging study. Biol Psychiatry 74:458–466 Available at: http://www.ncbi.nlm.nih.gov/pubmed/23746539 [Accessed January 30, 2018].

Laird AR, Riedel MC, Okoe M, Jianu R, Ray KL, Eickhoff SB, Smith SM, Fox PT, Sutherland MT (2017) Heterogeneous fractionation profiles of meta-analytic coactivation networks. Neuroimage 149:424–435.

Laird AR, Eickhoff SB, Fox PM, Uecker AM, Ray KL, Saenz JJ, McKay DR, Bzdok D, Laird RW, Robinson JL, Turner JA, Turkeltaub PE, Lancaster JL, Fox PT (2011a) The BrainMap strategy for standardization, sharing, and meta-analysis of neuroimaging data. BMC Res Notes 4:349 Available at: http://www.ncbi.nlm.nih.gov/pubmed/21906305 [Accessed July 19, 2016].

Laird AR, Fox PM, Eickhoff SB, Turner JA, Ray KL, McKay DR, Glahn DC, Beckmann CF, Smith SM, Fox PT (2011b) Behavioral interpretations of intrinsic connectivity networks. J Cogn Neurosci 23:4022–4037 Available at: http://www.ncbi.nlm.nih.gov/pubmed/21671731.

Laird AR, Fox PM, Price CJ, Glahn DC, Uecker AM, Lancaster JL, Turkeltaub PE, Kochunov P, Fox PT (2005a) ALE meta-analysis: Controlling the false discovery rate and performing statistical contrasts. Hum Brain Mapp 25:155–164 Available at: http://onlinelibrary.wiley.com.spot.lib.auburn.edu/doi/10.1002/hbm.20136/abstract.

Laird AR, Lancaster JL, Fox PT (2005b) BrainMap: the social evolution of a human brain mapping database. Neuroinformatics 3:65–78 Available at: http://www.ncbi.nlm.nih.gov/pubmed/15897617.

Laird AR, Riedel MC, Sutherland MT, Eickhoff SB, Ray KL, Uecker AM, Fox PM, Turner JA, Fox PT (2015) Neural architecture underlying classification of face perception paradigms. Neuroimage 119:70–80.

Lee T-H, Telzer EH (2016) Negative functional coupling between the right fronto-parietal and limbic resting state networks predicts increased self-control and later substance use onset in adolescence. Dev Cogn Neurosci 20:35–42 Available at: http://www.ncbi.nlm.nih.gov/pubmed/27344035 [Accessed January 31, 2018].

Lois G, Linke J, Wessa M (2014) Altered functional connectivity between emotional and cognitive resting state networks in euthymic bipolar I disorder patients. PLoS One 9:e107829 Available at: http://www.ncbi.nlm.nih.gov/pubmed/25343370 [Accessed January 31, 2018].

Madden DJ, Parks EL, Tallman CW, Boylan MA, Hoagey DA, Cocjin SB, Packard LE, Johnson MA, Chou Y, Potter GG, Chen N, Siciliano RE, Monge ZA, Honig JA, Diaz MT (2017) Sources of disconnection in neurocognitive aging: cerebral white-matter integrity, resting-state functional connectivity, and white-matter hyperintensity volume. Neurobiol Aging 54:199–213

Available at: https://www.sciencedirect.com/science/article/pii/S0197458017300969?via%3Dihub [Accessed January 31, 2018].

Meda SA, Gill A, Stevens MC, Lorenzoni RP, Glahn DC, Calhoun VD, Sweeney JA, Tamminga CA, Keshavan MS, Thaker G, Pearlson GD (2012) Differences in Resting-State Functional Magnetic Resonance Imaging Functional Network Connectivity Between Schizophrenia and Psychotic Bipolar Probands and Their Unaffected First-Degree Relatives. Biol Psychiatry 71:881–889 Available at: https://www.sciencedirect.com/science/article/pii/S0006322312000935?via%3Dihub [Accessed January 30, 2018].

Minzenberg MJ et al. (2009) Meta-analysis of 41 Functional Neuroimaging Studies of Executive Function in Schizophrenia. Arch Gen Psychiatry 66:811 Available at: http://archpsyc.jamanetwork.com/article.aspx?doi=10.1001/archgenpsychiatry.2009.91 [Accessed September 25, 2017].

Ochsner KN, Silvers JA, Buhle JT (2012) Functional imaging studies of emotion regulation: a synthetic review and evolving model of the cognitive control of emotion. Ann N Y Acad Sci 1251:E1–E24 Available at: http://doi.wiley.com/10.1111/j.1749-6632.2012.06751.x.

Poppe AB, Barch DM, Carter CS, Gold JM, Ragland JD, Silverstein SM, MacDonald AW, III (2016) Reduced Frontoparietal Activity in Schizophrenia Is Linked to a Specific Deficit in Goal Maintenance: A Multisite Functional Imaging Study. Schizophr Bull 42:1149–1157 Available at: http://www.ncbi.nlm.nih.gov/pubmed/27060129 [Accessed January 31, 2018].

Price CJ, Moore CJ, Friston KJ (1997) Subtractions, conjunctions, and interactions in experimental design of activation studies. In: Human Brain Mapping, pp. 264–272.

Putcha D, Ross RS, Cronin-Golomb A, Janes AC, Stern CE (2015) Altered intrinsic functional coupling between core neurocognitive networks in Parkinson's disease. NeuroImage Clin 7:449–455 Available at: http://www.ncbi.nlm.nih.gov/pubmed/25685711 [Accessed January 31, 2018].

Putcha D, Ross RS, Cronin-Golomb A, Janes AC, Stern CE (2016) Salience and Default Mode Network Coupling Predicts Cognition in Aging and Parkinson's Disease. J Int Neuropsychol Soc 22:205–215 Available at: http://www.ncbi.nlm.nih.gov/pubmed/26888617 [Accessed January 31, 2018].

Ray KL, McKay DR, Fox PM, Riedel MC, Uecker AM, Beckmann CF, Smith SM, Fox PT, Laird AR (2013) ICA model order selection of task co-activation networks. Front Neurosci 7:237 Available at: http://www.ncbi.nlm.nih.gov/pubmed/24339802.

Riedel MC, Yanes JA, Ray KL, Eickhoff SB, Fox PT, Sutherland MT, Laird AR (2018) DIssociable meta-analytic brain networks contribute to coordinated emotional processing. Hum Brain Mapp.

Servaas MN, Geerligs L, Renken RJ, Marsman J-BC, Ormel J, Riese H, Aleman A (2015) Connectomics and neuroticism: an altered functional network organization. Neuropsychopharmacology 40:296–304 Available at: http://www.ncbi.nlm.nih.gov/pubmed/25005250 [Accessed January 30, 2018].

Smith SM, Fox PT, Miller KL, Glahn DC, Fox PM, Mackay CE, Filippini N, Watkins KE, Toro R, Laird AR, Beckmann CF (2009) Correspondence of the brain's functional architecture during activation and rest. Proc Natl Acad Sci U S A 106:13040–13045 Available at: http://www.ncbi.nlm.nih.gov/pubmed/19620724.

Solé-Padullés C, Castro-Fornieles J, de la Serna E, Romero S, Calvo A, Sánchez-Gistau V, Padrós-Fornieles M, Baeza I, Bargalló N, Frangou S, Sugranyes G (2016) Altered Cortico-Striatal Connectivity in Offspring of Schizophrenia Patients Relative to Offspring of Bipolar Patients and Controls. PLoS One 11:e0148045 Available at: http://www.ncbi.nlm.nih.gov/pubmed/26885824 [Accessed January 31, 2018].

Spitzer M (1993) The psychopathology, neuropsychology, and neurobiology of associative and working memory in schizophrenia. Eur Arch Psychiatry Clin Neurosci 243:57–70.

Thomann AK, Griebe M, Thomann PA, Hirjak D, Ebert MP, Szabo K, Reindl W, Wolf RC (2017) Intrinsic neural network dysfunction in quiescent Crohn's Disease. Sci Rep 7:11579 Available at: http://www.ncbi.nlm.nih.gov/pubmed/28912568 [Accessed January 31, 2018].

Turkeltaub PE, Eden GF, Jones KM, Zeffiro TA (2002) Meta-Analysis of the Functional Neuroanatomy of Single-Word Reading: Method and Validation. Neuroimage 16:765–780 Available at: http://www.sciencedirect.com/science/article/pii/S1053811902911316.

Turkeltaub PE, Eickhoff SB, Laird AR, Fox M, Wiener M, Fox P (2012) Minimizing within-experiment and within-group effects in activation likelihood estimation meta-analyses. Hum Brain Mapp 33:1–13.

Van Horn JD, Toga AW (2014) Human neuroimaging as a "Big Data" science. Brain Imaging Behav 8:323–331 Available at: http://www.ncbi.nlm.nih.gov/pubmed/24113873 [Accessed September 20, 2017].

Wang H, Zeng L-L, Chen Y, Yin H, Tan Q, Hu D (2015a) Evidence of a dissociation pattern in default mode subnetwork functional connectivity in schizophrenia. Sci Rep 5:14655 Available at: http://www.ncbi.nlm.nih.gov/pubmed/26419213 [Accessed January 31, 2018].

Wang L, Chou Y-H, Potter GG, Steffens DC (2015b) Altered Synchronizations among Neural Networks in Geriatric Depression. Biomed Res Int 2015:343720 Available at: http://www.ncbi.nlm.nih.gov/pubmed/26180795 [Accessed January 31, 2018].

Weinberger DR, Berman KF (1996) Prefrontal function in schizophrenia: confounds and controversies. Philos Trans R Society Biol Sci 351:1495–1503.

Worhunsky PD, Dager AD, Meda SA, Khadka S, Stevens MC, Austad CS, Raskin SA, Tennen H, Wood RM, Fallahi CR, Potenza MN, Pearlson GD (2016) A Preliminary Prospective Study of an Escalation in "Maximum Daily Drinks", Fronto-Parietal Circuitry and Impulsivity-Related Domains in Young Adult Drinkers. Neuropsychopharmacology 41:1637–1647 Available at: http://www.ncbi.nlm.nih.gov/pubmed/26514582 [Accessed January 31, 2018].

Worhunsky PD, Potenza MN, Rogers RD (2017) Alterations in functional brain networks associated with loss-chasing in gambling disorder and cocaine-use disorder. Drug Alcohol Depend 178:363–371 Available at: https://www.sciencedirect.com/science/article/pii/S0376871617302995?via%3Dihub [Accessed January 31, 2018].

Yarkoni T, Poldrack RA, Nichols TE, Van Essen DC, Wager TD (2011) Large-scale automated synthesis of human functional neuroimaging data. Nat Methods 8:665–670 Available at: https://doi.org/10.1038/nmeth.1635 [Accessed March 24, 2015].

Network Modulation in Neuropsychiatric Disorders Using the Virtual Brain

Andrea B. Protzner, Sora An, and Viktor Jirsa

Introduction

Over the last few decades, neuroimaging has been extensively applied to healthy and diseased populations and has led to advances that profoundly shape our understanding of brain function. However, much of this previous work employs mass-univariate analyses to make inferences about the effects of interest at the voxel level, where differences are identified between groups or at the aggregate level across multiple subjects. These approaches are limited, as they do not capture linked relationships across voxels, nor do they capture linked relationships across different imaging modalities. These approaches also are not suited for classifying individuals into diagnostic groups nor for providing individualized prognostic information based on neuroimaging findings. As a consequence, neuroimaging has so far had relatively little practical clinical impact in shaping diagnosis and prognosis. This poor translational history has fostered recent interest in alternate approaches that can combine information in a principled way across voxels (connectivity) as well as from multiple imaging modalities, and can be applied to predict clinically relevant outcomes, such as diagnosis or treatment response, from individual patient neuroimaging measures.

Two dominant approaches in translational neuroscience use data in very different ways. One approach uses large patient datasets to identify biologically homogeneous subtypes that parse patients within the boundaries of phenotypic diagnoses (Cuthbert and Insel 2013). Such biomarker-defined subtypes identify clinically

A. B. Protzner
Department of Psychology, University of Calgary, Calgary, AB, Canada

Hotchkiss Brain Institute, Calgary, AB, Canada

S. An · V. Jirsa (✉)
INSERM, INS, Inst Neurosci Syst, Aix Marseille University, Marseille, France
e-mail: viktor.jirsa@univamu.fr

V. A. Diwadkar, S. B. Eickhoff (eds.), *Brain Network Dysfunction in Neuropsychiatric Illness*, https://doi.org/10.1007/978-3-030-59797-9_8

meaningful differences between patients and controls or within patient populations, to facilitate a deeper understanding of the underlying brain mechanisms. This research sets the stage for neuroimaging biomarkers that separate a broad-illness phenotype into a set of treatment-relevant subgroups (Trusheim et al. 2007). The second approach is to focus on individual patient brains and create personalized large-scale network models by fusing the individual's structural data and mathematical modeling of brain activations. These models represent an in silico platform where clinicians can test and develop personalized strategies toward therapy and intervention. Although we will cover research from both of these dominant approaches, our focus in this chapter is on the latter: the use of computational modeling as a rational way to generate new insights into brain function in health and disease.

When linking brain network modeling approaches to empirical brain signals, the current state-of-the-art simulation engine for brain network modeling is arguably The Virtual Brain (TVB), an international neuroscience project (see http://www.thevirtualbrain.org) and part of the European Flagship Human Brain Project (HBP). It promotes the large-scale network perspective of brain function and dysfunction, assembling neuroinformatics tools and workflows for the generation and simulation of personalized brain networks. Jirsa et al. (2002) proposed the use of connectivity (later referred to as the connectome, Sporns et al. 2005) derived from diffusion-weighted MRI (dMRI) to constrain large-scale brain network models. An upsurge of connectome-based model development followed with applications to the resting state (Deco et al. 2009; Ghosh et al. 2008; Honey et al. 2007), aging (Nakagawa et al. 2013), and pathologies such as epilepsy (Jirsa et al. 2017), schizophrenia (Deco and Kringelbach 2014), and stroke (Falcon et al. 2015). Although some efforts exist to link physiological mechanisms to brain imaging signals (Schirner et al. 2018), generally research using TVB aims to understanding the network mechanisms of the brain's structure–function relationship as it expresses itself in brain imaging. Much of the large-scale brain network modeling has largely focused on the resting-state activity, reproducing the set of functionally active links between brain areas (functional connectivity). This approach has been hampered by the stationary nature of most connectivity-based metrics applied to validate the models (Hansen et al. 2015). More recent novel approaches explore the highly nonstationary seizure propagation patterns in epilepsy (Jirsa et al. 2017; Proix et al. 2018) or hybrid approaches driving brain models by empirical functional data (Schirner et al. 2018). The former suffers from lack of generalizability to other brain diseases, the latter from high dimensionality and practical model nonidentifiability (Raue et al. 2009). Most meaningful situations and tasks in neuroscience, however, are nonstationary processes including the resting-state dynamics, as well as cognitive and motor behaviors (Allen et al. 2012; Hansen et al. 2015). It is for this reason why stimulation paradigms and nonstationary analyses are of importance.

A successful clinical translation using TVB should be rooted in a workflow leading from the imaging of a patient's brain through the virtualization and modeling of the patient's large-scale brain network to the prediction of a therapeutic intervention. The continuity of the workflow depends on three factors, emphasizing TVB's

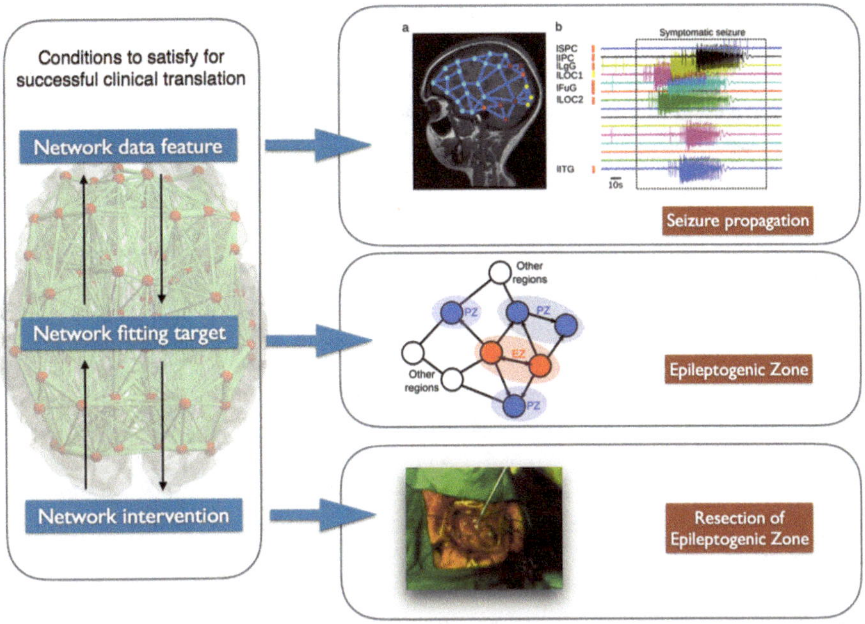

Fig. 1 Steps of successful translation of The Virtual Brain (TVB) applications into clinic. TVB requires the expression of data features in the brain signals that are informative about the brain network organization. These network mechanisms supporting these data features must be linkable to network parameters, serving as a fitting target and offering an entry point for therapeutic network modulation. For the example of epilepsy, the network data feature is the seizure propagation pattern, the network fitting target is the spatial organization of the epileptogenic zone, and the network intervention is the resection of the epileptogenic zone

network nature. These factors are illustrated in Fig. 1 and comprise (1) network data features, (2) network fitting target, and (3) network intervention. Data features extracted from the brain imaging signals should be informative regarding the spatiotemporal network organization. When fitting these data features to model parameters, the fitting target should be highly predictive of the subsequent network dynamics as expressed through the data feature. Finally, the data fitting target should be linkable to a therapeutic intervention in order to fully translate the approach into clinic. If only the first two criteria are satisfied, then the workflow can still generate biomarkers useful for patient stratification. The example of epilepsy illustrates the full workflow in Fig. 1. Here, the network data feature is the seizure propagation pattern, more precisely the envelope of a high-frequency band and its spatiotemporal pattern. The fitting target is the epileptogenic zone, as expressed as the epileptogenicity distributed over the brain regions in the network. Finally, the intervention is the resection of the brain regions with the highest epileptogenicity.

In the remainder of this chapter, we first provide a brief overview of TVB. We explain how the model links large-scale (macroscopic) brain dynamics with biophysical parameters at the meso- and microscopic levels, and we describe our inclusion of brain noise as being crucial for accurately modeling brain function. We then highlight how individual differences in neuroimaging measures can be used to characterize disease mechanisms and to predict the outcome of therapies and treatment approaches in psychiatric diseases. Lastly, we describe our first proof of concept for the successful application of TVB to deep brain stimulation for treatment-resistant depression (TRD).

The Virtual Brain (TVB)

TVB is a free open-source neuroinformatics tool. The first prototype was launched in October 2012 and has been curated and regularly updated since then. It provides the possibility to feed computational neuronal network models with information about structural and functional imaging data including population (stereotactic electroencephalography [sEEG], EEG, magnetoencephalography [MEG]) activity, spatially highly resolved whole-brain metabolic/vascular signals (functional magnetic resonance imaging [fMRI]), global measures of neuronal connections (diffusion tensor imaging [DTI]), and structure, from T1-weighted images. TVB is network node model agnostic and offers a wide range of neural population models to be used as network nodes. The software infrastructure of TVB is composed of a functional core running the large-scale brain simulations independently or in batch mode, a web-based interface to access the simulator, and a command-line interface to develop more extensive applications. All simulations may be performed on workstations and laptops, as well as on high-performance clusters. Manipulations of network parameters within TVB allow researchers and clinicians to test the effects of experimental paradigms, interventions (such as stimulation and surgery), and therapeutic strategies (such as pharmaceutical interventions targeting local regions). The computational environment allows the user to visualize the simulated data in 2D and 3D and perform data analyses in the same way as commonly performed with empirical data. Figure 2 shows a cartoon of a typical workflow using The Virtual Brain.

TVB enables the linkage of large-scale (macroscopic) brain dynamics with biophysical parameters at the meso- and microscopic levels. Mesoscopic dynamics describe the mean field activity of populations of neurons organized as cortical columns or subcortical nuclei (Fig. 3). Common assumptions in mean-field modeling are that explicit structural features or temporal details of neuronal networks (e.g., spiking dynamics of single neurons) are irrelevant for the analysis of complex mesoscopic dynamics, and the emergent collective behavior is only weakly sensitive to the details of individual neuron behavior (Breakspear and Jirsa 2007). Basic mean field models capture changes of the mean firing rate (Brunel and Wang 2003; Zerlaut et al. 2018), whereas other mean field models account for parameter dispersion in the neurons and the subsequent richer behavior of the mean field dynamics (Assisi et al. 2005; Stefanescu and Jirsa 2008, 2011; Jirsa and Stefanescu 2011). These approaches demonstrate the relatively new concept from statistical physics that macroscopic physical systems obey laws that are independent of the details of

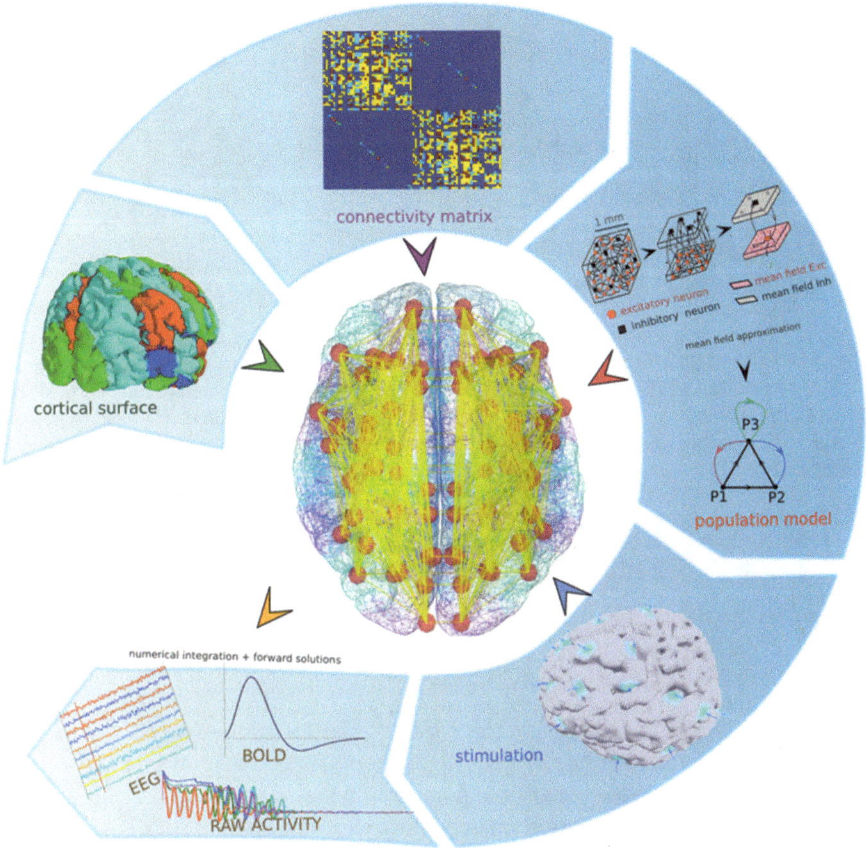

Fig. 2 Workflow in The Virtual Brain. (Clockwise) A template for the cortical surface defines three-dimensional geometry and parcellation. Methods of dMRI allow reconstructing connections and tract lengths of fiber bundles to define a connectivity matrix and time delays via signal transmission. Neural population models serve as network nodes generating dynamic signals due to their intrinsic dynamics and interactions with other nodes. Interventions such as lesioning of the connectivity or stimulating brain regions offer means to probe the brain dynamics in TVB. The resulting brain network dynamics is registered as raw local field potentials (LFPs), bold fMRI, EEG, and MEG signals

the microscopic constituents they are built of (Haken 1975). In TVB, our main interest lies in deriving the mesoscopic laws that drive the observed dynamical processes at the macroscopic large brain scale in a systematic manner.

When traversing the scale to the large-scale network, then each network node is governed by its own intrinsic dynamics in interaction with the dynamics of all other network nodes. This interaction happens through the connectivity matrix via specific connection weights and time delays due to signal transmission delays. The following (generic) evolution equation (Jirsa 2009) captures all the above features and underlies the emergence of the spatiotemporal network dynamics in TVB:

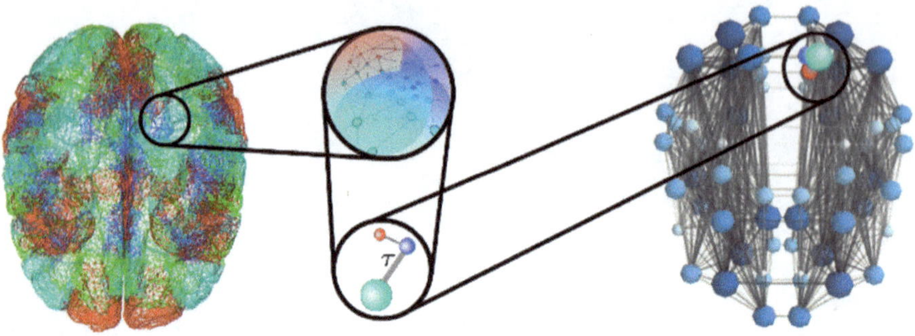

Fig. 3 Two forms of connectivity are implemented in large-scale brain models. Local connectivity (left) between near neighbors generates a dense mesh tracing out the geometry of the cortical surface in three dimensions. Global connectivity (right) representing the human connectome defines a large-scale connectivity skeleton. In TVB, the two connectivity patterns are simultaneously present (middle)

$$\psi(x,t) = N(\psi(x,t)) + \int_{\Gamma} g_{\text{local}}(x,x')\,S(\psi(x',t))\,dx'$$
$$+ \int_{\Gamma} g_{\text{local}}(x,x')\,S\left(\psi\left(x',t - \frac{|x-x'|}{v}\right)\right)dx' + I(x,t) + \xi(x,t)$$

The equation describes the stochastic differential equation of a network of connected neural populations. $\psi(x,t)$ is the neural population activity vector at the location x in 3D physical space and time point t. It has as many state variables as are defined by the neural population model, which is specified by $N(\psi(x,t))$. The connectivity distinguishes local and global connections, which are captured separately in two expressions. The local network connectivity $g_{\text{local}}(x,x')$ is described by connection weights between x and x', whereas global connectivity is defined by $g_{\text{global}}(x,x')$. The critical difference between the two types of connectivity is threefold: 1. Local connectivity is short range (on the order of 0.1 cm) and global connectivity is long range (on the order of 10 cm). 2. Signal transmission via local connections is instantaneous, but via global connections undergoes a time delay dependent on the distance $|x - x'|$ and the transmission speed v. 3. Local connectivity is typically spatially invariant (of course with variations from area to area, but generally it falls off with distance), and global connectivity is highly heterogeneous.

Stimuli of any form, in the context of normal brain function (e.g., perceptual, cognitive, or behavioral perturbations) and therapeutic intervention or brain damage (e.g., lesions, stimulation), are introduced into the Virtual Brain via the expression $I(x,t)$ and are defined over a location x with a particular time course.

Noise plays a crucial role for the brain dynamics and hence for brain function (see McIntosh et al. 2010). In TVB, it is introduced via the expression $\xi(x,t)$ where the type of noise and its spatial and temporal correlations can be specified independently.

Various numerical algorithms are available in TVB and can be coarsely categorized into deterministic (no noise) and stochastic (with noise). They include the Heun algorithm, Runge-Kutta of various orders, Euler Maruyama, and others. TVB performs mechanistic model simulations at every network node. To link to human brain imaging data, biologically realistic forward solutions are computed at the desired sensor locations. For scalp surface data (EEG, MEG), realistic head models are implemented, as well as the approximate spherical shell solutions. sEEG is directly computed from the Maxwell equations. fMRI signals are computed using the Balloon–Windkessel model.

The Influence of Individual Variability on the Outcome of Therapies and Treatment Approaches in Psychiatric Diseases

Neuropsychiatric disorders are devastating not only to individuals but are also a growing health burden for society. There exist challenges with reliable differential diagnosis and with predicting clinical trajectories and treatment success in individual patients (Kapur et al. 2012; Krystal and State 2014; Stephan et al. 2016). Disease classification is based on symptoms and signs, which are associated with varied causes and mechanisms across patients. The clinical categories proposed by these classifications, like depression or schizophrenia, do not define clinical trajectories and do not predict treatment efficacy for individual patients (Cuthbert and Insel 2013; Kapur et al. 2012). Without tests to guide personalized treatment choices, physicians select therapies based on trial-and-error and generally introduce several treatment adjustments before identifying an effective option.

To make progress in nosology and diagnostics, we need a better understanding of how normal brain function becomes unbalanced in neuropsychiatric disorders (Deco and Kringelbach 2014), and how these imbalances can be righted through different treatment options. We examine these concepts in the context of major depressive disorder (MDD), which is the most prevalent of all mental illnesses and is the leading source of disability and disease burden in the Western world (Kessler et al. 2003; WHO 2008). Clinically, MDD is a multidimensional syndrome, involving disruption of mood, cognition, sensorimotor functions, sleep, appetite, and libido. For those affected, it reduces quality of life, prevents social integration, and leads to suicide. MDD can be treated with medication, psychotherapy, or somatic treatments including electroconvulsive therapy (ECT), transcranial magnetic stimulation (TMS), and vagus nerve stimulation (VNS). However, treatment challenges are aptly illustrated by a study examining acute and long-term treatment outcomes in a sample of over 3500 patients with MDD, where even after four successive treatment adjustments, the cumulative remission rate was 67% (Rush et al. 2006). Those who remain severely depressed despite adequate, often aggressive, treatment are classified as having TRD (Holtzheimer and Mayberg 2011). Patients with TRD are underemployed and have frequent relapses and hospital admissions, placing a huge burden on the healthcare system and the economy (Fekadu et al. 2009; Crown et al. 2002). At present, there are no proven treatment options available for such intractable depression (Kennedy et al. 2009).

Neurostimulation therapies, such as TMS and deep brain stimulation (DBS), have taken-on an increasingly prominent role in the treatment of TRD patients (Mayberg et al. 2005; Holtzheimer et al. 2012; Lozano et al. 2012; Puigdemont et al. 2011; Ramasubbu et al. 2013). Critical to the development of these therapies has been the evolving understanding of the brain networks that mediate normal and abnormal mood states and the systematic characterization of changes in these networks that accompany successful treatment response. Much of this work is based on well-established treatments, including medication and psychotherapy (Mayberg 2009). Studies using diverse neuroimaging techniques such as positron emission tomography (PET; McGrath et al. 2014), fMRI (Fox et al. 2012), DTI (Riva-Posse et al. 2014; Lujan et al. 2012), and source-localized EEG power spectra (Saletu et al. 2010; Korb et al. 2012) have defined a putative mood regulation network mediating active symptoms and treatment effects. The most reproducible findings suggest that depression reflects dysregulation between hyperactive ventral paralimbic regions (bottom-up reactive regions, primarily medial prefrontal and subcallosal cingulate) and hypoactive cortical regions (top-down control regions, mainly dorsolateral frontal and dorsal anterior cingulate/dorsomedial prefrontal) (Mayberg 1997; Giacobbe et al. 2009). Importantly, abnormalities normalize with effective treatment (Mayberg et al. 2005; Kennedy et al. 2001; Quraan et al. 2014) and baseline differences have been shown to be predictive of treatment success (Jaworska and Protzner 2013; Jaworska et al. 2017; Mayberg 2003; Seminowicz et al. 2004).

Abnormal connectivity patterns in MDD are being leveraged to identify novel subtypes of depression with stronger neurobiological correlates that predict treatment responsiveness. Drysdale et al. (2017) developed a method for defining depression subtypes by clustering subjects according to distinct, whole-brain patterns of abnormal functional connectivity in resting-state networks, unbiased by assumptions about the involvement of particular brain regions, and tested it in a large, multisite data set. Their analyses revealed four biotypes that were defined by homogeneous patterns of dysfunctional connectivity in frontostriatal and limbic networks, and that could be diagnosed with high sensitivity and specificity in individual subjects. Importantly, these biotypes were also prognostically informative, predicting which patients responded to repetitive TMS. This research is an important step in identifying neuroimaging biomarkers that separate a broad-illness phenotype into a treatment-relevant subgroups (Trusheim et al. 2007).

These large datasets are not available for the most invasive stimulation strategy, DBS, which has revolutionized the treatment of several neurological conditions over the past 20 years. Based on extensive clinical investigations, DBS has become an established modality of treatment for advanced medication intractable movement disorders, like Parkinson's disease, essential tremor, and dystonia. DBS also is under investigation for conditions such as intractable epilepsy, and neurobehavioral and psychiatric disorders such as addiction, Alzheimer's disease, obesity, traumatic brain injury, obsessive compulsive disorders, and TRD. In all of these cases, the exact mechanism of action is the subject of ongoing investigation.

For TRD, the subcallosal cingulate is commonly used for DBS targeting, because it has monosynaptic connections with many regions involved in the mood regulation network and the antidepressant response (Kennedy et al. 2009; Riva-Posse et al. 2014; Mayberg 1997; Freedman et al. 2000). Several open-label studies of subcallosal cingulate DBS showed benefit and safety (Mayberg et al. 2005; Holtzheimer et al. 2012; Lozano et al. 2012; Puigdemont et al. 2011; Ramasubbu et al. 2013). However, only 40–60% of patients responded. Unfortunately, in the push to achieve FDA approval, an industry sponsored large randomized double-blind sham-controlled trial failed to show efficacy (Holtzheimer et al. 2017). Despite these setbacks, DBS patients continue to improve over time (Bari et al. 2018; Riva-Posse et al. 2014), suggesting that stimulation effects are gradual, and were likely missed by the industry study design. Given mixed results regarding outcome and the invasive nature of DBS, it is critical to identify predictors and mechanisms of treatment response and nonresponse.

To our knowledge, there are two papers that use computational modeling to inform mechanisms of action and treatment choices for individual patients in the context of DBS. Riva-Posse et al. (2014) used activation volume tractography modeling to identify patient-specific DBS target locations within the subcallosal cingulate and stimulation parameter settings that would activate critical tracts that mediate subcallosal cingulate DBS antidepressant response. This modeling technique is useful for identifying optimal anatomical location and stimulation parameters on an individual patient level. It does not, however, examine the influence of these parameters on large-scale network function. Ramirez-Mahaluf et al. (2017) created a parsimonious biophysical computational model of MDD network function, in part to examine downstream effects of changes in ventral anterior cingulate cortex activation. They modeled activity and connectivity in two regions, ventral anterior cingulate cortex and dorsolateral prefrontal cortex, which are critical for switching between emotional and cognitive processing. To mimic MDD-associated changes in this network, they slowed glutamate decay in ventral anterior cingulate cortex and generated sustained ventral anterior cingulate cortex activation. This ventral anterior cingulate cortex hyperactivity interfered with typical dorsolateral prefrontal cortex responses to cognitive signals, and thus simulated the neurocognitive dysfunction seen in MDD. Simulation of clinical treatments (SSRI and deep brain stimulation) counteracted aberrant ventral anterior cingulate cortex activity. The model shows how glutamate dysregulation can cause aberrant brain dynamics, and how treatments that alter ventral anterior cingulate cortex function can lead to downstream effects on the dorsolateral prefrontal cortex, and normalization in the affected network. In this case, a parsimonious model was chosen to explain several features of the pathophysiology and behavior associated with MDD and its treatment. However, these more simple models, comprised of the minimum number of brain regions that are necessary to simulate a given condition, may miss the contributions and constraints and context that the "rest of the brain" imposes on the simulated treatment and the sought-after mechanisms.

The Promise of a TVB Application to MDD

As neuropsychiatric disorders involve a large network disorganization, TVB bears the promise to be able to address some of the network reorganization following stimulation. Here, we lay out the technical steps involved and provide first proof of concept for the successful application of TVB in treatment-resistant depression. Our network simulations reproduce patient-specific brain activation patterns in response to stimulation at particular anatomical locations and stimulation parameters. In other words, our feasibility protocol models the stimulation-evoked potential, which is time locked to the stimulation pulse, as a method to observe patient-specific patterns of signal propagation from subcallosal cingulate DBS targets (Waters et al. 2018). The stimulation locally activates a specific regions within the subcallosal cingulate and is followed by propagation and sequential recruitment through the individual brain connectome, thereby generating a unique spatiotemporal response pattern.

As a first step, the cortical/subcortical surface and the structural connectivity are reconstructed from diffusion MRI scans and T1-weighted images using SCRIPTS pipeline (Proix et al. 2016, 2017; Jirsa et al. 2017). Each patient's brain is divided into 164 regions including 148 cortical regions (based on the Destrieux atlas, Destrieux et al. 2010) and 16 subcortical regions, and mean anatomical positions of nerve fibers connecting different brain regions (fiber tracts) are derived. Through the co-registration with T1-weighted images and CT scans, patient-specific electrode contact location is also calculated. Based on the structural brain reconstruction, individual brain network model is built. The reconstructed brain surface is composed of triangular meshes, and each vertex (brain node) is defined by neural mass model replicating dynamics of neuronal population (Here, the number of vertices is about 8000–10,000 depending on each patient's brain surface area). Each brain node interacts with other brain nodes according to homogeneous connectivity (intraregional connections) and heterogeneous connectivity (interregional connections) (Spiegler et al. 2016).

The finite difference method (or the finite element method) can be used to calculate voltage distribution within a particular boundary according to stimulation voltage and active contact position(s). According to the result, voltage distribution applied to each fiber tract (along the anatomical structure) is obtained, which allows to select the fiber tracts to be activated. Then, based on the location of the peak voltage in the selected fiber tract, stimulation is applied to the brain regions at both ends considering time delays.

Figure 4 illustrates the spatiotemporal response pattern when stimulation (single pulse) is applied to the optimal electrode contacts. It represents the normalized energy distribution over time, which shows that the stimulation activates the internal

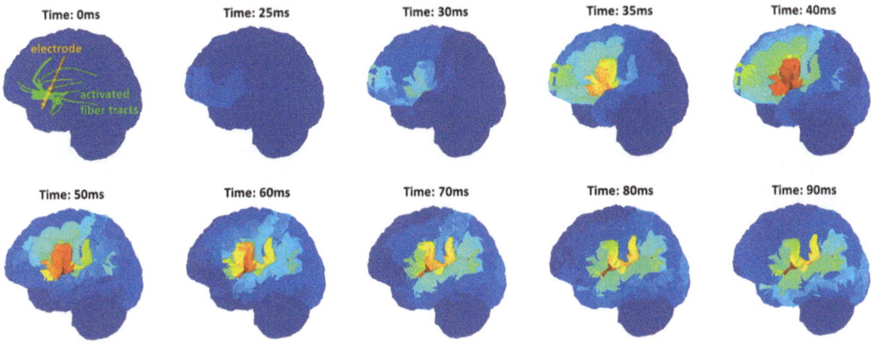

Fig. 4 Spatiotemporal brain activation pattern to subcallosal cingulate DBS. The green lines at 0 ms show the locations of the fiber tracts activated by stimulation. When a single pulse of 8 V is applied to the optimal electrode contacts, the internal structures including the subcallosal cingulate cortex are initially activated, and then prefrontal regions are recruited. After the entire frontal region is activated, the induced energy moves progressively to temporal and parietal regions

structures including the subcallosal cingulate cortex first, then the activities propagate to prefrontal regions, entire frontal regions, and temporal and parietal regions gradually. These response patterns can be simulated according to active contact position(s) and stimulation parameters, and the results can identify the distinct patterns each other. Moreover, the brain activation pattern depends not only on the anatomical location of the fiber tracts and electrode contacts but also on the individual brain connectome, so that a unique pattern can be derived for each patient.

These network simulation results can be validated by comparison with empirical data recorded from patients. In a recent paper, Waters and colleagues analyzed the propagation characteristics of event-related potentials (ERPs) due to electrical stimulation, based on recorded EEG data while applying a 2 Hz stimulation to a specific position (Waters et al. 2018). Integrative approaches involving these empirical studies are required to increase the reliability of the individual brain network model. The validated brain model has the potential to be used as a simulation tool to predict and optimize the DBS treatment effect in a patient-specific manner.

Conclusions

This chapter provides an overview of the use of computational modeling to interrogate neuroimaging data with the purpose of characterizing disease mechanisms and linking these mechanisms to treatment outcome. Our approach involves creating personalized brain models with TVB to support clinical decision-making in identifying interventions that not only alter the underlying disease state but prevent the brain from entering or remaining in that state at inappropriate times. Success will depend on close collaborations between clinical and computational scientists carrying out carefully designed prospective validation studies.

Acknowledgments We acknowledge support to ABP by the Canadian Institutes of Health Research (CIHR; grant number PJT-148762) and Natural Sciences and Engineering Council of Canada (NSERC; grant number 418454-2013), to SA by the SATT Sud-Est (TVB-Epilepsy), and to VJ by Fondation pour la Recherche Médicale (FRM) (grant number DIC20161236442), European Commission's Human Brain Project (grant agreement H2020-720270) and the FHU EPINEXT with the support of the A*MIDEX project (ANR-11-IDEX-0001-02) from the "Investissements d'Avenir" French Government program managed by the French National Research Agency (ANR).

References

Allen M, Dietz M, Blair KS, van Beek M, Rees G, Vestergaard-Poulsen P, et al. Cognitive-Affective Neural Plasticity following Active-Controlled Mindfulness Intervention, *J Neurosci.* 32[44], 15601-15610. 2012.

Assisi CG, Jirsa VK, Kelso JAS. Synchrony and clustering in heterogeneous networks with global coupling and parameter dispersion. *Phys. Rev. Lett.* 94, 018106. 2005.

Bari AA, Mikell CB, Abosch A, Ben-Haim S, Buchanan RJ, Burton AW, et al. Charting the road forward in psychiatric neurosurgery: proceedings of the 2016 American Society for Stereotactic and Functional Neurosurgery workshop on neuromodulation for psychiatric disorders. *J. Neurol. Neurosurg. Psychiatry.* 89[8]. 2018.

Breakspear M, Jirsa VK. Neuronal Dynamics and Brain Connectivity. In: Jirsa VK, McIntosh A (eds) Handbook of Brain Connectivity. Understanding Complex Systems. Springer, Berlin, Heidelberg. 2007.

Brunel N, Wang XJ. What determines the frequency of fast network oscillations with irregular neural discharges? I. synaptic dynamics and excitation-inhibition balance. *J Neurophysiol.* 90, 415-430. 2003.

Crown WH, Finkelstein S, Berndt ER, Ling D, Poret AW, Rush AJ, et al. The impact of treatment-resistant depression on health care utilization and costs. *J. Clin. Psychiatry.* 63[11], 963-971. 2002.

Cuthbert BN, Insel TR. Toward the future of psychiatric diagnosis: The seven pillars of RDoC. *BMC Medicine*. 11, 126. 2013.

Deco G, Jirsa V, McIntosh AR, Sporns O, Kötter R. Key role of coupling, delay, and noise in resting brain fluctuations. *Proc Natl Acad Sci USA*. 106[25], 10302-10307. 2009.

Deco G, Kringelbach ML. Great Expectations: Using whole-brain computational connectomics for understanding neuropsychiatric disorders. *Neuron.* 84[5], 892-905. 2014.

Destrieux C, Fischl B, Dale A, Halgren E. Automatic parcellation of human cortical gyri and sulci using standard anatomical nomenclature. *Neuroimage.* 53[1], 1-15. 2010.

Drysdale AT, Grosenick L, Downar J, Dunlop K, Mansouri F, Meng Y, et al. Resting-state connectivity biomarkers define neurophysiological subtypes of depression. *Nat. Med.* 23, 28-38. 2017.

Falcon MI, Riley JD, Jirsa V, McIntosh AR, Shereen AD, Chen EE, et al. The Virtual Brain: modeling biological correlates of recovery after chronic stroke. *Front Neurol.* 6[288], 1-13. 2015

Fekadu A, Wooderson SC, Markopoulo K, Donaldson C, Papadopoulos A, Cleare AJ. What happens to patients with treatment-resistant depression? A systematic review of medium to long term outcome studies. *J. Affect. Disord.* 116[1–2], 4-11. 2009.

Fox MD, Liu H, Pascual-Leone A. Identification of reproducible individualized targets for treatment of depression with TMS based on intrinsic connectivity. *Neuroimage*. 66, 151-160. 2012.

Freedman LJ, Insel TR, Smith Y. Subcortical projections of area 25 (subgenual cortex) of the macaque monkey. *J. Comp. Neurol.* 421, 172-188. 2000.

Ghosh A., Rho Y., McIntosh AR, Kötter R, Jirsa VK. Cortical network dynamics with time delays reveals functional connectivity in the resting brain. *Cogn Neurodyn.* 2, 115-120. 2008.

Giacobbe P, Mayberg HS, Lozano AM. Treatment resistant depression as a failure of brain homeostatic mechanisms: implications for deep brain stimulation. *Exp Neurol.* 219, 44-52. 2009.

Haken H. Statistical physics of bifurcation, spatial structures, and fluctuations of chemical reactions. *Zeitschrift für Physik B Condensed Matter*, 20[4], 413-420. 1975.

Hansen ECA, Battaglia D, Spiegler A, Deco G, Jirsa VK. Functional connectivity dynamics: Modeling the switching behavior of the resting state. *Neuroimage.* 105, 525-535. 2015.

Holtzheimer PE, Mayberg HS. Stuck in a rut: rethinking depression and its treatment. *Trends Neurosci.* 34[1], 1-9. 2011.

Holtzheimer PE, Kelley ME, Gross RE, Filkowski MM, Garlow SJ, Barrocas A, et al. Subcallosal cingulate deep brain stimulation for treatment-resistant unipolar and bipolar depression. *Arch. Gen. Psychiatry.* 69[2], 150-158. 2012.

Holtzheimer PE, Husain MM, Lisanby SH, Taylor SF, Whitworth LA, McClintock S, et al. Subcallosal cingulate deep brain stimulation for treatment-resistant depression: a multisite, randomised, sham-controlled trial. *The Lancet Psychiatry* 4[11], 839-849. 2017.

Honey CJ, Kötter R, Breakspear M, Sporns O. Network structure of cerebral cortex shapes functional connectivity on multiple time scales. *Proc Natl Acad Sci USA.* 104[24], 10240-10245. 2007.

Jaworska N, Protzner A. Electrocortical features of depression and their utility in assessing antidepressant treatment outcome. *Can. J. Psychiatry.* 58[9], 509–514. 2013.

Jaworska N, Wang H, Smith DM, Blier P, Knott V, Protzner AB. Pre-treatment EEG signal variability is associated with treatment success in depression. *Neuroimage Clin.* 17, 368-377. 2017.

Jirsa VK, Jantzen KJ, Fuchs A, Kelso JAS. Spatiotemporal forward solution of the EEG and MEG using network modeling. *IEEE Trans Med Imaging.* 21[5], 493-504. 2002.

Jirsa VK. Neural field dynamics with local and global connectivity and time delay. *Philos Trans A Math Phys Eng Sci.* 367[1891], 1131-1143. 2009.

Jirsa VK, Stefanescu RA. Neural population modes capture biologically realistic large scale network dynamics. *Bull Math Biol.* 73[2], 325-343. 2011.

Jirsa VK, Proix T, Perdikis D, Woodman MM, Wang H, Gonzalez-Martinez J, et al. The virtual epileptic patient: individualized whole-brain models of epilepsy spread. *Neuroimage.* 145, 377-388. 2017.

Kapur S, Phillips AG, Insel TR. Why has it taken so long for biological psychiatry to develop clinical tests and what to do about it? *Mol. Psychiatry.* 17, 1174–1179. 2012.

Kennedy SH, Evans KR, Kruger S, Mayberg HS, Meyer JH, McCann S, et al. Changes in regional brain glucose metabolism measured with positron emission tomography after paroxetine treatment of major depression. *Am. J. Psychiatry.* 158[6], 899-905. 2001.

Kennedy SH, Milev R, Giacobbe P, Ramasubbu R, Lam RW, Parikh SV, et al. Canadian Network for Mood and Anxiety Treatments (CANMAT) Clinical guidelines for the management of major depressive disorder in adults. IV. Neurostimulation therapies. *J. Affect. Disord.* 117[Suppl 1], S44-S53. 2009.

Kessler RC, Berglund P, Demler O, Jin R, Koretz D, Merikangas KR, et al. The epidemiology of major depressive disorder: results from the National Comorbidity Survey Replication (NCS-R). *JAMA.* 289[23], 3095-3105. 2003.

Korb AS, Hunter AM, Cook IA, Leuchter AF. Rostral anterior cingulate cortex theta current density and response to antidepressants and placebo in major depression. *Clin Neurophysiol.* 120, 1313–1319. 2012.

Krystal JH, State MW. Psychiatric disorders: Diagnosis to therapy. *Cell.* 157, 201–214. 2014.

Lozano AM, Giacobbe P, Hamani C, Rizvi SJ, Kennedy SH, Kolivakis TT, et al. A multicenter pilot study of subcallosal cingulate area deep brain stimulation for treatment-resistant depression. *J. Neurosurg.* 116[2], 315-322. 2012.

Lujan JL, Chaturvedi A, Malone DA, Rezai AR, Machado AG, McIntyre CC. Axonal pathways linked to therapeutic and nontherapeutic outcomes during psychiatric deep brain stimulation. *Hum. Brain Mapp.* 33, 958–968. 2012.

Mayberg HS. Limbic-cortical dysregulation: A proposed model of depression. *J. Neuropsychiatry Clin. Neurosci.* 9, 471-481. 1997.

Mayberg HS. Modulating dysfunctional limbic-cortical circuits in depression: towards development of brain-based algorithms for diagnosis and optimised treatment. *Br. Med. Bull.* 65, 193-207. 2003.

Mayberg HS, Lozano AM, Voon V, McNeely HE, Seminowicz D, Hamani C, et al. Deep brain stimulation for treatment-resistant depression. *Neuron* 45[5], 651-660. 2005.

Mayberg HS. Targeted electrode-based modulation of neural circuits for depression. *J. Clin. Invest.* 119, 717-725. 2009.

McGrath CL, Kelley ME, Dunlop BW, Holtzheimer PE, Craighead WE, Mayberg HS. Pretreatment brain states identify likely nonresponse to standard treatments for depression. *Biol. Psychiatry.* 76, 527-535. 2014.

McIntosh AR, Kovacevic N, Lippe S, Garrett D, Grady C, Jirsa V. The development of a noisy brain. *Archives Italiennes de Biologie.* 148, 323-337. 2010.

Nakagawa TT, Jirsa VK, Spiegler A, McIntosh AR, Deco G. Bottom up modeling of the connectome: Linking structure and function in the resting brain and their changes in aging. *NeuroImage.* 80, 318-329. 2013.

Proix T, Spiegler A, Schirner M, Rothmeier S, Ritter P, Jirsa VK. How do parcellation size and short-range connectivity affect dynamics in large-scale brain network models? *Neuroimage.* 142,135-149. 2016.

Proix T, Bartolomei F, Guye M, Jirsa VK. Individual brain structure and modelling predict seizure propagation. *Brain.* 140[3], 641-54. 2017.

Proix T, Jirsa VK, Bartolomei F, Guye M, Truccolo W. Predicting the spatiotemporal diversity of seizure propagation and termination in human focal epilepsy. *Nat. Commun.* 9, 1088. 2018.

Puigdemont D, Perez-Egea R, Portella MJ, Molet J, de Diego-Adelino J, Gironell A, et al. Deep brain stimulation of the subcallosal cingulate gyrus: further evidence in treatment-resistant major depression. *Int. J. Neuropsychopharmacol.* 15[1], 1-13. 2011.

Quraan M, Protzner AB, Daskalakis ZJ, Giacobbe P, Wei Tang C, Kennedy SH, et al. Power asymmetry and functional connectivity as a marker of treatment effectiveness in DBS surgery for depression. *Neuropsychopharmacology,* 39[5], 1270-1281. 2014.

Ramasubbu R, Anderson S, Haffenden A, Chavda S, Kiss ZH. Double-blind optimization of subcallosal cingulate deep brain stimulation for treatment-resistant depression: a pilot study. *J. Psychiatry Neurosci.* 38[5], 325-332. 2013.

Ramirez-Mahaluf JP, Roxin A, Mayberg HS, Compte A. A Computational Model of Major Depression: the Role of Glutamate Dysfunction on Cingulo-Frontal Network Dynamics. *Cereb Cortex.* 27[1], 660-679. 2017.

Raue A, Kreutz C, Maiwald T, Bachmann J, Schilling M, Klingmüller U, et al. Structural and practical identifiability analysis of partially observed dynamical models by exploiting the profile likelihood. *Bioinformatics.* 25[25], 1923-1929. 2009.

Riva-Posse P, Choi KS, Holtzheimer PE, McIntyre CC, Gross RE, Chaturvedi A, et al. Defining critical white matter pathways mediating successful subcallosal cingulate deep brain stimulation for treatment-resistant depression. *Biol. Psychiatry.* 76, 963-969. 2014.

Rush A, Trivedi M, Wisniewski S, Nierenberg A, Stewart J, Warden D, et al. Acute and longer-term outcomes in depressed outpatients requiring one or several treatment steps: A STAR*D report. *Am J Psychiatry.* 163, 1905-1917. 2006.

Saletu B, Anderer P, Saletu-Zyhlarz GM. EEG topography and tomography (LORETA) in diagnosis and pharmacotherapy of depression. *Clin. EEG Neurosci.* 41, 203-210. 2010.

Schirner M, McIntosh AR, Jirsa V, Deco G, Ritter P. Inferring multi-scale neural mechanisms with brain network modelling. *Elife.* 7. 2018.

Seminowicz DA, Mayberg HS, McIntosh AR, Goldapple K, Kennedy S, Segal Z, et al. Limbic-frontal circuitry in major depression: a path modeling metaanalysis. *Neuroimage.* 22[1], 409-418. 2004.

Spiegler A, Hansen EC, Bernard C, McIntosh AR, Jirsa VK. Selective activation of resting-state networks following focal stimulation in a connectome-based network model of the human brain. *eNeuro.* 3[5], ENEURO. 0068-16.2016. 2016.

Sporns O, Tononi G, Kötter R. The human connectome: a structural description of the human brain. *PLoS Comput Biol.* 1[4], e42. 2005.

Stefanescu RA, Jirsa VK. A low dimensional description of globally coupled heterogeneous neural networks of excitatory and inhibitory neurons. *PLoS Comput Biol.* 4[11], e1000219. 2008.

Stefanescu RA, Jirsa VK. Reduced representations of heterogeneous mixed neural networks with synaptic coupling. *Phys. Rev. E.* 83, 026204. 2011.

Stephan KE, Bach DR, Fletcher PC, Flint J, Frank MJ, Friston KJ, et al. Charting the landscape of priority problems in psychiatry, part 1: Classification and diagnosis. *Lancet Psychiatry.* 3, 77-83. 2016

Trusheim MR, Berndt ER, Douglas FL. Stratified medicine: strategic and economic implications of combining drugs and clinical biomarkers. *Nat. Rev. Drug Discov.* 6, 287-293. 2007.

Waters AC, Veerakumar A, Choi KS, Howell B, Tiruvadi V, Bijanki KR, et al. Test–retest reliability of a stimulation-locked evoked response to deep brain stimulation in subcallosal cingulate for treatment resistant depression. *Human Brain Mapping.* 39[12], 4844-4856. 2018.

World Health Organization, 2008.

Zerlaut Y, Chemla S, Chavane F, Destexhe A. Modeling mesoscopic cortical dynamics using a mean-field model of conductance-based networks of adaptive exponential integrate-and-fire neurons. *J Comput Neurosci.* 44[1], 45-61. 2018.

Part II
Applications

Networks-Mediated Spreading of Pathology in Neurodegenerative Diseases

Yasser Iturria-Medina and Alan C. Evans

Introduction

The last two decades have seen a tremendous growth in multimodal studies of brain connectivity or "connectomics." This mapping of brain networks initially focused on direct anatomical connections (Hagmann 2005; Sporns et al. 2005) but rapidly extended to include other modalities, such as functional, metabolic, and even vascular connectivity (Leergaard et al. 2012; Melie-García et al. 2013; Stam et al. 2009; Zuo et al. 2012). It also extended to the study of microscopic and mesoscopic brain subnetworks (Leergaard et al. 2012; Oh et al. 2014; Shimono and Beggs 2018). These studies have highlighted the importance of characterizing the brain's complex connectional architecture for understanding the principles underlying cognitive and behavioral function.

In addition to its significant contribution to our understanding of brain organization and information flow (Bullmore and Sporns 2009; Hagmann et al. 2010; Iturria-Medina et al. 2011; Sporns 2013), connectomics has also been helpful for clinical research on neurological and neuropsychiatric disorders. Brain disconnection and hyperconnection have now been reported for many diseases (Pievani et al. 2014; Rubinov and Bullmore 2013). The term "pathoconnectomics" has been coined for the study of abnormal brain connections and is associated with the postulate that major neuropsychiatric disorders, rather than be a consequence of some focal lesion, are abnormalities of brain networks (Rubinov and Bullmore 2013). Indeed, the fact that altered connectional patterns are significantly associated with cognitive and behavioral impairment (e.g., depression, schizophrenia) is not surprising. Any physical, biological, or social network with an abnormally wired substrate is likely to

Y. Iturria-Medina (✉) · A. C. Evans (✉)
McConnell Brain Imaging Center, Montreal Neurological Institute, Montreal, QC, Canada

Ludmer Centre for NeuroInformatics and Mental Health, Montreal, QC, Canada
e-mail: yasser.iturriamedina@mcgill.ca; alan.evans@mcgill.ca

V. A. Diwadkar, S. B. Eickhoff (eds.), *Brain Network Dysfunction in Neuropsychiatric Illness*, https://doi.org/10.1007/978-3-030-59797-9_9

develop abnormal function. This may well explain the great expectations that the scientific community has on pathoconnectomic analyses (Deco and Kringelbach 2014).

However, the connectomics of neurological and neuropsychiatric disorders is not limited to those cases where the brain networks are physically altered. A growing body of evidence supports the view that brain connections, independent of their anatomical and functional integrity, may also be a critical conduit for the spreading of pathological factors that cause disease (Iturria-Medina and Evans 2015). Consistent with this latter view, we provide a brief overview of the biological mechanisms implicated in both (i) intra-brain disease factor spreading and connectional dysfunction effects on neurodegenerative progression and (ii) how brain computational modeling can contribute to understand further and potentially control such pathological processes. The chapter is organized in three sections. The first offers a brief overview on prion-like propagation modeling. The second discusses the use of functional dynamic models to studying neural activity dysregulation associated with neurodegenerative progression. The third emphasizes the importance of considering multiple biological factors, instead of only one (e.g., misfolded protein (MP) status or neuronal activity), to inform multifactorial brain models for understanding neurological and neuropsychiatric disorders.

Misfolded Proteins Spreading Through Brain Connections

Neurodegenerative diseases affect millions of people worldwide. Although covering a wide spectrum of clinical presentation, these disorders are invariably associated with a gradual loss of neural cells, and their prevalence rises notably with age. Historically, diverse disease-specific hypotheses for the origin of the neurodegeneration have been proposed and are still the subject of intense scientific debate (Buckner et al. 2005; Frost et al. 2009; Iadecola 2013). Among these, the prion-like hypothesis implicates neurodegeneration with a cell-to-cell transmission of specific toxic proteins which fail to configure correctly and are known as misfolded proteins (MPs) (Brundin et al. 2010; Frost et al. 2009; Hallbeck et al. 2013). MPs are thought to be involved in the dysfunction of cells, tissues, and organs (Walker and Levine 2000a, b), and are hypothesized to play a causal role in multiple non-native protein conformational disorders, such as Prion Diseases, Alzheimer's disease (AD), Parkinson's disease (PD), Creutzfeldt-Jakob disease (CJD), and amyotrophic lateral sclerosis (ALS) (Braak et al. 2004; Braak 1991; Dobson 2002, 2003; Reynaud 2010). However, the basic MPs mechanisms (i.e., protein seeding, aggregation, inter-cellular propagation, and toxicity) are not totally understood and still constitute an active research field (Eisele and Duyckaerts 2016; Lewis and Dickson 2016; Uchihara and Giasson 2016).

Since the early 1990s, we have seen a growing interest in the application of mathematical models for understanding the pathologic processes associated with MP propagation. As stated in a comprehensive review (Carbonell et al. 2018), the proposed MP-based mathematical models can be categorized into two main classes. Firstly, there is a large class of models designed to reproduce molecular level processes (e.g., seeding, aggregation, short-range spatial spreading). Secondly and more recently, a relatively small class of models account for inter-regional macroscopic interactions (e.g., long-range MP propagation in the human brain). Owing to its direct link with macroscopic brain networks and pathological inter-regional relationships, this chapter will focus on the second class of models. Readers may refer to Carbonell et al. (2018) for a detailed description of molecular-level models, most with application in the chemical kinetics field.

Motivated by the prion-like hypothesis, some previous studies (Matthäus 2006, 2009) focused on the theoretical modeling of MP spreading among neurons interconnected at the short-scale level (e.g., within a cortical column). However, it was not until the last decade that reliable in vivo imaging evidence supported the macroscopic cross-network propagation of MPs in the human brain. In 2009, Seeley and colleagues showed that different neurodegenerative disorders are associated with spatially dissociable atrophy patterns observed with structural MRI, with each pattern corresponding to consistent structural covariance and functional brain subnetwork analysis (Seeley et al. 2009). The same research group (Zhou et al. 2012) also reported that the observed brain's regional structural atrophy levels will be significantly associated with each region's functional connectivity strength with disease-specific "epicenters." Buckner et al. (2009) reported a strong correspondence between regional deposition values of Aß (a toxic protein historically associated with AD) and functional connectivity levels. However, gray matter atrophy patterns may not be uniquely provoked by MP toxicity. Instead, other pathological factors (e.g., neuronal hyperactivity, metabolic, and vascular deregulations) could have been responsible for the observed atrophy. In another study, to describe the underlying large-scale spreading processes using a mathematical perspective, Raj et al. (2012), introduced a Diffusion Network Model (DNM) of intra-brain MP propagation. According to this model, the increase over time of the number of diseased afferents from a given brain region to any other region depends upon the disease concentration factor in both regions and upon the anatomical connection strength between them (Raj et al. 2012). Based on the analysis of an analytical expression relating MPs spreading and structural atrophy dynamics, the authors observed a significant correspondence between specific dissociable structural connectivity modules and the characteristic atrophy patterns of different neurodegenerative diseases (i.e., AD or behavioral variant frontotemporal dementia [bvFTD]). These results suggested that the structural atrophy patterns observed in aging could be the weighted combination of characteristic atrophy patterns from the most prevalent neurodegenerative diseases, that is, each disease-characteristic module weighted by the individual predisposition to express the corresponding disease.

Altogether, the previous studies demonstrated the critical link between the brain's functional and structural connectivity and the commonly observed atrophy patterns in prevalent neurodegenerative diseases. However, it still was unclear why the toxic proteins will propagate progressively through the direct connections, without any system resistance, to induce cell death and brain dysfunction. Why would the typical defense mechanisms of the brain (e.g., Aß degradation by macrophages and microglia cells) not control such pathologic spreading? To address this question, Iturria-Medina and colleagues introduced an Epidemic Spreading Model (ESM) to account for the brain's capacity to produce, spread, and/or clear the MPs. In practice, the ESM explained the changes in MP deposition at a given region as a function of the local MP production level, the incoming MP from other brain regions via the anatomical network, the local MP clearance capacity, and potential external inputs (e.g., drugs, diet, physical exercise). When applied to the study of AD using Aß PET imaging data (see Fig. 1; Iturria-Medina et al. 2014), the ESM approach reconstructed individual/group Aß deposition patterns, highlighting the strategic role of the Aß outbreak regions and their connectional architecture on the disease's temporal progression. Consistent with experimental results from radioimmunoassay studies (Mawuenyega et al. 2010; Patterson et al. 2015), this work revealed that it is not an *increase* in Aß production but mainly a *decrease* in Aß clearance processes and an early Aß onset age that gave rise to the increased Aß deposition pattern observed in AD (Fig. 1e–f). Importantly, additional ESM-based analysis clarified the impact of genetic and demographic properties on individual MP production and clearance levels, as well as on the resultant intra-brain Aß propagation.

Neuronal Dynamics on the Neurodegenerative Brain

Contrary to the prion-like hypothesis of neurodegeneration, which assumes that the MPs have a dominant pathological role, consistent evidence suggest that other different biological factors also influence the neurodegenerative progression, and even MP dynamics (Bero et al. 2011; Buckner et al. 2009; Cirrito et al. 2005, 2008; Kamenetz et al. 2003). For instance, it is known that regional manipulations of the basal neuronal activity levels can impact the local vulnerabilities to Aß presence and toxicity (Cirrito et al. 2005). Cirrito and colleagues showed that exogenous increases in neuronal and synaptic activity in the hippocampus, induced by electric stimulation, increase the extracellular Aß concentrations. Similarly, endogenous physiological neuronal activity fluctuations have a strong causal impact on Aß concentrations in the interstitial fluid (Bero et al. 2011). These findings are in line with the strong linear relationship observed between regional functional connectivity/hubness and Aß deposition (Buckner et al. 2008, 2009). Functional hyperactivity and hyperconnectivity are significantly associated with semantic memory deficits and neurodegenerative progression (Gardini et al. 2015). Taken together, these and other consistent observations have led to the *neuronal activity-dependent degeneration* hypothesis, which characterizes neurodegeneration as a consequence of neuronal/synaptic hyperactivity that expands a "toxic" effect on surrounding connected neurons/synapses (Buckner 2005; Buckner et al. 2008; de Haan et al. 2012).

Models of large-scale neuronal dynamics constitute a promising framework for clarifying pathological mechanisms underlying neuronal activity alterations (Breakspear 2017; Deco and Kringelbach 2014). Seeking to determine if aberrant

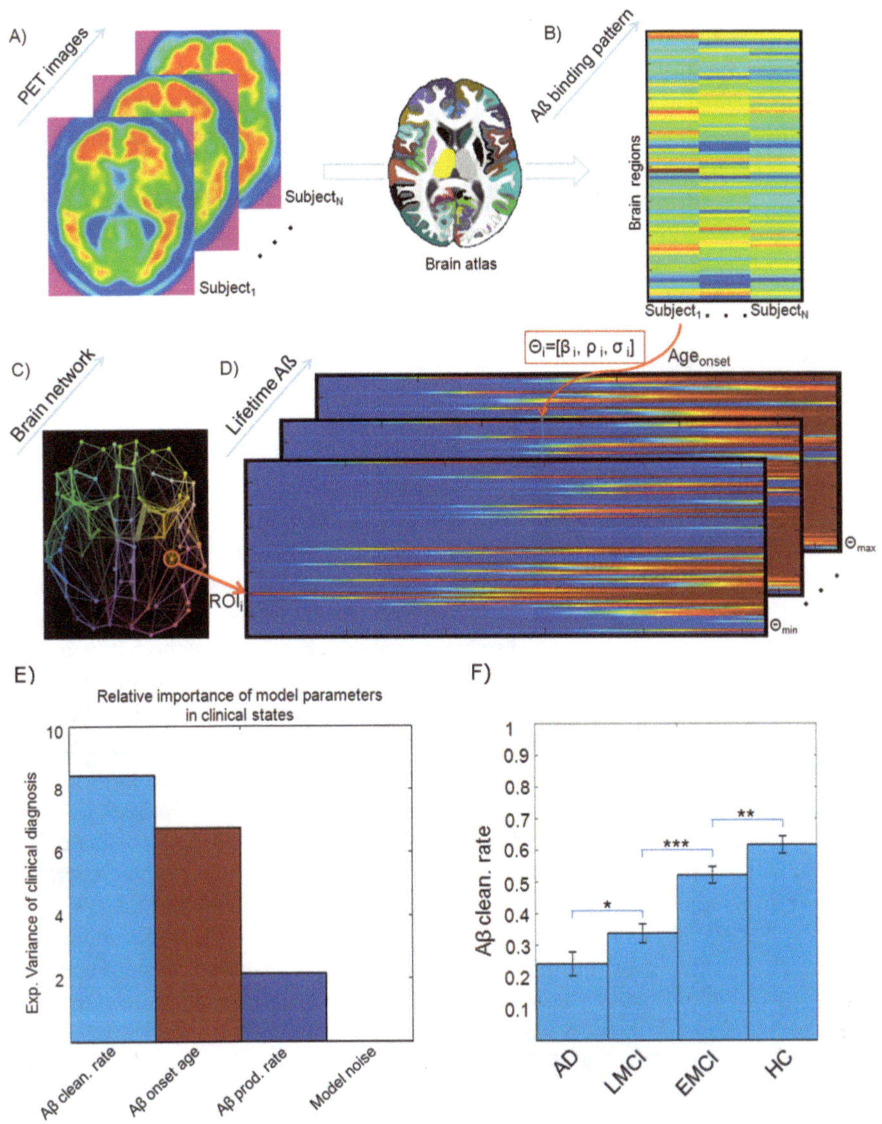

Fig. 1 Reconstruction of individual network-mediated Aß propagation/deposition histories using the ESM. (**a**) ^{18}F-AV-45 PET scans (N = 733) were used to calculate individual Aß deposition patterns for different regions covering all the brain's gray matter. (**b**) Anatomical connectome from a young healthy group, estimated via diffusion weighted MRI tractography (DW-MRI). (**d**) ESM-based generation of multiple hypothetical lifetime Aß propagation/deposition trajectories (each corresponding to an initial set of spreading seed regions and a different set of global model parameters, for example, Aß production and clearance rates and Aß onset age). An iterative algorithm estimated the individual model parameters that maximized the similarity between the generated and the reference Aß deposition pattern, characterizing each subject's Aß propagation/deposition history. (**e**) Explained variance of the clinical diagnoses by the different Aß propagation model parameters. (**f**) Mean Aß clearance rate (± standard error) for different clinical groups, that is, healthy controls (HC), mild cognitive impairment (EMCI), late cognitive impairment (LMCI), and AD. The values were adjusted for gender and educational level. $^*p < 0.05$, $^{**}p < 0.01$, $^{***}p < 10^{-5}$, Student's t-test. Figure adapted from Iturria-Medina et al. (2014), with permission

neuronal/synaptic activity could have an upstream role in neurodegenerative progression, de Haan et al. 2012, used neural mass modeling to explore local neuronal activity in relation to large-scale connectivity in normal and abnormal conditions. This model considered the average activity in relatively large groups of interacting excitatory and inhibitory neurons (i.e., neural masses), interconnecting the neural masses using the structural brain connectome from healthy subjects. In addition to simulating neural dynamics and obtaining an EEG-/MEG-like signal for each brain region, the authors induced progressive damage to the regions based on their level of activity. The results supported the notion that the highly structurally connected regions should be the most functionally active regions, with an excessive connectivity-dependent neuronal activity having a critical role in neurodegenerative progression (de Haan et al. 2012).

Recently, de Haan et al. (2017), used the same computational model to explore the effects of non-pharmacological therapeutic network interventions (e.g., deep brain stimulation or transcranial magnetic stimulation). For this, the authors considered six defense strategies that either enhanced or diminished neuronal excitability, targeting excitatory, and inhibitory neurons in combination or separately (de Haan et al. 2017). Notably, previous pharmacological interventions (Bakker et al. 2012; Sanchez et al. 2012) had shown that reducing the hyperactivation in both human and mice with neurodegeneration can partially reverse the synaptic and cognitive deficits. However, contrary to the idea that reducing neuronal activity would be the most effective activity-focused intervention, the results from de Haan et al. (2017) suggested that a selective stimulation of all excitatory neurons in the network will substantially preserve system integrity. In other words, the functional network damage due to pathological neuronal activity in neurodegeneration could be controlled by the selective increase of the neuronal excitability levels. These large-scale neuronal dynamic simulations also showed that reducing the hyperactivation would still be clinically beneficial but its positive effects would be less stable with time/aging. Based on these interesting results, the authors finally suggested that trying to oppose the degeneration mechanism may not be the optimum strategy, but to keep the network "alive" for as long as possible by stimulating the remaining neurons in the decaying system (de Haan et al. 2017).

Population heterogeneity is a crucial confounding factor undermining our understanding of the main neurodegenerative processes. Zimmermann et al. (2018) investigated if the previous neuronal activity-dependent degeneration observations varied in a heterogeneous aging and diseased population. These authors used a novel large-scale neuronal model for analyzing the resting state activity at the individual level and its relation with clinical phenotypes (see Fig. 2). Each brain region was modeled as a local neural mass comprising connected excitatory and inhibitory neurons linked with N-methyl-D-aspartate (NMDA) and gamma-aminobutyric acid (GABA) synapses, while the activity generated depended on intra-regional local parameters, excitatory and inhibitory inputs, and inter-regional structural connectivity (Deco et al. 2014a, b). For each subject, the biophysical parameters identification was achieved by iterative optimization until the similarity between measured and simulated region-region functional connectivity was maximized (Fig. 2a–d). These

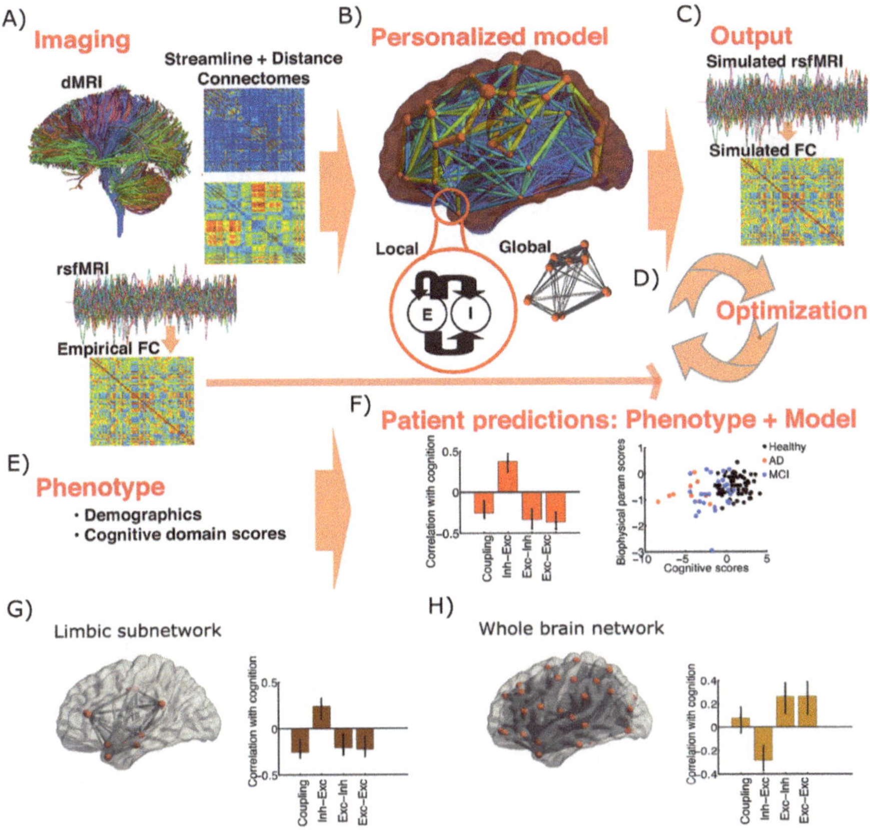

Fig. 2 Individual large-scale neuronal activity modeling in neurodegeneration. (**a**) Individual imaging data used by the model. The anatomical connectome is estimated using each participant's DW-MRI data, and the functional connectome is inferred from resting-state functional MRI (rs-fMRI). While the anatomical connectome is the input to the model, the functional connectivity is used for model fitting and optimization (i.e., aiming to maximize the similarity among the simulated and the real functional connectome patterns). (**b**) The whole brain neuronal activity simulated results from the integration of local and global dynamics, mediated by the brain's structural architecture. (**c**) This simulated activity is used then to calculate a functional connectome. (**d**) Both simulated and real individual connectomes are compared, providing a quantitative measure of the simulation's validity. (**e**) Individual patient demographics, cognitive, and behavioral scores are used to make (**f**) patient predictions by correlating these scores with the model's biological parameters. (**g**) Correlation between biophysical model parameters (global coupling, inhibition-excitation, excitation-inhibition, and excitation-excitation) and cognitive domain z-scores across all subjects for the limbic subnetwork and (**h**) the whole brain network. Figure adapted from Zimmermann et al. (2018), with permission

results emphasized the influence of the patient-specific neuronal inhibition/excitation and long-range coupling properties on the clinical phenotypes (Fig. 2g–h).

Interestingly, the authors found that the cognitive performances correlated both positively and negatively with the inhibition/excitation model parameters, depending on the choice of brain regions used for the neuronal activity modeling. Within a

Limbic system subnetwork, the individual excitation parameters correlated negatively with cognition (i.e., lower excitation implying better cognitive performance), while inhibition correlated positively with cognition. This result was consistent with the previous population-based findings, suggesting that an over-excitation promotes degeneration (Buckner 2005; Buckner et al. 2008; de Haan et al. 2012). However, when using a whole brain network for the neuronal activity generation, both the biophysical excitation and the inhibition parameters were positively correlated with the cognitive performance. These network-dependent results reflect the high complexity of brain dynamics, and how our modeling-based observations may vary at different scales. In line with the findings from de Haan et al. (2017), these recent results evidence the intricate, and still unpredictable, relationship among neuronal hyper-excitation and neurodegenerative evolution.

Multiple Interrelated Disease Factors in Neurodegeneration

Brain disorders are thought to be caused by multiple concomitant factors (Sheikh et al. 2012). Contrary to the Aß cascade and the neuronal activity dependent degeneration hypotheses, a large body of evidence supports the view that neither Aß nor neuronal hyperactivity is a unique dominant factor in neurodegenerative progression. Other complex multifactorial mechanisms are expected to present major pathological roles (Bero et al. 2011; Crossley et al. 2014; Iturria-Medina and Evans 2015; Schwarz et al. 2004; Wu et al. 2016). For example, in both health and disease, the major pathways of Aß clearance are associated with three different biological mechanisms (Qosa et al. 2014): (i) the transport across the blood brain barrier (BBB) (Deane et al. 2009), (ii) the degradation in the brain tissue (Iwata et al. 2000), and (iii) the clearance by the lymphatic clearance routes of cerebrospinal fluid (CSF) (Dreha-Kulaczewski et al. 2015; Louveau et al. 2015; Nedergaard and Goldman 2016). In another notable example, nerves and blood vessels present alike and interdependent branching patterns (Lewis 1902), a phenomenon named "neurovascular congruency" (Andreone et al. 2015; Martin 1989). The neurovascular cross-talk initiates early in life and sets the stage for a continuing relationship (Bautch and James 2009). In addition, neuronal structural degeneration can trigger functional hyperexcitability, being an important cellular mechanism underlying brain pathology and dysfunction (Šišková et al. 2014). Energy supply and neuronal function are strongly dependent on the cerebral blood flow (CBF) regulation (Andreone et al. 2015). Aβ increase negatively impacts mitochondrial function, which can alter brain metabolism, subsequently reducing mitochondrial Aβ clearance in a continuous feed-forward loop (Readnower et al. 2011). In addition to numerous factor-factor interactions, a varied spreading of factor-specific alterations is thought to be also a major feature in neurodegenerative progression (Cabral et al. 2012; Friston et al. 2014; Iturria-Medina et al. 2014, 2017; Iturria-Medina and Evans 2015; Sanz Leon et al. 2013). Concurrently with the conventional functional and Aß changes propagation (Brundin et al. 2010; de Haan et al. 2012), glucose metabolic impairment and structural degeneration spreading are well known pathological processes, the former being associated with cortical depression spreading (CSD) (Chang et al. 2013; Mergenthaler et al. 2013) and the latter with axonal degeneration (Lingor et al. 2012).

In summary, the concurrent multifactorial interactions and alterations spreading imply that the direct coupling among multiple pathological mechanisms is a key feature in neurodegeneration. Due to the marked complexity of the involved processes, only multifactorial causal analyses could lead to an integrated understanding of the neurodegenerative evolution, potentially clarifying the intrinsic multifactorial pathological interactions/spreading that take place at different spatiotemporal scales. Recently, Iturria-Medina and colleagues characterized these processes via a data-driven Multifactorial Causal Model (MCM) of Brain (dys) Organization (Iturria-Medina et al. 2017). The MCM combines control theory (Kalman 1963; Klickstein et al. 2016) and multimodal longitudinal data (e.g., PET, MRI, cognitive measures) in a multifactorial framework. It accounts for causal interactions among these multiple factors at the regional level, pathological propagation through axonal and vascular connectomes, impact on cognitive integrity, and assessment of single-target or combined therapeutic interventions. Specifically, this model (Fig. 3a–b) proposes that once a factor-specific event occur in a given brain region or set of regions, it directly interacts with other biological factors and alters their states. These new alterations may spread through anatomical and vascular connections to other brain areas, where similar factor-factor and propagation mechanisms may occur, in a continuous cycle.

Applied to the study of late-onset AD (LOAD), the MCM revealed that, at least at the population level, a vascular dysregulation may be the most likely initial pathologic event leading to this disorder (Iturria-Medina et al. 2017). The results also supported that LOAD is not caused by a unique dominant biological factor (e.g., Aβ, vascular or functional dysregulation) but by the complex interplay among multiple factor-factor interactions, revealing that each biological factor has a key role on disease progression. The authors then used the model as an in silico evaluator of interventional treatments (Fig. 3b), obtaining quantitative evidence for the crucial advantage of using combinatorial over single-target therapies and the critical need for considering multiple brain factors as endpoints, instead of merely cognitive biomarkers. These findings offered hope toward the assisted personalized intervention of brain pathologies, based on the computational characterization and controlling of individual multifactorial mechanisms at multiple spatiotemporal scales, from local causal interactions to networks-mediated propagation.

To further explore this idea, in a subsequent study (Iturria-Medina et al. 2018), the authors extended the MCM formulation to define personalized Therapeutic Intervention Fingerprints (pTIF). Based on the spatiotemporal analysis of individual multimodal imaging data, the pTIF values are a set of multivariate metrics that reflect the biological deformation required to stop the pathologic progression or restore the condition to normality (Iturria-Medina et al. 2018). In other words, the pTIF proposes to integrate large amount of data into a simplified individual patient profile of quantitative biological factor modifications needed to control disease

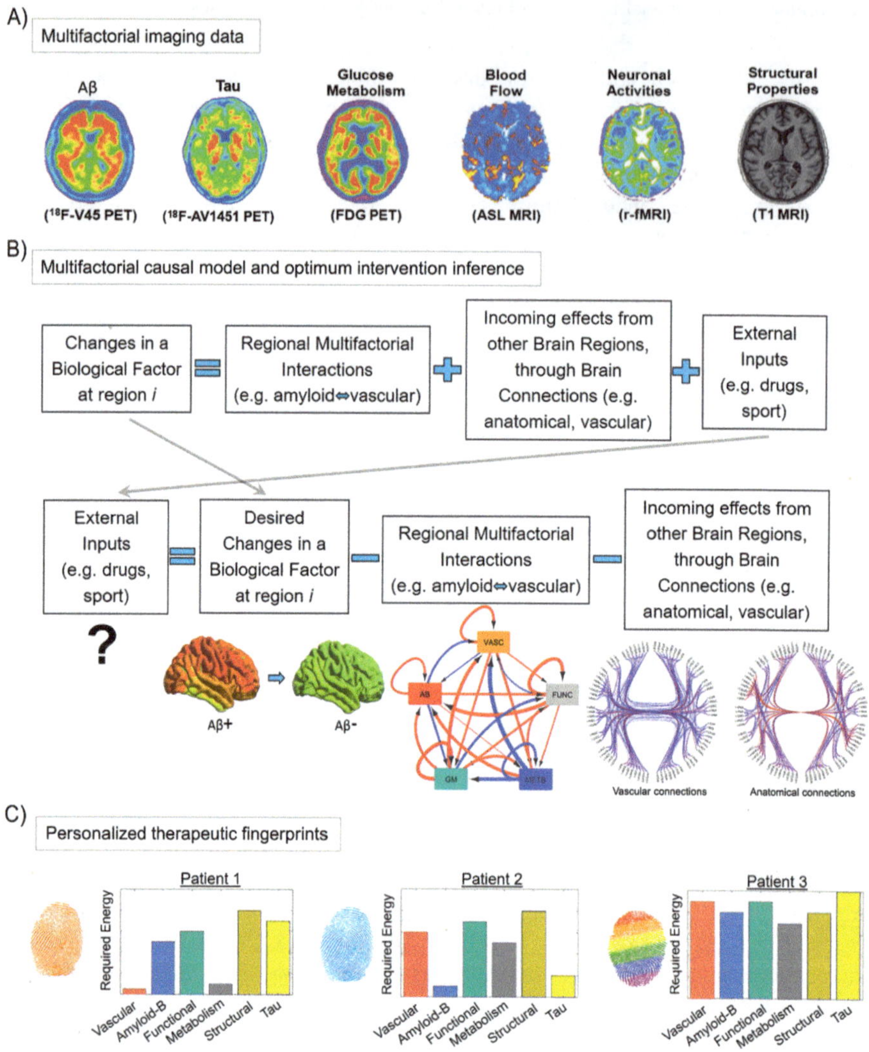

Fig. 3 Identifying multifactorial causal mechanisms and therapeutic needs. (**a**) Images are acquired using multimodal imaging modalities at different time-points, with each imaging modality associated with a given biological factor of interest: amyloid, Tau, glucose metabolism, vascular flow, functional activity at rest, and structural tissue, respectively. (**b**) The MCM approach allows characterization of the intra-brain direct factor-factor biological interactions and the multifactorial spreading mechanisms through vascular/anatomical networks. The inversion of the model's fundamental equation allows estimation of the input signals and associated cost-energy values required to produce a desired (e.g., healthy) state, through targeting individual biological factors or their combinations. (**c**) The pTIF is defined as the set of cost-energy values obtained for each patient, which are inversely proportional to the effectiveness of each intervention. Three dissimilar pTIF patterns, for three different participants in the same population, are presented. Patient 1 requires lower cost-energy values for vascular and metabolic interventions while Patient 2 requires lower cost-energy values for anti-Aß and anti-tau interventions. These patients may be benefited by specific single-target therapies (e.g., physical exercise and aducanumab, respectively). Contrary, Patient 3 requires high cost-energy for all the single-target interventions, suggesting that single-target treatments would not be beneficial and that combinatorial treatments may well be more positive. Figure adapted from Iturria-Medina et al. (2018), with permission

evolution (see Fig. 3c). When applied to an aging and neurodegenerative population (total N = 331), pTIF allows for the classification of individual subjects in terms of severity of pathological disruption in several brain variables (i.e., amyloid, tau, structural, metabolism, vascular, functional), providing a potential guide to optimal treatment interventions. The pTIF approach vastly outperforms cognitive/clinical evaluations in predicting individual gene expression (GE) profiles, providing biological validity to the model. Indeed, pTIF-based patients subgroups exhibited distinctively altered molecular pathway signatures. This leads naturally to stratification of dissimilar pathological subtypes and therapeutic needs. Although the therapeutic fingerprints concept/framework remains to be validated in the clinic, treatments that are targeted to an individual profile of brain disruptions have the potential for more effective medical care, decreased undesired secondary effects, and substantial reduction of pharmaceutical/clinical costs associated with clinical trials in neurodegeneration.

Conclusions

In this chapter, we have discussed the critical role of brain connections as a conduit for the spreading of pathological factors in neurodegeneration. This might have profound implications on our understanding of neurodegeneration, focusing on concurrent multifactorial causal interactions and networks-mediated disease propagation instead of focal alteration. The disease-focused studies described above force us to move from the classic "diseased vs healthy" brain comparisons to the characterization of the direct causal effects and pathology spreading. Although still needing further extension and validation, mathematical and computational brain models will meet the urgent need for identifying effective therapeutic interventions for neurodegeneration. These models are also promising tools toward a more effective clinical care. For instance, by estimating the most-likely initial triggering causal effects and the direct interactions among multiple relevant biological factors, it becomes possible to recover the multifactorial chronological stages that lead to the individual outcome of a complex disease. This allows us to characterize each patient's multifactorial dynamic brain changes in response to potential therapeutic inputs. A dynamic analysis of the consequent brain reconfiguration may provide robust prediction of treatment response and accelerate individual-level identification of effective disease-modifying agents.

References

Andreone, B.J., Lacoste, B., Gu, C., 2015a. Neuronal and vascular interactions. Neurosci Res. 38, 25–46. doi:https://doi.org/10.1146/annurev-neuro-071714-033835

Bakker, A., Krauss, G.L., Albert, M.S., Speck, C.L., Jones, L.R., Stark, C.E., Yassa, M.A., Bassett, S.S., Shelton, A.L., Gallagher, M., 2012. Reduction of Hippocampal Hyperactivity Improves

Cognition in Amnestic Mild Cognitive Impairment. Neuron 74, 467–474. doi:https://doi.org/10.1016/j.neuron.2012.03.023

Bautch, V.L., James, J.M., 2009. Neurovascular development: The beginning of a beautiful friendship. Cell Adhes. Migr. 3, 199–204. doi:https://doi.org/10.4161/cam.3.2.8397

Bero, A.W., Yan, P., Roh, J.H., Cirrito, J.R., Stewart, F.R., Raichle, M.E., Lee, J., Holtzman, D.M., 2011. Neuronal activity regulates the regional vulnerability to amyloid- b deposition 14, 5–7. doi:https://doi.org/10.1038/nn.2801

Braak H, Ghebremedhin E, Rub U, Bratzke H.D.T.K., 2004. Stages in the development of Parkinson's disease-related pathology. Cell Tissue Res. 318, 121–134.

Braak H.B.E., 1991. Neuropathological staging of Alzheimer-related changes. Acta Neuropathol. 82, 239–59.

Breakspear, M., 2017. Dynamic models of large-scale brain activity. Nat. Neurosci. 20, 340–352. doi:https://doi.org/10.1038/nn.4497

Brundin, P., Melki, R., Kopito, R., 2010. Prion-like transmission of protein aggregates in neurodegenerative diseases. Nat Rev. 11, 301–307.

Buckner, R., 2005. Molecular, Structural, and Functional Characterization of Alzheimer's Disease: Evidence for a Relationship between Default Activity, Amyloid, and Memory. J. Neurosci. 25, 7709–7717. doi:https://doi.org/10.1523/JNEUROSCI.2177-05.2005

Buckner, R., Andrews-Hanna, J., Schacter, D., 2008. The brain's default network: anatomy, function, and relevance to disease. Ann N Y Acad Sci. 1124, 1–38.

Buckner, R., Snyder, A., Shannon, B., LaRossa, G., Sachs, R., Fotenos, A., Sheline, Y., Klunk, W., Mathis, C., Morris, J., Mintun, M., 2005. Molecular, structural, and functional characterization of Alzheimer's disease: evidence for a relationship between default activity, amyloid, and memory. J. Neurosci. 25, 7709–7717.

Buckner, R.L., Sepulcre, J., Talukdar, T., Krienen, F.M., Liu, H., Hedden, T., Andrews-Hanna, J.R., Sperling, R. a, Johnson, K. a, 2009. Cortical hubs revealed by intrinsic functional connectivity: mapping, assessment of stability, and relation to Alzheimer's disease. J. Neurosci. 29, 1860–73. doi:https://doi.org/10.1523/JNEUROSCI.5062-08.2009

Bullmore, E., Sporns, O., 2009. Complex brain networks: graph theoretical analysis of structural and functional systems 10. doi:https://doi.org/10.1038/nrn2575

Cabral, J., Hugues, E., Kringelbach, M.L., Deco, G., 2012. Modeling the outcome of structural disconnection on resting-state functional connectivity. Neuroimage. doi:https://doi.org/10.1016/j.neuroimage.2012.06.007

Carbonell, F., Iturria-Medina, Y., Evans, A.C., 2018. Mathematical modeling of protein misfolding mechanisms in neurological diseases: A historical overview. Front. Neurol. 9. doi:https://doi.org/10.3389/fneur.2018.00037

Chang, J.C., Brennan, K.C., He, D., Huang, H., Miura, R.M., Wilson, P.L., Wylie, J.J., 2013. A Mathematical Model of the Metabolic and Perfusion Effects on Cortical Spreading Depression. PLoS One 8, 1–9. doi:https://doi.org/10.1371/journal.pone.0070469

Cirrito, J.R., Kang, J., Lee, J., Stewart, F.R., Verges, D.K., Silverio, L.M., Bu, G., Mennerick, S., Holtzman, D.M., 2008. Article Endocytosis Is Required for Synaptic Activity-Dependent Release of Amyloid- b In Vivo 42–51. doi:https://doi.org/10.1016/j.neuron.2008.02.003

Cirrito, J.R., Yamada, K.A., Finn, M.B., Sloviter, R.S., Bales, K.R., May, P.C., Schoepp, D.D., Paul, S.M., Mennerick, S., Holtzman, D.M., 2005. Synaptic Activity Regulates Interstitial Fluid Amyloid- b Levels In Vivo 48, 913–922. doi:https://doi.org/10.1016/j.neuron.2005.10.028

Crossley, N.A., Mechelli, A., Scott, J., Carletti, F., Fox, P.T., Mcguire, P., Bullmore, E.T., 2014. The hubs of the human connectome are generally implicated in the anatomy of brain disorders. Brain 137, 2382–2395. doi:https://doi.org/10.1093/brain/awu132

de Haan, W., Mott, K., van Straaten, E.C.W., Scheltens, P., Stam, C.J., 2012. Activity dependent degeneration explains hub vulnerability in Alzheimer's disease. PLoS Comput. Biol. 8, e1002582. doi:https://doi.org/10.1371/journal.pcbi.1002582

de Haan, W., van Straaten, E.C.W., Gouw, A.A., Stam, C.J., 2017. Altering neuronal excitability to preserve network connectivity in a computational model of Alzheimer's disease. PLOS Comput. Biol. 13, e1005707. doi:https://doi.org/10.1371/journal.pcbi.1005707

Deane, R., Bell, R., Sagare, A., Zlokovic, B., 2009. Clearance of amyloid-beta Alzheimer's, peptide across the blood-brain barrier: implication for therapies in disease. CNS Neurol. Disord. Drug. Targets 8, 16–30.

Deco, G., Kringelbach, M.L., 2014. Great expectations: Using whole-brain computational connectomics for understanding neuropsychiatric disorders. Neuron 84, 892–905. doi:https://doi.org/10.1016/j.neuron.2014.08.034

Deco, G., McIntosh, A.R., Shen, K., Hutchison, R.M., Menon, R.S., Everling, S., Hagmann, P., Jirsa, V.K., 2014a. Identification of Optimal Structural Connectivity Using Functional Connectivity and Neural Modeling. J. Neurosci. 34, 7910–7916. doi:https://doi.org/10.1523/JNEUROSCI.4423-13.2014

Deco, G., Ponce-Alvarez, A., Hagmann, P., Romani, G.L., Mantini, D., Corbetta, M., 2014b. How Local Excitation-Inhibition Ratio Impacts the Whole Brain Dynamics. J. Neurosci. 34, 7886–7898. doi:https://doi.org/10.1523/JNEUROSCI.5068-13.2014

Dobson, C.M., 2003. Protein folding and misfolding. Nature 426, 884–890. doi:https://doi.org/10.1038/nature02261

Dobson, C.M., 2002. Protein misfolding diseases: Getting out of shape. Nature 418, 729–730.

Dreha-Kulaczewski, S., Joseph, A. a, Merboldt, K.-D., Ludwig, H.-C., Gärtner, J., Frahm, J., 2015. Inspiration Is the Major Regulator of Human CSF Flow. J. Neurosci. 35, 2485–91. doi:https://doi.org/10.1523/JNEUROSCI.3246-14.2015

Eisele, Y.S., Duyckaerts, C., 2016. Propagation of Aß pathology: hypotheses, discoveries, and yet unresolved questions from experimental and human brain studies. Acta Neuropathol. 131, 5–25. doi:https://doi.org/10.1007/s00401-015-1516-y

Friston, K.J., Kahan, J., Biswal, B., Razi, A., 2014. A DCM for resting state fMRI. Neuroimage 94, 396–407. doi:https://doi.org/10.1016/j.neuroimage.2013.12.009

Frost, B., Jacks, R. & Diamond, M., 2009. Propagation of tau misfolding from the outside to the inside of a cell. J. Biol. Chem. 284, 12845–12852.

Gardini, S., Venneri, A., Sambataro, F., Cuetos, F., Fasano, F., Marchi, M., Crisi, G., Caffarra, P., 2015. Increased functional connectivity in Default Mode Network in Mild Cognitive Impairment: a maladaptive compensatory mechanism associated with poor semantic memory performance. J. Alzheimer's Dis. In press.

Hagmann, P., 2005. From diffusion MRI to brain connectomics. Lausanne: Ecole Polytechnique Fédérale de Lausanne (EPFL).

Hagmann, P., Sporns, O., Madan, N., Cammoun, L., Pienaar, R., Wedeen, V.J., Meuli, R., Thiran, J., 2010. White matter maturation reshapes structural connectivity in the late developing human brain 1–6. doi:https://doi.org/10.1073/pnas.1009073107

Hallbeck, M., Nath, S., Marcusson, J., 2013. Neuron-to-neuron transmission of neurodegenerative pathology. Neuroscientist 19, 560–6. doi:https://doi.org/10.1177/1073858413494270

Iadecola, C., 2013. Review The Pathobiology of Vascular Dementia. Neuron 80, 844–866. doi:https://doi.org/10.1016/j.neuron.2013.10.008

Iturria-Medina, Y., Carbonell, F.M., Evans, A.C., ADNI, 2018. Multimodal Imaging-based Therapeutic Fingerprints for Optimizing Personalized Interventions: Application to Neurodegeneration. Neuroimage 179, 40–50. doi:https://doi.org/10.1016/j.neuroimage.2018.06.028

Iturria-Medina, Y., Carbonell, F.M., Sotero, R.C., Chouinard-Decorte, F., Evans, A.C., 2017. Multifactorial causal model of brain (dis)organization and therapeutic intervention: Application to Alzheimer's disease. Neuroimage 152, 60–77. doi:https://doi.org/10.1016/j.neuroimage.2017.02.058

Iturria-Medina, Y., Evans, A.C., 2015. On the central role of brain connectivity in neurodegenerative disease progression. Front. Aging Neurosci. 7, 90. doi:https://doi.org/10.3389/fnagi.2015.00090

Iturria-Medina, Y., Pérez Fernández, A., Morris, D.M., Canales-Rodríguez, E.J., Haroon, H.A., García Pentón, L., Augath, M., Galán García, L., Logothetis, N., Parker, G.J.M., Melie-García, L., 2011. Brain hemispheric structural efficiency and interconnectivity rightward asymmetry in human and nonhuman primates. Cereb. Cortex 21, 56–67.

Iturria-Medina, Y., Sotero, R.C., Toussaint, P.J., Evans, A.C., 2014. Epidemic spreading model to characterize misfolded proteins propagation in aging and associated neurodegenerative disorders. PLoS Comput. Biol. 10, e1003956. doi:https://doi.org/10.1371/journal.pcbi.1003956

Iwata, N., Tsubuki, S., Takaki, Y., Watanabe, K., Sekiguchi, M., Hosoki, E., Kawashima-Morishima, M., Lee, H.J., Hama, E., Sekine-Aizawa, Y., Saido, T.C., 2000. Identification of the major Abeta1-42-degrading catabolic pathway in brain parenchyma: suppression leads to biochemical and pathological deposition. Nat. Med. 6, 143–150.

Kalman, R., 1963. Mathematical description of linear dynamical systems. J. Soc. Indus. Appl. Math. Ser. A 1, 152–192.

Kamenetz, F., Tomita, T., Hsieh, H., Seabrook, G., Borchelt, D., Iwatsubo, T., Sisodia, S., Malinow, R., Point, W., 2003. APP Processing and Synaptic Function State University of New York at Stony Brook 37, 925–937.

Klickstein, I.S., Shirin, A., Sorrentino, F., 2016. Optimal Target Control of Complex Networks.

Leergaard, T.B., Hilgetag, C.C., Sporns, O., 2012. Mapping the Connectome: Multi-Level Analysis of Brain Connectivity. Front. Neuroinform. 6, 1–6. doi:https://doi.org/10.3389/fninf.2012.00014

Lewis, J., Dickson, D.W., 2016. Propagation of tau pathology: hypotheses, discoveries, and yet unresolved questions from experimental and human brain studies. Acta Neuropathol. 131, 27–48. doi:https://doi.org/10.1007/s00401-015-1507-z

Lewis, W., 1902. The development of the arm in man. Am J Anat. 1, 145–185.

Lingor, P., Koch, J.C., Tönges, L., Bähr, M., 2012. Axonal degeneration as a therapeutic target in the CNS. Cell Tissue Res. 349, 289–311. doi:https://doi.org/10.1007/s00441-012-1362-3

Louveau, A., Smirnov, I., Keyes, T.J., Eccles, J.D., Rouhani, S.J., Peske, J.D., Derecki, N.C., Castle, D., Mandell, J.W., Lee, K.S., Harris, T.H., Kipnis, J., 2015. Structural and functional features of central nervous system lymphatic vessels. Nature 523, 337–341. doi:https://doi.org/10.1038/nature14432

Martin P.L.J., 1989. Origins of the neurovascular bundle: interactions between developing nerves and blood vessels in embryonic chick skin. Int J Dev Biol. 33, 379–387.

Matthäus, F., 2009. The spread of prion diseases in the brain – models of reaction and transport on networks. J. Biol. Syst. 17, 623–641. doi:https://doi.org/10.1142/S0218339009003010

Matthäus, F., 2006. Diffusion versus network models as descriptions for the spread of prion diseases in the brain. J. Theor. Biol. 240, 104–113. doi:https://doi.org/10.1016/j.jtbi.2005.08.030

Mawuenyega, K., Sigurdson, W., Ovod, V., Munsell, L., Kasten, T., Morris, J., Yarasheski, K., Bateman, R., 2010. Decreased clearance of CNS beta-amyloid in Alzheimer's disease. Science (80-.). 330(6012), 1774.

Melie-García, L., Sanabria-Diaz, G., Sánchez-Catasús, C., 2013. Studying the topological organization of the cerebral blood flow fluctuations in resting state. Neuroimage 64, 173–184.

Mergenthaler, P., Lindauer, U., Dienel, G., Meisel, A., 2013. Sugar for the brain: the role of glucose in physiological and pathological brain function. Trends Neurosci. 36, 587–597. doi:https://doi.org/10.1016/j.tins.2013.07.001.Sugar

Nedergaard, M., Goldman, S.A., 2016. Brain Drain. Sci. Am. 314, 342–347. doi:https://doi.org/10.1038/scientificamerican0316-44

Oh, S.W., Harris, J.A., Ng, L., Winslow, B., Cain, N., Mihalas, S., Wang, Q., Lau, C., Kuan, L., Henry, A.M., Mortrud, M.T., Ouellette, B., Nguyen, T.N., Sorensen, S.A., Slaughterbeck, C.R., Wakeman, W., Li, Y., Feng, D., Ho, A., Nicholas, E., Hirokawa, K.E., Bohn, P., Joines, K.M., Peng, H., Hawrylycz, M.J., Phillips, J.W., Hohmann, J.G., Wohnoutka, P., Gerfen, C.R., Koch, C., Bernard, A., Dang, C., Jones, A.R., Zeng, H., 2014. ARTICLE A mesoscale connectome of the mouse brain. Nature 508, 207–214. doi:https://doi.org/10.1038/nature13186

Patterson, B., Elbert, D., Mawuenyega, K., Kasten, T., 2015. Age and Amyloid Effects on Human CNS Amyloid-Beta Kinetics. Ann. Neurol. 78, 439–453. doi:https://doi.org/10.1161/CIRCRESAHA.116.303790.The

Pievani, M., Filippini, N., van den Heuvel, M.P., Cappa, S.F., Frisoni, G.B., 2014. Brain connectivity in neurodegenerative diseases-from phenotype to proteinopathy. Nat. Rev. Neurol. 10, 620–633. doi:https://doi.org/10.1038/nrneurol.2014.178

Qosa, H., Abuasal, B.S., Romero, I.A., Weksler, B., Couraud, P.-O., Keller, J.N., Kaddoumi, A., 2014. Differences in amyloid-β clearance across mouse and human blood–brain barrier models.pdf.

Raj, A., Kuceyeski, A., Weiner, M., 2012. A Network Diffusion Model of Disease Progression in Dementia. Neuron 73, 1204–1215. doi:https://doi.org/10.1016/j.neuron.2011.12.040

Readnower, R., Sauerbeck, A., Sullivan, P., 2011. Mitochondria, Amyloid β, and Alzheimer's Disease.pdf. doi:https://doi.org/10.4061/2011/104545

Reynaud, E., 2010. Protein Misfolding and Degenerative Diseases. Nat. Educ. 3(9), 28.

Rubinov, M., Bullmore, E., 2013. Fledgling pathoconnectomics of psychiatric disorders. Trends Cogn. Sci. 17, 641–647. doi:https://doi.org/10.1016/j.tics.2013.10.007

Sanchez, P.E., Zhu, L., Verret, L., Vossel, K.A., Orr, A.G., Cirrito, J.R., Devidze, N., Ho, K., Yu, G.-Q., Palop, J.J., Mucke, L., 2012. Levetiracetam suppresses neuronal network dysfunction and reverses synaptic and cognitive deficits in an Alzheimer's disease model. Proc. Natl. Acad. Sci. 109, E2895–E2903. doi:https://doi.org/10.1073/pnas.1121081109

Sanz Leon, P., Knock, S. a, Woodman, M.M., Domide, L., Mersmann, J., McIntosh, A.R., Jirsa, V., 2013. The Virtual Brain: a simulator of primate brain network dynamics. Front. Neuroinform. 7, 10. doi:https://doi.org/10.3389/fninf.2013.00010

Schwarz, Q., et al., 2004. Vascular endothelial growth factor controls neuronal migration and cooperates with Sema3A to pattern distinct compartments of the facial nerve. Genes Dev. 18, 2822–2834.

Seeley, W.W., Crawford, R.K., Zhou, J., Miller, B.L., Greicius, M.D., 2009. Neurodegenerative Diseases Target Large-scale Human Brain Networks. Neuron 62, 42–52. doi:https://doi.org/10.1016/j.neuron.2009.03.024

Sheikh, S., Haque, E., Mir, S., 2012. Neurodegenerative Diseases: Multifactorial Conformational Diseases and Their Therapeutic Interventions. J. Neurodegener. Dis. 2013, 8. doi:https://doi.org/10.1155/2013/563481\

Shimono, M., Beggs, J.M., 2018. Functional Clusters, Hubs, and Communities in the Cortical Microconnectome 3743–3757. doi:https://doi.org/10.1093/cercor/bhu252

Šišková, Z., Justus, D., Kaneko, H., Friedrichs, D., Henneberg, N., Beutel, T., Pitsch, J., Schoch, S., Becker, A., VonderKammer, H., Remy, S., 2014. Dendritic structural degeneration is functionally linked to cellular hyperexcitability in a mouse model of Alzheimer's disease. Neuron 84, 1023–1033. doi:https://doi.org/10.1016/j.neuron.2014.10.024

Sporns, O., 2013. Making sense of brain network data. Nat Methods 10, 491–493. doi:https://doi.org/10.1038/nmeth.2485

Sporns, O., Tononi, G., Kötter, R., 2005. The human connectome: A structural description of the human brain. PLoS Comput. Biol. 1e42.

Stam, C.J., de Haan, W., Daffertshofer, A, Jones, B.F., Manshanden, I., van Cappellen van Walsum, A.M., Montez, T., Verbunt, J.P.A., de Munck, J.C., van Dijk, B.W., Berendse, H.W., Scheltens, P., 2009. Graph theoretical analysis of magnetoencephalographic functional connectivity in Alzheimer's disease. Brain 132, 213–24. doi:https://doi.org/10.1093/brain/awn262

Uchihara, T., Giasson, B.I., 2016. Propagation of alpha-synuclein pathology: hypotheses, discoveries, and yet unresolved questions from experimental and human brain studies. Acta Neuropathol. 131, 49–73. doi:https://doi.org/10.1007/s00401-015-1485-1

Walker, L.C., Levine, H., 2000a. The cerebral proteopathies. Neurobiol. Aging 21, 559–561. doi:https://doi.org/10.1016/S0197-4580(00)00160-3

Walker, L.C., Levine, H., 2000b. The cerebral proteopathies. Mol. Neurobiol. 21, 83–95.

Wu, J.W., Hussaini, S.A., Bastille, I.M., Rodriguez, G.A., Mrejeru, A., Rilett, K., Sanders, D.W., Cook, C., Fu, H., Boonen, R.A.C.M., Herman, M., Nahmani, E., Emrani, S., Figueroa, Y.H., Diamond, M.I., Clelland, C.L., Wray, S., Duff, K.E., 2016. Neuronal activity enhances tau propagation and tau pathology in vivo. Nat. Neurosci. doi:https://doi.org/10.1038/nn.4328

Zhou, J., Gennatas, E., Kramer, J., Miller, B., Seeley, W., 2012. Predicting Regional Neurodegeneration from the Healthy Brain Functional Connectome. Neuron 73, 1216–1227. doi:https://doi.org/10.1016/j.neuron.2012.03.004

Zimmermann, J., Perry, A., Breakspear, M., Schirner, M., Sachdev, P., Wen, W., Kochan, N.A., Mapstone, M., Ritter, P., McIntosh, A.R., Solodkin, A., 2018. Differentiation of Alzheimer's disease based on local and global parameters in personalized Virtual Brain models. NeuroImage Clin. doi:https://doi.org/10.1016/j.nicl.2018.04.017

Zuo, X., Ehmke, R., Mennes, M., Imperati, D., Castellanos, F.X., Sporns, O., Milham, M.P., 2012. Network Centrality in the Human Functional Connectome 1862–1875. doi:https://doi.org/10.1093/cercor/bhr269

Resting-State Functional Network Disturbances in Schizophrenia

Qingbao Yu and Vince D. Calhoun

Introduction

Schizophrenia (SZ) is a severe mental illness characterized by symptoms, including delusions, hallucinations, apathy, social withdrawal, and deficits in cognitive functions such as attention and working memory (Marin 2012). Functional brain imaging studies about this disease support a leading hypothesis that SZ stems from dysconnectivity, that is, abnormal interactions between widespread brain networks (Andreasen and Pierson 2008; Friston and Frith 1995; He et al. 2013; Smieskova et al. 2011; Stephan et al. 2006; Williamson and Allman 2012). Resting-state functional magnetic resonance imaging (R-fMRI) is a powerful tool to examine the disturbances of brain networks in SZ, helping us to understand the underlying pathology of this brain disorder.

During resting-state experiments, participants are typically instructed to relax, stay awake, and not to think of anything in particular while their patterns of spontaneous brain activation were measured (van den Heuvel and Hulshoff Pol 2010). Functional connectivity, which can be defined as the coordination of activity across brain regions (Friston et al. 1993¸ 1996; Rosazza et al. 2012), is often estimated as the temporal correlations of low frequency (0.01–0.10 HZ) oscillations in the blood oxygen level-dependent (BOLD) signal between anatomically distinct brain areas

Q. Yu (✉)
The Mind Research Network, Albuquerque, NM, USA
e-mail: qyu@mrn.org

V. D. Calhoun (✉)
The Mind Research Network, Albuquerque, NM, USA

Department of Electrical and Computer Engineering, University of New Mexico, Albuquerque, NM, USA

Department of Psychiatry, Yale University, New Haven, CT, USA
e-mail: vcalhoun@unm.edu

V. A. Diwadkar, S. Eickhoff (eds.), *Brain Network Dysfunction in Neuropsychiatric Illness*, https://doi.org/10.1007/978-3-030-59797-9_10

(van den Heuvel and Hulshoff Pol 2010). During the resting state, brain regions tend to be functionally connected with somewhat prototypical patterns, revealing the intrinsic organization of so-called resting-state networks (RSNs). Research has demonstrated that RSNs are largely present in both rest and task, though they do change during tasks (Calhoun et al. 2008, 2011) and between eyes open and closed conditions (Wu et al. 2010; Zou et al. 2009). Using R-fMRI, the motor network (Biswal et al. 1995), default mode network (Raichle et al. 2001; Raichle and Snyder 2007), and several other RSNs can be readily identified (Deco and Corbetta 2011; Deco et al. 2011; Rosazza and Minati 2011).

There are a number of advantages of studying the functional organization of brain networks during the resting state, as opposed to during a task (Pearlson and Calhoun 2009). First, spontaneous activity accounts for the large majority of the brain's activity (Raichle and Mintun 2006), whereas task-related increases in neuronal metabolism are usually small (<5%) (Raichle 2010). It is important to assess the spontaneous activity of brain during resting state to fully understand brain function (Fox and Raichle 2007). Secondly, in practice, it is difficult to attain equivalent level of performance in a task for healthy controls and patients, whereas R-fMRI is largely free from this potentially confounding effect (Greicius 2008; Pearlson and Calhoun 2009). In addition, R-fMRI experiments are less prone to multi-site variability stemming from differences in experimental design or implementation, allowing for a wider range of patients to be included and considered in analyses (Fox and Greicius 2010). Finally, R-fMRI leads to less movement than task fMRI studies requiring subjects to respond with motor activity (Rosazza and Minati 2011).

R-fMRI data have typically been investigated using seed-based functional connectivity analysis (Biswal et al. 1995; Biswal 2012) or spatial independent component analysis (ICA) (Calhoun et al. 2001). In addition, graph theory-based complex network analysis methods were applied to further characterize the underlying networks in both seed/ROI-based and ICA (or network-based) analyses (Fornito et al. 2015). Moreover, amplitude of low-frequency fluctuation (ALFF) (Golestani et al. 2017; Zang et al. 2007) and regional homogeneity (ReHo) (Golestani et al. 2017; Zang et al. 2004) are two widely used measures to characterize R-fMRI data. Most recently, dynamic connectivity analysis was performed on R-fMRI data to estimate time-varying connectivity in human brain (Calhoun et al. 2014; Hutchison et al. 2013). In this chapter, we review and summarize multiple findings of brain dysconnectivity in SZ discovered using these approaches in R-fMRI data and discuss ongoing challenges and future work.

Dysconnectivity of SZ Identified by Seed-Based Analysis

In seed-based or region-of-interest (ROI)-based approaches to assess functional brain connectivity in fMRI data (Rosazza and Minati 2011), BOLD time series were first extracted from an ROI (or seed). Then, the temporal correlation between this extracted signal and the time series from all other brain voxels was determined

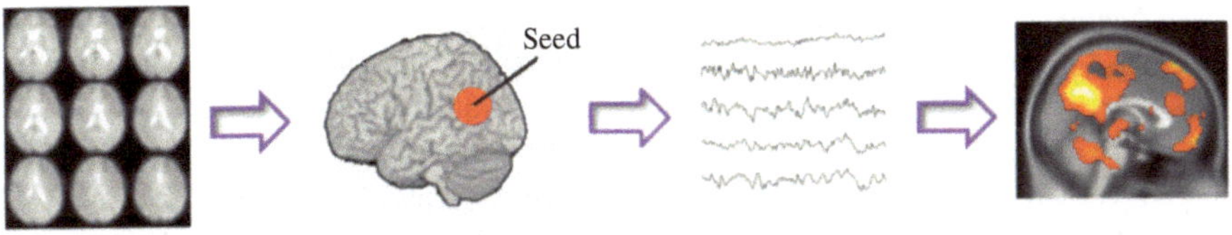

Fig. 1 Pipeline for seed-based functional connectivity analysis in R-fMRI studies. The first step was to select a brain region of interest (ROI) as a seed. Next, time courses were extracted from the seed and all other brain voxels. After preprocessing (e.g., band-pass filtering, autocorrelation correction, covarying for white matter variance or other confounding variables), a functional connectivity brain map showing the temporal correlation between the given seed and all other brain voxels was produced

after some preprocessing (e.g., band-pass filtering, autocorrelation correction, covarying for white matter variance or other confounding variables). For a schematic pipeline of this approach, see Fig. 1. This approach is very useful; however, it also has some disadvantages, such as dependence on a priori-defined seed region, a limited ability to study multiple systems simultaneously, and the possibility that the extracted waveform may not be a true independent variable when assessing statistical significance (Fox and Raichle 2007). The seed-based approach has been widely used due to its inherent simplicity, sensitivity, and ease of interpretation (Fox and Raichle 2007; Greicius 2008; Rosazza and Minati 2011; Rosazza et al. 2012). It also provides results largely consistent with ICA measures in many cases (Erhardt et al. 2011).

Findings from Single ROI Seeds

Other work has investigated brain connectivity in SZ using single seeds. A few studies defined seeds in the frontal lobe. For example, when using seeds located in bilateral dorsolateral prefrontal cortex (bDLPFC), the bDLPFC showed reduced functional connectivity (FC) to the parietal lobe, posterior cingulate cortex, thalamus and striatum, and the left DLPFC showed enhanced FC to the left mid-posterior temporal lobe and the paralimbic regions in patients with first-episode SZ (Zhou et al. 2007a). Another study which used DLPFC regions as seeds estimated brain connectivity in patients with SZ, unaffected first-degree relatives (UR), and healthy controls (HCs). Results showed that SZ and UR had impaired connectivity from DLPFC to its coordinated regions which were distributed in the bilateral caudate, left middle/inferior frontal gyrus, left precentral gyrus, and right cerebellum. Moreover, the individual FC strength between the left DLPFC and its coordinated regions was correlated with individual executive function performance among whole subjects (Su et al. 2013). When using the ventromedial prefrontal cortex (vMPFC) as a seed, hypoconnectivity between vMPFC and the medial frontal cortex, right middle temporal lobe (MTL), right hippocampus, parahippocampal cortex

(PHC), and amygdala was discovered in SZ. Also, the strength of the negative connectivity (or anticorrelation) between the vMPFC and the bDLPFC, and pre-supplementary motor areas was decreased in SZ. In addition, reduced vMPFC-DLPFC connectivity was positively correlated with positive symptoms on the positive and negative syndrome scales (PANSS), while vMPFC –right MTL/PHC/amygdala FC was positively correlated with the performance of emotional regulation in SZ patients (Fan et al. 2013).

Temporal areas have been used as seeds as well. When evaluating FC between bilateral planum temporal (PT) and other brain regions in SZ patients, unaffected first-degree relatives, and healthy controls, results demonstrated reduced FC between PT and temporal, parietal, limbic, and subcortical regions in SZ patients and relatives comparing with controls. Interestingly, altered FC was correlated with predisposition toward hallucinations in both patients and relatives. Reduced functional asymmetry of the superior temporal gyrus in patients and relatives was also observed, which correlated significantly with acute severity of hallucinations in the patient group (Oertel-Knochel et al. 2013).

Thalamus is another popular seed used to evaluate brain FC. SZ patients had exhibited enhanced thalamic connectivity with bilateral precentral gyrus, dorsal medial frontal gyrus, middle occipital gyrus, and lingual gyrus. And reduced thalamic connectivity in SZ was found in bilateral superior frontal gyrus, anterior cingulate cortex, inferior parietal lobe, and cerebellum (Wang et al. 2015b). Another study identified higher overall thalamocortical FC in SZ, which was mostly accounted for by differences in thalamic connections to right ventrolateral prefrontal and bilateral secondary motor and sensory (superior temporal and lateral occipital) cortical areas (Klingner et al. 2014).

Seeds located in many other brain areas have also been assessed. FC between bilateral hippocampi and brain regions including posterior cingulate cortex, extrastriate cortex, medial prefrontal cortex, and parahippocampus gyrus were found to be reduced in patients with paranoid SZ (Zhou et al. 2008). Reduced connectivity between the retrosplenial cortex and both the temporal lobe and regions of the default network in SZ had been reported (Bluhm et al. 2009). Decreased correlation between posterior cingulate and lateral parietal, medial prefrontal, and cerebellar regions has also been observed in SZ (Bluhm et al. 2007).

Findings from Seeds Defined with Subregions of Specific Brain Areas

Dividing a specific brain area into subregions was an interesting method for defining the seeds, which would estimate the particular contribution of each subregion to the dysconnectivity in SZ. The frontal pole (FP) was subdivided into the orbital (FPo), medial (FPm), and lateral (FPl) subregions as seeds to perform a seed-based brain connectivity analysis. Results revealed reduced FC between the bilateral FPl and several cognitive-related regions, including the dorsolateral prefrontal cortex, medial prefrontal cortex, anterior cingulate cortex, posterior cingulate cortex/precuneus, temporal cortex, and inferior parietal lobule in SZ. Although the FPl exhibited obvious atrophy, FC changes were unrelated to volumetric atrophy in the FPl, to duration of illness, and to antipsychotic medication dosage. Notably, no significant differences were observed in the FC of other FP subregions (Zhou et al. 2015).

When using two symmetric ACC parcellation regions derived from an anatomical template (left and right) cognitive division of anterior cingulate cortex (ACC-cd) as seeds, decreased positive connectivity with the bilateral putamen and caudate, increased negative connectivity with the left posterior cingulated cortex (PCC), increased asymmetry of connectivity strength with the contralateral inferior frontal gyrus (IFG) in SZ patients was observed (Yan et al. 2012).

The inferior parietal lobule (IPL) has been divided into six subregions to study FC between each IPL subregion and gray matter voxels throughout the brain. Among the six subregions of the left IPL and the five subregions of the right IPL, only the bilateral PFm (the magnocellular part of PF, a transition zone of the IPL) subregions exhibited abnormal FC in SZ. Specifically, the left PFm showed increased FC with the bilateral lingual gyri in SZ. The right PFm exhibited increased FC with the right lingual gyrus and inferior occipital gyrus, and bilateral mid-cingulate and sensorimotor cortices in SZ. These results suggest a selective FC abnormality in the IPL subregions in SZ (Liu et al. 2016).

The temporal pole (TP) was subdivided into dorsal (TPd) and ventral (TPv) subregions and results showed reduced FC between the right TPd and brain regions involved in language processing and multisensory integration in SZ, including the left superior temporal gyrus, left mid-cingulate cortex, and right insular cortex. FC changes of the right TPd in SZ were found to be independent of the gray matter reduction of this subregion. Moreover, these FC changes were unrelated to illness severity, duration of illness, and antipsychotic medication dosage. No significant group differences were observed in the FC of the left TPd and bilateral TPv subregions. These results suggested a selective (the right TPd) functional dysconnection of TP subregions in SZ (Xu et al. 2015).

When using three subregions of the insula (dorsal anterior, ventral anterior, and posterior insula) as seeds, decreased connectivity exclusively between the right insula and key hubs of central executive network (DLPFC) and DMN (PCC) was found (Moran et al. 2013).

Findings from Multiple Seeds

Multiple seeds have been used to investigate a wider network of dysconnectivity in SZ. When using six regions of interest including the prefrontal cortex, motor cortex/supplementary motor area, somatosensory cortex, temporal lobe, posterior parietal cortex, and occipital lobe as seeds to perform FC analysis, results showed reduced prefrontal–thalamic connectivity and increased motor/somatosensory–thalamic connectivity in schizophrenia (Woodward et al. 2012). When using seeds from the left crus I, right crus I, and lobule IX, which have previously been identified as being involved in the DMN, results revealed that, compared with the healthy controls, the SZ patients showed significantly reduced cerebellar functional connectivity with the thalamus and several frontal regions including the middle frontal gyrus, anterior cingulate cortex, and supplementary motor area. Moreover, the positive correlations between the strength of fronto-cerebellar and thalamo-cerebellar functional connectivity observed in the healthy subjects were diminished in the patients (Wang et al. 2014).

Findings from Connectivity of Whole-Brain ROIs

Dysconnectivity in SZ has also been estimated by computing connectivity among ROIs across the whole brain. When comparing the correlation of each pair of 116 ROIs of the whole brain, decreased connectivity was widely distributed throughout the entire brain in SZ (Liang et al. 2006). Another study computed the FC after dividing the whole brain into 90 ROIs; it was found that most of the altered FC was located within visual cortical, default mode, and sensorimotor networks. Furthermore, during reconstruction arithmetic, the fusiform gyrus exhibited the greatest weight. This study used classification method, and demonstrated that schizophrenia patients might be successfully differentiated from healthy subjects by using whole-brain R-fMRI, and the fusiform gyrus might play an important functional role in the physiological symptoms manifested by schizophrenic patients (Tang et al. 2012).

Whole-brain gray matter can be segmented into 77 anatomical ROIs using the semi-automatic FreeSurfer software and these have been used to estimate whole-brain connectivity. SZ patients exhibited increased FC between the medial parietal regions, including the posterior cingulate gyrus, and the frontal lobe (pars triangularis and opercularis of the inferior frontal gyrus and dorsolateral prefrontal cortex). SZ showed reduced FC between the same medial parietal region and the left temporal lobe (inferior temporal gyrus and amygdala). In addition, the interhemispheric connectivity between the left and the right temporal regions expressed reduced FC in the selected features analysis only. The functional connections that exhibited hyperconnectivity in the SZ group (i.e., between the medial parietal and inferior frontal and dorsolateral prefrontal regions) were negatively correlated with patients' scores on the negative and general subscales of the PANSS. Furthermore, the functional connections that exhibited hypoconnectivity in the schizophrenia group (i.e., between the medial parietal and temporal regions) were positively correlated with patients' scores on the positive subscales of the PANSS (Venkataraman et al. 2012).

Another study performed a voxel-level connectivity analysis. For each participant, the average Pearson correlation of each voxel in the scan volume with all gray matter voxels (from FreeSurfer parcellation) in a whole-brain mask was calculated. Results from this whole-brain connectivity analysis found 26 brain regions with decreased functional correlations in SZ. These regions were found to be organized into two function-related networks. The first with regions associated with social and higher level cognitive processing, and the second with regions involved in somatosensory and motor processing (Berman et al. 2016).

Symptom-Related Findings

Most interestingly, some findings of dysconnectivity in SZ were associated with patients' symptoms. For example, SZ patients with a history of auditory hallucinations had significantly reduced interhemispheric connectivity in primary and secondary auditory cortices when compared with SZ patients without auditory hallucinations and healthy controls (Gavrilescu et al. 2010). When using a bilateral amygdala seed, Hoptman et al. (2010) revealed reductions in connectivity between amygdala and ventral prefrontal cortex regions in patients with SZ or schizoaffective disorder. Furthermore, the lower connectivity in patients was associated with higher levels of self-rated aggression. When using bilateral temporo-parietal junction as seed regions, Vercammen et al. (2010) revealed reduced connectivity between left temporo-parietal junction and the right homotope of Broca's area in SZ patients with auditory–verbal hallucinations. More severe auditory–verbal hallucinations were associated with reduced correlation between left temporo-parietal junction and bilateral anterior cingulate as well as the bilateral amygdale.

Resting-state FC assessed in the meta-analytically defined "mirror neuron system (MNS)" and the "mentalizing network (MENT)" demonstrated decreased connectivity within the MNS and also the MENT in SZ patients. Notably, dysconnectivity of the MNS was related to symptom severity, while no such relationship was observed for the MENT. These findings demonstrated that differential patterns of dysconnectivity exist in SZ, which may contribute differently to the interpersonal difficulties commonly observed in the disorder (Schilbach et al. 2016).

Findings About Connectivity Between and Within RSNs

In terms of connectivity within and between resting-state brain networks (RSNs), Whitfield-Gabrieli et al. (2009) found abnormal high functional connectivity within the default mode network in SZ, consistent with Garrity et al. (2007). Another study (Woodward et al. 2011) investigated the FC in several RSNs including default mode, dorsal attention, executive control, and salience networks using seed-based method. A number of differences between SZ patients and controls were revealed in default mode, dorsal attention, and executive control networks. Specifically, patients demonstrated greater connectivity between the posterior cingulated cortex and the left inferior gyrus, left middle frontal gyrus, and left middle temporal gyrus. These regions were more strongly connected to the executive control network in healthy controls. Decreased connectivity in the executive control and dorsal attention networks in SZ was observed by calculating the correlation among clusters within each network. The findings suggested the process of functional specialization was altered in SZ. In another study, after identification of ROIs in "task-positive" network which was identified by using the right dorsolateral prefrontal cortex as seed, and ROIs in "task-negative" network which was identified by using posterior cingulated cortex/precuneus as seed, connectivity between each pair of these brain regions was evaluated. Results revealed abnormal connectivity with the bilateral dorsal medial prefrontal cortex, the lateral parietal region, the inferior temporal gyrus, the right dorsolateral prefrontal cortex, and the right dorsal premotor cortex in patients with paranoid SZ (Zhou et al. 2007b).

Dysconnectivity of SZ Identified by ICA

Spatial ICA is a technique to decompose the spatio-temporal BOLD signal into spatial brain components and their associated time courses (Beckmann et al. 2005; Calhoun et al. 2001; Lagioia et al. 2010; McKeown et al. 1998). Each component provides a grouping of brain activity into regions that share the same response pattern, thus providing a natural measure of functional connectivity (Calhoun et al. 2009b). Although the results are dependent on the number of components and the user must determine which components reflect noise and which components resemble neuro-anatomical systems (Fox and Raichle 2007), ICA is a data-driven approach which does not require a priori definition of seed regions to analyze the functional connectivity networks in human brain. Spatial ICA is a popular technique to identify temporally coherent networks and identify potential biomarkers in SZ by processing R-fMRI data (Calhoun et al. 2008).

Disrupted Spatial Brain Components

In line with studies using seed-based analysis, multiple dysconnected brain networks in SZ have been observed using ICA approaches. For example, Ongur et al. (2010) observed that SZ patients had greater recruitment (spatial extent) of the frontopolar cortex/basal ganglia and decreased recruitment in the dorsal anterior cingulate cortex, as well as reduced DMN connectivity in the medial prefrontal cortex. Healthy controls revealed stronger effects of ICA-derived connectivity measures in right and left dorsolateral prefrontal cortices, bilateral medial frontal cortex, left precuneus and left posterior lateral parietal cortex, while stronger effects in SZ patients were found in the right amygdala, left orbitofrontal cortex, right anterior cingulate and bilateral inferior temporal cortices. In patients, an inverse correlation of negative symptoms with right anterior prefrontal cortex activity at rest and negative symptoms was observed. These findings suggest that aberrant default mode network connectivity contributes to regional functional pathology in schizophrenia and bears significance for core symptoms (Mingoia et al. 2012). Another ICA study identified increased functional integration in the bilateral caudate nucleus in SZ. And the caudate nucleus in SZ showed altered functional connection with the prefrontal area and cerebellum. These results identified the importance of basal ganglia in SZ (Duan et al. 2015).

A number of dysconnectivity findings in SZ observed by ICA studies were associated with patients' symptoms. For example, Camchong et al. (2011) found lower FC in medial frontal gyrus and anterior cingulate gyrus in patients with chronic SZ. Notably, frontal connectivity in patients was positively associated with symptoms as well as with general cognitive ability measures. Wolf et al. (2011) evaluated the FC of some RSNs in patients with SZ who have auditory–verbal hallucinations (AVHs). Within a speech-related network, patients showed increased connectivity in bilateral temporal regions and decreased connectivity in the cingulate cortex. Within two additional RSNs associated with attention and executive control, respectively, patients exhibited less connectivity in the left precuneus and increased connectivity in the right lateral prefrontal areas. Importantly, correlations between measures of AVH severity and FC of the left anterior cingulate, left superior temporal gyrus, and right lateral prefrontal cortex were revealed. Another study (Sorg et al. 2012) investigated the intrinsic activity of striatum underlining different disorder states using ICA. During psychosis, coherent intrinsic activity of the striatum was increased in the dorsal part and correlated with positive symptoms. In psychotic remission of the same patients, activity of the ventral striatum was increased and correlated with negative symptoms. Rotarska-Jagiela et al. (2010) reported that, SZ patients showed decreased connectivity in posterior cingulate and hippocampus, which was correlated with severity of hallucinations and delusions, and decreased hemispheric separation of fronto-parietal activity, which was correlated with disorganization symptoms.

Disrupted Correlation Between Spatial Brain Components

Other interesting findings based on ICA include altered functional network connectivity (FNC) in SZ. FNC is the functional connectivity of ICA components by evaluating the temporal relationships among ICA time courses across different components, rather than within components (Calhoun et al. 2009a). Jafri et al. (2008) was the first study to evaluate the differences in FNC of resting-state brain components in SZ patients versus healthy controls. That work found higher correlation among most of the dominant RSNs in SZ (Fig. 2). Subsequent studies reported both increased and decreased FNC in SZ. For example, Lui et al. (2010) reported increased connectivity within the frontoparietal-temporal network and decreased connectivity within the temporal-ventromedial frontal network, DMN, and medial frontal areas in drug-naive first-episode SZ patients. Meda et al. (2012) found lower connectivity between fronto/occipital network and anterior default mode/prefrontal

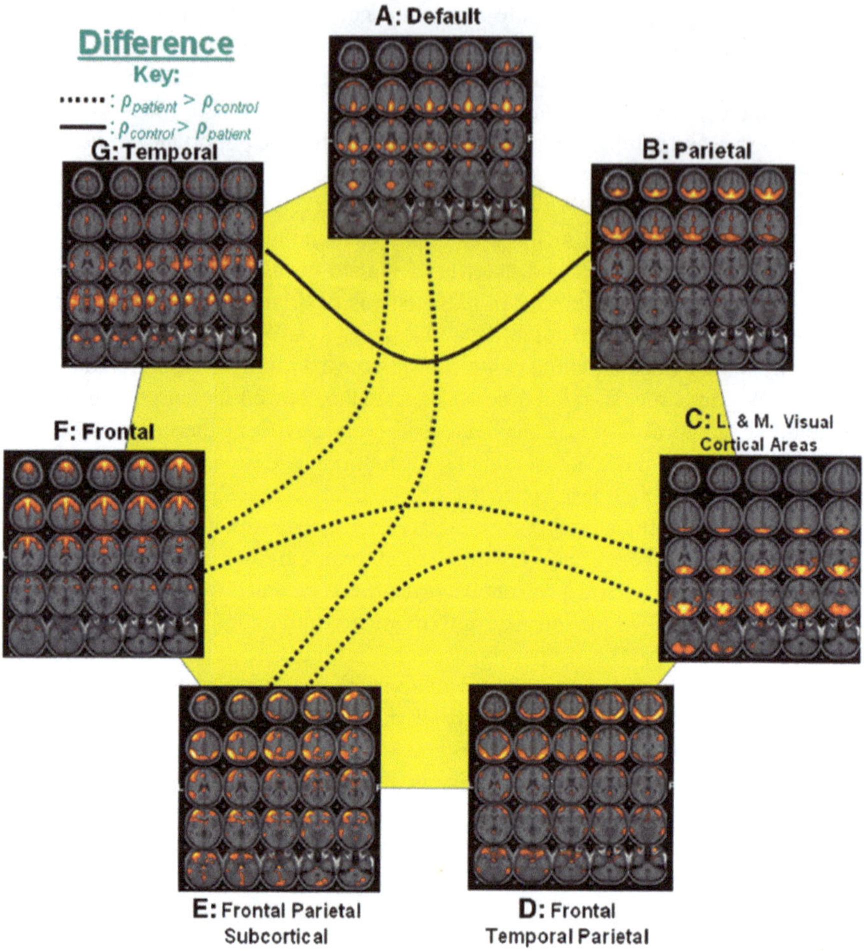

Fig. 2 Group differences of FNC [from (Jafri et al. 2008)]. Out of 21 possible correlation combinations between 7 components, only 5 combinations passed the two-sample t-test ($p < 0.01$). The solid line represents the significant connectivity where controls have higher mean correlation than patients, while dotted line represents connectivity where patients have higher mean correlation. Presence of dotted lines rejects the hypothesis that controls should have more correlation between two components than patients

network, and between meso/paralimbic network and sensory/motor network in SZ. FNC of DMN patterns has also been used to differentiate patients from controls with 76.9% accuracy (Wang et al. 2015a). Another machine learning study showed that classification with higher accuracy was achievable using simple nonlinear discriminative methods such as k-nearest neighbors (KNNs), which is very promising (Arbabshirani et al. 2013).

Another ICA study identified networks of the auditory cortex, left inferior frontal language regions and the anterior cingulate region, associated with attention. FNC analysis revealed decreased connectivity between the auditory and language networks in SZ. Conversely, SZ patients showed increased connectivity between the attention and language network compared to controls (Liemburg et al. 2012).

Connectivity between cerebellum and cerebrum in SZ had been investigated by combining seed-based analysis and ICA methods. Firstly, 12 ROIs of the cerebellum were selected from 6 brain components identified by ICA including the cingulo-opercular network (CON), the dorsal default-mode networks (dDMN), the ventral default-mode network (vDMN), the left frontoparietal networks (lFPN), the right frontoparietal network (rFPN), and the motor network (MOT). Then, FC from the 12 ROIs to all other brain voxels was computed. SZ patients showed a decrease in FC between the cerebellar ROIs and the thalamus in all networks except the MOT. The FC was decreased between cerebellar ROIs and the frontal cortex in the CON, rFPN, and MOT in SZ. However, the FC was increased with the precentral gyrus and postcentral gyrus for the CON, lFPN, and dDMN in SZ. An increased FC with the occipital fusiform gyrus and the temporal occipital fusiform gyrus was also noted in the dDMN, vDMN, and MOT in SZ. These results suggested that the dysconnection of the cerebellum in the SZ was network-specific. The cerebellar–thalamic dysconnection was the most prominent finding and was common to all cognitive-related networks, whereas the cortico-cerebellar connectivity involved both an increase and decrease in SZ, and depended more on the nature of the specific network (Chen et al. 2013b).

Altered spatial maps and FNC in SZ were analyzed in a same dataset to identify default mode network, central executive network, and anterior insular salience network (DMN, CEN, and SN) by ICA. Connectivity between networks and spatial maps of the networks were compared between controls and SZ patients. Spatial BOLD-synchronicity was altered in SZ in SN, DMN and CEN, including decreased activity in the right anterior insular (rAI). FC between DMN and CEN was increased and related to hallucinations severity in SZ. Patients' time-lagged connectivity between SN and DMN/CEN was reduced. And decreased rAI activity of the SN was associated with both hallucinations and increased FC between DMN and CEN (Manoliu et al. 2014).

To study brain FC in first-episode, treatment-naïve patients with SZ, 90 brain components which were extracted by ICA were ascribed to 10 previously well-defined brain networks/areas. Twenty-nine altered functional connections including 19 connections with increased connectivity and 10 connections with decreased connectivity in SZ were revealed after examining the connectivity between those brain components. Increased FC was mainly within the default mode network (DMN) and between the DMN and cognitive networks, whereas decreased connectivity was predominantly associated with sensory networks. Given the key roles of the DMN in internal mental processes and sensory networks in inputs from the external environment, these patterns of altered brain network connectivity suggested imbalanced neural processing of internal and external information in SZ (Li et al. 2015). Another study which investigated the FNC in SZ using rest and task fMRI data found stable group differences of heightened connectivity in SZ between multiple brain areas with thalamus across tasks which underscore the importance of the thalamus as a gateway to sensory input (Cetin et al. 2014).

Dysconnectivity of SZ Identified by Graph Theory-Based Analysis

Graph theory-based network analysis is another widely used method to describe R-fMRI data in order to study mind–brain mechanisms (Bassett and Gazzaniga 2011). Topological properties of brain networks described by nodes and edges can be investigated by graph theory-based analysis. In practice, graph measures such as clustering coefficient, degree, betweenness, path length, local efficiency, global efficiency, and modularity [for the definition of these metrics, see Rubinov and Sporns (Rubinov and Sporns 2010)] are often used to characterize system properties (Telesford et al. 2011). See Fig. 3 for a flowchart of graph theory-based analysis implemented in R-fMRI studies. Applications of this approach have demonstrated that the intrinsic activity of human brain tends to be organized as small-world, with significant modularity and highly connected hub regions which indicate high efficiency of brain systems (Bullmore and Sporns 2009; Wang et al. 2010). Recent studies implemented the graph-based analysis in brain disorders and found altered graph metrics in patients with SZ during resting state (Fornito et al. 2012).

Altered Graph Measures

Most of the topological metrics of small-world brain networks have been found to be altered in SZ. For example, when using atlas-based ROIs as brain graph nodes, Liu et al. (2008) observed lower degree, lower connectivity strength, decreased clustering coefficient, and longer characteristic path length in patients with SZ. The

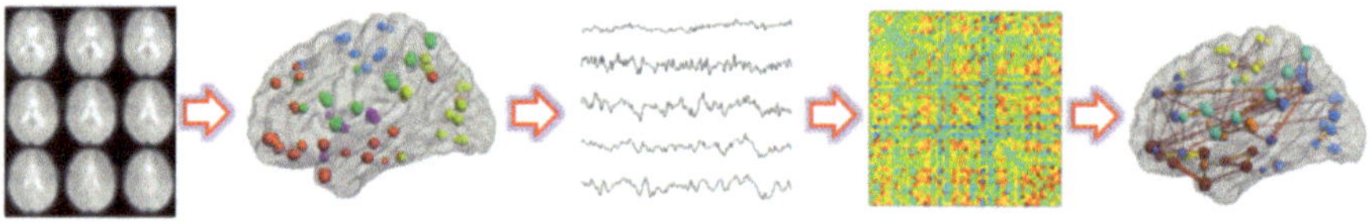

Fig. 3 A schematic flowchart for implementing graph theory-based analysis in R-fMRI data. The first step was defining nodes based on ROIs of brain atlas, voxels, or brain components. Secondly, the time courses of all nodes were extracted followed by preprocessing (e.g., band-pass filtering, autocorrelation correction, covarying for white matter variance, or other confounding variables). Subsequently, a correlation matrix representing the temporal correlation between any pair of nodes was computed. Next, a brain graph with nodes and edges could be built based on the correlation matrix. Then, network metrics of the graph such as clustering, efficiencies, and path length could be evaluated

properties of particular brain regions in the prefrontal, parietal and temporal lobes have been shown to be altered. Lynall et al. (2010) consistently found decreased connectivity strength, reduced clustering, and small-worldness. In addition, reduced degree and clustering were locally significant in medial parietal, premotor and cingulated, and right orbitofrontal areas. Alexander-Bloch et al. (2010) also revealed reduced connectivity strength, decreased clustering and local efficiency, but increased global efficiency and robustness in patients with childhood-onset SZ. Cortical regions with abnormally reduced local connectivity included the left and right superior temporal gyrus, left ventral occipital cortex, right cingulate, right insula, and right frontal operculum; subcortical decreases of clustering bilaterally in the thalamus, caudate, and accumbens; increased efficiency in the right inferior parietal lobule, left ventral temporal cortex, bilateral frontal operculum, and right planum polare. When using brain components as network nodes, Yu et al. (2011) discovered higher clustering coefficient, higher local efficiency, higher characteristic path length, and lower global efficiency in SZ. The topological measures were locally altered in brain components involving frontal, parietal, occipital, and cerebellar areas in SZ. Importantly, characteristic path length and global efficiency were correlated with negative scale of PANSS in patients.

Multi-level analysis can be performed by combining graph theory-based analysis and other techniques. For example, Bassett et al. (2012) quantitatively characterized the univariate wavelet entropy of regional activity, the bivariate pairwise functional connectivity between regions, and the multivariate network organization of connectivity patterns. The results indicated that univariate measures were less sensitive to disease state than higher level bivariate and multivariate measures. In addition, by considering the network structure as a function of correlation strength, network organization specifically of weak connections was strongly correlated with attention, memory, and negative symptom scores of patients with SZ, and displays potential as a clinical biomarker.

One study tested the hypothesis that brain network randomization is an endophenotype of schizophrenia and therefore evident also in nonpsychotic relatives of patients. R-fMRI data were acquired from patients with schizophrenia, first-degree relatives of patients, and healthy volunteers. Graphs were used to model FC as a set of edges between regional nodes. Topological efficiency, clustering, degree distribution, resilience, and connection distance of each functional graph were estimated. The schizophrenic group demonstrated significant randomization of global network metrics (reduced clustering, greater efficiency), a shift in the degree distribution to a more homogeneous form (fewer hubs), a shift in the distance distribution (proportionally more long-distance edges), and greater resilience to targeted attack on network hubs. The graphs of the relatives also demonstrated abnormal randomization and resilience compared with healthy controls, but they were typically less topologically abnormal than the patients' networks and did not have abnormal connection distances. These findings implied that schizophrenia was associated with replicable and convergent evidence for functional network randomization, and a similar topological profile was evident also in nonpsychotic relatives, suggesting that this is a systems-level endophenotype or marker of familial risk (Lo et al. 2015).

Graph analysis has been performed to assess functional integration and functional segregation of brain networks in unmedicated SZ patients. Graph nodes were defined by 279 ROIs. R-fMRI scans were obtained from healthy controls, and SZ patients while unmedicated, and again after 6 weeks of treatment with risperidone. Results showed that compared with healthy controls, unmedicated SZ patients showed reduced global efficiency and increased clustering coefficients. And this pattern of aberrant functional network integration and segregation was modulated with antipsychotic medications, but only in those who responded to treatment. These findings lend support to the concept of schizophrenia as a dysconnectivity syndrome, and suggested that faulty brain network topology in schizophrenia was modulated by antipsychotic medication as a function of treatment response (Hadley et al. 2016).

When using independent brain components detected by ICA as graph nodes, SZ patients show significantly less clustering (transitivity) among components than healthy controls with networks less likely to be connected, and also showed lower small-world connectivity than healthy controls (Anderson and Cohen 2013).

Brain graph measures have also been used to classify SZ patients and controls. Using 278 ROI nodes, whether the network measure of betweenness centrality could classify persons with SZ and normal controls was tested. The betweenness centrality of each node, or fraction of shortest paths that pass through it, was calculated in order to characterize the centrality of the different regions. The nodes with high betweenness centrality agreed well with hub nodes reported in previous studies of structural and functional networks. Using a linear support vector machine algorithm, the schizophrenia group was differentiated from non-psychiatric controls using the ten nodes with the highest betweenness centrality. The classification accuracy was around 80%, and stable against connectivity thresholding. Better performance was achieved when using the ranks as feature space as opposed to the actual values of betweenness centrality. Overall, these findings suggested that a specific network property, betweenness centrality, can classify persons with SZ with a high level of accuracy (Cheng et al. 2015).

Another study systematically explored relationships between functional connectivity, complex network topology, and anatomical (Euclidean) distance between connected brain regions, in R-fMRI data in patients with childhood-onset SZ (COS). Results showed that normal between-subject differences in average distance of connected edges in brain graphs were strongly associated with variation in topological properties of functional networks. In addition, a club or subset of connector hubs was identified, in lateral temporal, parietal, dorsal prefrontal, and medial prefrontal/cingulate cortical regions. In COS, there was reduced functional connectivity over short distances especially, and therefore, global mean connection distance of thresholded graphs was significantly greater than normal. As predicted from relationships between spatial and topological properties of normal networks, this disorder-related proportional increase in connection distance was associated with reduced clustering and modularity and increased global efficiency of COS networks. Between-group differences in connection distance were localized specifically to connector hubs of multimodal association cortex (Alexander-Bloch et al. 2013).

Altered Modular Organization

Modular community structure has been repeatedly demonstrated in resting-state functional brain connectivity networks (He et al. 2009; Meunier et al. 2009; Salvador et al. 2005). Modular organization of brain networks may play a critical role in its evolution and neurodevelopment (Bassett et al. 2011; Kashtan and Alon 2005; Meunier et al. 2010). Since it has been hypothesized that brain disorders such as SZ are associated with developmental factors (van Os et al. 2010), the examination of modular organization in SZ may provide biomarkers of altered brain development in this psychosis (Alexander-Bloch et al. 2010). Indeed, reduced modularity which implies proportionally less intra-modular edges and more inter-modular edges in patients with childhood-onset SZ were shown as well (Alexander-Bloch et al. 2010). When using brain components as network nodes to build weighted networks, Yu et al. (2012) found more and smaller modules with different hubs in SZ (see Fig. 4). Alexander-Bloch et al. (2012) also reported altered modular community structure in patients with childhood-onset SZ. The difference in the modular partition was most striking for a module that consistently includes brain regions around the right anterior insula in healthy controls. In contrast, these regions were distributed over a number of other modules in patient group.

Another study that used 264 ROIs as nodes sought to determine whether community structure alterations were present in an adult-onset schizophrenia cohort while stringently controlling for sources of imaging artifacts. Group-level average

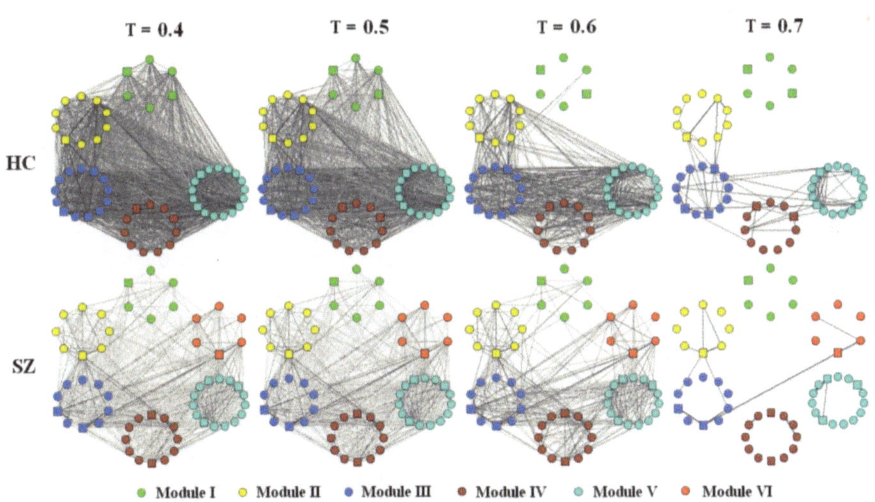

Fig. 4 Schematic modular architecture of FNC for healthy controls and patients with SZ [from (Yu et al. 2012)]. Nodes in a given module were placed in a small circle. To show how edges disappear with threshold (T) change, graphs (each consisted of 57 brain components) built by T at 0.4, 0.5, 0.6, and 0.7 were shown, respectively. An edge (either- or intra-module edge) was shown in the graph if its weight was bigger than that of T; the thicker and the darker an edge, the higher its weight. Hubs were indicated by square nodes

graphs in healthy controls and individuals with SZ exhibited visually similar network community structures and high amounts of normalized mutual information (NMI). However, testing of individual subject community structures identified small but significant alterations in community structure with alterations being driven by changes in node community membership in the somatosensory, auditory, default mode, salience, and subcortical networks (Lerman-Sinkoff and Barch 2016).

Altered Dynamic Functional Connectivity in SZ

While early studies focused on the static brain connectivity in R-fMRI data, temporal dynamic connectivity across seconds to minutes had been investigated by fMRI studies during the last few years (Calhoun et al. 2014; Sakoglu et al. 2010). Sliding time-windows is one of the most popular methods for assessing dynamic brain connectivity in resting-state fMRI data (Allen et al. 2014). Different connectivity states can be estimated by this approach. Recent studies reported disrupted dynamic brain connectivity in schizophrenia.

Altered Dynamic Connectivity States

By analyzing brain connectivity states using dynamic approach, SZ patients were observed to be making fewer transitions to some states, and showed differences from controls including both hyper- and hypoconnectivity in one common connectivity state. These results provided new information about that illness and strongly suggested that state-based analyses were critical to avoid averaging together important factors that can help distinguish these clinical groups (Rashid et al. 2014).

By applying temporal independent component analysis (tICA) to dynamic functional network connectivity (FNC) of the brain, it was revealed that SZ patients showed reduced FNC dynamism. (1) SZ patients exhibited diminished dynamic fluidity: (a) SZ patients switched less frequently between five-dimensional meta-states. (b) SZ patients occupied a smaller number of distinct meta-states during their scans. (2) SZ patients operated over a restricted dynamic range: (a) SZ patients remain trapped in a smaller radius hypersphere of the state space and (b) SZ patients covered less distance as they move through the state space (Miller et al. 2014, 2016).

In another study which performed both static and dynamic connectivity analysis, dynamic analysis suggested that (1), on average, SZ patients spend much less time than healthy controls in states typified by strong, large-scale connectivity and (2) abnormal connectivity patterns were more pronounced during these connectivity states. In particular, states exhibiting cortical–subcortical antagonism (anticorrelations) and strong positive connectivity between sensory networks showed group differences of thalamic hyperconnectivity and sensory hypoconnectivity. Group differences were weak or absent during other connectivity states. Dynamic analysis also revealed hypoconnectivity between the putamen and sensory networks during the same states of thalamic hyperconnectivity. Notably, this finding cannot be observed in the static connectivity analysis. In addition, post-hoc analyses observed that the relationships between subcortical low frequency power and connectivity with sensory networks were altered in patients, suggesting different functional interactions between subcortical nuclei and sensorimotor cortex during specific connectivity states (Damaraju et al. 2014).

When estimating dynamic connectivity within default mode network using a sliding time window method followed by K-means clustering, controls were found to show stronger inter-subsystem interaction than patients. Patients spent more time in the connectivity states with nodes sparsely connected. For each state, SZ patients presented relatively weaker connectivity strengths mainly in inter-subsystem. Patients also exhibited lower values in averaged node strength, clustering coefficient, global efficiency, and local efficiency than controls. These findings indicated that SZ patients had impaired interaction among DMN subsystems, with a reduced central role for posterior cingulate cortex (PCC) and anterior medial prefrontal cortex (aMPFC) hubs as well as weaker interaction between dorsal medial prefrontal cortex (dMPFC) subsystem and medial temporal lobe (MTL) subsystem (Du et al. 2016).

Other Altered Dynamic Properties

By computing the graph measures of dynamic brain connectivity, it had been shown that SZ patients have decreased variance in the dynamic graph metrics. And graph measures of connectivity states showed lower values in SZ (Yu et al. 2015). Another study also revealed lower variability in the default-mode network in SZ patients (Zhang et al. 2016).

A classification study showed that the dynamic FNC significantly outperformed the static FNC in terms of predictive accuracy, indicating that features from dynamic FNC have distinct advantages over static FNC for classification purposes. Moreover, combining static and dynamic FNC features slightly increase but do not significantly improve the classification performance over the dynamic FNC features alone, suggesting that static FNC does not add significant information beyond dynamic FNC for classification purposes (Rashid et al. 2016).

Altered ALFF in SZ

Amplitude of low-frequency fluctuation (ALFF) is defined as the sum of amplitude of each voxel's signal frequency spectrum within the low-frequency range and reflects the amplitude of spontaneous low-frequency fluctuations in the BOLD signal (Golestani et al. 2017). This mathematical measure of signal power has been widely applied to characterize resting-state fMRI data (Biswal et al. 1995; Zang et al. 2007). Altered ALFF in SZ has been shown in several studies.

A study investigated the ALFF and fractional ALFF (fALFF) between SZ patients and healthy controls. Results showed that regressing out motion and non-gray matter signals decreases fALFF throughout the brain as well as ALFF in the cortical edge, but increases ALFF in subcortical regions. Regression had little effect on site, age, and diagnosis effects on ALFF, other than to reduce diagnosis effects in subcortical regions. Controls showed greater ALFF in the occipital, posterior parietal, and superior temporal lobe, while SZ showed smaller clusters of greater ALFF in the frontal and temporal/insular regions as well as in the caudate, putamen, and hippocampus. Controls showed greater fALFF compared with patients in all regions, though subcortical differences were only significant for original fALFF (Turner et al. 2013). A few ICA studies have revealed lower ALFF in multiple brain components in SZ and bipolar disorder (Calhoun et al. 2011; Garrity et al. 2007).

Another study measured the ALFF and fALFF in three different frequency bands (slow-5: 0.01–0.027 Hz; slow-4: 0.027–0.08 Hz; and typical band: 0.01–0.08 Hz) of R-fMRI data in SZ patients and controls. Significant differences in ALFF/fALFF between the two bands (slow-5 and slow-4) in regions including basal ganglia, midbrain, and ventromedial prefrontal cortex were found. Importantly, significant interaction between frequency bands and groups in inferior occipital gyrus, precuneus, and thalamus was also identified. The results suggested that the abnormalities of low-frequency oscillations in schizophrenia was dependent on the frequency band and suggested that future studies should take the different frequency bands into account when measuring intrinsic brain activity (Yu et al. 2014).

Cognitive correlates of ALFF reductions were researched using R-fMRI data of SZ patients and healthy controls. Voxelwise fractional ALFF (fALFF) was regressed on neuropsychological measures of sustained attention and working memory. Poorer sustained attention was associated with smaller fALFF in the left superior frontal cortex and bilateral temporo-parietal junction in both groups, with additional relationships in bilateral posterior parietal, posterior cingulate, dorsal anterior cingulate (ACC), and right dorsolateral prefrontal cortex (DLPFC) evident only in SZ. Poorer working memory was associated with smaller fALFF in bilateral ACC/mPFC, DLPFC, and posterior parietal cortex in both groups. These findings suggest that smaller amplitudes of low-frequency BOLD oscillations during rest, measured by fALFF, were significantly associated with poorer cognitive performance, sometimes similarly in both groups and sometimes only in SZ, in regions known to subserve sustained attention and working memory. These data suggest that the magnitude of resting-state BOLD oscillations shows promise as a biomarker of cognitive function in health and disease (Fryer et al. 2015).

The abnormality of ALFF in adolescents with early-onset schizophrenia (EOS) has also been investigated. In a study, voxel-wise ALFF of R-fMRI data was compared between EOS and healthy controls. And seed-based functional connectivity between significantly disturbed ALFF regions and whole-brain voxels was computed. EOS exhibited increased ALFF in orbitofrontal cortex (OFC) and decreased ALFF in ventral precuneus compared with controls. Decreased ALFF in the precuneus of EOS showed a significant negative correlation with negative symptom scores on the positive and negative syndrome scale (PANSS). Disturbed FC mainly occurred between the orbitofrontal cortex and the temporal cortex in EOS (Zheng et al. 2016).

Dynamic ALFF has also been examined. One study calculated the ALFF of the time series of dynamic connectivity from the typical frequency band (0.01–0.08 Hz). Significant ALFF differences between SZ patients and controls were revealed by two-sample t-tests. The results show significant abnormality in the dynamics of internetwork FC in SZ, which contributes toward the characterization and differentiation of schizophrenic patients, and may be used as a potential biomarker (Shen et al. 2014). Fu et al. (2017) consistently showed altered dynamic ALFF and correlation between dynamic ALFF and dynamic brain connectivity in SZ.

Altered ReHo in SZ

Regional Homogeneity (ReHo) was defined as the Kendall's coefficient of concordance between a given voxel and its 27 neighboring voxels and represents the synchronization between the time series of a given voxel and its neighbors (Golestani et al. 2017; Zang et al. 2004).

In a study which analyzed the ReHo, 12 ROIs in the frontal and temporal lobes were generated and one-way ANOVA was performed to test the ReHo differences within these ROIs between three groups, healthy controls, patients with SZ, and patients with major depressive disorder (MDD). The ReHo values within ROIs were extracted to investigate whether a left–right asymmetry existed in a mental disorder. One-way ANOVA showed significant differences in ReHo in the right superior frontal gyrus and left superior temporal gyrus; post hoc analysis revealed that SZ patients had lower ReHo in the left superior temporal gyrus than either control subjects or patients with MDD. Increased ReHo was observed in the right superior frontal gyrus in SZ patients compared with control subjects, and a left-less-than-right asymmetry was also found in this region in SZ patients. This work suggested that the altered ReHo in the superior frontal and temporal gyrus may be specific for SZ rather than MDD. A left-less-than-right asymmetry activation pattern may exist in the resting-state superior frontal gyrus in SZ (Chen et al. 2013a).

Another study estimated ReHo in three different frequency bands (slow-5: 0.01–0.027 Hz; slow-4: 0.027–0.08 Hz; and typical band: 0.01–0.08 Hz) in SZ patients and healthy controls. Compared with controls, SZ patients exhibited decreased ReHo in the precentral gyrus, middle occipital gyrus, and posterior insula, and increased ReHo in the medial prefrontal cortex and anterior insula. Significant differences in ReHo between the two bands were found in fusiform gyrus and superior frontal gyrus (slow-4 > slow-5), and in basal ganglia, parahippocampus, and dorsal middle prefrontal gyrus (slow-5 > slow-4). Importantly, that work identified significant interaction between frequency bands and groups in the inferior occipital gyrus and caudate body. These findings suggested that ReHo changes in schizophrenia were widespread and frequency-dependent (Yu et al. 2013).

Voxel-wise ReHo has been calculated in patients with delusions of SZ who were rated with PANSS and characteristics of the delusional rating scale (CDRS). A two-sample t-test showed that ReHo of the right anterior cingulate gyrus and left medial superior frontal gyrus was higher in patients, and ReHo of the left superior occipital gyrus was lower, compared to healthy controls. Further, among patients, correlation analysis showed a significant difference between delusion scores of CDRS and ReHo of brain regions. ReHo of the left medial superior frontal gyrus was negatively correlated with patients' CDRS scores but not with delusional PANSS scores. These results suggested that altered local synchronization of spontaneous brain activity may be related to the pathophysiology of delusion in SZ (Gao et al. 2015).

Summary and Challenges in Future Works

In summary, ALFF, ReHo, and spatial maps of ICA are powerful ways to examine the disturbances of single brain regions' activation in SZ using R-fMRI data. Seed-based (and/or ROI-based) connectivity analysis and ICA-based FNC analysis are powerful to examine dysconnectivity between brain regions in SZ. Graph theory analysis is powerful to examine the disturbances of brain connectivity in SZ at both a global system level and local region level. These characterizations have been assessed from a dynamic view in addition to a static view. Generally, disturbances of above-mentioned features in SZ are revealed to be not only located at a few brain regions, but widespread across the whole brain. All of those findings suggest that both segregation and integration performance of the brain are disturbed in this mental illness.

The majority of seed-based studies showed decreased connectivity, while ICA studies reported both increased and decreased connectivity in SZ (in different networks). Medial frontal cortex and thalamic area were often identified to be involved in aberrant connectivity in SZ by both approaches. Findings revealed by graph theory-based analysis which can examine the properties (such as efficiency and modular organization) of the whole system are complementary to seed-based and ICA studies. Brain graphs in SZ were found to exhibit more randomness than healthy controls. Some studies showed relationships between dysconnectivity and typical symptoms of the disorder. The findings improved our understanding of the underlying deficits of brain networks in SZ. However, there were some limitations. For example, in most studies, medication effects were not distinguished from those related purely to the disease because there were very few studies of unmedicated subjects. One approach to clarifying this is to examine unaffected first-degree relatives of schizophrenia patients to see if abnormal connectivity is inherited as a marker of illness predisposition (i.e., an endophenotype) as opposed to a biological illness marker (Meda et al. 2012).

Specificity is another unresolved problem, that is, whether abnormal connectivity detected in schizophrenia patients is unique to that disorder or shared with other psychotic illnesses such as psychotic bipolar disorder. Again, relatively few studies have attempted to clarify this distinction (Meda et al. 2012).

Regarding the direction of alteration in connectivity, both decreased and increased connectivity in SZ had been reported, although most seed-based studies revealed reduced connectivity in SZ. Bi-directed topological metrics have also been revealed by graph theory-based analysis. The explanation of these inconsistencies in direction was not clear, though methodological differences among studies (such as seeds definition in seed-based analysis, number of components in FNC analysis, nodes definition in graph theory-based analysis) may play a role. Relationships between the alteration of ALFF, ReHo, and connectivity in SZ should be investigated in future studies (Fu et al. 2017).

In addition, all studies focused on dysconnectivity in SZ at a group level, and it is difficult to identify dysconnectivity at single subject level to be used in clinic. However, classification may be a potential approach to do discrimination on a case-by-case basis, because classification provides information for each individual subject (Du et al. 2012). Although it is not easy to design an accurate, robust classifier for classifying SZ patients based on a small number of training samples and high dimensional data, a few studies have tried to design objective prognostic/diagnostic tools using R-fMRI data. For example, based on functional connectivity between pairs of 116 ROIs, Shen et al. (2010) classified SZ patients from controls with a high accuracy (94% for patients and 75% for healthy controls). Arbabshirani and Calhoun (2011) also classified patients from healthy controls with high accuracy using FNC. These results suggested that machine learning applied to R-fMRI may be a useful tool for improving the current diagnosis of SZ.

There are some specific challenges for each technique. In studies using seed-based analysis, different seeds (or ROIs) are defined in different studies. A standard criterion for determining seed(s) should be developed. In studies doing ICA-based FNC analysis, different numbers of brain components involving different areas were selected. The number of brain components and which areas they include should be provided when implementing an FNC analysis. In graph theory-based analysis studies, nodes and edges are defined differently. Nodes are usually defined by atlas-based ROIs or ICA components (Yu et al. 2017). Edges are usually defined by partial correlation or Pearson correlation. The best method of building brain graphs to explore brain networks is not clear and can have a large impact on the results and conclusions. The above challenges should be considered in future studies.

References

Alexander-Bloch, A., Lambiotte, R., Roberts, B., Giedd, J., Gogtay, N., Bullmore, E., 2012. The discovery of population differences in network community structure: new methods and applications to brain functional networks in schizophrenia. Neuroimage 59, 3889–3900.

Alexander-Bloch, A.F., Gogtay, N., Meunier, D., Birn, R., Clasen, L., Lalonde, F., Lenroot, R., Giedd, J., Bullmore, E.T., 2010. Disrupted modularity and local connectivity of brain functional networks in childhood-onset schizophrenia. Front Syst Neurosci 4, 147.

Alexander-Bloch, A.F., Vertes, P.E., Stidd, R., Lalonde, F., Clasen, L., Rapoport, J., Giedd, J., Bullmore, E.T., Gogtay, N., 2013. The anatomical distance of functional connections predicts brain network topology in health and schizophrenia. Cereb Cortex 23, 127–138.

Allen, E.A., Damaraju, E., Plis, S.M., Erhardt, E.B., Eichele, T., Calhoun, V.D., 2014. Tracking whole-brain connectivity dynamics in the resting state. Cereb Cortex 24, 663–676.

Anderson, A., Cohen, M.S., 2013. Decreased small-world functional network connectivity and clustering across resting state networks in schizophrenia: an fMRI classification tutorial. Front Hum Neurosci 7, 520.

Andreasen, N.C., Pierson, R., 2008. The role of the cerebellum in schizophrenia. Biol Psychiatry 64, 81–88.

Arbabshirani, M.R., Calhoun, V.D., 2011. Functional network connectivity during rest and task: comparison of healthy controls and schizophrenic patients. Conference proceedings: ... Annual International Conference of the IEEE Engineering in Medicine and Biology Society. IEEE Engineering in Medicine and Biology Society. Conference 2011, 4418–4421.

Arbabshirani, M.R., Kiehl, K.A., Pearlson, G.D., Calhoun, V.D., 2013. Classification of schizophrenia patients based on resting-state functional network connectivity. Front Neurosci 7, 133.

Bassett, D.S., Gazzaniga, M.S., 2011. Understanding complexity in the human brain. Trends Cogn Sci 15, 200–209.

Bassett, D.S., Nelson, B.G., Mueller, B.A., Camchong, J., Lim, K.O., 2012. Altered resting state complexity in schizophrenia. Neuroimage 59, 2196–2207.

Bassett, D.S., Wymbs, N.F., Porter, M.A., Mucha, P.J., Carlson, J.M., Grafton, S.T., 2011. Dynamic reconfiguration of human brain networks during learning. Proc Natl Acad Sci U S A 108, 7641–7646.

Beckmann, C.F., DeLuca, M., Devlin, J.T., Smith, S.M., 2005. Investigations into resting-state connectivity using independent component analysis. Philosophical transactions of the Royal Society of London. Series B, Biological sciences 360, 1001–1013.

Berman, R.A., Gotts, S.J., McAdams, H.M., Greenstein, D., Lalonde, F., Clasen, L., Watsky, R.E., Shora, L., Ordonez, A.E., Raznahan, A., Martin, A., Gogtay, N., Rapoport, J., 2016. Disrupted sensorimotor and social-cognitive networks underlie symptoms in childhood-onset schizophrenia. Brain 139, 276–291.

Biswal, B., Yetkin, F.Z., Haughton, V.M., Hyde, J.S., 1995. Functional connectivity in the motor cortex of resting human brain using echo-planar MRI. Magn Reson Med 34, 537–541.

Biswal, B.B., 2012. Resting state fMRI: a personal history. Neuroimage 62, 938–944.

Bluhm, R.L., Miller, J., Lanius, R.A., Osuch, E.A., Boksman, K., Neufeld, R.W., Theberge, J., Schaefer, B., Williamson, P., 2007. Spontaneous low-frequency fluctuations in the BOLD signal in schizophrenic patients: anomalies in the default network. Schizophr Bull 33, 1004–1012.

Bluhm, R.L., Miller, J., Lanius, R.A., Osuch, E.A., Boksman, K., Neufeld, R.W., Theberge, J., Schaefer, B., Williamson, P.C., 2009. Retrosplenial cortex connectivity in schizophrenia. Psychiatry Res 174, 17–23.

Bullmore, E., Sporns, O., 2009. Complex brain networks: graph theoretical analysis of structural and functional systems. Nat Rev Neurosci 10, 186–198.

Calhoun, V.D., Adali, T., Pearlson, G.D., Pekar, J.J., 2001. A method for making group inferences from functional MRI data using independent component analysis. Hum Brain Mapp 14, 140–151.

Calhoun, V.D., Eichele, T., Pearlson, G., 2009a. Functional brain networks in schizophrenia: a review. Front Hum Neurosci 3, 17.
Calhoun, V.D., Kiehl, K.A., Pearlson, G.D., 2008. Modulation of temporally coherent brain networks estimated using ICA at rest and during cognitive tasks. Hum Brain Mapp 29, 828–838.
Calhoun, V.D., Liu, J., Adali, T., 2009b. A review of group ICA for fMRI data and ICA for joint inference of imaging, genetic, and ERP data. Neuroimage 45, S163–172.
Calhoun, V.D., Miller, R., Pearlson, G., Adali, T., 2014. The chronnectome: time-varying connectivity networks as the next frontier in fMRI data discovery. Neuron 84, 262–274.
Calhoun, V.D., Sui, J., Kiehl, K., Turner, J., Allen, E., Pearlson, G., 2011. Exploring the psychosis functional connectome: aberrant intrinsic networks in schizophrenia and bipolar disorder. Frontiers in psychiatry/Frontiers Research Foundation 2, 75.
Camchong, J., MacDonald, A.W., 3rd, Bell, C., Mueller, B.A., Lim, K.O., 2011. Altered functional and anatomical connectivity in schizophrenia. Schizophr Bull 37, 640–650.
Cetin, M.S., Christensen, F., Abbott, C.C., Stephen, J.M., Mayer, A.R., Canive, J.M., Bustillo, J.R., Pearlson, G.D., Calhoun, V.D., 2014. Thalamus and posterior temporal lobe show greater inter-network connectivity at rest and across sensory paradigms in schizophrenia. Neuroimage 97, 117–126.
Chen, J., Xu, Y., Zhang, K., Liu, Z., Xu, C., Shen, Y., Xu, Q., 2013a. Comparative study of regional homogeneity in schizophrenia and major depressive disorder. Am J Med Genet B Neuropsychiatr Genet 162B, 36–43.
Chen, Y.L., Tu, P.C., Lee, Y.C., Chen, Y.S., Li, C.T., Su, T.P., 2013b. Resting-state fMRI mapping of cerebellar functional dysconnections involving multiple large-scale networks in patients with schizophrenia. Schizophr Res 149, 26–34.
Cheng, H., Newman, S., Goni, J., Kent, J.S., Howell, J., Bolbecker, A., Puce, A., O'Donnell, B.F., Hetrick, W.P., 2015. Nodal centrality of functional network in the differentiation of schizophrenia. Schizophr Res 168, 345–352.
Damaraju, E., Allen, E.A., Belger, A., Ford, J.M., McEwen, S., Mathalon, D.H., Mueller, B.A., Pearlson, G.D., Potkin, S.G., Preda, A., Turner, J.A., Vaidya, J.G., van Erp, T.G., Calhoun, V.D., 2014. Dynamic functional connectivity analysis reveals transient states of dysconnectivity in schizophrenia. Neuroimage Clin 5, 298–308.
Deco, G., Corbetta, M., 2011. The dynamical balance of the brain at rest. Neuroscientist 17, 107–123.
Deco, G., Jirsa, V.K., McIntosh, A.R., 2011. Emerging concepts for the dynamical organization of resting-state activity in the brain. Nat Rev Neurosci 12, 43–56.
Du, W., Calhoun, V.D., Li, H., Ma, S., Eichele, T., Kiehl, K.A., Pearlson, G.D., Adali, T., 2012. High classification accuracy for schizophrenia with rest and task FMRI data. Front Hum Neurosci 6, 145.
Du, Y.H., Pearlson, G.D., Yu, Q.B., He, H., Lin, D.D., Sui, J., Wu, L., Calhoun, V.D., 2016. Interaction among subsystems within default mode network diminished in schizophrenia patients: A dynamic connectivity approach. Schizophrenia Research 170, 55–65.
Duan, M., Chen, X., He, H., Jiang, Y., Jiang, S., Xie, Q., Lai, Y., Luo, C., Yao, D., 2015. Altered Basal Ganglia Network Integration in Schizophrenia. Front Hum Neurosci 9, 561.
Erhardt, E.B., Allen, E.A., Damaraju, E., Calhoun, V.D., 2011. On network derivation, classification, and visualization: a response to Habeck and Moeller. Brain connectivity 1, 1–19.
Fan, F.M., Tan, S.P., Yang, F.D., Tan, Y.L., Zhao, Y.L., Chen, N., Li, B.B., Song, C.S., Wang, Y.H., Jin, Z., Zhou, D.F., Milham, M.P., Zou, Y.Z., Zuo, X.N., 2013. Ventral medial prefrontal functional connectivity and emotion regulation in chronic schizophrenia: a pilot study. Neuroscience bulletin 29, 59–74.
Fornito, A., Zalesky, A., Breakspear, M., 2015. The connectomics of brain disorders. Nat Rev Neurosci 16, 159–172.
Fornito, A., Zalesky, A., Pantelis, C., Bullmore, E.T., 2012. Schizophrenia, neuroimaging and connectomics. Neuroimage 62, 2296–2314.

Fox, M.D., Greicius, M., 2010. Clinical applications of resting state functional connectivity. Front Syst Neurosci 4, 19.

Fox, M.D., Raichle, M.E., 2007. Spontaneous fluctuations in brain activity observed with functional magnetic resonance imaging. Nat Rev Neurosci 8, 700–711.

Friston, K.J., Frith, C.D., 1995. Schizophrenia: a disconnection syndrome? Clin Neurosci 3, 89–97.

Friston, K.J., Frith, C.D., Fletcher, P., Liddle, P.F., Frackowiak, R.S., 1996. Functional topography: multidimensional scaling and functional connectivity in the brain. Cereb Cortex 6, 156–164.

Friston, K.J., Frith, C.D., Liddle, P.F., Frackowiak, R.S., 1993. Functional connectivity: the principal-component analysis of large (PET) data sets. Journal of cerebral blood flow and metabolism: official journal of the International Society of Cerebral Blood Flow and Metabolism 13, 5–14.

Fryer, S.L., Roach, B.J., Ford, J.M., Turner, J.A., van Erp, T.G., Voyvodic, J., Preda, A., Belger, A., Bustillo, J., O'Leary, D., Mueller, B.A., Lim, K.O., McEwen, S.C., Calhoun, V.D., Diaz, M., Glover, G., Greve, D., Wible, C.G., Vaidya, J., Potkin, S.G., Mathalon, D.H., 2015. Relating Intrinsic Low-Frequency BOLD Cortical Oscillations to Cognition in Schizophrenia. Neuropsychopharmacology 40, 2705–2714.

Fu, Z., Tu, Y., Di, X., Du, Y., Pearlson, G.D., Turner, J.A., Biswal, B.B., Zhang, Z., Calhoun, V.D., 2017. Characterizing dynamic amplitude of low-frequency fluctuation and its relationship with dynamic functional connectivity: An application to schizophrenia. Neuroimage.

Gao, B., Wang, Y., Liu, W., Chen, Z., Zhou, H., Yang, J., Cohen, Z., Zhu, Y., Zang, Y., 2015. Spontaneous Activity Associated with Delusions of Schizophrenia in the Left Medial Superior Frontal Gyrus: A Resting-State fMRI Study. PLoS One 10, e0133766.

Garrity, A.G., Pearlson, G.D., McKiernan, K., Lloyd, D., Kiehl, K.A., Calhoun, V.D., 2007. Aberrant "default mode" functional connectivity in schizophrenia. Am J Psychiatry 164, 450–457.

Gavrilescu, M., Rossell, S., Stuart, G.W., Shea, T.L., Innes-Brown, H., Henshall, K., McKay, C., Sergejew, A.A., Copolov, D., Egan, G.F., 2010. Reduced connectivity of the auditory cortex in patients with auditory hallucinations: a resting state functional magnetic resonance imaging study. Psychol Med 40, 1149–1158.

Golestani, A.M., Kwinta, J.B., Khatamian, Y.B., Chen, J.J., 2017. The Effect of Low-Frequency Physiological Correction on the Reproducibility and Specificity of Resting-State fMRI Metrics: Functional Connectivity, ALFF, and ReHo. Front Neurosci 11, 546.

Greicius, M., 2008. Resting-state functional connectivity in neuropsychiatric disorders. Curr Opin Neurol 21, 424–430.

Hadley, J.A., Kraguljac, N.V., White, D.M., Ver Hoef, L., Tabora, J., Lahti, A.C., 2016. Change in brain network topology as a function of treatment response in schizophrenia: a longitudinal resting-state fMRI study using graph theory. NPJ Schizophr 2, 16014.

He, Y., Wang, J., Wang, L., Chen, Z.J., Yan, C., Yang, H., Tang, H., Zhu, C., Gong, Q., Zang, Y., Evans, A.C., 2009. Uncovering intrinsic modular organization of spontaneous brain activity in humans. PLoS One 4, e5226.

He, Z., Deng, W., Li, M., Chen, Z., Jiang, L., Wang, Q., Huang, C., Collier, D.A., Gong, Q., Ma, X., Zhang, N., Li, T., 2013. Aberrant intrinsic brain activity and cognitive deficit in first-episode treatment-naive patients with schizophrenia. Psychol Med 43, 769–780.

Hoptman, M.J., D'Angelo, D., Catalano, D., Mauro, C.J., Shehzad, Z.E., Kelly, A.M., Castellanos, F.X., Javitt, D.C., Milham, M.P., 2010. Amygdalofrontal functional disconnectivity and aggression in schizophrenia. Schizophr Bull 36, 1020–1028.

Hutchison, R.M., Womelsdorf, T., Allen, E.A., Bandettini, P.A., Calhoun, V.D., Corbetta, M., Della Penna, S., Duyn, J.H., Glover, G.H., Gonzalez-Castillo, J., Handwerker, D.A., Keilholz, S., Kiviniemi, V., Leopold, D.A., de Pasquale, F., Sporns, O., Walter, M., Chang, C., 2013. Dynamic functional connectivity: promise, issues, and interpretations. Neuroimage 80, 360–378.

Jafri, M.J., Pearlson, G.D., Stevens, M., Calhoun, V.D., 2008. A method for functional network connectivity among spatially independent resting-state components in schizophrenia. Neuroimage 39, 1666–1681.

Kashtan, N., Alon, U., 2005. Spontaneous evolution of modularity and network motifs. Proc Natl Acad Sci U S A 102, 13773–13778.
Klingner, C.M., Langbein, K., Dietzek, M., Smesny, S., Witte, O.W., Sauer, H., Nenadic, I., 2014. Thalamocortical connectivity during resting state in schizophrenia. European archives of psychiatry and clinical neuroscience 264, 111–119.
Lagioia, A., Van De Ville, D., Debbane, M., Lazeyras, F., Eliez, S., 2010. Adolescent resting state networks and their associations with schizotypal trait expression. Frontiers in systems neuroscience 4.
Lerman-Sinkoff, D.B., Barch, D.M., 2016. Network community structure alterations in adult schizophrenia: identification and localization of alterations. Neuroimage Clin 10, 96–106.
Li, M.L., Deng, W., He, Z.L., Wang, Q., Huang, C.H., Jiang, L.J., Gong, Q.Y., Ziedonis, D.M., King, J.A., Ma, X.H., Zhang, N.Y., Li, T., 2015. A splitting brain: Imbalanced neural networks in schizophrenia. Psychiatry Research-Neuroimaging 232, 145–153.
Liang, M., Zhou, Y., Jiang, T., Liu, Z., Tian, L., Liu, H., Hao, Y., 2006. Widespread functional disconnectivity in schizophrenia with resting-state functional magnetic resonance imaging. Neuroreport 17, 209–213.
Liemburg, E.J., Vercammen, A., Ter Horst, G.J., Curcic-Blake, B., Knegtering, H., Aleman, A., 2012. Abnormal connectivity between attentional, language and auditory networks in schizophrenia. Schizophr Res 135, 15–22.
Liu, X., Zhuo, C., Qin, W., Zhu, J., Xu, L., Xu, Y., Yu, C., 2016. Selective functional connectivity abnormality of the transition zone of the inferior parietal lobule in schizophrenia. Neuroimage Clin 11, 789–795.
Liu, Y., Liang, M., Zhou, Y., He, Y., Hao, Y., Song, M., Yu, C., Liu, H., Liu, Z., Jiang, T., 2008. Disrupted small-world networks in schizophrenia. Brain 131, 945–961.
Lo, C.Y.Z., Su, T.W., Huang, C.C., Hung, C.C., Chen, W.L., Lan, T.H., Lin, C.P., Bullmore, E.T., 2015. Randomization and resilience of brain functional networks as systems-level endophenotypes of schizophrenia. Proceedings of the National Academy of Sciences of the United States of America 112, 9123–9128.
Lui, S., Li, T., Deng, W., Jiang, L., Wu, Q., Tang, H., Yue, Q., Huang, X., Chan, R.C., Collier, D.A., Meda, S.A., Pearlson, G., Mechelli, A., Sweeney, J.A., Gong, Q., 2010. Short-term effects of antipsychotic treatment on cerebral function in drug-naive first-episode schizophrenia revealed by "resting state" functional magnetic resonance imaging. Arch Gen Psychiatry 67, 783–792.
Lynall, M.E., Bassett, D.S., Kerwin, R., McKenna, P.J., Kitzbichler, M., Muller, U., Bullmore, E., 2010. Functional connectivity and brain networks in schizophrenia. J Neurosci 30, 9477–9487.
Manoliu, A., Riedl, V., Zherdin, A., Muhlau, M., Schwerthoffer, D., Scherr, M., Peters, H., Zimmer, C., Forstl, H., Bauml, J., Wohlschlager, A.M., Sorg, C., 2014. Aberrant dependence of default mode/central executive network interactions on anterior insular salience network activity in schizophrenia. Schizophr Bull 40, 428–437.
Marin, O., 2012. Interneuron dysfunction in psychiatric disorders. Nat Rev Neurosci 13, 107–120.
McKeown, M.J., Makeig, S., Brown, G.G., Jung, T.P., Kindermann, S.S., Bell, A.J., Sejnowski, T.J., 1998. Analysis of fMRI data by blind separation into independent spatial components. Hum Brain Mapp 6, 160–188.
Meda, S.A., Gill, A., Stevens, M.C., Lorenzoni, R.P., Glahn, D.C., Calhoun, V.D., Sweeney, J.A., Tamminga, C.A., Keshavan, M.S., Thaker, G., Pearlson, G.D., 2012. Differences in resting-state functional magnetic resonance imaging functional network connectivity between schizophrenia and psychotic bipolar probands and their unaffected first-degree relatives. Biological Psychiatry 71, 881–889.
Meunier, D., Lambiotte, R., Bullmore, E.T., 2010. Modular and hierarchically modular organization of brain networks. Front Neurosci 4, 200.
Meunier, D., Lambiotte, R., Fornito, A., Ersche, K.D., Bullmore, E.T., 2009. Hierarchical modularity in human brain functional networks. Front Neuroinform 3, 37.
Miller, R.L., Yaesoubi, M., Calhoun, V.D., 2014. Higher Dimensional Analysis Shows Reduced Dynamism of Time-Varying Network Connectivity in Schizophrenia Patients. 2014 36th

Annual International Conference of the Ieee Engineering in Medicine and Biology Society (Embc), 3837–3840.
Miller, R.L., Yaesoubi, M., Turner, J.A., Mathalon, D., Preda, A., Pearlson, G., Adali, T., Calhoun, V.D., 2016. Higher Dimensional Meta-State Analysis Reveals Reduced Resting fMRI Connectivity Dynamism in Schizophrenia Patients. PLoS One 11, e0149849.
Mingoia, G., Wagner, G., Langbein, K., Maitra, R., Smesny, S., Dietzek, M., Burmeister, H.P., Reichenbach, J.R., Schlosser, R.G., Gaser, C., Sauer, H., Nenadic, I., 2012. Default mode network activity in schizophrenia studied at resting state using probabilistic ICA. Schizophr Res 138, 143–149.
Moran, L.V., Tagamets, M.A., Sampath, H., O'Donnell, A., Stein, E.A., Kochunov, P., Hong, L.E., 2013. Disruption of anterior insula modulation of large-scale brain networks in schizophrenia. Biol Psychiatry 74, 467–474.
Oertel-Knochel, V., Knochel, C., Matura, S., Prvulovic, D., Linden, D.E., van de Ven, V., 2013. Reduced functional connectivity and asymmetry of the planum temporale in patients with schizophrenia and first-degree relatives. Schizophr Res 147, 331–338.
Ongur, D., Lundy, M., Greenhouse, I., Shinn, A.K., Menon, V., Cohen, B.M., Renshaw, P.F., 2010. Default mode network abnormalities in bipolar disorder and schizophrenia. Psychiatry Res 183, 59–68.
Pearlson, G.D., Calhoun, V.D., 2009. Convergent approaches for defining functional imaging endophenotypes in schizophrenia. Front Hum Neurosci 3, 37.
Raichle, M.E., 2010. Two views of brain function. Trends in cognitive sciences 14, 180–190.
Raichle, M.E., MacLeod, A.M., Snyder, A.Z., Powers, W.J., Gusnard, D.A., Shulman, G.L., 2001. A default mode of brain function. Proceedings of the National Academy of Sciences of the United States of America 98, 676–682.
Raichle, M.E., Mintun, M.A., 2006. Brain work and brain imaging. Annual review of neuroscience 29, 449–476.
Raichle, M.E., Snyder, A.Z., 2007. A default mode of brain function: a brief history of an evolving idea. Neuroimage 37, 1083–1090; discussion 1097–1089.
Rashid, B., Arbabshirani, M.R., Damaraju, E., Cetin, M.S., Miller, R., Pearlson, G.D., Calhoun, V.D., 2016. Classification of schizophrenia and bipolar patients using static and dynamic resting-state fMRI brain connectivity. Neuroimage 134, 645–657.
Rashid, B., Damaraju, E., Pearlson, G.D., Calhoun, V.D., 2014. Dynamic connectivity states estimated from resting fMRI Identify differences among Schizophrenia, bipolar disorder, and healthy control subjects. Front Hum Neurosci 8, 897.
Rosazza, C., Minati, L., 2011. Resting-state brain networks: literature review and clinical applications. Neurological sciences: official journal of the Italian Neurological Society and of the Italian Society of Clinical Neurophysiology 32, 773–785.
Rosazza, C., Minati, L., Ghielmetti, F., Mandelli, M.L., Bruzzone, M.G., 2012. Functional connectivity during resting-state functional MR imaging: study of the correspondence between independent component analysis and region-of-interest-based methods. AJNR. American journal of neuroradiology 33, 180–187.
Rotarska-Jagiela, A., van de Ven, V., Oertel-Knochel, V., Uhlhaas, P.J., Vogeley, K., Linden, D.E., 2010. Resting-state functional network correlates of psychotic symptoms in schizophrenia. Schizophr Res 117, 21–30.
Rubinov, M., Sporns, O., 2010. Complex network measures of brain connectivity: uses and interpretations. Neuroimage 52, 1059–1069.
Sakoglu, U., Pearlson, G.D., Kiehl, K.A., Wang, Y.M., Michael, A.M., Calhoun, V.D., 2010. A method for evaluating dynamic functional network connectivity and task-modulation: application to schizophrenia. MAGMA 23, 351–366.
Salvador, R., Suckling, J., Coleman, M.R., Pickard, J.D., Menon, D., Bullmore, E., 2005. Neurophysiological architecture of functional magnetic resonance images of human brain. Cereb Cortex 15, 1332-1342.

Schilbach, L., Derntl, B., Aleman, A., Caspers, S., Clos, M., Diederen, K.M., Gruber, O., Kogler, L., Liemburg, E.J., Sommer, I.E., Muller, V.I., Cieslik, E.C., Eickhoff, S.B., 2016. Differential Patterns of Dysconnectivity in Mirror Neuron and Mentalizing Networks in Schizophrenia. Schizophr Bull 42, 1135–1148.

Shen, H., Li, Z., Zeng, L.L., Yuan, L., Chen, F., Liu, Z., Hu, D., 2014. Internetwork dynamic connectivity effectively differentiates schizophrenic patients from healthy controls. Neuroreport 25, 1344–1349.

Shen, H., Wang, L.B., Liu, Y.D., Hu, D.W., 2010. Discriminative analysis of resting-state functional connectivity patterns of schizophrenia using low dimensional embedding of fMRI. Neuroimage 49, 3110–3121.

Smieskova, R., Allen, P., Simon, A., Aston, J., Bendfeldt, K., Drewe, J., Gruber, K., Gschwandtner, U., Klarhoefer, M., Lenz, C., Scheffler, K., Stieglitz, R.D., Radue, E.W., McGuire, P., Riecher-Rossler, A., Borgwardt, S.J., 2011. Different duration of at-risk mental state associated with neurofunctional abnormalities. A multimodal imaging study. human Brain Mapping, https://doi.org/10.1002/hbm.21360.

Sorg, C., Manoliu, A., Neufang, S., Myers, N., Peters, H., Schwerthoffer, D., Scherr, M., Muhlau, M., Zimmer, C., Drzezga, A., Forstl, H., Bauml, J., Eichele, T., Wohlschlager, A.M., Riedl, V., 2012. Increased Intrinsic Brain Activity in the Striatum Reflects Symptom Dimensions in Schizophrenia. Schizophrenia Bulletin, https://doi.org/10.1093/schbul/sbr1184.

Stephan, K.E., Baldeweg, T., Friston, K.J., 2006. Synaptic plasticity and dysconnection in schizophrenia. Biol Psychiatry 59, 929–939.

Su, T.W., Lan, T.H., Hsu, T.W., Biswal, B.B., Tsai, P.J., Lin, W.C., Lin, C.P., 2013. Reduced neuro-integration from the dorsolateral prefrontal cortex to the whole brain and executive dysfunction in schizophrenia patients and their relatives. Schizophr Res 148, 50–58.

Tang, Y., Wang, L.F., Cao, F., Tan, L.W., 2012. Identify schizophrenia using resting-state functional connectivity: an exploratory research and analysis. Biomedical engineering online 11.

Telesford, Q.K., Simpson, S.L., Burdette, J.H., Hayasaka, S., Laurienti, P.J., 2011. The brain as a complex system: using network science as a tool for understanding the brain. Brain connectivity 1, 295–308.

Turner, J.A., Damaraju, E., van Erp, T.G., Mathalon, D.H., Ford, J.M., Voyvodic, J., Mueller, B.A., Belger, A., Bustillo, J., McEwen, S., Potkin, S.G., Fbirn, Calhoun, V.D., 2013. A multi-site resting state fMRI study on the amplitude of low frequency fluctuations in schizophrenia. Front Neurosci 7, 137.

van den Heuvel, M.P., Hulshoff Pol, H.E., 2010. Exploring the brain network: a review on resting-state fMRI functional connectivity. Eur Neuropsychopharmacol 20, 519–534.

van Os, J., Kenis, G., Rutten, B.P., 2010. The environment and schizophrenia. Nature 468, 203–212.

Venkataraman, A., Whitford, T.J., Westin, C.F., Golland, P., Kubicki, M., 2012. Whole brain resting state functional connectivity abnormalities in schizophrenia. Schizophr Res 139, 7–12.

Vercammen, A., Knegtering, H., den Boer, J.A., Liemburg, E.J., Aleman, A., 2010. Auditory hallucinations in schizophrenia are associated with reduced functional connectivity of the temporo-parietal area. Biological Psychiatry 67, 912–918.

Wang, H., Zeng, L.L., Chen, Y., Yin, H., Tan, Q., Hu, D., 2015a. Evidence of a dissociation pattern in default mode subnetwork functional connectivity in schizophrenia. Sci Rep 5, 14655.

Wang, H.L., Rau, C.L., Li, Y.M., Chen, Y.P., Yu, R., 2015b. Disrupted thalamic resting-state functional networks in schizophrenia. Front Behav Neurosci 9, 45.

Wang, J., Zuo, X., He, Y., 2010. Graph-based network analysis of resting-state functional MRI. Front Syst Neurosci 4, 16.

Wang, L., Zou, F., Shao, Y., Ye, E., Jin, X., Tan, S., Hu, D., Yang, Z., 2014. Disruptive changes of cerebellar functional connectivity with the default mode network in schizophrenia. Schizophr Res 160, 67–72.

Whitfield-Gabrieli, S., Thermenos, H.W., Milanovic, S., Tsuang, M.T., Faraone, S.V., McCarley, R.W., Shenton, M.E., Green, A.I., Nieto-Castanon, A., LaViolette, P., Wojcik, J., Gabrieli, J.D., Seidman, L.J., 2009. Hyperactivity and hyperconnectivity of the default network in

schizophrenia and in first-degree relatives of persons with schizophrenia. Proceedings of the National Academy of Sciences of the United States of America 106, 1279–1284.
Williamson, P.C., Allman, J.M., 2012. A framework for interpreting functional networks in schizophrenia. Frontiers in human neuroscience 6, 184.
Wolf, N.D., Sambataro, F., Vasic, N., Frasch, K., Schmid, M., Schonfeldt-Lecuona, C., Thomann, P.A., Wolf, R.C., 2011. Dysconnectivity of multiple resting-state networks in patients with schizophrenia who have persistent auditory verbal hallucinations. Journal of psychiatry & neuroscience: JPN 36, 366–374.
Woodward, N.D., Karbasforoushan, H., Heckers, S., 2012. Thalamocortical dysconnectivity in schizophrenia. Am J Psychiatry 169, 1092–1099.
Woodward, N.D., Rogers, B., Heckers, S., 2011. Functional resting-state networks are differentially affected in schizophrenia. Schizophrenia Research 130, 86–93.
Wu, L., Eichele, T., Calhoun, V.D., 2010. Reactivity of hemodynamic responses and functional connectivity to different states of alpha synchrony: a concurrent EEG-fMRI study. Neuroimage 52, 1252–1260.
Xu, L., Qin, W., Zhuo, C., Zhu, J., Liu, H., Liu, X., Xu, Y., Yu, C., 2015. Selective Functional Disconnection of the Dorsal Subregion of the Temporal Pole in Schizophrenia. Sci Rep 5, 11258.
Yan, H., Tian, L., Yan, J., Sun, W., Liu, Q., Zhang, Y.B., Li, X.M., Zang, Y.F., Zhang, D., 2012. Functional and anatomical connectivity abnormalities in cognitive division of anterior cingulate cortex in schizophrenia. PLoS One 7, e45659.
Yu, Q., Du, Y., Chen, J., He, H., Sui, J., Pearlson, G., Calhoun, V.D., 2017. Comparing brain graphs in which nodes are regions of interest or independent components: a simulation study. Journal of neuroscience methods.
Yu, Q., Erhardt, E.B., Sui, J., Du, Y., He, H., Hjelm, D., Cetin, M.S., Rachakonda, S., Miller, R.L., Pearlson, G., Calhoun, V.D., 2015. Assessing dynamic brain graphs of time-varying connectivity in fMRI data: application to healthy controls and patients with schizophrenia. Neuroimage 107, 345–355.
Yu, Q., Plis, S.M., Erhardt, E.B., Allen, E.A., Sui, J., Kiehl, K.A., Pearlson, G., Calhoun, V.D., 2012. Modular Organization of Functional Network Connectivity in Healthy Controls and Patients with Schizophrenia during the Resting State. Front Syst Neurosci 5, 103.
Yu, Q., Sui, J., Rachakonda, S., He, H., Gruner, W., Pearlson, G., Kiehl, K.A., Calhoun, V.D., 2011. Altered topological properties of functional network connectivity in schizophrenia during resting state: a small-world brain network study. PLoS One 6, e25423.
Yu, R., Chien, Y.L., Wang, H.L., Liu, C.M., Liu, C.C., Hwang, T.J., Hsieh, M.H., Hwu, H.G., Tseng, W.Y., 2014. Frequency-specific alternations in the amplitude of low-frequency fluctuations in schizophrenia. Hum Brain Mapp 35, 627–637.
Yu, R., Hsieh, M.H., Wang, H.L., Liu, C.M., Liu, C.C., Hwang, T.J., Chien, Y.L., Hwu, H.G., Tseng, W.Y., 2013. Frequency dependent alterations in regional homogeneity of baseline brain activity in schizophrenia. PLoS One 8, e57516.
Zang, Y., Jiang, T., Lu, Y., He, Y., Tian, L., 2004. Regional homogeneity approach to fMRI data analysis. Neuroimage 22, 394–400.
Zang, Y.F., He, Y., Zhu, C.Z., Cao, Q.J., Sui, M.Q., Liang, M., Tian, L.X., Jiang, T.Z., Wang, Y.F., 2007. Altered baseline brain activity in children with ADHD revealed by resting-state functional MRI. Brain Dev 29, 83–91.
Zhang, J., Cheng, W., Liu, Z., Zhang, K., Lei, X., Yao, Y., Becker, B., Liu, Y., Kendrick, K.M., Lu, G., Feng, J., 2016. Neural, electrophysiological and anatomical basis of brain-network variability and its characteristic changes in mental disorders. Brain 139, 2307–2321.
Zheng, J., Zhang, Y., Guo, X., Duan, X., Zhang, J., Zhao, J., Chen, H., 2016. Disrupted amplitude of low-frequency fluctuations in antipsychotic-naive adolescents with early-onset schizophrenia. Psychiatry Res 249, 20–26.

Zhou, Y., Liang, M., Jiang, T., Tian, L., Liu, Y., Liu, Z., Liu, H., Kuang, F., 2007a. Functional dysconnectivity of the dorsolateral prefrontal cortex in first-episode schizophrenia using resting-state fMRI. Neuroscience letters 417, 297–302.

Zhou, Y., Liang, M., Tian, L., Wang, K., Hao, Y., Liu, H., Liu, Z., Jiang, T., 2007b. Functional disintegration in paranoid schizophrenia using resting-state fMRI. Schizophr Res 97, 194–205.

Zhou, Y., Ma, X., Wang, D., Qin, W., Zhu, J., Zhuo, C., Yu, C., 2015. The selective impairment of resting-state functional connectivity of the lateral subregion of the frontal pole in schizophrenia. PLoS One 10, e0119176.

Zhou, Y., Shu, N., Liu, Y., Song, M., Hao, Y., Liu, H., Yu, C., Liu, Z., Jiang, T., 2008. Altered resting-state functional connectivity and anatomical connectivity of hippocampus in schizophrenia. Schizophrenia Research 100, 120–132.

Zou, Q., Long, X., Zuo, X., Yan, C., Zhu, C., Yang, Y., Liu, D., He, Y., Zang, Y., 2009. Functional connectivity between the thalamus and visual cortex under eyes closed and eyes open conditions: a resting-state fMRI study. human Brain Mapping 30, 3066–3078.

Disturbed Brain Networks in the Psychosis High-Risk State?

André Schmidt and Stefan Borgwardt

Introduction

Early detection of emerging mental illnesses is decisive for the application of preventive interventions. Identification and intervention early in the course of the illness maximizes treatment effectiveness, clinical outcome and quality of life. A promising avenue to establish predictive markers for ongoing disease progression is research on high-risk states. However, the prognostic accuracy of clinical assessment instruments to predict risk propensity in individuals at increased risk for developing psychiatric disorders, in particular psychosis, is limited mainly because they do not capture relevant pathophysiological processes (Fusar-Poli et al. 2015; Schmidt et al. 2016a). This forms a major barrier to understanding disease mechanisms and developing more effective treatments. Interventions need to directly target the pathophysiological processes causing a disorder in a manner that enduringly modifies its progression (Millan et al. 2016). There is thus an urgent need to find pathophysiological data to improve risk prediction in people at high risk for developing psychiatric disorders.

Psychiatry has long been searching for biological test (Andreasen 1995) to validate and improve the limited predictive validity of the phenomenologically based definition of mental diseases (Cuthbert and Insel 2013). Like in other medical specialities, biological assays will unlikely replace clinical diagnosis but might help to supplement them (Kapur et al. 2012) by uncovering pathophysiological processes

A. Schmidt
Department of Psychiatry (UPK), University of Basel, Basel, Switzerland

S. Borgwardt (✉)
Department of Psychiatry (UPK), University of Basel, Basel, Switzerland

Department of Psychiatry, Psychosomatics and Psychotherapy, University of Lübeck, Lübeck, Germany
e-mail: stefan.borgwardt@web.de

V. A. Diwadkar, S. B. Eickhoff (eds.), *Brain Network Dysfunction in Neuropsychiatric Illness*, https://doi.org/10.1007/978-3-030-59797-9_11

of the disease. In search of potential biomarkers in psychiatry, brain imaging has emerged as a very powerful tool to map direct neurobiological processes associated with the development of the disorder. Numerous imaging studies have demonstrated that psychiatric disorders such as schizophrenia and major depressive disorder are associated with structural, functional and chemical abnormalities in many different brain regions (Drevets et al. 2008; Fusar-Poli et al. 2011; Grace 2016; Hamilton et al. 2012; Howes et al. 2012; Liao et al. 2012; Palaniyappan 2017; Radua et al. 2012, 2015; Russo and Nestler 2013). More recent connectome-wide association studies provide a more fine-grained interpretation by revealing that brain abnormalities in psychiatric disorders are not solely attributable to changes in local regions and connections but rather emerge from changes in the topology of the network as a whole, the connectome of the brain (Crossley et al. 2014, 2015; Filippi et al. 2013; Fornito et al. 2015, 2017; Rubinov and Bullmore 2013; van den Heuvel et al. 2010). The explicit linkage of human neuroimaging data with whole-brain computational modelling has shown great potential not only for a deeper understanding of the computational and biophysical mechanisms underlying healthy states, but also for the discovery of the causes of the breakdown in emerging psychiatric disorders. These network findings have been integrated in the US National Institute of Mental Health's Research Domain Criteria (RDoC), a research road map that focuses on dimensions and pathophysiology-driven classification of psychiatric disorders, with special attention to brain network dysfunction (Insel et al. 2010; Insel and Cuthbert 2015). The RDoC framework highlights the fact that many phenomena vary continuously within and between psychiatric patients and in the general population at large and become pathological only at the extremes of an otherwise normal distribution. The hope is that the study of disruptions to the brain connectome helps to elucidate the pathophysiological mechanisms underlying emerging psychiatric disorders (Castellanos et al. 2013; Deco and Kringelbach 2014) and ultimately to improve risk prediction and the development of preventive interventions in high-risk individuals.

In this chapter, we first briefly introduce the conceptualization of different high-risk samples for psychosis and concepts for the analysis of brain networks, in particular resting-state functional network analyses, graph theoretical approaches and model-based investigations into effective connectivity. We then present a selective review of research showing disrupted brain networks in individuals at high risk for developing psychosis. Finally, we suggest necessary developments for future research to approach a brain network-driven stratification of individual risk propensity in people with an increased risk for psychosis. We conclude that much more large-scale longitudinal studies using standardized clinical assessment tools, thoughtful hub-based network analyses and validated machine learning approaches are required to ultimately evaluate the clinical utility of brain network measures to individually predict the conversion of psychosis in individuals at increased risk.

Table 1 Conceptualization of different high-risk samples for psychosis

Phenotype	Characterization
Clinical high risk	Help-seeking people having one or more of the following signs/symptoms: Attenuated psychotic symptoms (APS) Brief limited intermittent psychotic symptoms (BLIPS) Genetic risk and deterioration syndrome (GRD) Basic symptoms (BS)
Genetic high risk	Monozygotic twins concordant or discordant for mental disorders People with non-clinical symptoms having first- or second-degree relatives with psychiatric disorders People with mutations in candidate genes related to psychiatric disorders. For instance, persons with a 22q11.2 deletion syndrome have a risk for developing schizophrenia spectrum disorders
Schizotypal traits	**Persons (including non-clinical healthy volunteers, CHR and GHR individuals or patients with schizotypal or schizoid personality disorder) fulfilling the following three factors:** **Positive/cognitive-perceptual factor: magical thinking, unusual perceptual experiences, ideas of reference and paranoia** **Disorganized factor: odd behaviour and odd speech** **Interpersonal factor: constricted affect, social anxiety, lack of close personal relationships and suspiciousness**

Conceptualization of High-Risk Samples for Psychosis

High-risk individuals for psychosis can be conceptualized according to their putative endophenotypes as genetic high-risk individuals (GHR) or based on the expression of their clinical symptoms or signs as clinical high-risk (CHR) individuals (Smieskova et al. 2012) and/or persons with schizotypal traits (Debbané and Barrantes-Vidal 2015; Debbané et al. 2015; Debbané and Mohr 2015; Nelson et al. 2013) (Table 1).

Currently, there are two complementary approaches for the characterization of the CHR state of psychoses: the ultra-high risk (UHR) and the basic symptoms criteria (Fusar-Poli et al. 2013; Schultze-Lutter et al. 2015). The UHR concept comprises the attenuated psychotic symptom (APS) criterion, the brief limited intermittent psychotic symptom (BLIPS) criterion, and the genetic risk and deterioration (GRD) criterion. It has been originally developed with the explicit aim of detecting persons at risk for developing a first episode within the next 12 months, with the ultimate goal to administer preventive interventions (Stafford et al. 2013; van der Gaag et al. 2013). In contrast to the UHR concept, the basic symptoms approach, consisting of the cognitive-perceptive basic symptoms (COPER) and the cognitive disturbances (COGDIS) criteria, was developed to detect the risk for psychosis as early as possible in the development of the illness, ideally before functional impairments appear (Klosterkötter et al. 2001). In UHR samples, the transition to psychosis in UHR individuals is most likely to occur within the first 2 years after presentation to clinical services, with 25% of transitions occurring by 106 days and 50% by 240 days (Fusar-Poli and Schultze-Lutter 2016; Kempton et al. 2015). The risk of transition to psychosis accumulates to 29% at 2 years and reaches approximately 35% after 10 years (Fusar-Poli et al. 2012a). This risk (29%) is significantly higher than the risk of psychosis observed in the general population (0.063%) (Kirkbride et al. 2012). It has also been shown that the BLIPS subgroup shows the highest risk of transition to psychosis, the GRD the lowest risk and the APS an intermediate risk and that the APS and the BLIPS subgroups of the UHR show an increased risk of psychosis as compared to subjects seeking help at clinical services but not meeting UHR criteria (Fusar-Poli et al. 2016). In COGDIS samples, the transition rates ranged from 25.3% at 1 year to 61.3% at > 4 years' follow-up (Schultze-Lutter et al. 2015). However, only two studies are available so far to estimate the conversion rates in COGDIS samples (Phillips et al. 2000; Riecher-Rössler et al. 2008).

On the other hand, GHR samples can consist of monozygotic twin pairs concordant or discordant for mental illnesses (Borgwardt et al. 2010; Ettinger et al. 2012; van Haren et al. 2004). In discordant pairs, one twin has the disorder while their co-twin does not; in concordant pairs, both twins have the disorder. Contrasting the unaffected members from discordant pairs and control pairs allows to test to what extent the genetic risk drives disorder-related changes (Ettinger et al. 2012). Studying individuals with first-degree (e.g. parents, siblings) or second-degree (e.g. grandparents, aunt/uncle) relatives of patients affected with a severe mental illness also offers insights into the genetic risk of developing the disorder (Dean et al. 2010; Gottesman et al. 2010; Whalley et al. 2004, 2005). First-degree relatives of schizophrenia patients have approximately a tenfold increased risk for developing psychosis over lifetime (Johnstone et al. 2002). Notably, although previous family high-risk studies have reported that familial risk was diagnosis specific, for example children of parents with schizophrenia were at increased risk for non-affective psychosis but not for mood disorders (Baron et al. 1985; Gershon et al. 1982; Goldstein et al. 2010), accumulating evidence indicates children of parents with severe mental illnesses also having increased risk for mental disorders other than the disorder diagnosed in the parent (Dean et al. 2010). In line with such a view, twin and molecular studies have reported that the risk for mood disorder and schizophrenia is associated with common polygenetic variation (Consortium et al. 2013; Lichtenstein et al. 2009; Purcell et al. 2009). Moreover, findings from a very recent meta-analysis of family high-risk studies confirmed that high-risk offspring had a significantly increased rate not only of the disorder present in the parent but also of other types of severe mental illnesses (Rasic et al. 2014). In particular, the risk of mood disorders was significantly increased among offspring of parents with schizophrenia and the risk of schizophrenia was significantly increased in offspring of parents with bipolar disorder. They concluded that offspring of parents with severe mental disorders are at increased risk for a range of psychiatric disorders and one-third of them may develop a severe mental illness by early adulthood (Rasic et al. 2014). This means that a transdiagnostic staging approach is going to be necessary (McGorry and Nelson 2016). Another approach to study genetic effects on the development of psychiatric disorders is to study candidate genes already associated with disease risk, as for instance persons with a deletion of a small piece of chromosome 22 (22q11.2 deletion syndrome). 30–40% of these individuals will develop schizophrenia spectrum disorders during adulthood and even more will experience psychotic symptoms during their lifetime (Monks et al. 2014; Schneider et al. 2014).

Finally, schizotypal traits also present a putative phenotypic marker of elevated risk for psychosis (Debbané et al. 2015; Nelson et al. 2013). Schizotypy includes personality traits such as odd or bizarre behaviour, strange speech, magical thinking, unusual perceptual experiences and social anhedonia. Although still a matter of debate (Mason and Claridge 2006), the contemporary understanding of schizotypy is that it is comprised of 3 factors, which broadly correspond to the positive, negative and disorganized dimensions of schizophrenia. The first factor is the 'cognitive-perceptual factor', which includes magical thinking, unusual perceptual experiences, ideas of reference and paranoia. The second is the 'disorganized factor', which includes odd behaviour and odd speech. The final 'interpersonal factor' resembles the negative dimension of schizophrenia and includes constricted affect, social anxiety, lack of close personal relationships and suspiciousness (Debbané et al. 2015; Nelson et al. 2013). It has been suggested that there might be a schizotypy–schizophrenia spectrum not only with regard to symptoms and clinical signs, but also common underlying biological factors (Barrantes-Vidal et al. 2015; Ettinger et al. 2014; Hazlett et al. 2012; Nelson et al. 2013; Raine 2006). This dimensional approach posits that the latent structure of schizotypy is on a continuum, ranging from low schizotypy in healthy non-clinical populations, to extremely high schizotypy and potential dysfunction in the form of psychosis (Nelson et al. 2013). While schizotypal traits per se are not sufficient to indicate risk for psychopathology, high levels of schizotypy combined with other aetiological risk factors increase individuals' susceptibility to schizophrenia and other psychotic disorders (Rawlings et al. 2008). In other words, unless high schizotypy is combined with other risk factors, it is considered neutral in regard to psychopathology (Rawlings et al. 2008).

Brief Summary of Brain Network Analyses

Resting-State Functional Connectivity Approaches

Resting-state functional connectivity (rsFC) has become popular over the last decade in psychiatric research because it involves no task demands and minimal compliance during a brief magnetic resonance imaging (MRI) scan, and does not have behavioural confounds. rsFC is based on the analysis of low-frequency fluctuations present in the blood oxygenation level-dependent (BOLD) signal obtained from functional magnetic resonance imaging (fMRI) (Biswal et al. 2010). These low-frequency fluctuations have been shown to be temporally correlated within spatially distinct but functionally related resting-state networks (RSN) (Fox et al. 2005), such as for instance the default mode network, sensorimotor network, executive control network or fronto-parietal network (Smith et al. 2009). Although conventional analysis of resting-state fMRI data does not allow studying topological properties of brain networks, evaluating rsFC can provide important information regarding inherent brain function that may help identify networks key to emerging psychiatric disorders, which may be diagnostically or therapeutically useful (Woodward and Cascio 2015).

The two most applied approaches for the evaluation of rsFC are seed-based correlation analysis (SCA) (Biswal et al. 1995) and independent components analysis (ICA) (Beckmann et al. 2005), but several other valuable methods exist as well including graph theoretical approaches (see below for more details) (Meszlényi et al. 2017; Smith et al. 2011; Smith et al. 2013). SCA (Biswal et al. 1995) is a univariate hypothesis-driven approach, whereby connectivity is calculated by correlating the time series from a priori defined seed voxel (or region of interest) with time series from all other voxel across the brain (or specific regions of interest). The result of SCA is a connectivity map showing correlation coefficient (or Z-scores) for each voxel indicating how well its time series correlates with the time series of the seed. ICA (McKeown et al. 1998; McKeown and Sejnowski 1998) is a multivariate data-driven technique, which identifies a predefined number of spatial templates and corresponding time courses to model the data: the activities of the individual voxel are a linear combination of these component time series (Meszlényi et al. 2017). Each independent source defines a component, therefore clusters arise not only from actual neural synchronization (i.e. RSN), but from other signal sources such as measurement noise and artefacts (Meszlényi et al. 2017). A considerable advantage of such exploratory techniques over more hypothesis-based SCA techniques is the ability to identify various types of signal fluctuations by virtue of their spatial and/or temporal characteristics without the need to specify an explicit temporal model. Such flexibility in data modelling is essential in cases where the effects of interest are not well understood and cannot be predicted accurately (Beckmann et al. 2005).

Model-Based Investigations of Effective Connectivity

A second approach constitutes model-based formulations of context-specific effective (directed) connectivity that are based on the biophysics of neuronal interactions, such as dynamic causal modelling (DCM) (Park and Friston 2013). DCM (Friston et al. 2003, 2013) is a generic Bayesian system identification technique, which allows for inference on 'hidden' neurophysiological mechanisms that generated observed measures, such as BOLD signal in fMRI or evoked responses measured with electroencephalography (EEG). The principle idea thereby is to formulate a simplified model of neuronal population responses and combine this with a modality-specific forward model such that one can predict the measurement that would arise from any particular neuronal circuit. DCM is particularly useful to test of a priori defined competing theories about neuronal circuits, which can be compared with model comparison procedures such as Bayesian model selection (Stephan et al. 2007b, 2009). In DCM for fMRI, the dynamics of the neural states underlying

BOLD responses are modelled by a bilinear differential equation that describes how the neural states (x) change over time (t) as a function of endogenous interregional connections (matrix A), as well as modulatory effects on these connections (matrix B), and driving inputs (matrix C) (see formula below) (Friston et al. 2003; Stephan et al. 2007a). The endogenous connections represent coupling strengths in the absence of input to the system, whereas the modulatory effects represent task-specific alterations in this connectivity.

$$f\left(x,u\right)=\frac{dx}{xt}=\left(A+\sum_{i=1}^{m}uB^{(j)}\right)x+Cu$$

Graph Theoretical Network Approaches

The most recently used approach to measure brain networks is based on mathematical graph theory describing the brain's network topology in terms of undirected connections (correlations) derived from structural or functional MRI data (Bullmore and Sporns 2009). Graph theory analysis has emerged as a very helpful approach to infer complex network properties of the brain from a more global perspective. Graph theory can be applied to data from many imaging modalities such as structural and functional MRI or DTI. The topological properties or patterns of connectivity can be expressed by a variety of parameters, including measures of segregation (e.g. clustering, transitivity, modularity), integration (e.g. path length, efficiency), centrality (e.g. edge and node betweenness) or resilience (e.g. assortativity) (Bullmore et al. 2009; Rubinov and Sporns 2010). One major contribution of graph theory to our understanding of psychiatric disorders has been in highlighting the important role of hubs, which are nodes of the network with an unusually high number or strength of connections (van den Heuvel and Sporns 2013a). They are therefore positioned to influence – and to be influenced by – many other network nodes (Fornito et al. 2017). It has been shown that some of these brain hubs tend to be more densely interconnected among themselves than would be expected solely from their high degree, forming together a 'rich-club' (van den Heuvel and Sporns 2011). This term is in analogy with social systems, where highly connected people (who are 'rich' in their social connections) form a highly interconnected club (Astikainen et al. 2004). These members of the brain's rich-club serve as a macroscopic anatomical substrate to cross-link functional networks and thus play an important role in the integration of information between segregated functional domains of the human cortex (van den Heuvel and Sporns 2013). The high level of centrality of brain hubs also renders them points of vulnerability that are susceptible to disconnection and dysfunction in psychiatric disorders (Crossley et al. 2014). These findings suggest that psychiatric disorders are associated with a selective breakdown of hub connectivity rather than global brain dysfunction (Fornito et al. 2017).

Disorganized Brain Networks in Subjects at High Risk for Psychosis?

Resting-State Functional Connectivity Approaches

Using diverse hypothesis-driven seed regions, several SCA have been conducted to study rsFC in high-risk samples for psychosis. For instance, a previous study showed that CHR subjects had significantly greater positive connectivity than did controls between the posterior cingulate cortex seed region and other areas in the bilateral anterior cingulate cortex, medial prefrontal cortex, precuneus and lateral parietal cortex (Shim et al. 2010). Also using the posterior cingulate cortex as seed region, it has been shown that patients with psychotic disorders and non-psychotic siblings had increased posterior cingulate cortex connectivity with the left inferior parietal lobule, the left precuneus and the right medial prefrontal cortex. Of note, the functional connectivity of siblings was intermediate to that of controls and patients (Peeters et al. 2015). Again, using the posterior cingulate cortex as the seed region to address default mode network (dys)connectivity, another investigation revealed that the default mode network in GHR individuals showed reduced functional connectivity in the prefrontal areas, posterior cingulate cortex and precuneus. The reduction in prefrontal connectivity was thereby related to the total and general scores on the Positive and Negative Syndrome Scale (Jang et al. 2011). Using an ICA approach to detect RSN, it has further been shown that subjects with a 22q11.2 depletion syndrome showed decreased functional connectivity within the default mode network (Debbané et al. 2012; Padula et al. 2015), which correlated with prodromal symptom intensity and neuropsychological performances (Debbané et al. 2012). In contrast, however, another study observed greater default mode network connectivity in 22q11.2 subjects (Mattiaccio et al. 2016). Another ICA study observed a significant correlation between dysfunctional connectivity in the self-referential RSN and schizotypal trait expression (odd speech and behaviour) in community-living and help-seeking adolescents (Lagioia et al. 2010).

In order to tackle deficits in verbal fluency, a further study focused on the rsFC of Broca's area. The authors reported that CHR subjects exhibited reduced functional connectivity of Broca's area to the right dorsolateral prefrontal cortex and left medial superior frontal cortex, but also increased functional connectivity of Broca's area to the left dorsolateral prefrontal cortex and right anterior insula (Jung et al. 2012). Fornito and colleagues investigated the rsFC of 6 different striatal seed regions, given that dysregulation of corticostriatal circuitry has long been thought to be critical in the aetiology of psychotic disorder (Fornito et al. 2013). They found that individuals with early psychosis and their unaffected relatives showed corticostriatal dysregulation, with dorsal striatal hypo-connectivity and ventral striatal hyperconnectivity with cortical targets (Fornito et al. 2013). Intriguingly, a very recent study showed that schizotypy scores, especially from the positive dimension, in participants from the general public were positively correlated with rsFC between the ventral striatum and frontal cortex and negatively associated with rsFC between the dorsal striatum and posterior cingulate (Wang et al. 2017). This study provides further neurobiological support for dimensionality from schizotypal personality to the clinical disorder. The most recent SCA study in 243 CHR subjects focused on thalamo-cortical connectivity (Anticevic et al. 2015) because this pathway has been previously implicated in established psychosis (Woodward et al. 2012). The findings revealed hypo-connectivity between the thalamus and prefrontal and cerebellar areas, as well as hyper-connectivity between the thalamus and sensory-motor regions. Both patterns were more prominent in CHR subjects who converted to psychosis and significantly correlated with prodromal symptom severity (Anticevic et al. 2015). The last SCA examples (Anticevic et al. 2015; Fornito et al. 2013; Wang et al. 2017) represent straightforward and promising strategies to find brain markers for the onset of psychosis because they systematically follow up replicated evidence from previous cross-sectional studies in chronic psychosis (Borgwardt and Schmidt 2016).

Several investigations have been conducted to examine inter-network RSN connectivity. This is important as, for example, anti-correlation between the default mode and fronto-parietal network is critical for the flexibility of normal cognitive control and its disruption under pathological conditions (Lawrence et al. 2003; McKiernan et al. 2003). Indeed, hyper-connectivity in the default mode network of CHR individuals was associated with reduced connectivity in the task-related network comprising fronto-parietal brain regions (Shim et al. 2010). This finding has been confirmed in other studies in CHR subjects (Fryer et al. 2013; Pelletier-Baldelli et al. 2015; Wotruba et al. 2013). Notably, the enhanced connectivity between the default mode and task-related network was related to poor performance in cognitive functions (Wotruba et al. 2013). In accordance with these studies in CHR subjects, the inverse correlation between the default mode network and the task-positive network is also diminished in GHR individuals (Whitfield-Gabrieli et al. 2009). Another study further showed reduced inter-network connectivity between the fronto-parietal network and the cingulo-opercular network and a cerebellar network in individuals with psychosis and their unaffected siblings and that these reductions were associated with both cognitive impairments and clinical symptoms (Repovs et al. 2011). This is important evidence, as it suggests that antagonism between the default mode and FP activity is mediated by the cingulo-opercular system (Bressler and Menon 2010).

Taken together, these rsFC studies demonstrate a variety of connectivity abnormalities in high-risk subjects for psychosis. The most consistent findings are hyper-connectivity in default mode network and diminished anti-correlation between the default mode and fronto-parietal system, as well as alterations within cortico-striatal-thalamic loops. These findings should guide future resting-state fMRI investigations in order to establish predictive brain marker for psychosis onset in high-risk individuals.

Model-Based Investigations into Effective Connectivity

Using DCM, several studies have investigated whether effective connectivity during working memory processing (N-back task) is impaired in CHR individuals (for a review, see (Dauvermann et al. 2014)). The purpose of these studies was based on evidence showing that working memory deficits are evident not only at the time of the first episode of the disease (Mesholam-Gately et al. 2009) but also in CHR individuals and may help to predict transition to psychosis (Fusar-Poli et al. 2012b). With a focus on endogenous connections (i.e. not modulated by a specific task condition), the first DCM study reported reduced connectivity from the superior temporal gyrus to the middle frontal gyrus in CHR subjects (Crossley et al. 2009). In line with findings in patients with schizophrenia (Deserno et al. 2012) and first-episode psychosis (FEP) (Nielsen et al. 2017), Schmidt and colleagues further found a significant decrease in working memory-induced modulation of fronto-parietal connectivity in CHR subjects (Schmidt et al. 2014b). Notably, the extent of the disruption of fronto-parietal connectivity was negatively related to the formation of psychotic symptoms in CHR subjects, suggesting a mechanistic relation between the degree of functional network integrity and the clinical expression of psychosis (Schmidt et al. 2014b). Moreover, the same group further demonstrated that fronto-parietal connectivity during working memory processing was progressively reduced from healthy controls to CHR participants further to patients with FEP, and that the abnormal modulation of connectivity in FEP patients was normalized by treatment with antipsychotics (Schmidt et al. 2013). The findings of this study suggest that abnormalities in working memory-induced modulation of fronto-parietal connectivity may separate healthy controls from CHR individuals and FEP patients and may thus serve as a physiologic marker that predates the onset of psychosis (Schmidt et al. 2014a).

There are several other DCM studies using different paradigms to address abnormal effective connectivity in high-risk cohorts. For example, it has been shown that CHR and FEP patients revealed decreased connectivity from the right posterior hippocampus to the right inferior frontal gyrus during a short-term memory task (delayed matching to sample task) (Benetti et al. 2009). Using the Hayling Sentence Completion Task (HSCT) to study altered effective connectivity during response initiation and suppression, Allen et al. further found increased endogenous connectivity between the anterior cingulate cortex and the middle temporal gyrus in CHR subjects (Allen et al. 2010). This study also explored task-induced modulation of connectivity but found no difference between healthy controls and CHR individuals. Using psychophysiological interaction to explore effective connectivity, this group also showed that CHR subjects who subsequently developed psychosis showed greater midbrain-to-prefrontal cortex connectivity during a verbal fluency task compared with those who did not become psychotic (Allen et al. 2012). Also during the HSCT, another study demonstrated that the connection strength with the nonlinear modulation between the mediodorsal thalamus and the inferior frontal gyrus was significantly reduced in GHR subjects with psychotic symptoms and in four ill subjects who subsequently developed schizophrenia in contrast to the healthy controls (Dauvermann et al. 2013). There was also a significant correlation between the individual connection strength and the presence of delusions in GHR subjects with psychotic symptoms (Dauvermann et al. 2013). Furthermore, Diwadkar and colleagues revealed disordered corticolimbic interactions during affective processing in children and adolescent at risk for schizophrenia. In particular, they found decreased driving inputs to the visual cortex, decreased intrinsic coupling, most robustly between fronto-limbic regions; and increased modulatory inhibition by negative valence of fronto-limbic connections (Diwadkar et al. 2012). The same group did also explore whether GHR subjects exhibit effective connectivity abnormalities during sustained attention (modified version of a continuous performance task) (Diwadkar et al. 2014). DCM revealed that GHR subjects were characterized by significant reductions in coupling across both fronto-striatal and fronto-parietal pathways (Diwadkar et al. 2014). Finally, informed by recent animal models (Lodge and Grace 2011a, b; Modinos et al. 2014), another study measured neural responses within a hippocampal–basal ganglia–midbrain network during reward, novelty and aversion processing in CHR subjects (Winton-Brown et al. 2017). DCM showed that reward-induced modulation of connectivity from the ventral striatum/pallidum to the midbrain was greater in CHR subjects than in controls, and that, in CHR subjects, the strength of connectivity in this pathway was correlated with the severity of their abnormal beliefs (Winton-Brown et al. 2017).

In summary, these studies demonstrate a broad spectrum of effective connectivity abnormalities in diverse brain circuits in high-risk subjects for psychosis. The strength of such model-based investigations is to test concrete a priori defined pathophysiological mechanisms underlying the transition to psychosis, leading to hypothesis-driven strategies rather than trying to find the needle in the haystack (Borgwardt et al. 2016). One of the most consistent findings is alteration in fronto-parietal connectivity (Schmidt et al. 2014a), which may indicate that abnormalities in this pathway particularly reflect a vulnerability marker for emerging psychosis.

Graph Theoretical Network Approaches

More recent graph theoretic approaches revealed that brain functional networks have efficient small-world properties, which support efficient parallel information transfer at a relatively low cost (Liu et al. 2008). It has been suggested that they live on 'the edge of chaos', supporting the kind of rapid formation, dissolution and adaptation of connectivity that is critical for mental activity (Bassett et al. 2006). Such brain 'hubs' represent specific regions that play a central role in the overall network organization, as indexed by a high degree, low clustering, short path length, high centrality and participation in multiple communities across the network (Sporns et al. 2007). It has been shown that schizophrenia patients displayed altered small-world network organization (Bassett et al. 2008; Liu et al. 2008; van den Heuvel et al. 2013; Zhang et al. 2012), characterized by an imbalance between integration and segregation processes. Another study revealed that the rich-club organization is significantly affected in patients with schizophrenia, together with a reduced density of rich-club connections predominantly comprising the white matter pathways that link the midline frontal, parietal and insular hub regions. This reduction in rich-club density was found to be associated with lower levels of global communication capacity (van den Heuvel et al. 2013). This study provides novel biological evidence that schizophrenia is characterized by a selective disruption of structural brain connectivity among central hub regions of the brain, potentially leading to reduced communication capacity and altered functional brain dynamics (van den Heuvel et al. 2013). Intriguingly, a previous DTI study showed that abnormal rich-club organization is already evident in unaffected siblings of schizophrenia patients if compared with healthy subjects, but less affected than in schizophrenia patients. This study suggests that impaired rich-club connectivity is related to familial, possibly reflecting genetic, vulnerability for schizophrenia (Collin et al. 2013). Moreover, reduced efficiency in central hub regions have also been found in CHR subjects for psychosis (Schmidt et al. 2016b) and individuals with subclinical psychotic experiences (Drakesmith et al. 2015) expressed by a more segregated and less integrated system. In accordance with these findings, impaired integration and efficiency of structural and functional hub regions have also been observed in individuals with a chromosome 22q11.2 deletion syndrome (Ottet et al. 2013) and individuals with high schizotypy (Wang et al. 2015). Furthermore, in line with schizophrenia patients and unaffected siblings, reduced anatomical rich-club connectivity and increased modularity is also evident in CHR subjects (Schmidt et al. 2016b). The disruption of rich-club organization was associated with the severity of negative psychotic symptoms and led to an elevated level of modularity in CHR subjects (Schmidt et al. 2016b). This study highlights the role of structural rich-club connectivity as the backbone for network-wide information integration (van den Heuvel et al. 2012, 2013) and shows a breakdown of this interplay in CHR subjects for psychosis. A very recent study found that schizophrenia offspring also exhibit connectivity deficits of the brain's central rich-club system relative to both control subjects and offspring of patients with bipolar disorder (Collin et al. 2017). Again, the disruption in anatomical rich-club connectivity in schizophrenia offspring was associated with increased modularity of the functional connectome (Collin et al. 2017). Further evidence comes from a study in individuals with a chromosome 22q11.2 deletion syndrome reporting reduced connectivity strength among A-core regions, which exhibited stronger-than-expected interconnectivity and thus resemble a weighted rich-club (Vása et al. 2016). These specific disturbances are associated with a rerouting of shortest network paths that circumvent the A-core in 22q11DS, 'de-centralizing' the network (Vása et al. 2016). These findings together support the notion that a breakdown of hub network connectivity may constitute a core pathoconnectomical hallmark of psychosis (Crossley et al. 2014; Rubinov et al. 2013) and already exists in CHR samples, indicating the potential of network connectivity measures to predict outcomes in psychosis (Borgwardt et al. 2016; McGuire et al. 2015; Schmidt et al. 2014a).

Although the number of graph-based network studies in high-risk is still limited, findings of altered structural and functional dysorganization of central hub regions, in particular rich-club regions across different high-risk samples for psychosis, are remarkably consistent. Graph theoretic measures of hub connectivity appear to be promising to improve individual prediction of conversion to psychosis in high-risk individuals. Such information from candidate brain networks might be more accurate for detecting neural changes in emerging psychiatric disorders than information from isolated brain regions (Schmidt et al. 2014a).

Towards a Brain Network-Based Prediction of Mental Illness

In this chapter, we provided a selective overview of disturbed brain networks in high-risk subjects for psychosis. The findings so far suggest that abnormalities in fronto-parietal networks, altered interactions between the default mode and fronto-parietal network, as well as disorganization of central hub regions may be most critical for emerging psychosis. Although these findings are promising, it has yet to be shown whether they allow individual prediction of risk propensity. To test the predictive potential of these brain network markers, the field has to overcome major hurdles (Borgwardt et al. 2016; Kapur et al. 2012; McGuire and Dazzan 2017).

The first issue is that most studies are clearly underpowered and of cross-sectional nature. Currently, ongoing multicentre studies – such as PRONIA (Personalised Prognostic Tools for Early Psychosis Management), PSYSCAN (Translating Neuroimaging Findings From Research Into Clinical Practice) and NAPLS (North American Prodrome Longitudinal Study) – will be able to address this issue by collecting large high-risk data samples. These samples should also contain a meaningful ratio of converters and non-converters to ensure robust statistical inferences. These datasets should then be analysed with standardized network approaches by focusing on previously established evidence.

Most studies to date have also used a single cross-sectional neuroimaging measure to detect brain network abnormalities. However, to establish predictive brain markers, it has to be disentangled how putative brain markers develop over time and if they are related to transition or remission. Only when such makers could be found, studies can then test if one-time neuroimaging assessments at baseline are sufficient to make robust prediction of personalized outcomes over time. Multi-site studies should also help to face this open challenge by providing longitudinal datasets. Another point of contention is the high clinical heterogeneity of high-risk samples, such as for instance in CHR subjects for psychosis. This is partly due to the different HR criteria used across centres (Fusar-Poli et al. 2013; Smieskova et al. 2010). An important next step is thus to agree on a standardized clinical instrument for the definition of the CHR state for psychosis. Furthermore, as indicated in the introduction, the risk of individuals is not necessarily specific to one certain disease but is rather related to a range of psychiatric disorders, indicating that a transdiagnostic staging approach is going to be necessary. Future research on high-risk states may better focus on mapping transdiagnostic dimensions of psychopathology to specific brain networks. Useful clinical predictions have to be made at the single subject level. An established method for this purpose is the application of pattern recognition techniques such as machine learning. In parallel with the rise of interest in brain networks, there has been an increase in the use and development of machine-learning techniques in clinical neuroscience (Pereira et al. 2009) and in particular in the field of individualized early psychosis prediction (Bendfeldt et al. 2015; de Wit et al. 2017; Koutsouleris et al. 2014; Mechelli et al. 2017; Salvador et al. 2017; Schreiner et al. 2017; Young et al. 2016). Future studies using standardized machine-learning approaches are urgently required to test the clinical utility of brain network markers to predict individual outcomes in high-risk subjects. Of uppermost importance, these models need then to be validated on independent datasets. As indicated in the introduction, brain network markers will unlikely replace any clinical assessment but may help to supplement them. A complementary and innovative approach has recently been suggested in the field of psychosis prediction. This study introduced a multistage sequential testing approach to improve risk prediction (Clark et al. 2015). Sequentially adding (biological) information such as brain network markers on top of the initial clinical assessment might indeed become an avenue for individual risk stratification of high-risk individuals and optimizing preventive interventions (Schmidt et al. 2016a).

In summary, this book chapter reviewed most recent research on disrupted brain network properties in individuals at high risk for developing psychosis. The findings of these studies clearly show that brain networks are already altered in these high-risk subjects and thus precede later chronic disease stages. The pattern of abnormalities is very diverse and more research is required to detect which of them are useful for clinical decision-making. We conclude that much more large-scale longitudinal studies using standardized clinical assessment tools, thoughtful hub-based network approaches and validated machine learning analyses are required to ultimately evaluate the clinical utility of brain network measures to predict the conversion to psychosis in individuals at increased risk.

References

Allen P, Luigjes J, Howes OD, Egerton A, Hirao K, Valli I, *et al* (2012). Transition to psychosis associated with prefrontal and subcortical dysfunction in ultra high-risk individuals. *Schizophr Bull* **38**(6): 1268–1276.

Allen P, Stephan KE, Mechelli A, Day F, Ward N, Dalton J, *et al* (2010). Cingulate activity and fronto-temporal connectivity in people with prodromal signs of psychosis. *Neuroimage* **49**(1): 947–955.

Andreasen NC (1995). The validation of psychiatric diagnosis: new models and approaches. *Am J Psychiatry* **152**(2): 161–162.

Anticevic A, Haut K, Murray JD, Repovs G, Yang GJ, Diehl C, *et al* (2015). Association of Thalamic Dysconnectivity and Conversion to Psychosis in Youth and Young Adults at Elevated Clinical Risk. *JAMA Psychiatry* **72**(9): 882–891.

Astikainen P, Ruusuvirta T, Wikgren J, Korhonen T (2004). The human brain processes visual changes that are not cued by attended auditory stimulation. *Neurosci Lett* **368**(2): 231–234.

Baron M, Gruen R, Rainer JD, Kane J, Asnis L, Lord S (1985). A family study of schizophrenic and normal control probands: implications for the spectrum concept of schizophrenia. *Am J Psychiatry* **142**(4): 447–455.

Barrantes-Vidal N, Grant P, Kwapil TR (2015). The role of schizotypy in the study of the etiology of schizophrenia spectrum disorders. *Schizophr Bull* **41 Suppl 2**: S408–416.

Bassett DS, Bullmore E, Verchinski BA, Mattay VS, Weinberger DR, Meyer-Lindenberg A (2008). Hierarchical organization of human cortical networks in health and schizophrenia. *J Neurosci* **28**(37): 9239–9248.

Bassett DS, Meyer-Lindenberg A, Achard S, Duke T, Bullmore E (2006). Adaptive reconfiguration of fractal small-world human brain functional networks. *Proc Natl Acad Sci U S A* **103**(51): 19518–19523.

Beckmann CF, DeLuca M, Devlin JT, Smith SM (2005). Investigations into resting-state connectivity using independent component analysis. *Philos Trans R Soc Lond B Biol Sci* **360**(1457): 1001–1013.

Bendfeldt K, Smieskova R, Koutsouleris N, Klöppel S, Schmidt A, Walter A, *et al* (2015). Classifying individuals at high-risk for psychosis based on functional brain activity during working memory processing. *Neuroimage Clin* **9**: 555–563.

Benetti S, Mechelli A, Picchioni M, Broome M, Williams S, McGuire P (2009). Functional integration between the posterior hippocampus and prefrontal cortex is impaired in both first episode schizophrenia and the at risk mental state. *Brain* **132**(Pt 9): 2426–2436.

Biswal B, Yetkin FZ, Haughton VM, Hyde JS (1995). Functional connectivity in the motor cortex of resting human brain using echo-planar MRI. *Magn Reson Med* **34**(4): 537–541.

Biswal BB, Mennes M, Zuo XN, Gohel S, Kelly C, Smith SM, *et al* (2010). Toward discovery science of human brain function. *Proc Natl Acad Sci U S A* **107**(10): 4734–4739.

Borgwardt S, Schmidt A (2016). Is neuroimaging clinically useful in subjects at high risk for psychosis? *World Psychiatry* **15**(2): 178–179.

Borgwardt SJ, Picchioni MM, Ettinger U, Toulopoulou T, Murray R, McGuire PK (2010). Regional gray matter volume in monozygotic twins concordant and discordant for schizophrenia. *Biol Psychiatry* **67**(10): 956–964.

Bressler SL, Menon V (2010). Large-scale brain networks in cognition: emerging methods and principles. *Trends Cogn Sci* **14**(6): 277–290.

Bullmore E, Sporns O (2009). Complex brain networks: graph theoretical analysis of structural and functional systems. *Nat Rev Neurosci* **10**(3): 186–198.

Castellanos FX, Di Martino A, Craddock RC, Mehta AD, Milham MP (2013). Clinical applications of the functional connectome. *Neuroimage* **80**: 527–540.

Clark SR, Schubert KO, Baune BT (2015). Towards indicated prevention of psychosis: using probabilistic assessments of transition risk in psychosis prodrome. *J Neural Transm (Vienna)* **122**(1): 155–169.

Collin G, Kahn RS, de Reus MA, Cahn W, van den Heuvel MP (2013). Impaired Rich Club Connectivity in Unaffected Siblings of Schizophrenia Patients. *Schizophr Bull.*

Collin G, Scholtens LH, Kahn RS, Hillegers MHJ, van den Heuvel MP (2017). Affected Anatomical Rich Club and Structural-Functional Coupling in Young Offspring of Schizophrenia and Bipolar Disorder Patients. *Biol Psychiatry.*

Consortium C-DGotPG, Smoller G, Craddock N, Kendler K (2013). Identification of risk loci with shared effects on five major psychiatric disorders: A genome-wide analysis. *Lancet* **381**: 1371–1379.

Crossley NA, Mechelli A, Fusar-Poli P, Broome MR, Matthiasson P, Johns LC, *et al* (2009). Superior temporal lobe dysfunction and frontotemporal dysconnectivity in subjects at risk of psychosis and in first-episode psychosis. *Hum Brain Mapp* **30**(12): 4129–4137.

Crossley NA, Mechelli A, Ginestet C, Rubinov M, Bullmore ET, McGuire P (2015). Altered Hub Functioning and Compensatory Activations in the Connectome: A Meta-Analysis of Functional Neuroimaging Studies in Schizophrenia. *Schizophr Bull.*

Crossley NA, Mechelli A, Scott J, Carletti F, Fox PT, McGuire P, *et al* (2014). The hubs of the human connectome are generally implicated in the anatomy of brain disorders. *Brain* **137**(Pt 8): 2382–2395.

Cuthbert BN, Insel TR (2013). Toward the future of psychiatric diagnosis: the seven pillars of RDoC. *BMC Med* **11**: 126.

Dauvermann MR, Whalley HC, Romaniuk L, Valton V, Owens DG, Johnstone EC, *et al* (2013). The application of nonlinear Dynamic Causal Modelling for fMRI in subjects at high genetic risk of schizophrenia. *Neuroimage* **73**: 16–29.

Dauvermann MR, Whalley HC, Schmidt A, Lee GL, Romaniuk L, Roberts N, *et al* (2014). Computational Neuropsychiatry – Schizophrenia as a Cognitive Brain Network Disorder. *Front Psychiatry* **5**: 30.

de Wit S, Ziermans TB, Nieuwenhuis M, Schothorst PF, van Engeland H, Kahn RS, *et al* (2017). Individual prediction of long-term outcome in adolescents at ultra-high risk for psychosis: Applying machine learning techniques to brain imaging data. *Hum Brain Mapp* **38**(2): 704–714.

Dean K, Stevens H, Mortensen PB, Murray RM, Walsh E, Pedersen CB (2010). Full spectrum of psychiatric outcomes among offspring with parental history of mental disorder. *Arch Gen Psychiatry* **67**(8): 822–829.

Debbané M, Barrantes-Vidal N (2015). Schizotypy from a developmental perspective. *Schizophr Bull* **41 Suppl 2**: S386–395.

Debbané M, Eliez S, Badoud D, Conus P, Flückiger R, Schultze-Lutter F (2015). Developing psychosis and its risk states through the lens of schizotypy. *Schizophr Bull* **41 Suppl 2**: S396–407.

Debbané M, Lazouret M, Lagioia A, Schneider M, Van De Ville D, Eliez S (2012). Resting-state networks in adolescents with 22q11.2 deletion syndrome: associations with prodromal symptoms and executive functions. *Schizophr Res* **139**(1-3): 33–39.

Debbané M, Mohr C (2015). Integration and development in schizotypy research: an introduction to the special supplement. *Schizophr Bull* **41 Suppl 2**: S363–365.

Deco G, Kringelbach ML (2014). Great expectations: using whole-brain computational connectomics for understanding neuropsychiatric disorders. *Neuron* **84**(5): 892–905.

Deserno L, Sterzer P, Wüstenberg T, Heinz A, Schlagenhauf F (2012). Reduced prefrontal-parietal effective connectivity and working memory deficits in schizophrenia. *J Neurosci* **32**(1): 12–20.

Diwadkar VA, Bakshi N, Gupta G, Pruitt P, White R, Eickhoff SB (2014). Dysfunction and Dysconnection in Cortical-Striatal Networks during Sustained Attention: Genetic Risk for Schizophrenia or Bipolar Disorder and its Impact on Brain Network Function. *Front Psychiatry* **5**: 50.

Diwadkar VA, Wadehra S, Pruitt P, Keshavan MS, Rajan U, Zajac-Benitez C, *et al* (2012). Disordered corticolimbic interactions during affective processing in children and adolescents at risk for schizophrenia revealed by functional magnetic resonance imaging and dynamic causal modeling. *Arch Gen Psychiatry* **69**(3): 231–242.

Drakesmith M, Caeyenberghs K, Dutt A, Zammit S, Evans CJ, Reichenberg A, *et al* (2015). Schizophrenia-like topological changes in the structural connectome of individuals with subclinical psychotic experiences. *Hum Brain Mapp* **36**(7): 2629–2643.

Drevets WC, Price JL, Furey ML (2008). Brain structural and functional abnormalities in mood disorders: implications for neurocircuitry models of depression. *Brain Struct Funct* **213**(1–2): 93–118.

Ettinger U, Meyhöfer I, Steffens M, Wagner M, Koutsouleris N (2014). Genetics, cognition, and neurobiology of schizotypal personality: a review of the overlap with schizophrenia. *Front Psychiatry* **5**: 18.

Ettinger U, Schmechtig A, Toulopoulou T, Borg C, Orrells C, Owens S, *et al* (2012). Prefrontal and striatal volumes in monozygotic twins concordant and discordant for schizophrenia. *Schizophr Bull* **38**(1): 192–203.

Filippi M, van den Heuvel MP, Fornito A, He Y, Hulshoff Pol HE, Agosta F, *et al* (2013). Assessment of system dysfunction in the brain through MRI-based connectomics. *Lancet Neurol* **12**(12): 1189–1199.

Fornito A, Bullmore ET, Zalesky A (2017). Opportunities and Challenges for Psychiatry in the Connectomic Era. *Biological Psychiatry: CNNI* **2**: 9–19.

Fornito A, Harrison BJ, Goodby E, Dean A, Ooi C, Nathan PJ, *et al* (2013). Functional dysconnectivity of corticostriatal circuitry as a risk phenotype for psychosis. *JAMA Psychiatry* **70**(11): 1143–1151.

Fornito A, Zalesky A, Breakspear M (2015). The connectomics of brain disorders. *Nat Rev Neurosci* **16**(3): 159–172.

Fox MD, Snyder AZ, Vincent JL, Corbetta M, Van Essen DC, Raichle ME (2005). The human brain is intrinsically organized into dynamic, anticorrelated functional networks. *Proc Natl Acad Sci U S A* **102**(27): 9673–9678.

Friston K, Moran R, Seth AK (2013). Analysing connectivity with Granger causality and dynamic causal modelling. *Curr Opin Neurobiol* **23**(2): 172–178.

Friston KJ, Harrison L, Penny W (2003). Dynamic causal modelling. *Neuroimage* **19**(4): 1273–1302.

Fryer SL, Woods SW, Kiehl KA, Calhoun VD, Pearlson GD, Roach BJ, *et al* (2013). Deficient Suppression of Default Mode Regions during Working Memory in Individuals with Early Psychosis and at Clinical High-Risk for Psychosis. *Front Psychiatry* **4**: 92.

Fusar-Poli P, Bonoldi I, Yung AR, Borgwardt S, Kempton MJ, Valmaggia L, *et al* (2012a). Predicting psychosis: meta-analysis of transition outcomes in individuals at high clinical risk. *Arch Gen Psychiatry* **69**(3): 220–229.

Fusar-Poli P, Borgwardt S, Bechdolf A, Addington J, Riecher-Rössler A, Schultze-Lutter F, *et al* (2013). The psychosis high-risk state: a comprehensive state-of-the-art review. *JAMA Psychiatry* **70**(1): 107–120.

Fusar-Poli P, Borgwardt S, Crescini A, Deste G, Kempton MJ, Lawrie S, *et al* (2011). Neuroanatomy of vulnerability to psychosis: a voxel-based meta-analysis. *Neurosci Biobehav Rev* **35**(5): 1175–1185.

Fusar-Poli P, Cappucciati M, Borgwardt S, Woods SW, Addington J, Nelson B, *et al* (2016). Heterogeneity of Psychosis Risk Within Individuals at Clinical High Risk: A Meta-analytical Stratification. *JAMA Psychiatry* **73**(2): 113–120.

Fusar-Poli P, Cappucciati M, Rutigliano G, Schultze-Lutter F, Bonoldi I, Borgwardt S, *et al* (2015). At risk or not at risk? A meta-analysis of the prognostic accuracy of psychometric interviews for psychosis prediction. *World Psychiatry* **14**(3): 322–332.

Fusar-Poli P, Deste G, Smieskova R, Barlati S, Yung AR, Howes O, *et al* (2012b). Cognitive Functioning in Prodromal Psychosis: A Meta-analysis Cognitive Functioning in Prodromal Psychosis. *Arch Gen Psychiatry* **69**(6): 562–571.

Fusar-Poli P, Schultze-Lutter F (2016). Predicting the onset of psychosis in patients at clinical high risk: practical guide to probabilistic prognostic reasoning. *Evid Based Ment Health* **19**(1): 10–15.

Gershon ES, Hamovit J, Guroff JJ, Dibble E, Leckman JF, Sceery W, *et al* (1982). A family study of schizoaffective, bipolar I, bipolar II, unipolar, and normal control probands. *Arch Gen Psychiatry* **39**(10): 1157–1167.

Goldstein JM, Buka SL, Seidman LJ, Tsuang MT (2010). Specificity of familial transmission of schizophrenia psychosis spectrum and affective psychoses in the New England family study's high-risk design. *Arch Gen Psychiatry* **67**(5): 458–467.

Gottesman II, Laursen TM, Bertelsen A, Mortensen PB (2010). Severe mental disorders in offspring with 2 psychiatrically ill parents. *Arch Gen Psychiatry* **67**(3): 252–257.

Grace AA (2016). Dysregulation of the dopamine system in the pathophysiology of schizophrenia and depression. *Nat Rev Neurosci*.

Hamilton JP, Etkin A, Furman DJ, Lemus MG, Johnson RF, Gotlib IH (2012). Functional neuroimaging of major depressive disorder: a meta-analysis and new integration of base line activation and neural response data. *Am J Psychiatry* **169**(7): 693–703.

Hazlett EA, Goldstein KE, Kolaitis JC (2012). A review of structural MRI and diffusion tensor imaging in schizotypal personality disorder. *Curr Psychiatry Rep* **14**(1): 70–78.

Howes OD, Kambeitz J, Kim E, Stahl D, Slifstein M, Abi-Dargham A, *et al* (2012). The nature of dopamine dysfunction in schizophrenia and what this means for treatment. *Arch Gen Psychiatry* **69**(8): 776–786.

Insel T, Cuthbert B, Garvey M, Heinssen R, Pine DS, Quinn K, *et al* (2010). Research domain criteria (RDoC): toward a new classification framework for research on mental disorders. *Am J Psychiatry* **167**(7): 748–751.

Insel TR, Cuthbert BN (2015). Medicine. Brain disorders? Precisely. *Science* **348**(6234): 499–500.

Jang JH, Jung WH, Choi JS, Choi CH, Kang DH, Shin NY, *et al* (2011). Reduced prefrontal functional connectivity in the default mode network is related to greater psychopathology in subjects with high genetic loading for schizophrenia. *Schizophr Res* **127**(1-3): 58–65.

Johnstone EC, Lawrie SM, Cosway R (2002). What does the Edinburgh high-risk study tell us about schizophrenia? *Am J Med Genet* **114**(8): 906–912.

Jung WH, Jang JH, Shin NY, Kim SN, Choi CH, An SK, *et al* (2012). Regional brain atrophy and functional disconnection in Broca's area in individuals at ultra-high risk for psychosis and schizophrenia. *PLoS One* **7**(12): e51975.

Kapur S, Phillips AG, Insel TR (2012). Why has it taken so long for biological psychiatry to develop clinical tests and what to do about it? *Mol Psychiatry* **17**(12): 1174–1179.

Kempton MJ, Bonoldi I, Valmaggia L, McGuire P, Fusar-Poli P (2015). Speed of Psychosis Progression in People at Ultra-High Clinical Risk: A Complementary Meta-analysis. *JAMA Psychiatry*.

Kirkbride JB, Errazuriz A, Croudace TJ, Morgan C, Jackson D, Boydell J, *et al* (2012). Incidence of schizophrenia and other psychoses in England, 1950-2009: a systematic review and meta-analyses. *PLoS One* **7**(3): e31660.

Klosterkötter J, Hellmich M, Steinmeyer EM, Schultze-Lutter F (2001). Diagnosing schizophrenia in the initial prodromal phase. *Arch Gen Psychiatry* **58**(2): 158–164.

Koutsouleris N, Riecher-Rössler A, Meisenzahl EM, Smieskova R, Studerus E, Kambeitz-Ilankovic L, *et al* (2014). Detecting the Psychosis Prodrome Across High-risk Populations Using Neuroanatomical Biomarkers. *Schizophr Bull.*

Lagioia A, Van De Ville D, Debbané M, Lazeyras F, Eliez S (2010). Adolescent resting state networks and their associations with schizotypal trait expression. *Front Syst Neurosci* **4**.

Lawrence NS, Ross TJ, Hoffmann R, Garavan H, Stein EA (2003). Multiple neuronal networks mediate sustained attention. *J Cogn Neurosci* **15**(7): 1028–1038.

Liao Y, Huang X, Wu Q, Yang C, Kuang W, Du M, *et al* (2012). Is depression a disconnection syndrome? Meta-analysis of diffusion tensor imaging studies in patients with MDD. *J Psychiatry Neurosci* **37**(4): 110180.

Lichtenstein P, Yip BH, Björk C, Pawitan Y, Cannon TD, Sullivan PF, *et al* (2009). Common genetic determinants of schizophrenia and bipolar disorder in Swedish families: a population-based study. *Lancet* **373**(9659): 234–239.

Liu Y, Liang M, Zhou Y, He Y, Hao Y, Song M, *et al* (2008). Disrupted small-world networks in schizophrenia. *Brain* **131**(Pt 4): 945–961.

Lodge DJ, Grace AA (2011a). Developmental pathology, dopamine, stress and schizophrenia. *Int J Dev Neurosci* **29**(3): 207–213.

Lodge DJ, Grace AA (2011b). Hippocampal dysregulation of dopamine system function and the pathophysiology of schizophrenia. *Trends Pharmacol Sci* **32**(9): 507–513.

Mason O, Claridge G (2006). The Oxford-Liverpool Inventory of Feelings and Experiences (O-LIFE): further description and extended norms. *Schizophr Res* **82**(2–3): 203–211.

Mattiaccio LM, Coman IL, Schreiner MJ, Antshel KM, Fremont WP, Bearden CE, *et al* (2016). Atypical functional connectivity in resting-state networks of individuals with 22q11.2

deletion syndrome: associations with neurocognitive and psychiatric functioning. *J Neurodev Disord* **8**: 2.

McGorry P, Nelson B (2016). Why We Need a Transdiagnostic Staging Approach to Emerging Psychopathology, Early Diagnosis, and Treatment. *JAMA Psychiatry* **73**(3): 191–192.

McGuire P, Dazzan P (2017). Does neuroimaging have a role in predicting outcomes in psychosis? *World Psychiatry* **16**(2): 209–210.

McGuire P, Sato JR, Mechelli A, Jackowski A, Bressan RA, Zugman A (2015). Can neuroimaging be used to predict the onset of psychosis? *Lancet Psychiatry* **2**(12): 1117–1122.

McKeown MJ, Makeig S, Brown GG, Jung TP, Kindermann SS, Bell AJ, *et al* (1998). Analysis of fMRI data by blind separation into independent spatial components. *Hum Brain Mapp* **6**(3): 160–188.

McKeown MJ, Sejnowski TJ (1998). Independent component analysis of fMRI data: examining the assumptions. *Hum Brain Mapp* **6**(5–6): 368–372.

McKiernan KA, Kaufman JN, Kucera-Thompson J, Binder JR (2003). A parametric manipulation of factors affecting task-induced deactivation in functional neuroimaging. *J Cogn Neurosci* **15**(3): 394–408.

Mechelli A, Lin A, Wood S, McGorry P, Amminger P, Tognin S, *et al* (2017). Using clinical information to make individualized prognostic predictions in people at ultra high risk for psychosis. *Schizophr Res* **184**: 32–38.

Mesholam-Gately RI, Giuliano AJ, Goff KP, Faraone SV, Seidman LJ (2009). Neurocognition in first-episode schizophrenia: a meta-analytic review. *Neuropsychology* **23**(3): 315–336.

Meszlényi RJ, Hermann P, Buza K, Gál V, Vidnyánszky Z (2017). Resting State fMRI Functional Connectivity Analysis Using Dynamic Time Warping. *Front Neurosci* **11**: 75.

Millan MJ, Andrieux A, Bartzokis G, Cadenhead K, Dazzan P, Fusar-Poli P, *et al* (2016). Altering the course of schizophrenia: progress and perspectives. *Nat Rev Drug Discov*.

Modinos G, Allen P, Grace AA, McGuire P (2014). Translating the MAM model of psychosis to humans. *Trends Neurosci*.

Monks S, Niarchou M, Davies AR, Walters JT, Williams N, Owen MJ, *et al* (2014). Further evidence for high rates of schizophrenia in 22q11.2 deletion syndrome. *Schizophr Res* **153**(1–3): 231–236.

Nelson MT, Seal ML, Pantelis C, Phillips LJ (2013). Evidence of a dimensional relationship between schizotypy and schizophrenia: a systematic review. *Neurosci Biobehav Rev* **37**(3): 317–327.

Nielsen JD, Madsen KH, Wang Z, Liu Z, Friston KJ, Zhou Y (2017). Working Memory Modulation of Frontoparietal Network Connectivity in First-Episode Schizophrenia. *Cereb Cortex* **27**(7): 3832–3841.

Ottet MC, Schaer M, Debbané M, Cammoun L, Thiran JP, Eliez S (2013). Graph theory reveals dysconnected hubs in 22q11DS and altered nodal efficiency in patients with hallucinations. *Front Hum Neurosci* **7**: 402.

Padula MC, Schaer M, Scariati E, Schneider M, Van De Ville D, Debbané M, *et al* (2015). Structural and functional connectivity in the default mode network in 22q11.2 deletion syndrome. *J Neurodev Disord* **7**(1): 23.

Palaniyappan L (2017). Progressive cortical reorganisation: A framework for investigating structural changes in schizophrenia. *Neurosci Biobehav Rev* **79**: 1–13.

Park HJ, Friston K (2013). Structural and functional brain networks: from connections to cognition. *Science* **342**(6158): 1238411.

Peeters SC, van de Ven V, Gronenschild EH, Patel AX, Habets P, Goebel R, *et al* (2015). Default mode network connectivity as a function of familial and environmental risk for psychotic disorder. *PLoS One* **10**(3): e0120030.

Pelletier-Baldelli A, Bernard JA, Mittal VA (2015). Intrinsic Functional Connectivity in Salience and Default Mode Networks and Aberrant Social Processes in Youth at Ultra-High Risk for Psychosis. *PLoS One* **10**(8): e0134936.

Pereira F, Mitchell T, Botvinick M (2009). Machine learning classifiers and fMRI: a tutorial overview. *Neuroimage* **45**(1 Suppl): S199–209.

Phillips LJ, Yung AR, McGorry PD (2000). Identification of young people at risk of psychosis: validation of Personal Assessment and Crisis Evaluation Clinic intake criteria. *Aust N Z J Psychiatry* **34 Suppl**: S164–169.

Purcell SM, Wray NR, Stone JL, Visscher PM, O'Donovan MC, Sullivan PF, *et al* (2009). Common polygenic variation contributes to risk of schizophrenia and bipolar disorder. *Nature* **460**(7256): 748–752.

Radua J, Borgwardt S, Crescini A, Mataix-Cols D, Meyer-Lindenberg A, McGuire PK, *et al* (2012). Multimodal meta-analysis of structural and functional brain changes in first episode psychosis and the effects of antipsychotic medication. *Neurosci Biobehav Rev.*

Radua J, Schmidt A, Borgwardt S, Heinz A, Schlagenhauf F, McGuire P, *et al* (2015). Ventral Striatal Activation During Reward Processing in Psychosis: A Neurofunctional Meta-Analysis. *JAMA Psychiatry* **72**(12): 1243–1251.

Raine A (2006). Schizotypal personality: neurodevelopmental and psychosocial trajectories. *Annu Rev Clin Psychol* **2**: 291–326.

Rasic D, Hajek T, Alda M, Uher R (2014). Risk of mental illness in offspring of parents with schizophrenia, bipolar disorder, and major depressive disorder: a meta-analysis of family high-risk studies. *Schizophr Bull* **40**(1): 28–38.

Rawlings D, Williams B, Haslam N, Claridge G (2008). Is schizotypy taxonic? Response to Beauchaine, Lenzenweger, and Waller. *Personality and Individual Differences* **44**: 1663–1672.

Repovs G, Csernansky JG, Barch DM (2011). Brain network connectivity in individuals with schizophrenia and their siblings. *Biol Psychiatry* **69**(10): 967–973.

Riecher-Rössler A, Aston J, Ventura J, Merlo M, Borgwardt S, Gschwandtner U, *et al* (2008). [The Basel Screening Instrument for Psychosis (BSIP): development, structure, reliability and validity]. *Fortschr Neurol Psychiatr* **76**(4): 207–216.

Rubinov M, Bullmore E (2013). Fledgling pathoconnectomics of psychiatric disorders. *Trends Cogn Sci* **17**(12): 641–647.

Rubinov M, Sporns O (2010). Complex network measures of brain connectivity: uses and interpretations. *Neuroimage* **52**(3): 1059–1069.

Russo SJ, Nestler EJ (2013). The brain reward circuitry in mood disorders. *Nat Rev Neurosci* **14**(9): 609–625.

Salvador R, Radua J, Canales-Rodríguez EJ, Solanes A, Sarró S, Goikolea JM, *et al* (2017). Evaluation of machine learning algorithms and structural features for optimal MRI-based diagnostic prediction in psychosis. *PLoS One* **12**(4): e0175683.

Schmidt A, Cappucciati M, Radua J, Rutigliano G, Rocchetti M, Dell'Osso L, *et al* (2016a). Improving Prognostic Accuracy in Subjects at Clinical High Risk for Psychosis: Systematic Review of Predictive Models and Meta-analytical Sequential Testing Simulation. *Schizophr Bull.*

Schmidt A, Crossley NA, Harrisberger F, Smieskova R, Lenz C, Riecher-Rössler A, *et al* (2016b). Structural Network Disorganization in Subjects at Clinical High Risk for Psychosis. *Schizophr Bull.*

Schmidt A, Diwadkar VA, Smieskova R, Harrisberger F, Lang UE, McGuire P, *et al* (2014a). Approaching a network connectivity-driven classification of the psychosis continuum: a selective review and suggestions for future research. *Front Hum Neurosci* **8**: 1047.

Schmidt A, Smieskova R, Aston J, Simon A, Allen P, Fusar-Poli P, *et al* (2013). Brain connectivity abnormalities predating the onset of psychosis: correlation with the effect of medication. *JAMA Psychiatry* **70**(9): 903–912.

Schmidt A, Smieskova R, Simon A, Allen P, Fusar-Poli P, McGuire PK, *et al* (2014b). Abnormal effective connectivity and psychopathological symptoms in the psychosis high-risk state. *J Psychiatry Neurosci* **39**(4): 239–248.

Schneider M, Debbané M, Bassett AS, Chow EW, Fung WL, van den Bree M, *et al* (2014). Psychiatric disorders from childhood to adulthood in 22q11.2 deletion syndrome: results from the International Consortium on Brain and Behavior in 22q11.2 Deletion Syndrome. *Am J Psychiatry* **171**(6): 627–639.

Schreiner M, Forsyth JK, Karlsgodt KH, Anderson AE, Hirsh N, Kushan L, *et al* (2017). Intrinsic Connectivity Network-Based Classification and Detection of Psychotic Symptoms in Youth With 22q11.2 Deletions. *Cereb Cortex* **27**(6): 3294–3306.
Schultze-Lutter F, Michel C, Schmidt SJ, Schimmelmann BG, Maric NP, Salokangas RK, *et al* (2015). EPA guidance on the early detection of clinical high risk states of psychoses. *Eur Psychiatry* **30**(3): 405–416.
Shim G, Oh JS, Jung WH, Jang JH, Choi CH, Kim E, *et al* (2010). Altered resting-state connectivity in subjects at ultra-high risk for psychosis: an fMRI study. *Behav Brain Funct* **6**: 58.
Smieskova R, Fusar-Poli P, Allen P, Bendfeldt K, Stieglitz RD, Drewe J, *et al* (2010). Neuroimaging predictors of transition to psychosis--a systematic review and meta-analysis. *Neurosci Biobehav Rev* **34**(8): 1207–1222.
Smieskova R, Marmy J, Schmidt A, Bendfeldt K, Riecher-Rössler A, Walter M, *et al* (2012). Do subjects at clinical high risk for psychosis differ from those with a genetic high risk? – A systematic review of structural and functional brain abnormalities. *Curr Med Chem.*
Smith SM, Fox PT, Miller KL, Glahn DC, Fox PM, Mackay CE, *et al* (2009). Correspondence of the brain's functional architecture during activation and rest. *Proc Natl Acad Sci U S A* **106**(31): 13040–13045.
Smith SM, Miller KL, Salimi-Khorshidi G, Webster M, Beckmann CF, Nichols TE, *et al* (2011). Network modelling methods for FMRI. *Neuroimage* **54**(2): 875–891.
Smith SM, Vidaurre D, Beckmann CF, Glasser MF, Jenkinson M, Miller KL, *et al* (2013). Functional connectomics from resting-state fMRI. *Trends Cogn Sci* **17**(12): 666–682.
Sporns O, Honey CJ, Kötter R (2007). Identification and classification of hubs in brain networks. *PLoS One* **2**(10): e1049.
Stafford MR, Jackson H, Mayo-Wilson E, Morrison AP, Kendall T (2013). Early interventions to prevent psychosis: systematic review and meta-analysis. *BMJ* **346**: f185.
Stephan KE, Harrison LM, Kiebel SJ, David O, Penny WD, Friston KJ (2007a). Dynamic causal models of neural system dynamics:current state and future extensions. *J Biosci* **32**(1): 129–144.
Stephan KE, Penny WD, Daunizeau J, Moran RJ, Friston KJ (2009). Bayesian model selection for group studies. *Neuroimage* **46**(4): 1004–1017.
Stephan KE, Weiskopf N, Drysdale PM, Robinson PA, Friston KJ (2007b). Comparing hemodynamic models with DCM. *Neuroimage* **38**(3): 387–401.
van den Heuvel MP, Kahn RS, Goñi J, Sporns O (2012). High-cost, high-capacity backbone for global brain communication. *Proc Natl Acad Sci U S A* **109**(28): 11372–11377.
van den Heuvel MP, Mandl RC, Stam CJ, Kahn RS, Hulshoff Pol HE (2010). Aberrant frontal and temporal complex network structure in schizophrenia: a graph theoretical analysis. *J Neurosci* **30**(47): 15915–15926.
van den Heuvel MP, Sporns O (2011). Rich-club organization of the human connectome. *J Neurosci* **31**(44): 15775–15786.
van den Heuvel MP, Sporns O (2013). An anatomical substrate for integration among functional networks in human cortex. *J Neurosci* **33**(36): 14489–14500.
van den Heuvel MP, Sporns O (2013a). Network hubs in the human brain. *Trends Cogn Sci.*
van den Heuvel MP, Sporns O, Collin G, Scheewe T, Mandl RC, Cahn W, *et al* (2013). Abnormal rich club organization and functional brain dynamics in schizophrenia. *JAMA Psychiatry* **70**(8): 783–792.
van der Gaag M, Smit F, Bechdolf A, French P, Linszen DH, Yung AR, *et al* (2013). Preventing a first episode of psychosis: meta-analysis of randomized controlled prevention trials of 12 month and longer-term follow-ups. *Schizophr Res* **149**(1-3): 56–62.
van Haren NE, Picchioni MM, McDonald C, Marshall N, Davis N, Ribchester T, *et al* (2004). A controlled study of brain structure in monozygotic twins concordant and discordant for schizophrenia. *Biol Psychiatry* **56**(6): 454–461.
Vása F, Griffa A, Scariati E, Schaer M, Urben S, Eliez S, *et al* (2016). An affected core drives network integration deficits of structural connectome in 22q11.2 deletion syndrome. *Neuroimage Clin* **10**: 239–249.

Wang Y, Ettinger U, Meindl T, Chan RCK (2017). Association of schizotypy with striatocortical functional connectivity and its asymmetry in healthy adults. *Hum Brain Mapp*.

Wang Y, Yan C, Yin DZ, Fan MX, Cheung EF, Pantelis C, *et al* (2015). Neurobiological changes of schizotypy: evidence from both volume-based morphometric analysis and resting-state functional connectivity. *Schizophr Bull* **41 Suppl 2**: S444–454.

Whalley HC, Simonotto E, Flett S, Marshall I, Ebmeier KP, Owens DG, *et al* (2004). fMRI correlates of state and trait effects in subjects at genetically enhanced risk of schizophrenia. *Brain* **127**(Pt 3): 478–490.

Whalley HC, Simonotto E, Marshall I, Owens DG, Goddard NH, Johnstone EC, *et al* (2005). Functional disconnectivity in subjects at high genetic risk of schizophrenia. *Brain* **128**(Pt 9): 2097–2108.

Whitfield-Gabrieli S, Thermenos HW, Milanovic S, Tsuang MT, Faraone SV, McCarley RW, *et al* (2009). Hyperactivity and hyperconnectivity of the default network in schizophrenia and in first-degree relatives of persons with schizophrenia. *Proc Natl Acad Sci U S A* **106**(4): 1279–1284.

Winton-Brown T, Schmidt A, Roiser JP, Howes OD, Egerton A, Fusar-Poli P, *et al* (2017). Altered activation and connectivity in a hippocampal-basal ganglia-midbrain circuit during salience processing in subjects at ultra high risk for psychosis. *Transl Psychiatry* **7**(10): e1245.

Woodward ND, Cascio CJ (2015). Resting-State Functional Connectivity in Psychiatric Disorders. *JAMA Psychiatry* **72**(8): 743–744.

Woodward ND, Karbasforoushan H, Heckers S (2012). Thalamocortical dysconnectivity in schizophrenia. *Am J Psychiatry* **169**(10): 1092–1099.

Wotruba D, Michels L, Buechler R, Metzler S, Theodoridou A, Gerstenberg M, *et al* (2013). Aberrant Coupling Within and Across the Default Mode, Task-Positive, and Salience Network in Subjects at Risk for Psychosis. *Schizophr Bull*.

Young J, Kempton MJ, McGuire P (2016). Using machine learning to predict outcomes in psychosis. *Lancet Psychiatry* **3**(10): 908–909.

Zhang Y, Lin L, Lin CP, Zhou Y, Chou KH, Lo CY, *et al* (2012). Abnormal topological organization of structural brain networks in schizophrenia. *Schizophr Res* **141**(2-3): 109–118.

Functional Connectivity in Autism Spectrum Disorders: Challenges and Perspectives

Ralph-Axel Müller and Annika Linke

Autism Spectrum Disorders

Autism spectrum disorders (ASDs) are neurodevelopmental disorders with core symptomatology including socio-communicative deficits and restricted and repetitive behaviors, often accompanied by sensory anomalies (American Psychiatric Association 2013). Although core symptomatology is defined predominantly in the socio-communicative domain, the overall neuropsychological profile is much broader, with impairment or abnormal function reported for all sensorimotor domains (Cascio et al. 2012; Chukoskie et al. 2013; Esposito and Pasca 2013; Gowen and Hamilton 2012; Marco et al. 2011; O'Connor 2012; Puts et al. 2014; Rozenkrantz et al. 2015; Schecklmann et al. 2013; Simmons et al. 2009), as well as attention (Keehn et al. 2013), language (Groen et al. 2008), and executive control (Craig et al. 2016; Hill 2004). With ever-increasing prevalence (Maenner et al. 2020), ASDs present a major public health challenge and are associated with enormous personal and societal burdens (Leigh and Du 2015). Advances toward fully targeted treatments and precision medicine that can reduce symptom load or even prevent the disorder before it emerges in the first years of life are therefore an urgent public health need. However, after decades of extensive genetic and neuroscientific research, mechanistic models that could broadly account for "idiopathic" (nonsyndromic) ASDs remain elusive. As described in detail later in this chapter, a crucial challenge relates to the use of the label "ASD" at different levels of clinical practice and scientific study.

Ralph-Axel Müller and Annika Linke contributed to this chapter.

R.-A. Müller (✉) · A. Linke
Brain Development Imaging Laboratories, Department of Psychology, San Diego State University, San Diego, CA, USA
e-mail: rmueller@sdsu.edu

V. A. Diwadkar, S. B. Eickhoff (eds.), *Brain Network Dysfunction in Neuropsychiatric Illness*, https://doi.org/10.1007/978-3-030-59797-9_12

By consensus, ASDs are considered neurological disorders. The neuroimaging literature on ASDs has grown exponentially over the past few decades, but only few areas of consensus have emerged. Among these are findings from multiple groups suggesting atypical brain growth patterns, with overgrowth in the first years of life, followed by flattened growth (Courchesne et al. 2001; Hazlett et al. 2011; Sparks et al. 2002). A caveat relates to abnormal volume of cerebrospinal fluid compartments whose relation to early overgrowth is not yet fully understood (Shen et al. 2013, 2017). Potentially precocious pseudo-maturation of white matter tracts seen in some diffusion weighted imaging (DWI) studies of infants and preschoolers (Andrews et al. 2019; Solso et al. 2015; Weinstein et al. 2011; Wolff et al. 2012) may be considered consistent with early anatomical overgrowth, although not all DWI findings are in clear agreement (Lewis et al. 2017). As a more general caveat, early brain overgrowth is detected statistically at the level of larger groups, and most children with ASDs do not actually show this feature (Amaral et al. 2017).

Another example of tentative consensus is agreement that ASDs cannot be tied to a single locus of brain anomaly, nor a single functional or neurotransmitter system, but instead requires investigation at the level of multiple distributed brain networks (Menon 2011). While brain network abnormality in ASDs is widely accepted, the exact patterns remain contentious, ranging from functional underconnectivity (Just et al. 2012) to massive overconnectivity (Supekar et al. 2013) with reduced network segregation or differentiation (Nebel et al. 2014; Rudie and Dapretto 2013; Rudie et al. 2012; Shih et al. 2011), and atypical between-network “cross-talk” (Fishman et al. 2014, 2015).

In this chapter, we will briefly describe some major themes inherent in the current evidence on functional connectivity in ASDs. We will then turn to the many sources of divergence in findings and will finally evaluate approaches to overcoming gaps and inconsistencies in our knowledge of brain network organization in ASDs.

Functional Connectivity

According to a commonly cited definition by Friston et al. (1993), *functional connectivity* describes “temporal correlations between spatially remote neurophysiological events.” A more recent and broader definition refers to “statistical dependence between remote neural processes” (Honey et al. 2007). Both definitions highlight the fact that the study of neural activity can only indirectly inform us about (anatomical) connections in the brain. In principle, two brain regions may show highly correlated activity changes without any direct anatomical connections. The term “connectivity” therefore has fundamentally different meanings in functional vs. anatomical studies. Within the realm of functional connectivity, the term is further applied with different meanings in studies testing hemodynamic vs. electrophysiological signals (Mash et al. 2018). Given the extensive methodological (and conceptual) challenges in functional MRI approaches to connectivity in ASDs, detailed discussion of anatomical and electrophysiological techniques is beyond the scope of this chapter.

Over the past 10–15 years, the technique of choice in the study of functional connectivity has been functional connectivity MRI (fcMRI). Its origins go back to Biswal et al. (1995), who showed that fluctuations in the blood oxygen-level-dependent (BOLD) signal are synchronized within distributed brain networks (such as the motor system). This finding is surprising given the low temporal resolution of fMRI, limited by both the sluggishness of the hemodynamic response and a low sampling rate (commonly about every 2 seconds). However, BOLD correlations in frequency bands below 0.1 Hz (Cordes et al. 2001)—or only slightly higher (Gohel and Biswal 2015)—have been shown to be linked to network-specific fluctuations in local field potentials, which reflect the summation of local neuronal activity (Leopold et al. 2003; Schölvinck et al. 2010). The most commonly implemented *intrinsic fcMRI* (iFC) approach uses resting-state fMRI data (or data from which task-induced signal changes have been regressed out (Fair et al. 2007)) to test for correlations in spontaneous, low-frequency BOLD fluctuations (Van Dijk et al. 2010). Despite challenges related to uncertainties about the "resting state" (Buckner et al. 2013), the iFC approach has been successfully applied in the study of numerous networks and circuits (Power et al. 2014b; Smith et al. 2013). The technique has also been applied in many recent ASD studies.

Findings in ASDs

The notion of the autistic brain being "underconnected" has gained much popularity (Hughes 2007), probably because of its intuitive simplicity. Reduced connectivity may equal reduced functionality and thus account for behavioral problems. From a neurodevelopmental perspective, this intuition is misguided, as selective and "regressive" processes (such as synaptic pruning and axonal loss) imply that excess connectivity in the first postnatal years is, on the contrary, linked to functional inefficiency (Johnson 2011; Kandel et al. 2000). However, there is a core of reasonably consistent evidence indicating that *network integration*, i.e., the integrity of connections within neurotypical networks (e.g., a language network connecting Broca's, Wernicke's, and other areas involved in language processing (Price 2010)) may indeed be impaired in ASDs. A large corpus of supportive evidence comes from studies testing task-induced BOLD signal correlations between regions of interest (ROIs) activated by a task in a domain of interest, such as speech comprehension (Just et al. 2004; Kana et al. 2006), verbal working memory (Koshino et al. 2005), response inhibition (Agam et al. 2010; Kana et al. 2007; Lee et al. 2009), problem solving (Just et al. 2007), attention orienting (Fitzgerald et al. 2014), cognitive control (Solomon et al. 2009), finger movement (Mostofsky et al. 2009), visuomotor coordination (Villalobos et al. 2005), face processing (Kleinhans et al. 2008b), self-representation (Lombardo et al. 2010; Mizuno et al. 2011), and theory of mind (Kana et al. 2009, 2012; Mason et al. 2008).

These studies differ from the common iFC approach described above, and several studies have indicated that a focus on stimulus or task-driven BOLD correlations tends to accentuate underconnectivity findings in ASDs (Jones et al. 2010; Müller et al. 2011; Nair et al. 2014), as described in more detail later. Nonetheless, there is some concordant evidence from iFC studies supporting underconnectivity within specialized networks in ASDs. Probably the most widely replicated finding concerns underconnectivity between the two main nodes of the default mode network (DMN) in medial prefrontal cortex and posterior cingulate gyrus/precuneus (Abbott et al. 2016; Assaf et al. 2010; Doyle-Thomas et al. 2015; Falahpour et al. 2016; Joshi et al. 2017; Jung et al. 2014; Monk et al. 2009; Starck et al. 2013; von dem Hagen et al. 2013; Washington et al. 2013; Yerys et al. 2015). These converging findings from multiple groups strongly indicate reduced network integration of the DMN in ASDs.

Interestingly, many of the studies cited above also reported *overconnectivity* findings between DMN and regions *outside* the DMN in ASD cohorts. For example, several groups testing independent samples of children and adolescents with ASDs observed atypically increased iFC between the DMN hub in posterior cingulate cortex and dorsolateral prefrontal task-positive regions, as well as temporo-occipital cortices (Abbott et al. 2016; Doyle-Thomas et al. 2015; Lynch et al. 2013; Yerys et al. 2015). This pattern of reduced within-network iFC accompanied by atypically increased out-of-network connectivity may reflect a dual impairment affecting both integration and segregation (or differentiation), which has been reported for networks other than the DMN as well and may reflect a general principle of atypical connectivity in ASDs (Keown et al. 2017; Rudie et al. 2012; Shih et al. 2011) (Fig. 1). Examples include social and imitation networks (Chien et al. 2015; Fishman et al. 2014, 2015) and somatotopic differentiation in primary motor cortex (Nebel et al. 2014). Such dual impairment in *both* integration and differentiation suggests that fundamental developmental principles of network formation that rely on the interplay of constructive (e.g., synaptogenesis, axonal myelination) and regressive mechanisms (e.g., synaptic pruning) (Johnson 2011) may be affected in ASDs.

However, the above attempt at a synthesis captures fcMRI findings only in the broadest of strokes and the specifics of anomalous connectivity in ASDs remain fraught with inconsistencies and non-replications. For but one example, several studies have tested iFC of posterior superior temporal sulcus (pSTS), given its known role in several domains implicated in ASDs, such as audiovisual integration, face processing, and social cognition (Hein and Knight 2008; Redcay 2008). Alaerts and colleagues (2015), examining multisite data from the Autism Brain Imaging Data Exchange (ABIDE) (Di Martino et al. 2014), found complex patterns of under- and overconnectivity for right pSTS, with linear and non-linear trajectories of iFC changes between ages 7 and 30 years. Abrams et al. (2013) studied a cohort of 20 children with ASDs in a much narrower age range around 10 years and observed underconnectivity between seeds in left and right pSTS and ventrolateral prefrontal cortices, for which no effects were reported in the larger sample tested by Alaerts and colleagues. Fishman et al. (2014), including 25 slightly older children (mean

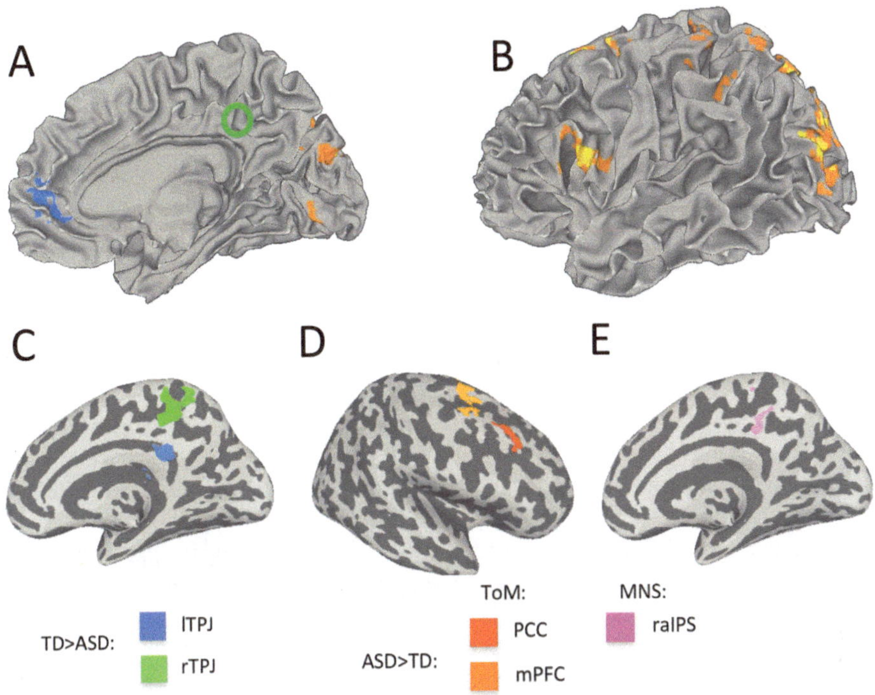

Fig. 1 Examples of dual connectivity impairment in children and adolescents with ASDs, compared to matched TD groups. (**A**) Default mode network (DMN) integration is reduced, as seen by blue cluster of reduced iFC in medial prefrontal DMN hub for a seed in posterior cingulate DMN hub (green circle). (**B**) However, numerous clusters of atypically increased iFC (red-yellow) can be seen in frontal and parieto-occipital regions, indicating reduced network segregation. Adapted from Abbott et al. (2016). (**C**) Integration within "theory-of-mind" (ToM) network in ASD group is reduced, with green and blue clusters indicating lower iFC for seed in left and right temporoparietal junction (TPJ). However, there is also overconnectivity, which manifests as "cross-talk" between ToM network and mirror neuron system (MNS). On the one hand (**D**), seeds belonging to the ToM network (posterior cingulate cortex [PCC], medial prefrontal cortex [mPFC]) are overconnected with premotor MNS regions; on the other (**E**), an MNS seed in right anterior intraparietal sulcus (raIPS) is overconnected with a ToM region in PCC. (Adapted from Fishman et al. 2014)

age 14.8 years) from an in-house cohort, did not replicate this finding either and instead observed extensive within-group iFC between left pSTS and ventrolateral prefrontal regions in their ASD group, with more modest effects in TD peers. In an earlier study, testing iFC after regressing out effects of a visual search task, Shih et al. (2011) even reported extensive overconnectivity of pSTS bilaterally that was associated with reduced differentiation (i.e., increased connectivity overlap) between three small subsections of pSTS. Divergent findings highlight the crucial impact of cohort demographics and methodological specifics on group findings in the ASD iFC literature, which we will consider later in this chapter, but also fundamental conceptual questions, to which we will turn now.

Conceptual Challenges

Many Findings, But No Single ASD "Biomarker"

The decades-old insight that autism is a neurological disorder – as important as it was when contrasted with theories based on parental behavior (Bettelheim, 1967) – has prompted the search for brain-based causes of the disorder that may be as crucial as it is futile. Although causation is no longer considered primarily at the behavioral level, the diagnostic definition of ASD (American Psychiatric Association 2013; Lord et al. 2012; Rutter et al. 2003) has remained exclusively behavioral. If one accepts the consensus of autism being a brain-based disorder, the prospect that "autism spectrum disorder" (in the singular of the DSM-5) may simply not exist at the neural level is jarring. However, it may be a necessary and in fact urgent logical conclusion. It has been argued that ASD lacks validity as a neurobiological construct and should therefore be abandoned (Lynn Waterhouse et al. 2016, 2017). While such a drastic step may not further the ultimate translational goals of imaging research (Müller 2017; Müller and Amaral 2017), the underlying insight is accurate. After more than three decades of neuroimaging in ASDs, there is still no *single* brain finding (or complex pattern of brain findings) that uniquely distinguishes individuals with ASDs from their TD peers. And correspondingly, there is no *single* generally accepted model of brain anomalies in ASDs, let alone a mechanistic model linking genetic (or other) causes, neurodevelopmental disturbances, and outcome brain-behavior anomalies. However, the words "no single" in these two statements is ambiguous. While it appears to mean "none," it could equally be replaced by "many." This ambiguity highlights the fact that the very same answer may be as frustrating in response to an ill-posed question ("What is the biomarker of ASD?"), as it is illuminating in response to a correctly posed question ("What are the many biomarkers of the many ASDs?"). Indeed, the number of brain findings distinguishing some individuals with an ASD (combined in a given study cohort) from TD peers is in the thousands. Numerous methodological factors (described later in this chapter) may affect differences in findings, but many of the differences likely reflect the differential composition of participants in one study versus another.

Of course, this is old news. The field has been aware of heterogeneity along the autism spectrum for many years (e.g., Happé et al. 2006; Rutter and Schopler 1987). However, imaging studies have continued to be dominated by conventional group comparison designs that neglect within-group variability as a statistical nuisance rather than an effect of interest. This is risky in common small sample studies with no more than 20–30 ASD participants, but may even be deceptive when applied to very large multisite samples, as those drawn from ABIDE (Di Martino et al. 2014, 2017). Experienced neuroimaging researchers in the field will have encountered the phenomenon of "intriguing findings" with strong effect sizes in an early preliminary sample that simply melt away with the acquisition of additional data. This does not imply that the early findings were meaningless but indicates that they were specific to some individuals with ASDs that happened to predominate in the preliminary sample.

Are Large Samples the Answer?

The conclusion that very large ASD samples (like the total of 1060 subjects in ABIDE-I and II; fcon_1000.projects.nitrc.org/indi/abide) provide an easy remedy and that any findings may be authoritative by virtue of statistical power alone is tempting, but misses the point. If we take seriously the suspected broad etiological heterogeneity of the disorder, what does it mean to have a finding in samples of 1000 ASD participants compared to 1000 matched TD peers? Sample size may be associated with impressively low p-values in some statistical tests. However, aside from the general misperception that low p-values indicate meaningful findings (Wasserstein 2016), closer inspection will often reveal very small effect sizes (e.g., van Rooij et al. 2017). If neurodevelopmental heterogeneity implies divergent profiles of brain anomalies, the question whether a specific brain anomaly present in 10 out of 1000 ASD participants may or may not be detected will depend on how many other ASD variants may share this anomaly, present no anomaly on this measure, or show anomaly with opposite polarity. While the three different scenarios will determine whether a finding will be detected with large or small effect size or not at all in a full sample, these differences in group-wise findings are entirely uninformative with respect to the mentioned 10 individuals with ASD: The brain anomaly may be crucially associated with their behavioral profiles and symptomatology, even if it has little or no impact on the group-wise findings. Simply put, etiological heterogeneity implies that conventional group-level comparisons may reveal little more than "smallest common denominators" across a large sample of individuals with ASDs, which may explain very little about the more complex profiles of brain anomalies accounting for core symptomatology and other behavioral problems in *any single* ASD participant.

This potential limitation of large sample findings (when combined with conventional group-level statistics) finds its analogy at the level of theory building. There can be no single explanatory model of causation and outcome brain anomalies in ASDs. The existence of numerous such models is *in principle* desirable because etiological heterogeneity implies the existence of multiple variants that necessitate multiple explanatory models. The proliferation of such models promoted by journals such as *Medical Hypotheses*—proposing that autism might be explained by mercury (Bernard et al. 2001), thimerosal (Blaxill et al. 2004), zinc (Johnson 2001), clostridium (Bolte 1998), opiates (White and White 1987), nicotinic receptors (Lippiello 2006), and so on and so forth—may be exasperating. However, while many of such proposals are empirically weak, a plurality of mechanistic models will ultimately indeed be necessary, given the likely plurality of etiological variants.

Imaging the Many ASDs

Heterogeneity in ASDs can be primarily approached in two ways. The traditional approach has been to test how the variability of imaging measures relates to behavioral variability. More recent approaches interrogate imaging data for patterns suggesting subgroups that may reflect variants of the disorder.

Variability and Brain-Behavior Links

A common assumption in the field is that an imaging finding distinguishing an ASD and a TD cohort becomes truly meaningful only if the measured variable correlates with symptom severity or some other behavioral variable in ASD participants. Indeed, the likelihood of a statistically significant imaging finding being spurious is lower if there is additional evidence of a behavioral association. Such additional evidence will yield strength to a brain finding, if and only if methodological caution is taken. For example, symptom severity or other behavioral measures may be correlated with head motion during scanning, and failure to control for motion may generate spurious behavioral correlations in tandem with spurious group differences. While this set of challenges can be addressed through good data processing practices, a deeper conceptual question arises.

For reasons of feasibility, the large majority of imaging studies of ASDs is conducted in school-age children, adolescents, or young adults. Aside from study-specific hypotheses, the underlying question—often not explicitly formulated—is that a better understanding of the neural causes of ASDs is needed for ultimate translational goals. This would imply that one is primarily interested in brain findings that reflect the developmental emergence of the disorder. From this perspective, experiential plasticity may be considered a nuisance as it adds variability to the causal developmental differences in brain organization we would like to detect in an imaging study. Because the impact of experience in a child with ASDs can probably not be operationalized comprehensively, any claim that an imaging finding *explains why* children in a cohort have ASDs is therefore preferably avoided with caution. Consider the example of the milestone study by Just et al. (2004), cited by almost 2000 other papers as support for an underconnectivity theory of ASDs. The authors relate their finding of reduced BOLD correlations between several ROI pairs during sentence comprehension in adults with ASDs to "maturational processes" and argue that "the explanation of the disturbance is likely to be found in developmental neurobiological processes" (p.1819). This may sound like an account of developmental causation of the disorder (or its associated language impairments). However, it is at least equally likely that reduced BOLD correlations resulted from differences in language experience. Language delay and atypical socio-communicative interaction are prevalent among children with ASDs and will likely have secondary effects on the maturing language networks. The functional connectivity differences may thus not explain the development of impaired or atypical language, but may rather *be explained by* it.

In the study by Just et al. (2004), the comparison was categorical (group difference), but analogous questions arise with regard to behavioral correlations. In fact, an imaging finding linked to symptom severity is even more likely to be "shallow" and uninformative of developmental causation. Considering the evidence of multiple divergent etiologies in ASDs discussed earlier, developmental trajectories in ASDs converge onto a comparably narrow set of diagnostic criteria (American Psychiatric Association 2013). This is the reason why we call children "autistic" while recognizing that the underlying neurodevelopmental causation may differ across individuals. However, with this insight in mind, a brain finding that correlates attractively with a symptom severity score on the Autism Diagnostic Observation Schedule (Lord et al. 2012) or the Social Responsiveness Scale (Constantino 2012) is unlikely to tell us much about the developmental causation of the disorder. It is an outcome brain feature that is linked to an outcome behavioral feature. It is "shallow" because it is probably not informative of etiological diversity. Or to make the complementary argument: One cannot expect that a brain finding tied to early causal neurodevelopmental disturbances would correlate with outcome symptom severity, once it is acknowledged that many *different* etiologies converge onto a relatively small set of behavioral outcome criteria under the clinical ASD umbrella. However, this does not mean that any group-level imaging finding that shows no symptom correlation would give us important insight into etiology. The problem remains that *any* neuroimaging finding in a child that has lived with an ASD for years will be affected by experiential plasticity.

Causation Equals Genes—Or Does It?

This chapter is about imaging, but in the context of the search for "deep" causal explanations of ASDs, a brief foray into genetics is suitable. Over decades, large-sample gene hunting in ASDs has identified many hundreds of risk loci (Oron and Elliott 2017; Vorstman et al. 2017). The lessons may appear analogous to those from imaging research, which has been similarly plagued by too many (often unreplicated) findings. Similar lessons in large part reflect the "ill-posed" questions behind the hunt for "autism genes" or "autism brain markers" discussed earlier. However, the genetic literature on ASDs has contributed fundamentally to our understanding of the disorder, albeit in ways not initially anticipated. Rather than "autism genes" (Myers et al. 2020), it has revealed numerous genetic loci that may each carry only small risk. In contrast to syndromic forms of ASDs (Sztainberg and Zoghbi 2016), the genetic literature on "idiopathic" ASDs has highlighted how intractably complex polygenic developmental disorders are. Increasing awareness of epigenetic factors beyond coding DNA, such as noncoding RNA (Gudenas et al. 2017), methylation and gene expression (Ansel et al. 2017; Bedrosian et al. 2018; Rylaarsdam and Guemez-Gamboa 2019), as well as environmental factors (in and ex utero (Gardner et al. 2020; Kim et al. 2017)), suggests that each case of nonsyndromic ASD may reflect one of possibly uncountable permutations of "multiple hits". Although some of these risk factors may converge onto more distinct etiologies (e.g., pathways affecting synapses and circuit formation (Velmeshev et al. 2019)), fully mechanistic models of causation and neurodevelopment for the large population of nonsyndromic ASDs probably remain in the distance. The question is: What does this imply for neuroimaging?

Shallow Neuroimaging and Translational Shortcuts

Earlier in this chapter, we argued that neuroimaging findings in children or adolescents may not be informative of developmental causation even when associated with symptom severity (or other behavioral outcome measures). However, this argument may partly miss the translational goals of neuroimaging. Even if fcMRI findings in a 10-year old, for example, may not be easily interpretable with respect to neurodevelopmental disturbances causing the emergence of ASD in the first years of life, such findings may still have great translational value. They may not indicate what would need to be done in infants (or prenatally) to prevent the onset of the disorder, but they may pinpoint what needs to be done to successfully treat the 10-year-old child in the study. In other words, while neuroimaging may not unequivocally reveal distal causes (i.e., how the disorder first came about), it can be most powerful in detecting *proximal causation* of the symptomatology that clinicians wish to improve.

How much do network differences between a cohort of school-age children with ASDs and TD peers in an fcMRI study reveal about distal primary causes of neurodevelopmental disturbances vs. proximal causation of outcome behavioral disturbances? The question is rarely posed explicitly in imaging investigations. However, the distinction is crucial, making the difference between likely failure and potential success. Although imaging features from children or adults may be considered “intermediate phenotypes” that are more informative of etiological subtypes of the disorder than behavioral measures alone (Hong et al. 2020; Rasetti and Weinberger 2011), it appears unlikely that even the most comprehensive multimodal imaging can generate a mechanistic understanding of developmental history. What outcome neuroimaging can contribute is to pinpoint variability and clusters of variants that may be useful to other fields (genetics, epigenetics, developmental neuroscience; animal models) in the pursuit of mechanistic models.

On the other hand, giving up on the often implicit assumption that imaging findings may inform about developmental causation is a first step toward a more feasible goal: To pinpoint what the crucial brain anomalies are that are linked to behavioral problems and cognitive limitations in the *outcome phenotype* of a given *individual* with ASDs. Even in the absence of a mechanistic understanding of distal primary causes, precise and individualized knowledge of proximal causation of specific behavioral anomalies will be an indispensable step toward tailored intervention. A whole array of noninvasive brain stimulation techniques, such as transcranial magnetic (Valero-Cabre et al. 2017) and direct cortical stimulation (Philip et al. 2017) or neurofeedback training (Sitaram et al. 2017), are under development. None of them is ready for anything other than exploratory interventions (e.g., Datko et al.

2017; Kang et al. 2019; Ramot et al. 2017); but understanding of how such techniques can be calibrated and optimized to exert a specific effect that may rectify or normalize a neurofunctional anomaly detected in an individual with ASD is slated to improve in coming years. Imaging techniques, at the same time, will need to improve in the detection of brain anomalies at the level of the single individual that may be targets for such interventions. As already mentioned (and further explained and described in Methodological Challenges and Recommendations below), the still common habit of performing *t*-tests comparing largest possible ASD and TD samples cannot reach this goal.

Methodological Challenges

In the previous section, we reviewed conceptual considerations that call for fundamental changes in the way neuroimaging needs to approach the questions of "autism biomarkers" and mechanistic models of ASDs. However, even if one accepts changes in perspective, away from group-level analyses and toward the detection of subtypes and individual brain patterns, methodological best practices remain indispensable. In this section, we will therefore review methodological issues that have contributed to the lack of coherent imaging findings, such as inconsistent reports of functional over- and underconnectivity described earlier, and will try to identify sources of variability between studies that might contribute to these divergent results. We will describe these in detail referring to data from ABIDE I and II (total $N = 2226$, ASD = 1060, TD = 1166 from 39 different data sets (Di Martino et al. 2014, 2017)) as illustrations whenever possible. We will then offer recommendations for how the challenges of interpreting and comparing results from heterogeneous cohorts and study protocols can be addressed in future research.

Sources of Variability

Cohort Characteristics

The vast majority of studies investigating functional connectivity differences in ASDs and typical development have focused on children and adolescents, with the average age of participants in fMRI studies submitted to ABIDE being 15.95 years (SD = 8.6). This age range marks a period of particularly rapid and extensive brain development, and differences in the developmental trajectory of functional connectivity in ASDs have been reported in multiple independent studies (Kelly et al. 2009; Lee et al. 2017; Nomi and Uddin 2015; Padmanabhan et al. 2013). For instance, one study reported atypically increased within-network connectivity in ASDs in early childhood (<11 years), but not in adolescence (Nomi and Uddin 2015). While ASD and TD groups are usually matched for age within a study, they are rarely matched well between studies. For example, the mean age varies from 10 to 34.4 years (SD = 6.77) in the 20 different data sets in ABIDE I and from 6.8 to 38.5 years (SD = 7.98) across the 19 data sets in ABIDE II. This renders comparisons of results obtained from cohorts with dissimilar age distributions is difficult. Including age as a covariate in statistical tests rectifies this problem to some extent but is not always practiced. Note also that chronological age matching does not guarantee matching for maturational stage, which is harder to determine objectively and—given atypical growth trajectories described earlier—may not transparently relate to chronological age in ASDs.

Furthermore, cohorts vary substantially in gender distribution. ASDs are more frequently diagnosed in males, with an estimated ratio between 4:1 (American Psychiatric Association 2013) and 3:1 (Loomes et al. 2017). This ratio tends to be even higher in fMRI cohorts, with an average percentage of 13.4% female participants in the ABIDE data sets (ABIDE I: range across sites 0–28.6%, SD = 9.69; ABIDE II: range across sites 0–53.6%, SD = 14.8). 9 data sets do not include any female participants. The discrepancy is explained by often lower general level of functioning and higher symptom severity in female variants of ASDs (Frazier et al. 2014), which makes acquisition of high-quality (low-motion) MRI data extremely challenging. Given findings of differences in brain organization between girls and boys diagnosed with ASDs (Calderoni et al. 2012; Irimia et al. 2017; Lai et al. 2013; Olson et al. 2020), varying gender distribution across studies could contribute to inconsistent findings. Furthermore, functional connectivity is likely affected by handedness (Pool et al. 2014). Differences in the number of left-handed participants should therefore be taken into account (ABIDE I left-handed: 8.69% average for 17 sites that report handedness "category," range across sites 0–17.9%, SD = 5.36, ambidextrous: 2.28%, range across sites 0–21.05%, SD = 5.42; ABIDE II left handed: 6.37%, range across sites 0–21.43%, SD = 7.03, ambidextrous: 6.26%, range across sites 0–28%, SD = 7.2).

In addition to basic demographic factors like age, gender, and handedness, autism severity and cognitive abilities differ across cohorts as well. Because fMRI studies require participants to be able to follow instructions and remain still for a prolonged period of time in the unusual environment of an MRI scanner and in the presence of loud and unfamiliar noises, children and adolescents with ASDs included in fMRI studies are predominantly high functioning (Nordahl et al. 2016). This is generally also true for the ABIDE data sets (with a mean full scale IQ of 106 among ASD participants, range 41–149, SD = 17). The inclusion of low-functioning individuals differs substantially between data sets submitted to ABIDE (ABIDE I: 10.1% ASD participants with FSIQ <85, range across sites 0–28.6%, SD = 9.7; ABIDE II: 12.8%, range across sites 0–47%, SD = 11.7). Comparing ASD cohorts with different proportions of low- and high-functioning individuals to matched TD groups might therefore also lead to seemingly inconsistent findings between studies. This problem is further complicated if TD groups include participants with below-average IQs (<85) to assure group matching (4 sites in ABIDE I, 3 sites in ABIDE II), as it is questionable whether such participants are truly "neurotypical."

Lastly, comorbidities and use of psychotropic medications are common in children and adolescents with ASDs. Spencer et al. (2013) reported that among 33,565 commercially insured children with ASDs in the United States, 64% had a prescription of at least one psychoactive medication, and 35% were prescribed more than one. These medications are intended to alleviate symptoms of ASDs or treat comorbid conditions such as anxiety or attention-deficit/hyperactivity disorder (ADHD), and many of the same medications have been shown to affect brain function in schizophrenia, bipolar disorder, depression, and ADHD (Abbott et al. 2013; Hafeman et al. 2012; Phillips et al. 2008; Rubia et al. 2014; Singh and Chang 2012). However, for studies of brain function in ASDs, medication status and history have rarely been included in cohort characteristics or taken into account in analyses. For example, medication status at the time of scanning is not reported for 11 out of 39 data sets in ABIDE (ABIDE I: 7, ABIDE II: 4). Similarly, 26 data sets do not have information on comorbidities (ABIDE I: 17, ABIDE II: 9). Additionally, some sites instructed participants to withhold stimulants for the day of the scan. However, even this precaution does not rule out confounds related to medication effects because withdrawal from a regularly taken medication may also affect brain function and connectivity in ways other than a simple return to a medication-naïve state (cf. Lanius et al. 2010).

A recent study by our group (Linke et al. 2017) showed different patterns of functional connectivity in a group of children and adolescents taking psychotropic medication at the time of the scan compared to those who were medication-naïve (Fig. 2). However, the wide range and different combinations of medications prescribed make it difficult—if not impossible—to replicate the exact findings in independent samples, highlighting the need to take cohort differences in medication status into consideration. Aside from current medication status, medication history with its likely effect on brain development needs to be considered as a source of variability. Analogously, behavioral interventions, their duration and the age at which they were started are likely to affect brain development. Because availability of and participation in interventions are highly dependent on socioeconomic status (Patten et al. 2013), cohort characteristics might differ substantially by study location and recruitment strategies.

The solution cannot be to simply exclude individuals with comorbidities and current or past medication use from ASD samples in neuroimaging studies because such samples would be grossly unrepresentative of the ASD population at large. More generally, there are so many factors of clinical, demographic, and treatment variability that creation of "clean" samples unaffected by any of them would be infeasible and probably impossible. On the other hand, documentation of comorbidities, past and present medication usage, interventions received, and socioeconomic status, should be provided in as much detail as possible and taken into account in data analysis and interpretation of findings. The commonly sparse coverage of these types of collateral data in the ASD imaging literature has made it harder to account for differences in findings.

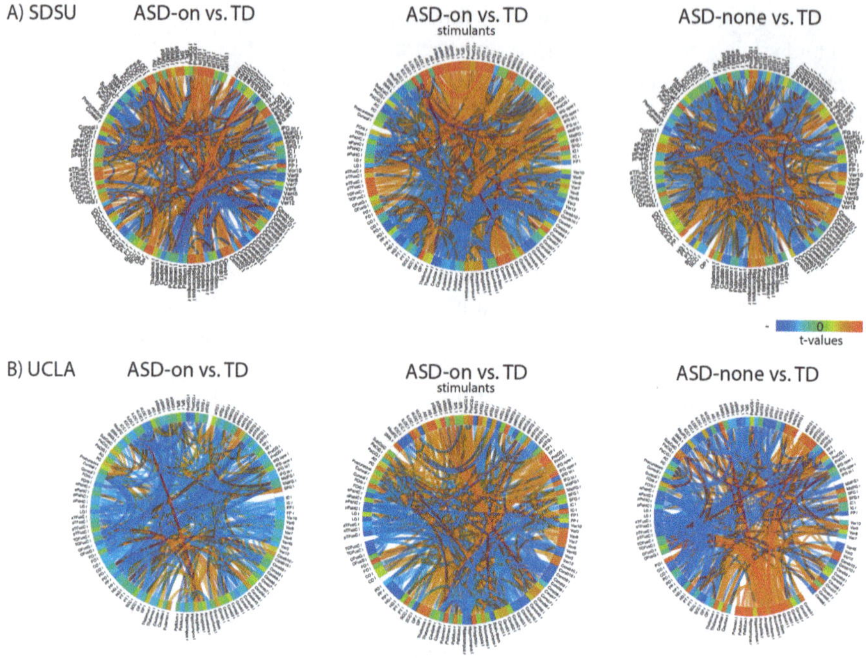

Fig. 2 (**a**) Differences in functional connectivity pattern between ASD participants taking psychotropic medication (ASD-on, $n = 25$) compared to a motion-, age-, nonverbal IQ-, and gender-matched TD group ($n = 50$), and compared to an equally matched medication-naive ASD group (ASD-none, $n = 24$). Blue colors indicate underconnectivity, and red colors show overconnectivity for t-tests of 132 regions of interest (ROIs) obtained from the FSL Harvard-Oxford cortical and subcortical atlas and the AAL cerebellar regions (as distributed with the CONN toolbox), thresholded at $p < 0.05$ (uncorrected) to illustrate broad pattern differences. (**b**) In comparison, the ASD-on ($n = 18$) and matched TD ($n = 32$) groups from the UCLA-1 and UCLA-2 cohorts available through ABIDE show a very different pattern of connectivity differences compared to the SDSU cohort. This is likely due to differences in psychotropic medications taken by the ASD participants in the two cohorts. Despite much smaller sample sizes, and polypharmacy with different drug classes, the pattern of ASD-on>TD connectivity differences becomes more similar when only comparing ASD participants taking stimulants (second column, SDSU: $n = 14$, UCLA: $n = 6$). This highlights the need to take psychotropic medication usage and cohort differences into account when comparing functional connectivity results reported in different studies. Figure adapted from (Linke et al. 2017)

Data Acquisition

MRI hardware and echo-planar imaging (EPI) sequences commonly used for fMRI data acquisition can substantially affect temporal signal-to-noise ratio (tSNR) and ultimately fMRI findings (Wald and Polimeni 2017). While these effects will usually be similar in ASD and TD groups compared at any given site, there may be differential interactions between imaging hardware and protocols, on the one hand, and demographic or diagnostic variables, on the other. For example, different sources of noise dominate at different field strengths, with physiological noise contributions increasing at higher field strength (Havsteen et al. 2017; Triantafyllou et al. 2005). Accelerated EPI sequences such as simultaneous multislice (or "multiband") EPI that allow for higher temporal resolution likewise seem to be impacted differently by physiological noise than standard 2D EPI sequences (Boubela et al. 2014; Golestani et al. 2017), as are multichannel array coils with higher numbers of channels (Faraji-Dana et al. 2016; Triantafyllou et al. 2011). Any etiology or cohort-specific factor linked to physiological differences (heart rate, respiration) such as age, medication usage, comorbidities (e.g., anxiety or ADHD), or prescan preparation (to mitigate risk of claustrophobia or anxiety) might interact with the way fMRI data are acquired and result in spurious functional connectivity differences.

Additionally, head coils differ in their tSNR profile across the brain. Generally, higher cortical tSNR is achieved with increasing numbers of channels, but this often comes at the expense of reduced subcortical tSNR, especially when coil fit is poor (e.g., in younger children for whom the head coil will fit less tightly). Especially when comparing subcortical connectivity findings between studies, it is therefore important to consider the potential impact of hardware and image acquisition sequences.

EPI sequences that are accelerated by means of parallel imaging (such as GRAPPA and SENSE on Siemens or ARC on GE MRI scanners) often suffer from a trade-off between faster image acquisition (i.e., higher temporal resolution) and overall reduced image quality and tSNR (Bhavsar et al. 2014; de Zwart et al. 2006). Current multiband-accelerated EPI sequences have been shown to be able to attain higher tSNR for acceleration factors up to around four simultaneously acquired slices, after which tSNR decreases (Feinberg and Setsompop 2013; Todd et al. 2016). Different EPI sequences may therefore affect the sensitivity to detect group differences.

Even when comparing data derived from similar hardware or in cohorts with similar demographic characteristics, participants may experience different states of vigilance during data acquisition, which might subsequently impact functional connectivity measures (Fukunaga et al. 2006; Larson-Prior et al. 2009; Wong et al. 2013a) (cf. Fig. 3). Vigilance may be affected by the anxiety-evoking noisy and constrained setting of an MRI session or in response to specific instructions given prior to a resting state. Many study protocols ask participants to keep their eyes open and fixated on a central point (e.g., a fixation cross on an otherwise blank screen) for the duration of the resting-state fMRI data collection, but only some groups monitor adherence using video recordings or eye tracking. Other groups instruct participants to close their eyes or give no explicit instructions at all. In the ABIDE data sets, eye status varies (548 participants with eyes closed, 1607 with eyes open). Eye status has been shown to affect functional connectivity measures (Patriat et al. 2013; Zou et al. 2015) and detected differences between ASD and TD groups (S. Nair et al. 2017). More generally, vigilance (which might vary depending on eye status, motivation of the participant, age, time of day, medications taken prior to the scan) is known to impact functional connectivity measures and global signal amplitude (Liu et al. 2017; Wang et al. 2015; Wong et al. 2013a, b).

Lastly, scan duration and the number of volumes acquired will affect the robustness of functional connectivity measures, with longer scans and a higher number of

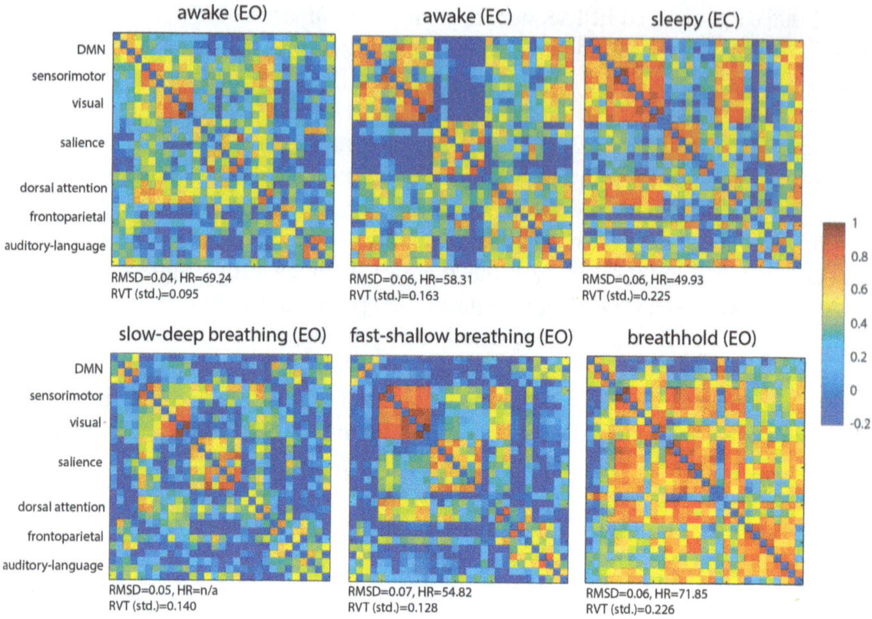

Fig. 3 Pattern of connectivity (Fisher z-transformed correlations) in one adult (32-year-old female) undergoing 6 min of resting-state fMRI in different physiological states on four separate days. The same standard preprocessing and denoising was applied to all data (including rigid-body realignment, 6 mm smoothing, censoring of volumes with >0.5 mm framewise displacement, 0.008–0.08 Hz bandpass filtering, and nuisance regression including white matter, CSF, and motion regressors, as well as their derivatives). Connectivity between 30 cortical ROIs distributed with the CONN toolbox, covering seven resting-state networks derived from ICA in 497 young adults who participated in the Human Connectome Project was estimated using Pearson correlations. The pattern of connectivity varies substantially, highlighting the importance of physiological state in standard correlation analyses used to estimate resting-state functional connectivity, and the importance of matching groups or correcting for individual physiological differences. RMSD, root mean square deviation; HR, heart rate (per minute); RVT, respiratory volume per time

volumes significantly improving reliability of findings (Birn et al. 2013; Noble et al. 2017). An earlier pseudo-consensus that 6 min of resting-state scanning was fully sufficient for reliable fcMRI measurements (e.g., Van Dijk et al. 2010) may have misled the ASD fcMRI field at large and resulted in acquisition of data with suboptimal reliability Noble et al. (2017). For example, in ABIDE I-II, the total duration of the rs-fMRI scan per participant across sites varies from 5 to over 10 min, with one to three separate scans acquired per participant. Hence, the duration of data acquisition and number of time points acquired should also be considered when comparing inconsistent functional connectivity findings.

Data Preprocessing and Analysis

As described earlier, inter-regional signal correlations in fcMRI are typically obtained from resting-state fMRI but can also be extracted from data collected during task performance. Both approaches have been broadly used in the ASD literature and have often been treated interchangeably. However, instructions (task vs. rest) given to participants differ vastly, with corresponding differences in mental states, and analysis strategies diverge greatly. Our group (Müller et al. 2011) previously found that many of the prominent underconnectivity results in ASDs were observed in task-related coactivation studies (e.g., (Just et al. 2004; Mostofsky et al. 2009; Villalobos et al. 2005; Kleinhans et al. 2008a; Kana et al. 2006; Koshino et al. 2005), while many findings inconsistent with a general underconnectivity theory were generated using resting-state fMRI scans (e.g., (Di Martino et al. 2011; Fishman et al. 2014, 2015; Khan et al. 2015; Lynch et al. 2013; Supekar et al. 2013). Notably, Nair and colleagues (Nair et al. 2014) showed that in the same task-fMRI data sets robust underconnectivity findings were obtained when task effects were left intact, while overconnectivity was found when task regression was performed prior to functional connectivity analyses. This suggests that the choice of analysis pipeline has a substantial effect on detected group differences.

Preprocessing methods may also account for some inconsistencies in findings. Motion is a likely confound when comparing patient populations to typically developing controls (Deen and Pelphrey 2012). Strategies to prevent motion vary, including mock scanning to allow participants to practice lying still in the MRI scanner and to familiarize themselves with the procedure and environment, as well as the use of head restraints ranging from foam cushions to bite bars and custom-printed head cases (e.g., CaseForge, https://caseforge.co/). The success of any procedure may vary across individuals, and some procedures (such as bite bars) may be impractical with many clinical or pediatric populations.

Any head motion during scanning is typically corrected using rigid-body realignment, from which six motion parameters are obtained. These are often used to identify participants that should be excluded from the analyses due to excessive motion during the scan. Frequently, individual volumes or chunks of volumes affected by motion are censored or "scrubbed," either by deletion or by interpolation (Power et al. 2012, 2015). The thresholds for participant exclusion and volume censoring, however, vary across studies. While censoring cutoffs $\leq$0.2 mm motion between time points are commendable, such strict criteria are often infeasible in the study of challenging populations (such as children with ASDs). Motion denoising procedures have therefore varied across ASD studies from such strict thresholds to lenient ones (e.g., 3.4 mm in Weng et al. 2010) and even complete absence of censoring (e.g., Lombardo et al. 2010). Such differences are likely to affect findings, in particular if ASD and TD comparison groups are not tightly matched for motion parameters.

Additional preprocessing steps likely to contribute to divergent findings include spatial smoothing, filtering (low-pass, high-pass, or bandpass), and nuisance regression. Spatial smoothing may be particularly relevant given the recent discovery that most common fMRI analysis software packages previously underestimated spatial autocorrelation when setting significance thresholds for cluster correction (Eklund et al. 2016). This also highlights the fact that different statistical thresholding methods chosen for multiple comparison correction might make it harder to detect similarities between studies—a problem that could be mitigated by routine submission of unthresholded maps and comparisons of effect sizes across the entire brain (as opposed to exclusive focus on regional clusters exceeding a corrected statistical "significance" threshold).

High-pass filtering of fMRI data is a relatively uncontroversial denoising step, implemented by most groups (Caballero-Gaudes and Reynolds 2017; Cole 2010). Low-pass or bandpass filtering is more contentious, as there is evidence that BOLD signal fluctuations may extend up to 0.8 Hz (Bright et al. 2017; Chen and Glover 2015; Della-Maggiore et al. 2002; Niazy et al. 2011; Skudlarski et al. 1999). While most intrinsic functional connectivity studies have implemented bandpass filters approximating $0.1 > f > .01$ Hz, others—especially coactivation fcMRI studies testing for task-related signal correlations—have included higher frequencies as well (e.g., Brieber et al. 2010; Kana et al. 2009; Kleinhans et al. 2008a, b; Lombardo et al. 2010). Importantly, underconnectivity findings in ASDs have been observed relatively frequently in studies applying no low-pass filter (Müller et al. 2011).

The sequence of preprocessing steps determining when exactly confound regressors are implemented in the pipeline and when data are filtered (prior to nuisance regression, simultaneously or subsequently) can also influence results (Caballero-Gaudes and Reynolds 2017; Ciric et al. 2017; Parkes et al. 2018; Power et al. 2014a). Prior to fcMRI analyses, nuisance regression is carried out to remove any remaining noise due to image artifacts, motion, physiological variables (cardiac and respiratory), or scanner drift. When fMRI data acquired during task performance are analyzed for intrinsic functional connectivity, signal changes related to task performance will also be regressed out (Fair et al. 2007). In addition to the motion parameters obtained from performing rigid-body-realignment, nuisance regressors can be derived from white matter and cerebrospinal fluid masks (to correct for motion and physiological noise) by calculating a mean time course within these masks or by performing principal components analysis (PCA, e.g., aCompCor; Behzadi et al. 2007), from physiological monitoring devices (such as a pulse-oximeter, ECG and/or a respiratory belt) worn by an individual during the scan, or by performing independent component analysis (ICA) and extracting time courses from components identified as noise. Frequently, derivatives of the various nuisance regressors are included as well (Power et al. 2014a).

Recent comparisons to identify the best combination of nuisance regressors for denoising of resting-state fMRI data have yielded inconsistent results (Caballero-Gaudes and Reynolds 2017; Ciric et al. 2017; Parkes et al. 2018; Power et al. 2014a). Nonetheless, there is some broad consensus on the benefit of most noise removal methods described above. This is not the case for global signal regression. The inclusion of a global signal regressor for the removal of global artifacts remains highly contentious. While considered to be the most efficient method of noise removal by some (see Liu et al. 2017 for a review; Power et al. 2014a), others have pointed out that global signal fluctuations partly reflect neuronal activity (Schölvinck et al. 2010) and their removal can alter between-group effects, for example, in comparisons between ASD and TD groups (Abbott et al. 2016; Gotts et al. 2013) (Fig. 4).

Recommendations

Multisite, Multimodal Data

Differences in cohort characteristics, data acquisition, and analysis methods make it difficult to integrate functional connectivity findings from different ASD studies. Recent efforts to pool data from many different sites and to make them openly available might help address some challenges of interpretation described in the previous sections. In addition to the very large repositories in ABIDE I and II described earlier, imaging data are openly available as part of the National Database for Autism Research (NDAR; $n = 9389$, $n = 1480$ and $n = 1753$ with anatomical MRI, fMRI, and DWI data, respectively, as of early 2018) and other data sharing initiatives (Loth et al. 2017). Using multisite data to relate atypical brain function in ASDs to neuropsychological measures or medical history is often difficult as the information collected and instruments used vary across sites. Some studies using ABIDE data suggest that it may be possible to identify common ASD brain markers across sites, despite heterogeneous study populations and protocols (e.g., Abraham et al. 2017). However, modest diagnostic prediction accuracies below 70% reflect not only the methodological sources of variability described in this section but also the critical conceptual challenges discussed earlier in this chapter. While large samples suffer the limitations of group-level studies (in the sense that they may only reveal "smallest common denominators"), they nonetheless open avenues for addressing the heterogeneity problem through data-driven clustering or unsupervised machine learning (Dwyer et al. 2018). These techniques can identify subgroups in an ASD cohort that share brain features (considered "intermediate phenotypes"; Rasetti and Weinberger 2011) and may therefore indicate the existence of brain-based variants of the disorder. In a recent example of such an approach including a subsample of 107 ASD data sets from ABIDE, Hong and colleagues (Hong et al. 2017) detected three ASD subgroups differentiated by several anatomical MRI features. Easson et al. (2017) detected two ASD subgroups (in a different ABIDE subsample of 163 ASD participants) using static and dynamic fcMRI features. Inclusion of larger sample sizes (if well controlled for the numerous sources of variability affecting multisite data) will make such approaches more powerful because less prevalent subtypes are more likely to be represented in larger samples. Inclusion of multimodal data (anatomical, functional, diffusion MRI, MEG, etc.) is expected to further improve the sensitivity of such data-driven approaches, as one type of data (e.g., from T1-weighted ana-

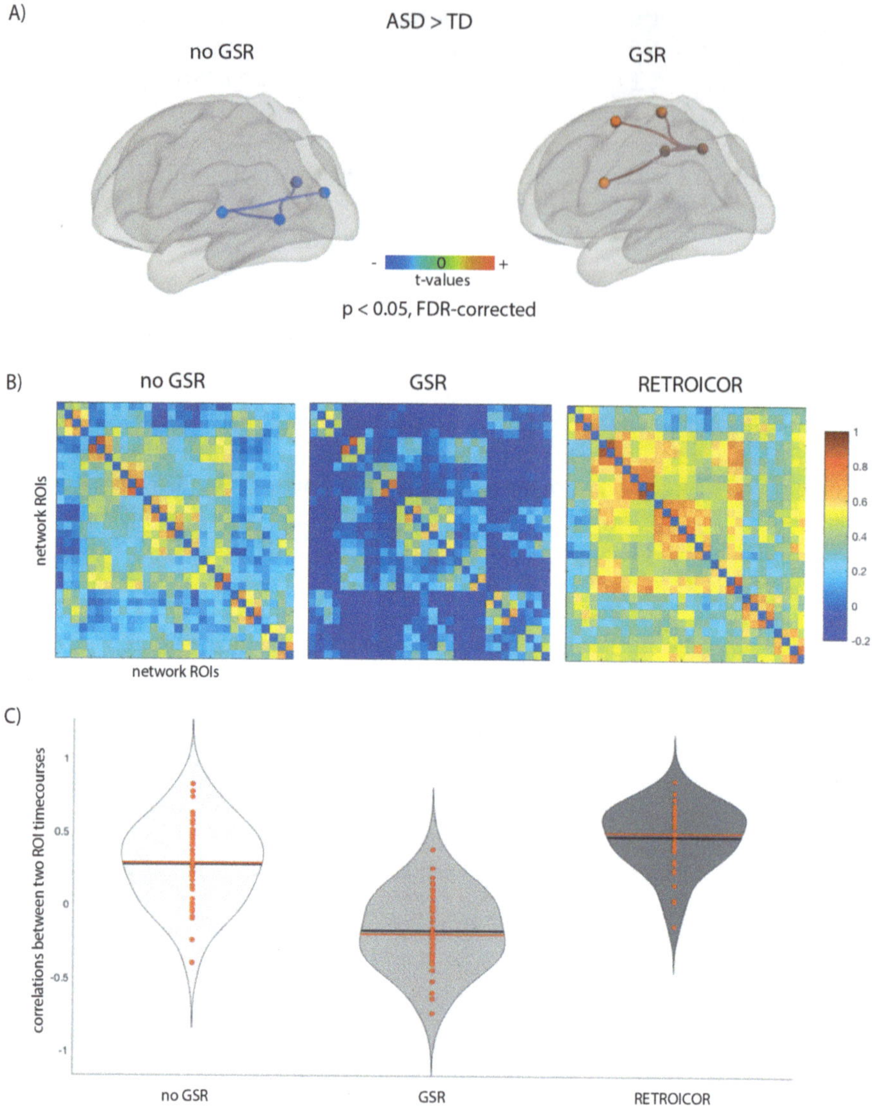

Fig. 4 (**a**) Multiple comparison corrected ($p < 0.05$, FDR) results of comparing ROI-ROI functional connectivity differences between age-, gender-, and motion-matched ASD and TD groups ($n = 18$ ASD, $n = 20$ TD; adults ages 40–65 years); 30 ROIs covering the whole brain (as in Fig. 3) were included. Significant group differences in connectivity vary, dependent on preprocessing method. Without GSR, only underconnectivity between several temporooccipital ROIs survive multiple comparison correction, while a strikingly different pattern of overconnectivity between several frontoparietal emerges when GSR is used. (**b**) Pattern of connectivity between 30 ROIs (as in Fig. 3) in the same 38 adults, after implementing different preprocessing methods. Preprocessing always included rigid-body realignment, 6 mm smoothing, censoring of volumes with >0.5 mm framewise displacement, 0.008–0.08 Hz band-pass filtering, and nuisance regression including white matter, CSF, and motion regressors, as well as their derivatives. In the middle panel, preprocessing also included GSR, in the right panel RETROICOR. (**c**) Correlation values for one representative ROI-ROI pair (a left visual and a left language ROI) when different preprocessing (no GSR, GSR, or RETROICOR) are applied. Importantly, the mean correlation value changes substantially depending on the preprocessing strategy chosen

tomical MRI, as in the study by Hong et al.) may not be sufficient for the detection of crucial brain differences between ASD variants.

For future studies, more standardized collection and reporting of medical and clinical history, including type, duration, and dosage of behavioral and pharmaceutical interventions, may elucidate the effects of different treatment strategies on functional brain development in ASDs. In the short-term, detailed summaries of available demographic and medical information and interpretation of findings with respect to differences in data acquisition and analysis, as well as cohort characteristics, sample, and effect sizes, may help account for seemingly inconsistent results.

Data Acquisition, Hardware, and Imaging Sequences

Advancements in MRI hardware make it easier to control and correct for motion. Real-time motion tracking systems (e.g., Kineticor, www.kineticor.com), imaging sequences that use navigators and can reacquire images impacted by motion (e.g., Tisdall et al. 2012; Watanabe et al. 2016; White et al. 2010), and quieter imaging (such as the "whisper mode" for structural scans on Siemens MRI scanners, or "Silent" structural sequences on GE scanners; e.g., Alibek et al. 2014) may increase data quality and allow inclusion of lower-functioning or younger participants. A child-friendly environment and mock sessions to prepare children or adults with neurodevelopmental disorders for their scan can further improve data quality (Nordahl et al. 2016). Acquiring physiological data (heart rate and respiration) during a resting-state scan and collecting a detailed medical history allow for assessment of potential physiological differences related to diagnostic group or age. If available, physiological noise regression is advisable to improve denoising of resting-state fMRI data (e.g., using RETROICOR, Glover et al. 2000). New sequences for fMRI image acquisition can also significantly increase SNR through improved noise detection. For example, multi-echo EPI (Kundu et al. 2011, 2013; Power et al. 2018) acquires images at multiple echo times, and ICA run separately on data acquired with different echo times can then be used to separate noise (common across echo times) from BOLD signal (detected only at specific echo times). With multiband accelerated EPI acquisition, it additionally becomes possible to run temporal ICA and use resulting time series for denoising (Glasser et al. 2017).

Natural Stimulation fMRI

Collecting functional MRI data during the resting state makes it possible to detect the intrinsic synchronization of BOLD fluctuations between different brain regions and to study brain function in populations who might not be able to follow task instructions. One downside of resting-state fMRI is that fluctuations of the BOLD signal are only an indirect measure of neural activity and reflect a complicated mixture of neuronal, physiological, and hemodynamic factors. In the absence of a task and corresponding model of the neural response, it is often difficult to discern signal from noise in resting-state fMRI data. Because there is little control over a participant's cognitive state during a resting-state fMRI scan (Buckner et al. 2013), assessment of individual differences is further complicated. Natural stimulation approaches in which participants watch the same movie or listen to an audio recording might offer an alternative to resting-state functional connectivity (Hasson and Honey 2012; Hasson et al. 2010; Malinen et al. 2007). With this approach, intersubject correlation analysis allows assessment of how similar responses to the same natural stimulus are across participants, and where they diverge, which could offer insights over how information processing is disrupted in participants with neurodevelopmental disorders such as autism and how these differences relate to symptomatology. As with resting-state functional connectivity, it also allows testing of a large range of hypotheses. First studies using natural stimulation in populations with ASDs have shown idiosyncratic brain responses to real-life scenarios and a lack of synchronization with activation patterns of TD participants (Hasson et al. 2009; Salmi et al. 2013; Wolf et al. 2010; Lyons et al. 2020). Some studies in typical controls have reported reduced motion when participants are watching a movie or are listening to a story compared to resting-state fMRI, potentially making this approach feasible even for lower-functioning children with ASDs. The collection of "movie fMRI" as part of large open-source projects such as the Nathan Kline Institute's (NKI) Healthy Brain Biobank (Alexander et al. 2017) further allows comparison of neural responses to a standardized natural stimulus in an individual to a large, heterogeneous sample. In smaller samples, movie fMRI with specific social interaction content has been implemented in adults with ASDs (Byrge et al. 2015).

New Analysis Approaches

Functional connectivity is typically assessed using Pearson correlation. This assumes a linear relationship of the BOLD signal in "functionally connected" regions of the brain. This assumption is too simplistic, as shown by a number of studies (Anzellotti and Coutanche 2018; Hlinka et al. 2011; Lahaye et al. 2003; Mitra et al. 2015). Even a slight lag in the BOLD signal between regions will result in low correlation magnitude, even if the two signals are otherwise identical. Additionally, Pearson correlation magnitude is highly dependent on the preprocessing methods chosen, easily affected by physiological, hemodynamic, and other confounds and has low test–retest reliability in the same individual (Birn et al. 2013; Noble et al. 2017; Patriat et al. 2013; Shah et al. 2016). As such, assigning functional meaning to the correlation magnitude between regions is difficult, and interpreting main effects in this context is not advisable as correlation magnitude is impacted by the choice of preprocessing methods. It is therefore questionable, for example, whether high-positive correlations between two regions in a TD group demonstrates "functional connectivity," whereas low correlation in an ASD group reflects absence of this connection. Instead, group differences may be more meaningfully interpreted as a difference in synchronization between two regions, without functional interpretation of the correlation magnitudes in the individual groups.

Other analysis approaches, such as comparing patterns of functional connectivity rather than individual ROI-ROI differences between participants, are more robust to the choice of preprocessing method chosen. Nonlinear methods for comparison of BOLD signal fluctuations between different brain regions, such as graph inference or dynamic time warping (Linke et al. 2020), also hold promise but have not yet been widely studied (see review in Anzellotti and Coutanche 2018).

Functional connectivity MRI in ASDs has been almost exclusively limited to static analyses, based on time series correlations across an entire run (of 5 or more minutes in duration). Dynamic connectivity analyses that assess changes in correlation over the course of a scan, especially when combined with multimodal approaches (e.g., through simultaneous collection of EEG and fMRI data) may offer a powerful way of detecting differences in functional brain dynamics in individuals with ASDs (reviewed in Mash et al. 2018).

Conclusion

The imaging literature on brain network connectivity in ASDs is vast and rapidly growing. Some gross patterns emerge, such as atypical brain growth trajectories especially early in life, which are well established, or a dual impairment of functional network integration and segregation, which remains under debate. However, firmly replicated findings remain largely elusive. The reasons are manifold. There are many methodological sources of variability, but steady improvement in preprocessing and analysis techniques and better agreement across research groups on "best practices" may mitigate these issues in the future. However, even optimal methods may not reveal a single "autism biomarker" (or a small set of such markers). The pursuit of such unique markers is questionable because ASDs comprise an unknown but possibly large number of disorders that likely differ with respect to primary causes and ensuing neurodevelopmental disturbances. Improved methodological approaches will therefore have to be dedicated to the detection of neurobiologically defined variants of the disorder.

Acknowledgments The authors are supported by the National Institutes of Health R01 MH101173, R01 DC017736, and R01 MH103494.

References

Abbott, A. E., Nair, A., Keown, C. L., Datko, M., Jahedi, A., Fishman, I., & Müller, R. A. (2016). Patterns of Atypical Functional Connectivity and Behavioral Links in Autism Differ Between Default, Salience, and Executive Networks. *Cereb Cortex, 26*(10), 4034–4045.

Abbott, C. C., Jaramillo, A., Wilcox, C. E., & Hamilton, D. A. (2013). Antipsychotic drug effects in schizophrenia: a review of longitudinal FMRI investigations and neural interpretations. *Current medicinal chemistry, 20*(3), 428–437.

Abraham, A., Milham, M. P., Di Martino, A., Craddock, R. C., Samaras, D., Thirion, B., & Varoquaux, G. (2017). Deriving reproducible biomarkers from multi-site resting-state data: An Autism-based example. *Neuroimage, 147*, 736–745.

Abrams, D. A., Lynch, C. J., Cheng, K. M., Phillips, J., Supekar, K., Ryali, S., . . . Menon, V. (2013). Underconnectivity between voice-selective cortex and reward circuitry in children with autism. *Proc Natl Acad Sci U S A, 110*(29), 12060–12065.

Agam, Y., Joseph, R. M., Barton, J. J., & Manoach, D. S. (2010). Reduced cognitive control of response inhibition by the anterior cingulate cortex in autism spectrum disorders. *Neuroimage, 52*(1), 336–347.

Alaerts, K., Nayar, K., Kelly, C., Raithel, J., Milham, M. P., & Di Martino, A. (2015). Age-related changes in intrinsic function of the superior temporal sulcus in autism spectrum disorders. *Soc Cogn Affect Neurosci.*

Alexander, L. M., Escalera, J., Ai, L., Andreotti, C., Febre, K., Mangone, A., . . . Milham, M. P. (2017). Data Descriptor: An open resource for transdiagnostic research in pediatric mental health and learning disorders. *Scientific Data, 4*, 170181.

Alibek, S., Vogel, M., Sun, W., Winkler, D., Baker, C. A., Burke, M., & Gloger, H. (2014). Acoustic noise reduction in MRI using Silent Scan: An initial experience. *Diagnostic and Interventional Radiology, 20*, 360–363.

Amaral, D. G., Li, D., Libero, L., Solomon, M., Van de Water, J., Mastergeorge, A.,. .. Wu Nordahl, C. (2017). In pursuit of neurophenotypes: The consequences of having autism and a big brain. *Autism Res, 10*(5), 711–722.

American Psychiatric Association. (2013). *Diagnostic and Statistical Manual of Mental Disorders - 5* (4th ed. Vol. http://www.dsm5.org/). Washington, DC: American Psychiatric Association.

Andrews, D. S., Lee, J. K., Solomon, M., Rogers, S. J., Amaral, D. G., & Nordahl, C. W. (2019). A diffusion-weighted imaging tract-based spatial statistics study of autism spectrum disorder in preschool-aged children. *J Neurodev Disord, 11* (1), 32.

Ansel, A., Rosenzweig, J. P., Zisman, P. D., Melamed, M., & Gesundheit, B. (2017). Variation in Gene Expression in Autism Spectrum Disorders: An Extensive Review of Transcriptomic Studies. *Front Neurosci, 10*, 601.

Anzellotti, S., & Coutanche, M. N. (2018). Beyond Functional Connectivity: Investigating Networks of Multivariate Representations *Trends in Cognitive Sciences* (Vol. 22, pp. 258–269): Elsevier.

Assaf, M., Jagannathan, K., Calhoun, V. D., Miller, L., Stevens, M. C., Sahl, R., . . . Pearlson, G. D. (2010). Abnormal functional connectivity of default mode sub-networks in autism spectrum disorder patients. *Neuroimage, 53*(1), 247–256.

Behzadi, Y., Restom, K., Liau, J., & Liu, T. T. (2007). A component based noise correction method (CompCor) for BOLD and perfusion based fMRI. *Neuroimage, 37*(1), 90–101.

Bedrosian, T. A., Quayle, C., Novaresi, N., & Gage, F. H. (2018). Early life experience drives structural variation of neural genomes in mice. *Science, 359*, 1395–1399.

Bernard, S., Enayati, A., Redwood, L., Roger, H., & Binstock, T. (2001). Autism: a novel form of mercury poisoning. *Med Hypotheses, 56*(4), 462–471.

Bettelheim, B. (1967). *The Empty Fortress: Infantile Autism and the Birth of the Self.* New York: Free Press.

Bhavsar, S., Zvyagintsev, M., & Mathiak, K. (2014). BOLD sensitivity and SNR characteristics of parallel imaging-accelerated single-shot multi-echo EPI for fMRI. *NeuroImage, 84*, 65–75.

Birn, R. M., Molloy, E. K., Patriat, R., Parker, T., Meier, T. B., Kirk, G. R., . . . Prabhakaran, V. (2013). The effect of scan length on the reliability of resting-state fMRI connectivity estimates. *NeuroImage, 83*, 550–558.

Biswal, B., Yetkin, F. Z., Haughton, V. M., & Hyde, J. S. (1995). Functional connectivity in the motor cortex of resting human brain using echo-planar MRI. *Magn Reson Med, 34*(4), 537–541.

Blaxill, M. F., Redwood, L., & Bernard, S. (2004). Thimerosal and autism? A plausible hypothesis that should not be dismissed. *Med Hypotheses, 62*(5), 788–794.

Bolte, E. R. (1998). Autism and Clostridium tetani. *Med Hypotheses, 51*(2), 133–144.

Boubela, R. N., Kalcher, K., Nasel, C., & Moser, E. (2014). Scanning fast and slow: current limitations of 3 Tesla functional MRI and future potential. *Frontiers in Physics, 2*, 1.
Brieber, S., Herpertz-Dahlmann, B., Fink, G. R., Kamp-Becker, I., Remschmidt, H., & Konrad, K. (2010). Coherent motion processing in autism spectrum disorder (ASD): an fMRI study. *Neuropsychologia, 48*(6), 1644–1651.
Bright, M. G., Tench, C. R., & Murphy, K. (2017). Potential pitfalls when denoising resting state fMRI data using nuisance regression. *NeuroImage, 154*, 159–168.
Buckner, R. L., Krienen, F. M., & Yeo, B. T. (2013). Opportunities and limitations of intrinsic functional connectivity MRI. *Nat Neurosci, 16*(7), 832–837.
Byrge, L., Dubois, J., Tyszka, J. M., Adolphs, R., & Kennedy, D. P. (2015). Idiosyncratic brain activation patterns are associated with poor social comprehension in autism. *J Neurosci, 35*(14), 5837–5850.
Caballero-Gaudes, C., & Reynolds, R. C. (2017). Methods for cleaning the BOLD fMRI signal. *NeuroImage, 154*, 128–149.
Calderoni, S., Retico, A., Biagi, L., Tancredi, R., Muratori, F., & Tosetti, M. (2012). Female children with autism spectrum disorder: an insight from mass-univariate and pattern classification analyses. *Neuroimage, 59*(2), 1013–1022.
Cascio, C. J., Moana-Filho, E. J., Guest, S., Nebel, M. B., Weisner, J., Baranek, G. T., & Essick, G. K. (2012). Perceptual and neural response to affective tactile texture stimulation in adults with autism spectrum disorders. *Autism Res, 5*(4), 231–244.
Chen, J. E., & Glover, G. H. (2015). BOLD fractional contribution to resting-state functional connectivity above 0.1Hz. *NeuroImage, 107*, 207–218.
Chien, H. Y., Lin, H. Y., Lai, M. C., Gau, S. S., & Tseng, W. Y. (2015). Hyperconnectivity of the Right Posterior Temporo-parietal Junction Predicts Social Difficulties in Boys with Autism Spectrum Disorder. *Autism Res, 8*(4), 427–441.
Chukoskie, L., Townsend, J., & Westerfield, M. (2013). Motor skill in autism spectrum disorders: a subcortical view. *Int Rev Neurobiol, 113*, 207–249.
Ciric, R., Wolf, D. H., Power, J. D., Roalf, D. R., Baum, G. L., Ruparel, K., . . . Satterthwaite, T. D. (2017). Benchmarking of participant-level confound regression strategies for the control of motion artifact in studies of functional connectivity. *NeuroImage, 154*, 174–187.
Cole. (2010). Advances and pitfalls in the analysis and interpretation of resting-state FMRI data. *Frontiers in Systems Neuroscience, 4*, 8.
Constantino, J. N. (2012). *Social Responsiveness Scale-2*. Los Angeles, CA: Western Psychological Services.
Cordes, D., Haughton, V. M., Arfanakis, K., Carew, J. D., Turski, P. A., Moritz, C. H., . . . Meyerand, M. E. (2001). Frequencies contributing to functional connectivity in the cerebral cortex in "resting-state" data. *AJNR Am J Neuroradiol, 22*(7), 1326–1333.
Courchesne, E., Karns, C. M., Davis, H. R., Ziccardi, R., Carper, R. A., Tigue, Z. D., . . . Courchesne, R. Y. (2001). Unusual brain growth patterns in early life in patients with autistic disorder: an MRI study. *Neurology, 57*(2), 245–254.
Craig, F., Margari, F., Legrottaglie, A. R., Palumbi, R., de Giambattista, C., & Margari, L. (2016). A review of executive function deficits in autism spectrum disorder and attention-deficit/hyperactivity disorder. *Neuropsychiatr Dis Treat, 12*, 1191–1202.
Datko, M., Pineda, J. A., & Müller, R.-A. (2017). Positive effects of neurofeedback on autism symptoms correlate with brain activation during imitation and observation. *European Journal of Neuroscience*, in press.
de Zwart, J. A., van Gelderen, P., Golay, X., Ikonomidou, V. N., & Duyn, J. H. (2006). Accelerated parallel imaging for functional imaging of the human brain. *NMR in Biomedicine, 19*, 342–351.
Deen, B., & Pelphrey, K. (2012). Perspective: Brain scans need a rethink. *Nature, 491*(7422), S20.
Della-Maggiore, V., Chau, W., Peres-Neto, P. R., & McIntosh, A. R. (2002). An empirical comparison of SPM preprocessing parameters to the analysis of fMRI data. *NeuroImage, 17*, 19–28.
Di Martino, A., Kelly, C., Grzadzinski, R., Zuo, X. N., Mennes, M., Mairena, M. A., . . . Milham, M. P. (2011). Aberrant striatal functional connectivity in children with autism. *Biological Psychiatry, 69*(9), 847–856.

Di Martino, A., O'Connor, D., Chen, B., Alaerts, K., Anderson, J. S., Assaf, M., . . . Milham, M. P. (2017). Enhancing studies of the connectome in autism using the autism brain imaging data exchange II. *Sci Data, 4*, 170010.

Di Martino, A., Yan, C. G., Li, Q., Denio, E., Castellanos, F. X., Alaerts, K., . . . Milham, M. P. (2014). The autism brain imaging data exchange: towards a large-scale evaluation of the intrinsic brain architecture in autism. *Mol Psychiatry, 19*(6), 659–667.

Doyle-Thomas, K. A., Lee, W., Foster, N. E., Tryfon, A., Ouimet, T., Hyde, K. L., . . . NeuroDevNet, A. S. D. I. G. (2015). Atypical functional brain connectivity during rest in autism spectrum disorders. *Ann Neurol, 77*(5), 866–876.

Dwyer, D. B., Falkai, P., & Koutsouleris, N. (2018). Machine Learning Approaches for Clinical Psychology and Psychiatry. *Annu Rev Clin Psychol.*

Easson, A. K., Fatima, Z., & McIntosh, A. R. (2017). Defining subtypes of autism spectrum disorder using static and dynamic functional connectivity. *bioRxiv*, 198093.

Eklund, A., Nichols, T. E., & Knutsson, H. (2016). Cluster failure: Why fMRI inferences for spatial extent have inflated false-positive rates. *Proceedings of the National Academy of Sciences of the United States of America.*

Esposito, G., & Pasca, S. P. (2013). Motor abnormalities as a putative endophenotype for Autism Spectrum Disorders. *Front Integr Neurosci, 7*, 43.

Fair, D. A., Schlaggar, B. L., Cohen, A. L., Miezin, F. M., Dosenbach, N. U., Wenger, K. K., . . . Petersen, S. E. (2007). A method for using blocked and event-related fMRI data to study "resting state" functional connectivity. *Neuroimage, 35*(1), 396–405.

Falahpour, M., Thompson, W. K., Abbott, A. E., Jahedi, A., Mulvey, M. E., Datko, M., . . . Müller, R. A. (2016). Underconnected, But Not Broken? Dynamic Functional Connectivity MRI Shows Underconnectivity in Autism Is Linked to Increased Intra-Individual Variability Across Time. *Brain Connect, 6*(5), 403–414.

Faraji-Dana, Z., Tam, F., Chen, J. J., & Graham, S. J. (2016). Interactions between head motion and coil sensitivity in accelerated fMRI. *Journal of Neuroscience Methods, 270*, 46–60.

Feinberg, D. A., & Setsompop, K. (2013). Ultra-fast MRI of the human brain with simultaneous multi-slice imaging. *Journal of Magnetic Resonance, 229*, 90–100.

Fishman, I., Datko, M., Cabrera, Y., Carper, R. A., & Müller, R. A. (2015). Reduced integration and differentiation of the imitation network in autism: A combined fcMRI and DWI study. *Ann Neurol, 78*(6), 958–969.

Fishman, I., Keown, C. L., Lincoln, A. J., Pineda, J. A., & Müller, R.-A. (2014). Atypical Cross Talk Between Mentalizing and Mirror Neuron Networks in Autism Spectrum Disorder. *JAMA Psychiatry, 71*(751–760).

Fitzgerald, J., Johnson, K., Kehoe, E., Bokde, A. L., Garavan, H., Gallagher, L., & McGrath, J. (2014). Disrupted Functional Connectivity in Dorsal and Ventral Attention Networks During Attention Orienting in Autism Spectrum Disorders. *Autism Res.*

Frazier, T. W., Georgiades, S., Bishop, S. L., & Hardan, A. Y. (2014). Behavioral and cognitive characteristics of females and males with autism in the Simons Simplex Collection. *J Am Acad Child Adolesc Psychiatry, 53*(3), 329–340 e321–323.

Friston, K. J., Frith, C. D., & Frackowiak, R. S. J. (1993). Time-dependent changes in effective connectivity measured with PET. *Human Brain Mapping, 1*, 69–79.

Fukunaga, M., Horovitz, S. G., van Gelderen, P., de Zwart, J. A., Jansma, J. M., Ikonomidou, V. N., . . . Duyn, J. H. (2006). Large-amplitude, spatially correlated fluctuations in BOLD fMRI signals during extended rest and early sleep stages. *Magnetic Resonance Imaging, 24*, 979–992.

Gardner, R. M., Lee, B. K., Brynge, M., Sjoqvist, H., Dalman, C., & Karlsson, H. (2020). Neonatal Levels of Acute Phase Proteins and Risk of Autism Spectrum Disorder. *Biol Psychiatry.*

Glasser, M. F., Coalson, T. S., Bijsterbosch, J. D., Harrison, S. J., Harms, M. P., Anticevic, A., . . . Smith, S. M. (2017). Using Temporal ICA to Selectively Remove Global Noise While Preserving Global Signal in Functional MRI Data. *bioRxiv*, 193862.

Glover, G. H., Li, T. Q., & Ress, D. (2000). Image-based method for retrospective correction of physiological motion effects in fMRI: RETROICOR. *Magnetic Resonance in Medicine, 44*, 162–167.

Gohel, S. R., & Biswal, B. B. (2015). Functional integration between brain regions at rest occurs in multiple-frequency bands. *Brain Connect, 5*(1), 23–34.
Golestani, A. M., Faraji-Dana, Z., Kayvanrad, M. A., Setsompop, K., Graham, S., & Chen, J. J. (2017). Simultaneous Multi-slice Resting-state fMRI at 3 Tesla: Slice-Acceleration Related Biases in Physiological Effects. *Brain Connectivity*, brain.2017.0491.
Gotts, S. J., Saad, Z. S., Jo, H. J., Wallace, G. L., Cox, R. W., & Martin, A. (2013). The perils of global signal regression for group comparisons: a case study of Autism Spectrum Disorders. *Front Hum Neurosci, 7*, 356.
Gowen, E., & Hamilton, A. (2012). Motor Abilities in Autism: A Review Using a Computational Context. *J Autism Dev Disord.*
Groen, W. B., Zwiers, M. P., van der Gaag, R. J., & Buitelaar, J. K. (2008). The phenotype and neural correlates of language in autism: an integrative review. *Neurosci Biobehav Rev, 32*(8), 1416–1425.
Gudenas, B. L., Srivastava, A. K., & Wang, L. (2017). Integrative genomic analyses for identification and prioritization of long non-coding RNAs associated with autism. *PLoS One, 12*(5), e0178532.
Hafeman, D. M., Chang, K. D., Garrett, A. S., Sanders, E. M., & Phillips, M. L. (2012). Effects of medication on neuroimaging findings in bipolar disorder: An updated review (Vol. 14, pp. 375–410).
Happé, F., Ronald, A., & Plomin, R. (2006). Time to give up on a single explanation for autism. *Nat Neurosci, 9*(10), 1218–1220.
Hasson, U., Avidan, G., Gelbard, H., Vallines, I., Harel, M., Minshew, N., & Behrmann, M. (2009). Shared and idiosyncratic cortical activation patterns in autism revealed under continuous real-life viewing conditions. *Autism Research, 2*, 220–231.
Hasson, U., & Honey, C. J. Future trends in Neuroimaging: Neural processes as expressed within real-life contexts, NeuroImage 62 1272–1278 (2012).
Hasson, U., & Malach, R., & Heeger, D. J. Reliability of cortical activity during natural stimulation, 14, NIH Public Access 40–48 (2010).
Havsteen, I., Ohlhues, A., Madsen, K. H., Nybing, J. D., Christensen, H., & Christensen, A. Are movement artifacts in magnetic resonance imaging a real problem?-a narrative review, 8, Frontiers 232 (2017).
Hazlett, H. C., Poe, M. D., Gerig, G., Styner, M., Chappell, C., Smith, R. G., . . . Piven, J. (2011). Early brain overgrowth in autism associated with an increase in cortical surface area before age 2 years. *Arch Gen Psychiatry, 68*(5), 467–476.
Hein, G., & Knight, R. T. (2008). Superior temporal sulcus--It's my area: or is it? *J Cogn Neurosci, 20*(12), 2125–2136.
Hill, E. L. (2004). Executive dysfunction in autism. *Trends Cogn Sci, 8*(1), 26–32.
Hlinka, J., Paluš, M., Vejmelka, M., Mantini, D., & Corbetta, M. (2011). Functional connectivity in resting-state fMRI: Is linear correlation sufficient? *NeuroImage, 54*, 2218–2225.
Honey, C. J., Kotter, R., Breakspear, M., & Sporns, O. (2007). Network structure of cerebral cortex shapes functional connectivity on multiple time scales. *Proc Natl Acad Sci U S A, 104*(24), 10240–10245.
Hong, S. J., Valk, S. L., Di Martino, A., Milham, M. P., & Bernhardt, B. C. (2017). Multidimensional Neuroanatomical Subtyping of Autism Spectrum Disorder. *Cereb Cortex*, 1–11.
Hong, S. J., Vogelstein, J. T., Gozzi, A., Bernhardt, B. C., Yeo, B. T. T., Milham, M. P., & Di Martino, A. (2020). Toward Neurosubtypes in Autism. *Biol Psychiatry, 88*(1), 111–128.
Hughes, J. R. (2007). Autism: the first firm finding = underconnectivity? *Epilepsy Behav, 11*(1), 20–24.
Irimia, A., Torgerson, C. M., Jacokes, Z. J., & Van Horn, J. D. (2017). The connectomes of males and females with autism spectrum disorder have significantly different white matter connectivity densities. *Sci Rep, 7*, 46401.
Johnson, M. H. (2011). Interactive specialization: A domain-general framework for human functional brain development. *Developmental Cognitive Neuroscience, 1*(1), 7–21.

Johnson, S. (2001). Micronutrient accumulation and depletion in schizophrenia, epilepsy, autism and Parkinson's disease? *Med Hypotheses, 56*(5), 641–645.

Jones, T. B., Bandettini, P. A., Kenworthy, L., Case, L. K., Milleville, S. C., Martin, A., & Birn, R. M. (2010). Sources of group differences in functional connectivity: an investigation applied to autism spectrum disorder. *Neuroimage, 49*(1), 401–414.

Joshi, G., Arnold Anteraper, S., Patil, K. R., Semwal, M., Goldin, R. L., Furtak, S. L., . . . Whitfield-Gabrieli, S. (2017). Integration and Segregation of Default Mode Network Resting-State Functional Connectivity in Transition-Age Males with High-Functioning Autism Spectrum Disorder: A Proof-of-Concept Study. *Brain Connect, 7*(9), 558–573.

Jung, M., Kosaka, H., Saito, D. N., Ishitobi, M., Morita, T., Inohara, K., . . . Iidaka, T. (2014). Default mode network in young male adults with autism spectrum disorder: relationship with autism spectrum traits. *Mol Autism, 5*, 35.

Just, M. A., Cherkassky, V. L., Keller, T. A., Kana, R. K., & Minshew, N. J. (2007). Functional and anatomical cortical underconnectivity in autism: evidence from an FMRI study of an executive function task and corpus callosum morphometry. *Cereb Cortex, 17*(4), 951–961.

Just, M. A., Cherkassky, V. L., Keller, T. A., & Minshew, N. J. (2004). Cortical activation and synchronization during sentence comprehension in high-functioning autism: evidence of underconnectivity. *Brain, 127*(Pt 8), 1811–1821.

Just, M. A., Keller, T. A., Malave, V. L., Kana, R. K., & Varma, S. (2012). Autism as a neural systems disorder: A theory of frontal-posterior underconnectivity. *Neurosci Biobehav Rev, 36*(4), 1292–1313.

Kana, R. K., Keller, T. A., Cherkassky, V. L., Minshew, N. J., & Adam Just, M. (2009). Atypical frontal-posterior synchronization of Theory of Mind regions in autism during mental state attribution. *Soc Neurosci, 4*(2), 135–152.

Kana, R. K., Keller, T. A., Cherkassky, V. L., Minshew, N. J., & Just, M. A. (2006). Sentence comprehension in autism: thinking in pictures with decreased functional connectivity. *Brain, 129*(Pt 9), 2484–2493.

Kana, R. K., Keller, T. A., Minshew, N. J., & Just, M. A. (2007). Inhibitory control in high-functioning autism: decreased activation and underconnectivity in inhibition networks. *Biol Psychiatry, 62*(3), 198–206.

Kana, R. K., Libero, L. E., Hu, C. P., Deshpande, H. D., & Colburn, J. S. (2012). Functional Brain Networks and White Matter Underlying Theory-of-Mind in Autism. *Soc Cogn Affect Neurosci.*

Kandel, E. R., Jessell, T. M., & Sanes, J. R. (2000). Sensory experience and the fine tuning of synaptic connections. In E. R. Kandel, J. H. Schwartz & T. M. Jessell (Eds.), *Principles of Neural Science* (4th ed., pp. 1115–1130). New York: Elsevier.

Kang, J. N., Song, J. J., Casanova, M. F., Sokhadze, E. M., & Li, X. L. (2019). Effects of repetitive transcranial magnetic stimulation on children with low-function autism. *CNS Neurosci Ther, 25*(11), 1254–1261.

Keehn, B., Müller, R.-A., & Townsend, J. (2013). Atypical attentional networks and the emergence of autism. *Neurosci Biobehav Rev, 37*(2), 164–183.

Kelly, A. M. C., Di Martino, A., Uddin, L. Q., Shehzad, Z., Gee, D. G., Reiss, P. T., . . . Milham, M. P. (2009). Development of Anterior Cingulate Functional Connectivity from Late Childhood to Early Adulthood. *Cerebral Cortex, 19*(3), 640–657.

Keown, C. L., Datko, M. C., Chen, C. P., Maximo, J. O., Jahedi, A., & Müller, R.-A. (2017). Network Organization Is Globally Atypical in Autism: A Graph Theory Study of Intrinsic Functional Connectivity. *Biological Psychiatry: Cognitive Neuroscience and Neuroimaging, 2*(1), 66–75.

Khan, A. J., Nair, A., Keown, C. L., Datko, M. C., Lincoln, A. J., & Müller, R. A. (2015). Cerebro-cerebellar Resting-State Functional Connectivity in Children and Adolescents with Autism Spectrum Disorder. *Biol Psychiatry, 78*(9), 625–634.

Kim, D., Volk, H., Girirajan, S., Pendergrass, S., Hall, M. A., Verma, S. S., . . . Selleck, S. B. (2017). The joint effect of air pollution exposure and copy number variation on risk for autism. *Autism Res, 10*(9), 1470–1480.

Kleinhans, N. M., Müller, R.-A., Cohen, D. N., & Courchesne, E. (2008a). Atypical functional lateralization of language in autism spectrum disorders. *Brain Research, 1221*, 115–125.

Kleinhans, N. M., Richards, T., Sterling, L., Stegbauer, K. C., Mahurin, R., Johnson, L. C., . . . Aylward, E. (2008b). Abnormal functional connectivity in autism spectrum disorders during face processing. *Brain, 131*(Pt 4), 1000–1012.

Koshino, H., Carpenter, P. A., Minshew, N. J., Cherkassky, V. L., Keller, T. A., & Just, M. A. (2005). Functional connectivity in an fMRI working memory task in high-functioning autism. *Neuroimage, 24*(3), 810–821.

Kundu, P., Brenowitz, N. D., Voon, V., Worbe, Y., Vertes, P. E., Inati, S. J., . . . Bullmore, E. T. (2013). Integrated strategy for improving functional connectivity mapping using multiecho fMRI. *Proceedings of the National Academy of Sciences, 110*, 16187–16192.

Kundu, P., Inati, S. J., Evans, J. W., Luh, W.-M., & Bandettini, P. a. (2011). Differentiating BOLD and non-BOLD signals in fMRI time series using multi-echo EPI. *NeuroImage.*

Lahaye, P. J., Poline, J. B., Flandin, G., Dodel, S., & Garnero, L. (2003). Functional connectivity: Studying nonlinear, delayed interactions between BOLD signals. *NeuroImage, 20*, 962–974.

Lai, M. C., Lombardo, M. V., Suckling, J., Ruigrok, A. N., Chakrabarti, B., Ecker, C., . . . Baron-Cohen, S. (2013). Biological sex affects the neurobiology of autism. *Brain.*

Lanius, R. A., Brewin, C. R., Bremner, J. D., Daniels, J. K., Friedman, M. J., Liberzon, I., . . . Vermetten, E. (2010). Does neuroimaging research examining the pathophysiology of posttraumatic stress disorder require medication-free patients? *J Psychiatry Neurosci, 35*(2), 80–89.

Larson-Prior, L. J., Zempel, J. M., Nolan, T. S., Prior, F. W., Snyder, A. Z., & Raichle, M. E. (2009). Cortical network functional connectivity in the descent to sleep. *Proc Natl Acad Sci U S A, 106*, 4489–4494.

Lee, P. S., Yerys, B. E., Della Rosa, A., Foss-Feig, J., Barnes, K. A., James, J. D., . . . Kenworthy, L. E. (2009). Functional connectivity of the inferior frontal cortex changes with age in children with autism spectrum disorders: a fcMRI study of response inhibition. *Cereb Cortex, 19*(8), 1787–1794.

Lee, Y., Park, B.-Y., James, O., Kim, S.-G., & Park, H. (2017). Autism Spectrum Disorder Related Functional Connectivity Changes in the Language Network in Children, Adolescents and Adults. *Frontiers in Human Neuroscience, 11*, 418–418.

Leigh, J. P., & Du, J. (2015). Brief Report: Forecasting the Economic Burden of Autism in 2015 and 2025 in the United States. *J Autism Dev Disord.*

Leopold, D. A., Murayama, Y., & Logothetis, N. K. (2003). Very slow activity fluctuations in monkey visual cortex: implications for functional brain imaging. *Cereb Cortex, 13*(4), 422–433.

Lewis, J. D., Evans, A. C., Pruett, J. R., Jr., Botteron, K. N., McKinstry, R. C., Zwaigenbaum, L., . . . Infant Brain Imaging Study, N. (2017). The Emergence of Network Inefficiencies in Infants With Autism Spectrum Disorder. *Biol Psychiatry.*

Linke, A. C., Olson, L., Gao, Y., Fishman, I., & Müller, R. A. (2017). Psychotropic Medication Use in Autism Spectrum Disorders May Affect Functional Brain Connectivity. *Biological Psychiatry: Cognitive Neuroscience and Neuroimaging.*

Linke, A. C., Mash, L. E., Fong, C. H., Kinnear, M. K.,Kohli, J. S.,Wilkinson, M.,Tung, R., Keehn, R. J. J., Carper, R. A., Fishman, I., & Müller, R. A. (2020). Dynamic time warping outperforms Pearson correlation in detecting atypical functional connectivity in autism *Neuroimage*, 117383.

Lippiello, P. M. (2006). Nicotinic cholinergic antagonists: a novel approach for the treatment of autism. *Med Hypotheses, 66*(5), 985–990.

Liu, T. T., Nalci, A., & Falahpour, M. (2017). The global signal in fMRI: Nuisance or Information? *NeuroImage, 150*, 213–229.

Linke, A. C., Mash, L. E., Fong, C. H., Kinnear, M.K.,Kohli, J. S.,Wilkinson, M.,Tung, R., Keehn, R. J. J., Carper, R. A., Fishman, I., & Müller, R.A. (2020). Dynamic time warping outperforms Pearson correlation in detecting atypical functional connectivity in autism. *Neuroimage*, 117383.

Lombardo, M. V., Chakrabarti, B., Bullmore, E. T., Sadek, S. A., Pasco, G., Wheelwright, S. J., . . . Baron-Cohen, S. (2010). Atypical neural self-representation in autism. *Brain, 133*(Pt 2), 611–624.

Loomes, R., Hull, L., & Mandy, W. P. L. (2017). What Is the Male-to-Female Ratio in Autism Spectrum Disorder? A Systematic Review and Meta-Analysis. *J Am Acad Child Adolesc Psychiatry, 56*(6), 466–474.

Lord, C., Rutter, M., DiLavore, P., Risi, S., Gotham, K., & Bishop, S. L. (2012). *Autism Diagnostic Observation Schedule - 2* (2nd ed.). Torrance (CA): Western Psychological Services.

Loth, E., Charman, T., Mason, L., Tillmann, J., Jones, E. J. H., Wooldridge, C., . . . Buitelaar, J. K. (2017). The EU-AIMS Longitudinal European Autism Project (LEAP): design and methodologies to identify and validate stratification biomarkers for autism spectrum disorders. *Mol Autism, 8*, 24.

Lynch, C. J., Uddin, L. Q., Supekar, K., Khouzam, A., Phillips, J., & Menon, V. (2013). Default Mode Network in Childhood Autism: Posteromedial Cortex Heterogeneity and Relationship with Social Deficits. *Biol Psychiatry*.

Lyon, K. M., Stevenson, R. A., Owen, A. M., Stojanoski, B. (2020). Examining the relationship between measures of autistic traits and neural synchrony during movies in children with and without autism. *Neuroimage: Clinical*, 102477.

Maenner, M. J., Shaw, K. A., Baio, J., EdS, Washington, A., Patrick, M., . . . Dietz, P. M. (2020). Prevalence of Autism Spectrum Disorder Among Children Aged 8 Years - Autism and Developmental Disabilities Monitoring Network, 11 Sites, United States, 2016. *MMWR Surveill Summ, 69*(4), 1–12.

Malinen, S., Hlushchuk, Y., & Hari, R. (2007). Towards natural stimulation in fMRI-Issues of data analysis. *NeuroImage, 35*, 131–139.

Marco, E. J., Hinkley, L. B., Hill, S. S., & Nagarajan, S. S. (2011). Sensory processing in autism: a review of neurophysiologic findings. *Pediatr Res, 69*(5 Pt 2), 48R-54R.

Mash, L. E., Reiter, M. A., Linke, A. C., Townsend, J., & Müller, R.-A. (2018). Multimodal Approaches to Functional Connectivity in Autism Spectrum Disorders: An Integrative Perspective. *Developmental Neurobiology, in press*.

Mason, R. A., Williams, D. L., Kana, R. K., Minshew, N., & Just, M. A. (2008). Theory of Mind disruption and recruitment of the right hemisphere during narrative comprehension in autism. *Neuropsychologia, 46*(1), 269–280.

Menon, V. (2011). Large-scale brain networks and psychopathology: a unifying triple network model. *Trends Cogn Sci, 15*(10), 483–506.

Mitra, A., Snyder, A. Z., Blazey, T., & Raichle, M. E. (2015). Lag threads organize the brain's intrinsic activity. *Proceedings of the National Academy of Sciences, 112*, E2235-E2244.

Mizuno, A., Liu, Y., Williams, D. L., Keller, T. A., Minshew, N. J., & Just, M. A. (2011). The neural basis of deictic shifting in linguistic perspective-taking in high-functioning autism. *Brain, 134*(Pt 8), 2422–2435.

Monk, C. S., Peltier, S. J., Wiggins, J. L., Weng, S. J., Carrasco, M., Risi, S., & Lord, C. (2009). Abnormalities of intrinsic functional connectivity in autism spectrum disorders. *Neuroimage, 47*(2), 764–772.

Mostofsky, S. H., Powell, S. K., Simmonds, D. J., Goldberg, M. C., Caffo, B., & Pekar, J. J. (2009). Decreased connectivity and cerebellar activity in autism during motor task performance. *Brain, 132*(Pt 9), 2413–2425.

Müller, R.-A. (2017). Abandoning ASD? A response to Waterhouse, London, and Gillberg. *Autism Res, 10*(7), 1183.

Müller, R.-A., & Amaral, D. G. (2017). Editorial: Time to give up on Autism Spectrum Disorder? *Autism Res, 10*(1), 10–14.

Müller, R.-A., Shih, P., Keehn, B., Deyoe, J. R., Leyden, K. M., & Shukla, D. K. (2011). Underconnected, but how? A survey of functional connectivity MRI studies in autism spectrum disorders. *Cerebral Cortex, 21*(10), 2233–2243.

Myers, S. M., Challman, T. D., Bernier, R., Bourgeron, T., Chung, W. K., Constantino, J. N., . . . Ledbetter, D. H. (2020). Insufficient Evidence for "Autism-Specific" Genes. *Am J Hum Genet, 106*(5), 587–595.

Nair, A., Keown, C. L., Datko, M., Shih, P., Keehn, B., & Müller, R. A. (2014). Impact of methodological variables on functional connectivity findings in autism spectrum disorders. *Hum Brain Mapp, 35*(8), 4035–4048.

Nair, S., Jao Keehn, R. J., Berkebile, M. M., Maximo, J. O., Witkowska, N., & Müller, R.-A. (2017). Local resting state functional connectivity in autism: site and cohort variability and the effect of eye status. *Brain Imaging Behav*, E-pub.

Nebel, M. B., Joel, S. E., Muschelli, J., Barber, A. D., Caffo, B. S., Pekar, J. J., & Mostofsky, S. H. (2014). Disruption of functional organization within the primary motor cortex in children with autism. *Hum Brain Mapp, 35*(2), 567–580.

Niazy, R. K., Xie, J., Miller, K., Beckmann, C. F., & Smith, S. M. (2011). Spectral characteristics of resting state networks. *Progress in Brain Research, 193*, 259–276.

Noble, S., Spann, M. N., Tokoglu, F., Shen, X., Constable, R. T., & Scheinost, D. (2017). Influences on the Test-Retest Reliability of Functional Connectivity MRI and its Relationship with Behavioral Utility. *Cerebral Cortex, 27*, 5415–5429.

Nomi, J. S., & Uddin, L. Q. (2015). Developmental changes in large-scale network connectivity in autism. *NeuroImage: Clinical, 7*, 732–741.

Nordahl, C. W., Mello, M., Shen, A. M., Shen, M. D., Vismara, L. A., Li, D., . . . Amaral, D. G. (2016). Methods for acquiring MRI data in children with autism spectrum disorder and intellectual impairment without the use of sedation. *J Neurodev Disord, 8*, 20.

O'Connor, K. (2012). Auditory processing in autism spectrum disorder: a review. *Neuroscience and Biobehavioral Reviews, 36*(2), 836–854.

Olson, L. A., Mash, L. E., Linke, A. C., Fong, C. H., Müller, R. -A., & Fishman, I. (2020). Sex-related patterns of intrinsic functional connectivity in children and adolescents with autism spectrum disorders. *Autism, 24*(8), 2190–2201.

Oron, O., & Elliott, E. (2017). Delineating the Common Biological Pathways Perturbed by ASD's Genetic Etiology: Lessons from Network-Based Studies. *Int J Mol Sci, 18*(4).

Padmanabhan, A., Lynn, A., Foran, W., Luna, B., & O'Hearn, K. (2013). Age related changes in striatal resting state functional connectivity in autism. *Frontiers in Human Neuroscience, 7*, 814–814.

Parkes, L., Fulcher, B., Yücel, M., & Fornito, A. (2018). An evaluation of the efficacy, reliability, and sensitivity of motion correction strategies for resting-state functional MRI. *NeuroImage, 171*, 415–436.

Patriat, R., Molloy, E. K., Meier, T. B., Kirk, G. R., Nair, V. A., Meyerand, M. E., . . . Birn, R. M. (2013). The effect of resting condition on resting-state fMRI reliability and consistency: A comparison between resting with eyes open, closed, and fixated. *NeuroImage, 78*, 463–473.

Patten, E., Baranek, G. T., Watson, L. R., & Schultz, B. (2013). Child and family characteristics influencing intervention choices in autism spectrum disorders. *Focus Autism Other Dev Disabl, 28*(3), 138–146.

Philip, N. S., Nelson, B. G., Frohlich, F., Lim, K. O., Widge, A. S., & Carpenter, L. L. (2017). Low-Intensity Transcranial Current Stimulation in Psychiatry. *Am J Psychiatry, 174*(7), 628–639.

Phillips, M. L., Travis, M. J., Fagiolini, A., & Kupfer, D. J. (2008). Medication effects in neuroimaging studies of bipolar disorder (Vol. 165, pp. 313–320).

Pool, E. M., Rehme, A. K., Fink, G. R., Eickhoff, S. B., & Grefkes, C. (2014). Handedness and effective connectivity of the motor system. *NeuroImage, 99*, 451–460.

Power, J. D., Barnes, K. A., Snyder, A. Z., Schlaggar, B. L., & Petersen, S. E. (2012). Spurious but systematic correlations in functional connectivity MRI networks arise from subject motion. *Neuroimage, 59*(3), 2142–2154.

Power, J. D., Mitra, A., Laumann, T. O., Snyder, A. Z., Schlaggar, B. L., & Petersen, S. E. (2014a). Methods to detect, characterize, and remove motion artifact in resting state fMRI. *Neuroimage, 84*, 320–341.

Power, J. D., Plitt, M., Gotts, S. J., Kundu, P., Voon, V., Bandettini, P. A., & Martin, A. (2018). Ridding fMRI data of motion-related influences: Removal of signals with distinct spatial and physical bases in multiecho data. *Proceedings of the National Academy of Sciences, 115*, 201720985.

Power, J. D., Schlaggar, B. L., & Petersen, S. E. (2014b). Studying brain organization via spontaneous fMRI signal. *Neuron, 84*(4), 681–696.

Power, J. D., Schlaggar, B. L., & Petersen, S. E. (2015). Recent progress and outstanding issues in motion correction in resting state fMRI. *Neuroimage, 105C*, 536–551.

Price, C. J. (2010). The anatomy of language: a review of 100 fMRI studies published in 2009. *Ann N Y Acad Sci, 1191*, 62–88.

Puts, N. A., Wodka, E. L., Tommerdahl, M., Mostofsky, S. H., & Edden, R. A. (2014). Impaired tactile processing in children with autism spectrum disorder. *J Neurophysiol, 111*(9), 1803–1811.

Ramot, M., Kimmich, S., Gonzalez-Castillo, J., Roopchansingh, V., Popal, H., White, E., . . . Martin, A. (2017). Direct modulation of aberrant brain network connectivity through real-time NeuroFeedback. *Elife, 6*.

Rasetti, R., & Weinberger, D. R. (2011). Intermediate phenotypes in psychiatric disorders. *Curr Opin Genet Dev, 21*(3), 340–348.

Redcay, E. (2008). The superior temporal sulcus performs a common function for social and speech perception: implications for the emergence of autism. *Neurosci Biobehav Rev, 32*(1), 123–142.

Rozenkrantz, L., Zachor, D., Heller, I., Plotkin, A., Weissbrod, A., Snitz, K., . . . Sobel, N. (2015). A Mechanistic Link between Olfaction and Autism Spectrum Disorder. *Curr Biol, 25*(14), 1904–1910.

Rubia, K., Alegria, A. A., Cubillo, A. I., Smith, A. B., Brammer, M. J., & Radua, J. (2014). Effects of stimulants on brain function in attention-deficit/hyperactivity disorder: A systematic review and meta-analysis. *Biological Psychiatry, 76*(8), 616–628.

Rudie, J. D., & Dapretto, M. (2013). Convergent evidence of brain overconnectivity in children with autism? *Cell Rep, 5*(3), 565–566.

Rudie, J. D., Shehzad, Z., Hernandez, L. M., Colich, N. L., Bookheimer, S. Y., Iacoboni, M., & Dapretto, M. (2012). Reduced functional integration and segregation of distributed neural systems underlying social and emotional information processing in autism spectrum disorders. *Cereb Cortex, 22*(5), 1025–1037.

Rutter, M., LeCouteur, A., & Lord, C. (2003). *Autism Diagnostic Interview - R*. Los Angeles (CA): Wester Psychological Services.

Rutter, M., & Schopler, E. (1987). Autism and pervasive developmental disorders: concepts and diagnostic issues. *J Autism Dev Disord, 17*(2), 159–186.

Rylaarsdam, L., & Guemez-Gamboa, A. (2019). Genetic Causes and Modifiers of Autism Spectrum Disorder. *Front Cell Neurosci, 13*, 385.

Salmi, J., Roine, U., Glerean, E., Lahnakoski, J., Nieminen-Von Wendt, T., Tani, P., . . . Sams, M. (2013). The brains of high functioning autistic individuals do not synchronize with those of others. *NeuroImage: Clinical, 3*, 489–497.

Schecklmann, M., Schwenck, C., Taurines, R., Freitag, C., Warnke, A., Gerlach, M., & Romanos, M. (2013). A systematic review on olfaction in child and adolescent psychiatric disorders. *J Neural Transm, 120*(1), 121–130.

Schölvinck, M. L., Maier, A., Ye, F. Q., Duyn, J. H., & Leopold, D. A. (2010). Neural basis of global resting-state fMRI activity. *Proc Natl Acad Sci U S A, 107*(22), 10238–10243.

Shah, L. M., Cramer, J. A., Ferguson, M. A., Birn, R. M., & Anderson, J. S. (2016). Reliability and reproducibility of individual differences in functional connectivity acquired during task and resting state. *Brain and behavior, 6*(5), e00456-e00456.

Shen, M. D., Kim, S. H., McKinstry, R. C., Gu, H., Hazlett, H. C., Nordahl, C. W., . . . Gu, H. (2017). Increased Extra-axial Cerebrospinal Fluid in High-Risk Infants Who Later Develop Autism. *Biol Psychiatry, 82*(3), 186–193.

Shen, M. D., Nordahl, C. W., Young, G. S., Wootton-Gorges, S. L., Lee, A., Liston, S. E., . . . Amaral, D. G. (2013). Early brain enlargement and elevated extra-axial fluid in infants who develop autism spectrum disorder. *Brain, 136*(Pt 9), 2825–2835.

Shih, P., Keehn, B., Oram, J. K., Leyden, K. M., Keown, C. L., & Müller, R.-A. (2011). Functional differentiation of posterior superior temporal sulcus in autism: A functional connectivity magnetic resonance imaging study. *Biological Psychiatry, 70*(3), 270–277.

Simmons, D. R., Robertson, A. E., McKay, L. S., Toal, E., McAleer, P., & Pollick, F. E. (2009). Vision in autism spectrum disorders. *Vision Res, 49*(22), 2705–2739.

Singh, M. K., & Chang, K. D. (2012). The Neural Effects of Psychotropic Medications in Children and Adolescents (Vol. 21, pp. 753–771): NIH Public Access.

Sitaram, R., Ros, T., Stoeckel, L., Haller, S., Scharnowski, F., Lewis-Peacock, J., . . . Sulzer, J. (2017). Closed-loop brain training: the science of neurofeedback. *Nat Rev Neurosci, 18*(2), 86–100.

Skudlarski, P., Constable, R. T., & Gore, J. C. (1999). ROC analysis of statistical methods used in functional MRI: Individual subjects. *NeuroImage, 9*, 311–329.

Smith, S. M., Vidaurre, D., Beckmann, C. F., Glasser, M. F., Jenkinson, M., Miller, K. L., . . . Van Essen, D. C. (2013). Functional connectomics from resting-state fMRI. *Trends Cogn Sci, 17*(12), 666–682.

Solomon, M., Ozonoff, S. J., Ursu, S., Ravizza, S., Cummings, N., Ly, S., & Carter, C. S. (2009). The neural substrates of cognitive control deficits in autism spectrum disorders. *Neuropsychologia, 47*(12), 2515–2526.

Solso, S., Xu, R., Proudfoot, J., Hagler, D. J., Jr., Campbell, K., Venkatraman, V., . . . Courchesne, E. (2015). Diffusion Tensor Imaging Provides Evidence of Possible Axonal Overconnectivity in Frontal Lobes in Autism Spectrum Disorder Toddlers. *Biol Psychiatry*.

Sparks, B. F., Friedman, S. D., Shaw, D. W., Aylward, E. H., Echelard, D., Artru, A. A., . . . Dager, S. R. (2002). Brain structural abnormalities in young children with autism spectrum disorder. *Neurology, 59*(2), 184–192.

Spencer, D., Marshall, J., Post, B., Kulakodlu, M., Newschaffer, C., Dennen, T., . . . Jain, A. (2013). Psychotropic medication use and polypharmacy in children with autism spectrum disorders. *Pediatrics, 132*(5), 833–840.

Starck, T., Nikkinen, J., Rahko, J., Remes, J., Hurtig, T., Haapsamo, H., . . . Kiviniemi, V. J. (2013). Resting state fMRI reveals a default mode dissociation between retrosplenial and medial prefrontal subnetworks in ASD despite motion scrubbing. *Front Hum Neurosci, 7*, 802.

Supekar, K., Uddin, L. Q., Khouzam, A., Phillips, J., Gaillard, W. D., Kenworthy, L. E., . . . Menon, V. (2013). Brain hyperconnectivity in children with autism and its links to social deficits. *Cell Rep, 5*(3), 738–747.

Sztainberg, Y., & Zoghbi, H. Y. (2016). Lessons learned from studying syndromic autism spectrum disorders. *Nat Neurosci, 19*(11), 1408–1417.

Tisdall, M. D., Hess, A. T., Reuter, M., Meintjes, E. M., Fischl, B., van der Kouwe, A. J. W. (2012). Volumetric navigators for prospective motion correction and selective reacquisition in neuroanatomical MRI, 68, NIH Public Access 389–399.

Todd, N., Moeller, S., Auerbach, E. J., Yacoub, E., Flandin, G., & Weiskopf, N. (2016). Evaluation of 2D multiband EPI imaging for high-resolution, whole-brain, task-based fMRI studies at 3T: Sensitivity and slice leakage artifacts. *NeuroImage, 124*, 32–42.

Triantafyllou, C., Hoge, R. D., Krueger, G., Wiggins, C. J., Potthast, A., Wiggins, G. C., & Wald, L. L. (2005). Comparison of physiological noise at 1.5 T, 3 T and 7 T and optimization of fMRI acquisition parameters. *NeuroImage, 26*, 243–250.

Triantafyllou, C., Polimeni, J. R., & Wald, L. L. (2011). Physiological noise and signal-to-noise ratio in fMRI with multi- channel array coils. *Neuroimage, 15*, 597–606.

Valero-Cabre, A., Amengual, J. L., Stengel, C., Pascual-Leone, A., & Coubard, O. A. (2017). Transcranial magnetic stimulation in basic and clinical neuroscience: A comprehensive review of fundamental principles and novel insights. *Neurosci Biobehav Rev, 83*, 381–404.

Van Dijk, K. R., Hedden, T., Venkataraman, A., Evans, K. C., Lazar, S. W., & Buckner, R. L. (2010). Intrinsic functional connectivity as a tool for human connectomics: theory, properties, and optimization. *J Neurophysiol, 103*(1), 29–321.

van Rooij, D., Anagnostou, E., Arango, C., Auzias, G., Behrmann, M., Busatto, G. F., . . . Buitelaar, J. K. (2017). Cortical and Subcortical Brain Morphometry Differences Between Patients With Autism Spectrum Disorder and Healthy Individuals Across the Lifespan: Results From the ENIGMA ASD Working Group. *Am J Psychiatry*, appiajp201717010100.

Velmeshev, D., Schirmer, L., Jung, D., Haeussler, M., Perez, Y., Mayer, S., . . . Kriegstein, A. R. (2019). Single-cell genomics identifies cell type-specific molecular changes in autism. *Science, 364*(6441), 685–689.

Villalobos, M. E., Mizuno, A., Dahl, B. C., Kemmotsu, N., & Müller, R.-A. (2005). Reduced functional connectivity between V1 and inferior frontal cortex associated with visuomotor performance in autism. *Neuroimage, 25*(3), 916–925.

von dem Hagen, E. A., Stoyanova, R. S., Baron-Cohen, S., & Calder, A. J. (2013). Reduced functional connectivity within and between 'social' resting state networks in autism spectrum conditions. *Soc Cogn Affect Neurosci, 8*(6), 694–701.

Vorstman, J. A. S., Parr, J. R., Moreno-De-Luca, D., Anney, R. J. L., Nurnberger, J. I., Jr., & Hallmayer, J. F. (2017). Autism genetics: opportunities and challenges for clinical translation. *Nat Rev Genet, 18*(6), 362–376.

Wald, L. L., & Polimeni, J. R. (2017). Impacting the effect of fMRI noise through hardware and acquisition choices – Implications for controlling false positive rates. *NeuroImage, 154*, 15–22.

Wang, X. H., Li, L., Xu, T., & Ding, Z. (2015). Investigating the Temporal Patterns within and between Intrinsic Connectivity Networks under Eyes-Open and Eyes-Closed Resting States: A Dynamical Functional Connectivity Study Based on Phase Synchronization. *PLoS One, 10*(10), e0140300-e0140300.

Washington, S. D., Gordon, E. M., Brar, J., Warburton, S., Sawyer, A. T., Wolfe, A., . . . Vanmeter, J. W. (2013). Dysmaturation of the default mode network in autism. *Hum Brain Mapp.*

Wasserstein, R. L. (2016). ASA statement on statistical significance and P-values. *The American Statistician, 70*(2), 129–133.

Watanabe, K., Kakeda, S., Igata, N., Watanabe, R., Narimatsu, H., Nozaki, A., . . . Korogi, Y. (2016). Utility of real-time prospective motion correction (PROMO) on 3D T1-weighted imaging in automated brain structure measurements. *Scientific Reports, 6*, 38366.

Waterhouse, L., London, E., & Gillberg, C. (2016). ASD Validity. *Review Journal of Autism and Developmental Disorders*, 1–28.

Waterhouse, L., London, E., & Gillberg, C. (2017). The ASD diagnosis has blocked the discovery of valid biological variation in neurodevelopmental social impairment. *Autism Res, 10*(7), 1182.

Weinstein, M., Ben-Sira, L., Levy, Y., Zachor, D. A., Ben Itzhak, E., Artzi, M., . . . Ben Bashat, D. (2011). Abnormal white matter integrity in young children with autism. *Hum Brain Mapp, 32*(4), 534–543.

Weng, S. J., Wiggins, J. L., Peltier, S. J., Carrasco, M., Risi, S., Lord, C., & Monk, C. S. (2010). Alterations of resting state functional connectivity in the default network in adolescents with autism spectrum disorders. *Brain Res, 1313*, 202–214.

White, B. B., & White, M. S. (1987). Autism from the inside. *Med Hypotheses, 24*(3), 223–229.

White, N., Roddey, C., Shankaranarayanan, A., Han, E., Rettmann, D., Santos, J., . . . Dale, A. (2010). PROMO: Real-time prospective motion correction in MRI using image-based tracking. *Magnetic Resonance in Medicine, 63*, 91–105.

Wolf, I., Dziobek, I., & Heekeren, H. R. (2010). Neural correlates of social cognition in naturalistic settings: A model-free analysis approach. *NeuroImage, 49*, 894–904.

Wolff, J. J., Gu, H., Gerig, G., Elison, J. T., Styner, M., Gouttard, S., . . . Piven, J. (2012). Differences in white matter fiber tract development present from 6 to 24 months in infants with autism. *American Journal of Psychiatry, 169*(6), 589–600.

Wong, C. W., Olafsson, V., Tal, O., & Liu, T. T. (2013a). The amplitude of the resting-state fMRI global signal is related to EEG vigilance measures. *NeuroImage, 83*, 983–990.

Wong, C. W., Olafsson, V., Tal, O., & Liu, T. T. (2013b). Caffeine-induced reductions in the resting-state fMRI global signal reflect increases in EEG vigilance measures. *ISMRM*, 33.

Yerys, B. E., Gordon, E. M., Abrams, D. N., Satterthwaite, T. D., Weinblatt, R., Jankowski, K. F., . . . Vaidya, C. J. (2015). Default mode network segregation and social deficits in autism spectrum disorder: Evidence from non-medicated children. *Neuroimage Clin, 9*, 223–232.

Zou, Q., Miao, X., Liu, D., Wang, D. J., Zhuo, Y., & Gao, J. H. (2015). Reliability comparison of spontaneous brain activities between BOLD and CBF contrasts in eyes-open and eyes-closed resting states. *Neuroimage, 121*, 91–105.

Functional Resting-State Network Disturbances in Bipolar Disorder

Gwladys Rey, Camille Piguet, and Patrik Vuilleumier

Introduction

Brain research on bipolar disorders (BD) is increasingly seeking to understand how disturbances in the dynamical organization of multiple brain areas within large-scale networks contribute to the cognitive, sensorimotor, and emotional symptoms observed in patients. Conceptual and methodological advances in resting-state functional magnetic resonance imaging (rs-fMRI) data analysis and their application to clinical populations have provided new opportunities to better characterize these disturbances, yielding a growing amount of literature in recent years, but with a considerable variety of results concerning the neural underpinnings of BD. In this chapter, we review neuroimaging findings on this topic and highlight how they may offer new insights into the bipolar brain. We will not review the equally large and important field of investigations examining brain function in active cognitive or affective tasks in BD patients.

In terms of methodological approaches, most of the studies investigating brain function at rest in BD have focused on a few selective regions of interest (ROIs) – informed by current knowledge on their putative role and previous task-related results in BD – with the aim of either defining seed-based connectivity maps for these ROIs or examining specific connections within particular networks. Other studies included a large number of ROIs (e.g., 90 or 122) covering almost the entire brain, so as to assess more global large-scale network organization. Among the resting-state investigations employing a single ROI methodology, the amygdala has been the most studied region (i.e., taken as a seed) due to its important role in

G. Rey · P. Vuilleumier (✉)
Department of Basic Neurosciences, University of Geneva, Geneva, Switzerland
e-mail: gwladys.rey@unige.ch; patrik.vuilleumier@unige.ch

C. Piguet
Faculty of Medicine, University of Geneva, Geneva, Switzerland
e-mail: camille.piguet@unige.ch

V. A. Diwadkar, S. B. Eickhoff (eds.), *Brain Network Dysfunction in Neuropsychiatric Illness*, https://doi.org/10.1007/978-3-030-59797-9_13

emotion processing (Anand et al. 2009; Anticevic et al. 2013; Brady et al. 2016, 2017a; Chai et al. 2011; Favre et al. 2014; Li et al. 2015a, b; Liu et al. 2014; Rey et al. 2016; Singh et al. 2014, 2015; Souza-Queiroz et al. 2016; Torrisi et al. 2013), followed by the pregenual anterior cingulate cortex (pgACC) or rostromedial prefrontal cortex (rMPFC) (Anticevic et al. 2015; Brady et al. 2016; Magioncalda et al. 2014; Martino et al. 2016a; Rey et al. 2016). Still another commonly used method is the independent component analysis (ICA), which aims at separating the fMRI signal into multivariate subcomponents that are statistically independent from each other. These independent components (ICs), whose number (generally chosen a priori) can vary a lot depending on the studies (e.g., n = 20 to 75), are sometimes grouped into networks based on analogies with other higher-order models of brain architecture. The ICs generally identified as non-artifactual independent resting-state networks (RSN) include the default mode network (DMN), fronto-parietal attention network (FPN), central executive network (CEN), cingulo-opercular network (CON), visuospatial network, saliency network (SN), as well as the sensorimotor network (SMN), visual, and auditory networks. Recently, new methods have also been developed to further explore different aspects of resting-state functional connectivity (FC) in BD, such as global connectivity (Stoddard et al. 2016), homotopic connectivity (Wang et al. 2015), regional homogeneity (Liang et al. 2013), graph theory metrics (Spielberg et al. 2016), or dynamic connectivity (Rashid et al. 2014). All techniques, however, are generally based on measures of the temporal covariance of neural signals within or between multiple spatially distinct brain regions or large-scale networks.

Another important methodological point in BD studies is the type of populations recruited. Most of the previous investigations enrolled not just one but two clinical populations, either (i) testing for disease-specificity or non-specificity, by including also patients with unipolar depression disorder, schizophrenia, schizoaffective disorder, or borderline personality disorder, (ii) including developmental control groups with healthy offspring of BD probands, or (iii) comparing different subgroups of BD patients based on the symptomatology, like psychotic status or mood state. Contrary to task-based literature, there is, to our knowledge, no comparison of BD type I vs type II during resting state. Moreover, in terms of study design, resting-state fMRI studies in BD adults include almost exclusively cross-sectional investigations, that is, testing each patient in each relevant group once, while follow-up fMRI assessment has been exceptionally reported (Brady et al. 2017a; Martino et al. 2016b).

In the next section, before addressing functional connectivity per se, we will briefly overview studies that aim at characterizing the neural signal at rest in individual areas, independently of the temporal synchronization of these areas. We then review resting-state FC studies in BD, grouped around changes in the default-mode network, amygdala, salience and meso/paralimbic networks, cognitive control, as well as sensorimotor networks.

Brain Activity at Rest in Bipolar Disorder Patients

In the late 1990s/early 2000, studies using position emission tomography (PET), single-photon emission computed tomography (SPECT), and arterial spin-labeling fMRI (ASL) provided the first measurements of perfusion, regional cerebral blood flow (CBF), and metabolism (glucose intake) related to regional neural activity at rest in BD patients. Unlike the standard BOLD contrast in fMRI which is a relative measure, these methods provide absolute values without contrasting different conditions, and thus are more applicable during rest. Most often enrolling modest-sized groups of BD patients ($n < 20$), these early investigations observed quite consistent changes in CBF and metabolism in limbic, paralimbic, and striatal regions in euthymic patients, depressed patients, and also but less clearly in manic/hypomanic patients. Specifically, increased activity was reported in the amygdala, accumbens nucleus, caudate nucleus, putamen, parahippocampal cortex, and subgenual ACC, while decreased activity was reported in dorsolateral prefrontal cortex (DLPFC) (Brooks et al. 2009a, b, c, 2010; Ketter et al. 2001; Krüger et al. 2003; Li et al. 2015a; Mah et al. 2007).

More recently developed, other methods have allowed exploring brain activity level based on the BOLD signal derived from MRI measurements. For instance, the amplitude of low-frequency fluctuations (ALFF) measures the power of the BOLD signal in a given frequency range, typically between 0.01 Hz and 0.08 Hz. While these frequencies are filtered out in task-related fMRI analyses, at rest such oscillations are thought to relate to spontaneous neuronal activity fluctuations. Moreover, while the ALFF indexes the amplitude of oscillations, the fractional ALFF (fALFF) can be calculated as the ALFF value normalized by the power in the entire frequency range to assess the contribution of the low-frequency oscillations to the whole spectrum (Zuo et al. 2010). A few studies measured ALFF from the BOLD signal in BD patients, respectively, enrolling 26 depressed BD patients (Liu et al. 2012a), 44 (Jiang et al. 2017) and 29 (Xu et al. 2014) patients with various mood states (mixed samples), and also 57 (Lui et al. 2015) and 180 (Meda et al. 2015) psychotic patients.[1] These studies reported *increased* ALFF or fALFF in the striatum, thalamus and subgenual ACC/ventromedial prefrontal cortex (VMPFC), insula, as well as prefrontal areas in inferior frontal gyrus (IFG), dorsolateral prefrontal cortex (DPFC) and frontal eye field, plus inferior and middle temporal areas. On the other hand, *decreased* ALFF/ fALFF was reported in the pregenual ACC, dorsomedial prefrontal cortex/anterior middle cingulate cortex (aMCC), posterior cingulate cortex (PCC), precuneus, precentral gyrus (precG), postcentral gyrus (postcG), inferior and superior parietal cortex, and lingual gyrus. While increases in ALFF were replicated for insula and striatum (in depressed, psychotic, and mixed samples), and for the subgenual ACC (in psychotic and mixed samples), different changes were observed in the hippocampus/parahippocampal cortex, with an

[1] By psychotic patients, we refer to BD patients with a history of psychotic symptoms. Strictly, current psychotic symptoms can be observed only during depression or mania.

increase in psychotic patients and mixed mood states (Meda et al. 2015; Jiang et al. 2017) but a decrease in depressed patients (Liu et al. 2012b).

Another way of studying activity oscillations at rest is to measure the variability (SD) of the BOLD signal or the normalized, fractional SD (fSD). Using this measure, Martino et al. (2016b) investigated the balance between the DMN and three other large-scale networks in 20 euthymic, 20 depressed, and 20 manic BD patients at rest. After excluding global changes in variability, they analyzed fSD in 4 networks: DMN, CEN, SN, and SMN by averaging 5 to 9 nodes for each of the networks. They computed different ratio in two frequency ranges, 0.01–0.027 Hz (called Slow5) and 0.027–0.073 Hz. The Slow5 range showed an increased fSD DMN/SMN ratio in depressed patients compared with manic and euthymic patients (and a trend decrease for depressed patients compared with controls) while manic patients showed decreased fSD as compared to controls. Importantly, the same patterns of results were replicated in an independent group of 23 patients (9 depressed, 8 euthymic, 6 manic) and 18 HC. The authors interpreted these findings as being consistent with symptoms of excessive self-focus and psychomotor inhibition during depression, and, on the contrary, enhanced externally focused attention, perceptual distraction, and motor excitation during mania.

Thus, ALFF and fSD investigations support limbic/paralimbic and subcortical hyperactivity at rest in BD patients, in line with previous investigations of CBF and metabolism. However, the effects of mood state and psychotic symptoms on neural oscillations across the brain need to be further investigated. The appealing hypothesis of an imbalance between DMN and SMN favoring the DMN during depression and favoring the SMN during mania also warrants further research.

Default Mode Network

Many findings derived from different methodologies, that is, ICA, seed-based connectivity, and graph theory-based metrics, argue for functional connectivity alterations within the DMN and its interactions with other networks.

First, changes within the DMN were repeatedly observed with ICA in BD. Decreased coherence was reported within anterior and posterior components of the DMN in BD patients with psychotic history (Khadka et al. 2013), in currently psychotic patients (n = 300; Meda et al. 2014), and during acute manic episodes (Ongür et al. 2010). In the study by Meda et al., which showed decreased connectivity in both schizophrenia and bipolar disorder patients, these changes were reported to be associated with poorer social functioning and greater negative and positive symptoms, but not with mania and depression scores (Meda et al. 2014). Together, these data point to a relationship between decreased DMN coherence and current psychotic symptoms, which however needs more direct confirmation. Interestingly, using an atlas-based correlation approach, Brady et al. (2017b) demonstrated hypoconnectivity between dorsal frontal nodes and the rest of the DMN in euthymic patients (n = 24), while manic patients (n = 23) were similar to controls.

However, other findings from ICA studies seem to contradict reports of decreased DMN coherence in psychotic or euthymic patients. For instance, Lois et al. (2014) found no difference in DMN coherence (either anterior or posterior) when testing euthymic patients and no difference between patients with and without a history of psychosis (Lois et al. 2014). In addition, Du et al. (2015) applied group information-guided (GIG) ICA to characterize 12 independent networks identified in psychotic BD patients, patients with schizophrenia or schizoaffective disorder, and healthy controls. Among regions classified as the most discriminative between groups (with Support Vector Machine), two (slightly ventral) MPFC clusters showed the highest degree of coherence with one of the three DMN components in the BD group. Finally, in the study by Das et al. (2014), euthymic patients exhibited increased connectivity between two subcomponents of the DMN, a precuneus network (or posterior DMN) and a more global DMN component.

Various alterations in DMN connectivity have also been reported during depression. In 26 depressed BD patients, Liu et al. (2012b) investigated local synchronization – or local FC – within the DMN by measuring regional homogeneity (ReHo) in ICA-identified networks and found greater local synchronization in dorsomedial prefrontal cortex and left inferior parietal cortex in patients relative to controls. In another group of 13 unmedicated, mostly depressed BD patients (He et al. 2016), graph theory methodology revealed that some nodes putatively belonging to the DMN (including DLPFC/ventrolateral prefrontal cortex (VLPFC), superior frontal gyrus (SFG), and ACC) exhibited stronger local functional interconnections (strength, clustering coefficient, and local coefficient) than in controls, suggesting closer interactions between nearby regions and stronger topological structures. However, the conclusion of hyperconnected DMN in depression may be too simplistic. In a recent study, Wang et al. (2017) enrolled 31 depressed, unmedicated BD patients to examine network modularity with graph theory metrics as well, and their results suggest that the DMN (and the limbic network as well) show disrupted effective segregation and integration of information processing across the network. This was supported by changes in several global parameters: decreased clustering coefficient, local efficiency and global efficiency, and increased characteristic path lengths (Wang et al. 2017).

Studies focusing more specifically on central nodes in the DMN, that is, MPFC and PCC, also indicate dysfunction in BD. For instance, the MPFC shows weaker interactions with the rest of the brain, at least in BD patients with psychosis history. In a study computing connections of the MPFC with every other voxels in a gray matter mask, Anticevic et al. (2013) demonstrated decreased connectivity in 61 remitted BD patients compared to controls. A comparison between patient subtypes clearly indicated that this effect was driven by psychosis history. Likewise, applying wavelet transformation to signal extracted from 266 ROIs, Argyelan et al. (2014) showed lower global connectivity strength of MPFC with all other regions in 19 BD patients – among which 16 had a history of psychosis. This result was confirmed by a complementary voxel-wise analysis, that is, by computing global connectivity strength for each voxel rather than for each ROI (Argyelan et al. 2014). A global connectivity study in 37 unmedicated, depressed BD patients (but psychosis status unknown) reported decreased global connectivity for some regions of the DMN, including dorsomedial PFC but also the precuneus, PCC, and middle temporal gyrus (Wang et al. 2016), whereas another study in 26 of these patients reported decreased interhemispheric connectivity in the MPFC in a slightly more ventral area (Wang et al. 2015). It should, however, be underscored that a decreased global connectivity has limited interpretability as it indicates that *in average* the MPFC (or any relevant area) has fewer interactions with the whole brain, including structures both inside and outside the DMN.

Regarding the PCC, one study reported increases in global PCC connectivity in a group of 47 BD patients including both euthymic and non-euthymic patients (Stoddard et al. 2016). In manic patients, decreased FC has been reported between PCC and pregenual ACC compared with both depressed patients and controls, whose magnitude correlated with mania scores (Martino et al. 2016a). In this study, euthymic patients had intermediate, not significantly different FC values. However, another study reported a significant decrease in FC between the PCC and subgenual ACC in euthymic patients compared to controls (Rey et al. 2016).

Some studies suggest that, in BD, DMN or its main nodes may differentially interact with other specific brain areas outside the DMN or with other networks. For instance, in euthymic patients compared to controls, MPFC is negatively coupled to DLPFC (Favre et al. 2014), but less connected with the left temporo-parietal junction (Martino et al. 2016a). Also in euthymic patients, the MPFC has been suggested to mediate an increased connectivity between amygdala and VLPFC (Torrisi et al. 2013). In addition, increased (positive instead of negative) MPFC-left insula and decreased MPFC-right insula were reported in manic patients (Chai et al. 2011; Martino et al. 2016a).

Likewise, ICA-based studies probing temporal correlations between network time-series found increased connection of DMN components with the cognitive control network in depressed patients (He et al. 2016), and increased DMN connectivity with the visual network in a mixed sample (Meda et al. 2012). In addition, a recent graph theory study in 78 remitted patients showed that two core regions of the DMN, ventromedial prefrontal cortex (VMPFC) and hippocampus, had lower coefficients of between-networks integration (Doucet et al. 2017).

In sum, the findings reported here tend to support decreased connectivity within the DMN plus decreased MPFC global connectivity in BD patients, generally reflecting the presence of psychosis. During depression, changes may involve more local increases for some prefrontal DMN nodes, and perhaps more global changes as well, whereas euthymia and mania are more frequently associated with hypoconnectivity between specific areas. Together, these changes suggest various losses in the integrative role of the DMN and its functional coupling with other networks.

Amygdala

Amygdala Interactions with MPFC and PCC

The amygdala–ACC/MPFC couple is probably the most documented connection in BD.

Studies typically define a seed in the amygdala and/or in a midline region within the prefrontal cortex (most often centered on the pregenual region of the cingulate gyrus or the medial superior frontal gyrus). For simplicity, we will refer to this later region as MPFC. Using such an approach, Favre et al. (2014) reported that FC between these regions was *increased* in a group of 20 euthymic patients and correlated with their illness duration. Anticevic et al. (2013) also reported an increased amygdala–MPFC FC in 63 BD patients but only in a subgroup who had a history of psychosis. On the contrary, *decreased* FC was reported in high-risk (healthy offspring of BD patients) compared to low-risk subjects (Singh et al. 2014), but also in manic and depressed patients (altogether) compared to control (Anand et al. 2009), and in manic compared to euthymic patients (Brady et al. 2016). The latter result was replicated in a subgroup of patients tested longitudinally (Brady et al. 2017a). However, other investigations did not report any difference in manic and euthymic patients in terms of amygdala–MPFC coupling (Chai et al. 2011; Rey et al. 2016).

Another study specifically compared connectivity from basolateral, centromedial, and superficial amygdala subdivisions in 18 BD patients. FC was decreased between each of these subregions and different parts of the ACC/MPFC along the rostral–ventral gradient (Liu et al. 2014). However, the clinical description of patients in this study did not allow drawing any conclusion with regard to specific symptoms or mood. Interestingly, Souza-Queiroz et al. (2016) reported that decreases in VMPFC–amygdala and VMPFC–hippocampus FC were associated with higher rate of childhood trauma (physical and emotional neglect) in euthymic BD patients.

Anatomical and functional interactions between MPFC and amygdala are known to play a key role in emotion regulation in healthy subjects (Delgado et al. 2008; Johansen-Berg et al. 2008; Pezawas et al. 2005). Although clarifications about the respective effects of mood and psychosis are necessary, the reported findings support an altered connectivity between these regions in BD patients, as a central component of current neurophysiological models of BD (Strakowski 2012). These findings are also consistent with alterations of the structural integrity of the pregenual ACC–amygdala white-matter connections that have been repeatedly shown in BD populations (Savitz et al. 2014).

While these results certainly indicate alterations in the functional feedback circuitry that regulates amygdala activity, they also raise questions about more general interactions between amygdala and the DMN at rest. FC changes have been reported between amygdala and DMN nodes. For instance, FC between amygdala and PCC/ventral precuneus was increased in non-euthymic patients (Rey et al. 2016; Singh et al. 2015). Interestingly, PCC-left amygdala FC was found to be increased and correlated with greater rumination tendencies in remitted depressed adolescent (Peters et al. 2016). Thus, amygdala coupling with PCC may be preferentially increased during mood episode, when coupling with MPFC is generally decreased instead.

Amygdala Interactions with VLPFC and DLPFC

Alterations of amygdala FC may also affect the VLPFC, another key player in neurophysiological models of BD (Chase and Phillips 2016; Strakowski 2012). Such alterations have been observed at rest, although much less often than during tasks (e.g., Cerullo et al. 2012; Foland et al. 2008). In healthy controls, amygdala–VLPFC FC is generally positive during rest (Roy et al. 2009). FC was reported to be increased in euthymic patients compared to controls (mediated by ACC; Torrisi et al. 2013) but decreased in manic and depressed patients compared to controls (Li et al. 2015b). If confirmed, these latter results may indicate that disruption of VLPFC–amygdala connectivity is involved in acute mood dysregulation. Possible compensatory (or over-compensatory) FC increases in VLPFC–amygdala may help restore emotional homeostasis during euthymia, in line with current models of emotion regulation (e.g., Ochsner et al. 2012). Besides, heightened coupling of subgenual ACC–VLPFC circuits in euthymic patients and its link with decreased rumination tendency (Rey et al. 2016) could be another facet of compensatory connectivity between prefrontal and over-active or hyper-responsive limbic areas.

Lastly, amygdala–DLPFC connectivity is negative at rest in controls (Roy et al. 2009), but less so in different groups of euthymic patients compared with controls (Li et al. 2015a; Torrisi et al. 2013). This is paralleled by increased amygdala metabolism and decreased DLPFC metabolism as measured by PET (Li et al. 2015a). Moreover, amygdala–DLPFC connectivity is more negative in psychotic than both non-psychotic BD patients and controls, while these two latter do not differ (Anticevic et al. 2013). If confirmed, a loss of negative connectivity between DLPFC and amygdala in euthymic patients could be interpreted as a residual lack of top-down prefrontal control on amygdala activity, which may contribute to the risk of relapse.

Graph Theory Highlights a Central Role of the Amygdala

Skåtun et al. (2016) derived global functional connectivity using voxel-wise eigenvector centrality mapping in 43 BD and 71 schizophrenia patients plus 196 controls. In this approach, a high value is attributed to nodes (here, voxels) that are strongly correlated to many other nodes that are themselves central (i.e., highly connected) within the network. In BD patients, the authors found decreased centrality in four (high-centrality) subcortical clusters, including putamen and hippocampus/

amygdala bilaterally. On the contrary, they reported increased centrality in (already high-centrality) right DLPFC and (low-centrality) left inferior parietal lobe (IPL). In another recent graph theory study, Spielberg et al. (2016) explored 181 ROIs (whole-brain except cerebellum) in 30 moderately depressed and 30 hypomanic (all unmedicated) BD patients, plus 30 controls, and found lower global efficiency (i.e., less efficient network communication or integration) in the patients. However, connectivity was increased in a network in which the right amygdala had the largest number of connections (42%), essentially toward sensory areas, together with a higher clustering coefficient for the amygdala.[2] Interestingly, after isolating variance associated with each set of symptoms, the authors found that mania, not depression, was correlated with hyperconnectivity in a network involving the amygdala, posterior SFG, and midbrain, including the ventral tegmental area (VTA), substantia nigra, and red nucleus. As the network around the amygdala involved many sensory areas, the authors postulated "a form of hypersalience, wherein affective significance is inappropriately overlaid on perception." Moreover, examining relationships between significant networks and each of three scores derived from a mania scale (based on Hanwella and de Silva 2011), they reported that the "elated mania" score (including elevated mood, language abnormalities/thought disorder, increased sexual interest and poor insight) uniquely predicted hyperconnectivity of amygdala (notably toward sensory areas) and hyperconnectivity of VTA and substantia nigra, which are well-known dopaminergic brain areas involved in reward processing. The "irritable mania" score (including irritability, increased motor activity/energy, and disruptive aggressive behavior) predicted posterior SFG hyperconnectivity (especially to association cortices) which may tentatively be related to goal-striving and frustration. On the other hand, depression scores were associated not only with hyperconnectivity in a network in which the posterior-middle orbitofrontal cortex (pmOFC) was the major hub (e.g., with central operculum, temporal regions, and sensory processing areas), but also weaker global network resilience and greater clustering for the pmOFC. These findings suggest that, during bipolar depression, the OFC may exert a greater influence on a widespread network within which information processing is generally more vulnerable to disruption. The authors speculated this may reflect effortful value-related processing to compensate a difficulty in discriminating rewards.

Thus, in keeping with the early neuroimaging studies and models of mood disorders (Kennedy et al. 2007; Mayberg 2009; Phillips et al. 2008; Price and Drevets 2010), the amygdala is increasingly ascribed a pivotal role in the pathophysiology of BD, with some modulations depending on mood state. Speculatively, changes in amygdala connectivity with sensory processing areas may be more specifically

---

[2] As the clustering coefficient of a node is the fraction of its neighbors that are also connected to each other, higher clustering coefficient for the amygdala in BD actually does not contradict previous reports of decreased (high-) centrality for this structure. Hub nodes may even be expected to have low clustering coefficient (see Chap. 5. Centrality and hubs; Fornito et al. 2016). Here, these combined results may indicate that the amygdala loses his function of "hub" and connects with regions that are connected together in a larger extent than it does in controls.

related to mania, while changes in connections involving the OFC are related to depression. However, depressed patients may also show decreased global connectivity of the amygdala (n = 37; Wang et al. 2016).

Possible Laterality Effect in Amygdala Connectivity Changes

Regarding laterality, there are some suggestions that right and left amygdala may be differentially affected in BD at rest and that right amygdala connectivity may be more sensitive to mood states. For instance, right amygdala connectivity with ACC was decreased in non-euthymic patients, not in euthymic, compared to controls (subgenual part; Rey et al. 2016) and in manic compared to euthymic patients (pregenual part; Brady et al. 2016). Additionally, connectivity between the right amygdala and right hippocampus was increased in manic patients but decreased in depressed patients (Li et al. 2015b). Other studies reported significant FC changes involving the right, not left amygdala (Favre et al. 2014; Spielberg et al. 2016; Torrisi et al. 2013; however, see Anand et al. 2009; Chepenik et al. 2010; Li et al. 2015a). Because many experimental factors may influence fMRI signal asymmetry in the amygdala (Sergerie et al. 2008), further research is needed to confirm and explain laterality effects.

Subcortical Circuits: Striatum, Thalamus, and Midbrain

Interconnecting the MPFC, striatum, thalamus, and also the amygdala, the cortico-striato-thalamic loops have been proposed to "form the core of the neural system of mood disorders" (Price and Drevets 2012) and are frequently invoked in models specifically addressing BD (Strakowski 2012). Several findings from rs-fMRI support alterations in these loops.

Enrolling 15 BD patients and 16 controls, Teng et al. (2013) parceled their brains into 90 ROIs, and then defined the most discriminative connections between groups with clustering analysis. These connections all concerned the basal ganglia. In a subsequent reanalysis, Teng et al. (2014) selected two "modules" or clusters of connections significant in each group, one involving 11 striatal and thalamic regions, the other involving 19 regions labeled as "DMN." They confirmed significant decreases in striatum/thalamus, striatum/MCC-PCC, and thalamus/MCC functional coupling in the patients compared to controls. These findings are consistent with previous reports of decreased thalamus connectivity with pregenual ACC in a group of manic and depressed BD patients (Anand et al. 2009) and decreased global connectivity for the thalamus in patients with psychosis history (Argyelan et al. 2014).

In more recent work on reward processing, 23 bipolar and 22 unipolar depressed patients underwent a monetary reward task followed by a resting-state scanning session (Satterthwaite et al. 2015). During the task, depression severity (independently of diagnosis) was associated with a blunted response of the reward system. At rest, depression severity was associated with robust and selective decrease in connectivity strength of the ventral striatum with thalamus, and VTA with VMPFC, among several ROIs selected a priori. This effect was similar across unipolar and bipolar groups, suggesting a direct negative relationship between depressed mood and low connectivity within the reward network. Conversely, another study in 30 manic and 30 depressed unmedicated patients also tested FC from several seeds in basal ganglia, and found increased connectivity between ventral striatum and bilateral thalamus during mania, consistent with higher reward-seeking behavior in this condition (Altinay et al. 2016). Manic patients also showed increased FC between dorsal caudate and midbrain/substantia nigra, while depressed patients showed increased connectivity between putamen (somatosensory and associative subdivisions) and temporal cortex/posterior insula (Altinay et al. 2016). Moreover, the VTA–substantia nigra connectivity in BD patients as a whole (depressed and manic) was found to be positively correlated with the "elation" factor derived from mania scores (Spielberg et al. 2016). Taken together, these results support the idea that the reward-seeking behavior in mania may reflect hyperactivity or hyper-responsiveness of the mesocortical/mesolimbic dopaminergic pathways (see also Ashok et al. 2017). Note that the amygdala also interacts with the striato-thalamic loops (Price and Drevets 2012), and was found to exhibit altered FC with reward areas in putamen and orbital IFG in BD (18 manic and 12 depressed patients; see Li et al. (2015b)).

To conclude, there is growing evidence for functional changes of cortico-striato-thalamic circuits in BD patients at rest, in agreement with anatomical alterations observed in WM tracts between striatum and thalamus (Lu et al. 2012; Sussmann et al. 2009). Such changes are likely to be related to emotional and cognitive regulation difficulties, in particular in reward-related processing – for instance, between ventral striatum and thalamus whose connectivity is increased during mania but decreased during depression, concomitant to changes in motivation and reward mechanisms.

Salience and Meso/Paralimbic Networks

Aberrant engagement of the salience network has been observed in several neuropsychiatric disorders (Menon 2011). The SN is formed by the dorsal ACC, together with anterior insula and adjacent inferior frontal gyrus, and plays a pivotal role in detection and attention to salient events and the coordination of neural resources mediated by other large-scale networks dynamics like the DMN and CEN (Menon 2015; Sridharan et al. 2008; Uddin 2015). The SN is also thought to include subcortical structures such as the amygdala, the ventral striatum, thalamus and hypothalamus, substantia nigra, and VTA (Seeley et al. 2007), providing access to the emotional and reward value of any salient stimulus. Therefore, functional connectivity changes in amygdala and striato-thalamic loops as described in the previous section may also partly reflect abnormal SN function in BD. However, other more specific changes in the salience and meso/paralimbic networks have also been identified.

For instance, Yip et al. (2014) investigated rs-fMRI in 15 euthymic, unmedicated, and young BD patients using ICA and reported increased coherence in a temporo-insular network that could be an aggregate of (or overlapping with) auditory/language/salience networks. This effect concerned bilateral insula but also IFG, supplementary motor area, superior temporal gyrus, and subcortical areas (putamen, caudate). Moreover, one of the most discriminating regions between BD, other psychotic disorders, and controls, was found in the insula (Du et al. 2015). This region showed higher degree of coherence with the auditory network (not the SN) in 20 BD patients. These findings converge with the report of increased FC between the ventral anterior insula and perigenual ACC in a recent investigation specifically assessing insula FC in 46 remitted BD patients (Li et al. 2017).

Other findings argue for altered interactions of the SN with other networks. Sixty-four patients with psychotic symptoms (current for half of them) showed increased FC between the fronto-temporal/paralimbic network – that is, the salience network similar to Yip et al. (2014) – and the meso-paralimbic network (MPN), consisting primarily of hippocampal complex, amygdala, and mesial temporal cortex (Meda et al. 2012). Moreover, this increase in BD patients was associated with greater negative symptoms on the PANSS (positive and negative syndrome scale). Within-network FC explorations on the same data additionally showed group differences in the MPN, not the SN, with patients showing increased FC in the superior temporal gyrus compared to controls (Khadka et al. 2013).

In another study enrolling 16 euthymic patients, ICA generated a "social salience" network including the dACC and orbital fronto-insular cortices, plus the inferior frontal gyrus, posterior superior temporal sulcus, IPL, and temporo-parietal junction. Coupling between this "social salience" network and the ventromedial network was increased in patients (although the effect did not survive correction for multiple comparisons) and was associated with lack of emotional clarity in psychometric tests (Das et al. 2014).

In sum, according to these few studies, the SN may show higher within-network connectivity during euthymia, but greater coupling with other limbic and paralimbic structures may in turn account for affective biases in detection and processing of information. Again, the respective role of mood state and psychosis need to be investigated.

Cognitive Control/Attention Networks

Networks implicated in cognitive control and attention such as the FPN and dorsal attention network (DAN) have been found to be modified in several studies. Baker et al. (2014) recruited 100 psychotic patients with schizophrenia, schizoaffective disorders, or bipolar disorder, most of whom had current psychotic symptoms (77% of the 40 BD patients). The authors investigated FC across 122 regions clustered into either 7 or 17 networks. In both cases, they identified reduced connectivity patterns in the FPN that concerned the DLPFC, posterior MPFC, lateral parietal cortex, and posterior temporal cortex, among both BD and schizophrenia groups. Decreased coherence in the FPN, centered on the left VLPFC, was also reported in high-risk subjects, that is, healthy offspring of BD patients, compared to low-risk subjects (Singh et al. 2014). In another study enrolling mainly depressed BD patients, Goya-Maldonado et al. (2016) compared FC within 3 networks, that is, DMN, FPN, and CON. Unlike in the above-mentioned studies, depressed patients showed increased coherence of the left DLPFC/VLPFC and IPL with the FPN network.

Abnormal interactions at rest between the FPN and other networks have been also reported. In a study of 30 euthymic patients, Lois et al. (2014) did not find any changes for within-network FC but instead found a positive correlation between MPN and right FPN in the patients (unlike controls). In depressed patients, on the other side, nodal metrics applied to 48 independent components arranged in 7 networks (visual, auditory, cognitive control, DMN, SMN, cerebellar and subcortical networks) suggested that the cognitive control network – roughly encompassing the FPN and CON – had higher FC with both DMN and SMN (He et al. 2016).

Finally, regarding the DAN, Du et al. (2015) reported that the IPL had the most discriminative connectivity change in 20 BD patients compared to other groups with psychosis history. Another recent investigation reported that posterior areas of the DAN show hypoconnectivity during euthymia but hyperconnectivity during mania (Brady et al. 2017b). However, changes were also observed for interhemispheric connectivity in anterior sectors of the DAN, such as the middle frontal gyrus (MFG) and precG, in remitted BD patients (Zhao et al. 2017).

In sum, cognitive control networks may show within-network connectivity changes depending on mood state (e.g., FPN increase during depression) and psychosis status (e.g., FPN decrease with current psychosis). Such alterations are generally compatible with task-based studies in BD patients, showing deficits in executive functions and abnormal brain activation patterns during the performance of attentional tasks (see following chapter). In addition, a stronger coupling of control networks with DMN and SMN during depression, but with MPN during euthymia, might reflect changes in attentional focus and cognitive biases directed to internal vs external representations.

Sensorimotor Network

Besides expected changes in emotion and cognition-related systems of the brain, there is growing evidence that the sensorimotor network is also affected in BD. For instance, abnormal connectivity has repeatedly been reported for precG and postcG, that is, primary motor and somatosensory areas forming major nodes of the SMN. In patients with a history of psychotic symptoms, Lui et al. (2015) showed decreased FC of these regions with the parahippocampal cortex, and increased FC with the thalamus, insula, and SFG, while increases with caudate were found in their first-degree relatives. ICA analyses also showed increased coherence of the SFG/MFG with the SMN in psychotic patients (50% with current symptoms) (Khadka et al. 2013).

In a study by Stoddard et al. (2016), conducted in 37 BD patients (27 euthymic, 10 non-euthymic patients, 2 unknown) and 196 controls with a voxel-wise global connectivity approach, 4 clusters with heightened FC were identified in patients compared to controls: right IPL, right PCC, left superior temporal gyrus, and left precG. By then using these regions as seeds and now applying k-means clustering and multidimensional scaling to all detected regions, the authors could demonstrate hyperconnectivity within and between two specific networks: one involving cortico-striatal areas associated with fronto-parietal, dorsal/ventral attention and SMN (putamen, inferior parietal lobe, precG, IFG/insula, DLPFC), and the other involving temporo-parietal areas associated with visual networks and SMN. Importantly, this pattern remained unchanged after excluding non-euthymic patients. On the contrary, in fully remitted patients (n = 71) (and unaffected siblings as well), a study using graph theory reported a loss of connectivity inside the SMN combined with increased interactions with regions from other networks (Doucet et al. 2017).

There is also evidence that the SMN may differentially interact with other networks during mood episodes. As previously mentioned, Spielberg et al. (2016) demonstrated that a large proportion of FC differences observed in BD compared to controls concerned hyperconnectivity between the amygdala and sensory processing areas, in association with mania scores. Consistent with this, manic BD patients show increased FC of the precG with both hippocampus and amygdala (Singh et al. 2015), and of the supplementary motor area with amygdala (Brady et al. 2017a). Conversely, in depressed patients, components of the SMN show higher connectivity with components of the cognitive control network (ACC-IFG) and visual network (postcG-cuneus) (He et al. 2016), while OFC-sensory areas connectivity increases were found to correlate with depression severity (Spielberg et al. 2016).

Lastly, innovative approaches in fMRI analysis have begun to explore moment-to-moment fluctuations of functional connectivity at rest in BD patients. Calhoun's team applied static and dynamic FC analysis methods in one study enrolling 38 BD patients, 60 schizophrenia or schizo-affective disorders patients, and 70 healthy controls (Rashid et al. 2014). Using group ICA, they first identified 49 ICs grouped into 6 large-scale networks (cognitive control, default mode, sensorimotor, visual, auditory, and subcortical). No difference in static FC was found between the groups. However, when then analyzing dynamic FC using sliding windows and k-means clustering, they identified 5 re-occurring patterns of FC states. In one particular state, BD patients had lower FC between paracentral gyrus (putatively from the SMN) and superior parietal lobe (cognitive control) than healthy participants, and greater FC between bilateral fusiform (visual) and left supramarginal cortex (SMN). In a follow-up work (Rashid et al. 2016), the authors extracted most discriminative features from both static and dynamic FC analyses (notably SMN components) and used them to classify data from each group (based on support vector machine). Results clearly showed that using static plus dynamic features or dynamic features alone led to better classification results than using static features only.

Although no definite conclusion can be drawn at this point, recent findings suggest that the SMN show decreased within-network connectivity during euthymia, and may also display differential interactions with other networks depending on current mood state in BD patients. Higher coupling with cognitive control and visual networks during euthymia may, for instance, result from compensatory mechanisms to maintain emotional homeostasis. On the other side, as suggested by Spielberg et al. (2016), sensorimotor coupling with limbic networks may account for hypersalience and hypersensitivity to sensory stimulation. More studies are needed to compare mood episodes, and clarify such changes of the SMN during depression versus mania.

Conclusion

The variety of approaches in the study of bipolar disorders has brought about a rich and interesting literature on functional connectivity in BD, but also highlights a large diversity of findings and frequent lack of reproducibility. By systematically discussing and organizing the existing studies in terms of specific networks, our review attempts to underscore the major functional patterns emerging from this work (Fig. 1) and tentatively draws some preliminary conclusions or hypothesis concerning cognitive-affective mechanisms at play in BD. These data may help guiding future research on this topic and motivate further studies addressing more specific hypothesis concerning particular clinical dimensions and particular functional network properties.

Overall, findings regarding the DMN suggest decreased within-network connectivity in BD, which could be linked to the psychotic dimension more than to the affective dimension. In parallel, both current psychosis and depression symptoms may impact on control functions of the FPN including not only within-network connectivity changes (decreased or increased, respectively), but also differential interactions with the DMN, SMN, and mesoparalimbic networks depending on mood state.

In general, cortico-striato-thalamic loop and reward-related circuits associated with affect and motivation demonstrate increased functional connectivity in manic patients and decreased connectivity in depressed patients, which can meaningfully be related to clinical symptoms. The amygdala constitutes a major hub for alterations, showing strengthened connections with MPFC and VLPFC during euthymia but possible decoupling during mood episodes. Its connectivity with sensory processing areas appears stronger during mania, whereas depression may be associated with hyperconnectivity of OFC with sensory areas and hypoconnectivity among several reward-related structures. In the same vein, the salience network is likely to be altered as well, in terms of both abnormal connectivity with amygdala and striatum as individual seeds and more global functional coupling of SN–limbic/paralimbic structures.

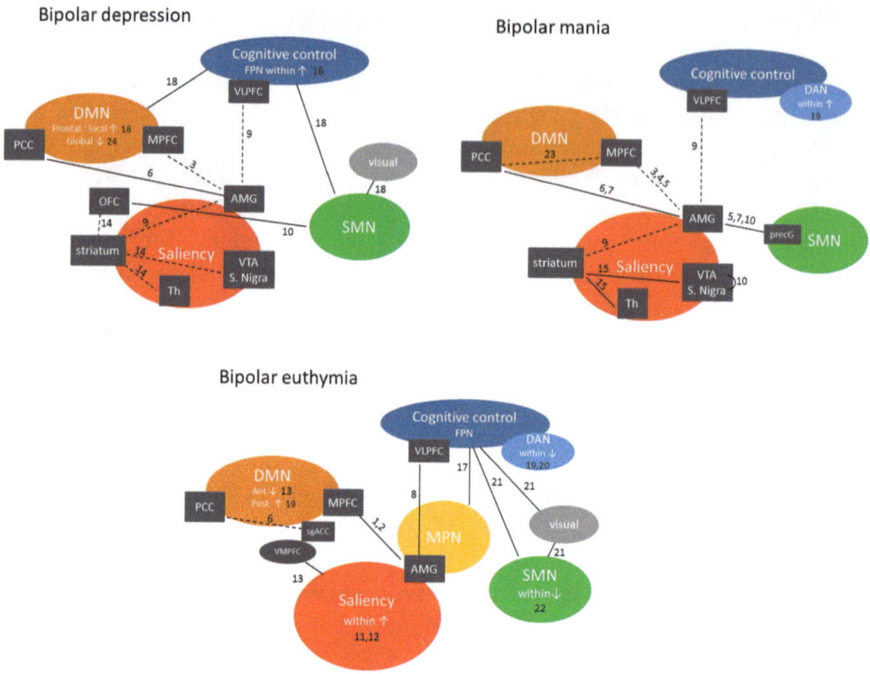

Fig. 1 Schematics representation of functional connectivity changes at rest in bipolar disorder patients during euthymia, depression, and mania, compared to healthy controls. Plain lines represent increased connectivity while dashed lines represent decreased connectivity between networks (colored bubbles) or brain areas (gray squares). AMG: amygdala; DAN: dorsal attention network; DMN: default mode network; MPFC: medial prefrontal cortex; MPN: mesoparalimbic network; VLPFC: ventrolateral prefrontal cortex; OFC: orbitofrontal cortex; PCC: posterior cingulate cortex; SMN: sensorimotor; Th: thalamus; VTA: ventral tegmental area
References are indicated by numbers: 1.(Favre et al. 2014); 2.(Anticevic et al. 2013); 3.(Anand et al. 2009); 4. (Brady et al. 2016); 5.(Brady et al. 2017a); 6.(Rey et al. 2016); 7.(Singh et al. 2015); 8.(Torrisi et al. 2013); 9.(Li et al. 2015b); 10.(Spielberg et al. 2016); 11.(Yip et al. 2014); 12.(Li et al. 2017); 13.(Das et al. 2014); 14.(Satterthwaite et al. 2015); 15.(Altinay et al. 2016); 16.(Goya-Maldonado et al. 2016); 17.(Lois et al. 2014); 18.(He et al. 2016); 19.(Brady et al. 2017b); 20.(Zhao et al. 2017); 21.(Stoddard et al. 2016); 22.(Doucet et al. 2017); 23.(Martino et al. 2016a); 24.(Wang et al. 2017)

Finally, recent findings highlighted FC changes in the sensorimotor network, both within the network and in its interactions with visual and cognitive control networks during euthymia, and with limbic networks during mania as previously stated.

Taken together, functional connectivity measures offered by rs-fMRI accord with earlier reports of changes in blood flow and glucose metabolism in several subcortical and limbic areas in BD and mood disorders in general. They may also open new perspectives in terms of better defining risk factors prognosis, or treatment efficacy. However, as underlined in our review, research in this field would clearly benefit from a better clinical description of the patients, primarily in terms of mood and psychosis, but also concerning other symptom dimensions and illness history (i.e., recent episodes, etc.). The proportion of euthymic, depressed, and manic or hypomanic patients in studies including mixed mood samples, as well as correlational analysis of mood effects, should help interpret results and clarify discrepancies. An optimal mood-comparison design would be made possible by following-up the same patients over time (e.g., Rey et al. 2014). Since psychosis has been shown to strongly impact resting-state connectivity, precise information should be systematically provided concerning both past history and current psychosis symptomatology. Better description of positive and negative connectivity patterns for the relevant networks should be reported, not only group comparisons, as this would help interpret functional significance of the connections and group differences. Moreover, larger samples of participants should be recruited to reach stronger statistical confidence, as well as allow better characterization of subtypes. Finally, some findings suggest that resting-state connectivity patterns might also be usefully considered in relation to task-related activation and behavioral performance in cognitive tests, in order to yield a better understanding of spontaneous/baseline alterations observed at rest. Variability in resting-state conditions might also be valuably constrained by measuring rs-fMRI connectivity subsequent to standardized tasks (Eryilmaz et al. 2011, 2014), for instance, to unveil dynamic changes in emotion regulation and homeostatic processes that may vary as a function of individual traits. Through the convergence of these approaches, our measures of brain activity changes at rest in patients suffering from bipolar disorder should continue to illuminate the pathophysiology of this disease and ultimately improve our diagnosis and care of individual patients.

References

Altinay, M.I., Hulvershorn, L.A., Karne, H., Beall, E.B., and Anand, A. (2016). Differential Resting-State Functional Connectivity of Striatal Subregions in Bipolar Depression and Hypomania. Brain Connect. *6,* 255–265.

Anand, A., Li, Y., Wang, Y., Lowe, M.J., and Dzemidzic, M. (2009). Resting state corticolimbic connectivity abnormalities in unmedicated bipolar disorder and unipolar depression. Psychiatry Res. *171*, 189–198.

Anticevic, A., Brumbaugh, M.S., Winkler, A.M., Lombardo, L.E., Barrett, J., Corlett, P.R., Kober, H., Gruber, J., Repovs, G., Cole, M.W., et al. (2013). Global prefrontal and fronto-amygdala dysconnectivity in bipolar I disorder with psychosis history. Biol. Psychiatry *73*, 565–573.

Anticevic, A., Savic, A., Repovs, G., Yang, G., McKay, D.R., Sprooten, E., Knowles, E.E., Krystal, J.H., Pearlson, G.D., and Glahn, D.C. (2015). Ventral anterior cingulate connectivity distinguished nonpsychotic bipolar illness from psychotic bipolar disorder and schizophrenia. Schizophr. Bull. *41*, 133–143.

Argyelan, M., Ikuta, T., DeRosse, P., Braga, R.J., Burdick, K.E., John, M., Kingsley, P.B., Malhotra, A.K., and Szeszko, P.R. (2014). Resting-state fMRI connectivity impairment in schizophrenia and bipolar disorder. Schizophr. Bull. *40*, 100–110.

Ashok, A.H., Marques, T.R., Jauhar, S., Nour, M.M., Goodwin, G.M., Young, A.H., and Howes, O.D. (2017). The dopamine hypothesis of bipolar affective disorder: the state of the art and implications for treatment. Mol. Psychiatry. *22*, 666–679.

Baker, J.T., Holmes, A.J., Masters, G.A., Yeo, B.T.T., Krienen, F., Buckner, R.L., and Öngür, D. (2014). Disruption of cortical association networks in schizophrenia and psychotic bipolar disorder. JAMA Psychiatry *71*, 109–118.

Brady, R.O., Masters, G.A., Mathew, I.T., Margolis, A., Cohen, B.M., Öngür, D., and Keshavan, M. (2016). State dependent cortico-amygdala circuit dysfunction in bipolar disorder. J. Affect. Disord. *201*, 79–87.

Brady, R.O., Margolis, A., Masters, G.A., Keshavan, M., and Öngür, D. (2017a). Bipolar mood state reflected in cortico-amygdala resting state connectivity: A cohort and longitudinal study. J. Affect. Disord. *217*, 205–209.

Brady, R.O., Tandon, N., Masters, G.A., Margolis, A., Cohen, B.M., Keshavan, M., and Öngür, D. (2017b). Differential brain network activity across mood states in bipolar disorder. J. Affect. Disord. *207*, 367–376.

Brooks, J.O., Hoblyn, J.C., Woodard, S.A., Rosen, A.C., and Ketter, T.A. (2009a). Corticolimbic metabolic dysregulation in euthymic older adults with bipolar disorder. J. Psychiatr. Res. *43*, 497–502.

Brooks, J.O., Wang, P.W., Bonner, J.C., Rosen, A.C., Hoblyn, J.C., Hill, S.J., and Ketter, T.A. (2009c). Decreased prefrontal, anterior cingulate, insula, and ventral striatal metabolism in medication-free depressed outpatients with bipolar disorder. J. Psychiatr. Res. *43*, 181–188.

Brooks, J.O., Rosen, A.C., Hoblyn, J.C., Woodard, S.A., Krasnykh, O., and Ketter, T.A. (2009b). Resting prefrontal hypometabolism and paralimbic hypermetabolism related to verbal recall deficits in euthymic older adults with bipolar disorder. Am. J. Geriatr. Psychiatry Off. J. Am. Assoc. Geriatr. Psychiatry *17*, 1022–1029.

Brooks, J.O., Hoblyn, J.C., and Ketter, T.A. (2010). Metabolic evidence of corticolimbic dysregulation in bipolar mania. Psychiatry Res. *181*, 136–140.

Cerullo, M.A., Fleck, D.E., Eliassen, J.C., Smith, M.S., DelBello, M.P., Adler, C.M., and Strakowski, S.M. (2012). A longitudinal functional connectivity analysis of the amygdala in bipolar I disorder across mood states. Bipolar Disord. *14*, 175–184.

Chai, X.J., Whitfield-Gabrieli, S., Shinn, A.K., Gabrieli, J.D.E., Nieto Castañón, A., McCarthy, J.M., Cohen, B.M., and Ongür, D. (2011). Abnormal medial prefrontal cortex resting-state connectivity in bipolar disorder and schizophrenia. Neuropsychopharmacol. Off. Publ. Am. Coll. Neuropsychopharmacol. *36*, 2009–2017.

Chase, H.W., and Phillips, M.L. (2016). Elucidating neural network functional connectivity abnormalities in bipolar disorder: toward a harmonized methodological approach. Biol. Psychiatry Cogn. Neurosci. Neuroimaging *1*, 288–298.

Chepenik, L.G., Raffo, M., Hampson, M., Lacadie, C., Wang, F., Jones, M.M., Pittman, B., Skudlarski, P., and Blumberg, H.P. (2010). Functional connectivity between ventral prefrontal cortex and amygdala at low frequency in the resting state in bipolar disorder. Psychiatry Res. *182*, 207–210.

Das, P., Calhoun, V., and Malhi, G.S. (2014). Bipolar and borderline patients display differential patterns of functional connectivity among resting state networks. NeuroImage *98*, 73–81.

Delgado, M.R., Nearing, K.I., Ledoux, J.E., and Phelps, E.A. (2008). Neural circuitry underlying the regulation of conditioned fear and its relation to extinction. Neuron *59*, 829–838.

Doucet, G.E., Bassett, D.S., Yao, N., Glahn, D.C., and Frangou, S. (2017). The Role of Intrinsic Brain Functional Connectivity in Vulnerability and Resilience to Bipolar Disorder. Am. J. Psychiatry *174*, 1214–1222.

Du, Y., Pearlson, G.D., Liu, J., Sui, J., Yu, Q., He, H., Castro, E., and Calhoun, V.D. (2015). A group ICA based framework for evaluating resting fMRI markers when disease categories are unclear: application to schizophrenia, bipolar, and schizoaffective disorders. NeuroImage *122*, 272–280.

Eryilmaz, H., Van De Ville, D., Schwartz, S., and Vuilleumier, P. (2011). Impact of transient emotions on functional connectivity during subsequent resting state: a wavelet correlation approach. NeuroImage *54*, 2481–2491.

Eryilmaz, H., Van De Ville, D., Schwartz, S., and Vuilleumier, P. (2014). Lasting impact of regret and gratification on resting brain activity and its relation to depressive traits. J. Neurosci. *34*, 7825–7835.

Favre, P., Baciu, M., Pichat, C., Bougerol, T., and Polosan, M. (2014). fMRI evidence for abnormal resting-state functional connectivity in euthymic bipolar patients. J. Affect. Disord. *165*, 182–189.

Foland, L.C., Altshuler, L.L., Bookheimer, S.Y., Eisenberger, N., Townsend, J., and Thompson, P.M. (2008). Evidence for deficient modulation of amygdala response by prefrontal cortex in bipolar mania. Psychiatry Res. *162*, 27–37.

Fornito, A., Zalesky, A., and Bullmore, E. (2016). Fundamentals of Brain Network Analysis (Academic Press).

Goya-Maldonado, R., Brodmann, K., Keil, M., Trost, S., Dechent, P., and Gruber, O. (2016). Differentiating unipolar and bipolar depression by alterations in large-scale brain networks. Hum. Brain Mapp. *37*, 808–818.

Hanwella, R., and de Silva, V.A. (2011). Signs and symptoms of acute mania: a factor analysis. BMC Psychiatry *11*, 137.

He, H., Yu, Q., Du, Y., Vergara, V., Victor, T.A., Drevets, W.C., Savitz, J.B., Jiang, T., Sui, J., and Calhoun, V.D. (2016). Resting-state functional network connectivity in prefrontal regions differs between unmedicated patients with bipolar and major depressive disorders. J. Affect. Disord. *190*, 483–493.

Jiang, X., Dai, X., Kale Edmiston, E., Zhou, Q., Xu, K., Zhou, Y., Wu, F., Kong, L., Wei, S., Zhou, Y., et al. (2017). Alteration of cortico-limbic-striatal neural system in major depressive disorder and bipolar disorder. J. Affect. Disord. *221*, 297–303.

Johansen-Berg, H., Gutman, D.A., Behrens, T.E.J., Matthews, P.M., Rushworth, M.F.S., Katz, E., Lozano, A.M., and Mayberg, H.S. (2008). Anatomical connectivity of the subgenual cingulate region targeted with deep brain stimulation for treatment-resistant depression. Cereb. Cortex *18*, 1374–1383.

Kennedy, S.H., Konarski, J.Z., Segal, Z.V., Lau, M.A., Bieling, P.J., McIntyre, R.S., and Mayberg, H.S. (2007). Differences in brain glucose metabolism between responders to CBT and venlafaxine in a 16-week randomized controlled trial. Am. J. Psychiatry *164*, 778–788.

Ketter, T.A., Kimbrell, T.A., George, M.S., Dunn, R.T., Speer, A.M., Benson, B.E., Willis, M.W., Danielson, A., Frye, M.A., Herscovitch, P., et al. (2001). Effects of mood and subtype on cerebral glucose metabolism in treatment-resistant bipolar disorder. Biol. Psychiatry *49*, 97–109.

Khadka, S., Meda, S.A., Stevens, M.C., Glahn, D.C., Calhoun, V.D., Sweeney, J.A., Tamminga, C.A., Keshavan, M.S., O'Neil, K., Schretlen, D., et al. (2013). Is aberrant functional connectivity a psychosis endophenotype? A resting state functional magnetic resonance imaging study. Biol. Psychiatry *74*, 458–466.

Krüger, S., Seminowicz, D., Goldapple, K., Kennedy, S.H., and Mayberg, H.S. (2003). State and trait influences on mood regulation in bipolar disorder: blood flow differences with an acute mood challenge. Biol. Psychiatry *54*, 1274–1283.

Li, C.-T., Tu, P.-C., Hsieh, J.-C., Lee, H.-C., Bai, Y.-M., Tsai, C.-F., Wang, S.-J., Hsu, J.-W., Huang, K.-L., Hong, C.-J., et al. (2015a). Functional dysconnection in the prefrontal-amygdala circuitry in unaffected siblings of patients with bipolar I disorder. Bipolar Disord. *17*, 626–635.

Li, J., Tang, Y., Womer, F., Fan, G., Zhou, Q., Sun, W., Xu, K., and Wang, F. (2017). Two patterns of anterior insular cortex functional connectivity in bipolar disorder and schizophrenia. World J. Biol. Psychiatry Off. J. World Fed. Soc. Biol. Psychiatry *19*, S115–S123.

Li, M., Huang, C., Deng, W., Ma, X., Han, Y., Wang, Q., Li, Z., Guo, W., Li, Y., Jiang, L., et al. (2015b). Contrasting and convergent patterns of amygdala connectivity in mania and depression: a resting-state study. J. Affect. Disord. *173*, 53–58.

Liang, M.-J., Zhou, Q., Yang, K.-R., Yang, X.-L., Fang, J., Chen, W.-L., and Huang, Z. (2013). Identify changes of brain regional homogeneity in bipolar disorder and unipolar depression using resting-state FMRI. PloS One *8*, e79999.

Liu, C.-H., Li, F., Li, S.-F., Wang, Y.-J., Tie, C.-L., Wu, H.-Y., Zhou, Z., Zhang, D., Dong, J., Yang, Z., et al. (2012a). Abnormal baseline brain activity in bipolar depression: A resting state functional magnetic resonance imaging study. Psychiatry Res. Neuroimaging *203*, 175–179.

Liu, C.-H., Ma, X., Li, F., Wang, Y.-J., Tie, C.-L., Li, S.-F., Chen, T.-L., Fan, T., Zhang, Y., Dong, J., et al. (2012b). Regional Homogeneity within the Default Mode Network in Bipolar Depression: A Resting-State Functional Magnetic Resonance Imaging Study. PLOS ONE *7*, e48181.

Liu, H., Tang, Y., Womer, F., Fan, G., Lu, T., Driesen, N., Ren, L., Wang, Y., He, Y., Blumberg, H.P., et al. (2014). Differentiating patterns of amygdala-frontal functional connectivity in schizophrenia and bipolar disorder. Schizophr. Bull. *40*, 469–477.

Lois, G., Linke, J., and Wessa, M. (2014). Altered Functional Connectivity between Emotional and Cognitive Resting State Networks in Euthymic Bipolar I Disorder Patients. PloS One *9*, e107829.

Lu, L.H., Zhou, X.J., Fitzgerald, J., Keedy, S.K., Reilly, J.L., Passarotti, A.M., Sweeney, J.A., and Pavuluri, M. (2012). Microstructural abnormalities of white matter differentiate pediatric and adult-onset bipolar disorder. Bipolar Disord. *14*, 597–606.

Lui, S., Yao, L., Xiao, Y., Keedy, S.K., Reilly, J.L., Keefe, R.S., Tamminga, C.A., Keshavan, M.S., Pearlson, G.D., Gong, Q., et al. (2015). Resting-state brain function in schizophrenia and psychotic bipolar probands and their first-degree relatives. Psychol. Med. *45*, 97–108.

Magioncalda, P., Martino, M., Conio, B., Escelsior, A., Piaggio, N., Presta, A., Marozzi, V., Rocchi, G., Anastasio, L., Vassallo, L., et al. (2015). Functional connectivity and neuronal variability of resting state activity in bipolar disorder-reduction and decoupling in anterior cortical midline structures. Hum. Brain Mapp. *36*, 666-682.

Mah, L., Zarate, C.A., Singh, J., Duan, Y.-F., Luckenbaugh, D.A., Manji, H.K., and Drevets, W.C. (2007). Regional cerebral glucose metabolic abnormalities in bipolar II depression. Biol. Psychiatry *61*, 765–775.

Martino, M., Magioncalda, P., Saiote, C., Conio, B., Escelsior, A., Rocchi, G., Piaggio, N., Marozzi, V., Huang, Z., Ferri, F., et al. (2016a). Abnormal functional-structural cingulum connectivity in mania: combined functional magnetic resonance imaging-diffusion tensor imaging investigation in different phases of bipolar disorder. Acta Psychiatr. Scand. *134*, 339–349.

Martino, M., Magioncalda, P., Huang, Z., Conio, B., Piaggio, N., Duncan, N.W., Rocchi, G., Escelsior, A., Marozzi, V., Wolff, A., et al. (2016b). Contrasting variability patterns in the default mode and sensorimotor networks balance in bipolar depression and mania. Proc. Natl. Acad. Sci. *113*, 4824–4829.

Mayberg, H.S. (2009). Targeted electrode-based modulation of neural circuits for depression. J. Clin. Invest. *119*, 717–725.

Meda, S.A., Gill, A., Stevens, M.C., Lorenzoni, R.P., Glahn, D.C., Calhoun, V.D., Sweeney, J.A., Tamminga, C.A., Keshavan, M.S., Thaker, G., et al. (2012). Differences in resting-state functional magnetic resonance imaging functional network connectivity between schizophrenia and psychotic bipolar probands and their unaffected first-degree relatives. Biol. Psychiatry *71*, 881–889.

Meda, S.A., Ruaño, G., Windemuth, A., O'Neil, K., Berwise, C., Dunn, S.M., Boccaccio, L.E., Narayanan, B., Kocherla, M., Sprooten, E., et al. (2014). Multivariate analysis reveals genetic associations of the resting default mode network in psychotic bipolar disorder and schizophrenia. Proc. Natl. Acad. Sci. *111*, E2066–E2075.

Meda, S.A., Wang, Z., Ivleva, E.I., Poudyal, G., Keshavan, M.S., Tamminga, C.A., Sweeney, J.A., Clementz, B.A., Schretlen, D.J., Calhoun, V.D., et al. (2015). Frequency-Specific Neural Signatures of Spontaneous Low-Frequency Resting State Fluctuations in Psychosis: Evidence From Bipolar-Schizophrenia Network on Intermediate Phenotypes (B-SNIP) Consortium. Schizophr. Bull. *41*, 1336–1348.

Menon, V. (2011). Large-scale brain networks and psychopathology: a unifying triple network model. Trends Cogn. Sci. *15*, 483–506.

Menon, V. (2015). Salience Network A2 - Toga, Arthur W. In Brain Mapping, (Waltham: Academic Press), pp. 597–611.

Ochsner, K.N., Silvers, J.A., and Buhle, J.T. (2012). Functional imaging studies of emotion regulation: a synthetic review and evolving model of the cognitive control of emotion. Ann. N. Y. Acad. Sci. *1251*, E1–24.

Ongür, D., Lundy, M., Greenhouse, I., Shinn, A.K., Menon, V., Cohen, B.M., and Renshaw, P.F. (2010). Default mode network abnormalities in bipolar disorder and schizophrenia. Psychiatry Res. *183*, 59–68.

Peters, A.T., Burkhouse, K., Feldhaus, C.C., Langenecker, S.A., and Jacobs, R.H. (2016). Aberrant resting-state functional connectivity in limbic and cognitive control networks relates to depressive rumination and mindfulness: A pilot study among adolescents with a history of depression. J. Affect. Disord. *200*, 178–181.

Pezawas, L., Meyer-Lindenberg, A., Drabant, E.M., Verchinski, B.A., Munoz, K.E., Kolachana, B.S., Egan, M.F., Mattay, V.S., Hariri, A.R., and Weinberger, D.R. (2005). 5-HTTLPR polymorphism impacts human cingulate-amygdala interactions: a genetic susceptibility mechanism for depression. Nat. Neurosci. *8*, 828–834.

Phillips, M.L., Ladouceur, C.D., and Drevets, W.C. (2008). A neural model of voluntary and automatic emotion regulation: implications for understanding the pathophysiology and neurodevelopment of bipolar disorder. Mol. Psychiatry *13*, 829, 833–857.

Price, J.L., and Drevets, W.C. (2010). Neurocircuitry of mood disorders. Neuropsychopharmacol. Off. Publ. Am. Coll. Neuropsychopharmacol. *35*, 192–216.

Price, J.L., and Drevets, W.C. (2012). Neural circuits underlying the pathophysiology of mood disorders. Trends Cogn. Sci. *16*, 61–71.

Rashid, B., Damaraju, E., Pearlson, G.D., and Calhoun, V.D. (2014). Dynamic connectivity states estimated from resting fMRI Identify differences among Schizophrenia, bipolar disorder, and healthy control subjects. Front. Hum. Neurosci. *8*, 897.

Rashid, B., Arbabshirani, M.R., Damaraju, E., Cetin, M.S., Miller, R., Pearlson, G.D., and Calhoun, V.D. (2016). Classification of schizophrenia and bipolar patients using static and dynamic resting-state fMRI brain connectivity. NeuroImage *134*, 645–657.

Rey, G., Desseilles, M., Favre, S., Dayer, A., Piguet, C., Aubry, J.-M., and Vuilleumier, P. (2014). Modulation of brain response to emotional conflict as a function of current mood in bipolar disorder: preliminary findings from a follow-up state-based fMRI study. Psychiatry Res. *223*, 84–93.

Rey, G., Piguet, C., Benders, A., Favre, S., Eickhoff, S.B., Aubry, J.-M., and Vuilleumier, P. (2016). Resting-state functional connectivity of emotion regulation networks in euthymic and non-euthymic bipolar disorder patients. Eur. Psychiatry J. Assoc. Eur. Psychiatr. *34*, 56–63.

Roy, A.K., Shehzad, Z., Margulies, D.S., Kelly, A.M.C., Uddin, L.Q., Gotimer, K., Biswal, B.B., Castellanos, F.X., and Milham, M.P. (2009). Functional connectivity of the human amygdala using resting state fMRI. NeuroImage *45*, 614–626.

Satterthwaite, T.D., Kable, J.W., Vandekar, L., Katchmar, N., Bassett, D.S., Baldassano, C.F., Ruparel, K., Elliott, M.A., Sheline, Y.I., Gur, R.C., et al. (2015). Common and Dissociable Dysfunction of the Reward System in Bipolar and Unipolar Depression. Neuropsychopharmacol. Off. Publ. Am. Coll. Neuropsychopharmacol. *40*, 2258–2268.

Savitz, J.B., Price, J.L., and Drevets, W.C. (2014). Neuropathological and neuromorphometric abnormalities in bipolar disorder: view from the medial prefrontal cortical network. Neurosci. Biobehav. Rev. *42*, 132–147.

Seeley, W.W., Menon, V., Schatzberg, A.F., Keller, J., Glover, G.H., Kenna, H., Reiss, A.L., and Greicius, M.D. (2007). Dissociable intrinsic connectivity networks for salience processing and executive control. J. Neurosci. Off. J. Soc. Neurosci. *27*, 2349–2356.

Sergerie, K., Chochol, C., and Armony, J.L. (2008). The role of the amygdala in emotional processing: a quantitative meta-analysis of functional neuroimaging studies. Neurosci. Biobehav. Rev. *32*, 811–830.

Singh, M.K., Chang, K.D., Kelley, R.G., Saggar, M., L Reiss, A., and Gotlib, I.H. (2014). Early signs of anomalous neural functional connectivity in healthy offspring of parents with bipolar disorder. Bipolar Disord. *16*, 678–689.

Singh, M.K., Kelley, R.G., Chang, K.D., and Gotlib, I.H. (2015). Intrinsic Amygdala Functional Connectivity in Youth With Bipolar I Disorder. J. Am. Acad. Child Adolesc. Psychiatry *54*, 763–770.

Skåtun, K.C., Kaufmann, T., Tønnesen, S., Biele, G., Melle, I., Agartz, I., Alnæs, D., Andreassen, O.A., and Westlye, L.T. (2016). Global brain connectivity alterations in patients with schizophrenia and bipolar spectrum disorders. J. Psychiatry Neurosci. JPN *41*, 150159.

Souza-Queiroz, J., Boisgontier, J., Etain, B., Poupon, C., Duclap, D., d'Albis, M.-A., Daban, C., Hamdani, N., Le Corvoisier, P., Delavest, M., et al. (2016). Childhood trauma and the limbic network: a multimodal MRI study in patients with bipolar disorder and controls. J. Affect. Disord. *200*, 159–164.

Spielberg, J.M., Beall, E.B., Hulvershorn, L.A., Altinay, M., Karne, H., and Anand, A. (2016). Resting State Brain Network Disturbances Related to Hypomania & Depression in Medication-Free Bipolar Disorder. Neuropsychopharmacol. *41*, 3016–3024.

Sridharan, D., Levitin, D.J., and Menon, V. (2008). A critical role for the right fronto-insular cortex in switching between central-executive and default-mode networks. Proc. Natl. Acad. Sci. U. S. A. *105*, 12569–12574.

Stoddard, J., Gotts, S.J., Brotman, M.A., Lever, S., Hsu, D., Zarate, C., Ernst, M., Pine, D.S., and Leibenluft, E. (2016). Aberrant intrinsic functional connectivity within and between corticostriatal and temporal-parietal networks in adults and youth with bipolar disorder. Psychol. Med. *46*, 1509.

Strakowski, S.M. (2012). Integration and consolidation – a neurophysiological model of bipolar disorder. In The Bipolar Brain: Integrating Neuroimaging and Genetics, (New York: Oxford University Press), pp. 253–274.

Sussmann, J.E., Lymer, G.K.S., McKirdy, J., Moorhead, T.W.J., Muñoz Maniega, S., Job, D., Hall, J., Bastin, M.E., Johnstone, E.C., Lawrie, S.M., et al. (2009). White matter abnormalities in bipolar disorder and schizophrenia detected using diffusion tensor magnetic resonance imaging. Bipolar Disord. *11*, 11–18.

Teng, S., Lu, C.-F., Wang, P.-S., Hung, C.-I., Li, C.-T., Tu, P.-C., Su, T.-P., and Wu, Y.-T. (2013). Classification of bipolar disorder using basal-ganglia-related functional connectivity in the resting state. 35th Annual International Conference of the IEEE Engineering in Medicine and Biology Society *2013*, 1057–1060.

Teng, S., Lu, C.-F., Wang, P.-S., Li, C.-T., Tu, P.-C., Hung, C.-I., Su, T.-P., and Wu, Y.-T. (2014). Altered resting-state functional connectivity of striatal-thalamic circuit in bipolar disorder. PloS One *9*, e96422.

Torrisi, S., Moody, T.D., Vizueta, N., Thomason, M.E., Monti, M.M., Townsend, J.D., Bookheimer, S.Y., and Altshuler, L.L. (2013). Differences in resting corticolimbic functional connectivity in bipolar I euthymia. Bipolar Disord. *15*, 156–166.

Uddin, L.Q. (2015). Salience processing and insular cortical function and dysfunction. Nat. Rev. Neurosci. *16*, 55–61.

Wang, Y., Zhong, S., Jia, Y., Zhou, Z., Zhou, Q., and Huang, L. (2015). Reduced interhemispheric resting-state functional connectivity in unmedicated bipolar II disorder. Acta psychiatr. Scand. *132*, 400–407.

Wang, Y., Zhong, S., Jia, Y., Sun, Y., Wang, B., Liu, T., Pan, J., and Huang, L. (2016). Disrupted Resting-State Functional Connectivity in Nonmedicated Bipolar Disorder. Radiology 151641.

Wang, Y., Wang, J., Jia, Y., Zhong, S., Zhong, M., Sun, Y., Niu, M., Zhao, L., Zhao, L., Pan, J., et al. (2017). Topologically convergent and divergent functional connectivity patterns in unmedicated unipolar depression and bipolar disorder. Transl. Psychiatry *7*, e1165.

Xu, K., Liu, H., Li, H., Tang, Y., Womer, F., Jiang, X., Chen, K., Zhou, Y., Jiang, W., Luo, X., et al. (2014). Amplitude of low-frequency fluctuations in bipolar disorder: a resting state fMRI study. J. Affect. Disord. *152–154*, 237–242.

Yip, S.W., Mackay, C.E., and Goodwin, G.M. (2014). Increased temporo-insular engagement in unmedicated bipolar II disorder: an exploratory resting state study using independent component analysis. Bipolar Disord. *16*, 748–755.

Zhao, L., Wang, Y., Jia, Y., Zhong, S., Sun, Y., Qi, Z., Zhang, Z., and Huang, L. (2017). Altered interhemispheric functional connectivity in remitted bipolar disorder: A Resting State fMRI Study. Sci. Rep. *7*, 4698.

Zuo, X.-N., Di Martino, A., Kelly, C., Shehzad, Z.E., Gee, D.G., Klein, D.F., Castellanos, F.X., Biswal, B.B., and Milham, M.P. (2010). The oscillating brain: complex and reliable. NeuroImage *49*, 1432–1445.

An Overview of Resting State Functional Connectivity Studies of Major Depressive Disorder

Henry W. Chase

Richard Feynman's famous attributed quotation (Goodstein and Goodstein 1996) – that a topic isn't well understood unless it can be described to an undergraduate (specifically, freshman) audience – probably applies to the difficult undertaking of presenting the literature on resting-state connectivity in major depression. The word "probably" is used, as perhaps a scientist of Feynman's capabilities could make an adequate fist of it. My overall aim for this chapter is to demonstrate that it is by no means difficult to derive hypotheses to account for alterations in functional connectivity, but providing adequate evidence to support any of them – enough to convince an undergraduate or any other kind of reader – is in fact more difficult.

The success of the resting-state paradigm is remarkable (Buckner et al. 2013). Participants are asked to lie in an MRI scanner, either with their eyes open or closed, while functional images are collected usually for around 5–10 min. At the very least, this method took a radically different approach from the typical task-based fMRI acquisition. While task-based fMRI analysis has been dominated by general linear model (GLM)-based methods (Poline and Brett 2012), a vast number of methods have been applied to resting fMRI, many making markedly different assumptions about the nature of the underlying signal. Moreover, with respect to psychiatry in particular, the large samples have been collected facilitate data-driven analysis, for example, by affording out-of-sample validation. These large samples are possible in part due to the practicality of the method.

The method has also faced several challenges. These arise from the fact that many of the analysis methods employed rest on correlations of activity between brain regions. Given that fMRI images are made up of a combination of BOLD signal and artifacts caused by motion, physiological noise, and global signal fluctuations, there has been disagreement about the nature and extent of confounding induced by these artifacts. Whereas in task-based fMRI these artifacts are mostly

H. W. Chase (✉)
Department of Psychiatry, University of Pittsburgh, Pittsburgh, PA, USA
e-mail: chaseh@upmc.edu

V. A. Diwadkar, S. B. Eickhoff (eds.), *Brain Network Dysfunction in Neuropsychiatric Illness*, https://doi.org/10.1007/978-3-030-59797-9_14

assumed to go into the residuals of the GLM and be uncorrelated with the task manipulation, this cannot be assumed in rsfMRI. In practice, most rsMRI analysis methods try to estimate and remove artifact components before computing correlations between regional activities (e.g., Satterthwaite et al. 2017).

Of the variety of methods that have been applied to rsfMRI, most provide overall support for a broad framework which includes prototypical large-scale networks (that are also seen in task-based paradigms (Smith et al. 2009) and a modular architecture (Meunier et al. 2009). Finally, these networks are reproducible both across individuals, but also the estimated strength of a connection can show trait-like properties. This latter point can be demonstrated in several ways, but primarily using intraclass correlation (ICC: Shrout and Fleiss 1979). This statistic summarizes the extent to which between-participant variance can be replicated between testing sessions. A high ICC would therefore demonstrate that between-participant variance, presumably relating to a trait of some kind, can be seen at subsequent testing sessions. A low ICC would demonstrate that between-participant variance is seen irregularly across testing sessions, and is likely – though not necessarily (Chase et al. 2015b) – related to physiological or thermal noise. In general, rsfMRI-derived functional connectivity can show moderate/good ICCs (Braun et al. 2012; Geerligs et al. 2017; Zuo and Xing 2014). Furthermore, a connectivity "fingerprint" that characterizes individuals can be seen across different paradigms (e.g., emotion and working memory paradigms) as well as the resting state (Finn et al. 2015). These demonstrations provide a vital basis for the investigation of individual differences – in the same way as adequate reliability is a prerequisite for valid questionnaire measures.

Yet, a lingering question emerges at this point – what kinds of traits might variation in these connections reflect? With conventional task-related fMRI, task-evoked activation can be described in terms of a percentage change in the Blood Oxygenation Level-Dependent (BOLD) signal. In other words, manipulation "X" causes a percentage change of BOLD of "Y"%. This is then assumed to be driven by underlying physiological processes including neural activity (broadly defined for the present purposes), and a highly predictable effect on oxygen consumption and blood flow (Goense et al. 2016). Between-subject variation is then naturally interpreted in terms of a capacity to evoke such activation, even though other possible explanations may need to be addressed (e.g., Murphy et al. 2006). Functional connectivity relies on a similar set of assumptions to describe correlated neural activity across brain regions. Thus, on a trait level, individuals may vary in the capacity for the synchronization of a particular anatomical connection or circuit. What is not often discussed however, are the supposed underlying constructs that this variation reflects. Put simply, what does it mean to have a functional connection between two regions with a value of $Z = 1$, compared to $Z = 0.5$? In other areas of individual differences research, questionnaires or behavioral measures can be normed against known population statistics, and usually have some construct validity insofar as they are associated statistically with scales or measures that reflect the same construct. This would mean that a score of $Z = 1$ on a depression scale would reasonably be interpreted as reflecting the individual showing more severe depression symptoms than an individual with a score of $Z = 0.5$ (particularly if the scale has

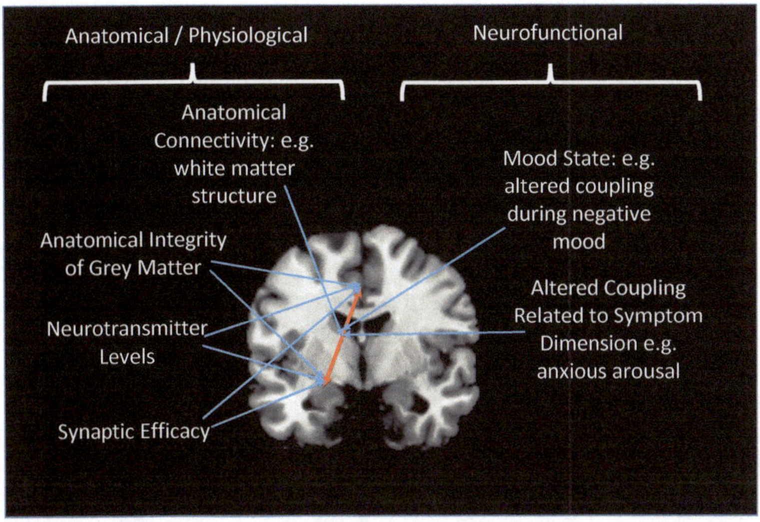

Fig. 1 Various types of explanation that might be invoked to describe alteration in the functional connectivity between, in this case, the amygdala and cingulate cortex. These might be broadly clustered into "anatomical/physiological" to describe the neural underpinnings of altered connectivity, or "neurofunctional" to describe the association between functional connectivity and symptom profiles or mood state

good psychometric properties). But for rsfMRI, this translation is not so simple. While the rsfMRI method can be broadly validated with respect to anatomical connectivity (indeed, we have deployed such arguments, see Chase et al. 2015a), this type of validation seems much stronger for the population mean of two regions' connectivity than the between-subject variance. In other words, although some kind of anatomical connection, be it mono- or polysynaptic, is a necessary prerequisite for nonzero functional connectivity, it is by no means clear how measurable differences in anatomy or physiology (e.g., more/richer neuronal connectivity between regions) might be reflected in functional connectivity. In the coming sections, I will offer some suggestions as to what between-subject variability in resting connectivity might reflect (see Fig. 1), and discuss whether these can be used as plausible interpretations of the findings of rsfMRI studies of Major Depressive Disorder (MDD) and related psychiatric illnesses.

Structure/Function Relationships

The most obvious place to look for an explanation is the relationship between structural and functional connectivity. Within this category, there are two obvious explanations for altered connectivity. First, if a given region shows reduced or absent activity because it has atrophied, or its internal activity is otherwise directly reduced, we would expect it to show reduced correlated activity with other regions that are within the same network. Second, if critical white matter pathways that connect two regions are disrupted, we might expect disrupted correlated activity between two regions. In other words, both these possibilities reflect changes in rsfMRI connectivity that would be expected from lesions, to the gray or white matter respectively. Other theoretical work (e.g., Fornito et al. 2015) describes how the effect of lesions might impact functional connectivity through less direct means.

Before addressing the rsfMRI literature, the reliability of evidence of structural deficits in MDD should be assessed. Meta-analytic studies emphasize regions of the cingulate and insula (Schmaal et al. 2017; Wise et al. 2017) as showing reduced gray matter concentration in MDD, a finding which is observed across a variety of psychiatric disorders (Goodkind et al. 2015). Other findings include alterations in the temporal and frontal lobes (dorsolateral, superior frontal, and orbitofrontal: Schmaal et al. 2017; Wise et al. 2017), as well as small increases in the occipital lobe (Wise et al. 2017). Focused analysis on the hippocampus has also revealed reduced volume (Schmaal et al. 2016; see also Wise et al. 2017). Importantly, the effect sizes for these differences are not large: probably around d = 0.1–0.2 (Schmaal et al. 2017; Schmaal et al. 2016). Thus, the overall effect size of gray matter differences may not be sufficient to account for the differences in rsfMRI connectivity which have been observed: even a relatively large study with patient/control cohorts of 50 would yield a power of around 0.2 at best. Similarly, modest effect sizes have been estimated for white matter alterations in MDD (Wise et al. 2016) to the cingulate bundle and corpus callosum. Examination of white matter hyperintensities in depression has also yielded modest effect sizes, as well as the suggestion of substantial between-cohort heterogeneity (Wang et al. 2014). There are a variety of possible mechanisms underlying this heterogeneity. For example, individuals showing high levels of oxidative stress may show distinct patterns of white matter alteration (see Versace et al. 2014 for a study in bipolar disorders). Of course, there is another possibility – that there is no obvious structural lesion or deficit that could be measured with MRI but the activity in a given region is compromised in MDD due to an otherwise undetected anatomical alteration of some kind.

The major assumption behind this line of argument is that gray or white matter in a critical pathway would be linearly related to functional connectivity. Clearly, in the extreme case, the presence of an anatomical pathway predicates functional connectivity. But, regardless of moderators, would a small reduction in gray or white matter be expected to be associated with a small reduction in functional connectivity? The literature is somewhat mixed on this point, with some studies demonstrating linear between-subject structure/function relationships (van den Heuvel et al. 2008), some arguing for state/context-dependent relationships (Baria et al. 2013) and others demonstrating that structural and functional connectivity are largely unrelated (Tsang et al. 2017). Our own work in suicidal individuals suggests that some of these relationships might be somewhat complex (Chase et al. 2017). We expected a simple reduction in the functional connectivity between anterior cingulate cortex (ACC) and posterior cingulate cortex cortices (PCC) in individuals with reduced white matter integrity in the cingulum bundle (e.g., van den Heuvel et al.

2008). In fact, a marginal relationship between cingulum bundle radial diffusivity and dorsal/ventral PCC connectivity with the ACC was observed, which was largely related to the shared impact of group. Moreover, stronger findings were seen by examining relationships between low frequency BOLD in the PCC and cingulum radial diffusivity/fractional anisotropy, which differed across groups. These low-frequency BOLD measures reflect the power of low-frequency oscillations relative to broadband BOLD signal (e.g., Baria et al. 2013), and were themselves related to connectivity measures, as would be expected. Overall, these findings suggest that functional connectivity might be an emergent property of oscillatory activity within the cingulate (c.f. Cabral et al. 2011), and can be influenced by white matter connectivity and group membership. It also suggests that the simple relationships we expect between structure and function may in fact be complex and/or nonlinear.

In the absence of striking structural differences, another straightforward interpretation of trait-level differences in functional connectivity might be in terms of Hebbian plasticity. Specifically, if two anatomically connected regions are repetitively co-activated, a consequent strengthening of synapses that connected them might subsequently be expected to lead to increases in firing correlations between regions (Fregnac 2003). The relevance for psychiatric disorders might be dependent on the region and disorder in question, but it is possible to imagine that over-activation of the amygdala for example might lead to altered correlations with distal, connected regions. Another relevant point is that plasticity itself is thought to be altered in MDD (Normann et al. 2007). In fact, direct examinations of Hebbian plasticity using rsfMRI have yielded quite complex data (Johnen et al. 2015), in which paired transcranial magnetic stimulation (TMS) to the motor and premotor cortex did not increase resting functional connectivity, but did alter connectivity during grasping. Surprisingly, paired TMS also affected functional connectivity between the motor cortex and unstimulated areas including the parietal lobe at rest.

When taking these various strands together therefore, it is difficult to trace a link between basic anatomical (e.g., structural) and physiological (i.e., synaptic potentiation) processes, and altered patterns of resting connectivity in MDD. In many ways, this difficulty is reflected in the inconsistency of the extant literature: while numerous studies of resting-state connectivity with the amygdala have been performed, very few consistent findings have emerged. Given the uncertain mechanisms at play, perhaps the most promising way forward would be to embrace the multivariate nature of both structure and function, and explore multivariate relationships across individuals. It may be that more sophisticated and detailed modeling of biophysical generators of BOLD may allow the crucial linear structure/function relationships, and synaptic mechanisms to be explored in more detail. However, such methods are presumably reliant on having established a reliable marker using conventional methods, and for this, it may be that more exploratory studies are necessary. I will address these in the next section.

Mood State/Heterogeneity

The first section described examples of *physiologically* plausible mechanisms for variation in functional connectivity. But those examples do not directly deal with what we are perhaps more interested in, namely, a *psychologically* relevant mechanism. Here, we might consider the wider literature on emotion representation in the brain, and perhaps generate hypotheses based on how these regions might interact in different mood states. In practice, this is trickier than it sounds. First, despite numerous neuroimaging studies of emotion, there still remains considerable uncertainty about how emotions are represented as a distributed code or in terms of interactions between specialized, localized systems (see e.g., Chang et al. 2015; Kragel et al. 2016). Thus, although we might suppose that the functional connectivity of classical "emotion" or "regulation"-related regions such as the amygdala or prefrontal cortex might be altered in MDD, these hypotheses do not take us as far as we might like. In terms of mood disorders more widely, a detailed reading of the literature suggests that connectivity between the amygdala and ACC/medial prefrontal cortex might be altered by mood state across individuals with bipolar disorders (Chase and Phillips 2016), and there is evidence that there are similar alterations in MDD (Anand et al. 2005). In addition, interactions between different prefrontal cortex regions involved in regulating the amygdala (i.e., ventrolateral prefrontal cortex, medial prefrontal cortex) may also be altered in mood disorders (Chase and Phillips 2016; Scheinost et al. 2017). Nevertheless, this is a far from consistent observation, and it is difficult even to argue that the effect of mood would correspond to hyper- or hypoconnectivity between these structures. This may be partly due to the inconsistencies caused by different patterns of medial prefrontal functional connectivity with different amygdala sub-regions (Kerestes et al. 2017).

It has still been possible to make some progress along these lines. This is in part because some rsfMRI datasets are so substantial that they allow validation of data-driven analyses by out-of-sample replications. A good example of this is the recent work by Drysdale and colleagues (Drysdale et al. 2017): using a combination of canonical correlation analysis (CCA) and hierarchical clustering within a very large cohort of MDD patients, they identified four clusters of patients ("biotypes"), each showing distinct patterns of whole brain connectivity. The patterns of connectivity associated with these clusters could be accurately replicated at a second testing session, and a similar clustering pattern could be obtained within a second independent dataset. The four biotypes were associated with different levels of anhedonia and anxiety, while the overall depression severity was largely matched between the groups. The connectivity patterns tended to differ along several dimensions: reduced fronto-amygdala connectivity, enhanced thalamic/fronto-striatal connectivity, reduced anterior cingulate, and orbitofrontal cortex connectivity.

Other approaches have identified significant findings while concurrently testing large numbers of (nondirectional) hypotheses. Such techniques require very large effect sizes to be present to appropriately correct for multiple significance testing. Nevertheless, there are examples where this approach has been successful. For example, Satterthwaite and colleagues identified distinct effects of depression and anxiety severity on amygdala connectivity across cohorts of MDD and post-traumatic stress disorder (PTSD) patients (Satterthwaite et al. 2016). This study employed a method known as multivariate distance-based matrix regression (MDMR), which identified the amygdala as showing a global, or at least nonspecific, alteration in functional connectivity with the rest of the brain which was related to depression severity. A follow-up, seed-based analysis with the amygdala revealed that depression severity was associated with decoupling of the structure with the dorsolateral prefrontal cortex, the anterior cingulate cortex, and the insula, while anxiety was associated with increased connectivity between the amygdala and the ventromedial prefrontal cortex.

A key similarity of the Satterthwaite and Drysdale studies is that for the most part, both studies were relatively hypothesis-free. Although a focus on the amygdala and prefrontal cortex is supported by considerable prior work (e.g., Phillips et al. 2008), few other constraints were imposed or anticipated, and indeed, these regions were identified in a data-driven fashion. Broadly then, the strength of these studies depends largely on the magnitude of the underlying statistical effect sizes – as well as the methods needed to identify them. As described above, these exploratory studies are a necessary part of a longer process of uncovering the physiological mechanisms underlying these psychological traits. Further work is needed to explore the generality of the findings across cohorts, and to establish an underlying mechanism for the brain/behavior relationships of interest.

A more important point is that both studies directly addressed the heterogeneity within MDD. It is very likely that much of the complexity and inconsistency within the extant literature is related to the heterogeneity of the participant cohorts themselves – a point which is reflected in the RDoC framework (Insel et al. 2010). These findings confirm that modeling individual differences in symptomatology is a promising approach, and thus, it may yet replace the conventional patient/control analysis for statistical analysis. Although heterogeneity of MDD may be associated with differences in subjective mood state, this line of analysis does seem to move the discussion away from functional connectivity patterns that might reflect negative mood, and toward a picture in which various neurobiological traits reflected in functional connectivity may combine to influence the clinical phenotype. Of course, these relationships are largely correlational, so a number of state/trait relationships are possible. Ideally, we would want to be in a position where an established trait-related alteration of functional connectivity in MDD might be modulated by mood induction. Indeed, there are several smaller studies where mood induction on a depressed cohort has been performed (e.g., Admon and Pizzagalli 2015; Price et al. 2017; Renner et al. 2017).

Arousal and the Orchestration of Large-Scale Networks

An issue with working back from data-driven analyses, particularly regarding rsfMRI, is that the method provides no control over the psychological state of the participant. At the very least, this provides an explanatory gap. Although it may well be plausible to fill this gap with hypotheses about the interactions of different brain regions, there are also explanations that may have nothing to do with the core symptoms of MDD or mood state. One account that is difficult to discount pertains to the arousal level of an individual. Although arousal has a rather unfortunate "catch-all" quality, clearly not operating a unitary construct (Satpute et al. 2018; Trofimova and Robbins 2016), the term nevertheless conveys something useful in this context. Individuals enter the scanner at different times of day, during different daily schedules, after different levels of caffeine or perhaps nicotine intake, and find the experience of being in an MRI scanner somewhat different. On the one hand, we can imagine an individual who is nervous about the scan and has drunk a coffee or two before entering the scanner. On the other, we can imagine an individual, perhaps a medical student or nurse, trying to fit in an evening scan after a long day at work. The latter may be familiar with the scanner, and find the whole procedure unremarkable. Both individuals may nominally be "controls" and show little variation in demographic or other measures that would plausibly capture some relevant individual differences. However, the rsfMRI method may be uniquely sensitive to the differences described: while task-related fMRI may engage both participants, attracting their attention to the task and encouraging them to deploy the necessary cognitive resources, the rsfMRI paradigm makes no such requirement.

We then return to the concept of arousal – or at least, a concept that might help to describe the variation described above. Empirically, there are a few options for describing this variation. One is to ask participants about how they were feeling in the scanner. This has been a successful approach, and there have been a variety of findings describing brain (connectivity) – behavior relationships in the scanner (e.g., Andrews-Hanna et al. 2010). A second is to monitor eye movements and pupil dilation in the scanner (Wang et al. 2016): pupil dilation is a useful proxy of wakefulness or arousal, so periods in which the pupil is narrow versus when it is wide are related to differences in resting connectivity. The canonical effect observed (see also De Havas et al. 2012) is that reduced wakefulness/arousal is associated with increased connectivity between networks which are typically uncoupled or anticorrelated. Examples include the salience/task-positive networks and the default mode network. Finally, we might assume that arousal may change slowly through the scan, gradually decreasing as familiarity with the environment, and perhaps boredom or fatigue increase. Interestingly, there are several recent observations of BOLD signal/connectivity changes that follow this pattern (Abrol et al. 2016; Bijsterbosch et al. 2017).

As the effect of arousal becomes better understood, it may be possible to decode this effect directly from images themselves: this could then be modeled and thereby corrected for in subsequent between- and within-subject analyses. In the absence of this, there remains the lingering concern, as with the effect of motion or global signal, that group differences may reflect confounding by arousal rather than an independent group difference. Indeed, a meta-analysis which compiled the existing rsfMRI literature on MDD identified some patterns of altered connectivity that might be consistent with large-scale network decoupling or reduced anticorrelation (Kaiser et al. 2015). Further work is thus probably needed to correct for the nonspecific influence of arousal or wakefulness.

Alterations in Neurotransmitter Levels

Another way forward might be to consider how neurotransmitter levels, either within particular circuits or more widely, might exert effects on functional connectivity. This idea has been developed in part from resting-state fMRI studies of schizophrenia, where global alterations in functional connectivity may reflect the balance of excitation (i.e., Glutamate) and inhibition (i.e., GABA: see Foss-Feig et al. 2017). Specifically, manipulations of neural excitability by ketamine have been shown to affect neural connectivity in a manner that corresponds to the predictions of a recurrent neural network model which includes excitatory (e.g., glutamatergic) and inhibitory (e.g., GABAergic) components. This body of work may be relevant for MDD for several reasons, but in this context, because alterations of GABA levels have been observed in MDD using magnetic resonance spectroscopy (MRS: e.g., Bhagwagar et al. 2008). One study to have examined MRS and resting fMRI connectivity in the same sample was conducted by Horn and colleagues (Horn et al. 2010), who found that functional connectivity between the pre-genual cingulate cortex and anterior insula was reduced in MDD patients with greater levels of glutamate and glutamine in their pre-genual cingulate. Overall, their findings suggested that the most severely depressed patients might show abnormalities in glutamate/glutamine in this region, and consequent alterations (increases) in functional connectivity.

Alterations in signaling by other neurotransmitters in depression may also be manifest in terms of altered functional connectivity. In particular serotoninergic function is classically related to MDD (Cowen 2008), and manipulations of this neurotransmitter system have been studied using rsfMRI. One particularly novel area of study has been the examination of the effect of psychedelic drugs such as psilocybin on the PCC/default mode network (DMN: Carhart-Harris et al. 2012). Other studies have examined DMN connectivity within the context of antidepressant treatment studies. For example, Posner and colleagues showed that antidepressant (duloxetine) treatment reduced hyperconnectivity within the DMN that was seen pretreatment in dysthymic individuals (Posner et al. 2013). Similar reductions in DMN connectivity by duloxetine were also seen by van Wingen and others in healthy individuals (van Wingen et al. 2014). Antidepressants can also alter the functional connectivity between the amygdala and prefrontal cortex (McCabe and Mishor 2011).

Dopamine has been less widely studied with neuroimaging in the context of MDD, but there are numerous studies reporting alterations in the functional connectivity of dopamine-rich regions including the striatum (Gabbay et al. 2013; Leaver et al. 2016; Meng et al. 2014). It seems likely that dopamine is of specific relevance to anhedonic or apathetic symptoms (Treadway and Zald 2011), but to the best of our knowledge, targeted examination of the effect of dopamine manipulations on striatal connectivity of anhedonic individuals has not been performed. One study which has assessed a dopamine manipulation (haloperidol, a D2 receptor antagonist) on amygdala connectivity in healthy individuals (Haaker et al. 2016) identified a decoupling of the amygdala from the posterior cingulate cortex. Interestingly, this observation is similar to a finding from our group in postpartum depression (PPD: Chase et al. 2013), but is opposite to a finding in a dysthymic population (Posner et al. 2013). Again, beyond the shared depressive symptomatology, dysthymia and PPD are clinically distinct in numerous ways, and it is an intriguing possibility that dopaminergic probes may help to identify the underlying neurobiological differences.

There are a variety of other pharmacological avenues that have been investigated, in an effort to study the complexity of molecular signaling in depression and in particular the stress-related components of mood disorders. These include norepinephrine (McCabe and Mishor 2011), hydrocortisone (Henckens et al. 2012), inflammation (Felger et al. 2016), and goserelin (a gonadotropin-releasing hormone agonist: Fisher et al. 2017). Again, amygdala connectivity is a common theme throughout many of these studies, but considerable further work is needed to develop a coherent picture from these individual pieces of evidence.

Summary

Despite the numerous resting-state fMRI studies that have been performed on patients with MDD, it is difficult to see a dominant theoretical and empirical picture emerging from the extant literature. Overall, to my knowledge, it has not been possible to connect a physiologically plausible mechanism with a psychological account of MDD using resting fMRI, and to provide replicable evidence of this mechanism.

In reality, these are the first few steps of a longer scientific process, in which various sources of variability need to be controlled before a clear picture emerges. There has been some success in the field of schizophrenia in replicating altered functional connectivity of the thalamus (Murray and Anticevic 2017), so we should be cautiously optimistic that this might be possible for MDD too. Notably, these challenges are not restricted to resting functional connectivity, but are arguably also present for task-based studies of MDD (Muller et al. 2017).

One prominent area is methodological heterogeneity between studies. One of the features of the rsfMRI field is the vast array of data analysis methods that have been developed for this work. Ideally, the differences between methods should not matter: aside from some arbitrary scaling or noise reduction differences, robust group differences – if they exist – should emerge regardless of the processing methods. In practice, however, some of the decisions do seem to make a difference. Global signal correction is commonly performed as it is particularly efficacious for removing artifacts and provides some convenient normalization of connectivity scores, but if groups differ in their global signal (Yang et al. 2014), it is likely that unstable and biased between-group differences will be observed if this confound is not appropriately modeled (Murphy et al. 2013). As a more general point, there are numerous methods for removing and correcting for noise in rsfMRI, and many have slightly different impact on test-retest reliability and the magnitude of the remaining signal (e.g., Birn et al. 2014; Varikuti et al. 2017). Selecting between these different methods has been difficult, and while there is an agreement that some nuisance removal is needed, proposed group differences would ideally be confirmed across methods. A final relevant and emerging methodological area is the location and resolution of the seed regions used for rsfMRI analysis (Cole et al. 2010). A second area is the mood and arousal state of the participant, discussed in Sect. 3. We anticipate that the methods described in that section may be employed more widely. A third area is an exploration of the relationships between particular symptoms, and functional connectivity as discussed in Sect. 2. In this context, state measures may provide a kind of mediating role between traits and estimate connectivity. Fourth, as discussed in Sect. 1, structure/function relationships may be a source of further insight. Finally, multimodal analysis of other physiological measures, including measurement of inflammation, neurotransmitters using MRS, or autonomic function may provide further capacity to explain differences between patients, groups, and/or studies.

Overall, our understanding of resting-state functional connectivity in MDD does not lead to a mechanistic account which can link physiologically grounded measurements of functional connectivity with the psychological abnormalities that characterize the disorder. Nevertheless, there are various appealing hypotheses, related to plasticity of relevant networks (e.g., ACC, amygdala) that can make contact with translational studies (e.g., Klavir et al. 2013). But these are generally at quite an early stage. We have described pieces of the progress that has been made, and possible strategies to assist reaching this goal. Whether an account of the extant literature on this topic makes good content for an undergraduate lecture may perhaps be dependent on whether one would focus on the vast technical, statistical, and logistic challenges that have been overcome to obtain our half-full literature, or whether one would focus on our half-empty understanding of this complex problem.

References

Abrol, A., Chaze, C., Damaraju, E., & Calhoun, V. D. (2016). The chronnectome: Evaluating replicability of dynamic connectivity patterns in 7500 resting fMRI datasets. *Conf Proc IEEE Eng Med Biol Soc, 2016*, 5571–5574. doi: https://doi.org/10.1109/EMBC.2016.7591989

Admon, R., & Pizzagalli, D. A. (2015). Corticostriatal pathways contribute to the natural time course of positive mood. *Nat Commun, 6*, 10065. doi: https://doi.org/10.1038/ncomms10065

Anand, A., Li, Y., Wang, Y., Wu, J., Gao, S., Bukhari, L., . . . Lowe, M. J. (2005). Activity and connectivity of brain mood regulating circuit in depression: a functional magnetic resonance study. *Biol Psychiatry, 57*, 1079–1088. doi: https://doi.org/10.1016/j.biopsych.2005.02.021

Andrews-Hanna, J. R., Reidler, J. S., Huang, C., & Buckner, R. L. (2010). Evidence for the default network's role in spontaneous cognition. *J Neurophysiol, 104*, 322–335. doi: https://doi.org/10.1152/jn.00830.2009 jn.00830.2009 [pii]

Baria, A. T., Mansour, A., Huang, L., Baliki, M. N., Cecchi, G. A., Mesulam, M. M., & Apkarian, A. V. (2013). Linking human brain local activity fluctuations to structural and functional network architectures. *Neuroimage, 73*, 144–155. doi: https://doi.org/10.1016/j.neuroimage.2013.01.072 S1053-8119(13)00119-5 [pii]

Bhagwagar, Z., Wylezinska, M., Jezzard, P., Evans, J., Boorman, E., P, M. M., & P, J. C. (2008). Low GABA concentrations in occipital cortex and anterior cingulate cortex in medication-free, recovered depressed patients. *Int J Neuropsychopharmacol, 11*, 255–260. doi: https://doi.org/10.1017/S1461145707007924

Bijsterbosch, J., Harrison, S., Duff, E., Alfaro-Almagro, F., Woolrich, M., & Smith, S. (2017). Investigations into within- and between-subject resting-state amplitude variations. *Neuroimage, 159*, 57–69. doi: https://doi.org/10.1016/j.neuroimage.2017.07.014

Birn, R. M., Cornejo, M. D., Molloy, E. K., Patriat, R., Meier, T. B., Kirk, G. R., . . . Prabhakaran, V. (2014). The influence of physiological noise correction on test-retest reliability of resting-state functional connectivity. *Brain Connect, 4*, 511–522. doi: https://doi.org/10.1089/brain.2014.0284

Braun, U., Plichta, M. M., Esslinger, C., Sauer, C., Haddad, L., Grimm, O., . . . Meyer-Lindenberg, A. (2012). Test-retest reliability of resting-state connectivity network characteristics using fMRI and graph theoretical measures. *Neuroimage, 59*, 1404–1412. doi: https://doi.org/10.1016/j.neuroimage.2011.08.044

Buckner, R. L., Krienen, F. M., & Yeo, B. T. (2013). Opportunities and limitations of intrinsic functional connectivity MRI. *Nat Neurosci, 16*, 832–837. doi: https://doi.org/10.1038/nn.3423

Cabral, J., Hugues, E., Sporns, O., & Deco, G. (2011). Role of local network oscillations in resting-state functional connectivity. *Neuroimage, 57*, 130–139. doi: https://doi.org/10.1016/j.neuroimage.2011.04.010

Carhart-Harris, R. L., Erritzoe, D., Williams, T., Stone, J. M., Reed, L. J., Colasanti, A., . . . Nutt, D. J. (2012). Neural correlates of the psychedelic state as determined by fMRI studies with psilocybin. *Proc Natl Acad Sci U S A, 109*, 2138–2143. doi: https://doi.org/10.1073/pnas.1119598109

Chang, L. J., Gianaros, P. J., Manuck, S. B., Krishnan, A., & Wager, T. D. (2015). A Sensitive and Specific Neural Signature for Picture-Induced Negative Affect. *PLoS Biol, 13*, e1002180. doi: https://doi.org/10.1371/journal.pbio.1002180

Chase, H. W., Clos, M., Dibble, S., Fox, P., Grace, A. A., Phillips, M. L., & Eickhoff, S. B. (2015a). Evidence for an anterior-posterior differentiation in the human hippocampal formation revealed by meta-analytic parcellation of fMRI coordinate maps: focus on the subiculum. *Neuroimage, 113*, 44–60. doi: https://doi.org/10.1016/j.neuroimage.2015.02.069

Chase, H. W., Fournier, J. C., Greenberg, T., Almeida, J. R., Stiffler, R., Zevallos, C. R., . . . Phillips, M. L. (2015b). Accounting for Dynamic Fluctuations across Time when Examining fMRI Test-Retest Reliability: Analysis of a Reward Paradigm in the EMBARC Study. *PLoS One, 10*, e0126326. doi: https://doi.org/10.1371/journal.pone.0126326

Chase, H. W., Moses-Kolko, E. L., Zevallos, C., Wisner, K. L., & Phillips, M. L. (2013). Disrupted posterior cingulate-amygdala connectivity in postpartum depressed women as measured with resting BOLD fMRI. *Soc Cogn Affect Neurosci*. doi: nst083 [pii] 191093/scan/nst083

Chase, H. W., & Phillips, M. L. (2016). Elucidating neural network functional connectivity abnormalities in bipolar disorder: toward a harmonized methodological approach. *Biol Psychiatry Cogn Neurosci Neuroimaging, 1*, 288–298. doi: https://doi.org/10.1016/j.bpsc.2015.12.006

Chase, H. W., Segreti, A. M., Keller, T. A., Cherkassky, V. L., Just, M. A., Pan, L. A., & Brent, D. A. (2017). Alterations of functional connectivity and intrinsic activity within the cingulate cortex of suicidal ideators. *J Affect Disord, 212*, 78–85. doi: https://doi.org/10.1016/j.jad.2017.01.013

Cole, D. M., Smith, S. M., & Beckmann, C. F. (2010). Advances and pitfalls in the analysis and interpretation of resting-state FMRI data. *Front Syst Neurosci, 4*, 8. doi: https://doi.org/10.3389/fnsys.2010.00008

Cowen, P. J. (2008). Serotonin and depression: pathophysiological mechanism or marketing myth? *Trends Pharmacol Sci.* doi: S0165–6147(08)00126-0 [pii] 241016/j.tips.2008.05.004

De Havas, J. A., Parimal, S., Soon, C. S., & Chee, M. W. (2012). Sleep deprivation reduces default mode network connectivity and anti-correlation during rest and task performance. *Neuroimage, 59*, 1745–1751. doi: https://doi.org/10.1016/j.neuroimage.2011.08.026

Drysdale, A. T., Grosenick, L., Downar, J., Dunlop, K., Mansouri, F., Meng, Y., . . . Liston, C. (2017). Resting-state connectivity biomarkers define neurophysiological subtypes of depression. *Nat Med, 23*, 28–38. doi: https://doi.org/10.1038/nm.4246

Felger, J. C., Li, Z., Haroon, E., Woolwine, B. J., Jung, M. Y., Hu, X., & Miller, A. H. (2016). Inflammation is associated with decreased functional connectivity within corticostriatal reward circuitry in depression. *Mol Psychiatry, 21*, 1358–1365. doi: https://doi.org/10.1038/mp.2015.168

Finn, E. S., Shen, X., Scheinost, D., Rosenberg, M. D., Huang, J., Chun, M. M., . . . Constable, R. T. (2015). Functional connectome fingerprinting: identifying individuals using patterns of brain connectivity. *Nat Neurosci, 18*, 1664–1671. doi: https://doi.org/10.1038/nn.4135

Fisher, P. M., Larsen, C. B., Beliveau, V., Henningsson, S., Pinborg, A., Holst, K. K., . . . Frokjaer, V. G. (2017). Pharmacologically Induced Sex Hormone Fluctuation Effects on Resting-State Functional Connectivity in a Risk Model for Depression: A Randomized Trial. *Neuropsychopharmacology, 42*, 446–453. doi: https://doi.org/10.1038/npp.2016.208

Fornito, A., Zalesky, A., & Breakspear, M. (2015). The connectomics of brain disorders. *Nat Rev Neurosci, 16*, 159–172. doi: https://doi.org/10.1038/nrn3901

Foss-Feig, J. H., Adkinson, B. D., Ji, J. L., Yang, G., Srihari, V. H., McPartland, J. C., . . . Anticevic, A. (2017). Searching for Cross-Diagnostic Convergence: Neural Mechanisms Governing Excitation and Inhibition Balance in Schizophrenia and Autism Spectrum Disorders. *Biol Psychiatry, 81*, 848–861. doi: https://doi.org/10.1016/j.biopsych.2017.03.005

Fregnac, Y. (2003). Hebbian Synaptic Plasticity. In M. A. Arbib (Ed.), *The Handbook of Brain Theory and Neural Networks* (Second Edition ed., pp. 515–522). Cambridge, MA: The MIT Press.

Gabbay, V., Ely, B. A., Li, Q., Bangaru, S. D., Panzer, A. M., Alonso, C. M., . . . Milham, M. P. (2013). Striatum-based circuitry of adolescent depression and anhedonia. *J Am Acad Child Adolesc Psychiatry, 52*, 628–641 e613. doi: https://doi.org/10.1016/j.jaac.2013.04.003

Geerligs, L., Tsvetanov, K. A., Cam, C., & Henson, R. N. (2017). Challenges in measuring individual differences in functional connectivity using fMRI: The case of healthy aging. *Hum Brain Mapp, 38*, 4125–4156. doi: https://doi.org/10.1002/hbm.23653

Goense, J., Bohraus, Y., & Logothetis, N. K. (2016). fMRI at High Spatial Resolution: Implications for BOLD-Models. *Front Comput Neurosci, 10*, 66. doi: https://doi.org/10.3389/fncom.2016.00066

Goodkind, M., Eickhoff, S. B., Oathes, D. J., Jiang, Y., Chang, A., Jones-Hagata, L. B., . . . Etkin, A. (2015). Identification of a common neurobiological substrate for mental illness. *JAMA Psychiatry, 72*, 305–315. doi: https://doi.org/10.1001/jamapsychiatry.2014.2206

Goodstein, D. L., & Goodstein, J. R. (1996). *Feynman's lost lecture : the motion of planets around the sun* (1st ed.). New York: Norton.

Haaker, J., Menz, M. M., Fadai, T., Eippert, F., & Buchel, C. (2016). Dopaminergic receptor blockade changes a functional connectivity network centred on the amygdala. *Hum Brain Mapp, 37*, 4148–4157. doi: https://doi.org/10.1002/hbm.23302

Henckens, M. J., van Wingen, G. A., Joels, M., & Fernandez, G. (2012). Corticosteroid induced decoupling of the amygdala in men. *Cereb Cortex, 22*, 2336–2345. doi: https://doi.org/10.1093/cercor/bhr313

Horn, D. I., Yu, C., Steiner, J., Buchmann, J., Kaufmann, J., Osoba, A., . . . Walter, M. (2010). Glutamatergic and resting-state functional connectivity correlates of severity in major depression - the role of pregenual anterior cingulate cortex and anterior insula. *Front Syst Neurosci, 4*. doi: https://doi.org/10.3389/fnsys.2010.00033

Insel, T., Cuthbert, B., Garvey, M., Heinssen, R., Pine, D. S., Quinn, K., . . . Wang, P. (2010). Research domain criteria (RDoC): toward a new classification framework for research on mental disorders. *Am J Psychiatry, 167*, 748–751. doi: 167/7/748 [pii] 421176/appi.ajp.2010.09091379

Johnen, V. M., Neubert, F. X., Buch, E. R., Verhagen, L., O'Reilly, J. X., Mars, R. B., & Rushworth, M. F. (2015). Causal manipulation of functional connectivity in a specific neural pathway during behaviour and at rest. *Elife, 4*. doi: https://doi.org/10.7554/eLife.04585

Kaiser, R. H., Andrews-Hanna, J. R., Wager, T. D., & Pizzagalli, D. A. (2015). Large-Scale Network Dysfunction in Major Depressive Disorder: A Meta-analysis of Resting-State Functional Connectivity. *JAMA Psychiatry, 72*, 603–611. doi: https://doi.org/10.1001/jamapsychiatry.2015.0071

Kerestes, R., Chase, H. W., Phillips, M. L., Ladouceur, C. D., & Eickhoff, S. B. (2017). Multimodal evaluation of the amygdala's functional connectivity. *Neuroimage, 148*, 219–229. doi: https://doi.org/10.1016/j.neuroimage.2016.12.023

Klavir, O., Genud-Gabai, R., & Paz, R. (2013). Functional connectivity between amygdala and cingulate cortex for adaptive aversive learning. *Neuron, 80*, 1290–1300. doi: https://doi.org/10.1016/j.neuron.2013.09.035

Kragel, P. A., Knodt, A. R., Hariri, A. R., & LaBar, K. S. (2016). Decoding Spontaneous Emotional States in the Human Brain. *PLoS Biol, 14*, e2000106. doi: https://doi.org/10.1371/journal.pbio.2000106

Leaver, A. M., Espinoza, R., Joshi, S. H., Vasavada, M., Njau, S., Woods, R. P., & Narr, K. L. (2016). Desynchronization and Plasticity of Striato-frontal Connectivity in Major Depressive Disorder. *Cereb Cortex, 26*, 4337–4346. doi: https://doi.org/10.1093/cercor/bhv207

McCabe, C., & Mishor, Z. (2011). Antidepressant medications reduce subcortical-cortical resting-state functional connectivity in healthy volunteers. *Neuroimage, 57*, 1317–1323. doi: https://doi.org/10.1016/j.neuroimage.2011.05.051

Meng, C., Brandl, F., Tahmasian, M., Shao, J., Manoliu, A., Scherr, M., . . . Sorg, C. (2014). Aberrant topology of striatum's connectivity is associated with the number of episodes in depression. *Brain, 137*, 598–609. doi: https://doi.org/10.1093/brain/awt290

Meunier, D., Lambiotte, R., Fornito, A., Ersche, K. D., & Bullmore, E. T. (2009). Hierarchical modularity in human brain functional networks. *Front Neuroinform, 3*, 37. doi: https://doi.org/10.3389/neuro.11.037.2009

Muller, V. I., Cieslik, E. C., Serbanescu, I., Laird, A. R., Fox, P. T., & Eickhoff, S. B. (2017). Altered Brain Activity in Unipolar Depression Revisited: Meta-analyses of Neuroimaging Studies. *JAMA Psychiatry, 74*, 47–55. doi: https://doi.org/10.1001/jamapsychiatry.2016.2783

Murphy, K., Birn, R. M., & Bandettini, P. A. (2013). Resting-state fMRI confounds and cleanup. *Neuroimage, 80*, 349–359. doi: https://doi.org/10.1016/j.neuroimage.2013.04.001

Murphy, K., Dixon, V., LaGrave, K., Kaufman, J., Risinger, R., Bloom, A., & Garavan, H. (2006). A validation of event-related FMRI comparisons between users of cocaine, nicotine, or cannabis and control subjects. *Am J Psychiatry, 163*, 1245–1251. doi: https://doi.org/10.1176/appi.ajp.163.7.1245

Murray, J. D., & Anticevic, A. (2017). Toward understanding thalamocortical dysfunction in schizophrenia through computational models of neural circuit dynamics. *Schizophr Res, 180*, 70–77. doi: https://doi.org/10.1016/j.schres.2016.10.021

Normann, C., Schmitz, D., Furmaier, A., Doing, C., & Bach, M. (2007). Long-term plasticity of visually evoked potentials in humans is altered in major depression. *Biol Psychiatry, 62*, 373–380. doi: https://doi.org/10.1016/j.biopsych.2006.10.006

Phillips, M. L., Ladouceur, C. D., & Drevets, W. C. (2008). A neural model of voluntary and automatic emotion regulation: implications for understanding the pathophysiology and neurodevelopment of bipolar disorder. *Mol Psychiatry, 13*, 829, 833–857. doi: https://doi.org/10.1038/mp.2008.65

Poline, J. B., & Brett, M. (2012). The general linear model and fMRI: does love last forever? *Neuroimage, 62*, 871–880. doi: https://doi.org/10.1016/j.neuroimage.2012.01.133

Posner, J., Hellerstein, D. J., Gat, I., Mechling, A., Klahr, K., Wang, Z., . . . Peterson, B. S. (2013). Antidepressants normalize the default mode network in patients with dysthymia. *JAMA Psychiatry, 70*, 373–382. doi: https://doi.org/10.1001/jamapsychiatry.2013.455

Price, R. B., Lane, S., Gates, K., Kraynak, T. E., Horner, M. S., Thase, M. E., & Siegle, G. J. (2017). Parsing Heterogeneity in the Brain Connectivity of Depressed and Healthy Adults During Positive Mood. *Biol Psychiatry, 81*, 347–357. doi: https://doi.org/10.1016/j.biopsych.2016.06.023

Renner, F., Siep, N., Arntz, A., van de Ven, V., Peeters, F., Quaedflieg, C., & Huibers, M. J. H. (2017). Negative mood-induction modulates default mode network resting-state functional connectivity in chronic depression. *J Affect Disord, 208*, 590–596. doi: https://doi.org/10.1016/j.jad.2016.10.022

Satpute, A. B., Kragel, P. A., Barrett, L. F., Wager, T. D., & Bianciardi, M. (2018). Deconstructing arousal into wakeful, autonomic and affective varieties. *Neurosci Lett*. doi: https://doi.org/10.1016/j.neulet.2018.01.042

Satterthwaite, T. D., Ciric, R., Roalf, D. R., Davatzikos, C., Bassett, D. S., & Wolf, D. H. (2017). Motion artifact in studies of functional connectivity: Characteristics and mitigation strategies. *Hum Brain Mapp*. doi: https://doi.org/10.1002/hbm.23665

Satterthwaite, T. D., Cook, P. A., Bruce, S. E., Conway, C., Mikkelsen, E., Satchell, E., . . . Sheline, Y. I. (2016). Dimensional depression severity in women with major depression and post-traumatic stress disorder correlates with fronto-amygdalar hypoconnectivity. *Mol Psychiatry, 21*, 894–902. doi: https://doi.org/10.1038/mp.2015.149

Scheinost, D., Holmes, S. E., DellaGioia, N., Schleifer, C., Matuskey, D., Abdallah, C. G., . . . Esterlis, I. (2017). Multimodal Investigation of Network Level Effects Using Intrinsic Functional Connectivity, Anatomical Covariance, and Structure-to-Function Correlations in Unmedicated Major Depressive Disorder. *Neuropsychopharmacology*. doi: https://doi.org/10.1038/npp.2017.229

Schmaal, L., Hibar, D. P., Samann, P. G., Hall, G. B., Baune, B. T., Jahanshad, N., . . . Veltman, D. J. (2017). Cortical abnormalities in adults and adolescents with major depression based on brain scans from 20 cohorts worldwide in the ENIGMA Major Depressive Disorder Working Group. *Mol Psychiatry, 22*, 900–909. doi: https://doi.org/10.1038/mp.2016.60

Schmaal, L., Veltman, D. J., van Erp, T. G., Samann, P. G., Frodl, T., Jahanshad, N., . . . Hibar, D. P. (2016). Subcortical brain alterations in major depressive disorder: findings from the ENIGMA Major Depressive Disorder working group. *Mol Psychiatry, 21*, 806–812. doi: https://doi.org/10.1038/mp.2015.69

Shrout, P. E., & Fleiss, J. L. (1979). Intraclass correlations: uses in assessing rater reliability. *Psychol Bull, 86*, 420–428.

Smith, S. M., Fox, P. T., Miller, K. L., Glahn, D. C., Fox, P. M., Mackay, C. E., . . . Beckmann, C. F. (2009). Correspondence of the brain's functional architecture during activation and rest. *Proc Natl Acad Sci U S A, 106*, 13040–13045. doi: https://doi.org/10.1073/pnas.0905267106

Treadway, M. T., & Zald, D. H. (2011). Reconsidering anhedonia in depression: lessons from translational neuroscience. *Neurosci Biobehav Rev, 35*, 537–555. doi: https://doi.org/10.1016/j.neubiorev.2010.06.006

Trofimova, I., & Robbins, T. W. (2016). Temperament and arousal systems: A new synthesis of differential psychology and functional neurochemistry. *Neurosci Biobehav Rev, 64*, 382–402. doi: https://doi.org/10.1016/j.neubiorev.2016.03.008

Tsang, A., Lebel, C. A., Bray, S. L., Goodyear, B. G., Hafeez, M., Sotero, R. C., . . . Frayne, R. (2017). White Matter Structural Connectivity Is Not Correlated to Cortical Resting-State Functional Connectivity over the Healthy Adult Lifespan. *Front Aging Neurosci, 9*, 144. doi: https://doi.org/10.3389/fnagi.2017.00144

van den Heuvel, M., Mandl, R., Luigjes, J., & Hulshoff Pol, H. (2008). Microstructural organization of the cingulum tract and the level of default mode functional connectivity. *J Neurosci, 28*, 10844–10851. doi: https://doi.org/10.1523/JNEUROSCI.2964-08.2008

van Wingen, G. A., Tendolkar, I., Urner, M., van Marle, H. J., Denys, D., Verkes, R. J., & Fernandez, G. (2014). Short-term antidepressant administration reduces default mode and task-positive network connectivity in healthy individuals during rest. *Neuroimage, 88*, 47–53. doi: https://doi.org/10.1016/j.neuroimage.2013.11.022

Varikuti, D. P., Hoffstaedter, F., Genon, S., Schwender, H., Reid, A. T., & Eickhoff, S. B. (2017). Resting-state test-retest reliability of a priori defined canonical networks over different preprocessing steps. *Brain Struct Funct, 222*, 1447–1468. doi: https://doi.org/10.1007/s00429-016-1286-x

Versace, A., Andreazza, A. C., Young, L. T., Fournier, J. C., Almeida, J. R., Stiffler, R. S., . . . Phillips, M. L. (2014). Elevated serum measures of lipid peroxidation and abnormal prefrontal white matter in euthymic bipolar adults: toward peripheral biomarkers of bipolar disorder. *Mol Psychiatry, 19*, 200–208. doi: https://doi.org/10.1038/mp.2012.188

Wang, C., Ong, J. L., Patanaik, A., Zhou, J., & Chee, M. W. (2016). Spontaneous eyelid closures link vigilance fluctuation with fMRI dynamic connectivity states. *Proc Natl Acad Sci U S A, 113*, 9653–9658. doi: https://doi.org/10.1073/pnas.1523980113

Wang, L., Leonards, C. O., Sterzer, P., & Ebinger, M. (2014). White matter lesions and depression: a systematic review and meta-analysis. *J Psychiatr Res, 56*, 56–64. doi: https://doi.org/10.1016/j.jpsychires.2014.05.005

Wise, T., Radua, J., Nortje, G., Cleare, A. J., Young, A. H., & Arnone, D. (2016). Voxel-Based Meta-Analytical Evidence of Structural Disconnectivity in Major Depression and Bipolar Disorder. *Biol Psychiatry, 79*, 293–302. doi: https://doi.org/10.1016/j.biopsych.2015.03.004

Wise, T., Radua, J., Via, E., Cardoner, N., Abe, O., Adams, T. M., . . . Arnone, D. (2017). Common and distinct patterns of grey-matter volume alteration in major depression and bipolar disorder: evidence from voxel-based meta-analysis. *Mol Psychiatry, 22*, 1455–1463. doi: https://doi.org/10.1038/mp.2016.72

Yang, G. J., Murray, J. D., Repovs, G., Cole, M. W., Savic, A., Glasser, M. F., . . . Anticevic, A. (2014). Altered global brain signal in schizophrenia. *Proc Natl Acad Sci U S A, 111*, 7438–7443. doi: https://doi.org/10.1073/pnas.1405289111

Zuo, X. N., & Xing, X. X. (2014). Test-retest reliabilities of resting-state FMRI measurements in human brain functional connectomics: a systems neuroscience perspective. *Neurosci Biobehav Rev, 45*, 100–118. doi: https://doi.org/10.1016/j.neubiorev.2014.05.009

Brain Network Dysfunction in Bipolar Disorder: Evidence from Structural and Functional MRI Studies

Giuseppe Delvecchio, Eleonora Maggioni, Letizia Squarcina, and Paolo Brambilla

Introduction

Bipolar disorder (BD) is a chronic and disabling neuropsychiatric condition associated with increased rates of comorbidity and mortality. With an estimated lifetime prevalence of 1.5–2% in EU countries, it represents a significant burden to public health and society (Whiteford et al. 2013). This complex illness is characterized by recurrent changes of mood, thought, and behavior, being associated with depressive, hypomanic or manic, mixed, and euthymic phases. In addition to showing neurobiological and symptom fluctuations (Wessa et al. 2015), BD exhibits heterogeneous clinical features and variable degrees of cognitive deficits, which influence the patients' clinical and functional outcome (Martino et al. 2008; Phillips and Kupfer 2013). Recent evidence also suggests that BD may have a progressive course, impacting long-term morbidity and determining unfavorable outcomes (Vieta et al. 2018).

The identification of early symptoms of BD has relevant treatment implications but is made difficult by the presence of overlapping symptom profiles with other mood disorders, especially major depressive disorder (MDD), and major psychoses, especially schizophrenia (SCZ). Indeed, up to 60% of patients affected by BD are

G. Delvecchio · L. Squarcina
Department of Pathophysiology and Transplantation, University of Milan, Milan, Italy

E. Maggioni
Department of Neurosciences and Mental Health, Fondazione IRCCS Ca' Granda, Ospedale Maggiore Policlinico, Milan, Italy

P. Brambilla (✉)
Department of Pathophysiology and Transplantation, University of Milan, Milan, Italy

Department of Neurosciences and Mental Health, Fondazione IRCCS Ca' Granda, Ospedale Maggiore Policlinico, Milan, Italy
e-mail: paolo.brambilla1@unimi.it

V. A. Diwadkar, S. B. Eickhoff (eds.), *Brain Network Dysfunction in Neuropsychiatric Illness*, https://doi.org/10.1007/978-3-030-59797-9_15

initially misdiagnosed as MDD, and only 20% of them receive the correct diagnosis in the first year from the illness onset (Hirschfeld et al. 2003). In this context, neurobiological markers hold great promise as objective tools to (1) identify MDD subjects at risk of conversion to BD, allowing early diagnosis and prodromal intervention, (2) predict the rate of disease progression, and (3) characterize the likelihood of treatment response.

Over the recent decades, sophisticated noninvasive brain imaging techniques have increasingly been employed in the study of the neurobiological bases of psychiatric illnesses. Among them, magnetic resonance imaging (MRI) has revolutionized our knowledge of the pathophysiology of BD, being able to noninvasively provide comprehensive information on brain anatomy and function with an unprecedented level of detail (Birur et al. 2017). Until now, the results of structural and functional MRI studies on BD have been inconsistent, leaving the neurobiological determinants of this complex illness largely unknown. Nonetheless, neuroimaging-based markers of BD have the potential to exert an impact on the official classificatory systems for mental disorders (Altamura et al. 2018; Maggioni et al. 2017b; Vieta et al. 2018).

In the present chapter, we will provide an overview of the most relevant findings of the MRI studies on BD. We will focus on both neuroanatomical alterations – in terms of gray matter volume and white matter diffusion – and neurofunctional deficits, linked to emotional processing, associated with BD.

Neuroanatomical Profile of Bipolar Disorder

In recent years, several structural MRI (sMRI) studies have inspected brain morphology in individuals with BD in comparison with healthy subjects, suggesting a link between specific gray matter (GM) abnormalities and this complex disorder (Arnone et al. 2009; Brambilla et al. 2005; Brambilla and Tansella 2009; Hibar et al. 2016, 2018). Nonetheless, there is still uncertainty on the brain areas, and corresponding circuitries, which are primarily involved in BD pathophysiology. Overall agreement among the neuroanatomical findings on BD is limited, at least in part, by methodological differences, heterogeneity of symptom profiles and cycles, as well as modest sample size characterizing most MRI studies.

The statistical power to detect diagnosis-related morphological alterations was recently enhanced by meta-analyses, which quantitatively combined the results from multiple studies on BD (Arnone et al. 2009; Kempton et al. 2008; McDonald et al. 2004; Selvaraj et al. 2012). This integrated approach usually accounts for potential confounding sources but still has limited capability in controlling study-specific random effects. This issue was partially addressed by mega-analytic techniques, which pooled relevant individual patient data across multiple centers and jointly analyzed them to extract significant characteristics. The largest mega-analytic studies to date were performed by the ENIGMA BD consortium, which coordinated the analyses of 20 and more sites and analyzed the resulting neuroanatomical parameters in a unique design (Hibar et al. 2016, 2018). To reduce post-processing biases, other studies have collected post hoc subject-level raw imaging data from multiple sites and centralized all the analyses (Ivleva et al. 2013; Maggioni et al. 2017b).

In this section, we provide a critical appraisal of the most robust knowledge of brain morphological correlates of BD by reporting the relevant findings from meta- and mega-analyses as well as large sample sMRI studies involving BD patients and healthy controls (HC). Evidence from voxel-based and region-based studies, concerning both subcortical and cortical regions, will be summarized. Possible effects of medication, duration of illness, and symptom dimensions on brain morphology will also be briefly discussed.

Regarding global brain measures, most of the studies found no significant differences in total cerebral volume in BD patients compared to HC (Hallahan et al. 2011; Hibar et al. 2016; Kempton et al. 2008; McDonald et al. 2004), confirming previous literature (Brambilla et al. 2005). An exception is the meta-analysis of Arnone et al. which described a global reduction of brain volume in BD (Arnone et al. 2009). This discrepancy may be explained in terms of lithium treatment and the stage of illness heterogeneity across studies. Indeed, Hallahan et al. (2011) showed that BD patients not taking lithium had brain volume deficits compared to lithium-treated patients and HC; conversely, cerebral volume was negatively correlated with duration of illness, suggesting that BD course is accompanied by neurodegenerative processes (Hallahan et al. 2011). Specifically, there are suggestions that illness progression mainly affects total GM volume, possibly reflecting abnormal cellular plasticity (Frey et al. 2008). Results from twin studies similarly suggest that GM volume changes are related to environmental or disease factors rather than being genetically determined (Squarcina et al. 2016).

With respect to subcortical structures, lateral ventricular enlargement has often been associated with BD (Hibar et al. 2016; Kempton et al. 2008; Rimol et al. 2010). Specifically, increased volume in the right lateral ventricle was repeatedly reported (Hallahan et al. 2011; Kempton et al. 2008; McDonald et al. 2004). The enlargement of the contralateral ventricle was found only in the study carried out by Arnone et al. (2009), whereas increased third ventricle size emerged in two investigations (Kempton et al. 2008; Rimol et al. 2010). Findings on the pituitary gland from original studies are variable, showing either increased (Takahashi et al. 2009) or decreased (Delvecchio et al. 2018; Sassi et al. 2001) volumes associated with BD. Overall, although the meta-analysis from Arnone et al. found no significant alterations (Arnone et al. 2009), a recent review from our group suggested a pituitary enlargement in BD (Delvecchio et al. 2017). Such heterogeneity of results may be related to gender, a factor that has been repeatedly shown to affect pituitary size (Arnone et al. 2009; Delvecchio et al. 2018; Takahashi et al. 2009).

Posterior fossa structures were suggested to be implicated in BD, but cerebellar and vermal structures have been poorly investigated (Brambilla et al. 2001a; Mills et al. 2005), and results are not universal, with big sample studies reporting reduced cerebellar volume (Rimol et al. 2010) and meta-analyses reporting no abnormalities (Kempton et al. 2008). Studies of corpus callosum size agreed on the presence of reduced callosal area in BD patients compared to HC (Brambilla et al. 2003; Arnone et al. 2009; Kempton et al. 2008). However, evidence from pediatric studies suggests that callosal deficits may be progressive during the illness course, being more pronounced in adult BD patients (Baloch et al. 2009).

The few studies on basal ganglia volumes in BD (Brambilla et al. 2001b; Hwang et al. 2006; Womer et al. 2014) suggested an involvement of lenticular nuclei (i.e., putamen and globus pallidus) in BD pathophysiology. Indeed, collaborative studies have shown increased volume in bilateral globi pallidi (Arnone et al. 2009) and right putamen (Hallahan et al. 2011) in BD patients compared to HC. This evidence, however, was not confirmed by the subcortical mega-analysis from the ENIGMA BD consortium (BD, n=1710; HC, n=2594) (Hibar et al. 2016). Consistently, no differences in caudate volume were described (Arnone et al. 2009; Hallahan et al. 2011; Hibar et al. 2016; Kempton et al. 2008; Rimol et al. 2010).

A larger number of research works investigated limbic structures, including hippocampus and amygdala, due to their key role in mood and emotion modulation and their possible involvement in BD symptomatology (Brambilla et al. 2008; Houenou et al. 2012). Although the results on the amygdala from single MRI studies are heterogeneous, both meta-analyses and multicenter studies agreed on the absence of significant volumetric alterations in this region in BD patients compared to HC (Arnone et al. 2009; Hallahan et al. 2011; Kempton et al. 2008; Maggioni et al. 2017b; Selvaraj et al. 2012). Only the ENIGMA study found a trending significant reduction in the amygdala of BD patients compared to HC (Hibar et al. 2016). The results on the hippocampus are more conflicting, with some studies reporting hippocampal deficits (Hibar et al. 2016; Rimol et al. 2010) but others showing no volumetric differences in this region (Arnone et al. 2009; Hallahan et al. 2011; Ivleva et al. 2013; Kempton et al. 2008; Maggioni et al. 2017b; McDonald et al. 2004; Selvaraj et al. 2012). Interestingly, a positive relation between amygdala and hippocampal volumes and lithium treatment has been documented, suggesting that the possible neuroprotective effects of lithium may have masked volume deficits in these limbic structures that could be related to BD pathogenesis (Hallahan et al. 2011).

Only a few studies have showed thalamic deficits, either bilateral (Hibar et al. 2016) or in the left hemisphere (Rimol et al. 2010) in BD patients compared to HC, whereas most have shown no significant abnormalities in thalamic volume (Arnone et al. 2009; Hallahan et al. 2011; Ivleva et al. 2013; Kempton et al. 2008; McDonald et al. 2004). This heterogeneity may be explained by the sex-by-diagnosis interactions emerged in this region, with female BD patients having increased thalamus volume (Hibar et al. 2016), and by the reported link between earlier age of onset and deficits in the left thalamus (Hallahan et al. 2011).

The numerous studies on cortical GM morphology in BD led to results that are of difficult harmonization. The recent cortical mega-analysis from the ENIGMA BD group (BD, n=2447; HC, n=4056) showed cortical thinning in pars opercularis, fusiform gyrus, and rostral middle frontal cortex of the left hemisphere. Interestingly, BD was characterized by a broad pattern of cortical thinning, exacerbated by illness duration, but not of surface area deficits (Hibar et al. 2018). Widespread cortical thinning was also described by Rimol et al. (2010), specifically in the bilateral frontal lobe, as well as in left superior temporal gyrus, left inferior parietal gyrus, and right fusiform gyrus.

The results concerning cortical GM volume are partially consistent, confirming a possible role of prefrontal and temporal regions in BD pathophysiology. Region-based studies highlighted either no significant differences (Kempton et al. 2008; McDonald et al. 2004) or temporal lobe deficits, especially on the left side (Hallahan et al. 2011), in BD patients compared to HC. Conversely, meta-analyses of voxel-based morphometry (VBM) studies showed diagnosis-specific deficits in clusters within anterior cingulate/medial frontal cortices (Bora et al. 2012), insula/inferior frontal region (Bora et al. 2012; Selvaraj et al. 2012), as well as temporal gyrus and claustrum (Selvaraj et al. 2012). In a recent multicenter VBM study conducted by our group, BD patients exhibited lower GM volume in anterior cingulate, superior temporal, and calcarine cortices of the right hemisphere (Maggioni et al. 2017b). Volume deficits in frontal, cingulate, insular, and temporal regions were also described in a psychotic BD sample (Ivleva et al. 2013). In this respect, lifetime psychosis seems to be associated with lower GM volume in the right temporoparietal cortex and with more extended and enhanced age-related GM volume deficits (Altamura et al. 2018).

Taken together, these results corroborate the hypothesis that BD is characterized by brain morphology alterations, with ventricular enlargement, callosal, prefrontal, and temporal deficits emerging as the most consistently reported findings (Fig. 1). Nonetheless, the evidence is still too heterogeneous and reflects the complexity of this disorder, the criteria for which may need to be revised based on pathophysiology in the framework of Research Domain Criteria (RDoC) (Insel et al. 2010). The RDoC proposes an innovative conceptualization of psychiatric diseases based not only on symptoms but also on biological and psychosocial features (Insel 2014). Thus, the classification of BD as one diagnostic category should be reviewed, because its phenotypic variability may reflect distinct subtypes of the illness, each with a specific neurobiological architecture. Recent interest has been driven toward the psychotic dimension, which is common to BD and SCZ and may represent an intermediate phenotype across multiple categorical diagnoses (Altamura et al. 2018; Maggioni et al. 2017a, b). Future confirmatory studies on more homogeneous samples are needed, together with multimodal analytical approaches that can translate the evidence of local regional abnormalities into a more complex and integrated framework of brain network dysfunction.

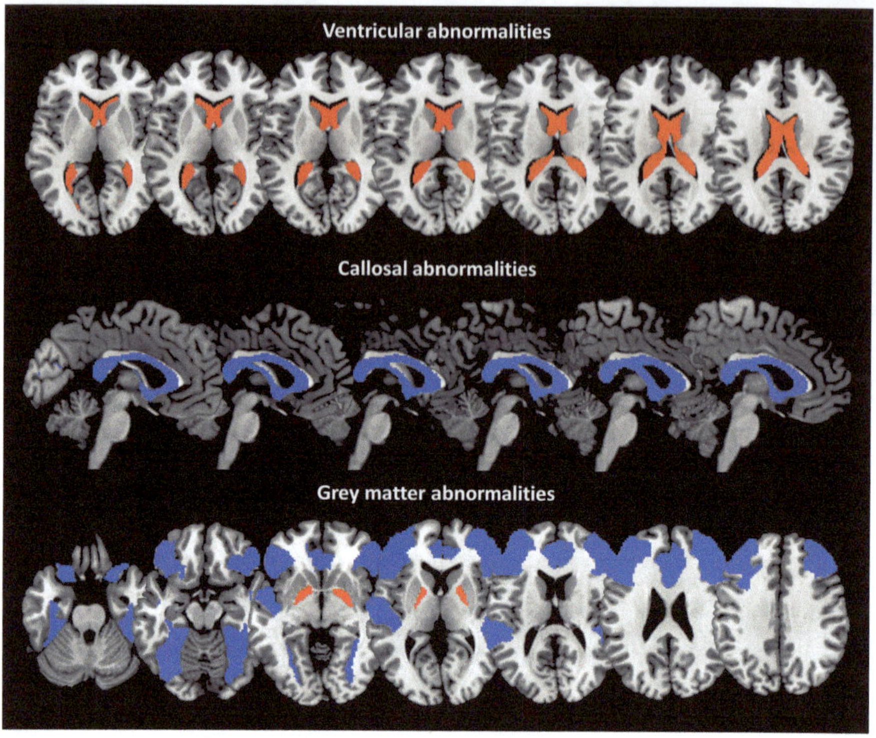

Fig. 1 Representation of the most consistent structural abnormalities in BD. Red and blue colors indicate regions with volumetric increase and decrease, respectively. Upper panel: lateral ventricles. Central panel: corpus callosum. Bottom panel: subcortical and cortical gray matter structures, highlighting (1) deficits in ventral and dorsolateral prefrontal cortex, anterior cingulate cortex, superior temporal, and fusiform gyri and (2) globus pallidus enlargement.

White Matter Alterations and Disconnectivity in Bipolar Disorder Investigated with Diffusion Tensor Imaging

With diffusion tensor imaging (DTI), it is possible to noninvasively investigate brain microstructural integrity, based on the behavior of water movement in tissues (Le Bihan et al. 1986). The organization and orientation of white matter tracts can be inferred through the estimation of specific diffusion indexes interconnected with each other (Basser and Pierpaoli 1996), thanks to a specific MRI pulse sequence sensitive to water motion. Indeed, free water moves without a preferential angle, in an isotropic manner, while when diffusion is hindered by obstacles, water motion occurs along definite directions: in this case, the diffusion is defined as anisotropic (Basser and Pierpaoli 1996). Diffusion is anisotropic in brain white matter, occurring faster along the direction of fibers than across them; the degree of anisotropy can be measured with parameters extracted from the diffusion tensor. Specifically, fractional anisotropy (FA) measures how much one direction is prevalent on the others, whereas mean diffusivity (MD) represents the magnitude of the diffusion. Axial (AD) and radial diffusivity (RD) measure the amount of diffusion that occurs along the principal axis of the diffusion tensor and perpendicularly to it (Alexander et al. 2007).

It has been demonstrated that, along with brain structure, brain tissue microstructure is affected in BD (Nortje et al. 2013). Because of the intrinsic characteristics of diffusion, DTI studies generally focus on white matter. Changes in diffusion reflect modifications in tissue integrity: fiber degradation as a result of trauma, demyelination, or Wallerian degeneration results in modified water diffusion. For example, Wallerian degeneration, characterized by myelin disruption, shows a reduction in measured FA, caused by a decrease of diffusion parallel to the fiber and an increase of perpendicular diffusion (Beaulieu 2002). Studies investigating BD consistently showed reduced FA and increased MD, widespread in all the three categories of tracts (commissural, association, and projection) (Nortje et al. 2013). Although findings are not always in agreement (Bellani et al. 2009), a majority of DTI studies detected a decrease in FA, which could be related to disruption of white matter. In fact, high FA values in white matter reflect highly organized and normally myelinated axons, while lower FA reflect probable damage to axonal membrane, demyelination, or reduced amount of intra-axonal structures (Beaulieu 2002).

DTI studies are conducted using either circumscribed regions of interest (ROIs), the whole brain, or white matter tracts. The first DTI studies were carried out at a ROI level, considering frontal and prefrontal areas (Adler et al. 2004; Beyer et al. 2005). The ROI approach considers a specific brain region, or a set of definite regions, in which DTI properties are evaluated, disregarding the rest of the brain. White matter alterations were found in various white matter areas, covering broadly the whole brain. For example, reduced FA was reported in frontal white matter tracts (Adler et al. 2006), corpus callosum (Oertel-Knöchel et al. 2014; Bellani et al. 2009), fornix (Oertel-Knöchel et al. 2014), corona radiata (Bauer et al. 2015), internal capsule, and uncinate fasciculi (Sussmann et al. 2009). The restriction of analyses to a region has the advantage of reducing noise, for example, constraining the region in the white matter, where the DTI signal is more reliable, and of exploiting information that is known a priori. Thus, focusing the analyses on corpus callosum characteristics allows testing hypotheses regarding disconnection in interhemispheric communication. However, the analysis of the whole brain has soon become prevalent as it promotes the search for possible brain alterations without restricting analyses to a priori hypotheses (Smith et al. 2006). The standard way to operate on whole brain level is voxel-wise statistical analysis, where data are registered to a common template and compared with the aim of identifying correlations with covariates of interest (e.g., disability scores, group). Diffusion data are usually analyzed using voxel-based analysis (VBA), after corrections for multiple comparisons and noise, and Tract-Based Spatial Statistics (TBSS, Smith et al. 2006), which overcomes limitations due to misalignment, partial volume, and registration. Using TBSS, only the tract centers are considered, using a fully automated method, which creates a template of the fiber bundles common to the subjects participating in a study. Then, the analyses are focused in the centers of the tracts, shared by the whole sample.

Studies performed using whole brain techniques showed an involvement of all the major white matter tracts. Decreased FA and increased MD in BD were found in frontal (Oertel-Knöchel et al. 2014; Yip et al. 2013), parietal, and occipital white matter (Kumar et al. 2015), in the corpus callosum (Bellani et al. 2009; Oertel-Knöchel et al. 2014; Yip et al. 2013), in the fornix (Oertel-Knöchel et al. 2014; Kurumaji et al. 2017), and in the corona radiata and internal capsule (Bauer et al. 2015). Interestingly, a minority of studies (Wessa et al. 2009; Bruno et al. 2008) found increased FA in frontal, parietal, temporal, and occipital areas, suggesting an increased coherence of fibers during remission or induced by medication. The degree of microstructure alteration has been shown to relate with the status of patients, with a gradient from the euthymic to the manic phase (Magioncalda et al. 2016), with cognitive symptoms and performance, and the use of medication (Haarman et al. 2016; Bauer et al. 2015). Notably, white matter alterations found in BD reveal a white matter disconnectivity, which is similarly observed in SCZ (Kumar et al. 2015; Squarcina et al. 2017a), identifying common neural factors distinctive of psychosis.

DTI also allows to track white matter fibers, following their pathways in the brain, with the possibility of evaluating fiber bundles specific for each subject participating in a study. Investigating their connections and shape can shed light on brain structure and connectivity. For example, Sun et al. (2017) found significant differences in the shape of cingulum, uncinate, and arcuate fasciculi in BD compared to controls, after reconstructing tracts. Other studies focused on DTI indexes obtained inside the reconstructed tracts (Sarrazin et al. 2014; Sprooten et al. 2016; Toteja et al. 2015) and typically found decreased FA and increased MD, confirming findings obtained at a whole brain level. Overall, studies using DTI showed an involvement of white matter in BD, with a correlation between tissue alterations and clinical features. The whole brain is affected in terms of white matter integrity, with

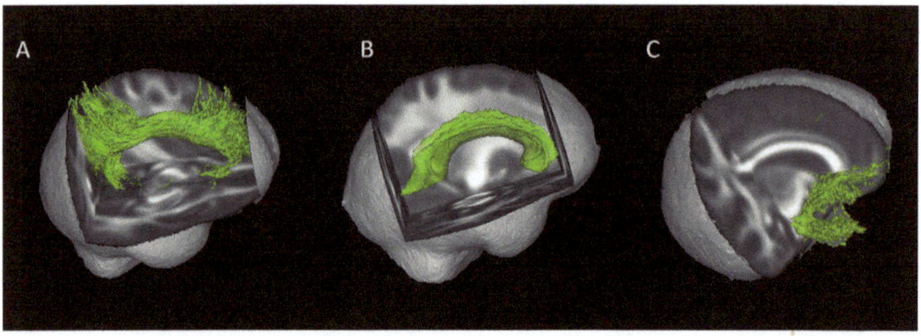

Fig. 2 Three-dimensional representation of the brain FA values and the tracts reconstructed using probabilistic tractography. Diffusion tensor imaging techniques allow the identification of disrupted white matter regions and tracts. Among the regions and tracts whose integrity results most altered are part of the limbic circuitry, as the arcuate fasciculus (panel A), the corpus callosum (panel B) and the prefrontal tracts, and the uncinate fasciculus (panel C)

predominant involvement of callosal, limbic, and prefrontal-limbic tracts, such as cingulum and uncinate (Houenou et al. 2015), indicating a disruptive effect of BD on tissue organization, highlighting the presence of anatomic disconnectivity induced by this disease (Fig. 2).

Facial Emotional Processing in Bipolar Disorder: Evidence from Functional Magnetic Resonance Imaging Studies in both Adults and Pediatric Populations

In recent decades, the study of emotional processing in mood disorders, including BD, has been a target of investigation for several functional magnetic imaging (fMRI) studies (Delvecchio et al. 2012). Specifically for BD, this is not surprising especially because BD is characterized by severe mood swings, which are often accompanied by cognitive, behavioral, and social disturbances. Interestingly, the majority of fMRI studies that investigated the emotional processing in BD patients used facial affect recognition tasks, especially because emotional facial expressions are important for social communication and interpersonal interactions. Therefore, in this section we will provide a critical appraisal of the major findings observed to date in patients with BD while performing facial emotional processing tasks, in both pediatric and adult populations.

FMRI Evidence in Adults BD Patients

In recent decades, fMRI studies investigating facial emotional processing in adults BD patients showed the presence of selective brain dysfunctions in several brain regions within subcortical-prefrontal networks and associated limbic modulating regions, regardless of mood phase, compared to healthy controls (HC) (Townsend and Altshuler 2012). Interestingly, these structures have been also found to be engaged in normal populations (Fusar-Poli et al. 2009), mainly because they modulate adaptive responses to shifts in emotional stimuli. However, there has been a persisting debate about whether these brain regions, alone or in combination, may have different roles in determining the type or symptoms associated with the various phases of BD. Specifically for manic symptoms, the majority of fMRI studies reported decreased activation of limbic regions while processing negative faces. These effects were observed mainly in the amygdala and ventral anterior cingulate cortex (vACC), in complement with hyper-activation of prefrontal regions, including dorsolateral prefrontal cortex (DLPFC), superior frontal gyrus, middle temporal gyrus, insula, striatum, as well as posterior and anterior cingulate cortex while processing of negative faces (Lennox et al. 2004; Chen et al. 2006; Hulvershorn et al. 2012). An exception to the latter set of results was a study by Liu et al. (2012), who reported decreased activation in DLPFC in manic BD patients compared to HC. Additionally, in manic BD patients, Chen et al. (2006) found increased activation in fusiform gyrus while processing of sad facial expressions.

With regard to depressed BD patients, it has been reported that these patients showed hyper-activity in fronto-striato-thalamic systems, which mainly include dorsal/ventral prefrontal, putamen, and thalamic regions (Chen et al. 2006), as well as amygdala (Grotegerd et al. 2014) in response to happy faces. Interestingly, amygdala (Almeida et al. 2010; Hulvershorn et al. 2012) and ventrolateral prefrontal (VLPFC)/orbitofrontal (Liu et al. 2012) hyper-activation has also been observed during the processing of negative valenced emotions. Moreover, Altshuler et al. (2008) suggested reduced activations in the VLPFC and DLPFC as well as increased activation in the lateral orbitofrontal cortex in response to faces with negative emotions. Interestingly, another study from the same research group also reported decreased activity in VLPFC, amygdala, and striatal regions in depressed BD patients compared to HC while viewing match emotions versus match forms (Vizueta et al. 2012). Finally, Lawrence et al. (2004) reported a positive correlation with the degree of depression and increased activation in the hippocampus while viewing sad facial affects.

For euthymic BD patients, a mixed picture of hypo- and hyper-activation in selective brain regions has been observed in response to either positive or negative emotions. Specifically, decreased activation in DLPFC, precuneus, and ACC in response to negative and positive emotions (Hassel et al. 2008; Dima et al. 2016; Tesli et al. 2015) as well as increased superior temporal gyrus and inferior parietal gyrus in response to fearful faces (Malhi et al. 2007) has been reported. Additionally, Surguladze et al. (2010) suggested increased medial prefrontal cortex activation in response to both fearful and happy faces as well as increased activation of the amygdala in response to happy faces. In contrast, Foland-Ross et al. (2012) reported that euthymic patients showed a significant decrease in activation of the VLPFC, insula, putamen, and lingual gyrus compared to HC during emotion labeling. Further, three studies also found hyper-activation in striatal regions in response to either positive or negative emotions (Surguladze et al. 2010; Hassel et al. 2008; Hulvershorn et al. 2012). Interestingly, fMRI studies investigating the impact of selective risk genes on emotional processing showed some interesting results. Specifically, Dima et al. (2013) reported that BD carriers of either the ANK3 or CACNA1C risk variants showed hypo-activation of VLPFC while processing of negative valenced emotions. Finally, Tesli et al. (2015) showed that the polygenic risk scores were positively correlated with increased right inferior frontal gyrus activation during negative face processing.

In conclusion, these sampled fMRI studies confirm the presence of a selective disruption of brain regions within the cortico-limbic network not only in BD patients during acute manic or depressed symptomatology but also during remission. Specifically, most of the studies reported increased activation of prefrontal regions and decreased activation of limbic areas, mainly the amygdala, in BD patients, regardless of mood phase, compared to HC (Fig. 3). However, future studies are

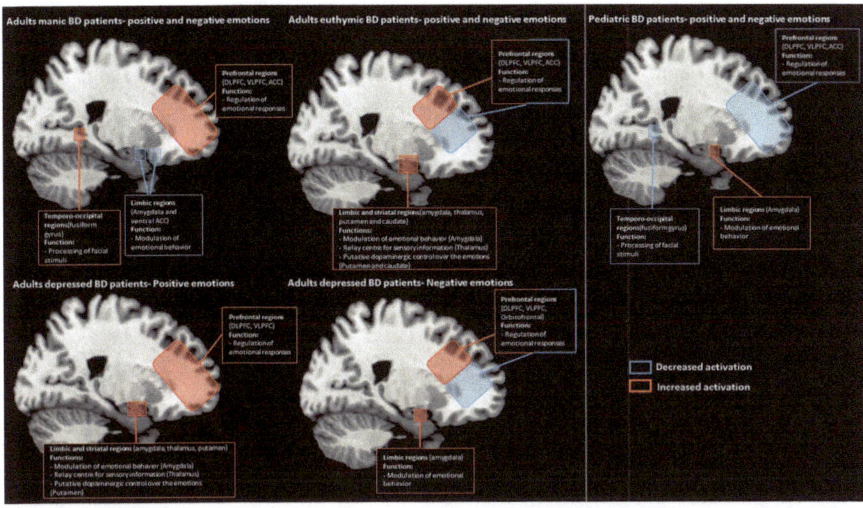

Fig. 3 Representation of the most consistent functional alterations found in adult (in euthymic, manic, and depressive phases) and pediatric BD patients during processing of facial emotional processing tasks

warranted to further investigate and clarify the variation in the pattern of activation associated with the specific emotional valance and different mood states.

FMRI Evidence in Pediatric BD Patients

Several fMRI studies explored emotional processing deficits in youth with BD. Overall these studies are of paramount importance for identifying biomarkers that might predict the development and the clinical trajectories of BD. Interestingly, the majority of the fMRI studies in pediatric cohorts reported increased activation in amygdala while processing fearful or angry faces (Kim et al. 2012; Pavuluri et al. 2007; Brotman et al. 2014) and happy faces (Ladouceur et al., 2011). In contrast, one study carried out by Deveney et al. (2014) found that BD patients failed to recruit the amygdala in response to faces with increasing expressions of anger. Furthermore, another consistent finding that emerged from the available evidence in pediatric BD patients is the hypo-activation of brain regions within the prefrontal cortex, mainly the DLPFC and VLPFC, while processing of happy or fearful faces (Ladouceur et al. 2011; Brotman et al. 2014; Diler et al. 2013; Hafeman et al. 2014; Pavuluri et al. 2007). Finally, one study also showed hypo-activation of the fusiform gyrus, especially while processing angry faces, in pediatric BD patients (Perlman et al. 2013).

This evidence, consistent with the sizable volume of results observed in adult BD patients, shows the presence of deficits within the top-down emotional control system, with increased activation in subcortical regions, mainly the amygdala, and decreased activation of prefrontal areas, ultimately suggesting that deficits within the affective circuits are already present early in the pathogenesis of bipolar illness and might be the results of developmental perturbations occurring early in life (Fig. 3).

The evidence sampled thus far firmly establishes the presence of functional abnormalities underlying affective dysregulation in BD. Indeed, the high degree of convergence across studies, despite the presence of sample and methodological differences, further supports the validity of the evidence to date. Specifically, the available studies produced significant results pointing toward the presence of fronto-limbic neural dysfunctions in both adults and pediatric BD patients, during the symptomatic phases of the illness and persisting also during remission. However, it is important to point out that in these studies, especially those involving adults, BD patients were taking different pharmacological medications, which may have influenced changes in regional activity and might explain some of the variability observed. Indeed, a growing body of evidence suggests that chronic antidepressant, antipsychotic, or mood stabilizer administration may result in visible functional changes. Therefore, future studies with more sophisticated design and homogenous samples are needed to generate a more valid neurobiological model of emotions in BD. This is because fMRI evidence provides valuable insight into the pathophysiology of this disabling illness, which may be potentially beneficial for earlier disease detection, more accurate diagnosis, and more targeted pharmacological interventions.

Conclusions

The morpho-functional and connectivity evidence reported above points toward the hypothesis that BD is a brain disorder characterized by deficits in selective brain networks, which may ultimately explain the specific clinical and cognitive alterations of this debilitating illness. Indeed, it is now evident that the roots of BD phenotypes lie in selective neurobiological disruptions, including emotional and impulse dyscontrol (Wessa et al. 2015; Swann et al. 2009). Moreover, several strands of evidence reported the presence of cognitive deficits in BD, regardless of clinical subtypes (Bora 2018) and of illness phase (Cipriani et al. 2017). Notably, a majority of the cognitive domains often found to be altered in BD, namely, attention, executive functions, and learning (Martínez-Arán et al. 2004; Brambilla et al. 2011, 2013) as well as language (Perlini et al. 2012; Raucher-Chéné et al. 2017), are strictly linked to prefronto-temporal functioning, which has consistently been found to be impaired in BD both by structural and functional neuroimaging studies.

Nonetheless, despite a tentative neurobiological model of BD being delineated, the mechanisms underpinning this illness have not been fully elucidated yet. Therefore, in recent years, the scientific community has driven its attention toward the investigation of the association between specific genetic variants and brain integrity in BD (Mandolini et al. 2018; Dima et al. 2016; Squarcina et al. 2017b) and in twin studies (Fagnani et al. 2014; Sugihara et al. 2017), trying to better understand gene-environment (GxE) interactions in the pathophysiology of BD. However, it is important to point out that more homogenous sample and methodological approaches are needed to better delineate the neurobiological signatures of BD (Dusi et al. 2018; Houenou et al. 2015). For instance, the majority of neuroimaging studies have enrolled BD patients with chronic pharmacological treatments, which may in part explain the heterogeneity observed in the findings across individual studies. Indeed, it has consistently been suggested that medication might influence neuroimaging results, especially for sMRI investigations (Hafeman et al. 2012), given evidence that lithium is associated with increased GM volumes in key regions involved in mood regulation (Hafeman et al. 2012; Sassi et al. 2002), supporting its role as neuroprotective agent (Hajek and Weiner 2016). Clearly structural and functional imaging has done much to elucidate the network underpinnings of BD, but much work still remains.

References

Altamura, A.C., Maggioni, E., Dhanoa, T., Ciappolino, V., Paoli, R.A., Cremaschi, L., Prunas, C., Orsenigo, G., Caletti, E., Cinnante, C.M., Triulzi, F.M., Dell'Osso, B., Yatham, L., Brambilla, P., 2018. The impact of psychosis on brain anatomy in bipolar disorder: A structural MRI study. J Affect Disord 233, 100–109.

Alexander, A.L., Lee, J.E., Lazar, M., Field, A,S., 2007. Diffusion tensor imaging of the brain. Neurotherapeutics. 4(3), 316–329. https://doi.org/10.1016/j.nurt.2007.05.011. PMID: 17599699; PMCID: PMC2041910.

Arnone, D., Cavanagh, J., Gerber, D., Lawrie, S.M., Ebmeier, K.P., McIntosh, A.M., 2009. Magnetic resonance imaging studies in bipolar disorder and schizophrenia: meta-analysis. Br J Psychiatry 195, 194–201.

Baloch, H.A., Brambilla, P., Soares, J.C., 2009. Corpus callosum abnormalities in pediatric bipolar disorder. Expert Rev Neurother 9, 949–955.

Birur, B., Kraguljac, N.V., Shelton, R.C., Lahti, A.C., 2017. Brain structure, function, and neurochemistry in schizophrenia and bipolar disorder-a systematic review of the magnetic resonance neuroimaging literature. NPJ Schizophr 3, 15.

Bora, E., Fornito, A., Yucel, M., Pantelis, C., 2012. The effects of gender on grey matter abnormalities in major psychoses: a comparative voxelwise meta-analysis of schizophrenia and bipolar disorder. Psychol Med 42, 295–307.

Brambilla, P., Glahn, D.C., Balestrieri, M., Soares, J.C., 2005. Magnetic resonance findings in bipolar disorder. Psychiatr Clin North Am 28, 443–467.

Brambilla, P., Harenski, K., Nicoletti, M., Mallinger, A.G., Frank, E., Kupfer, D.J., Keshavan, M.S., Soares, J.C., 2001a. MRI study of posterior fossa structures and brain ventricles in bipolar patients. J Psychiatr Res 35, 313–322.

Brambilla, P., Harenski, K., Nicoletti, M.A., Mallinger, A.G., Frank, E., Kupfer, D.J., Keshavan, M.S., Soares, J.C., 2001b. Anatomical MRI study of basal ganglia in bipolar disorder patients. Psychiatry Res 106, 65–80.

Brambilla, P., Hatch, J.P., Soares, J.C., 2008. Limbic changes identified by imaging in bipolar patients. Curr Psychiatry Rep 10, 505–509.

Brambilla P, Nicoletti MA, Sassi RB, Mallinger AG, Frank E, Kupfer DJ, Keshavan MS, Soares JC. Magnetic resonance imaging study of corpus callosum abnormalities in patients with bipolar disorder. Biol Psychiatry. 2003 Dec 1;54(11):1294–7.

Brambilla, P., Tansella, M., 2009. What are the perspectives of human brain mapping in the field of bipolar disorder? Int Rev Psychiatry 21, 295–296.

Delvecchio, G., Altamura, A.C., Soares, J.C., Brambilla, P., 2017. Pituitary gland in Bipolar Disorder and Major Depression: Evidence from structural MRI studies. J Affect Disord 218, 446–450.

Delvecchio, G., Mandolini, G.M., Perlini, C., Barillari, M., Marinelli, V., Ruggeri, M., Altamura, A.C., Bellani, M., Brambilla, P., 2018. Pituitary gland shrinkage in bipolar disorder: The role of gender. Compr Psychiatry 82, 95–99.

Frey, B.N., Zunta-Soares, G.B., Caetano, S.C., Nicoletti, M.A., Hatch, J.P., Brambilla, P., Mallinger, A.G., Soares, J.C., 2008. Illness duration and total brain gray matter in bipolar disorder: evidence for neurodegeneration? Eur Neuropsychopharmacol 18, 717–722.

Hallahan, B., Newell, J., Soares, J.C., Brambilla, P., Strakowski, S.M., Fleck, D.E., Kieseppa, T., Altshuler, L.L., Fornito, A., Malhi, G.S., McIntosh, A.M., Yurgelun-Todd, D.A., Labar, K.S., Sharma, V., MacQueen, G.M., Murray, R.M., McDonald, C., 2011. Structural magnetic resonance imaging in bipolar disorder: an international collaborative mega-analysis of individual adult patient data. Biol Psychiatry 69, 326–335.

Hibar, D.P., Westlye, L.T., Doan, N.T., Jahanshad, N., Cheung, J.W., Ching, C.R.K., Versace, A., Bilderbeck, A.C., Uhlmann, A., Mwangi, B., Kramer, B., Overs, B., Hartberg, C.B., Abe, C., Dima, D., Grotegerd, D., Sprooten, E., Boen, E., Jimenez, E., Howells, F.M., Delvecchio, G., Temmingh, H., Starke, J., Almeida, J.R.C., Goikolea, J.M., Houenou, J., Beard, L.M., Rauer, L., Abramovic, L., Bonnin, M., Ponteduro, M.F., Keil, M., Rive, M.M., Yao, N., Yalin, N., Najt, P., Rosa, P.G., Redlich, R., Trost, S., Hagenaars, S., Fears, S.C., Alonso-Lana, S., van Erp, T.G.M., Nickson, T., Chaim-Avancini, T.M., Meier, T.B., Elvsashagen, T., Haukvik, U.K., Lee, W.H., Schene, A.H., Lloyd, A.J., Young, A.H., Nugent, A., Dale, A.M., Pfennig, A., McIntosh, A.M., Lafer, B., Baune, B.T., Ekman, C.J., Zarate, C.A., Bearden, C.E., Henry, C., Simhandl, C., McDonald, C., Bourne, C., Stein, D.J., Wolf, D.H., Cannon, D.M., Glahn, D.C., Veltman, D.J., Pomarol-Clotet, E., Vieta, E., Canales-Rodriguez, E.J., Nery, F.G., Duran, F.L.S., Busatto, G.F., Roberts, G., Pearlson, G.D., Goodwin, G.M., Kugel, H., Whalley, H.C., Ruhe, H.G., Soares, J.C., Fullerton, J.M., Rybakowski, J.K., Savitz, J., Chaim, K.T., Fatjo-Vilas, M., Soeiro-de-Souza, M.G., Boks, M.P., Zanetti, M.V., Otaduy, M.C.G., Schaufelberger, M.S., Alda, M., Ingvar, M., Phillips, M.L., Kempton, M.J., Bauer, M., Landen, M., Lawrence, N.S., van Haren, N.E.M., Horn, N.R., Freimer, N.B., Gruber, O., Schofield, P.R., Mitchell, P.B., Kahn, R.S., Lenroot, R., Machado-Vieira, R., Ophoff, R.A., Sarro, S., Frangou, S., Satterthwaite, T.D., Hajek, T., Dannlowski, U., Malt, U.F., Arolt, V., Gattaz, W.F., Drevets, W.C., Caseras, X., Agartz, I., Thompson, P.M., Andreassen, O.A., 2018. Cortical abnormalities in bipolar disorder: an MRI analysis of 6503 individuals from the ENIGMA Bipolar Disorder Working Group. Mol Psychiatry 23, 932-942.

Hibar, D.P., Westlye, L.T., van Erp, T.G., Rasmussen, J., Leonardo, C.D., Faskowitz, J., Haukvik, U.K., Hartberg, C.B., Doan, N.T., Agartz, I., Dale, A.M., Gruber, O., Kramer, B., Trost, S., Liberg, B., Abe, C., Ekman, C.J., Ingvar, M., Landen, M., Fears, S.C., Freimer, N.B., Bearden, C.E., Costa Rica/Colombia Consortium for Genetic Investigation of Bipolar, E., Sprooten, E., Glahn, D.C., Pearlson, G.D., Emsell, L., Kenney, J., Scanlon, C., McDonald, C., Cannon, D.M., Almeida, J., Versace, A., Caseras, X., Lawrence, N.S., Phillips, M.L., Dima, D., Delvecchio, G., Frangou, S., Satterthwaite, T.D., Wolf, D., Houenou, J., Henry, C., Malt, U.F., Boen, E., Elvsashagen, T., Young, A.H., Lloyd, A.J., Goodwin, G.M., Mackay, C.E., Bourne, C., Bilderbeck, A., Abramovic, L., Boks, M.P., van Haren, N.E., Ophoff, R.A., Kahn, R.S., Bauer, M., Pfennig, A., Alda, M., Hajek, T., Mwangi, B., Soares, J.C., Nickson, T., Dimitrova, R., Sussmann, J.E., Hagenaars, S., Whalley, H.C., McIntosh, A.M., Thompson, P.M., Andreassen, O.A., 2016. Subcortical volumetric abnormalities in bipolar disorder. Mol Psychiatry 21, 1710-1716.

Hirschfeld, R.M., Lewis, L., Vornik, L.A., 2003. Perceptions and impact of bipolar disorder: how far have we really come? Results of the national depressive and manic-depressive association 2000 survey of individuals with bipolar disorder. J Clin Psychiatry 64, 161-174.

Houenou, J., d'Albis, M.A., Vederine, F.E., Henry, C., Leboyer, M., Wessa, M., 2012. Neuroimaging biomarkers in bipolar disorder. Front Biosci (Elite Ed) 4, 593-606.

Houenou, J., Perlini, C., Brambilla, P., 2015. Epidemiological and clinical aspects will guide the neuroimaging research in bipolar disorder. Epidemiol Psychiatr Sci 24(2), 117-120. https://doi.org/10.1017/S2045796014000766. Epub 2015 Jan 16. PMID: 25592435; PMCID: PMC6998109.

Hwang, J., Lyoo, I.K., Dager, S.R., Friedman, S.D., Oh, J.S., Lee, J.Y., Kim, S.J., Dunner, D.L., Renshaw, P.F., 2006. Basal ganglia shape alterations in bipolar disorder. Am J Psychiatry 163, 276-285.

Insel, T., Cuthbert, B., Garvey, M., Heinssen, R., Pine, D.S., Quinn, K., Sanislow, C., Wang, P., 2010. Research domain criteria (RDoC): toward a new classification framework for research on mental disorders. Am J Psychiatry 167, 748-751.

Insel, T.R., 2014. The NIMH Research Domain Criteria (RDoC) Project: precision medicine for psychiatry. Am J Psychiatry 171, 395-397.

Ivleva, E.I., Bidesi, A.S., Keshavan, M.S., Pearlson, G.D., Meda, S.A., Dodig, D., Moates, A.F., Lu, H., Francis, A.N., Tandon, N., Schretlen, D.J., Sweeney, J.A., Clementz, B.A., Tamminga, C.A., 2013. Gray matter volume as an intermediate phenotype for psychosis: Bipolar-Schizophrenia Network on Intermediate Phenotypes (B-SNIP). Am J Psychiatry 170, 1285-1296.

Kempton, M.J., Geddes, J.R., Ettinger, U., Williams, S.C., Grasby, P.M., 2008. Meta-analysis, database, and meta-regression of 98 structural imaging studies in bipolar disorder. Arch Gen Psychiatry 65, 1017-1032.

Maggioni, E., Altamura, A.C., Brambilla, P., 2017a. Exploring the neuroanatomical bases of psychotic features in bipolar disorder. Epidemiol Psychiatr Sci 26, 358-363.

Maggioni, E., Crespo-Facorro, B., Nenadic, I., Benedetti, F., Gaser, C., Sauer, H., Roiz-Santianez, R., Poletti, S., Marinelli, V., Bellani, M., Perlini, C., Ruggeri, M., Altamura, A.C., Diwadkar, V.A., Brambilla, P., group, E., 2017b. Common and distinct structural features of schizophrenia and bipolar disorder: The European Network on Psychosis, Affective disorders and Cognitive Trajectory (ENPACT) study. PLoS One 12, e0188000.

Martino, D.J., Strejilevich, S.A., Scapola, M., Igoa, A., Marengo, E., Ais, E.D., Perinot, L., 2008. Heterogeneity in cognitive functioning among patients with bipolar disorder. J Affect Disord 109, 149-156.

McDonald, C., Zanelli, J., Rabe-Hesketh, S., Ellison-Wright, I., Sham, P., Kalidindi, S., Murray, R.M., Kennedy, N., 2004. Meta-analysis of magnetic resonance imaging brain morphometry studies in bipolar disorder. Biol Psychiatry 56, 411-417.

Mills, N.P., Delbello, M.P., Adler, C.M., Strakowski, S.M., 2005. MRI analysis of cerebellar vermal abnormalities in bipolar disorder. Am J Psychiatry 162, 1530-1532.

Phillips, M.L., Kupfer, D.J., 2013. Bipolar disorder diagnosis: challenges and future directions. Lancet 381, 1663-1671.

Rimol, L.M., Hartberg, C.B., Nesvag, R., Fennema-Notestine, C., Hagler, D.J., Jr., Pung, C.J., Jennings, R.G., Haukvik, U.K., Lange, E., Nakstad, P.H., Melle, I., Andreassen, O.A., Dale, A.M., Agartz, I., 2010. Cortical thickness and subcortical volumes in schizophrenia and bipolar disorder. Biol Psychiatry 68, 41-50.

Sassi, R.B., Nicoletti, M., Brambilla, P., Harenski, K., Mallinger, A.G., Frank, E., Kupfer, D.J., Keshavan, M.S., Soares, J.C., 2001. Decreased pituitary volume in patients with bipolar disorder. Biol Psychiatry 50, 271-280.

Selvaraj, S., Arnone, D., Job, D., Stanfield, A., Farrow, T.F., Nugent, A.C., Scherk, H., Gruber, O., Chen, X., Sachdev, P.S., Dickstein, D.P., Malhi, G.S., Ha, T.H., Ha, K., Phillips, M.L., McIntosh, A.M., 2012. Grey matter differences in bipolar disorder: a meta-analysis of voxel-based morphometry studies. Bipolar Disord 14, 135-145.

Squarcina, L., Fagnani, C., Bellani, M., Altamura, C.A., Brambilla, P., 2016. Twin studies for the investigation of the relationships between genetic factors and brain abnormalities in bipolar disorder. Epidemiol Psychiatr Sci 25, 515-520.

Takahashi, T., Malhi, G.S., Wood, S.J., Walterfang, M., Yucel, M., Lorenzetti, V., Soulsby, B., Suzuki, M., Velakoulis, D., Pantelis, C., 2009. Increased pituitary volume in patients with established bipolar affective disorder. Prog Neuropsychopharmacol Biol Psychiatry 33, 1245-1249.

Vieta, E., Salagre, E., Grande, I., Carvalho, A.F., Fernandes, B.S., Berk, M., Birmaher, B., Tohen, M., Suppes, T., 2018. Early Intervention in Bipolar Disorder. Am J Psychiatry 175, 411-426.

Wessa, M., Perlini, C., Brambilla, P., 2015. Neuropsychological underpinnings of the dynamics of bipolar disorder. Epidemiol Psychiatr Sci 24, 479-483.

Whiteford, H.A., Degenhardt, L., Rehm, J., Baxter, A.J., Ferrari, A.J., Erskine, H.E., Charlson, F.J., Norman, R.E., Flaxman, A.D., Johns, N., Burstein, R., Murray, C.J., Vos, T., 2013. Global burden of disease attributable to mental and substance use disorders: findings from the Global Burden of Disease Study 2010. Lancet 382, 1575-1586.

Womer, F.Y., Wang, L., Alpert, K.I., Smith, M.J., Csernansky, J.G., Barch, D.M., Mamah, D., 2014. Basal ganglia and thalamic morphology in schizophrenia and bipolar disorder. Psychiatry Res 223, 75-83.

Adler, Caleb M, John Adams, Ba P Melissa DelBello, Scott K Holland, Vince Schmithorst, Ari Levine, BS Kelly, Jarvis, and Bse M Stephen Strakowski. 2006. "Evidence of White Matter Pathology in Bipolar Disorder Adolescents Experiencing Their First Episode of Mania: A Diffusion Tensor Imaging Study." *Am J Psychiatry*. Vol. 163.

Adler, Caleb M, Scott K Holland, Vince Schmithorst, Marko Wilke, Kenneth L Weiss, Hai Pan, and Stephen M Strakowski. 2004. "Abnormal Frontal White Matter Tracts in Bipolar Disorder: A Diffusion Tensor Imaging Study." *Bipolar Disorders* 6 (3):197–203.

Basser, Peter J., and Carlo Pierpaoli. 1996. "Microstructural and Physiological Features of Tissues Elucidated by Quantitative-Diffusion-Tensor MRI." *Journal of Magnetic Resonance – Series B* 111 (3):209–19.

Bauer, Isabelle E, Austin Ouyang, Benson Mwangi, Marsal Sanches, Giovana B Zunta-Soares, Richard S E Keefe, Hao Huang, and Jair C Soares. 2015. "Reduced White Matter Integrity and Verbal Fluency Impairment in Young Adults with Bipolar Disorder: A Diffusion Tensor Imaging Study." *Journal of Psychiatric Research* 62 (March). NIH Public Access:115–22.

Beaulieu, Christian. 2002. "The Basis of Anisotropic Water Diffusion in the Nervous System – A Technical Review." *NMR in Biomedicine* 15 (7–8):435–55.

Bellani, Marcella, Ping-Hong Yeh, Michele Tansella, Matteo Balestrieri, Jair C Soares, and Paolo Brambilla. 2009. "DTI Studies of Corpus Callosum in Bipolar Disorder." *Biochem Soc Trans* 37:1096–98.

Beyer, John L, Warren D Taylor, James R MacFall, Maragatha Kuchibhatla, Martha E Payne, James M Provenzale, Frederick Cassidy, and K Ranga R Krishnan. 2005. "Cortical White Matter Microstructural Abnormalities in Bipolar Disorder." *Neuropsychopharmacology* 30 (12):2225–29.

Bihan, D Le, Eric Breton, D Lallemand, P Grenier, E Cabanis, and M Laval-Jeantet. 1986. "MR Imaging of Intravoxel Incoherent Motions: Application to Diffusion and Perfusion in Neurologic Disorders." *Radiology* 161:401–7.

Bruno, S, M Cercignani, and M A Ron. 2008. "White Matter Abnormalities in Bipolar Disorder: A Voxel-Based Diffusion Tensor Imaging Study." *Bipolar Disorders* 10 (4):460–68.

Haarman, Bartholomeus C.M. ('Benno'), Rixt F. Riemersma – Van der Lek, Huibert Burger, Jan Cees de Groot, Hemmo A. Drexhage, Willem A. Nolen, and Leonardo Cerliani. 2016. "Diffusion Tensor Imaging in Euthymic Bipolar Disorder – A Tract-Based Spatial Statistics Study." *Journal of Affective Disorders* 203:281–91.

Kumar, J, S Iwabuchi, S Oowise, V Balain, L Palaniyappan, and P F Liddle. 2015. "Shared White-Matter Dysconnectivity in Schizophrenia and Bipolar Disorder with Psychosis." *Psychological Medicine* 45 (4). Europe PMC Funders:759–70.

Kurumaji, Akeo, Michio Itasaka, Akihito Uezato, Kazuo Takiguchi, Daisuke Jitoku, Mizue Hobo, and Toru Nishikawa. 2017. "A Distinctive Abnormality of Diffusion Tensor Imaging Parameters in the Fornix of Patients with Bipolar II Disorder." *Psychiatry Research: Neuroimaging* 266 (August):66–72.

Magioncalda, Paola, Matteo Martino, Benedetta Conio, Niccolò Piaggio, Roxana Teodorescu, Andrea Escelsior, Valentina Marozzi, et al. 2016. "Patterns of Microstructural White Matter Abnormalities and Their Impact on Cognitive Dysfunction in the Various Phases of Type I Bipolar Disorder." *Journal of Affective Disorders* 193:39–50.

Nortje, Gareth, Dan J. Stein, Joaquim Radua, David Mataix-Cols, and Neil Horn. 2013. "Systematic Review and Voxel-Based Meta-Analysis of Diffusion Tensor Imaging Studies in Bipolar Disorder." *Journal of Affective Disorders* 150 (2):192–200.

Oertel-Knöchel, Viola, Britta Reinke, Gilberto Alves, Alina Jurcoane, Sofia Wenzler, David Prvulovic, David Linden, and Christian Knöchel. 2014. "Frontal White Matter Alterations Are Associated with Executive Cognitive Function in Euthymic Bipolar Patients." *Journal of Affective Disorders* 155:223–33.

Sarrazin, Samuel, Cyril Poupon, Julia Linke, Michèle Wessa, Mary Phillips, Marine Delavest, Amelia Versace, et al. 2014. "A Multicenter Tractography Study of Deep White Matter Tracts in Bipolar I Disorder." *JAMA Psychiatry* 71 (4):388.

Smith, Stephen M, Mark Jenkinson, Heidi Johansen-Berg, Daniel Rueckert, Thomas E Nichols, Clare E Mackay, Kate E Watkins, et al. 2006. "Tract-Based Spatial Statistics: Voxelwise Analysis of Multi-Subject Diffusion Data."

Sprooten, Emma, Jennifer Barrett, D Reese McKay, Emma E Knowles, Samuel R Mathias, Anderson M Winkler, Margaret S Brumbaugh, et al. 2016. "A Comprehensive Tractography Study of Patients with Bipolar Disorder and Their Unaffected Siblings." *Human Brain Mapping* 37 (10):3474–85.

Squarcina, L., M. Bellani, M.G. Rossetti, C. Perlini, G. Delvecchio, N. Dusi, M. Barillari, et al. 2017a. "Similar White Matter Changes in Schizophrenia and Bipolar Disorder: A Tract-Based Spatial Statistics Study." *PLoS ONE* 12 (6).

Sun, Zhong Yi, Josselin Houenou, Delphine Duclap, Samuel Sarrazin, Julia Linke, Claire Daban, Nora Hamdani, et al. 2017. "Shape Analysis of the Cingulum, Uncinate and Arcuate Fasciculi in Patients with Bipolar Disorder." *Journal of Psychiatry & Neuroscience : JPN* 42 (1):27–36.

Sussmann, Jessika E., G. Katherine S. Lymer, James Mckirdy, T. William J. Moorhead, Susana Muñoz Maniega, Dominic Job, Jeremy Hall, et al. 2009. "White Matter Abnormalities in Bipolar Disorder and Schizophrenia Detected Using Diffusion Tensor Magnetic Resonance Imaging." *Bipolar Disorders* 11 (1):11–18.

Toteja, Nitin, Perihan Guvenek-Cokol, Toshikazu Ikuta, Vivian Kafantaris, Bart D Peters, Katherine E Burdick, Majnu John, Anil K Malhotra, and Philip R Szeszko. 2015. "Age-Associated Alterations in Corpus Callosum White Matter Integrity in Bipolar Disorder Assessed Using Probabilistic Tractography." *Bipolar Disorders* 17 (4):381–91.

Wessa, Michèle, Josselin Houenou, Marion Leboyer, Sandra Chanraud, Cyril Poupon, Jean Luc Martinot, and Marie Laure Paillère-Martinot. 2009. "Microstructural White Matter Changes in Euthymic Bipolar Patients: A Whole-Brain Diffusion Tensor Imaging Study." *Bipolar Disorders* 11 (5):504–14.

Yip, Sarah W, Rebecca A Chandler, Robert D Rogers, Clare E Mackay, and Guy M Goodwin. 2013. "White Matter Alterations in Antipsychotic- and Mood Stabilizer-Naïve Individuals with Bipolar II/NOS Disorder." *NeuroImage. Clinical* 3. Elsevier:271–78.

Almeida JR, Versace A, Hassel S, Kupfer DJ, Phillips ML. Elevated amygdala activity to sad facial expressions: a state marker of bipolar but not unipolar depression. Biol Psychiatry. 2010 Mar 1;67(5):414-21.

Altshuler L, Bookheimer S, Townsend J, Proenza MA, Sabb F, Mintz J, Cohen MS. Regional brain changes in bipolar I depression: a functional magnetic resonance imaging study. Bipolar Disord 2008; 10: 708-717.

Bora E. Neurocognitive features in clinical subgroups of bipolar disorder: A meta-analysis. J Affect Disord. 2018 Mar 15; 229:125-134.

Brambilla P, Cerruti S, Bellani M, Perlini C, Ferro A, Marinelli V, Giusto D, Tomelleri L, Rambaldelli G, Tansella M, Diwadkar VA. Shared impairment in associative learning in schizophrenia and bipolar disorder. Prog Neuropsychopharmacol Biol Psychiatry. 2011 Jun 1;35(4):1093-9.

Brambilla P, Perlini C, Bellani M, Tomelleri L, Ferro A, Cerruti S, Marinelli V, Rambaldelli G, Christodoulou T, Jogia J, Dima D, Tansella M, Balestrieri M, Frangou S. Increased salience of gains versus decreased associative learning differentiate bipolar disorder from schizophrenia during incentive decision making. Psychol Med. 2013 Mar;43(3):571-80.

Brotman MA, Tseng WL, Olsavsky AK, Fromm SJ, Muhrer EJ, Rutenberg JG, Deveney CM, Adleman NE, Zarate CA, Pine DS, Leibenluft E. Fronto-limbic-striatal dysfunction in pediatric and adult patients with bipolar disorder: impact of face emotion and attentional demands. Psychol Med. 2014 Jun;44(8):1639-51.

Chen CH, Lennox B, Jacob R, Calder A, Lupson V, Bisbrown-Chippendale R, Suckling J, Bullmore E. Explicit and implicit facial affect recognition in manic and depressed States of bipolar disorder: a functional magnetic resonance imaging study. Biol Psychiatry 2006; 59: 31-39.

Cipriani G, Danti S, Carlesi C, Cammisuli DM, Di Fiorino M. Bipolar Disorder and Cognitive Dysfunction: A Complex Link. J Nerv Ment Dis. 2017 Oct;205(10):743-756.

Delvecchio G, Fossati P, Boyer P, Brambilla P, Falkai P, Gruber O, Hietala J, Lawrie SM, Martinot JL, McIntosh AM, Meisenzahl E, Frangou S. Common and distinct neural correlates of emotional processing in Bipolar Disorder and Major Depressive Disorder: a voxel-based meta-analysis of functional magnetic resonance imaging studies. Eur Neuropsychopharmacol. 2012 Feb;22(2):100-13.

Deveney CM, Brotman MA, Thomas LA, Hinton KE, Muhrer EM, Reynolds RC, Adleman NE, Zarate CA Jr, Pine DS, Leibenluft E. Neural response during explicit and implicit face processing varies developmentally in bipolar disorder. Soc Cogn Affect Neurosci. 2014 Dec;9(12):1984-92.

Diler RS, de Almeida JR, Ladouceur C, Birmaher B, Axelson D, Phillips M. Neural activity to intense positive versus negative stimuli can help differentiate bipolar disorder from unipolar major depressive disorder in depressed adolescents: a pilot fMRI study. Psychiatry Res. 2013 Dec 30;214(3):277-84.

Dima D, de Jong S, Breen G, Frangou S. The polygenic risk for bipolar disorder influences brain regional function relating to visual and default state processing of emotional information. Neuroimage Clin. 2016 Nov 1;12:838-844.

Dima D, Jogia J, Collier D, Vassos E, Burdick KE, Frangou S. Independent modulation of engagement and connectivity of the facial network during affect processing by CACNA1C and ANK3 risk genes for bipolar disorder. JAMA Psychiatry. 2013 Dec;70(12):1303-11.

Dusi N, De Carlo V, Delvecchio G, Bellani M, Soares JC, Brambilla P. MRI features of clinical outcome in bipolar disorder: A selected review. J Affect Disord. 2018 Jun 1. pii: S0165-0327(18)30913-3.

Fagnani C, Bellani M, Soares JC, Stazi MA, Brambilla P. Discordant twins as a tool to unravel the aetiology of bipolar disorder. Epidemiol Psychiatr Sci. 2014 Jun;23(2):137–40.

Foland-Ross LC, Bookheimer SY, Lieberman MD, Sugar CA, Townsend JD, Fischer J, Torrisi S, Penfold C, Madsen SK, Thompson PM, Altshuler LL. Normal amygdala activation but deficient ventrolateral prefrontal activation in adults with bipolar disorder during euthymia. Neuroimage. 2012 Jan 2;59(1):738–44.

Fusar-Poli P, Placentino A, Carletti F, Landi P, Allen P, Surguladze S, Benedetti F, Abbamonte M, Gasparotti R, Barale F, Perez J, McGuire P, Politi P. Functional atlas of emotional faces processing: A voxelbased meta-analysis of 105 functional magnetic resonance imaging studies. J Psychiatry Neurosci 2009; 34: 418–432.

Grotegerd D, Stuhrmann A, Kugel H, Schmidt S, Redlich R, Zwanzger P, Rauch AV, Heindel W, Zwitserlood P, Arolt V, Suslow T, Dannlowski U. Amygdala excitability to subliminally presented emotional faces distinguishes unipolar and bipolar depression: an fMRI and pattern classification study. Hum Brain Mapp. 2014 Jul;35(7):2995–3007.

Hafeman DM, Chang KD, Garrett AS, Sanders EM, Phillips ML. Effects of medication on neuroimaging findings in bipolar disorder: an updated review. Bipolar Disord. 2012 Jun;14(4):375–410.

Hafeman DM, Bebko G, Bertocci MA, Fournier JC, Bonar L, Perlman SB, Travis M, Gill MK, Diwadkar VA, Sunshine JL, Holland SK, Kowatch RA, Birmaher B, Axelson D, Horwitz SM, Arnold LE, Fristad MA, Frazier TW, Youngstrom EA, Findling RL, Drevets W, Phillips ML. Abnormal deactivation of the inferior frontal gyrus during implicit emotion processing in youth with bipolar disorder: attenuated by medication. J Psychiatr Res. 2014 Nov;58:129–36.

Hajek T, Weiner MW. Neuroprotective Effects of Lithium in Human Brain? Food for Thought. Curr Alzheimer Res. 2016;13(8):862–72.

Hassel S, Almeida JR, Kerr N, Nau S, Ladouceur CD, Fissell K, Kupfer DJ, Phillips ML. Elevated striatal and decreased dorsolateral prefrontal cortical activity in response to emotional stimuli in euthymic bipolar disorder: no associations with psychotropic medication load. Bipolar Disord 2008; 10: 916–927.

Hulvershorn LA, Karne H, Gunn AD, Hartwick SL, Wang Y, Hummer TA, Anand A.Neural activation during facial emotion processing in unmedicated bipolar depression, euthymia, and mania. Biol Psychiatry. 2012 Apr 1;71(7):603–10.

Kim P, Thomas LA, Rosen BH, Moscicki AM, Brotman MA, Zarate CA Jr, Blair RJ, Pine DS, Leibenluft E. Differing amygdala responses to facial expressions in children and adults with bipolar disorder. Am J Psychiatry. 2012 Jun;169(6):642–9.

Ladouceur CD, Farchione T, Diwadkar V, Pruitt P, Radwan J, Axelson DA, Birmaher B, Phillips ML. Differential patterns of abnormal activity and connectivity in the amygdala-prefrontal circuitry in bipolar-I and bipolar-NOS youth. J Am Acad Child Adolesc Psychiatry. 2011 Dec;50(12):1275–89.e2.

Lawrence NS, Williams AM, Surguladze S, Giampietro V, Brammer MJ, Andrew C, Frangou S, Ecker C, Phillips ML. Subcortical and ventral prefrontal cortical neural responses to facial expressions distinguish patients with bipolar disorder and major depression. Biol Psychiatry 2004; 55: 578–587.

Lennox BR, Jacob R, Calder AJ, Lupson V, Bullmore ET. Behavioural and neurocognitive responses to sad facial affect are attenuated in patients with mania. Psychol Med 2004; 34: 795–802.

Liu J, Blond BN, van Dyck LI, Spencer L, Wang F, Blumberg HP: Trait and state corticostriatal dysfunction in bipolar disorder during emotional face processing. Bipolar Disord 2012; 14: 432–441

Malhi GS, Lagopoulos J, Sachdev PS, Ivanovski B, Shnier R, Ketter T. Is a lack of disgust something to fear? A functional magnetic resonance imaging facial emotion recognition study in euthymic bipolar disorder patients. Bipolar Disord 2007;9: 345–357.

Mandolini GM, Lazzaretti M, Pigoni A, Delvecchio G, Soares JC, Brambilla P. The impact of BDNF Val66Met polymorphism on cognition in Bipolar Disorder: A review. J Affect Disord. 2018 Jul 24. pii: S0165-0327(18)30909-1.

Martínez-Arán A, Vieta E, Reinares M, Colom F, Torrent C, Sánchez-Moreno J, Benabarre A, Goikolea JM, Comes M, Salamero M. Cognitive function across manic or hypomanic, depressed, and euthymic states in bipolar disorder. Am J Psychiatry. 2004 Feb;161(2):262–70.

Pavuluri MN, O'Connor MM, Harral E, Sweeney JA. Affective neural circuitry during facial emotion processing in pediatric bipolar disorder. Biol Psychiatry. 2007 Jul 15;62(2):158–67.

Perlini C, Marini A, Garzitto M, Isola M, Cerruti S, Marinelli V, Rambaldelli G, Ferro A, Tomelleri L, Dusi N, Bellani M, Tansella M, Fabbro F, Brambilla P. Linguistic production and syntactic comprehension in schizophrenia and bipolar disorder. Acta Psychiatr Scand. 2012 Nov;126(5):363–76.

Perlman SB, Fournier JC, Bebko G, Bertocci MA, Hinze AK, Bonar L, Almeida JR, Versace A, Schirda C, Travis M, Gill MK, Demeter C, Diwadkar VA, Sunshine JL, Holland SK, Kowatch RA, Birmaher B, Axelson D, Horwitz SM, Arnold LE, Fristad MA, Youngstrom EA,

Findling RL, Phillips ML. Emotional face processing in pediatric bipolar disorder: evidence for functional impairments in the fusiform gyrus. J Am Acad Child Adolesc Psychiatry. 2013 Dec;52(12):1314–1325.e3.

Raucher-Chéné D, Achim AM, Kaladjian A, Besche-Richard C. Verbal fluency in bipolar disorders: A systematic review and meta-analysis. J Affect Disord. 2017 Jan 1;207:359–366.

Sassi RB, Nicoletti M, Brambilla P, Mallinger AG, Frank E, Kupfer DJ, Keshavan MS, Soares JC. Increased gray matter volume in lithium-treated bipolar disorder patients. Neurosci Lett. 2002 Aug 30;329(2):243–5.

Squarcina L, Houenou J, Altamura AC, Soares J, Brambilla P. Association of increased genotypes risk for bipolar disorder with brain white matter integrity investigated with tract-based spatial statistics. J Affect Disord. 2017b Oct 15;221:312–317.

Sugihara G, Kane F, Picchioni MM, Chaddock CA, Kravariti E, Kalidindi S, Rijsdijk F, Toulopoulou T, Curtis VA, McDonald C, Murray RM, McGuire P. Effects of risk for bipolar disorder on brain function: A twin and family study. Eur Neuropsychopharmacol. 2017 May;27(5):494–503.

Surguladze SA, Marshall N, Schulze K, Hall MH, Walshe M, Bramon E, Phillips ML, Murray RM, McDonald C. Exaggerated neural response to emotional faces in patients with bipolar disorder and their first-degree relatives. Neuroimage. 2010; 53: 58–64.

Swann AC, Lijffijt M, Lane SD, Steinberg JL, Moeller FG. Severity of bipolar disorder is associated with impairment of response inhibition. J Affect Disord. 2009 Jul;116(1–2):30–6.

Tesli M, Kauppi K, Bettella F, Brandt CL, Kaufmann T, Espeseth T, Mattingsdal M, Agartz I, Melle I, Djurovic S, Westlye LT, Andreassen OA. Altered Brain Activation during Emotional Face Processing in Relation to Both Diagnosis and Polygenic Risk of Bipolar Disorder. PLoS One. 2015 Jul 29;10(7):e0134202.

Townsend J, Altshuler LL. Emotion processing and regulation in bipolar disorder: a review. Bipolar Disord 2012; 14: 326–339.

Vizueta N, Rudie JD, Townsend JD, Torrisi S, Moody TD, Bookheimer SY, Altshuler LL: Regional fMRI hypoactivation and altered functional connectivity during emotion processing in non-medicated depressed patients with bipolar II disorder. Am J Psychiatry 2012; 169:831–840

Wessa M, Perlini C, Brambilla P. Neuropsychological underpinnings of the dynamics of bipolar disorder. Epidemiol Psychiatr Sci 2015b Dec;24(6):479–83.

Alexander AL, Lee JE, Lazar M, Field AS. Diffusion tensor imaging of the brain. Neurotherapeutics. 2007 Jul;4(3):316-29. doi:10.1016/j.nurt.2007.05.011. PMID:17599699; PMCID: PMC2041910.

Houenou J, Perlini C, Brambilla P. Epidemiological and clinical aspects will guide the neuroimaging research in bipolar disorder. Epidemiol Psychiatr Sci. 2015 Apr;24(2):117-20. doi: 10.1017/S2045796014000766. Epub 2015 Jan 16. PMID: 25592435; PMCID: PMC6998109.

Understanding the Network Bases of ADHD: An Overview of the fMRI Evidence

Samuele Cortese

Introduction

Attention-deficit hyperactivity disorder (ADHD) is the most common neurodevelopmental disorder. According to the Diagnostic and Statistical Manual of Mental Disorders, Fifth Edition, DSM-5 (American Psychiatric Association 2013), ADHD is characterized by a persistent and impairing pattern of inattention and/or hyperactivity/impulsivity. Based on the type of symptoms, three presentations are possible: predominantly inattentive, predominantly hyperactive/impulsive, and combined. Hyperkinetic disorder (HKD), defined in the International Classification of Diseases, Tenth Edition (WHO 1992), is a narrower diagnostic category, requiring both symptoms of inattention and hyperactivity/impulsivity and including individuals diagnosed with the combined presentation ADHD as per DSM-5 (American Psychiatric Association 2013). A large body of evidence shows that ADHD is often comorbid with other psychiatric conditions, such as oppositional defiant disorder/conduct disorder, specific learning disorders, mood and anxiety disorders, and sleep disturbances (Biederman and Faraone 2005; Cortese et al. 2013).

ADHD is a major public health issue. Its worldwide-pooled prevalence is estimated at about 5% in school-age children (Polanczyk et al. 2007). Impairing symptoms of ADHD persist in adulthood in up to 65% of cases (Faraone et al. 2006), with

S. Cortese (✉)
Center for Innovation in Mental Health, School of Psychology and Clinical and Experimental Sciences (CNS and Psychiatry), Faculty of Medicine, University of Southampton, Southampton, UK

Solent NHS Trust, Southampton, UK

Hassenfeld Children's Hospital at NYU Langone, New York, New York, NY, USA

Division of Psychiatry and Applied Psychology, School of Medicine, University of Nottingham, Nottingham, UK
e-mail: samuele.cortese@soton.ac.uk

V. A. Diwadkar, S. B. Eickhoff (eds.), *Brain Network Dysfunction in Neuropsychiatric Illness*, https://doi.org/10.1007/978-3-030-59797-9_16

a pooled prevalence of adulthood ADHD ~ 2.5% (Simon et al. 2009). Because of its core symptoms and comorbid disorders, ADHD imposes an enormous burden on society in terms of psychological dysfunction, adverse vocational outcomes, stress on families, and societal financial costs. Average annual incremental costs of ADHD have been recently estimated at $143–$266 billion in the United States (Doshi et al. 2012). In Europe, annual national costs have been calculated between €1,041 and €1,529 million (Le et al. 2013).

Available treatments for ADHD include pharmacological (psychostimulant and non-psychostimulants) and non-pharmacological options (e.g., parent training, cognitive training).

Similarly to other mental health conditions, the paradigm shift from alterations in individual brain regions to dysfunction in brain networks has contributed to reshape our understanding of the pathophysiology of ADHD. The chapter is intended to be a critical, rather than a systematic, overview of the available fMRI evidence on the brain network bases of ADHD across the life span. The bulk of the chapter will focus on evidence from resting state fMRI (R-fMRI) studies. However, before moving to R-fMRI studies, the conclusions of two meta-analyses (in children and adults, respectively) of task-based fMRI studies, interpreted in the light of a brain network perspective, are presented, since they are informative in terms of our understating of the brain network bases of ADHD.

Meta-analyses of Task-Based fMRI Studies of ADHD from a Brain Network Perspective

In 2012, Cortese et al. (2012) published a meta-analysis of 55 task-based fMRI studies in children/adolescents and adults with ADHD. Studies were included if they were based on DSM/ICD categorical diagnoses of ADHD, used between-group (e.g., ADHD *vs.* controls) contrasts, and reported results as three-dimensional coordinates in stereotactic space at the whole brain level. Meta-analytic estimates were conducted using likelihood estimation meta-analysis using GingerALE, version 2.1.1 (www.brainmap.org/ale/). The meta-analysis of the 39 studies in children across tasks revealed significant ADHD-related hypoactivation in frontal regions and putamen bilaterally and in right parietal and right temporal regions. ADHD-related hyperactivation was found in the right angular gyrus, middle occipital gyrus, posterior cingulate cortex, and midcingulate cortex. The meta-analysis of the 16 studies in adults showed significant hypoactivation relative to comparison subjects in the right central sulcus, precentral gyrus, and middle frontal gyrus. ADHD-related hyperactivation was observed in a region with a peak in the right angular and middle occipital gyri. In order to help the interpretation of the results, Cortese et al. related each of the ADHD hypo- and hyperactivated regions from the above meta-analysis to seven reference networks defined by Yeo et al. (2011) on the basis of a data-driven analysis of resting state, which included the following networks: visual, somatomotor, dorsal attention, ventral attention, limbic, frontoparietal, and default. Considering studies both in children and adolescents, the majority of the ADHD-related hypoactivated areas fell within the ventral attention (41% of all the ADHD-related hypoactivated areas) and the frontoparietal (30%) networks. By contrast, the majority of ADHD-related hyperactivated areas fell within the default mode network (39%). Furthermore, 9% of the hyperactivated areas were within the visual network. The authors interpreted the results as an empirical support to the hypothesis that attentional lapses that characterize ADHD result from an inappropriate intrusion of the default network in the activity of task-positive networks frontoparietal, ventral, or dorsal attention networks, consistent with the default network hypothesis of ADHD put forward by Sonuga-Barke and Castellanos (2007). Furthermore, the hypoactivation of areas in the visual system was interpreted as a possible compensatory mechanism that individuals with ADHD would use to overcome executive dysfunctions. In this regard, it is interesting to note that, whereas in children 4% of the ADHD-related hyperactivated areas fell within the visual system, this percentage increased to 41% in adults, suggesting that such compensatory mechanism may develop over the life span.

In another meta-analysis of 24 task-based fMRI studies limited to adults, Cortese et al. (2016) first identified the following as ADHD-related hypoactivated regions consistently across studies: left putamen, right caudate, left inferior frontal gyrus (IFG) (pars opercularis), and left temporal pole. Then, the authors clustered the ADHD-related hypoactivated bran regions (resulting from the activation likelihood estimation meta-analysis) according to their functional coactivation profiles across all the experiments stored in BrainMap (no ADHD-related hyperactivation was found after using the stringent threshold of a more advanced activation likelihood estimation model). This procedure is referred to as meta-analytic connectivity modeling. On the basis of their meta-analytical coactivation, three regions (left putamen [medial], left temporal pole, and left putamen) clustered closely, a fourth one was isolated (left IFG, pars opercularis), and a fifth one (right caudate) was intermediate between the first three and the fourth. Leveraging on more than 7500 fMRI experiments in the BrainMap database, the authors also coded the function of each brain region resulting as significant form the ALE meta-analysis (functional decoding). Across tasks, the ADHD-related hypoactivated location in the left putamen was associated with domains related to cognitive aspects of music (e.g., music comprehension and production, tone monitor and discrimination), perception and audition, and speech and language. The ADHD-related hypoactivation in the left putamen was again related to cognitive aspects of music as well as language and speech. The ADHD-related hypoactivated region in the left IFG (pars opercularis) was associated with domains related to action or execution. The ADHD-related hypoactivation in the left temporal pole was associated with domains related to cognitive aspects of language and speech. Finally, the ADHD-related hypoactivated region in the right caudate was associated with domains related to action and execution, perception and somesthesis, and cognitive aspects of music. Interestingly, it could be concluded that the clustering of regions reflected their function detected with functional decoding. In particular, the clustering of the two regions in the putamen and temporal pole could reflect the fact that their functions revealed by functional decoding (music-related processes and semantic aspect of language, respectively) are linked to both timing and rhythm processing. Therefore, in addition to supporting models positing deficits in typical executive functions in ADHD, these results suggest that impairments in processes related to music perception/reproduction and language semantics, which have been so far downplayed in ADHD research, may be relevant for the pathophysiology of ADHD.

Overall, the two meta-analyses are an example of how evidence from task-based fMRI studies can be used to test or generate theoretical models about the brain network bases of ADHD. In the next section, R-fMRI studies of ADHD will be overviewed. The section will first highlight how current theories/models of ADHD have guided the search of the network bases underpinning ADHD. It will then focus on findings from theory-free, data-driven approaches. Following the delineation of the networks that are abnormal in ADHD, an overview of the genetic underpinnings of these networks will be provided. This will be followed by a section on the effects of ADHD treatments on brain networks. The final part of the chapter will discuss to which extent system neuroscience of ADHD is ready for clinical translation.

Theories/Models of ADHD Inspiring R-fMRI Studies

As mentioned in the previous section, one of the most recent and popular hypotheses put forward to gain insight into the pathophysiology of ADHD was proposed by Sonuga-Barke and Castellanos (2007). They posited that spontaneous patterns of very low-frequency (< 0.1 Hz) coherence within the default network persist or intrude into active task processing carried out by task-positive brain networks. Rather than providing a comprehensive model accounting for the classic triad of ADHD symptoms (namely, inattention, hyperactivity, and impulsivity), the default network hypothesis was intended to specifically address the performance variability associated with ADHD. Following the proposal of this model, a number of R-fMRI studies have been carried out to test abnormal connectivity of and within the default network. In a seminal study in 20 adults with ADHD and 20 controls, Castellanos et al. (2008) used a seed-based correlation approach to assess the strength of the antiphasic (negative) correlation between dorsal anterior cingulate cortex and default mode network components. A significant decrease of the negative resting state functional correlation (RSFC) was found in adults with ADHD compared to controls, thus providing initial support to the default network hypothesis. Furthermore, secondary analyses revealed significant attenuation in positive correlation strength between anterior (ventromedial prefrontal cortex, PFC) and posterior default mode network (DMN) components. These initial findings were substantially replicated in a sample of 19 boys with ADHD and 23 controls by Sun et al. (2012), who found that, compared with normal controls, boys with ADHD showed a significantly decreased negative RSFC between the dorsal anterior cingulate (dACC) with the DMN, including the dorsomedial prefrontal cortex and the posterior cingulate cortex. Furthermore, the strength of the RSFC between the dACC and the posterior cingulate cortex showed a significantly negative correlation with age in normal controls, but not in boys with ADHD. Another replication came from the study by Mattfeld et al. (2014), who measured RSFC in individuals who had a persistent diagnosis of ADHD in adulthood (n = 13), in those who met diagnosis in childhood but not in adulthood (n = 22), and in a control group without ADHD (n = 17). They found that a positive functional correlation between posterior cingulate and medial prefrontal cortices was significantly reduced only in patients whose diagnosis persisted into adulthood. However, a negative functional correlation between medial and dorsolateral prefrontal cortices was reduced in both persistent and remitted patients. An extension of the findings by Castellanos et al. was provided by Sripada et al. (2014) who, analyzing RSFC among 907 seeds in a subsample (133 patients with ADHD and 288 controls) of the ADHD-200 dataset, found significantly RSFC within the default network and between the default network and ventral attention, frontoparietal, and visual networks. They also found significantly reduced functional connectivity between ventral attention and frontoparietal networks and between ventral attention nodes and the default network, replicating the findings by Castellanos and extending them to anterior insula. Finally, it is worthy to mention a study by Kucyi et al. (2015), one of the few to assess cerebellar components of the default network, which is relevant since the cerebellum has been involved in the pathophysiology of ADHD in previous structural MRI studies (e.g., (Berquin et al. 1998)). Contrasting 23 adults with ADHD and 23 controls, the authors found that, in ADHD, the cerebellar components of the default network showed a positive, rather than a negative (as in controls), correlation with regions of salience, dorsal attention, and sensorimotor networks, as well as with frontoparietal and visual network regions.

Therefore, overall, there is mounting evidence, although results are not always consistent across studies, of altered RSFC between the default network and task-positive regions, as well as within the default network. However, it appeared soon clear in the field that this was just one locus of dysfunction in RSFC, and other circuits have been investigated since, under the influence of prominent alternative theoretical models of ADHD. One of these relates to dysfunctions in reward processing and altered delay discounting in ADHD. Accordingly, a number of R-fMRI studies have focused on this process. For instance, Costa Dias et al. (2013) examined differences in functional connectivity between the nucleus accumbens, a key region in reward processing, between 35 children with ADHD and 64 controls, and assessed to which extent the functional connectivity of the nucleus accumbens correlated with the scores on a delay discounting task. The authors found significant differences between the functional connectivity of the nucleus accumbens in children with ADHD and controls; furthermore, the ADHD-related increased connectivity between the nucleus accumbens and the prefrontal cortex was associated with steeper delayed reward discounting. However, findings have not been consistent so far. For example, Oldehinkel et al. (2016) assessed 169 children with ADHD and 122 controls from the NeuroIMAGE cohort. They failed to find significant alterations, related to ADHD, in the functional connectivity of the salience network, which includes key brain regions involved in reward processes. Therefore, further RSFC studies targeting reward circuits are needed in the field.

Beyond specific networks, another model of ADHD is based on a delay in maturation of the structure and function of key brain areas. This model was initially put forward and tested with structural MRI by the NIMH group (Dr. Rapoport, Castellanos, Shaw, and colleagues) (2007). The maturational lag hypothesis has been tested more recently using graph theory methods. For example, Marcos-Vidal et al. (2018) compared local and distant connectivity levels in 120 children with ADHD and local and distant connectivity levels in 120 children with ADHD and 120 age-matched controls. They found that, compared to children without ADHD, those with ADHD presented with more local connectivity in regions within the default mode, frontoparietal and ventral attentional networks, thus suggesting an immaturity of several networks. Longer follow-up studies have contributed to lend support to the maturational dysfunction hypothesis. For instance, Sudre et al. (2017a) found that symptoms of inattention persisting in adulthood in 205 individuals with childhood ADHD were associated with a loss of balance in the connections within the default mode as well as the connection between the default mode and the networks underpinning attention and cognitive control. No differences were found in resting state measures between adults with childhood inattentive symptoms in remission and controls.

Theory-Free, Data-Driven Approaches

This section briefly reports key findings across a number of analytic methods.

As for amplitude of low-frequency fluctuations (ALFF), the largest study found decreased ALFF in ventral prefrontal cortex and orbitofrontal cortex as well as increased ALFF in the pallidum and dorsal prefrontal cortex (Li et al. 2014). Regional homogeneity (ReHo) findings have been conflicting in the lingual gyrus but have consistently reported decreased ReHo in the prefrontal cortex (Castellanos and Aoki 2016). As for entropy, on the one hand, sample entropy was found to be reduced in the anterior cingulate cortex, precuneus and cuneus, and superior frontal gyrus, in adults with ADHD, overall indicating lower complexity of the networks (Sokunbi et al. 2013). On the other hand, graph spectral entropy was found to be increased in the pre- and postcentral gyri, superior temporal gyrus, and inferior frontal gyrus, overall suggesting an abnormal structure of the networks (Sato et al. 2013). On a related vein, graph theory-based methods have found decreased global efficiency (Wang et al. 2009; Lin et al. 2014). Interestingly, a 15% higher local functional connectivity in regions associated with reward processes, such as the orbitofrontal cortex, ventral striatum, and superior fontal cortex, and a 33% reduction in long distance functional connectivity density in the superior parietal cortex and the default mode have been reported (Tomasi and Volkow 2012).

Therefore, it is safe to conclude that a more comprehensive exploration of the various RSFC metrics is needed in the field.

Genetic Aspects Underpinning Resting-State MRI Abnormalities in ADHD

Recent fascinating work has shed light on the heritability of abnormal brain networks in ADHD. Sudre et al. (2017b) assessed 24 extended multigenerational families enriched for ADHD and confirmed the findings in 52 nuclear families using DTI and R-fMRI. They detected heritable patterns of functional connectivity within the default mode (cluster level significance, $p < 0.002$), cognitive control ($p< 0.002$), and ventral attention networks ($p < 0.002$). Of note, more symptoms of hyperactivity/impulsivity and inattention were associated with a decreased functional connectivity in the default mode network. Interestingly, a genetic shared cross-modal correlation was detected between the ventral attention network, identified with R-fMRI, and one DTI parameter (radial diffusivity) of the right inferior fronto-occipital fasciculus. Recent studies have also focused on specific genetic variants. Mizuno et al. (2017) assessed the impact of catechol-O-methyltransferase (COMT) polymorphism on the cortico-cerebellar executive function network. In a sample of 31 children with ADHD and 30 age- and IQ-matched controls, they found that Met carriers had significantly lower functional connectivity than the Val/Val carriers, suggesting variations in COMT genetic effects on the cortico-cerebellar executive function network. Additionally, Qian et al. (2018) assessed the effects of dopamine D4 receptor (DRD4) variants on ReHo and functional connectivity in a sample of 49 children with ADHD (n= 30 with 4-repeat/4-repeat (4R/4R) allele and n= 19 with DRD4 2-repeat (2R) allele). Participants with the DRD4 2R allele showed a significantly decreased functional connectivity to the left angular gyrus in the left striatum, right inferior frontal gyrus, and bilateral lobes of the cerebellum compared to those with the DRD4 4R/4R allele. They also showed increased left superior frontal gyrus, medial frontal gyrus, and rectus gyrus. Overall, the results show that DRD4 polymorphisms are associated with abnormalities in the frontal-striatal-cerebellar loop.

Effects of ADHD Treatment on fMRI-Defined Brain Networks

Recently, the effects of ADHD treatments on brain networks found as abnormal in ADHD via R-fMRI have been addressed. This is highly relevant from a scientific and clinical standpoint. While the body of evidence is still limited to be amenable to meta-analytic synthesis, key studies on the effects of pharmacological treatment and on the impact of non-pharmacological interventions are reported here. As for the pharmacological treatment, Yoo et al. (2018) assessed the effects of 12 weeks methylphenidate in a sample of 20 children with ADHD. The authors found an increase in large-scale connectivity between the left frontoparietal network and the left insula cortex and the right frontoparietal and the brainstem. ALFF was also increased in the bilateral superior parietal cortex, while it was decreased in the right inferior frontotemporal area. Overall, these results suggest that methylphenidate leads to alteration in a broad array of regions and impacts on the functional segregation of the large-scale intrinsic functional networks. In another study, Shang et al. (2016) set out to explore the differential effects of methylphenidate and atomoxetine (first- and second-line pharmacological treatment, respectively, for ADHD) in a sample of 46 children with ADHD. They found that the two medications acted on different networks: while reductions in inattentive symptoms were related to increased fALFF in the left superior temporal gyrus and left inferior parietal lobule in children treated with methylphenidate, those treated with atomoxetine showed reductions in the left lingual gyrus and left inferior occipital gyrus. As for adults, Yang et al. (2016) showed that 3 weeks of stimulant treatment decreased the positive functional connectivity between dorsolateral prefrontal cortex and insula, dorsolateral prefrontal cortex and anterior cingulate cortex, and medial prefrontal cortex and insula. Of note, these changes were associated with symptom improvement.

In relation to non-pharmacological treatment, in a small study (Wang et al. 2016) (20 participants) in adults, regional functional connectivity strength after 12 sessions of CBT was increased in the frontoparietal network and cerebellum, suggesting that CBT and pharmacological treatment may share common brain targets.

Therefore, the effects of pharmacological and non-pharmacological treatments of ADHD on brain networks start being elucidated against the common thinking that "we do not know what effect ADHD treatments have on the brain."

Clinical Translations

Machine learning approaches, including support vector machine, using neuroimaging data, including R-fMRI, have opened up the possibility to use neuroimaging data for prediction at the individual level, rather than group level, thus contributing to the use of neuroimaging findings which will be useful for the daily clinical practice. However, as highlighted in a thoughtful recent review (Arbabshirani et al. 2017), a number of methodological and conceptual issues need to be addressed. Indeed, an increasing number of R-fMRI studies (at least 11 in their review) have been published with the aim of predicting diagnostic status based on machine learning approaches. However, the reported accuracy in predicting the diagnosis of ADHD decreased with study sample sizes, which casts doubts on the generalizability of many of the small studies in ADHD. Beyond small sample sizes, more general issues in the literature on machine learning approaches for prediction at the individual level include feature selection bias, overfitting, and hyperparameter optimization. Additionally, many clinicians would argue that rather than prediction of diagnosis, prediction of response to treatment or prognosis may be more meaningful. While some studies using neuroimaging data to predict response to treatment have been published [(e.g., (Kim et al. 2015))], this needs to be further developed, and the contribution of the knowledge of abnormal brain networks may provide useful insights.

Conclusions

It is fair to state that the science of brain networks in ADHD defined by R-fMRI studies is still in its infancy. Study results are mostly based on small samples, and replicability has not been achieved in many areas. Whereas some results seem to be fairly consistent replicate din theory-driven studies (e.g., abnormalities within the DMN and in its connectivity), data-driven approaches need to be expanded further to gain comprehensive, unbiased insight into the brain networks bases of ADHD. While we start appreciating the genetic underpinnings of these networks and the effects of treatment, larger studies are needed. Finally, whereas the use of support vector machines has the potential to translate research findings on brain networks in clinicallyrelevant applications, methodological concerns need to be addressed. Large-scale collaborations and free data sharing will be essential to address these concens.

References

American Psychiatric Association. Diagnostic and Statistical Manual of Mental Disorders, Fifth Edition, DSM-5. American Psychiatric Publishing ed. 2013.

WHO. The ICD-10 classification of mental and behavioral disorders: clinical descriptions and diagnostic guidelines 1992; diagnostic criteria for research 1993. 1992. Geneva, WHO.

Biederman J, Faraone SV. Attention-deficit hyperactivity disorder. *Lancet* 2005; 366(9481):237-248.

Cortese S, Brown TE, Corkum P, Gruber R, O'Brien LM, Stein M et al. Assessment and management of sleep problems in youths with attention-deficit/hyperactivity disorder. *J Am Acad Child Adolesc Psychiatry* 2013; 52(8):784-796.

Polanczyk G, de Lima MS, Horta BL, Biederman J, Rohde LA. The worldwide prevalence of ADHD: a systematic review and metaregression analysis. *Am J Psychiatry* 2007; 164(6):942-948.

Faraone SV, Biederman J, Mick E. The age-dependent decline of attention deficit hyperactivity disorder: a meta-analysis of follow-up studies. *Psychol Med* 2006; 36(2):159-165.

Simon V, Czobor P, Balint S, Meszaros A, Bitter I. Prevalence and correlates of adult attention-deficit hyperactivity disorder: meta-analysis. *Br J Psychiatry* 2009; 194(3):204-211.

Doshi JA, Hodgkins P, Kahle J, Sikirica V, Cangelosi MJ, Setyawan J et al. Economic impact of childhood and adult attention-deficit/hyperactivity disorder in the United States. *J Am Acad Child Adolesc Psychiatry* 2012; 51(10):990-1002.

Le HH, Hodgkins P, Postma MJ, Kahle J, Sikirica V, Setyawan J et al. Economic impact of childhood/adolescent ADHD in a European setting: the Netherlands as a reference case. *Eur Child Adolesc Psychiatry* 2013.

Cortese S, Kelly C, Chabernaud C, Proal E, Di MA, Milham MP et al. Toward systems neuroscience of ADHD: a meta-analysis of 55 fMRI studies. *Am J Psychiatry* 2012; 169(10):1038-1055.

Yeo BT, Krienen FM, Sepulcre J, Sabuncu MR, Lashkari D, Hollinshead M et al. The organization of the human cerebral cortex estimated by intrinsic functional connectivity. *J Neurophysiol* 2011; 106(3):1125-1165.

Sonuga-Barke EJ, Castellanos FX. Spontaneous attentional fluctuations in impaired states and pathological conditions: a neurobiological hypothesis. *Neurosci Biobehav Rev* 2007; 31(7):977-986.

Cortese S, Castellanos FX, Eickhoff CR, D'Acunto G, Masi G, Fox PT, Laird AR, Eickhoff SB. Functional Decoding and Meta-analytic Connectivity Modeling in AdultAttention-Deficit/Hyperactivity Disorder. Biol Psychiatry. 2016 Dec15;80(12):896-904.

Castellanos FX, Margulies DS, Kelly C, Uddin LQ, Ghaffari M, Kirsch A et al. Cingulate-precuneus interactions: a new locus of dysfunction in adult attention-deficit/hyperactivity disorder. *Biol Psychiatry* 2008; 63(3):332-337.

Sun L, Cao Q, Long X, Sui M, Cao X, Zhu C et al. Abnormal functional connectivity between the anterior cingulate and the default mode network in drug-naive boys with attention deficit hyperactivity disorder. *Psychiatry Res* 2012; 201(2):120-127.

Mattfeld AT, Gabrieli JD, Biederman J, Spencer T, Brown A, Kotte A et al. Brain differences between persistent and remitted attention deficit hyperactivity disorder. *Brain* 2014; 137(Pt 9):2423-2428.

Sripada C, Kessler D, Fang Y, Welsh RC, Prem KK, Angstadt M. Disrupted network architecture of the resting brain in attention-deficit/hyperactivity disorder. *Hum Brain Mapp* 2014; 35(9):4693-4705.

Kucyi A, Hove MJ, Biederman J, Van Dijk KR, Valera EM. Disrupted functional connectivity of cerebellar default network areas inattention-deficit/hyperactivity disorder. Hum Brain Mapp. 2015 Sep;36(9):3373-86.

Berquin PC, Giedd JN, Jacobsen LK, Hamburger SD, Krain AL, Rapoport JL et al. Cerebellum in attention-deficit hyperactivity disorder: a morphometric MRI study. *Neurology* 1998; 50(4):1087-1093.

Costa Dias TG, Wilson VB, Bathula DR, Iyer SP, Mills KL, Thurlow BL et al. Reward circuit connectivity relates to delay discounting in children with attention-deficit/hyperactivity disorder. *Eur Neuropsychopharmacol* 2013; 23(1):33-45.

Oldehinkel M, Beckmann CF, Franke B, Hartman CA, Hoekstra PJ, Oosterlaan J et al. Functional connectivity in cortico-subcortical brain networks underlying reward processing in attention-deficit/hyperactivity disorder. *Neuroimage Clin* 2016; 12:796-805.

Shaw P, Eckstrand K, Sharp W, Blumenthal J, Lerch JP, Greenstein D et al. Attention-deficit/hyperactivity disorder is characterized by a delay in cortical maturation. *Proc Natl Acad Sci U S A* 2007; 104(49):19649-19654.

Marcos-Vidal L, Martínez-García M, Pretus C, Garcia-Garcia D, Martínez K, Janssen J, Vilarroya O, Castellanos FX, Desco M, Sepulcre J, Carmona S. Local functional connectivity suggests functional immaturity in children with attention-deficit/hyperactivity disorder. Hum Brain Mapp. 2018 Feb 22. https://doi.org/10.1002/hbm.24013. [Epub ahead of print]

Sudre G, Szekely E, Sharp W, Kasparek S, Shaw P. Multimodal mapping of the brain's functional connectivity and the adult outcome of attention deficit hyperactivity disorder. Proc Natl Acad Sci U S A. 2017a Oct31;114(44):11787-11792.

Li F, He N, Li Y, Chen L, Huang X, Lui S, et al. Intrinsic brain abnormalities in attention deficit hyperactivity disorder: a resting-state functional MR imaging study. Radiology. 2014; 272:514– 523.

Castellanos FX, Aoki Y. Intrinsic Functional Connectivity in Attention-Deficit/Hyperactivity Disorder: A Science in Development. Biol Psychiatry Cogn Neurosci Neuroimaging. 2016 May;1(3):253-261.

Sokunbi MO, Fung W, Sawlani V, Choppin S, Linden DE, Thome J. Resting state fMRI entropy probes complexity of brain activity in adults with ADHD. Psychiatry Res. 2013; 214:341–348.

Sato JR, Takahashi DY, Hoexter MQ, Massirer KB, Fujita A. Measuring network's entropy in ADHD: a new approach to investigate neuropsychiatric disorders. Neuroimage. 2013; 77:44–51.

Wang L, Zhu C, He Y, Zang Y, Cao Q, Zhang H, et al. Altered small-world brain functional networks in children with attention-deficit/hyperactivity disorder. Hum Brain Mapp. 2009; 30:638–649.

Lin P, Sun J, Yu G, Wu Y, Yang Y, Liang M, et al. Global and local brain network reorganization in attention-deficit/hyperactivity disorder. Brain Imaging Behav. 2014; 8:558–569.

Tomasi D, Volkow ND. Abnormal functional connectivity in children with attention-deficit/ hyperactivity disorder. Biol Psychiatry. 2012; 71:443–450.

Sudre G, Choudhuri S, Szekely E, Bonner T, Goduni E, Sharp W, Shaw P. Estimating the Heritability of Structural and Functional Brain Connectivity in Families Affected by Attention-Deficit/Hyperactivity Disorder. JAMA Psychiatry.2017b Jan 1;74(1):76-84.

Mizuno Y, Jung M, Fujisawa TX, Takiguchi S, Shimada K, Saito DN, Kosaka H, Tomoda A. Catechol-O-methyltransferase polymorphism is associated with the cortico-cerebellar functional connectivity of executive function in children with attention-deficit/hyperactivity disorder. Sci Rep. 2017 Jul 7;7(1):4850.

Qian A, Wang X, Liu H, Tao J, Zhou J, Ye Q, Li J, Yang C, Cheng J, Zhao K, Wang M. Dopamine D4 Receptor Gene Associated with the Frontal-Striatal-Cerebellar Loop in Children with ADHD: A Resting-State fMRI Study. Neurosci Bull. 2018 Mar 21. https://doi.org/10.1007/s12264-018-0217-7

Yoo JH, Kim D, Choi J, Jeong B. Treatment effect of methylphenidate on intrinsic functional brain network in medication-naïve ADHD children: A multivariate analysis. Brain Imaging Behav. 2018 Apr;12(2):518-531

Shang CY, Yan CG, Lin HY, Tseng WY, Castellanos FX, Gau SS. Differential effects of methylphenidate and atomoxetine on intrinsic brain activity in children with attention deficit hyperactivity disorder. Psychol Med. 2016Nov;46(15):3173-3185.

Yang Z, Kelly C, Castellanos FX, Leon T, Milham MP, Adler LA. Neural Correlates of Symptom Improvement Following Stimulant Treatment in Adults with Attention-Deficit/Hyperactivity Disorder. J Child Adolesc Psychopharmacol. 2016Aug;26(6):527-36.

Wang X, Cao Q, Wang J, Wu Z, Wang P, Sun L, Cai T, Wang Y. The effects of cognitive-behavioral therapy on intrinsic functional brain networks in adults with attention-deficit/hyperactivity disorder. Behav Res Ther. 2016 Jan;76:32-9.

Arbabshirani MR, Plis S, Sui J, Calhoun VD. Single subject prediction of brain disorders in neuroimaging: Promises and pitfalls. Neuroimage. 2017 Jan 15;145(Pt B):137-165.

Kim JW, Sharma V, Ryan ND. Predicting Methylphenidate Response in ADHD Using Machine Learning Approaches. Int J Neuropsychopharmacol. 2015 May 10;18(11):pyv052.

Cortical-Limbic and Default-Mode Networks in Borderline Personality Disorder

Annegret Krause-Utz and Christian Schmahl

Borderline personality disorder (BPD) is a severe mental disorder, characterized by a pronounced instability in affect, cognition, interpersonal relationships, and behavior (American Psychiatric Association 2013). Currently, the etiology of BPD seems to be best explained by a diathesis-stress model, assuming an interaction of genetic, neurobiological vulnerability and traumatic stress such as severe childhood abuse and neglect (Crowell et al. 2009; Golier et al. 2003; Leichsenring et al. 2011; Schmahl et al. 2014; Skodol et al. 2002; Wingenfeld and Wolf 2014).

Over the last decade, the understanding of possible neurobiological underpinnings of BPD has rapidly increased, with more and more neuroimaging studies providing deeper insight into structural and functional brain alterations in individuals with the disorder compared to controls. A growing number of neuroimaging studies in BPD has revealed altered (co)activity and structure in brain regions, including the amygdala, hippocampus, insula, anterior cingulate cortex (ACC), orbitofrontal cortex (OFC), and dorsolateral prefrontal cortex (dlPFC) (see, e.g., Krause-Utz et al. 2014a; van Zypthen et al. 2015).

There is growing consensus that mental disorders might be better understood in terms of network dysfunction rather than alterations in localized brain regions. Therefore, an increasing number of studies in BPD has focused on large-scale intrinsic connectivity networks, such as the default-mode network, salience network, and central executive network. These networks are characterized by relatively

A. Krause-Utz
Institute of Clinical Psychology, Leiden University, Leiden, The Netherlands

Leiden Institute for Brain and Cognition (LIBC), Leiden, The Netherlands

C. Schmahl (✉)
Department of Psychosomatic Medicine and Psychotherapy, Central Institute of Mental Health, Mannheim, Germany

Medical Faculty, University of Heidelberg, Mannheim, Germany
e-mail: Christian.schmahl@zi-mannheim.de

V. A. Diwadkar, S. B. Eickhoff (eds.), *Brain Network Dysfunction in Neuropsychiatric Illness*, https://doi.org/10.1007/978-3-030-59797-9_17

robust blood-oxygen-level-dependent (BOLD) co-activity among brain areas (Event-Menon 2011; Fox and Raichle 2007; Menon and Uddin 2010; Zuo et al. 2010), showing high consistency over time, across different sample populations, and across varying physical and experimental (psychological) conditions (Damoiseaux et al. 2006; Fox and Raichle 2007; Greicius et al. 2003; Margulies et al. 2007; Raichle et al. 2001; Seeley et al. 2007). Furthermore, these networks were successfully mapped both during experimental challenge and under resting state, widely corresponding to functional connectivity patterns during experimental tasks (Biswal et al. 1995; Broyd et al. 2009; Lowe et al. 1998; Shehzad et al. 2009; Smith et al. 2009).

In this chapter, we aim to give an overview over recent neuroimaging research in BPD, focusing on magnetic resonance imaging (MRI) studies that applied functional connectivity approaches and structural network analyses (for more extensive reviews on previous neuroimaging research, see, e.g., Krause-Utz et al. 2014a; van Zypthen et al. 2015). At first, central features of BPD psychopathology and brain networks of particular relevance to BPD psychopathology are introduced.

Psychopathology of BPD

Emotion dysregulation, involving a heightened sensitivity toward emotional stimuli, intense and instable emotions, and a lack of functional coping skills, is considered to be at the core of BPD (Carpenter and Trull 2013; Crowell et al. 2009; Gilbert et al. 2009; Glenn and Klonsky 2009; Leichsenring et al. 2011; Lynch et al. 2006; Rosenthal et al. 2008; Schmahl et al. 2014). Being one of the most detrimental features of the disorder, emotion dysregulation is closely linked to other BPD symptoms such as nonsuicidal self-injury (NSSI), cognitive disturbances including dissociation, self-harming impulsive behavior, and social disturbances. NSSI, such as cutting or burning, has been associated with heightened arousal, dissociative symptoms, and a reduced pain sensitivity, which is possibly mediated by emotion regulation processes or an altered appraisal of pain (Bohus et al. 2000; Cardenas-Morales et al. 2011; Kemperman et al. 1997; Kleindienst et al. 2008; Klonsky & Muehlenkamp 2007; Ludäscher et al. 2007; McCown et al. 1993; Niedtfeld et al. 2010; Russ et al. 1992; Schmahl et al. 2004, 2006, 2010). Stress-related dissociation is experienced by a high percentage (75–80%) of individuals with BPD Korzekwa et al. 2009a, b; Zanarini et al. 2000), including disturbances or disruptions in the usually integrated functions of consciousness, perception, identity, memory, and affect (American Psychiatric Association 2013). Dissociative symptoms, such as derealization, depersonalization, and emotional numbing (feelings of being detached from one's body or from the own person), may be understood as an attempt to create an inner distance to stressful events, for example, to cope with otherwise overwhelming emotions during traumatic situations (Korzekwa et al. 2009; Stiglmayr et al. 2008; Spiegel and Cardena 1991; Vermetten and Spiegel 2014). Aside from dissociation-related cognitive dysfunctions (e.g., amnesia), individuals with BPD may experience deficits in inhibitory control (especially of emotional material), disturbed self-referential processing, and an instable self-image (American Psychiatric Association 2013).

Impulsivity and impulsive aggression are another hallmark of the disorder (American Psychiatric Association 2013; Ende et al. 2016; McCloskey et al. 2009; New et al. 2008; Sebastian et al. 2014). Typical impulsive behaviors such as substance abuse, binge eating, high-risk behavior, aggressive outbursts, or sudden relationship breakups primarily occur under distress (Cackowski et al. 2014; Krause-Utz et al., 2013; Krause-Utz et al. in press) or in negative emotional contexts (Krause-Utz et al. 2012; Holtmann et al. 2013; Jacob et al. 2013; Prehn et al. 2013; Silbersweig et al. 2007).

Individuals with BPD further suffer from interpersonal disturbances, including difficulties in developing trust in others, heightened rejection sensitivity, and instable relationships (e.g., alternating between a strong need for belonging and need for autonomy), as well as alterations in social cognition and theory of mind (King-Casas et al. 2008; Lis and Bohus 2013; Staebler et al. 2010).

Brain Networks of Particular Relevance to BPD Psychopathology

A growing body of neuroimaging studies has aimed at linking BPD psychopathology to alterations in brain structure and function. There is growing evidence that a dysfunction in large-scale networks, most prominently in the default-mode network, the salience network, and a corticolimbic network, involving limbic and prefrontal regions, may underlie clinical expressions of BPD.

The default-mode network is assumed to play an important role in self-referential processes such as autobiographical memories, rumination, theory of mind (mentalizing), social cognition, moral decisions, and mind wandering. Key nodes of this network involve the posterior cingulate cortex (PCC), precuneus, medial prefrontal cortex/frontopolar cortex, temporoparietal junction, posterior inferior parietal lobe, and angular gyrus (Buckner et al. 2008; Buckner and Vincent 2007; Laird et al. 2011). While activity within the default-mode network has mainly (but not only) been observed during rest, "task-positive" networks such as the salience network and central executive network have been found to be mostly active during mentally demanding tasks (Buckner and Vincent 2007). The salience network, involving the dorsal ACC and insula, has been related to attention modulation, salience detection, encoding of negative emotions, interoceptive awareness, and pain processing (Craig 2011; Critchley et al. 2001; Dosenbach et al. 2006; Lee and Siegle 2012; Margulies et al. 2007; Menon 2011; Menon and Uddin 2010; Seeley et al. 2007). This network may further play a crucial role in switching between networks, for example, recruiting cognitive resources while initiating goal-directed behavior (Menon 2011; Sridharan et al. 2008). The central executive network, comprising frontoparietal regions such as the dlPFC and posterior parietal cortex, has been critically implicated in goal-directed behavior and cognitive control (e.g., attention, working memory, interference inhibition) (Banich et al. 2009; Menon 2011; Miller 2000; Owen et al. 2005; Smith and Jonides 1999). Cortical and corticostriatal pathways also play an important role in regulating impulsive urges and inhibiting motor responses (Bush et al. 2000; Coccaro et al. 2011; Fineberg et al. 2010; Schoenbaum et al. 2009; Leyton et al. 2001) and in the down-regulation of emotions (Anderson et al. 2004; Banich et al. 2009; Blair et al. 2007; Bush et al. 2000; Drevets and Raichle 1998; Etkin et al. 2011; Phillips et al. 2003; McRae et al. 2010; Ochsner and Gross 2007; Pessoa et al. 2012; Stein et al. 2007).

With respect to BPD symptoms such as emotion dysregulation, a network involving limbic(-related) regions (e.g., amygdala, hippocampus, ACC, and insula) and prefrontal cortical regions has gained more and more attention. Emotion regulation is assumed to involve dynamic interactions between an "affective-driven" bottom-up appraisal system (including the basal ganglia, nucleus accumbens, insula, and amygdala) and a "cognitive-driven" top-down system (e.g., ACC, OFC, dlPFC, dorso-/ventromedial prefrontal cortex, and ventrolateral cortex), which may modulate activation in limbic and subcortical areas (Ochsner et al. 2002; Ochsner and Gross 2007). Better emotion down-regulation capacities have been associated with a stronger coupling and inhibitory interactions between amygdala and prefrontal regions including the dorsomedial/dorsolateral PFC and ventromedial/ventrolateral PFC (Banks et al. 2007; Johnstone et al. 2007; Townsend et al. 2013). Stress seems to aggravate imbalance in this corticolimbic system (Oei et al. 2012; Veer et al. 2011).

Given the high relevance of the abovementioned networks to emotion regulation, salience processing, self-referential processing, impulse control, dissociation, and pain perception, more and more neuroimaging studies have aimed at investigating how alterations in these networks may be linked to BPD psychopathology. In the following, we first discuss recent resting-state and task-related functional connectivity studies before turning to research on structural networks in BPD.

Resting-State Functional Connectivity Studies in BPD

Studying how brain regions may interact during rest, in terms of synchronized BOLD signal fluctuations in the absence of experimental tasks, has become increasingly important for understanding the dynamic neural architecture of mental disorders (Biswal et al. 1995; Cole et al. 2010; Hampson et al. 2002, 2004).

In BPD, there is evidence for altered resting-state functional connectivity (RSFC), most prominently in default mode and corticolimbic regions (Doll et al. 2013; Krause-Utz et al. 2014b; Salvador et al. 2016; Visintin et al. 2016; Wolf et al. 2011; Xu et al. 2016).

According to a recent meta-analysis by Visintin et al. (2016), including seven fMRI studies (152 BPD patients, 147 controls), RSFC was most consistently altered in the default-mode network. More specifically, BPD patients showed increased

RSFC in the anterior and posterior midline hubs (including precuneus), while RSFC was decreased in the dorsal part (right lateral temporal and orbitofrontal cortex) of the network compared to healthy controls.

Aside from default-mode network and task-positive network abnormalities, Doll et al. (2013) observed inter-network abnormalities, more specifically a shift from RSFC in the central executive network to the salience network. In another RS fMRI study, BPD patients exhibited diminished anti-correlations between the posterior cingulate (a node of the default-mode network) and dorsal ACC (a node of the salience network), while healthy participants showed strong anti-correlations between those regions (Krause-Utz et al. 2014b), suggesting an inflexibility to switch between large-scale networks during rest.

Of note, previous resting-state fMRI studies in BPD differ substantially regarding methodological aspects such as medication status, inclusion/exclusion criteria, and sample characteristics. Sample sizes were mostly relatively small and BPD samples differed with respect to comorbidities (e.g., major depressive disorder or posttraumatic stress disorder (PTSD)), raising the question of how specific findings are for BPD. In a study by Das and colleagues (Das et al. 2014), BPD was specifically associated with decreased RSFC between "social salience" network regions and the precuneus as well as frontoparietal networks, which distinguished this patient group from bipolar disorder patients and healthy controls. To clarify the specificity of RSFC alterations for BPD, more studies including larger sample sizes of patients and clinical controls are needed.

The interpretation of previous resting-state fMRI studies in BPD is further complicated by the use of different analysis strategies (e.g., independent component analysis (ICA), seed-based correlations). While seed-based correlations detect differences in the co-activation (correlation) of an a priori selected region of interest and other regions across the brain (Friston et al. 1997; O'Reilly et al. 2012), data-driven approaches such as ICA are used to identify large-scale networks by differentiating a signal into nonoverlapping spatial and temporal components and by removing noise components (e.g., scanner drift and motion artefacts) in a data set. Combining different methodological approaches and techniques might help to gain a more global look at resting-state function in BPD (Krause-Utz and Schmahl 2016).

In a recent RS-fMRI study, Salvador et al. (2016) combined different methodological approaches including amplitude of low-frequency fluctuations (ALFF), calculation of global brain connectivity maps, and seed-based correlations (as well as diffusion tensor imaging, see section on structural networks). Compared to healthy controls, patients showed reduced connectivity in the right temporal lobe and altered RSFC in the perigenual ACC. Furthermore, patients demonstrated increased ALFF (suggesting increased resting-state activity) in the left amygdala and hippocampus, which was positively correlated with symptom severity. A follow-up seed-based analysis indicated that RS activity in this amygdala/hippocampus cluster was more strongly coupled with the ACC in patients than in healthy controls. The finding of altered (pre)frontal–amygdala connectivity is in line with previous observations (Krause-Utz et al. 2014b; New et al. 2007).

Symptoms such as dissociation and altered pain perception may modulate amygdala connectivity in BPD. For instance, self-reported trait dissociation modulated amygdala connectivity in BPD by positively predicting amygdala RSFC with dlPFC and negatively predicting amygdala RSFC with parts of the visual cortex (Krause-Utz et al. 2014b). In a study by Reitz et al. (2015), RSFC was assessed after an experimental stress induction (by the Montreal Imaging Stress Task), followed by either painful tissue damage (an incision into the forearm) compared to a sham condition. Incision into the forearm led to a decrease in self-reported arousal and heart rate, a decrease in amygdala activity, and normalized functional connectivity of amygdala and superior frontal gyrus in the patient group, suggesting a stress-reducing effect of tissue damage in BPD (as discussed in more detail below).

Aiming at investigating the topological organizations of brain networks in BPD, Xu et al. (2016) used graph-theory-based complex network analysis, creating frequency-specific functional brain networks based on the correlation of wavelet-filtered fMRI signals from 82 cortical and subcortical regions. Compared to healthy participants, BPD patients showed significantly larger measures of global network topology (clustering coefficient, small-worldness, size of largest connected graph component, and local efficiency), suggesting increased local cliquishness, within a network in the 0.03–0.06 Hz frequency band. They further demonstrated lower nodal centrality at several hub nodes, while exhibiting greater centrality at several nonhub nodes in the network. In an interconnected subnetwork in 0.03- to 0.06-Hz frequency band, BPD patients showed significantly lower mean connectivity, which was negatively correlated with increased global topology measures. Key network measures were further correlated to clinical measures, highly accurately classifying BPD patients against healthy controls in a linear discriminant analysis.

In sum, there is evidence for aberrant RSFC in default mode and corticolimbic regions as well as imbalanced inter-network RSFC and altered topological organizations of brain networks in BPD. More studies using larger samples sizes, repeated scanning sessions, and clinical control groups are needed to increase the understanding of RS alterations that are robust and specific for BPD. Assessing RS-fMRI before and after experimental tasks or interventions might further help to identify potential targets for treatment and to better relate neuroimaging findings to clinical expressions such as emotion dysregulation, hypervigilance to salient events, increased interoceptive awareness, and difficulties in switching between states of alertness and states of rest.

Symptom Provocation Studies

A growing number of studies has aimed at linking brain network function to clinical expressions of BPD by experimentally provoking symptoms during a scan. Early neuroimaging research that investigated neural responding to emotionally arousing stimuli (e.g., negative pictures, autobiographical scripts of negative life events) pointed to a crucial role of the amygdala in disturbed emotion processing in BPD:

Individuals with the disorder exhibited increased activity (Donegan et al. 2003; Herpertz et al. 2001; Koenigsberg et al. 2009; Krause-Utz et al. 2012; Minzenberg et al. 2007; Niedtfeld et al. 2010; Schulze et al. 2011) and prolonged activity in the amygdala, suggesting a failure to habituate to the repeated presentation of negative pictures (Hazlett et al. 2012), instructed fear conditions (Kamphausen et al. 2012), or fear conditioning (Krause-Utz et al., 2016) compared to healthy controls. The amygdala plays a critical role in emotion processing and in the initiation of stress and fear responses (LeDoux 1992; Davis and Whalen 2001; Davidson 2002; Phan, Wager, Taylor, & Liberzon, Phan et al. 2005). Along with the "medial temporal lobe memory system" (involving the hippocampus, parahippocampal gyrus, and retrosplenial cortex), the amygdala has also been implicated in emotional memory (Qin et al. 2016; Roy et al. 2009; Veer et al. 2011). Although contradictory findings were reported as well (Ruocco et al. 2013), the most recent meta-analysis on emotion processing in BPD suggests that amygdala hyperreactivity is a rather robust finding in medication-free patient samples (Schulze et al. 2016). Other earlier studies on emotional responding in BPD highlighted the role of the insula (Beblo et al. 2006; Krause-Utz et al. 2012, Under review; Niedtfeld et al. 2010; Prehn et al. 2013; Ruocco et al. 2013; Schulze et al. 2011), which has been critically implicated in the encoding of unpleasant feelings, awareness of one's own bodily experiences, social exclusion, and pain perception (Damasio et al. 2000; Menon and Uddin 2010). Aside from hyperreactivity of limbic regions, BPD patients demonstrated altered frontal activity during emotional challenge, such as blunted activity in the bilateral dlPFC (see Schulze et al. 2016) and in ventrolateral prefrontal cortices, while instructed to cognitively distance themselves from negative stimuli (Koenigsberg et al. 2009). Likewise, the use of cognitive reappraisal was associated with decreased recruitment of the left OFC along with increased insula activity in BPD (Schulze et al. 2011). Diminished recruitment or a failure to activate the ACC and medial PFC was further observed in response to fearful faces (Minzenberg et al. 2007), personalized auditory scripts of abandonment situations (Schmahl et al. 2003a), self-reported traumatic memories (Schmahl et al. 2004b), and an emotional Go/NoGo task (Silbersweig et al. 2007) in BPD patients compared to healthy controls.

As pointed out before, the presence of symptoms such as stress-related dissociation or altered pain processing may modulate activity within the corticolimbic network (Krystal et al. 1998; Schmahl and Bohus 2007). With respect to *dissociation*, it has been proposed that dissociative symptoms are associated with a "reversed" pattern of brain activation (i.e., increased recruitment of dorsal/rostral ACC and mPFC, dampened activity in amygdala and insula) relative to the neural pattern associated with emotion dysregulation (i.e., diminished ACC and mPFC activity, limbic hyperreactivity) (Lanius et al. 2010; Sierra and Berrios 1998). In their "corticolimbic-disconnection model," Sierra and Berrios (1998) propose that symptoms of depersonalization may lead to increased recruitment of prefrontal regions (ACC, mPFC, and dlPFC) and dampened amygdala activity, comparable to "shutting down the affective system." In line with these models, BPD patients with higher self-reported dissociation previously showed stronger inferior frontal gyrus activity (Ludäscher et al. 2010; Winter et al. 2015; Krause-Utz et al. Under review) and lower amygdala reactivity during presentation of highly arousing (trauma-related) pictures (Hazlett et al. 2012; Krause-Utz et al. 2012; Krause-Utz et al. Under review), although studies using other modalities than aversive images (e.g., words, pain stimulation) did not observe a dampening effect of dissociation on amygdala activity (Kraus et al. 2009; Winter et al. 2015).

With regard to altered *pain processing*, individuals with BPD showed increased activity in the dlPFC (Kraus et al. 2009; Schmahl et al. 2006) and diminished activity in the posterior parietal cortex, along with a deactivation in the amygdala and the perigenual ACC during painful heat stimulation (Schmahl et al. 2006), possibly due to an increased top-down regulation of emotional components of pain or an altered affective appraisal of pain (Niedtfeld et al. 2010; Pessoa et al. 2002; Ochsner et al. 2004).

With regard to *social perception*, patients with BPD showed exaggerated amygdala responses and altered frontotemporal activity (including subgenual ACC, inferior frontal gyrus, and dlPFC) to facial expressions and interpersonal scenes, suggesting a hypervigilance toward salient social information (Donegan et al. 2003; Frick et al. 2012; Guitart-Masip et al. 2009; Holtmann et al. 2013; Koenigsberg et al. 2009; Krause-Utz et al. 2012; Minzenberg et al. 2007; Niedtfeld et al. 2010; Prehn et al. 2013; Radaelli et al. 2012). During social cognition or empathy tasks, increased activity in left temporal pole, middle temporal gyrus, and medial frontal gyrus (Frick et al. 2012) and heightened activity in the middle and posterior insula (Dziobek et al. 2011) was observed. In experimental social exclusion conditions, BPD patients further exhibited increased activity in medial prefrontal cortex (Ruocco et al. 2010; Domsalla et al. 2014), demonstrating stronger activity in salience network regions (including dorsal ACC) already during a social inclusion condition (Domsalla et al. 2014).

To sum up so far, there is evidence for imbalanced activity within a corticolimbic system, including a hyperactivity of "bottom-up emotion-generating" limbic regions (e.g., amygdala and insula) and a diminished recruitment of prefrontal "cognitive control areas," which may underlie symptoms such as emotion dysregulation in BPD. In addition, there is evidence that patients with BPD show altered recruitment of regions that are critically involved in social cognition processes, when viewing social material or performing theory of mind tasks, which might underlie interpersonal disturbances in the disorder. Altered social perception (e.g., rejection hypersensitivity, attentional bias toward negative social information) may in turn contribute to emotion dysregulation, for example, by increasing emotional vulnerability (e.g., limbic hyperreactivity).

Task-Related Functional Connectivity Studies As previously mentioned, the complexity of BPD symptoms might be better understood by looking at dynamic interactions within networks rather than localized activation patterns. Therefore, more and more fMRI studies have used functional connectivity approaches such as ICA (see above) or PPI to investigate brain network dysfunction in BPD. PPI detects changes in the correlation of time series in a seed region and other areas dependent on an experimental condition. The principle underlying this approach is that if two brain areas are interacting in a task-dependent manner, time courses of activity in these areas will be correlated (Friston et al. 1997, 1998; O'Reilly et al. 2012).

There is evidence for a "hyper-connectivity" within the corticolimbic network, in terms of a stronger coupling between the amygdala and (pre)frontal brain regions during emotional challenge in BPD. Cullen et al. (2011) presented either neutral, overt fear, or masked far faces during fMRI, observing increased connectivity of the amygdala (seed region) with rostral ACC during overt fear and with bilateral thalamus as well as right caudate during the masked fear scan in BPD. Under neutral conditions, BPD patients showed diminished amygdala functional connectivity with mid-cingulate cortex compared to healthy individuals.

In another fMRI study, Kamphausen et al. (2012) instructed their participants that during a later fMRI scan, one of two visual stimuli might potentially indicate an aversive event (electrodermal stimulation), whereas the other stimulus would represent safety. In contrast to healthy controls, under instructed fear BPD patients did not show a habituation effect (deactivation) in the amygdala or an increase of vmPFC activity over time. Furthermore, they exhibited increased amygdala connectivity with vmPFC and a diminished coupling between subgenual and dorsal ACC (Kamphausen et al. 2012). Further evidence for a failure to effectively engage emotional habituation processes stems from a study by Koenigsberg et al. (2014). In this study, BPD patients showed a lack of increase in dorsal ACC activity along with smaller increases in insula-amygdala functional connectivity while being exposed to repeatedly presented (versus novel) negative pictures.

Krause-Utz et al. (Krause-Utz et al. 2014c) used PPI to examine functional connectivity of the bilateral amygdala (seed region) with other areas across the brain during emotional distraction (negative vs. neutral interpersonal pictures, presented as distractors in the context of a working memory task). BPD patients showed increased connectivity within the medial temporal lobe system, in terms of a stronger coupling of the amygdala with hippocampus, as well as increased amygdala–dorsomedial–PFC connectivity. These findings may point to enhanced retrieval of autobiographical events in individuals with BPD while being distracted by negative interpersonal pictures. In addition, alterations in salience network regions were observed: Compared to healthy controls, patients with BPD demonstrated a stronger coupling between the dorsal ACC (seed region) and left insula, posterior cingulate, superior temporal gyrus, and occipital gyrus during emotional distraction. BPD patients further showed a stronger coupling of both the dACC seed and amygdala seed with a cluster in the right dmPFC, which was positively correlated to reaction times during emotional distraction, suggesting difficulties in shifting attention away from task irrelevant but potentially self-relevant information. Interestingly, during emotional distraction, BPD patients with higher state dissociation showed stronger bilateral amygdala functional connectivity with right thalamus, right ACC, left insula, and left precentral gyrus, regions implicated in arousal modulation, body movements, attention, and filtering of sensory information.

In a more recent study (Krause-Utz et al. Under review), dissociation was experimentally induced using script-driven imagery before participants performed the emotional working memory task. In the script-driven imagery paradigm, a narrative of an autobiographical situation involving dissociative experiences (*dissociation script*, as compared to a neutral script) is created together with each participant. During fMRI, these scripts are presented and participants are instructed to recall the specific autobiographical situation as vividly as possible (Lanius et al. 2012; Ludäscher et al. 2010; Winter et al. 2015). Patients who had been exposed to a dissociation script showed diminished activity in the amygdala, left cuneus, lingual gyrus, and posterior cingulate, while BPD without dissociation showed amygdala hyperactivity to emotional distracting pictures (relative to healthy controls). For BPD patients exposed to the dissociation script, a significantly stronger positive coupling between activity in the amygdala and activity in the left inferior parietal lobule, right superior/middle temporal gyrus, right middle occipital gyrus, and left claustrum was observed during emotional distraction compared to nondissociative patients and healthy controls (Krause-Utz et al. Under review).

Other task-related functional connectivity studies in BPD aimed at investigating emotion processing in relation to altered pain perception in BPD. Niedtfeld et al. (2012) used PPI to study functional connectivity between corticolimbic regions during painful heat stimulation (vs. warmth perception) after presentation of arousing pictures in BPD. Painful sensory stimuli, in contrast to warmth perception, resulted in enhanced negative coupling between (para-)limbic and prefrontal structures in BPD, possibly reflecting an inhibition of limbic arousal. Additionally, BPD patients showed a stronger coupling of the dlPFC with posterior insula, suggesting altered salience and pain processing (Rainville 2002; Treede et al. 2000). Further evidence for altered functional connectivity during pain processing in BPD stems from an fMRI study by Kluetsch et al. (2012). In this study, patients with BPD exhibited lower functional connectivity between posterior cingulate cortex and dlPFC when exposed to painful heat rather than neutral temperature stimulation (Kluetsch et al. 2012). As previously mentioned, these regions are crucial nodes of the default mode network. Therefore, results of the study by Kluetsch et al. (2012)) may point to an altered cognitive and affective appraisal of pain (e.g., as being less aversive and self-relevant) in BPD. In sum, pain (and self-injury) may have a soothing effect in BPD (Reitz et al. 2012, 2015), which might be mediated by emotion regulation processes (attentional shift and altered appraisal of pain) (Niedtfeld et al. 2010).

Albein-Urios et al. (2013) investigated functional connectivity during an emotion regulation task (reappraisal versus maintenance of negative emotions) in highly impulsive patients with Cluster B personality disorders (including BPD) and comorbid cocaine use. Compared to cocaine users without cluster-B personality disorders, patients with a comorbid diagnosis displayed a significantly diminished coupling of the subgenual cingulate with the amygdala and the fusiform gyri as well as a stronger coupling with the anterior insula during the maintenance of emotions. During the reappraisal condition, they demonstrated significantly increased OFC–amygdala and diminished OFC–dorsal striatum functional connectivity. These findings may be related to an altered reappraisal of negative emotions, which might in turn contribute to impulsivity and dysfunctional beliefs in this highly impulsive patient group.

In another fMRI study, O'Neill et al. (2015) aimed to investigate links and differences between brain networks involved in emotion regulation and social cognition (e.g., theory of mind, ToM) in BPD. Compared to healthy participants, BPD patients showed significantly decreased ACC functional connectivity with three brain areas involved in ToM processes: the right supramarginal/inferior parietal lobes, the left superior temporal lobe, and right middle cingulate cortex. During rest, BPD patients further showed an increased coupling between the precuneus (default mode network seed) and the left inferior frontal lobe, left precentral/middle frontal, and left middle occipital/superior parietal lobes. Findings of this study (O'Neill et al. 2015) extend (the abovementioned) previous observations of altered limbic and fronto-temporal activity during presentation of social stimuli in BPD (Domsalla et al. 2014; Donegan et al. 2003; Dziobek et al. 2011; Frick et al. 2012; Guitart-Masip et al. 2009; Holtmann et al. 2013; King-Casas et al. 2008; Mier et al. 2013; Minzenberg et al. 2007; Prehn et al. 2013; Ruocco et al. 2010).

Possible Treatment Effects

It remains an interesting topic for future research to investigate possible effects of treatment interventions on brain networks in BPD. There is preliminary evidence that dialectical behavior therapy (DBT), which involves teaching of emotion regulation skills, has an impact on brain activity during emotion processing (Schnell and Herpertz 2007; Goodman et al. 2014). Recently, Winter et al. (2015) found decreased activity in the right inferior parietal lobe/supramarginal gyrus during distraction from negative vs. neutral stimuli, which significantly correlated to symptom improvement after a 12-week residential DBT-based treatment program. When viewing negative (versus neutral) pictures, patients who responded to DBT further showed a decrease in right perigenual ACC activity, suggesting lower emotional susceptibility. Schmitt et al. (in press) found decreased insula and anterior cingulate cortex activity during reappraisal in patients with BPD. Anterior cingulate connectivity to medial and superior frontal gyrus, superior temporal gyrus, and inferior parietal cortices increased after DBT. Treatment responders exhibited reduced activation in amygdala, ACC, OFC, and dlPFC together with increased connectivity within a limbic-prefrontal network during the reappraisal of negative stimuli after psychotherapy.

Paret et al. (2016) examined changes in amygdala–prefrontal connectivity after real-time fMRI neurofeedback (NF) in BPD. During four sessions of NF, patients with BPD were instructed to view aversive pictures, either with or without the instruction to down-regulate their emotions while receiving feedback on the amygdala BOLD signal from a thermometer display. The NF training increased amygdala–ventromedial PFC connectivity when patients were instructed to regulate (versus passive viewing of) pictures. Furthermore, across the four sessions, a significant linear increase of left amygdala RSFC with right dlPFC, increased right amygdala RSFC with the precentral and paracentral gyrus, and a linear decrease in

RSFC for left amygdala with a hippocampal–parahippocampal–thalamic cluster were observed. More studies including randomized-control trials are needed to clarify whether amygdala-target NF might strengthen the coupling of regions implicated in emotion down-regulation (i.e., the coupling between the amygdala and cortical "cognitive control" regions) during emotion processing, while reducing RSFC within a medial temporal lobe network implicated in self-referential processing in BPD.

Structural Network Studies in BPD

In addition to research on functional connectivity, an increasing number of studies has investigated structural network abnormalities in BPD. Earlier structural MRI research in patients with the disorder pointed to structural abnormalities in the amygdala, hippocampus, ACC, and insula (Brambilla et al., 2004; Driessen et al. 2000; Goodman et al. 2011; Hazlett et al. 2005; Minzenberg et al. 2008; Rüsch et al. 2003; Schmahl et al. 2009, 2003b; Soloff et al. 2008; Tebartz van Elst et al. 2003; Zetzsche et al. 2006), most prominently reduced volumes in hippocampus and amygdala (for a meta-analysis see Nunes et al. 2009).

Rather than focusing on localized (most prominently limbic) brain regions, more recent studies turned to whole-brain analyses of gray matter volume (GMV) (Irle et al. 2005, 2007; Lyoo et al. 1998; Soloff et al. 2008, 2012; Völlm et al. 2009). Reduced GMV was found in the OFC (Chanen et al. 2008; Soloff et al. 2012; Tebartz van Elst et al. 2003; Völlm et al. 2009), dlPFC (Brunner et al. 2010; Sala et al. 2011), ventromedial PFC (Bertsch et al. 2013), and ACC (Goodman et al. 2011; Hazlett et al. 2005; Soloff et al. 2008; Whittle et al. 2009), as well as hypothalamus (Kuhlmann et al. 2013) in BPD patients compared to healthy controls.

Interpretation of earlier structural studies is complicated, however, due to methodological aspects such as small sample sizes, treatment with psychotropic medication, trauma history, and comorbidities, most prominently with major depression and PTSD (Krause-Utz and Schmahl 2010; O'Neill and Frodl 2012; Schmahl and Bremner 2006). For instance, reductions in limbic areas such as the hippocampus may rather stem from a history of childhood trauma or comorbid PTSD, respectively (Dannlowski et al. 2012; Karl et al. 2006; Krause-Utz and Schmahl 2010; Lieb et al. 2004; Rodrigues et al. 2011; Stein et al. 1997; Woon and Hedges 2009). Niedtfeld et al. (2013) compared BPD patients with comorbid PTSD to BPD patients without PTSD using voxel-based morphometry. In this study, patients with BPD + PTSD specifically showed increased GMV in superior temporal gyrus and dlPFC. More recently, Depping et al. (2016) compared structural networks in patients with BPD to patients with major depressive disorder (MDD). Source-based morphometry (SBM) was used to group patterns of GMV alterations. For both patient groups (BPD, MDD), volume reductions in cingulate and lateral prefrontal regions were observed. Compared to both healthy controls and MDD patients, BPD patients exhibited significant reductions in the medial temporal/medial frontal network, which were associated with BPD symptomatology and impulsivity. In contrast, MDD patients showed volume reductions in a frontostriatal network bilaterally compared to the other groups. Findings of this study point to common and distinct structural network abnormalities in these two disorders, which show high comorbidity and overlap of affective symptoms.

Further evidence for a disturbed frontolimbic circuitry in BPD stems from studies using diffusion tensor imaging (DTI) to investigate white matter connections (Carrasco et al. 2012; Gan et al. 2016; Grant et al. 2007; Lee et al. 2016; Lischke et al. 2015; Rüsch et al. 2010; Rüsch et al. 2007; Maier-Hein et al. 2014; New et al. 2013; Whalley et al. 2015).

Lischke et al. (2015) used DTI to investigate fractional anisotropy (FA), axonal anisotropy (AD), and radial diffusivity (RD) in the uncinate fasciculus, a white-matter tract connecting (para-)limbic and prefrontal regions in BPD. Patients demonstrated lower FA and higher RD in the uncinate fasciculus than healthy controls, while AD in the uncinate fasciculus did not differ between the groups. In another DTI study by Walley and colleagues (2015), BPD patients showed decreased white matter integrity in the cingulum and fornix – white matter tracts connecting frontal and limbic regions. While no significant relationships with measures of childhood trauma could be found, there were significant associations between FA in the cingulum and fornix and measures of clinical symptoms (anger, affective instability, and avoidance of abandonment) in the patient group.

Extending previous observations of corpus callosum abnormalities (Rüsch et al. 2007; Carrasco et al. 2012) and corticolimbic dysconnectivity in BPD, Gan et al. (2016) observed significantly lower FA in the genu and body of the corpus callosum, right superior corona radiate, right anterior corona radiate, and higher RD in the left anterior thalamic radiation in young adults with BPD compared to healthy controls.

In another DTI study in patients with intermittent explosive disorder, impulsive aggression and BPD but not psychopathy or antisocial personality disorder was associated with lower FA in two clusters located in the superior longitudinal fasciculus, a large bundle of association fibers in the white matter, which connects frontal cortices with parietal, occipital, and temporal regions (Lee et al. 2016).

In the aforementioned RS-fMRI study by Salvador et al. (2016), measures of functional connectivity were compared to diffusion MRI data, aiming at a broader picture of brain alterations in BPD. Using this multimodal whole-brain approach, the authors detected convergence of alterations for genual and perigenual structures, with frontal white matter FA abnormalities partially encircling areas of increased mean diffusivity and global brain connectivity. These findings provide first support for both structural and functional abnormalities involving temporolimbic and frontomedial structures in BPD.

Summary

In sum, there is converging evidence for disturbances in default mode and salience network regions and in a corticolimbic circuitry involving the amygdala, hippocampus, insula, ACC, OFC, and medial PFC in BPD. These alterations may underlie clinical symptoms such as the marked expressions of emotion dysregulation, impulsivity, and interpersonal disturbances in individuals with the disorder. It remains an important topic for future research to clarify whether these structural and functional abnormalities are specific to BPD or related to other factors, such as a history of traumatic events in childhood, as similar alterations were also found in anxiety disorders (Killgore et al. 2014), major depression (Klauser et al. 2015; Wang et al. 2008), PTSD (Morey et al. 2009), trauma exposure in general (Lang et al. 2012), or psychotic disorder (Mukherjee et al. 2016).

Moreover, given the complexity of BPD psychopathology, a combination of multiple techniques capturing different levels (e.g., subjective reports of clinical symptoms, behavioral data from experimental tasks, and neurobiological measures including neuroimaging) may help to enhance the understanding of possible underlying mechanisms related to this severe disorder.

References

Albein-Urios, N., Verdejo-Román, J., Soriano-Mas, C., Asensio, S., Martínez-González, J.M., & Verdejo-García, A. (2013). Cocaine users with comorbid Cluster B personality disorders show dysfunctional brain activation and connectivity in the emotional regulation networks during negative emotion maintenance and reappraisal. European Neuropsychopharmacology, 23(12), 1698–1707. https://doi.org/10.1016/j.euroneuro.2013.04.012.

American Psychiatric Association (APA) (2013). Diagnostic and Statistical Manual of Mental Disorders (5th ed.). Arlington, VA: American Psychiatric Association. https://doi.org/10.1176/appi.books.9780890425596.744053.

Anderson, M. C., Ochsner, K. N., Kuhl, B., Cooper, J., Robertson, E., Gabrieli, S. W., ... Gabrieli, J. D. (2004). Neural systems underlying the suppression of unwanted memories. *Science, 303*(5655), 232–235. https://doi.org/10.1126/science.1089504.

Banich, M. T., Mackiewicz, K. L., Depue, B. E., Whitmer, A., Miller G. A. & Heller, W. (2009). Control mechanisms, emotion & memory: a neural perspective with implications for psychopathology. *Neuroscience and Biobehavioral Reviews, 33*(5), 613–630. https://doi.org/10.1016/jneubiorev.2008.09.010.

Banks, S. J., Eddy, K. T., Angstadt, M., Nathan, P. J. & Phan, K. L. (2007). Amygdala-frontal connectivity during emotion regulation. *Social Cognitive and Affective Neuroscience, 2*(4), 303–312. https://doi.org/10.1093/scan/nsm029.

Beblo, T., Driessen, M., Mertens, M., Wingenfeld, K., Piefke, M., Rullkoetter, N., . Woermann, F. G. (2006). Functional MRI correlates of the recall of unresolved life events in borderline personality disorder. Psychological medicine, 36(6), 845–856.

Bertsch, K., Grothe, M., Prehn, K., Vohs, K., Berger, C., Hauenstein, K., Keiper, P., Domes, G., Teipel, S., Herpertz, S. C. (2013). Brain volumes differ between diagnostic groups of violent criminal offenders. European Archives of Psychiatry and Clinical Neuroscience, 263 (7), 593–606.

Biswal, B., Yetkin, F. Z., Haughton, V. M., Hyde, J. S. (1995). Functional connectivity in the motor cortex of resting human brain using echo-planar MRI. *Magnetic Resonance in Medicine, 34*(4), 537–541. https://doi.org/10.1002/mrm.1910340409.

Blair, K. S., Smith, B. W., Mitchell, D. G., Morton, J., Vythilingam, M., Pessoa, L., … Blair, R. J. R. (2007). Modulation of emotion by cognition and cognition by emotion. *NeuroImage, 35*(1), 430–440. https://doi.org/10.1016/j.neuroimage.2006.11.048.

Bohus, M., Limberger, M., Ebner, U., Glocker, F. X., Schwarz, B., Wernz, M., & Lieb, K. (2000). Pain perception during self-reported distress and calmness in patients with borderline personality disorder and self-mutilating behavior. *Psychiatry research, 95*(3), 251–260.

Brambilla, P., Soloff, P. H., Sala, M., Nicoletti, M. A., Keshavan, M. S., & Soares, J. C. (2004). Anatomical MRI study of borderline personality disorder patients. *Psychiatry research, 131*(2), 125–133. https://doi.org/10.1016/j.pscychresns.2004.04.003

Broyd, S. J., Demanuele, C., Debener, S., Helps, S. K., James, C. J., & Sonuga-Barke, E. J. (2009). Default-mode brain dysfunction in mental disorders: a systematic review. Neuroscience and biobehavioral reviews, 33(3), 279–296. https://doi.org/10.1016/j.neubiorev.2008.09.002

Brunner, R., Henze, R., Parzer, P., Kramer, J., Feigl, N., Lutz, K., . Stieltjes, B. (2010). Reduced prefrontal and orbitofrontal gray matter in female adolescents with borderline personality disorder: is it disorder specific? NeuroImage, 49(1), 114–120.

Buckner, R., Andrews-Hanna, J., & Schacter, D. (2008). The brain's default network: anatomy, function, and relevance to disease. Annals of the New York Academy of Sciences, 1124, 1-38. https://doi.org/10.1196/annals.1440.011

Buckner, R. L. & Vincent, J. L. (2007). Unrest at rest: default activity and spontaneous network correlations. NeuroImage, 37(4), 1091–1096. discussion 1097–1099. https://doi.org/10.1016/j.neuroimage.2007.01.010

Bush, G., Luu, P., & Posner, M. I. (2000). Cognitive and emotional influences in anterior cingulate cortex. Trends Cogn Sci, 4(6), 215–222.

Cackowski, S., Reitz, A. C., Ende, G., Kleindienst, N., Bohus, M., Schmahl, C., Krause-Utz, A. (2014). Impact of stress on different components of impulsivity in borderline personality disorder. Psychol Med, 44(15), 3329–3340.

Cardenas-Morales, L., Fladung, A. K., Kammer, T., Schmahl, C., Plener, P. L., Connemann, B. J., & Schonfeldt-Lecuona, C. (2011). Exploring the affective component of pain perception during aversive stimulation in borderline personality disorder. Psychiatry Research, 186(2-3), 458–460.

Carpenter, R. W., & Trull, T. J. (2013). Components of emotion dysregulation in borderline personality disorder: a review. Curr Psychiatry Rep, 15(1), 335.

Carrasco, J. L., Tajima-Pozo, K., Diaz-Marsa, M., Casado, A., Lopez-Ibor, J. J., Arrazola, J. & Yus, M. (2012). Microstructural white matter damage at orbitofrontal areas in borderline personality disorder. Journal of Affective Disorders, 139 (2), 149–153.

Chanen, A. M., Velakoulis, D., Carison, K., Gaunson, K., Wood, S. J., Yuen, H. P., . Pantelis, C. (2008). Orbitofrontal, amygdala and hippocampal volumes in teenagers with first-presentation borderline personality disorder. Psychiatry research, 163(2), 116–125.

Coccaro, E. F., Sripada, C. S., Yanowitch, R. N., & Phan, K. L. (2011). Corticolimbic function in impulsive aggressive behavior. Biol Psychiatry, 69(12), 1153–1159.

Cole, D. M., Smith, S. M., & Beckmann, C. F. (2010). Advances and pitfalls in the analysis and interpretation of resting-state FMRI data. *Frontiers in Systems Neuroscience, 6*, 4–8. https://doi.org/10.3389/fnsys.2010.00008.

Craig, A. D. (2011). Significance of the insula for the evolution of human awareness of feelings from the body. *Annals of the New York Academy of Science, 1225*, 72–82. https://doi.org/10.1111/j.17496632.2011.05990.x.

Critchley, H. D., Mathias, C. J., & Dolan, R. J. (2001). Neural activity in the human brain relating to uncertainty and arousal during anticipation. *Neuron, 29*(2), 537–545. https://doi.org/10.1016/S0896-6273(01)00225-2.

Crowell, S. E., Beauchaine, T. P., & Linehan, M. M. (2009). A biosocial developmental model of borderline personality: Elaborating and extending Linehan's theory. Psychological bulletin, 135(3), 495–510.

Cullen, K. R., Vizueta, N., Thomas, K. M., Han, G. J., Lim, K. O., Camchong, J., . Schulz, S. C. (2011). Amygdala functional connectivity in young women with borderline personality disorder. . Brain Connect, 1(1), 61–71.
Damasio, A. R., Grabowski, T. J., Bechara, A., Damasio, H., Ponto, L. L. B., Parvizi, J., & Hichwa, R. D. (2000). Subcortical and cortical brain activity during the feeling of self-generated emotions. Nature Neuroscience, 3(10), 1049–1056.
Damoiseaux, J. S., Rombouts, S. A., Barkhof, F., Scheltens, P., Stam, C. J., Smith, S. M., & Beckmann, C. F. (2006). Consistent resting-state networks across healthy subjects. Proceedings of the National Academy of Sciences of the United States of America, 103(37), 13848–13853. https://doi.org/10.1073/pnas.0601417103
Dannlowski, U., Stuhrmann, A., Beutelmann, V., Zwanzger, P., Lenzen, T., Grotegerd, D., . Kugel, H. (2012). Limbic scars: long-term consequences of childhood maltreatment revealed by functional and structural magnetic resonance imaging. . Biological Psychiatry, 71(4), 286–293.
Das, P., Calhoun, V., & Malhi, G.S. (2014). Bipolar and borderline patients display differential patterns of functional connectivity among resting state networks. Neuroimage, 98, 73–81. https://doi.org/10.1016/j.neuroimage.2014.04.062.
Davidson, R. J. (2002). Anxiety and affective style: role of prefrontal cortex and amygdala. *Biological Psychiatry, 51*(1), 68–80. https://doi.org/10.1016/S0006-3223(01)01328-2.
Davis, M. & Whalen, P. J. (2001). The amygdala: vigilance and emotion. *Molecular Psychiatry, 6*(1), 13–34. https://doi.org/10.1038/sj.mp.4000812.
Depping, M. S., Wolf, N. D., Vasic, N., Sambataro, F., Thomann, P. A., & Wolf, R. C. (2016). Common and distinct structural network abnormalities in major depressive disorder and borderline personality disorder. Progress in Neuropsychopharmacology & Biological Psychiatry, 65, 127–133. https://doi.org/10.1016/j.pnpbp.2015.09.007.
Doll, A., Sorg, C., Manoliu, A., Woller, A., Meng, C., Forstl, H., Zimmer, C., Wohlschlager, A. M., Riedl, V. (2013). Shifted intrinsic connectivity of central executive and salience network in borderline personality disorder. Front Hum Neurosci, 7, 727. https://doi.org/10.3389/fnhum.2013.00727
Domsalla, M., Koppe, G., Niedtfeld, I., Vollstadt-Klein, S., Schmahl, C:, Bohus, M. & Lis, S. (2014). Cerebral processing of social rejection in patients with borderline personality disorder. Soc Cogn Affect Neurosci, 9 (11), 1789–1797.
Donegan, N. H., Sanislow, C. A., Blumberg, H. P., Fulbright, R. K., Lacadie, C., Skudlarski, P., . Wexler, B. E. (2003). Amygdala hyperreactivity in borderline personality disorder: implications for emotional dysregulation. Biological psychiatry, 54(11), 1284–1293.
Dosenbach, N. U., Visscher, K. M., Palmer, E. D., Miezin, F. M., Wenger, K. K., Kang, H. C., … Petersen, S. E. (2006). A core system for the implementation of task sets. *Neuron, 50*(5), 799–812. https://doi.org/10.1016/j.neuron.2006.04.031.
Drevets, W. C. & Raichle, M. E. (1998). Reciprocal suppression of regional cerebral blood flow during emotional versus higher cognitive processes: implications for interactions between emotion and cognition. *Cognition and Emotion, 12*(3), 353–385. https://doi.org/10.1080/026999398379646.
Driessen, M., Herrmann, J., Stahl, K., Zwaan, M., Meier, S., Hill, A., . Petersen, D. (2000). Magnetic resonance imaging volumes of the hippocampus and the amygdala in women with borderline personality disorder and early traumatization. Archives of General Psychiatry, 57(12), 1115–1122.
Dziobek, I., Preissler, S., Grozdanovic, Z., Heuser, I., Heekeren, H. R., & Roepke, S. (2011). Neuronal correlates of altered empathy and social cognition in borderline personality disorder. . Neuroimage, 57(2), 539–548.
Ende, G., Cackowski, S., VanEijk, J., Sack, M., Bohus, M., Sobanski, E., Krause-Utz, A., Schmahl, C. (2016). Impulsivity and aggression are associated with anterior cingulate glutamate and GABA concentrations in BPD and ADHD. Neuropsychopharmacology.
Etkin, A., Egner, T. & Kalisch, R. (2011). Emotional processing in anterior cingulate and medial prefrontal cortex. *Trends in Cognitive Science, 15*(2), 85–93. https://doi.org/10.1016/j.tics.2010.11.004.

Fineberg, N. A., Potenza, M. N., Chamberlain, S. R., Berlin, H. A., Menzies, L., Bechara, A., . Hollander, E. (2010). Probing compulsive and impulsive behaviors, from animal models to endophenotypes: a narrative review. Neuropsychopharmacology, 35(3), 591–604.

Frick, C., Lang, S., Kotchoubey, B., Sieswerda, S., Dinu-Biringer, R., Berger, M., . Barnow, S. (2012). Hypersensitivity in borderline personality disorder during mindreading. PLoS One, 7(8), e41650.

Friston, K. J., Büchel, C., Fink, G. R., Morris, J., Rolls, E. & Dolan, R. J. (1997). Psychophysiological and modulatory interactions in neuroimaging. Neuroimage, 6(3), 218–229. https://doi.org/10.1006/nimg.1997.0291.

Friston, K. J., Fletcher, P., Josephs, O., Holmes, A., Rugg, M. D. & Turner, R. (1998). Event-related fMRI: characterizing differential responses. Neuroimage, 7(1), 30–40. https://doi.org/10.1006/nimg.1997.0306.

Fox, M. D., & Raichle, M. E. (2007). Spontaneous fluctuations in brain activity observed with functional magnetic resonance imaging. Nature reviews. Neuroscience, 8(9), 700–711. https://doi.org/10.1038/nrn2201

Gan, J., Yi, J., Zhong, M., Cao, X., Jin, X., Liu, W., & Zhu, X. (in press). Abnormal white matter structural connectivity in treatment-naïve young adults with borderline personality disorder. Acta Psychiatrica Scandinavica, 2016 Sep 9. doi: 10.1111/acps.12640. [Epub ahead of print]

Gilbert, R., Widom, C. S., Browne, K., Fergusson, D., Webb, E., & Janson, S. (2009). Burden and consequences of child maltreatment in high-income countries. Lancet, 373(9657), 68–81.

Glenn, C. R., & Klonsky, E. D. (2009). Emotion dysregulation as a core feature of borderline personality disorder. Journal of Personality Disorders, 23(1), 20–28.

Golier, J. A., Yehuda, R., Bierer, L. M., Mitropoulou, V., New, A. S., Schmeidler, J., . Siever, L. J. (2003). The relationship of borderline personality disorder to posttraumatic stress disorder and traumatic events. Am J Psychiatry, 160(11), 2018–2024.

Goodman, M., Carpenter, D., Tang, C. Y., Goldstein, K. E., Avedon, J., Fernandez, N., Mascitelli, K. A., Blair, N. J., New, A. S., Triebwasser, J., Siever, L. J., Hazlett, E. A. (2014). Dialectical behavior therapy alters emotion regulation and amygdala activity inpatients with borderline personality disorder. J Psychiatr Res, 57, 108–116.

Goodman, M., Hazlett, E. A., Avedon, J. B., Siever, D. R., Chu, K. W., & New, A. S. (2011). Anterior cingulate volume reduction in adolescents with borderline personality disorder and co-morbid major depression. . Journal of psychiatric research, 45(6), 803–807.

Grant, J. E., Correia, S., Brennan-Krohn, T., Malloy, P. F., Laidlaw, D. H., & Schulz, S. C. (2007). Frontal white matter integrity in borderline personality disorder with self-injurious behavior. Journal of Neuropsychiatry and Clinical Neurosciences, 19(4), 383–390.

Greicius, M. D., Krasnow, B., Reiss, A. L., & Menon, V. (2003). Functional connectivity in the resting brain: a network analysis of the default mode hypothesis. Proceedings of the National Academy of Sciences of the United States of America, 100(1), 253–258. https://doi.org/10.1073/pnas.0135058100

Guitart-Masip, M., Pascual, J. C., Carmona, S., Hoekzema, E., Berge, D., Perez, V., . Vilarroya, O. (2009). Neural correlates of impaired emotional discrimination in borderline personality disorder: an fMRI study. Prog Neuropsychopharmacol Biol Psychiatry, 33(8), 1537–1545.

Hampson, M., Olson, I. R., Leung, H. C., Skudlarski, P., & Gore, J. C. (2004). Changes in functional connectivity of human MT/V5 with visual motion input. *Neuroreport, 15*(8), 1315–1319. https://doi.org/10.1097/01.wnr.0000129997.95055.15.

Hampson, M., Peterson, B. S., Skudlarski, P., Gatenby, J. C., & Gore, J. C. (2002). Detection of functional connectivity using temporal correlations in MR images. *Human Brain Mapping, 15*(4), 247–262. https://doi.org/10.1002/hbm.10022.

Hazlett, E. A., New, A. S., Newmark, R., Haznedar, M. M., Lo, J. N., Speiser, L. J., . Buchsbaum, M. S. (2005). Reduced anterior and posterior cingulate gray matter in borderline personality disorder. Biological Psychiatry, 58(8), 614–623.

Hazlett, E. A., Zhang, J., New, A.S., Zelmanova, Y., Goldstein, K. E., Haznedar, M. M., Meyerson, D., Goodman, M., Siever, L. J., Chu, K. W. (2012). Potentiated amygdala response to repeated emotional pictures in borderline personality disorder. Biol Psychiatry, 72(6), 448–456.

Herpertz, S. C., Dietrich, T. M., Wenning, B., Krings, T., Erberich, S. G., Willmes, K., . Sass, H. (2001). Evidence of abnormal amygdala functioning in borderline personality disorder: a functional MRI study. Biological psychiatry, 50(4), 292–298. Archives of General Psychiatry, 67(9), 946–954

Holtmann, J., Herbort, M. C., Wustenberg, T., Soch, J., Richter, S., Walter, H., . Schott, B. H. (2013). Trait anxiety modulates fronto-limbic processing of emotional interference in borderline personality disorder. Front Hum Neurosci, 7, 54. https://doi.org/10.3389/fnhum.2013.00054

Irle, E., Lange, C., & Sachsse, U. (2005). Reduced size and abnormal asymmetry of parietal cortex in women with borderline personality disorder. Biological psychiatry, 57(2), 173–182.

Irle, E., Lange, C., Weniger, G., & Sachsse, U. (2007). Size abnormalities of the superior parietal cortices are related to dissociation in borderline personality disorder. Psychiatry research, 156(2), 139–149.

Jacob, G. A., Zvonik, K., Kamphausen, S., Sebastian, A., Maier, S., Philipsen, A., . Tuscher, O. (2013). Emotional modulation of motor response inhibition in women with borderline personality disorder: an fMRI study. J Psychiatry Neurosci, 38(3), 164–172.

Johnstone, T., van Reekum, C. M., Urry, H. L., Kalin, N. H. & Davidson, R. J. (2007). Failure to regulate: counterproductive recruitment of top-down prefrontal-subcortical circuitry in major depression. *Journal of Neuroscience, 27*(33), 8877–8884. https://doi.org/10.1523/JNEUROSCI.2063-07.2007.

Kamphausen, S., Schroder, P., Maier, S., Bader, K., Feige, B., Kaller, C. P., . Tuscher, O. (2012). Medial prefrontal dysfunction and prolonged amygdala response during instructed fear processing in borderline personality disorder. World J Biol Psychiatry. https://doi.org/10.3109/15622975.2012.665174

Karl, A., Schaefer, M., Malta, L. S., Dorfel, D., Rohleder, N., & Werner, A. (2006). A meta-analysis of structural brain abnormalities in PTSD. Neuroscience and biobehavioral reviews, 30(7), 1004–1031.

Kemperman, I., Russ, M. J., & Shearin, E. N. (1997). Self-injurious behavior and mood regulation in borderline patients. Journal of Personality Disorders, 11(2), 146–157.

Killgore, W. D., Britton, J. C., Schwab, Z. J., Price, L. M., Weiner, M. R., Gold, A. L., Rosso, I. M., Simon, N. M., Pollack, M. H., Rauch, S. L. (2014). Cortico-limbic responses to masked affective faces across PTSD, panic disorder, and specific phobia. Depress Anxiety, 31(2), 150–159.

King-Casas, B., Sharp, C., Lomax-Bream, L., Lohrenz, T., Fonagy, P., & Montague, P. R. (2008). The rupture and repair of cooperation in borderline personality disorder. Science, 321(5890), 806–810.

Klauser, P., Fornito, A., Lorenzetti, V., Davey, C. G., Dwyer, D. B., Allen, N. B., & Yücel, M. (2015). Cortico-limbic network abnormalities in individuals with current and past major depressive disorder. Journal of affective disorders, 173, 45–52. https://doi.org/10.1016/j.jad.2014.10.041

Kleindienst, N., Bohus, M., Ludascher, P., Limberger, M. F., Kuenkele, K., Ebner-Priemer, U. W., … Schmahl, C. (2008). Motives for nonsuicidal self-injury among women with borderline personality disorder. The Journal of nervous and mental disease, 196(3), 230–236.

Kluetsch, R. C., Schmahl, C., Niedtfeld, I., Densmore, M., Calhoun, V. D., Daniels, J., . Lanius, R. A. (2012). Alterations in Default Mode Network Connectivity During Pain Processing in Borderline Personality Disorder. Archives of General Psychiatry, 1–11.

Klonsky, E. D., & Muehlenkamp, J. J. (2007). Self-injury: a research review for the practitioner. Journal of clinical psychology, 63(11), 1045–1056. https://doi.org/10.1002/jclp.20412.

Koenigsberg, H. W., Fan, J., Ochsner, K. N., Liu, X., Guise, K. G., Pizzarello, S., . Siever, L. J. (2009). Neural correlates of the use of psychological distancing to regulate responses to negative social cues: a study of patients with borderline personality disorder. Biological psychiatry, 66(9), 854–863.

Koenigsberg, H. W., Denny, B. T., Fan, J., Liu, X., Guerreri, S., Mayson, S. J., Rimsky, L., New, A. S., Goodman, M., Siever, L. J. (2014). The neural correlates of anomalous habituation to negative emotional pictures in borderline and avoidant personality disorder patients. Am J Psychiatry, 171(1), 82–90.

Korzekwa, M. I., Dell, P. F., Links, P. S., Thabane, L., & Fougere, P. (2009). Dissociation in borderline personality disorder: a detailed look. J Trauma Dissociation, 10(3), 346–367.
Korzekwa, M. I., Dell, P. F., & Pain, C. (2009a). Dissociation and borderline personality disorder: an update for clinicians. Current psychiatry reports, 11(1), 82–88. https://doi.org/10.1007/s11920-009-0013-1.
Korzekwa, M. I., Dell, P. F., Links, P. S., Thabane, L., & Fougere, P. (2009b). Dissociation in borderline personality disorder: a detailed look. Journal of trauma & dissociation : the official journal of the International Society for the Study of Dissociation (ISSD), 10(3), 346–367. https://doi.org/10.1080/15299730902956838.
Kraus, A., Esposito, F., Seifritz, E., Di Salle, F., Ruf, M., Valerius, G., Schmahl, C. (2009). Amygdala deactivation as a neural correlate of pain processing in patients with borderline personality disorder and co-occurrent posttraumatic stress disorder. . Biological psychiatry, 65(9), 819–822.
Krause-Utz, A., Cackowski, S., Daffner, S., Sobanski, E., Plichta, M. M., Bohus, M., Ende, G., & Schmahl, C. (in press). Delay discounting and response disinhibition under acute experimental stress in women with borderline personality disorder and adult attention deficit hyperactivity disorder. Psychol Med. 2016 Aug 30, 1–13. [Epub ahead of print] PubMed PMID: 27572473.
Krause-Utz, A., Elzinga, B. M., Oei, N. Y. L., Paret, C., Niedtfeld, I., Spinhoven, P., Bohus, M., Schmahl, C. (2014c). Amygdala and dorsal anterior cingulate connectivity during an emotional working memory task in borderline personality disorder patients with interpersonal trauma history. Front Hum Neurosci, 8, 848. https://doi.org/10.3389/fnhum.2014.00848.
Krause-Utz, A., Frost, R., Winter, D., & Elzinga, B. M. Dissociation and Alterations in Brain Function and Structure: Implications for Borderline Personality Disorder. Current Psychiatry Reports. Manuscript accepted for publication.
Krause-Utz, A., Keibel-Mauchnik, J., Ebner-Priemer, U., Bohus, M., Schmahl, C. (in press). Classical conditioning in borderline personality disorder: an fMRI study. Eur Arch Psychiatry Clin Neurosci, 2015 Mar 27. [Epub ahead of print] PubMed PMID: 25814470.
Krause-Utz, A., Oei, N. Y., Niedtfeld, I., Bohus, M., Spinhoven, P., Schmahl, C., & Elzinga, B. M. (2012). Influence of emotional distraction on working memory performance in borderline personality disorder. Psychol Med, 42(10), 2181–2192.
Krause-Utz, A., & Schmahl, C. (2010). Neurobiological Differentiation Between Borderline Patients With and Without Post-traumatic Stress Disorder. European Psychiatric Review 3(2), 63–68.
Krause-Utz, A., Sobanski, E., Alm, B., Valerius, G., Kleindienst, N., Bohus, M., & Schmahl, C. (2013). Impulsivity in relation to stress in patients with borderline personality disorder with and without co-occurring attention-deficit/hyperactivity disorder: an exploratory study. The Journal of nervous and mental disease, 201(2), 116–123. https://doi.org/10.1097/NMD.0b013e31827f6462.
Krause-Utz, A. & Schmahl, C. (2016). A More Global Look at Altered Neural Structure and Resting-State Function in Borderline Personality Disorder. Biological Psychiatry, 79(2), 76–77. https://doi.org/10.1016/j.biopsych.2015.10.011.
Krause-Utz, A., Veer, I. M., Rombouts, S. A. R. B., Bohus, M., Schmahl, C., Elzinga, B. M. (2014b). Amygdala and anterior cingulate resting-state functional connectivity in borderline personality disorder patients with a history of interpersonal trauma. Psychol Med, 44(13), 2889–2901
Krause-Utz, A., Winter, D., Niedtfeld, I., & Schmahl, C. (2014a). The latest neuroimaging findings in borderline personality disorder. Curr Psychiatry Rep, 16(3), 438. https://doi.org/10.1007/s11920-014-0438-z.
Krause-Utz A, Winter D, Schriner F, Chiu C-D, Lis S, Spinhoven P, et al. Reduced Amygdala Activity and Emotional Distractibility during Dissociation in Borderline Personality Disorder. Under review
Krystal, J. H., Bremner, J. D., Southwick, S. M., & Charney, D. S. (1998). The emerging neurobiology of dissociation: implications for treatment of posttraumatic stress disorder. In J. D. Bremner & C. R. Marmar (Eds.), (Vol. 1, pp. 321–363). Washington DC: American Psychiatric Press.

Kuhlmann, A., Bertsch, K., Schmidinger, I., Thomann, P. A. & Herpertz, S. C. (2013). Morphometric differences in central stress-regulating structures between women with and without borderline personality disorder. J Psychiatry Neurosci, 38 (2), 129–137.
Laird, A. R., Fox, P. M., Eickhoff, S. B., Turner, J. A., Ray, K. L., Mckay, D. R., … Fox, P. T. (2011). Behavioral interpretations of intrinsic connectivity networks. Journal of Cognitive Neuroscience, 23, 4022–4037. https://doi.org/10.1162/jocn_a_00077.
Lang, S., Kotchoubey, B., Frick, C., Spitzer, C., Grabe, H. J., & Barnow, S. (2012). Cognitive reappraisal in trauma-exposed women with borderline personality disorder. Neuroimage, 59(2), 1727–1734.
Lanius, R. A., Brand, B., Vermetten, E., Frewen, P. A., & Spiegel, D. (2012). The dissociative subtype of posttraumatic stress disorder: rationale, clinical and neurobiological evidence, and implications. [Review]. Depress Anxiety, 29(8), 701–708.
Lanius, R. A., Vermetten, E., Loewenstein, R. J., Brand, B., Schmahl, C., Bremner, J. D., & Spiegel, D. (2010). Emotion modulation in PTSD: Clinical and neurobiological evidence for a dissociative subtype. American Journal of Psychiatry, 167(6), 640–647.
LeDoux, J. E. (1992). Brain mechanisms of emotion and emotional learning. *Current Opinions in Neurobiology, 2*(2), 191–197. https://doi.org/10.1016/0959-4388(92)90011-9.
Lee, R., Arfanakis, K., Evia, A. M., Fanning, J., Keedy, S., & Coccaro, E. F. (2016). White Matter Integrity Reductions in Intermittent Explosive Disorder. Neuropsychopharmacology, 41(11), 2697–2703. https://doi.org/10.1038/npp.2016.74.
Leichsenring, F., Leibing, E., Kruse, J., New, A. S., & Leweke, F. (2011). Borderline personality disorder. Lancet, 377(9759), 74–84.
Leyton, M., Okazawa, H., Diksic, M., Paris, J., Rosa, P., Mzengeza, S., . Benkelfat, C. (2001). Brain Regional alpha-[11C]methyl-L-tryptophan trapping in impulsive subjects with borderline personality disorder. American Journal of Psychiatry, 158(5), 775–782.
Lieb, K., Zanarini, M. C., Schmahl, C., Linehan, M. M., & Bohus, M. (2004). Borderline personality disorder. Lancet, 364(9432), 453–461. https://doi.org/10.1016/s0140-6736(04)16770-6
Lis, S., & Bohus, M. (2013). Social interaction in borderline personality disorder. Curr Psychiatry Rep, 15(2), 338. https://doi.org/10.1007/s11920-012-0338-z
Lischke, A., Domin, M., Freyberger, H. J., Grabe, H. J., Mentel, R., Bernheim, D., & Lotze, M. (2015). Structural alterations in white-matter tracts connecting (para-)limbic and prefrontal brain regions in borderline personality disorder. Psychological Medicine, 45(15), 3171–3180. https://doi.org/10.1017/S0033291715001142.
Lowe, M. J., Mock, B. J., & Sorenson, J. A. (1998). Functional connectivity in single and multislice echoplanar imaging using resting-state fluctuations. NeuroImage, 7(2), 119–132. https://doi.org/10.1006/nimg.1997.0315.
Ludäscher, P., Bohus, M., Lieb, K., Philipsen, A., & Schmahl, C. (2007). Elevated pain thresholds correlate with dissociation and aversive arousal in patients with borderline personality disorder. Psychiatry Research, 149(1–3), 291–296.
Ludäscher, P., Valerius, G., Stiglmayr, C., Mauchnik, J., Lanius, R., Bohus, M., & Schmahl, C. (2010). Pain Sensitivity and Neural Processing during Dissociative States in Patients with Borderline Personality Disorder with and without comorbid PTSD - a Pilot Study. J Psychiatry Neuroscience 35(3):177–84.
Lynch, T. R., Rosenthal, M. Z., Kosson, D. S., Cheavens, J. S., Lejuez, C. W., & Blair, R. J. (2006). Heightened sensitivity to facial expressions of emotion in borderline personality disorder. . Emotion, 6(4), 647–655. d
Lyoo, I. K., Han, M. H., & Cho, D. Y. (1998). A brain MRI study in subjects with borderline personality disorder. Journal of Affective Disorders, 50(2–3), 235–243.
Maier-Hein, K. H., Brunner, R., Lutz, K., Henze, R., Parzer, P., Feigl, N., Kramer, J., Meinzer, H. P., Resch, F., Stieltjes, B. (2014). Disorder-specific white matter alterations in adolescent borderline personality disorder. Biol Psychiatry, 75 (1), 81–88.
Margulies, D. S., Kelly, A. M., Uddin, L. Q., Biswal, B. B., Castellanos, F. X. & Milham, M. P. (2007). Mapping the functional connectivity of anterior cingulate cortex. *NeuroImage, 37*(2), 579–588. https://doi.org/10.1016/j.neuroimage.2007.05.019.

McCloskey, M. S., New, A. S., Siever, L. J., Goodman, M., Koenigsberg, H. W., Flory, J. D., & Coccaro, E. F. (2009). Evaluation of behavioral impulsivity and aggression tasks as endophenotypes for borderline personality disorder. J Psychiatr Res, 43(12), 1036–1048.

McCown, W., Galina, H., Johnson, J., de Simone, P. A., & Posa, J. (1993). Borderline personality disorder and laboratory-induced cold pressor pain: Evidence of stress-induced analgesia. Journal of Psychopathology and Behavioral Assessment, 15(2), 87–95.

McRae, K., Hughes, B., Chopra, S., Gabrieli, J. D., Gross, J. J. & Ochsner, K. N. (2010). The neural bases of distraction and reappraisal. *Journal of Cognitive Neuroscience, 22*(2), 248–262. https://doi.org/10.1162/jocn.2009.21243.

Menon, V. (2011). Large-scale brain networks and psychopathology: a unifying triple network model. *Trends in Cognitive Science, 15*(10): 483–506. https://doi.org/10.1016/j.tics.2011.08.003.

Menon, V., & Uddin, L. Q. (2010). Saliency, switching, attention and control: a network model of insula function. Brain Struct Funct, 214(5–6), 655–667.

Mier, D., Lis, S., Esslinger, C., Sauer, C., Hagenhoff, M., Ulferts, J., . Kirsch, P. (2013). Neuronal correlates of social cognition in borderline personality disorder. Soc Cogn Affect Neurosci, 8(5), 531–537.

Miller, E. K. (2000). The prefrontal cortex and cognitive control. *Nature Reviews. Neuroscience, 1*(1), 59–65. https://doi.org/10.1038/35044578.

Minzenberg, M. J., Fan, J., New, A. S., Tang, C. Y., & Siever, L. J. (2007). Fronto-limbic dysfunction in response to facial emotion in borderline personality disorder: an event-related fMRI study. Psychiatry research, 155(3), 231–243.

Minzenberg, M. J., Fan, J., New, A. S., Tang, C. Y., & Siever, L. J. (2008). Frontolimbic structural changes in borderline personality disorder. Journal of psychiatric research, 42(9), 727–733.

Morey, R. A., Dolcos, F., Petty, C. M., Cooper, D. A., Hayes, J. P., LaBar, K. S., & McCarthy, G. (2009). The role of trauma-related distractors on neural systems for working memory and emotion processing in posttraumatic stress disorder. Journal of psychiatric research, 43(8), 809–817. https://doi.org/10.1016/j.jpsychires.2008.10.014.

Mukherjee, P., Sabharwal, A., Kotov, R., Szekely, A., Parsey, R., Barch, D. M., & Mohanty, A. (2016). Disconnection Between Amygdala and Medial Prefrontal Cortex in Psychotic Disorders. Schizophrenia Bulletin, 42(4), 1056–1067. https://doi.org/10.1093/schbul/sbw012.

New, A. S., Goodman, M., Triebwasser, J., & Siever, L. J. (2008). Recent advances in the biological study of personality disorders. Psychiatr Clin North Am, 31(3), 441–461.

New, A. S., Hazlett, E. A., Buchsbaum, M. S., Goodman, M., Mitelman, S. A., Newmark, R., ... Siever, L. J. (2007). Amygdala-prefrontal disconnection in borderline personality disorder. Neuropsychopharmacology, 32(7), 1629–1640.

New, A. S., Carpenter, D. M., Perez-Rodriguez, M. M., Ripoll, L. H., Avedon, J., Patil, U., …, Goodman, M. (2013). Developmental differences in diffusion tensor imaging parameters in borderline personality disorder. J Psychiatr Res, 47(8), 1101–1109.

Niedtfeld, I., Kirsch, P., Schulze, L., Herpertz, S. C., Bohus, M., & Schmahl, C. (2012). Functional Connectivity of Pain-Mediated Affect Regulation in Borderline Personality Disorder. PloS one, 7(3), e33293.

Niedtfeld, I., Schulze, L., Kirsch, P., Herpertz, S. C., Bohus, M., & Schmahl, C. (2010). Affect regulation and pain in borderline personality disorder: a possible link to the understanding of self-injury. Biological psychiatry, 68(4), 383–391.

Niedtfeld, I., Schulze, L., Krause-Utz, A., Demirakca, T., Bohus, M., & Schmahl, C. (2013). Voxel-based morphometry in women with borderline personality disorder with and without comorbid posttraumatic stress disorder. PloS one, 8(6), e65824. https://doi.org/10.1371/journal.pone.0065824.

Nunes, P. M., Wenzel, A., Borges, K. T., Porto, C. R., Caminha, R. M., & de Oliveira, I. R. (2009). Volumes of the hippocampus and amygdala in patients with borderline personality disorder: a meta-analysis. J Pers Disorders, 23(4), 333–345.

O'Neill, A., D'Souza, A., Samson, A., Carballedo, A., Kerskens, C., & Frodl, T. (2015). Dysregulation between emotion and theory of mind networks in borderline personality disorder. Psychiatry Research, 231(1), 25–32. https://doi.org/10.1016/j.pscychresns.2014.11.002.

O'Neill, A., & Frodl, T. (2012). Brain structure and function in borderline personality disorder. Brain Struct Funct, 217(4), 767–782.

Ochsner, K. N., Bunge, S. A., Gross, J. J., & Gabrieli, J. D. (2002). Rethinking feelings: an FMRI study of the cognitive regulation of emotion. J Cogn Neurosci 14(8), 1215–1229.

Ochsner, K. N., & Gross, J. J. (2007). The Neural Architecture of Emotion Regulation. In J. J. Gross (Ed.), Handbook of emotion regulation (p. 87–109). The Guilford Press.

Ochsner, K. N., Ray, R. D., Cooper, J. C., Robertson, E. R., Chopra, S., Gabrieli, J. D., & Gross, J. J. (2004). For better or for worse: neural systems supporting the cognitive down- and up-regulation of negative emotion. NeuroImage, 23(2), 483–499.

Oei, N. Y., Veer, I. M., Wolf, O. T., Spinhoven, P., Rombouts, S. A. & Elzinga, B. M. (2012). Stress shifts brain activation towards ventral "affective" areas during emotional distraction. *Social Cognitive and Affective Neuroscience, 7*(4), 403–412. https://doi.org/10.1093/scan/nsr024.

O'Reilly, J. X., Woolrich, M. W., Behrens, T. E., Smith, S. M. & Johansen-Berg, H. (2012). Tools of the trade: psychophysiological interactions and functional connectivity. Social Cognitive and Affective Neuroscience, 7(5), 604–609. https://doi.org/10.1093/scan/nss055.

Owen, A. M., McMillan, K. M., Laird, A. R., & Bullmore, E. (2005). N-back working memory paradigm: a meta-analysis of normative functional neuroimaging studies. *Human Brain Mapping, 25*(1), 46–59. https://doi.org/10.1002/hbm.20131.

Paret, C., Kluetsch, R., Zaehringer, J., Ruf, M., Demirakca, T., Bohus, M., …, Schmahl, C. (2016). Alterations of amygdala-prefrontal connectivity with real-time fMRI neurofeedback in BPD patients. Social Cognitive Affective Neuroscience, 11(6), 952–960. https://doi.org/10.1093/scan/nsw016.

Pessoa, L., McKenna, M., Gutierrez, E., & Ungerleider, L. G. (2002). Neural processing of emotional faces requires attention. Proc Natl Acad Sci U S A, 99(17), 11458–11463.

Pessoa, L., Padmala, S., Kenzer, A., & Bauer, A. (2012). Interactions between cognition and emotion during response inhibition. Emotion, 12(1), 192–197.

Phan, K. L., Fitzgerald, D. A., Nathan, P. J., Moore, G. J., Uhde, T. W. & Tancer, M. E. (2005). Neural substrates for voluntary suppression of negative affect: a functional magnetic resonance imaging study. *Biological Psychiatry, 57*(3), 210–219. https://doi.org/10.1016/j.biopsych.2004.10.030.

Phillips, M. L., Drevets, W. C., Rauch, S. L. & Lane, R. (2003). Neurobiology of emotion perception I: The neural basis of normal emotion perception. *Biological Psychiatry, 54*(5), 504–514. https://doi.org/10.1016/S0006-3223(03)00171-9.

Prehn, K., Schulze, L., Rossmann, S., Berger, C., Vohs, K., Fleischer, M., . Herpertz, S. C. (2013). Effects of emotional stimuli on working memory processes in male criminal offenders with borderline and antisocial personality disorder. World J Biol Psychiatry, 14(1), 71–78.

Qin, S., Duan, X., Supekar, K., Chen, H., Chen, T., & Menon, V. (2016). Large-scale intrinsic functional network organization along the long axis of the human medial temporal lobe. *Brain Structure and Function, 221*(6), 3237–58. https://doi.org/10.1007/s00429-015-1098-4.

Radaelli, D., Poletti, S., Dallaspezia, S., Colombo, C., Smeraldi, E., & Benedetti, F. (2012). Neural responses to emotional stimuli in comorbid borderline personality disorder and bipolar depression. Psychiatry Res, 203(1), 61–66.

Raichle, M. E., MacLeod, A. M., Snyder, A. Z., Powers, W. J., Gusnard, D. A., & Shulman, G. L. (2001). A default mode of brain function. Proceedings of the National Academy of Sciences of the United States of America, 98(2), 676–682. https://doi.org/10.1073/pnas.98.2.676.

Rainville, P. (2002). Brain mechanisms of pain affect and pain modulation. Curr Opin Neurobiol, 12(2), 195–204.

Reitz, S., Kluetsch, R., Niedtfeld, I., Knorz, T., Lis, S., Paret, C., …, Schmahl, C. (2015). Incision and stress regulation in borderline personality disorder: neurobiological mechanisms of self-injurious behaviour. British Journal of Psychiatry, 207(2), 165–172. https://doi.org/10.1192/bjp.bp.114.153379.

Reitz, S., Krause-Utz, A., Pogatzki-Zahn, E. M., Ebner-Priemer, U., Bohus, M., & Schmahl, C. (2012). Stress regulation and incision in borderline personality disorder – a pilot study modeling cutting behavior. J Pers Disord, 26, 605–615.

Rodrigues, E., Wenzel, A., Ribeiro, M. P., Quarantini, L. C., Miranda-Scippa, A., de Sena, E. P., & de Oliveira, I. R. (2011). Hippocampal volume in borderline personality disorder with and without comorbid posttraumatic stress disorder: a meta-analysis. Eur Psychiatry, 26(7), 452–456.

Rosenthal, M. Z., Gratz, K. L., Kosson, D. S., Cheavens, J. S., Lejuez, C. W., & Lynch, T. R. (2008). Borderline personality disorder and emotional responding: a review of the research literature. [Review]. Clinical psychology review, 28(1), 75–91.

Roy, A. K., Shehzad, Z., Margulies, D. S., Kelly, A. M., Uddin, L. Q., Gotimer, K., … Milham, M. P. (2009). Functional connectivity of the human amygdala using resting state fMRI. *NeuroImage, 45*(2), 614–626. https://doi.org/10.1016/j.neuroimage.2008.11.030.

Ruocco, A. C., Amirthavasagam, S., Choi-Kain, L. W., & McMain, S. F. (2013). Neural correlates of negative emotionality in borderline personality disorder: an activation-likelihood-estimation meta-analysis. Biol Psychiatry, 73(2), 153–160.

Ruocco, A. C., Medaglia, J. D., Tinker, J. R., Ayaz, H., Forman, E. M., Newman, C. F., . Chute, D. L. (2010). Medial prefrontal cortex hyperactivation during social exclusion in borderline personality disorder. Psychiatry Res, 181(3), 233–236.

Rüsch, N., Bracht, T., Kreher, B.W., Schnell, S., Glauche, V., Il'yasov, K. A., …, van Elst, L. T. (2010). Reduced interhemispheric structural connectivity between anterior cingulate cortices in borderline personality disorder. Psychiatry Res, 181(2), 151-154. https://doi.org/10.1016/j.pscychresns.2009.08.004.

Rüsch, N., van Elst, L. T., Ludaescher, P., Wilke, M., Huppertz, H. J., Thiel, T., … Ebert, D. (2003). A voxel-based morphometric MRI study in female patients with borderline personality disorder. NeuroImage, 20(1), 385–392.

Rüsch, N., Weber, M., Il'yasov, K. A., Lieb, K., Ebert, D., Hennig, J., & van Elst, L. T. (2007). Inferior frontal white matter microstructure and patterns of psychopathology in women with borderline personality disorder and comorbid attention-deficit hyperactivity disorder. . NeuroImage, 35(2), 738–747.

Russ, M. J., Roth, S. D., Lerman, A., Kakuma, T., Harrison, K., Shindledecker, R. D., ... Mattis, S. (1992). Pain perception in self-injurious patients with borderline personality disorder. Biological psychiatry, 32(6), 501–511.

Sala, M., Caverzasi, E., Lazzaretti, M., Morandotti, N., De Vidovich, G., Marraffini, E., . . . Brambilla, P. (2011). Dorsolateral prefrontal cortex and hippocampus sustain impulsivity and aggressiveness in borderline personality disorder. J Affect Disord, 131 (1–3), 417–421.

Salvador, R., Vega, D., Pascual, J. C., Marco, J., Canales-Rodríguez, E. J., Aguilar, S., …, Pomarol-Clotet, E. (2016).Converging Medial Frontal Resting State and Diffusion-Based Abnormalities in Borderline Personality Disorder. Biological Psychiatry, 79(2), 107–116. https://doi.org/10.1016/j.biopsych.2014.08.026.

Schmahl, C., Berne, K., Krause, A., Kleindienst, N., Valerius, G., Vermetten, E., & Bohus, M. (2009). Hippocampus and amygdala volumes in patients with borderline personality disorder with or without posttraumatic stress disorder. J Psychiatry Neurosci 34(4), 289–295.

Schmahl, C., & Bohus, M. (2007). Translational research issues in dissociation. In E. Vermetten, M. J. Dorahy & D. Spiegel (Eds.), Traumatic Dissociation (pp. 121–139). Washington DC: American Psychiatric Publishing.

Schmahl, C., Bohus, M., Esposito, F., Treede, R. D., Di Salle, F., Greffrath, W., . Seifritz, E. (2006). Neural correlates of antinociception in borderline personality disorder. Archives of General Psychiatry, 63(6), 659–667.

Schmahl, C., & Bremner, J. D. (2006). Neuroimaging in borderline personality disorder. Journal of Psychiatric Research 40(5), 419–427.

Schmahl, C., Meinzer, M., Zeuch, A., Fichter, M., Cebulla, M., Kleindienst, N., ... Bohus, M. (2010). Pain sensitivity is reduced in borderline personality disorder, but not in posttraumatic stress disorder and bulimia nervosa. The world journal of biological psychiatry, 364–371.

Schmahl, C. G., Elzinga, B. M., Vermetten, E., Sanislow, C., McGlashan, T. H., & Bremner, J. D. (2003a). Neural correlates of memories of abandonment in women with and without borderline personality disorder. Biological psychiatry, 54(2), 142–151.

Schmahl, C. G., Vermetten, E., Elzinga, B. M., & Douglas Bremner, J. (2003b). Magnetic resonance imaging of hippocampal and amygdala volume in women with childhood abuse and borderline personality disorder. Psychiatry Res, 122(3), 193–198.

Schmahl, C., Greffrath, W., Baumgartner, U., Schlereth, T., Magerl, W., Philipsen, A., … Treede, R. D. (2004a). Differential nociceptive deficits in patients with borderline personality disorder and self-injurious behavior: laser-evoked potentials, spatial discrimination of noxious stimuli, and pain ratings. Pain, 110(1-2), 470–479.

Schmahl, C. G., Vermetten, E., Elzinga, B. M., & Bremner, J. D. (2004b). A positron emission tomography study of memories of childhood abuse in borderline personality disorder. Biological psychiatry, 55(7), 759–765.

Schmahl, C., Herpertz, S. C., Bertsch, K., Ende, G., Flor, H., Kirsch, P., Lis, S., Meyer-Lindenberg, A., Rietschel, M., Schneider, M., Spanagel, R., Treede, R. D., & Bohus, M. (2014). Mechanisms of disturbed emotion processing and social interaction in borderline personality disorder: state of knowledge and research agenda of the German Clinical Research Unit. Borderline personality disorder and emotion dysregulation, 1, 12. https://doi.org/10.1186/2051-6673-1-12.

Schmitt, R., Winter, D., Niedtfeld, I., Herpertz, S., Schmahl, C.: Effects of psychotherapy on neuronal correlates of reappraisal in female patients with borderline personality disorder. Biological Psychiatry: Cognitive Neuroscience and Neuroimaging (in press)

Schnell, K., & Herpertz, S. C. (2007). Effects of dialectic-behavioral-therapy on the neural correlates of affective hyperarousal in borderline personality disorder. Journal of psychiatric research, 41(10), 837–847.

Schoenbaum, G., Roesch, M. R., Stalnaker, T. A., & Takahashi, Y. K. (2009). A new perspective on the role of the orbitofrontal cortex in adaptive behaviour. Nat Rev Neurosci, 10(12), 885-892.

Schulze, L., Domes, G., Kruger, A., Berger, C., Fleischer, M., Prehn, K., ... Herpertz, S. C. (2011). Neuronal correlates of cognitive reappraisal in borderline patients with affective instability. . Biological psychiatry, 69(6), 564–573.

Schulze, L., Schmahl, C., Niedtfeld, I (2016). Neural Correlates of Disturbed Emotion Processing in Borderline Personality Disorder: A Multimodal Meta-Analysis. Biol Psychiatry 79(2), 97–106. https://doi.org/10.1016/j.biopsych.2015.03.027.

Sebastian, A., Jung, P., Krause-Utz, A., Lieb, K., Schmahl, C., Tüscher, O. (2014). Frontal dysfunctions of impulse control - a systematic review in borderline personality disorder and attention-deficit/hyperactivity disorder. Front Hum Neurosci, 8, 698. https://doi.org/10.3389/fnhum.2014.00698.

Seeley, W. W., Menon, V., Schatzberg, A. F., Keller, J., Glover, G. H., Kenna, H., … Greicius, M. D. (2007). Dissociable intrinsic connectivity networks for salience processing and executive control. *Journal of Neuroscience, 27*(9), 2349–2356. https://doi.org/10.1523/JNEUROSCI.5587-06.2007.

Shehzad, Z., Kelly, A. M., Reiss, P. T., Gee, D. G., Gotimer, K., Uddin, L. Q., Lee, S. H., Margulies, D. S., Roy, A. K., Biswal, B. B., Petkova, E., Castellanos, F. X., & Milham, M. P. (2009). The resting brain: unconstrained yet reliable. Cerebral cortex (New York, N.Y. : 1991), 19(10), 2209–2229. https://doi.org/10.1093/cercor/bhn256.

Sierra, M., & Berrios, G. E. (1998). Depersonalization: neurobiological perspectives. Biological psychiatry, 44(9), 898–908.

Silbersweig, D., Clarkin, J. F., Goldstein, M., Kernberg, O. F., Tuescher, O., Levy, K. N., ... Stern, E. (2007). Failure of frontolimbic inhibitory function in the context of negative emotion in borderline personality disorder. American Journal of Psychiatry, 164(12), 1832–1841.

Skodol, A. E., Siever, L. J., Livesley, W. J., Gunderson, J. G., Pfohl, B., & Widiger, T. A. (2002). The borderline diagnosis II: biology, genetics, and clinical course. Biological psychiatry, 51(12), 951–963.

Smith, E. E. & Jonides, J. (1999). Storage and executive processes in the frontal lobes. *Science, 283*(5408), 1657–1661. https://doi.org/10.1126/science.238.5408.1657.

Smith, S. M., Fox, P. T., Miller, K. L., Glahn, D. C., Fox, P. M., Mackay, C. E., Filippini, N., Watkins, K. E., Toro, R., Laird, A. R., & Beckmann, C. F. (2009). Correspondence of the brain's functional architecture during activation and rest. Proceedings of the National Academy

of Sciences of the United States of America, 106(31), 13040–13045. https://doi.org/10.1073/pnas.0905267106.
Soloff, P., Nutche, J., Goradia, D., & Diwadkar, V. (2008). Structural brain abnormalities in borderline personality disorder: a voxel-based morphometry study. . Psychiatry Res, 164(3), 223–236.
Soloff, P. H., Pruitt, P., Sharma, M., Radwan, J., White, R., & Diwadkar, V. A. (2012). Structural brain abnormalities and suicidal behavior in borderline personality disorder. Journal of Psychiatric Research, 46(4), 516–525.
Sridharan, D., Levitin, D. J. & Menon, V. (2008). A critical role for the right fronto-insular cortex in switching between central-executive and default-mode networks. *Proceedings of the National Academy of Sciences USA, 105*(34), 12569–12574. https://doi.org/10.1073/pnas.0800005105.
Spiegel, D., & Cardeña, E. (1991) Disintegrated experience: The dissociative disorders revisited. Journal of Abnormal Psychology, 100(3), 366–378.
Staebler, K., Helbing, E., Rosenbach, C., & Renneberg, B. (2010). Rejection sensitivity and borderline personality disorder. Clin Psychol Psychotherapy 18(4):275–283.
Stein, J. L., Wiedholz, L. M., Bassett, D. S., Weinberger, D. R., Zink, C. F., Mattay, V. S., & Meyer-Lindenberg, A. (2007). A validated network of effective amygdala connectivity. NeuroImage, 36(3), 736–745. https://doi.org/10.1016/j.neuroimage.2007.03.022.
Stein, M. B., Koverola, C., Hanna, C., Torchia, M. G., & McClarty, B. (1997). Hippocampal volume in women victimized by childhood sexual abuse. Psych Med 27(4), 951–959.
Stiglmayr, C. E., Ebner-Priemer, U. W., Bretz, J., Behm, R., Mohse, M., Lammers, C. H., . Bohus, M. (2008). Dissociative symptoms are positively related to stress in borderline personality disorder. Acta psychiatr Scand, 117(2), 139-147.
Tebartz van Elst, L., Hesslinger, B., Thiel, T., Geiger, E., Haegele, K., Lemieux, L., ... Ebert, D. (2003). Frontolimbic brain abnormalities in patients with borderline personality disorder: a volumetric magnetic resonance imaging study Biol Psychiatry, 54(2), 163–171.
Townsend, J. D., Torrisi, S. J., Lieberman, M. D., Sugar, C. A., Bookheimer, S. Y. & Altshuler, L. L. (2013). Frontal-amygdala connectivity alterations during emotion downregulation in bipolar I disorder. *Biological Psychiatry, 73*(2), 127–135. https://doi.org/10.1016/j.biopsych.2012.06.030.
Treede, R.-D., Apkarian, A. V., Bromm, B., Greenspan, J. D., & Lenz, F. A. (2000). Cortical representation of pain: functional characterization of nociceptive areas near the lateral sulcus. Pain, 87(2), 113–119.
Veer, I. M., Oei, N. Y., Spinhoven, P., van Buchem, M. A., Elzinga, B. M., & Rombouts, S. A. (2011). Beyond acute social stress: increased functional connectivity between amygdala and cortical midline structures. *NeuroImage, 57*(4), 1534–1541. https://doi.org/10.1016/j.neuroimage.2011.05.074.
Vermetten, E., & Spiegel, D. (2014). Trauma and dissociation: implications for borderline personality disorder. Current psychiatry reports, 16(2), 434. https://doi.org/10.1007/s11920-013-0434-8.
Visintin, E., De Panfilis., Amore, M., Balestrieri, M., Wolf, R. C., & Sambataro, F. (2016). Mapping the brain correlates of borderline personality disorder: A functional neuroimaging meta-analysis of resting state studies. Journal of Affective Disorders, 204, 262–269. https://doi.org/10.1016/j.jad.2016.07.025.
Völlm, B. A., Zhao, L., Richardson, P., Clark, L., Deakin, J. F., Williams, S., & Dolan, M. C. (2009). A voxel-based morphometric MRI study in men with borderline personality disorder: preliminary findings. Crim Behav Ment Health, 19(1), 64–72.
Wang, L., Krishnan, K. R., Steffens, D. C., Potter, G. G., Dolcos, F., & McCarthy, G. (2008). Depressive state- and disease-related alterations in neural responses to affective and executive challenges in geriatric depression. The American journal of psychiatry, 165(7), 863–871. https://doi.org/10.1176/appi.ajp.2008.07101590.
Whalley, H. C., Nickson, T., Pope, M., Nicol, K., Romaniuk, L., Bastin, M. E., …, Hall, J. (2015). White matter integrity and its association with affective and interpersonal symptoms in borderline personality disorder. Neuroimage Clinical, 7, 476–481. https://doi.org/10.1016/j.nicl.2015.01.016.

Whittle, S., Chanen, A. M., Fornito, A., McGorry, P. D., Pantelis, C., & Yucel, M. (2009). Anterior cingulate volume in adolescents with first-presentation borderline personality disorder. Psychiatry research, 172(2), 155–160.
Wingenfeld, K. & Wolf, O. T. (2014). Effects of cortisol on cognition in major depressive disorder, posttraumatic stress disorder and borderline personality disorder. Psychoneuroendocrinology, 51C, 282–295.
Winter, D., Krause-Utz, A., Lis, S., Chiu, C. D., Lanius, R. A., Schriner, F., Bohus, M., & Schmahl, C. (2015). Dissociation in borderline personality disorder: Disturbed cognitive and emotional inhibition and its neural correlates. Psychiatry research, 233(3), 339–351. https://doi.org/10.1016/j.pscychresns.2015.05.018.
Wolf, R. C., Sambataro, F., Vasic, N., Schmid, M., Thomann, P. A., Bienentreu, S. D., & Wolf, N. D. (2011). Aberrant connectivity of resting-state networks in borderline personality disorder. Journal of psychiatry & neuroscience: JPN, 36(2), 100150.
Woon, F. L., & Hedges, D. W. (2009). Amygdala volume in adults with posttraumatic stress disorder: a meta-analysis. Journal of Neuropsychiatry Clinical Neuroscience, 21(1), 5–12.
Xu, T., Cullen, K. R., Mueller, B., Schreiner, M. W., Lim, K. O., Schulz, S. C., & Parhi, K. K. (2016) Network analysis of functional brain connectivity in borderline personality disorder using resting-state fMRI. Neuroimage Clinical, 11, 302-315. https://doi.org/10.1016/j.nicl.2016.02.006.
Zanarini M. C. (2000). Childhood experiences associated with the development of borderline personality disorder. The Psychiatric clinics of North America, 23(1), 89–101. https://doi.org/10.1016/s0193-953x(05)70145-3.
Zetzsche, T., Frodl, T., Preuss, U. W., Schmitt, G., Seifert, D., Leinsinger, G., Meisenzahl, E. M. (2006). Amygdala volume and depressive symptoms in patients with borderline personality disorder. Biol Psychiatry, 60(3), 302–310.
Zuo, X. N., Kelly, C., Adelstein, J. S., Klein, D. F., Castellanos, F. X., & Milham, M. P. (2010). Reliable intrinsic connectivity networks: test-retest evaluation using ICA and dual regression approach. NeuroImage, 49(3), 2163–2177. https://doi.org/10.1016/j.neuroimage.2009.10.080.
Zypthen van, L., Siep, N., Jacob, G. A., Goebel, R., & Arntz, A. (2015). Emotional sensitivity, emotion regulation and impulsivity in borderline personality disorder: a critical review of fMRI studies. Neuroscience and biobehavioral reviews, 51, 64–76. https://doi.org/10.1016/j.neubiorev.2015.01.001.

Structural and Functional Connectivity Changes Following Cognitive Remediation: A Systematic Review

H. New Fei, Jordon X. J. Tng, June Su Tan, and Kang Sim

Introduction

Cognitive dysfunction is a core feature of schizophrenia that significantly accounts for poor functional outcomes and quality of life (Green 1996; Green et al. 2000). The spectrum of neurocognitive deficits in schizophrenia span domains such as attention, processing speed, working memory, language processing, sensory processing, executive functioning (Nuechterlein et al. 2004), as well as social cognitive domains of emotion recognition, empathy/perspective-taking, social context appraisal, and understanding intentions (Green et al. 2015). Not only do some of these cognitive deficits precede onset of psychosis (Seidman and Mirsky 2017), they are more enduring than the positive symptoms of hallucinations and delusions (Hill et al. 2004). Cognitive deficits also impede treatment adherence and the ability of patients to receive psychosocial interventions. However, present medications cannot adequately treat cognitive deficits (Davidson et al. 2009).

Cognitive remediation (CR) is a behavioral-based intervention rooted in the principles of neuroplasticity that aims to target the cognitive deficits in schizophrenia (Medalia and Choi 2009). Approaches to CR include practice sessions to work on or get around deficits in neurocognition, social cognition, or a mixture of both domains. Typically, sessions are conducted a few times a week over a few weeks to a few months. While traditionally CR sessions are overseen by therapists in a clinical setting who provide instructions, supervise progress, and help relate the transfer of cognitive skills to everyday situations, initiatives are underway to explore remote CR through online exercises with minimal therapist intervention (Saperstein and Kurtz 2013).

Burgeoning studies in schizophrenia have linked abnormalities in functional and structural networks to dysfunction in different cognitive domains (Barch 2013).

H. New Fei · J. X. J. Tng · J. S. Tan · K. Sim (✉)
Research Division, Institute of Mental Health, Singapore, Singapore
e-mail: new_fei_ho@imh.com.sg; jordon_xinjie@imh.com.sg; kang_sim@imh.com.sg

V. A. Diwadkar, S. B. Eickhoff (eds.), *Brain Network Dysfunction in Neuropsychiatric Illness*, https://doi.org/10.1007/978-3-030-59797-9_18

A meta-analysis of task-based MRI studies show decreased activation of the left prefrontal cortex and right ventrolateral prefrontal cortex during tasks requiring executive control (McTeague et al. 2017). Resting-state functional connectivity show that abnormal dorsolateral prefrontal cortex functional connectivity with the rest of the brain is associated with weaker executive function in schizophrenia (Anticevic et al. 2015). Tractography studies show that the alterations of right cingulum bundle is associated with poorer working memory and processing speed in early course schizophrenia (Seitz et al. 2016). In a large-scale study of psychotic spectrum disorders, reduced global efficiency in the salience/cingulo-operculum/ventral attention network is associated with poorer generalized cognitive deficit, with group differences observed between schizophrenia and healthy controls (Sheffield et al. 2017).

By contrast, studies showing how interventions can restore these network abnormalities are far less. Pioneering work on CR-led brain changes focused on localized brain activity using single positron emission computed tomography (Penades et al. 2000; Wykes 1998). As task-based fMRI gained popularity with increasing availability of scanners, subsequent work were generally conducted on participants performing working memory tasks during the scan session. Investigators focused on frontal brain regions and demonstrated that CR led to task-related brain activation changes (Haut et al. 2010; Wexler et al. 2000; Wykes et al. 2002). These studies provide important proof-of-concept on brain changes associated with CR interventions, but the generalizability of findings are hampered by the lack of patient placebo controls and small sample sizes ($n < 10$ in each group), which significantly inflates the effect size of fMRI data (Yarkoni 2009).

Since the publication of a few earlier reviews of neuroimaging changes following CR (Isaac and Januel 2016; Penades et al. 2017; Thorsen et al. 2014), cross-MRI modality studies examining CR-mediated intrinsic brain activity have emerged (Donohoe et al. 2018; Eack et al. 2016; Fan et al. 2017; Ramsay et al. 2017). These task-free fMRI studies provide important complementary information on the slow and synchronous brain fluctuations that shape neural activity during task performance (Mennes et al. 2010). The increasing sophistication of analytical techniques also now permits examination of coherence of spatially distributed brain regions; recent studies have now begun to look into task-induced functional connectivity changes after CR (Keshavan et al. 2017).

The identification of CR-led changes in neural networks not only provide biological validation of the CR methods but may also serve as potential neuroplastic markers of CR response. Hence, in this study, we aim to provide an up-to-date systematic review of multimodal MRI findings of CR in schizophrenia, by focusing on studies with a sample size of at least 10 subjects per group. We discuss whether common findings of CR-mediated brain changes emerge from the studies hitherto, the limitations of existing literature and suggest future directions to address the gaps in the field.

Methods

Study Search and Selection

Two authors (JT and JS) independently searched and evaluated the articles for review. Disagreements were resolved upon discussion with a third author (NFH).

The systematic search to identify relevant articles on MEDLINE/PubMed and PsycInfo were based on the Preferred Reporting Items for Systematic Reviews and Meta-Analyses (PRISMA) guidelines. The search was conducted up till the end of September 2018. Keywords used were ("cogniti*" and "remediation" or "training" or "enhancement") and ("fMRI" or "MRI" or "DTI" or "connectivity") and "schizophrenia".

Studies were considered for inclusion if they were original articles, written in English, had pre- and post-MRI scans, involved patients with schizophrenia diagnosed using DSM or ICD-10, and involved comparison groups for patients receiving CR that can include treatment-as-usual, placebo treatment, and/or healthy individuals. Studies were excluded if they were not peer reviewed, had fewer than 10 participants in any of the groups, used non-MRI modalities and overlapping datasets with similar outcomes or articles in which data contributed to similar publications.

Recorded Variables

Details regarding the study cohort (sample size, gender, age, diagnosis), type of CR treatment (neurocognitive versus social cognitive and their subconstructs; presence of therapist-led sessions and frequency), MRI modality, analysis approach were extracted. Findings at baseline and after CR, and correlations between brain imaging measures and cognitive/behavioral scores were also extracted.

Results

The literature search outlined in the PRISMA flowchart in Fig. 1 identified 11 articles, with 311 individuals across nine study sites, which fulfilled the inclusion criteria. Details of the included studies such as the socio-demographic factors, CR approach, neuroimaging modality, and analytic approach, findings of baseline group differences, group-by-time interaction effects, and brain–behavioral correlations are provided in Table 1.

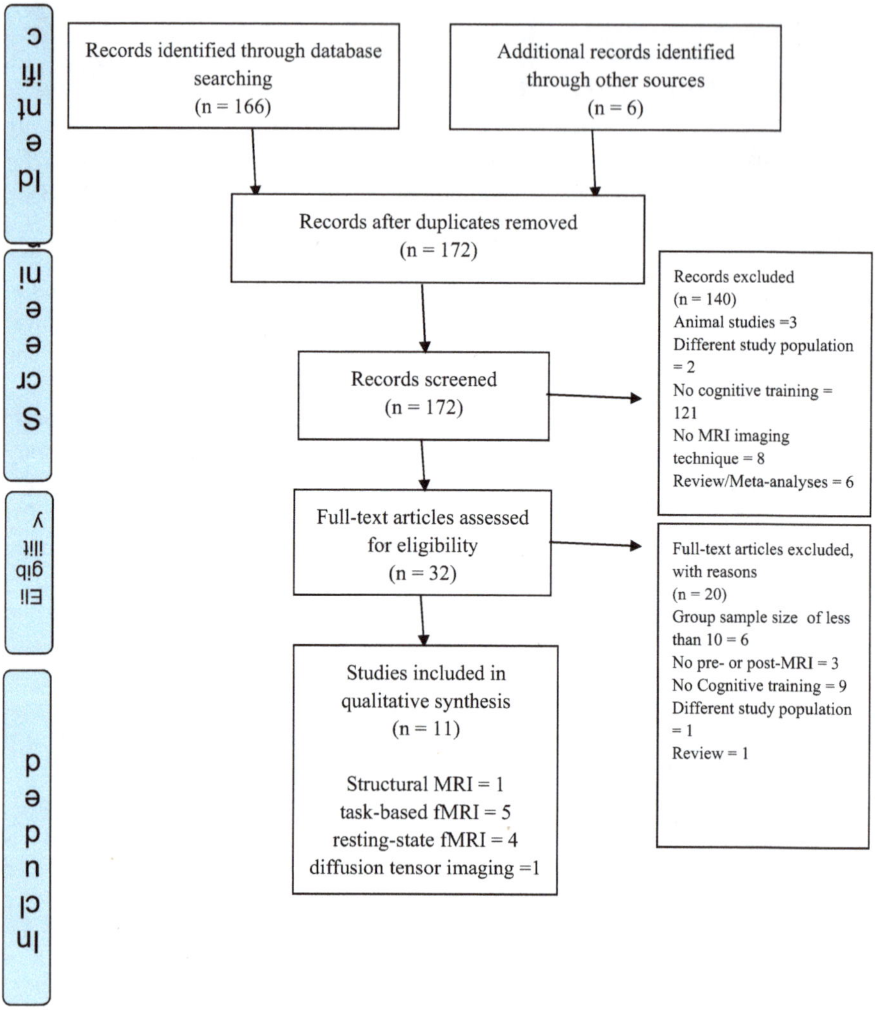

Fig. 1 PRISMA flowchart

Characteristics of Selected Studies

Study Design All the studies reviewed here were randomized controlled trials. Three independent study cohorts across four studies reviewed here included healthy controls as part of the study design (Habel et al. 2010; Penadés et al. 2013; Subramaniam et al. 2012, 2014). Active patient controls across nine studies received placebo interventions that were therapist-led sessions on psychoeducation and illness management (Eack et al. 2016; Keshavan et al. 2017; Penadés et al. 2013), open topic discussions with the therapist (Donohoe et al. 2018), computer skills training (Ramsay et al. 2017), computer games matched for CR treatment duration

Table 1 Summary of main characteristics of included studies

Author, year; site	Population	N (M/F), Age(SD)	CR approach (target cognitive domains); frequency	Imaging modality	Analysis approach (Whole-brain versus ROI/Seed); Software	Baseline group differences	Treatment effects (Group x Time)	Brain-cognition correlation
Ramsay et al. (2018); UCSF	Early-course schizoaffective, schizophrenia	CR: 22 (14 M/8F) 23.3 (4.6) Non-CR/ Placebo: 22(15 M/7F) 21.3 (3.7)	Neurocognitive (Auditory processing and WM exercises) Restorative Little/no therapist intervention 40 h/8 weeks	Structural MRI	ROI: Volume (thalamus) Free surfer v5.1	No patient group differences	No group x time effects	Baseline Both groups: left thalamic volume correlated with global cognitive performance Post-treatment CR: ↑ left thalamus volume positively correlated with cognitive improvement
Donohoe et al. (2018); Dublin, Wicklow and Galway	Schizophrenia	CR: 48 (31 M/17F) 43.5 (11.5) Non-CR/ Placebo: 42 (23 M/19F) 43.1 (11.0)	Neurocognitive (Auditory and visual WM, includes N-back training) Restorative Therapist (structured motivational interview) – 45 min/weekly 27 h/8 weeks to be completed over a 12 week window	Task-fMRI (N-back) CR:15 Non-CR:15	Whole-brain, Voxel based SPM 8	No patient group differences	No group x time effects	–
				Rs-fMRI CR:14 Non-CR:15	ROI (Mothersill et al. 2017)[a]: Left precuneus (DMN) Right precuneus (DMN) Left anterior cingulate cortex (AN) Left temporo-parietal junction (VAN) CONN v15	–	CR vs placebo: ↑ functional connectivity between the right precuneus and left inferior parietal lobule	–

(continued)

Table 1 (continued)

Author, year; site	Population	N (M/F), Age(SD)	CR approach (target cognitive domains); frequency	Imaging modality	Analysis approach (Whole-brain versus ROI/Seed); Software	Baseline group differences	Treatment effects (Group x Time)	Brain-cognition correlation
Fan et al. (2017); Beijing Huilongguan Hospital	Schizophrenia	CR: 12 (7 M/5F) 39.7 (6.2) Non-CR/ TAU: 11 (5 M/6F) 41.5 (3.5)	Neurocognitive + Social Cognitive (WM, EF and social function) Restorative Little therapist intervention 40 sessions/8 weeks	Rs-fMRI (ALFF)	ROI (defined postanalysis)[b]: mPFC/ACC Olfactory cortex STG Left IFG Insula MTG ITG Temporal pole Brainstem SPM5, DPARSF	Not analyzed	CR vs Placebo: CR: ↑ ALFF in the mPFC/ACC, CR: ↓ ALFF in the brainstem	Post-treatment CR + TAU: ↑ ALFF in the mPFC/ACC correlated with maze score changes
Ramsay et al. (2017); University of Minnesota	Schizophrenia	CR: 14 42.9 (10.6) Non-CR/ Placebo: 12 45.8 (7.7)	Neurocognitive (WM, includes N-back training) Restorative Therapist-led bridging sessions for both CR and non-CR group 48 h/16 weeks	Task-fMRI (Word and picture N-back)	Voxel-wise small-volume, FC of ROIs: Thalamus MFG FSL	No patient group differences	CR vs Placebo: CR: ↑FC in thalamus-ACC CR: ↑FC in thalamus-left MFG	No correlation found
				Rs-fMRI	FC of the ROIs: Thalamus MFG ACC FSL	No patient group differences	CR vs Placebo: CR: ↑FC in thalamus-right MFG CR: ↑FC in thalamus-ACC	Post-treatment CR: ↑ thalamus-right MFG FC positively correlated with MCCB change

Keshavan et al. (2017); Western Psychiatric Institute and Clinic in Pittsburgh, Pennsylvania, and surrounding community clinics.	Early-course schizophrenia, schizoaffective	CR: 25 (14 M/11F) 24.95 (5.16) Non-CR/ Placebo: 16 (12 M/4F) 26.13 (5.29)	Neurocognitive (Attention, memory, problem-solving) + Social Cognitive Restorative and strategy-based Therapist-led group sessions (for social cognition) 60 h/12 weeks for neurocognition and 45 weekly social cognitive sessions Total treatment duration 2 years	Task-fMRI (Preparing to Overcome Prepotency)	ROI-based BOLD activity and FC DLPFC ACC Superior PFC Medial PFC SPM8	CR vs Placebo: CR: less BOLD activity in right DLPFC	CR vs Placebo: CR:↑ BOLD activity in right DLPFC CR: ↓FC in right DLPFC-ACC CR: ↓FC in right DLPFC-left precentral gyrus CR: ↑FC in DLPFC-PCC	Post-treatment CR: ↑BOLD activity correlated with neurocognitive improvements CR: ↓FC in right DLPFC-right ACC correlated with neurocognitive improvements CR: ↓FC in right DLPFC-left precentral gyrus correlated with neurocognitive and social-cognitive improvements
								CR: ↑FC in DLPFC-PCC correlated with social-cognitive improvements

(continued)

Table 1 (continued)

Author, year; site	Population	N (M/F), Age(SD)	CR approach (target cognitive domains); frequency	Imaging modality	Analysis approach (Whole-brain versus ROI/Seed); Software	Baseline group differences	Treatment effects (Group x Time)	Brain-cognition correlation
Eack et al. (2016); Western Psychiatric Institute and Clinic in Pittsburgh, Pennsylvania, and surrounding community clinics.				Pseudo Rs-fMRI; aggregate of the rest blocks during the Preparing to Overcome Prepotency task	Seed to voxel FC analyses Seed: PCC Target ROIs: Lateral PFC Orbital PFC Medial PFC Insula Hippocampus	Not analyzed	CR vs Placebo: CR: less ↓FC in PCC-left DLPFC CR ↑FC in PCC-right insula	Post-treatment CR: less ↓ FC in PCC-left DLPFC correlated with emotion regulation improvements CR:↑FC in PCC-right insula correlated with emotion perception improvements
					Parahippocampal gyrus Amygdala Fusiform gyrus STG Functional connectivity toolbox in SPM8			

Subramaniam et al. (2014); UCSF	Schizophrenia	CR: 16 (12 M/4F) 40.69 (12.7) Non-CR/ Placebo: 14 (10 M/4F) 41.21 (9.48) HC: 15 (11 M/4F) 44.27 (11.2)	Neurocognitive + Social Cognitive (Auditory/visual processing and emotion identification) Restorative Little/no therapist intervention 80 h/16 weeks	Task-fMRI (N-back)	Seed-based (Perlstein et al. 2003): Seed (Talairach coordinates) Right MFG (38, 27, 24) Left MFG (−36, 38, 23) Right IFG (49,10, 23) Left IFG (−47, 7, 23) Right Insula (35, 15, 8) Left Insula (−40, 10, 10) 10 mm radius SPM2	HC: ↑activation in bilateral MFG SZ (CR and non-CR) vs HC: hypoactivation in left MFG	CR vs Placebo vs HC CR: ↑ activation in left MFG	Baseline: HC: ↑bilateral MFG activation correlated with task performance Post-treatment: CR: Right MFG activation correlated with task performance in CR 6 months follow-up: CR: right MFG activation at post-treatment correlated with letter-number span performance; bilateral MFG activation correlated with QLS scores

(continued)

Table 1 (continued)

Author, year; site	Population	N (M/F), Age(SD)	CR approach (target cognitive domains); frequency	Imaging modality	Analysis approach (Whole-brain versus ROI/Seed); Software	Baseline group differences	Treatment effects (Group x Time)	Brain-cognition correlation
Penades et al. (2013); Schizophrenia Unit at the Hospital Clinic, Barcelona	Schizophrenia	CR: 17 (12 M/5F) 36.35 (13.16) Non-CR/ Placebo: 18 (15 M/3F) 37.56 (8.99) HC: 15 (10 M/5F) 34.75 (3.14)	Neurocognitive (Flexibility, WM, planning) Strategy-based Therapist-led 40 h/16 weeks	Task-fMRI (N-back)	Tensorial independent component analysis (TICA) Multivariate Exploratory Linear Decomposition into Independent Components (MELODIC v3.1)	CR vs HC CR: increased activation in regions related to executive control network (f) CR vs Placebo CR: increased activation in regions related to the default mode network (ACC, MTG, cingulate gyrus, precuneus, inferior occipital gyrus, lingual gyrus)	CR vs Placebo: CR: ↓ activation in overactivated regions at baseline (Executive control: left superior parietal lobule and bilateral middle frontal gyri; default network: left precuneus and middle frontal gyrus)	Post-treatment: CR: ↓overactivation correlated with cognitive improvements CR: ↑FA correlated with cognitive improvements CR: ↑FA correlated with ↓overactivation during task

				DTI – FA	Tract Based Spatial Statistics (TBSS) TBSS v1.2	No healthy control or patient group differences	CR vs Placebo CR: ↑FA in the anterior genu of CC, right posterior thalamic radiation Placebo: decrease in bilateral superior longitudinal fasciculus and left inferior longitudinal fasiculus	
Hooker et al. (2013); UCSF; UC Berkeley	Schizoaffective, schizophrenia	CR: 11 (10 M/1F) 51.2 (5.8) Non-CR/ Placebo: 11 (8 M/3F) 41.0 (8.4)	Neurocognitive + Social Cognitive (Auditory and verbal info processing) Little/no therapist interaction 60 h/10 weeks	Task-fMRI (Emotion recognition; event related design) Events: Happy, surprise, fear; Angry, Sad, disgust	ROI Amygdala STC SRC SPM8	Not analyzed	CR vs Placebo: ↑ activity in bilateral amygdala, right putamen, and right MPFC	Post-treatment CR + Placebo:↑ activity in right amygdala, right putamen, and right MPFC correlated with improvements in perceiving emotions.
								CR: ↑right amygdala activity correlated with improvements in emotion perception

(continued)

Table 1 (continued)

Author, year; site	Population	N (M/F), Age(SD)	CR approach (target cognitive domains); frequency	Imaging modality	Analysis approach (Whole-brain versus ROI/Seed); Software	Baseline group differences	Treatment effects (Group x Time)	Brain-cognition correlation
Subramaniam et al. (2012): University of California, San Francisco (UCSF)	Schizophrenia	CR: 16 Non-CR/ Placebo: 15 HC: 16 Mean Age of SZ: 40 (1.17) Mean Age of HC: 45 (11.6)	Neurocognitive (Auditory and visual processing exercises) + Social Cognitive (emotion recognition) Restorative Little/no therapist intervention 90 h/16 weeks	Task-fMRI (Reality monitoring)	Seed-based: mPFC (Talairach coordinates −4, 52,8; 20 mm radius) (Cabeza et al. 2004) SPM2	CR vs Placebo vs HC HC: mPFC activation during task No patient group differences	CR vs Placebo vs HC CR:↑ mPFC activation	Baseline: HC: mPFC activation correlated with task accuracy CR group (6 months follow-up): mPFC activation correlated with improved social functioning
Habel et al. (2010); University of Dusseldorf	Schizophrenia	CR: 10 31.4 (7.8) Non-CR/ TAU: 10 33.7 (10.7) HC: 10 31.6 (8.8) (All male)	Social Cognitive (Facial affect recognition) Restorative Therapist-led 12 h/6 weeks	Task-fMRI (Emotion-discrimination: affect recognition and age discrimination)	Whole-brain (Regions showing changes over time in HC were used as a mask) SPM2	Not analyzed	CR vs Placebo CR: Affect recognition: ↑ activation in inferior frontal, left occipital and right parietal regions CR: Age discrimination: ↑	Post-treatment CR: Brain activation changes correlated positively with affect discrimination changes

							activation in superior temporal, precentral, postcentral, posterior cingulate and parahippocampal gyri, thalamus, basal ganglia	

Abbreviations: *ACC* anterior cingulate cortex, *ALFF* Amplitude of low-frequency fluctuations, *AN* affective network, *BOLD* blood-oxygenated level dependent, *CCN* cognitive control network, *CR* cognitive remediation, *DMN* default mode network, *DPARSF* data processing assistant for resting-state fMRI, *EF* executive function, *F* female, *FC* functional connectivity, *FSL* FMRIB software library, *HC* healthy controls, *IFG* inferior frontal gyrus, *ITG* inferior temporal gyrus, *M* male, *MFG* middle frontal gyrus, *mPFC* medial prefrontal cortex, *MTG* middle temporal gyrus, *PFC* prefrontal cortex, *SPM* Statistical Parametric Mapping, *SRC* somatosensory-related cortex, *STG* superior temporal gyrus, *STC* superior temporal cortex, *TAU* treatment-as-usual, *UCSF* University of California, San Francisco, *VAN* ventral attention network, *WM* working memory

[a]ROI derived from findings of differences between schizophrenia and healthy controls in authors' other study on resting-state functional connectivity, with seed regions denoting different networks

[b]Regions exhibiting within-group CR changes

and exposure (Hooker et al. 2012, 2013; Ramsay et al. 2018; Subramaniam et al. 2012, 2014). Two studies contrasted CR with treatment-as-usual (Fan et al. 2017; Habel et al. 2010).

Characteristics of Study Cohort There were nine unique study cohorts across the 11 different studies. In two studies, the N-back task fMRI and pseudo resting-state fMRI data were collected from the same study cohort of early course schizophrenia and schizoaffective disorder subjects (Eack et al. 2016; Keshavan et al. 2017). In another two studies, the N-back task fMRI and reality monitoring task fMRI data were acquired from chronic schizophrenia patients (Subramaniam et al. 2012, 2014). The majority of the studies were on patients who have been ill for some time, with the mean age close to 40–50 years of age. Three studies comprised patient participants that were ill for less than 5 years (Eack et al. 2016; Keshavan et al. 2017; Ramsay et al. 2018).

CR Approach Most studies reviewed here examined a combined CR approach of neurocognitive and social cognitive training (Eack et al. 2016; Fan et al. 2017; Hooker et al. 2012, 2013; Keshavan et al. 2017; Subramaniam et al. 2012, 2014). Some studies examined an exclusive neurocognitive approach to CR (Donohoe et al. 2018; Penadés et al. 2013; Ramsay et al. 2017, 2018) while one study examined social cognitive training (Habel et al. 2010). While in some studies the active ingredients of the training were distilled through computerized adaptive exercises with no or little therapist intervention (Fan et al. 2017; Hooker et al. 2012, 2013; Ramsay et al. 2018; Subramaniam et al. 2012, 2014), the CR in two studies were therapist-led (Habel et al. 2010; Penadés et al. 2013). The CR in other studies involved a mixture of computerized exercises and group sessions with therapists (Donohoe et al. 2018; Eack et al. 2016; Keshavan et al. 2017; Ramsay et al. 2017). The duration of CR across the studies ranged from 6 weeks (Habel et al. 2010) to 2 years (Keshavan et al. 2017). The number of hours of CR also varied, from 12 h (Habel et al. 2010) to 90 h (Subramaniam et al. 2012, 2014). Intensity of treatment, especially neurocognitive training, was typically 4–6 h per week.

MRI Modality The majority of the studies reviewed were task-based fMRI; four were working memory tasks (N-back) (Donohoe et al. 2018; Penadés et al. 2013; Ramsay et al. 2017; Subramaniam et al. 2014), one was a cognitive control task (Keshavan et al. 2017), and two tasks involved facial emotion recognition (Habel et al. 2010; Hooker et al. 2013). Four studies were task-free fMRI, of which two studies examined the resting-state functional connectivity of seed regions (Donohoe et al. 2018; Ramsay et al. 2017), one examined "pseudo" resting-state fMRI (merging of blocks of rest during an fMRI task) functional connectivity of a seed to target regions (Eack et al. 2016), and one centered on localized resting-state activity using the Amplitude of Low Frequency Fluctuations (ALFF) as a measure (Fan et al. 2017). One was a structural connectivity study, using diffusion tensor imaging (DTI) metrics of fractional anisotropy (FA) to quantify changes in white matter microstructure (Penadés et al. 2013), and only one volumetric MRI study met our selection criteria for review (Ramsay et al. 2018). Among these studies, three were multimodal: Donohoe and Ramsay et al. investigated both task-based and task-free fMRI (Donohoe et al. 2018; Ramsay et al. 2017) and Penades investigated task-based and DTI measures of CR changes (Penadés et al. 2013).

Findings of the Selected Studies

Here, we qualitatively review the studies by way of CR approach—namely social cognitive training, mixed neuro- and social cognitive training, and neurocognitive training—on the premise that different forms of learning would elicit neuroplasticity at different brain regions and highlight the neuroimaging modality used to examine the neuroplastic changes.

1. ***Social Cognitive Training.***

Task-Based fMRI One study in the present review focused exclusively on social cognitive training, specifically the recognition of basic facial emotions (Habel et al. 2010). CR was led by the therapist and was acute—around 12 h over 6 weeks. The study included healthy controls; brain regions showing effect of time in the healthy controls were masked out when performing patient group comparisons. Compared with patients receiving treatment-as-usual, patients who received social cognitive training showed activation in the left superior and middle occipital lobe, right inferior and superior parietal lobe, and bilateral inferior frontal cortex, correlating with performance in emotion identification. The same regions were activated in healthy controls during the emotion identification task, suggestive of restoration of cognitive function after CR.

2. **Mixed Neurocognitive and Social Cognitive Training**

   (a) **Cognitive Enhancement Therapy**

Cognitive enhancement therapy is a relatively long-term therapy of ~2 years that starts with 3 months of neurocognitive training through computerized exercises of attention, memory, and problem-solving, followed by therapist-led group social cognitive training sessions of emotion regulation, perspective-taking, and understanding social norms (Eack et al. 2016; Hogarty et al. 2004). Early-course schizophrenia patients assigned to receive CR were compared with patients assigned to receive enriched supportive therapy, an intervention of psychoeducation, illness management and relaxation training over the same period of time (Eack et al. 2016; Keshavan et al. 2017).

Pseudo Resting-State fMRI Keshavan and co-workers examined the pseudo resting state data, that is, the aggregate of rest blocks during a task-based fMRI study (Keshavan et al. 2017); the task measured attentional control and the results are described in the following paragraph. The outcome measure was the functional connectivity of the seed region, the posterior cingulate cortex—selected for its role as a hub in the default mode network—to a priori fronto-temporal regions such as the lateral, orbital, and medial prefrontal cortices; insula; hippocampus; parahippocampal gyrus; amygdala; fusiform gyrus; and superior temporal gyrus. After intervention, the functional connectivity between the posterior cingulate cortex and left dorsolateral prefrontal cortex functional connectivity was decreased in the placebo group and marginally increased in the CR group. A similar pattern of posterior cingulate cortex-insular functional connectivity reduction in placebo and enhancement in CR group that correlated with emotion perception scores was also observed (Eack et al. 2016).

Task-Based fMRI The ensuing study examined the group-by-time effects of the attentional control task (Preparing to overcome Prepotency) performed within the scanner and the BOLD activity of selected regions thought to be involved, namely the dorsolateral, superior, and medial prefrontal cortex and the anterior cingulate cortex. Group differences were observed at baseline. The BOLD activity in right dorsolateral prefrontal cortex was less in CR individuals than placebo; this increased after the intervention and correlated with improved neurocognition but not social cognition or symptomatic change in the CR group (Keshavan et al. 2017). After the intervention, the functional connectivity between the right dorsolateral prefrontal cortex and right anterior cingulate cortex, as well as left precentral gyrus, was found to be decreased. By contrast, the functional connectivity between the right dorsolateral prefrontal and posterior cingulate cortex increased. Nonetheless, both decreases and increases in functional connectivity of the dorsolateral prefrontal cortex were associated with cognitive improvement (Keshavan et al. 2017).

(b) **Auditory, Visual, Verbal, and Social Cognitive Working Memory**

Vinogradov and co-workers lead the field in the number of CR-related neuroimaging studies (Vinogradov et al. 2012). They advocate for restorative CR, a drill-and-practice approach that first targets lower-level neurocognitive function of perceptual processing and attention yoked to social cognitive processes of emotion recognition, followed by training in higher-level cognitive operations of working memory. Minimal therapist intervention is involved.

Task-Based fMRI The first study in the series compared chronic schizophrenia patients randomized to receive computerized CR for 90 h spread out over a course of 16 weeks versus patients randomized to play computer games over the same period (Subramaniam et al. 2012). Inside the scanner, the participants performed a reality monitoring task requiring them to separate words made up by themselves from the words shown to them. After the intervention period, the CR group showed more improvements at the task compared with healthy controls that correlated with increased activation in a predefined brain region–the medial prefrontal cortex that is known to be active during self-related tasks. No change in task performance or medial prefrontal cortex was found in in the placebo group. When the patients returned for a self-reported assessment of quality of life 6 months later, it was found that the post-treatment medial prefrontal cortex activation in the CR group correlated with increased well-being, despite no between-group differences in social functioning (Subramaniam et al. 2012). The findings are suggestive of the potential of medial prefrontal activity in prognostication of functional outcomes.

The same investigators proceeded to examine the effects of restorative CR on the N-back, a working memory task graded to increasing levels of difficulties, on the same study cohort (Subramaniam et al. 2014). A region-of-interest approach was similarly adopted, centering on the dorsolateral (middle frontal gyrus), ventrolateral (inferior frontal gyrus), and insula regions. At the start, the healthy controls were the only group to show task-related activation of bilateral middle frontal gyrus and that coincided with task accuracy. After the treatment, the CR group showed task-related activation in the left middle frontal gyrus and left inferior frontal gyrus, which correlated with task improvements. This post-treatment brain activations in the right middle frontal gyrus and bilateral middle frontal gyrus predicted generalized working memory performances and sense of well-being when the patients came back for their 6-month follow-up.

A concurrent trial used the same restorative and minimal therapist intervention CR approach, but with a stronger emphasis on social cognitive training (Hooker et al. 2012, 2013). Because there were two similar emotion recognition task-based fMRI studies on the same study cohort (the earlier was a block design and the latter event-related design), we report findings of the later study. No baseline group differences were investigated. Assessment of CR-related changes over time was focused on regions thought to be involved in facial emotion recognition, such as the superior temporal cortex, somatosensory-related cortices especially the amygdala. Tasks within the scanner were grouped under "Open-mouth" expressions of happy, surprise, and fear, which is thought to be harder to differentiate versus the other grouped expressions of angry, disgust, and sad. Compared with the placebo group, the CR group performed better in the MSCEIT perceiving emotion subscale but showed no improvements in global cognition or the fMRI task of facial emotion recognition. Also relative to the placebo group, the CR group showed post-training increase in activation of amygdala, putamen, and right medial prefrontal cortex during the open-mouth expressions, which was related to better MSCEIT performances. Also, the post-training enhanced activation of the amygdala related to accuracy of the behavioral task within the scanner.

(c) **Acute Cognitive Remediation**

ALFF ALFF measures the power of the low frequency signals extracted during resting-state fMRI, providing a measure of local intrinsic neural activity. The CR intervention in this study was acute and intensive: 40 h over 8 weeks and also involved nominal therapist intervention. The investigators examined within-group CR changes in ALFF using paired t-tests and used the brain regions exhibiting changes as ROIs for evaluating group-by-time interactions. Compared with baseline and treatment-as-usual, the CR group showed increased ALFF in the medial prefrontal cortex/anterior cingulate cortex and decreased ALFF in the brainstem. The ALFF within the medial prefrontal cortex/anterior cingulate cortex correlated with changes in problem-solving tasks in the combined two groups (CR and treatment-as-usual) (Fan et al. 2017).

3. **Neurocognition**

Task-Based fMRI and DTI Penades et al. (2013) examined the structural and task-positive functional connectivity effects of a therapist-led, strategy-based CR program carried out over 16 weeks side by side with active controls taught on illness management. Network based evaluation of the N-back task-fMRI data revealed differences in brain activity of certain networks between the groups at baseline: the CR group showed increased activation in regions related to the executive function network when compared with healthy controls and increased activation in regions related to the default mode network when compared with the placebo group. CR subsequently led to decreases in specific brain regions of the executive function and default network that were found to be increased at baseline, suggesting normalizing of abnormal networks. While no baseline differences were observed for the DTI fractional anisotropy metrics (a white matter microstructure diffusivity index), CR led to increases in fractional anisotropy of the anterior genu of the corpus callosum and right posterior thalamic radiation. The changes in the task-based networks and FA metrics in the CR group correlated with cognitive improvements (Penadés et al. 2013).

Task-Based fMRI and Resting-State fMRI Ramsay et al. (2017) investigated the prefrontal-thalamic functional connectivity between patients allocated to undergo a working memory-based restorative CR and patients allotted to computer skills training. The ROIs specified for both the N-back task-based and resting-state fMRI analyses were the thalamus, middle frontal gyrus, and anterior cingulate cortex. The resting-state functional connectivity of the thalamus to the anterior cingulate cortex was increased after CR relative to placebo intervention. Similarly, the resting-state functional connectivity of the thalamus to right middle frontal gyrus was found to be increased after CR and was associated with improvements in scores of a comprehensive battery of cognition. During N-back task engagement, analogous increases in the thalamus connectivity to the anterior cingulate cortex and left middle frontal gyrus was found post-CR relative to the active control group although no neurocognitive correlations were observed (Ramsay et al. 2017).

Remote CR CR that can be delivered at home, without having to travel to the clinics, reaches out to a greater number of patients in the community. The following two trials investigated the brain structure and function effects of portable CR with low therapist support.

Volume Leveraging on the positive findings of perceptual processing and working memory training in the abovementioned studies (Subramaniam et al. 2012, 2014), the investigators examined the effects of online auditory training exercises—without social cognitive training this time round—over 40 h of CR for an hour a day over 8–10 weeks in early-course schizophrenia (Ramsay et al. 2018). The investigators focused on volume of the thalamus as an outcome measure. Although no within- and between- group global cognitive improvements and the thalamic volume were found, the investigators observed a relationship between thalamic volume and global cognition across both groups at baseline. In the CR group, changes in cognition over time was positively associated with changes in left thalamic volume and negatively associated in the placebo group. Within the CR group, less severe negative and general psychopathology symptoms at baseline was associated with left thalamic volume change (Ramsay et al. 2018).

Task-Based and Resting-State fMRI A separate CR trial involved online exercises exclusively centered on working memory training and meeting with a therapist once a week for structured motivational interviewing over the course of 8 weeks (Donohoe et al. 2018). The placebo group met with the therapist with the same frequency and amount of time over the same period but discussed about open-ended topics during the session. A priori seeds were defined based on another study by the same group, and were regions which showed cluster-wise differences in schizophrenia patients relative to healthy controls (Mothersill et al. 2017). No group-by-time effect was found in the N-back based fMRI, but was found in the resting-state functional connectivity analyses on two of the six seed regions: the right precuneus and left inferior parietal lobule, and left anterior cingulate and right mid cingulate gyrus. No brain-behavior correlations were reported (Donohoe et al. 2018).

Discussion

The study of brain plasticity in CR of schizophrenia is still a nascent field. At this point, the identification of consistent pattern changes in brain networks following CR is still premature, owing to the small number of studies, heterogeneity in CR approaches, use of multimodal neuroimaging tools, and differences in analytical techniques. The recent findings covered in the present review show that CR-led brain changes in schizophrenia are not only constrained to the frontal regions but are widespread across the brain, corresponding to spatially distributed cognitive networks that are impaired in schizophrenia. Also, the largely positive findings suggest that brain circuits in schizophrenia patients, regardless of whether they are in the early or later stages of the illness, could still be remediated through CR interventions.

Among the task-based studies, the N-back was the most prevalent task used to study the effects of CR (Donohoe et al. 2018; Penadés et al. 2013; Ramsay et al. 2017; Subramaniam et al. 2014). The N-back measures working memory, loosely described as the storage and manipulation of short-term information (Baddeley 1986). The variability in analysis methods across the studies, however, complicates the synthesis of findings. Donohoe et al. (2018) took a whole-brain strategy and not find any CR related changes in BOLD response. Penades et al. (2013) used a data-driven approach to identify the task-evoked central executive networks (including the middle and inferior frontal gyrus) and default mode network (including the anterior cingulate cortex) and found hyperactivation within the executive control networks in patients that decreased following CR. Subramaniam et al. (2014) used predefined ROIs i.e., brain regions shown to elicit a BOLD response in an independent study—middle frontal gyrus/dorsolateral prefrontal cortex, inferior frontal gyrus/ventrolateral prefrontal cortex, and insula—found activation in the middle frontal gyrus in the healthy control sample and hypoactivation in schizophrenia patients that increased after CR. This appeared to be an opposing trend to what Penades et al. (2013) observed but in both studies the CR brain changes corresponded to cognitive improvements. Ramsay et al. (2017) focused on the functional connectivity of the BOLD responses of three thalamocortical regions, and showed between patient-group differences in thalamo-middle frontal gyrus and thalamo-anterior cingulate cortex BOLD functional connectivity following CR; this pattern was mirrored in the task-free intrinsic functional connectivity in the same study. Keshavan et al. (2017) studied a higher-level thought process of cognitive control, but also observed a similar pattern of CR-mediated change in the dorsolateral prefrontal cortical BOLD response, and changes in its functional connectivity with distributed brain regions of the anterior and posterior cingulate cortices and sensorimotor regions.

Resting-state fMRI have emerged as a popular tool in the study of clinical populations including schizophrenia because of its sensitivity to disease states and changes in physiological conditions. In Donohoe et al. (2018) study, CR-led changes in the intrinsic functional connectivity of the precuneus (a hub of the default mode network) and inferior parietal lobe (executive control network) was found, despite the lack of task-induced BOLD response. Another study showed that a short 8-week CR intervention program sufficed to elicit changes in ALFF, a measure of the magnitude of local intrinsic brain activity. Taken together, and that the observation that the bulk of energy consumed by the brain is during "rest" or not doing a task (Fox and Raichle 2007), the findings imply that intrinsic brain fluctuations may be a more sensitive measure of the effects of CR than task-based fMRI. Besides its sensitivity, resting-state fMRI has a shorter acquisition time relative to task-based fMRI. Performing a task inside a scanner may sometimes be an onerous affair for clinical population such as schizophrenia and is highly state dependent. Because of these reasons, we foresee a leap in the number of CR-related resting-state studies in future.

In contrast with studies of brain function, there was only one study examining the effects of CR on brain structure, which was focused on volume of the thalamus (Ramsay et al. 2018). Thalamic volume was not altered in the CR group when compared with the placebo group over time, although a within-group relationship between improvements in cognitive performances and thalamic volume increase was found. Also, subsidiary analyses found no treatment effect on the volumes of other brain regions thought to be implicated in cognitive remediation in schizophrenia (Ramsay et al. 2018). When juxtaposed with studies of other neuroimaging modalities, this may suggest that the neuroplastic effects of CR do not occur on a scale that can be captured by gross morphometry measures.

While the neuroimaging findings of CR intervention were largely encouraging—with nine out of the eleven studies here finding a group-by-time effect and an equivalent number of studies finding a relationship between brain changes and cognitive improvements—the heterogeneous nature of behavior-based interventions such as

CR makes it a challenge to pinpoint the exact neural networks underlying the therapeutic change. There is high variability across the CR therapies reviewed here, for instance, the types of CR therapy (drill-and-practice versus strategy-based), differences in duration and frequency of CR and the dependence on therapist skills, or the absence of therapist, in some studies. The inconsistency among the CR therapies make it hard to parse out the neurobiology underlying the active ingredients of treatment response.

Several limitations were common to studies reviewed here. All of them had modest sample sizes that ranged from 10 to 25 within a group, which increase the probability of Type I error. The studies were also largely hypothesis-driven. Other possible CR-mediated brain changes will not be picked up if the focus is on regions-of-interests. Other limitations pertain to specific studies. First, parts of the CR training in two studies involved tasks that were identical to the outcome measures (Donohoe et al. 2018; Ramsay et al. 2017) and the findings may be a result of practice effect and not a reflection of the active ingredients of CR. Second, many studies did not analyze baseline differences (Eack et al. 2016; Habel et al. 2010; Hooker et al. 2012, 2013) and hence could not rule out possible initial preintervention intra-individual differences that could influence the treatment-by-time effects. In the same vein, only two studies collected data from healthy controls, which could help determine whether CR normalizes overactive networks or restore weakened networks, as opposed to eliciting compensatory networks.

Future study designs should factor in a healthy control group. Theoretically, differences in neural circuitries between the healthy controls and schizophrenia (regardless of intervention) at baseline would suggest that these networks are abnormal in schizophrenia, and the associations of abnormalities in the neural circuitries with cognitive performances would suggest that the abnormal circuitries underlie cognitive deficits in schizophrenia. One could make the assumption that CR has restored the weakened brain function when the measures of the implicated neural circuitries moves toward the levels of those in healthy individuals after CR and the same effect is not observed in the placebo group. The conjecture that CR has led to compensatory mechanisms could be made when changes in neural circuitries, not found to be abnormal at baseline, are observed and coincide with cognitive improvements after CR intervention.

Future studies could also investigate the durability of CR-led brain changes. CR for now remains a one-off intervention in most clinical settings, and little is known about its long-term effects on brain functioning and cognition. Such findings may guide prescription such as the requirement of booster sessions or policy changes in the form of continued community interventions.

Finally, a multimodal neuroimaging approach could help capture the spatio-temporal nuances of CR-induced brain changes. The latest segmentation tools now permit automated labeling of certain gross subcortical structures into finer-grained subregions (Iglesias et al. 2015). Cognitive studies of schizophrenia are starting to embrace network science which affords more sensitive brain network metrics (Medaglia et al. 2015). The clinical utility of using neuroimaging markers to predict CR treatment response will largely depend on ongoing work to identify sensitive neuroimaging tools that can visualize robust and replicable brain structural or functional networks underlying the active ingredients of CR.

References

Anticevic A, Hu X, Xiao Y, Hu J, Li F, Bi F, *et al* (2015). Early-course unmedicated schizophrenia patients exhibit elevated prefrontal connectivity associated with longitudinal change. *J Neurosci* **35**(1): 267–286.

Baddeley AD (1986). *Working memory* Clarendon Press; Oxford University Press: Oxford Oxfordshire; New York, xi, 289 pp.

Barch DM (2013). Brain network interactions in health and disease. *Trends Cogn Sci* **17**(12): 603–605.

Cabeza R, Prince SE, Daselaar SM, Greenberg DL, Budde M, Dolcos F, *et al* (2004). Brain activity during episodic retrieval of autobiographical and laboratory events: an fMRI study using a novel photo paradigm. *J Cogn Neurosci* **16**(9): 1583-1594.

Davidson M, Galderisi S, Weiser M, Werbeloff N, Fleischhacker WW, Keefe RS, *et al* (2009). Cognitive effects of antipsychotic drugs in first-episode schizophrenia and schizophreniform disorder: a randomized, open-label clinical trial (EUFEST). *Am J Psychiatry* **166**(6): 675–682.

Donohoe G, Dillon R, Hargreaves A, Mothersill O, Castorina M, Furey E, *et al* (2018). Effectiveness of a low support, remotely accessible, cognitive remediation training programme for chronic psychosis: cognitive, functional and cortical outcomes from a single blind randomised controlled trial. *Psychol Med* **48**(5): 751–764.

Eack SM, Newhill CE, Keshavan MS (2016). Cognitive Enhancement Therapy Improves Resting-State Functional Connectivity in Early Course Schizophrenia. *J Soc Social Work Res* **7**(2): 211–230.

Fan F, Zou Y, Tan Y, Hong LE, Tan S (2017). Computerized cognitive remediation therapy effects on resting state brain activity and cognition in schizophrenia. *Sci Rep* **7**(1): 4758.

Fox MD, Raichle ME (2007). Spontaneous fluctuations in brain activity observed with functional magnetic resonance imaging. *Nat Rev Neurosci* **8**(9): 700–711.

Green MF (1996). What Are the Functional Consequences of Neurocognitive Deficits in Schizophrenia? *The American Journal of Psychiatry* **153**(3): 321.

Green MF, Horan WP, Lee J (2015). Social cognition in schizophrenia. *Nat Rev Neurosci* **16**(10): 620–631.

Green MF, Kern RS, Braff DL, Mintz J (2000). Neurocognitive Deficits and Functional Outcome in Schizophrenia: Are We Measuring the "Right Stuff"? *Schizophr Bulletin* **26**(1): 119–136.

Habel U, Koch K, Kellermann T, Reske M, Frommann N, Wölwer W, *et al* (2010). Training of affect recognition in schizophrenia: Neurobiological correlates. *Social Neuroscience* **5**(1): 92–104.

Haut KM, Lim KO, MacDonald A, 3rd (2010). Prefrontal cortical changes following cognitive training in patients with chronic schizophrenia: effects of practice, generalization, and specificity. *Neuropsychopharmacology* **35**(9): 1850–1859.

Hill SK, Schuepbach D, Herbener ES, Keshavan MS, Sweeney JA (2004). Pretreatment and longitudinal studies of neuropsychological deficits in antipsychotic-naive patients with schizophrenia. *Schizophr Res* **68**(1): 49–63.

Hogarty GE, Flesher S, Ulrich R, Carter M, Greenwald D, Pogue-Geile M, *et al* (2004). Cognitive enhancement therapy for schizophrenia: effects of a 2-year randomized trial on cognition and behavior. *Arch Gen Psychiatry* **61**(9): 866–876.

Hooker CI, Bruce L, Fisher M, Verosky SC, Miyakawa A, D'Esposito M, *et al* (2013). The influence of combined cognitive plus social-cognitive training on amygdala response during face emotion recognition in schizophrenia. *Psychiatry Research: Neuroimaging* **213**(2): 99–107.

Hooker CI, Bruce L, Fisher M, Verosky SC, Miyakawa A, Vinogradov S (2012). Neural activity during emotion recognition after combined cognitive plus social cognitive training in schizophrenia. *Schizophrenia Research* **139**(1-3): 53–59.

Iglesias JE, Augustinack JC, Nguyen K, Player CM, Player A, Wright M, *et al* (2015). A computational atlas of the hippocampal formation using ex vivo, ultra-high resolution MRI: Application to adaptive segmentation of in vivo MRI. *Neuroimage* **115**: 117–137.

Isaac C, Januel D (2016). Neural correlates of cognitive improvements following cognitive remediation in schizophrenia: a systematic review of randomized trials. *Socioaffect Neurosci Psychol* **6**: 30054.

Keshavan MS, Eack SM, Prasad KM, Haller CS, Cho RY (2017). Longitudinal functional brain imaging study in early course schizophrenia before and after cognitive enhancement therapy. *Neuroimage* **151**: 55–64.

McTeague LM, Huemer J, Carreon DM, Jiang Y, Eickhoff SB, Etkin A (2017). Identification of Common Neural Circuit Disruptions in Cognitive Control Across Psychiatric Disorders. *Am J Psychiatry* **174**(7): 676–685.

Medaglia JD, Lynall ME, Bassett DS (2015). Cognitive network neuroscience. *J Cogn Neurosci* **27**(8): 1471–1491.

Medalia A, Choi J (2009). Cognitive remediation in schizophrenia. *Neuropsychol Rev* **19**(3): 353–364.

Mennes M, Kelly C, Zuo XN, Di Martino A, Biswal BB, Castellanos FX, *et al* (2010). Inter-individual differences in resting-state functional connectivity predict task-induced BOLD activity. *Neuroimage* **50**(4): 1690–1701.

Mothersill O, Tangney N, Morris DW, McCarthy H, Frodl T, Gill M, *et al* (2017). Further evidence of alerted default network connectivity and association with theory of mind ability in schizophrenia. *Schizophr Res* **184**: 52–58.

Nuechterlein KH, Barch DM, Gold JM, Goldberg TE, Green MF, Heaton RK (2004). Identification of separable cognitive factors in schizophrenia. *Schizophr Res* **72**(1): 29–39.

Penades R, Boget T, Lomena F, Bernardo M, Mateos JJ, Laterza C, *et al* (2000). Brain perfusion and neuropsychological changes in schizophrenic patients after cognitive rehabilitation. *Psychiatry Res* **98**(2): 127–132.

Penades R, Gonzalez-Rodriguez A, Catalan R, Segura B, Bernardo M, Junque C (2017). Neuroimaging studies of cognitive remediation in schizophrenia: A systematic and critical review. *World J Psychiatry* **7**(1): 34–43.

Penadés R, Pujol N, Catalán R, Massana G, Rametti G, García-Rizo C, *et al* (2013). Brain Effects of Cognitive Remediation Therapy in Schizophrenia: A Structural and Functional Neuroimaging Study. *Biological Psychiatry* **73**(10): 1015–1023.

Perlstein WM, Dixit NK, Carter CS, Noll DC, Cohen JD (2003). Prefrontal cortex dysfunction mediates deficits in working memory and prepotent responding in schizophrenia. *Biol Psychiatry* **53**(1): 25–38.

Ramsay IS, Fryer S, Boos A, Roach BJ, Fisher M, Loewy R, *et al* (2018). Response to Targeted Cognitive Training Correlates with Change in Thalamic Volume in a Randomized Trial for Early Schizophrenia. *Neuropsychopharmacology* **43**(3): 590–597.

Ramsay IS, Nienow TM, MacDonald AW, 3rd (2017). Increases in Intrinsic Thalamocortical Connectivity and Overall Cognition Following Cognitive Remediation in Chronic Schizophrenia. *Biol Psychiatry Cogn Neurosci Neuroimaging* **2**(4): 355–362.

Saperstein AM, Kurtz MM (2013). Current trends in the empirical study of cognitive remediation for schizophrenia. *Can J Psychiatry* **58**(6): 311–318.

Seidman LJ, Mirsky AF (2017). Evolving Notions of Schizophrenia as a Developmental Neurocognitive Disorder. *J Int Neuropsychol Soc* **23**(9-10): 881–892.

Seitz J, Zuo JX, Lyall AE, Makris N, Kikinis Z, Bouix S, *et al* (2016). Tractography Analysis of 5 White Matter Bundles and Their Clinical and Cognitive Correlates in Early-Course Schizophrenia. *Schizophr Bull* **42**(3): 762–771.

Sheffield JM, Kandala S, Tamminga CA, Pearlson GD, Keshavan MS, Sweeney JA, *et al* (2017). Transdiagnostic Associations Between Functional Brain Network Integrity and Cognition. *JAMA Psychiatry* **74**(6): 605-613.

Subramaniam K, Luks Tracy L, Fisher M, Simpson Gregory V, Nagarajan S, Vinogradov S (2012). Computerized Cognitive Training Restores Neural Activity within the Reality Monitoring Network in Schizophrenia. *Neuron* **73**(4): 842–853.

Subramaniam K, Luks TL, Garrett C, Chung C, Fisher M, Nagarajan S, *et al* (2014). Intensive cognitive training in schizophrenia enhances working memory and associated prefrontal cortical efficiency in a manner that drives long-term functional gains. *NeuroImage* **99**: 281–292.

Thorsen AL, Johansson K, Loberg EM (2014). Neurobiology of cognitive remediation therapy for schizophrenia: a systematic review. *Front Psychiatry* **5**: 103.

Vinogradov S, Fisher M, de Villers-Sidani E (2012). Cognitive training for impaired neural systems in neuropsychiatric illness. *Neuropsychopharmacology* **37**(1): 43–76.

Wexler BE, Anderson M, Fulbright RK, Gore JC (2000). Preliminary evidence of improved verbal working memory performance and normalization of task-related frontal lobe activation in schizophrenia following cognitive exercises. *Am J Psychiatry* **157**(10): 1694–1697.

Wykes T (1998). What are we changing with neurocognitive rehabilitation? Illustrations from two single cases of changes in neuropsychological performance and brain systems as measured by SPECT. *Schizophr Res* **34**(1-2): 77–86.

Wykes T, Brammer M, Mellers J, Bray P, Reeder C, Williams C, *et al* (2002). Effects on the brain of a psychological treatment: cognitive remediation therapy: functional magnetic resonance imaging in schizophrenia. *Br J Psychiatry* **181**: 144–152.

Yarkoni T. Big Correlations in Little Studies: Inflated fMRI Correlations Reflect Low Statistical Power-Commentary on Vul et al. (2009). *Perspect Psychol Sci*. 2009 May;4. (3):294–8.

Part III
Implications

Single-subject Prediction: A Statistical Paradigm for Precision Psychiatry

Danilo Bzdok and Teresa M. Karrer

predictions, but not inferences, forecast what will happen.
White (1971)

Introduction

Psychiatric patients are treated every day by medical doctors based on clinical guidelines grounded in group definitions. Therapeutic intervention for a particular patient, however, frequently follows a trial-and-error path (cf. Rush et al. 2006). On average, only about 50% of patients benefit from a specific psychotropic drug therapy (Wong et al. 2010). Similar failure rates apply to common psychotherapeutic treatments (Hofmann et al. 2012). To render clinical care more effective, brain-imaging and genomics are among the most promising, yet expensive avenues. In research on the neural and genetic basis of psychiatric disease, the prevailing research ideology aims to discover new pathophysiological mechanisms as a stepping-stone to then reduce suffering of psychiatric patients (Insel and Cuthbert 2015). Instead of trying to exploit newly discovered disease mechanisms toward novel treatments that help patient groups on average, an alternative research agenda is coming into reach over recent years. A fast and cost-effective strategy is to accurately predict which of the currently existing treatment options is likely to work best for one particular patient (Bzdok and Meyer-Lindenberg 2018; Perna and Nemeroff 2017; Stephan et al. 2017b). Increasing individualization of therapeutic intervention

D. Bzdok (✉)
Department of Psychiatry, Psychotherapy and Psychosomatics, RWTH Aachen University, Aachen, Germany

JARA, Translational Brain Medicine, Aachen, Germany

Parietal Team, INRIA, Neurospin, Gif-sur-Yvette, France

T. M. Karrer
Department of Psychiatry, Psychotherapy and Psychosomatics, RWTH Aachen University, Aachen, Germany

V. A. Diwadkar, S. B. Eickhoff (eds.), *Brain Network Dysfunction in Neuropsychiatric Illness*, https://doi.org/10.1007/978-3-030-59797-9_19

in precision psychiatry would also open up the opportunity to automatically derive the diagnosis or expected disease course in single patients from data. Because psychiatric disorders result from disturbed brain biology, quantitative measures from in-vivo brain-imaging is ready for such predictive modeling approaches because structural, functional, or diffusion magnetic resonance imaging (sMRI, fMRI, dMRI) and sometimes positron emission tomography (PET) are already available in most modern psychiatric hospitals. Moreover, such spatially or temporally highly resolved imaging techniques produce data of sufficient quantity and information granularity on which to apply data-driven statistical techniques (Eyre et al. 2016).

Examples of Brain Network Phenotyping to Improve Predictions in Single Individuals

Several neuroscience studies have already demonstrated the clinical potential of combining extensive brain-imaging data with predictive data analytics toward important goals of precision medicine in psychiatry. As a first example, Drysdale et al. (2017) were able to forecast the treatment response in a large fMRI study on patients suffering from depression ($n = 1188$). The patients were automatically subdivided into four groups by applying data-driven clustering algorithms (hierarchical ward clustering) to their limbic and frontostriatal resting-state network data. Each of the four imaging-derived depression types was characterized by a particular pattern of neural dysfunction and could be linked to a distinct symptom constellation. The neurobiologically defined depression categories were showed to be clinically relevant as the subgroups differed in symptom reduction (according to Hamilton rating scale for depression) after a 5-week treatment with repetitive transcranial magnetic stimulation of the dorsomedial prefrontal cortex (Fig. 1). This neurostimulation method modulates connectivity of brain networks and is usually applied in refractory depression. Pattern-prediction algorithms (support vector machines) were built to predict individual treatment response based on brain connectivity data. The efficacy of treatment was correctly forecasted in nine out of ten new patients that were not involved in the previous model building procedure. Hence, a proof of concept is provided by a seminal stimulation-MRI investigation in major depressive disorder that brain-derived types of mental disease may enable the selection of the most promising treatment option in each particular patient.

This brute-force and ad hoc character of many successful pattern-prediction approaches was shown in a neuroimaging study in aphasia to, in certain cases, also allow for insights into disease pathophysiology. Brodersen et al. (2011) acquired experimental fMRI data of aphasic patients and healthy controls who listened to speech and time-reversed speech stimuli ($n = 37$). To estimate task-induced changes in directional neural coupling mechanisms, the investigators built a dynamic causal model of relevant auditory regions and the connections between these brain areas. The task-evoked thalamotemporal connectivity modulations were fed into a machine

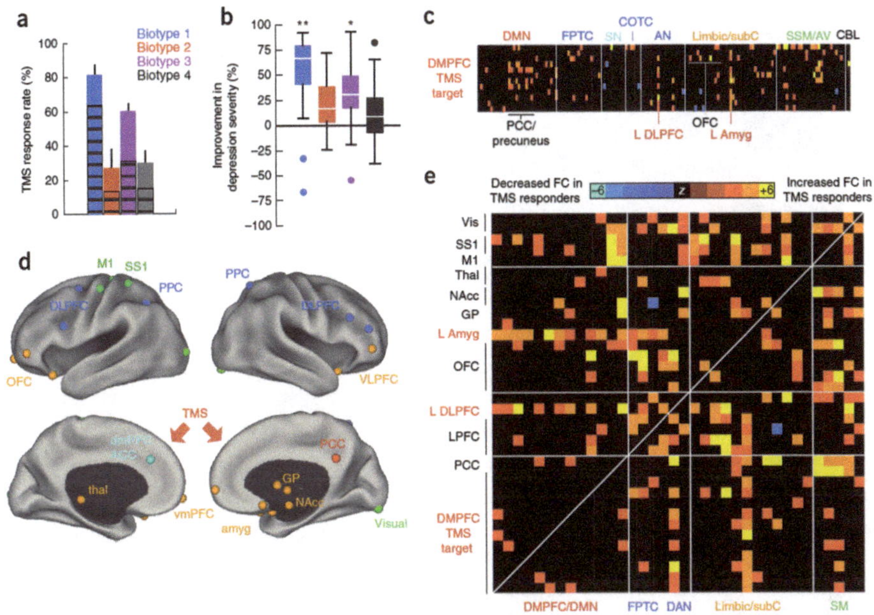

Fig. 1 Brain connectivity can predict treatment response in major depressive disorder. (**a**) Subgroups of patients clustered by resting-state connectivity profile differed in their clinical response to repetitive transcranial stimulation (rTMS) intervention on the dorsomedial prefrontal cortex (dmPFC). Treatment response rate corresponds to percentage of patients whose symptom severity was reduced by at least 25% as measured by the Hamilton rating scale for depression (HAMD). (**b**) Boxplot describing the distribution of symptom severity reduction in four depression biotypes. Percent improvement in symptom severity was computed as difference in total HAMD score before and after rTMS application. $^{**}P = 0.00001$–0.002 (Mann–Whitney), marks significant increase compared to biotypes 2–4; $^{*}P = 0.007$ (Mann–Whitney), marks significant increase compared to biotype 4. (**c**) Treatment responders differed in their functional coupling of dmPFC target area with other brain regions compared to non-responders: warm colors = increased connectivity, cold colors = decreased connectivity (Wilcoxon rank–sum tests, thresholded at $P < 0.005$). (**d**) Neuroanatomical locations of 25 brain regions (top 10%) which expose most discriminative connectivity features allowing detection of treatment responders versus non-responders. Red arrows point at rTMS target. (**e**) Heat maps illustrate how responders to rTMS differed in their functional connectivity profile compared to non-responders. (**c-e**) Brain regions colored by corresponding functional networks. Figure reused with permission from Drysdale et al. (2017)

learning algorithm (support vector classification) that automatically derived rules how to discriminate disease state. The classification algorithm was able to tell apart new patients and controls almost perfectly (98% prediction accuracy). Additionally, the predictive modeling approach was directly mechanistically interpretable by revealing disease-specific brain connections in patients with aphasia (Fig. 2). This second example demonstrates that often heuristic machine-learning methods can be grounded in mechanistically meaningful neurobiological quantities to offer means for a more complete understanding of pathophysiology and predictability of brain disorders.

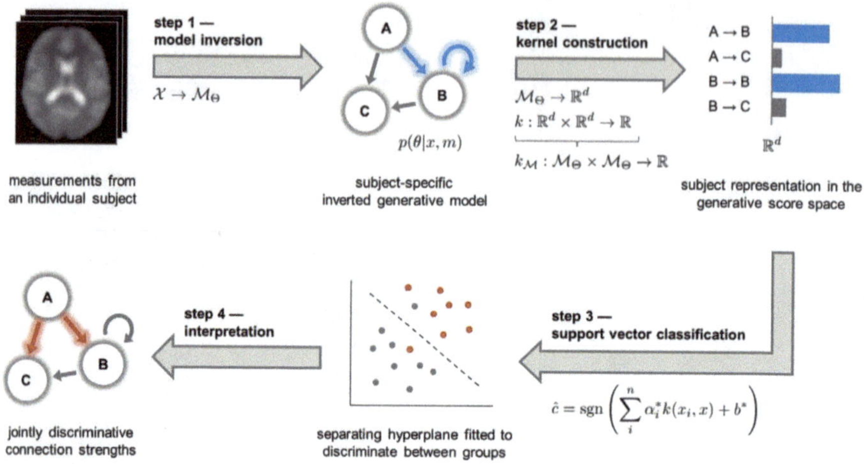

Fig. 2 Study rationale for discovering disease-relevant functional connections in aphasia disease. Brain activity as measured by the blood oxygen level dependent (BOLD) signal was acquired in patients with aphasia and healthy controls during a speech-processing task. In the first step, the functional data of each subject were used to estimate the parameters of a dynamic causal model (DCM). The ensuing directional connectivity estimates were supposed to capture quantities describing functional brain mechanisms. In the second analysis step, a kernel function was built to i a similarity metric between the fitted models from two subjects. The constructed kernel instantiated a model-based space of variables. In this exemplary subject, the influence of region A on region B and the self-connection of region B were particularly strong. In a third step, a machine-learning algorithm (support vector machine) was trained to disambiguate patients with aphasia and healthy controls exclusively based on connectivity strengths. In the last step, the relevance of functional coupling of brain regions A–C for discrimination of disease state could be interpreted mechanistically. In this example, the joint influences of region A on region B and C were clinically most relevant to tell apart patients and controls that used during model building (cross-validation). Figure reused with permission from Brodersen et al. (2011)

Our last example revisited a long-standing speculation about accelerated brain aging in schizophrenia by quantifying a scientifically valuable and clinically exploitable brain phenotype. Kraepelin himself has already suspected a neurodegenerative character of schizophrenia which is why he called the disease "dementia praecox" (Kraepelin 1899). More recently, Koutsouleris et al. (2014) employed predictive learning algorithms (support vector regression) to estimate brain age from the structural MRI scans of healthy subjects ($n = 800$). The obtained trained algorithm was then used to predict brain age from brain structure of at-risk individuals who exposed subthreshold psychotic symptoms and patients diagnosed with schizophrenia ($n = 230$). The discrepancy between model-derived brain age and the actual age of the subjects increased from at-risk over recent onset to recurrent disease states. Moreover, the brain phenotype of a prematurely aged brain in schizophrenia was predictive of disease status in specific individuals (Fig. 3). The predictive modeling strategy could hence quantitatively formalize accelerated aging effects for successful

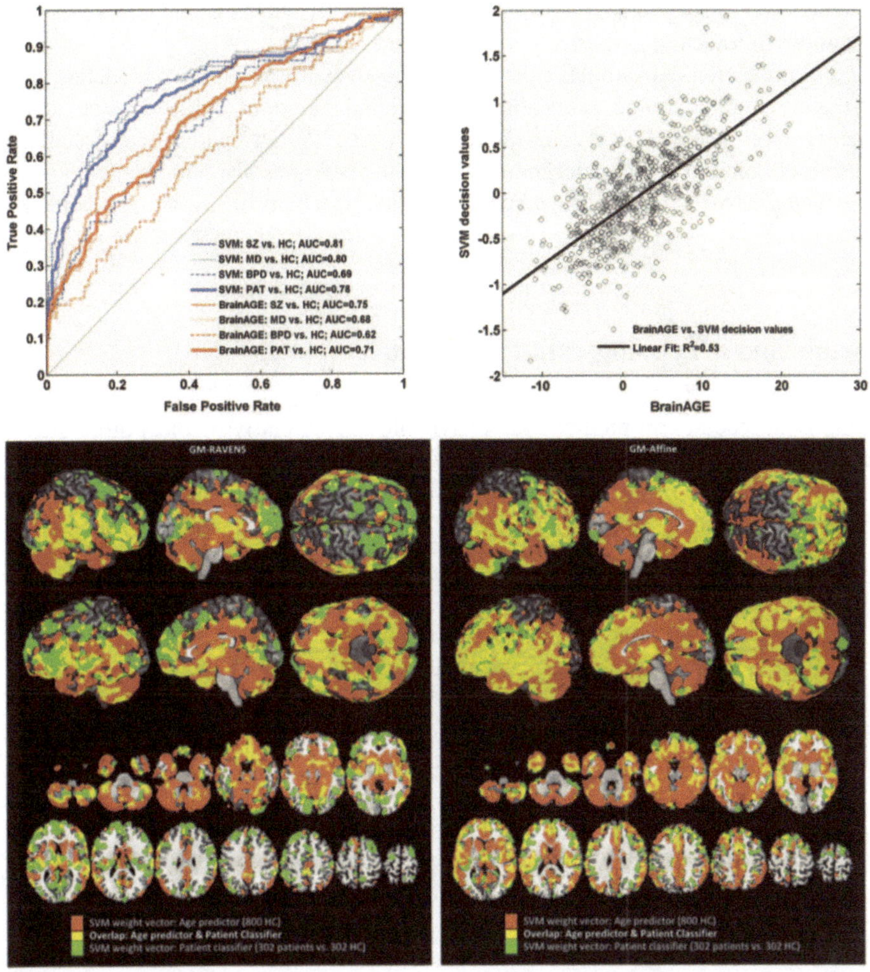

Fig. 3 Relationship between accelerated brain aging indexed by brain structure and classification of patients with schizophrenia (SCZ) versus healthy controls (HC). Tests evaluated whether the difference between actual and brain-derived age predicts disease discriminability, based on machine learning (support vector machines) in volumetric MRI data. Upper left: ROC curve of the disease classifier (blue line) and the brain age gap prediction (red line) to evaluate their performance in telling apart patients and controls not used for predictive model building. Upper right: continuous prediction (support vector regression) of patient classification values and brain age gap estimation. Both measures were highly correlated ($R2 = 0.53$; $T = 26.1$; $P < 0.001$). Lower panel: neuroanatomical overlap (yellow) of the brain regions relevant for age prediction (red) and disease state classification (green). Pattern-learning algorithms were fed by two structural brain maps: (i) gray matter-Regional Analysis of brain Volumes in Normalized Space (GM-RAVENS; left) and (ii) affinely registered gray matter data (right). Shared variation of patient- and age-predictive brain patterns might be an explanation for the observed correlation between brain aging and disease classification. Figure reused with permission from Koutsouleris et al. (2014)

prediction of disease state on a single-subject level, which substantiated a classical pathophysiological hypothesis in schizophrenia research.

The collection of examples exposes how the alliance of predictive modeling analytics and brain scanning can enable the individual prediction of disease state, treatment response, and clinical course in brain disorders. Against widespread belief, the fusion of "big data" and emerging statistical technologies also has the potential to open alternative windows on pathophysiological mechanisms in major psychiatric disorders.

Proposing a Typology of "Prediction"

Empirical research in medicine and psychology frequently invokes three different notions of how findings may allow for statements about the future of a particular individual (Bzdok 2017a; Casella and Berger 2002; Gabrieli et al. 2015; Woo et al. 2017):

(i) *Correlation*, such as Pearson or Spearman's rank correlation, computes a simple similarity metric between two series of measurements. For instance, the correlation between amygdala activity changes measured in an fMRI experiment and overall disease severity could be computed in a group of patients with schizophrenia. A high correlation between the two variables could motivate a deeper examination of the relationship between amygdala fMRI activity and disease severity. In this data-exploration setting, the goal is not to extract a model that embodies the discovered statistical relationship. As no model parameters are being estimated in common correlation analysis, the pattern interrogated in the data cannot be directly shipped to other researchers or clinicians for application to a new individual.

(ii) *Linear regression approaches*, such as ANOVA, ANCOVA, and MANOVA, estimates model parameters that encapsulate the mapping from explanatory (independent) input variables to explained (dependent) output variables. For instance, the linear relationship of how fMRI activity changes in the amygdala relate to the presence of schizophrenia symptoms can be extracted in a patient group. The obtained model embodies how the symptom severity can be expected to increase or decrease with a higher or lower fMRI activity in the amygdala in the subject sample at hand. The relevance of a particular explanatory variable can be quantified by comparison to the estimated importance of other explanatory variables, such as the measured fMRI activity in the prefrontal cortex, auditory cortex, fusiform gyrus, or other candidate brain regions. In this setting, a model has been generated that could be shipped to other researchers and psychiatrists for reuse in new individuals. However, for many linear-regression analyses there is a small tradition to explicitly ascertain that the isolated pattern still holds in subjects who the model has not yet seen.

(iii) *Single-subject prediction*, such as routinely performed with support vector machines, random forests, or artificial neural-network algorithms, is typically achieved by identifying relationships in one set of subjects as a function of how these patterns persist in other individuals from a different set of subjects. Here, model parameters are typically estimated on some data while the emerging model is explicitly put to the test in some independent data from unseen individuals (Shalev-Shwartz and Ben-David 2014). Such primary emphasis on model performance on single individuals can sometimes have the side effect of hindering understanding how the output variable is expected to change as specific input variables increase or decrease (Goodfellow et al. 2016; Hastie et al. 2001). For instance, pattern-recognition algorithms could possibly be built to derive optimal drug dosage suggestions for patients with schizophrenia from whole-brain fMRI activity, as measured from potentially tens of thousands of local brain measurements being the input variables for a given patient. In this setting, the investigator typically pursues the main goal to achieve the best-possible model accuracy for the optimal drug dosage in any single patient, even if this objective may in certain cases come at the expense of some reduced model interpretability.

In our opinion, predictions of type (iii) may be particularly tuned to the ambitions of precision medicine in psychiatry. This analysis paradigm, routinely practiced in many applications of pattern-recognition algorithms, is centered around evaluating the capacity of already extracted models to derive quantities of interest from new, potentially later encountered individuals. If an already extracted model embodying an identified relationship, reflected in the estimated parameters, is assessed in new individuals whose data were not used to estimate the parameters, the statistical analysis can be said to be an *out-of-sample prediction*. This form of building models from data has been explicitly optimized for and is naturally applicable to a single data point, such as one whole-brain scan or one sequenced genome of a particular individual. Whether an obtained model is useful in practice is judged based on its performance in achieving accurate predictions in independent individuals.[1] One may view these evaluation practices as more conservative measures when the goal is reliable single-subject predictions in patients admitted to a psychiatry hospital in the future (Bzdok and Meyer-Lindenberg 2018; James et al. 2013). In contrast, most correlation- and linear-regression-type analyses (i and ii) reflect *in-sample predictions* where model estimation and evaluation are typically performed in the same subject sample. In this analysis paradigm, the ensuing conclusions could be closely adapted to the particular subject sample at hand. Please also note that out-of-sample prediction is distinct from a *replication* study where the same statistical approach is entirely repeated on two different subject samples, rather than extracting a model from a first set of subjects while evaluating the performance of that model in an independent second set of subjects.

[1] Niels Bohr put this point in the following words: "Prediction is very difficult, especially if it's about the future."

Single-subject Prediction Is Important for Personalized Medicine

We invite the reader to imagine a future of centralized medical care, where comprehensive health-related information is available on a majority of the citizens, ignoring important technical and ethical challenges for a moment (Leonelli 2016; O'Neil 2016). *Scientific discovery* in such rich data resources has been a primary focus of the statistical methodology traditionally used in empirical research in medicine and psychology (Bzdok 2017b; Efron and Hastie 2016; James et al. 2013). This modeling goal is for instance especially suited to ask, "Which gene locations *contribute to* or *are associated* with schizophrenia?" Perhaps counterintuitively, thus identified genetic risk variants may not in all cases serve best to detect *whether* a given individual is affected by schizophrenia or is healthy (Shmueli 2010). This is because modeling for prediction typically asks a different kind of research question: "Which gene locations are *useful* to *distinguish* schizophrenic versus healthy individuals?" Finding such answers follows the heuristic agenda of prioritizing successful pattern-recognition to identify any relationship in the data that is able to derive the specified outcome in independent data, but may put smaller emphasis on *scientific insight into the neurobiological underpinnings of the schizophrenia disease*. As further example, scientific discovery inquires which genes provide mechanistic insight into explaining *why* a drug treatment for schizophrenia works in patients on average (James et al. 2013). Yet, pattern-recognition algorithms prioritize predictive genes that can reliably disambiguate *whether* a drug treatment will work in one specific patient with schizophrenia.

The data-driven identification of predictive principles from complex information has been a dominant focus in the machine-learning community (Breiman 2001; Goodfellow et al. 2016). Pattern-recognition algorithms have frequently been employed with the goal to achieve the best possible outcome detection (Hastie et al. 2001). Typically based on modest a priori knowledge (Abu-Mostafa et al. 2012), many algorithms in machine learning are estimated to deduce an output variable (e.g., presence of schizophrenia) from potentially many input variables (e.g., lifestyle and demographic indicators, brain function, neuropsychological tests). For qualitative output variables (e.g., healthy group versus schizophrenic group or drug responder versus non-responder), the modeling process can perform *classification*. For quantitative response variables (e.g., schizophrenia severity or probability of responding to a candidate drug), pattern-recognition algorithms are built for *regression* from potentially highly resolved input data from possibly different sources (Hastie et al. 2001). In the following, several aspects are highlighted that characterize methods embraced by machine learning with the prominent differences to tools often used in classical statistics. In particular, the distinct properties of both modeling approaches shed light on why the practices of pattern-learning methods are well suited to achieving accurate predictions *at the single-subject level*, whereas many traditional statistical approaches are frequently used in medicine and psychology to discern the "trueness" of an effect *at the group level*.

The *prediction paradigm*, such as practiced in machine learning community, departs in important ways from the *inference paradigm* grounded in classical null-hypothesis testing in types of conclusions that are to be drawn from data (Breiman 2001; Bzdok 2017a; Efron and Hastie 2016). The statistical paradigms anchored at inference or prediction are common in trying to evaluate whether an effect found in some data extrapolates to another sample of observations drawn from the same underlying population (Casella and Berger 2002; Efron 2012). To draw statistical inference, mainstream statistics has put a major emphasis on a framework revolving around rejection of a null-hypothesis in favor of an alternative hypothesis that is contradicting the status quo. Instead, many machine-learning applications have mainly focused on building pattern-recognition algorithms that predict the discovered patterns in independent, new data that did not influence model estimation. In classical statistics, inferential conclusions are drawn by formally testing for the existence of an effect expressed under the null-hypothesis (e.g., a gene is not associated with schizophrenia) in opposition to the alternative hypothesis (e.g., a gene is associated with schizophrenia). The ensuing *p*-value indicates whether data from the subject sample at hand are too extreme to occur under the null hypothesis.

While null-hypothesis testing typically takes the form of a single-step approach, the success of precision psychiatry will probably depend on predictive models that can be extracted and then "shipped" in a two-step approach. In classical null-hypothesis testing, the p-value is computed on the *entire* data from a particular subject sample in a single process. P-values are commonly obtained from all examined individuals (in-sample) and this quantitative outcome can usually not be used to test for the *same* statistical relationship in a later encountered single individual. In contrast, methods common in machine learning can quantify the prediction performance of a previously built algorithm applied to untapped data, such as from a new incoming patient, as a performance metric and immediate practical usefulness. This process of evaluating the prediction performance of learning algorithms is typically performed by a two-step procedure called *cross-validation* (Shalev-Shwartz and Ben-David 2014). In a first step, the machine-learning algorithm is built on a larger part of the dataset. In a second step, emerging candidate algorithms are evaluated and selected on unused data (Hastie et al. 2001). Because all conditions for independent, identically distributed observations are usually met for the left-out data, the out-of-sample prediction performance on the testing data samples can quantify how likely the same pattern could be detected in future, not yet seen patients.

It is this two-step nature of ensuring model generalization by means of cross-validation procedures that attempts to certify predictive models to be "shippable" to other mental health institutions for detection of a previously discovered pattern for a single individual (Arbabshirani et al. 2017; Stephan et al. 2017a). The option for single-subject prediction is especially relevant in precision psychiatry: the brain scans or genetic profile of a new patient can be fed into the (previously built) pattern-recognition algorithm to estimate a clinical outcome variable (Woo et al. 2017). Concretely, an association between a gene and a psychiatric disorder like schizophrenia with a statistically significant p-value does not necessarily imply that the same gene will be the best choice to successfully predict whether a given individual is affected by schizophrenia. Conversely, an effect that has been empirically shown to be highly predictive of schizophrenia disease based on cross-validation in independent individuals does not always go hand-in-hand with classical statistical tests evaluated to a significant p-value (Bzdok 2017a; Shmueli 2010). For these reasons, *cross-validated machine-learning algorithms and more traditional tools for null-hypothesis testing can sometimes lead to diverging conclusions in certain practical analysis settings (see* Fig. 4 *for an example).*

Moreover, many tools frequently used in machine learning may also be especially suited to achieve the goals of precision psychiatry because they are naturally capable of handling hundreds or thousands of outcomes at once (Bzdok and Meyer-Lindenberg 2018; Caruana 1998; Rahim et al. 2017). Classical null-hypothesis testing in medicine usually compares two possible output states that are expressed in the null and alternative hypothesis (Wasserstein and Lazar 2016) – the non-preferred null-hypothesis and the alternative hypothesis posited by the investigator – with limited scaling to more possible options (Efron 2012). The challenging question in precision psychiatry is usually not if a patient suffers from a psychiatric disorder or not. Instead, the psychiatrist rather decides which specific psychiatric disorder the patient is suffering from. Analogously, it may be clinically more relevant which treatment option a particular patient should be assigned to rather than asking whether a given patient needs a therapy or not. In fact, even when comparing a number of disease groups based on ANOVA, the null-hypothesis commonly being

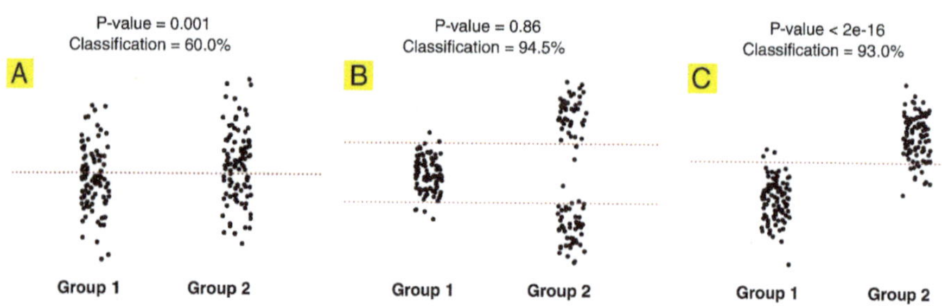

Fig. 4 Classical null-hypothesis testing and machine-learning algorithms can lead to diverging conclusions. The toy data distributions may reflect two groups (e.g., patients with schizophrenia versus healthy controls) as assessed by differences in a particular brain measurement (e.g., amygdala activity). Group differences are evaluated by (i) null-hypothesis testing using two-sample t-tests ("*P*-value") and (ii) machine-learning using a classification algorithm to predict which group each brain data point belongs to as indicated by dotted rule ("Classification"). In three cases with different brain data distributions, (**a**) t-test was statistically significant, while classification accuracy was poor. (**b**) Based on this data scenario, t-test was not statistically significant, while classification accuracy was high. (**c**) In yet another point distribution, both t-test was statistically significant and classification accuracy was high. This artificial example illustrates that null-hypothesis testing and machine-learning algorithms constitute two different statistical cultures that do not necessarily judge data distributions by the same aspects of evidence. Hence, group effects as assessed by significant p-values do not always entail a high prediction performance, and vice versa. Figure reused with permission from Arbabshirani et al. (2017)

tested is whether all groups are equal (null-hypothesis) or not (alternative hypothesis) (Casella and Berger 2002). For instance, one disease group being different from nine other ones or ten disease groups being different in every possible pair equally lead to rejection of the null-hypothesis of no difference. For these predictions of several diagnoses or treatment options in parallel, machine learning is especially suitable. This is because many machine-learning algorithms can be easily extended to the prediction of a large array of different outcomes in the same algorithm building process (Breiman and Friedman 1997; Caruana 1998; Rahim et al. 2017). In this way, several machine-learning approaches readily offer the opportunity to predict many clinical end-points in a single patient. For instance, a machine-learning algorithm could be built to derive i) which disease the patient is suffering from (e.g., schizophrenia versus bipolar disorder versus major depression), ii) which disease course can be expected (e.g., single episode versus recurrent versus chronic disease course), and iii) which treatment options will be most effective (e.g., antipsychotic drugs versus antidepressant drugs versus cognitive behavioral therapy versus their combinations).

Challenges to Overcome

Forecasting diagnosis, disease course, and effectiveness of treatment options in incoming patients offers a promising opportunity for improving medical care for psychiatric patients. However, several challenges still need to be overcome: From a technical point of view, prediction-focused statistical approaches typically require large amounts of data (Henke et al. 2016; Jordan et al. 2013). So called "big data" demand special data management skills and infrastructure that are today seldom available at clinical and research institutions (Bzdok and Meyer-Lindenberg 2018). To further mature training of pattern-recognition algorithm, predictable patterns have been advocated to fulfill several criteria before translation into clinical practice (Woo and Wager 2015). These authors suggested that an ideal predictive model should be (i) highly successful in diagnostic performance including both sensitivity and specificity, (ii) useful in terms of neuroscientific research, (iii) composed of precisely defined algorithms, and (iv) generalizable across different clinical settings (e.g., data acquisition means, geographic locations, and patient populations). Additionally, the application of predictive models must be checked for the economic efficiency, especially in light of the considerable costs of brain-imaging modalities such as Smri, Fmri, Dmri, and PET (Gabrieli et al. 2015).

Furthermore, ethical and societal aspects will probably become an important recurring theme in discussing the future of precision medicine (Gabrieli et al. 2015; O'Neil 2016).

Classical Null-Hypothesis Testing Is Not Obsolete

Tools for statistical hypothesis testing and more recently emerged machine learning techniques can be used to draw different types of conclusions from data. Whereas the core interest of machine-learning applications is to *predict* future events on the basis of patterns observed in data, classical statistics applications are probably more often used to *infer* scientific insight from the effects observed in data (White 1971). Both modeling paradigms can serve distinct statistical purposes in improving psychiatric practice based on brain-imaging and genetics (Bzdok and Yeo 2017; Shmueli 2010). Depending on the ultimate clinical or research question, a different set of statistical tools may suggest itself as more appropriate (James et al. 2013). It is therefore important for investigators and psychiatrists to acknowledge the partly diverging modeling goals and scopes of interpretation of these two distinct statistical cultures (Breiman 2001; Bzdok 2017a).

Traditional null-hypothesis testing emerged in the early twentieth century. This was a time in history when data were rare and expensive to acquire (Efron and Hastie 2016; Gigerenzer 1993). Well-controlled research experiments were carefully designed in advance. Nowadays, such datasets with few measured variables are still the norm in much research in psychology and medicine. Many early statistical tools were especially developed for such settings aiming at understanding the relationship between a few variables. If the goal is to examine whether an effect exists or which specific input variables have most impact on an output variable, classical statistics based on null-hypothesis testing is arguably still among the best tools. In practice, the focus routinely relies on the statistical analyses of few variables that tend to yield high interpretability, rather than perusing data for complex patterns that are predictive. Ideally of course, one would hope to achieve both interpretability and predictability. Several recent investigations have successfully combined "black-box" pattern-recognition analyses and model components that can be readily introspected for scientific understanding (cf. Brodersen et al. 2011).

Today, single-subject prediction becomes always more feasible due to the recent co-occurrence in data availability, computing power, and cheaper data storage (Goodfellow et al. 2016; Manyika et al. 2011). Brain-scanning and genetic measurements in psychiatry produce massive amounts of data at high granularity that classical statistical tools have not initially been invented to tackle (Efron 2012). In contrast, machine learning was designed to extract patterns from such observational data that was frequently acquired outside of a carefully controlled experimental context. Additionally, many machine-learning approaches specifically motivated for achieving prediction at scale, such as in thousands of individual subjects or for hundreds of outcomes, as well as when outcome variables are hard or expensive to collect. In precision psychiatry for instance, the accurate prediction of a psychiatric disease, the disease course, or efficacy of treatment options in individual patients is the relevant research goal.

However, it is important to appreciate that the potential immediate gains of the pragmatic goal to identify patterns useful to predict clinical end-points in complex data does not preclude the longer-term urge for *understanding* the biological nature underlying psychiatric diseases like schizophrenia. Carefully designed, meticulously conducted, and logistically expansive experiments to confirm or reject a priori verbalized research hypotheses in animals and humans will probably remain a cornerstone to generate neuroscientific insight into mental illness. As one of many potential scenarios for a happy cohabitation of the inference and prediction paradigms, pattern-recognition algorithms with successful prediction performance will probably need to undergo carefully controlled randomized clinical trials based on traditional null-hypothesis testing before regulatory stakeholders authorize translation into clinical practice.

Conclusions

The current diagnosis systems in psychiatry that are pervasively used every day to clinically diagnose thousands of patients based on their symptoms and to provide medical care were established largely based on expert opinion. Diagnosis *groups* have served well in clinical practice and scientific investigation to impose structure on the evasive conglomerate of mental health disorders. The now rapidly increasing amount, detail, and quality of health-related information on each given individual are heralding a major turning point in the history of psychiatry. We may increasingly abandon predefining disease concepts to group psychiatry patients based on observed symptoms for clinical investigation of brain dysfunction. Instead, it will be increasingly possible to first quantitatively derive disease stratifications directly from brain measurements in a data-guided fashion to then capitalize on the discovered brain-based phenotypes for patient-tailored monitoring, risk assessment, and therapeutic intervention.

Approaching this future of psychiatric diagnosis systems rooted in brain biology will, in our opinion, necessitate awareness of precise notions of *prediction* and the statistical techniques with the honest capacity for deriving rigorous statements on single individuals. Perhaps the most important point we tried to make in this crash course is this: medical research has had a long-standing focus on scientific discovery revolving around understanding disease mechanisms by statistical methods targeted at establishing *statistical inference.* This statistical goal is in many cases incompatible with the pragmatic wish to somewhat blindly exploit the quantifiable consequences of brain pathophysiology to achieve most accurate *predictions* about the future of individuals based on diverse and rich biological information. Appreciation of this *inference-prediction divergence* will probably be a necessary milestone in personalized medicine research, which will ultimately benefit the well-being of suffering psychiatric patients.

References

Abu-Mostafa, Y.S., Magdon-Ismail, M., Lin, H.T., 2012. Learning from data. AMLBook, California.

Arbabshirani, M.R., Plis, S., Sui, J., Calhoun, V.D., 2017. Single subject prediction of brain disorders in neuroimaging: Promises and pitfalls. Neuroimage 145, 137–165.

Breiman, L., 2001. Statistical Modeling: The Two Cultures. Statistical Science 16, 199–231.

Breiman, L., Friedman, J.H., 1997. Predicting multivariate responses in multiple linear regression. Journal of the Royal Statistical Society: Series B (Statistical Methodology) 59, 3–54.

Brodersen, K.H., Schofield, T.M., Leff, A.P., Ong, C.S., Lomakina, E.I., Buhmann, J.M., Stephan, K.E., 2011. Generative embedding for model-based classification of fMRI data. PLoS Comput Biol 7, e1002079.

Bzdok, D., 2017a. Classical Statistics and Statistical Learning in Imaging Neuroscience. Front Neurosci.

Bzdok, D., 2017b. Classical Statistics and Statistical Learning in Imaging Neuroscience. Frontiers in Neuroscience 11, 543.

Bzdok, D., Meyer-Lindenberg, A., 2018. Machine learning for precision psychiatry: Opportunities and challenges. Biological Psychiatry: CNNI, in press.

Bzdok, D., Yeo, B.T.T., 2017. Inference in the age of big data: Future perspectives on neuroscience. Neuroimage 14, 549–564.

Caruana, R., 1998. Multitask learning. Learning to learn. Springer, pp. 95–133.

Casella, G., Berger, R.L., 2002. Statistical inference. Duxbury Pacific Grove, CA.

Drysdale, A.T., Grosenick, L., Downar, J., Dunlop, K., Mansouri, F., Meng, Y., Fetcho, R.N., Zebley, B., Oathes, D.J., Etkin, A., Schatzberg, A.F., Sudheimer, K., Keller, J., Mayberg, H.S., Gunning, F.M., Alexopoulos, G.S., Fox, M.D., Pascual-Leone, A., Voss, H.U., Casey, B.J., Dubin, M.J., Liston, C., 2017. Resting-state connectivity biomarkers define neurophysiological subtypes of depression. Nature Medicine 23, 28–38.

Efron, B., 2012. Large-scale inference: empirical Bayes methods for estimation, testing, and prediction. Cambridge University Press.

Efron, B., Hastie, T., 2016. Computer-Age Statistical Inference. Cambridge University Press.

Eyre, H.A., Singh, A.B., Reynolds, C., 2016. Tech giants enter mental health. World Psychiatry 15, 21–22.

Gabrieli, J.D., Ghosh, S.S., Whitfield-Gabrieli, S., 2015. Prediction as a humanitarian and pragmatic contribution from human cognitive neuroscience. Neuron 85, 11–26.

Gigerenzer, G., 1993. The superego, the ego, and the id in statistical reasoning. A handbook for data analysis in the behavioral sciences: Methodological issues, 311–339.

Goodfellow, I.J., Bengio, Y., Courville, A., 2016. Deep learning. MIT Press, USA.

Hastie, T., Tibshirani, R., Friedman, J., 2001. The Elements of Statistical Learning. Springer Series in Statistics, Heidelberg, Germany.

Henke, N., Bughin, J., Chui, M., Manyika, J., Saleh, T., Wiseman, B., Sethupathy, G., 2016. The age of analytics: Competing in a data-driven world. Technical report, McKinsey Global Institute.

Hofmann, S.G., Asnaani, A., Vonk, I.J., Sawyer, A.T., Fang, A., 2012. The Efficacy of Cognitive Behavioral Therapy: A Review of Meta-analyses. Cognitive Therapy and Research 36, 427–440.

Insel, T.R., Cuthbert, B.N., 2015. Brain disorders? Precisely. Science 348, 499–500.

James, G., Witten, D., Hastie, T., Tibshirani, R., 2013. An introduction to statistical learning. Springer.

Jordan, M.I., Committee on the Analysis of Massive Data, Committee on Applied and Theoretical Statistics, Board on Mathematical Sciences and Their Applications, Division on Engineering and Physical Sciences, National Research Council, 2013. Frontiers in Massive Data Analysis. The National Academies Press, Washington, D.C.

Koutsouleris, N., Davatzikos, C., Borgwardt, S., Gaser, C., Bottlender, R., Frodl, T., Falkai, P., Riecher-Rossler, A., Moller, H.J., Reiser, M., Pantelis, C., Meisenzahl, E., 2014. Accelerated

brain aging in schizophrenia and beyond: a neuroanatomical marker of psychiatric disorders. Schizophrenia Bulletin 40, 1140–1153.
Kraepelin, E., 1899. Psychiatrie. Ein Lehrbuch für Studierende und Ärzte, 6th ed. Barth, Leipzig.
Leonelli, S., 2016. Data-centric biology: a philosophical study. University of Chicago Press.
Manyika, J., Chui, M., Brown, B., Bughin, J., Dobbs, R., Roxburgh, C., Byers, A., 2011. Big data: The next frontier for innovation, competition, and productivity. Technical report, McKinsey Global Institute.
O'Neil, C., 2016. Weapons of Math Destruction. How Big Data Increases Inequality and Threatens Democracy, New York: Crown.
Perna, G., Nemeroff, C.B., 2017. Personalized Medicine in Psychiatry: Back to the Future. Personalized Medicine in Psychiatry 1, 1.
Rahim, M., Thirion, B., Bzdok, D., Buvat, I., Varoquaux, G., 2017. Joint prediction of multiple scores captures better individual traits from brain images. Neuroimage 158, 145–154.
Rush, A.J., Trivedi, M.H., Wisniewski, S.R., Nierenberg, A.A., Stewart, J.W., Warden, D., Niederehe, G., Thase, M.E., Lavori, P.W., Lebowitz, B.D., 2006. Acute and longer-term outcomes in depressed outpatients requiring one or several treatment steps: a STAR* D report. American Journal of Psychiatry 163, 1905–1917.
Shalev-Shwartz, S., Ben-David, S., 2014. Understanding machine learning: From theory to algorithms. Cambridge University Press.
Shmueli, G., 2010. To explain or to predict? Statistical science, 289–310.
Stephan, K.E., Schlagenhauf, F., Huys, Q.J.M., Raman, S., Aponte, E.A., Brodersen, K.H., Rigoux, L., Moran, R.J., Daunizeau, J., Dolan, R.J., 2017a. Computational neuroimaging strategies for single patient predictions. Neuroimage 145, 180–199.
Stephan, K.E., Schlagenhauf, F., Huys, Q.J.M., Raman, S., Aponte, E.A., Brodersen, K.H., Rigoux, L., Moran, R.J., Daunizeau, J., Dolan, R.J., 2017b. Computational neuroimaging strategies for single patient predictions. Neuroimage.
Wasserstein, R.L., Lazar, N.A., 2016. The ASA's statement on p-values: context, process, and purpose. Am Stat 70, 129–133.
White, A.R., 1971. Inference. The Philosophical Quarterly 21, 289–302.
Wong, E.H.F., Yocca, F., Smith, M.A., Lee, C.-M., 2010. Challenges and opportunities for drug discovery in psychiatric disorders: the drug hunters' perspective. International Journal of Neuropsychopharmacology 13, 1269–1284.
Woo, C.-W., Chang, L.J., Lindquist, M.A., Wager, T.D., 2017. Building better biomarkers: brain models in translational neuroimaging. Nature Neuroscience 20, 365–377.
Woo, C.W., Wager, T.D., 2015. Neuroimaging-based biomarker discovery and validation. Pain 156, 1379–1381.

Genetic Imaging: Promises and Pitfalls

Thomas Nickl-Jockschat and Tom Wassink

Genetics of Neuropsychiatric Disorders

While heritability rates differ significantly between disease entities, ranging from below .4 in major depressive disorder (Otte et al. 2016) to .8 in schizophrenia (Sullivan et al. 2003) and .9 in bipolar disorder (McGuffin et al. 2003; Kieseppä et al. 2004), it is commonly acknowledged that genetic factors play an important role in the etiopathogenesis of neuropsychiatric disorders. Given this, it takes little wonder that considerable efforts have been undertaken to identify gene variants associated with these disorders and to subsequently characterize their biological functions from a molecular to a neural systems level.

Since its introduction in the late 1990s (Wassink et al. 1999), Genetic Imaging – that is, the investigation of neuroimaging correlates for genetic variation – has become highly popular. Under this umbrella, several basic approaches have been used:

(a) Classical imaging genetic approaches that study variation in brain structure and/or function associated with a given genetic variant of interest (usually a single-nucleotide polymorphism (SNP) or haplotype)
(b) Studies that utilize neuroimaging markers to discover associated genetic variants (again usually SNPs or haplotypes)
(c) Studies on carriers of rare larger genetic variants, mainly copy number variations (CNVs)
(d) Imaging studies on genetic animal models (which are rather new to the field)

Generally speaking, studies have so far identified two major types of genetic variations associated with neuropsychiatric disorders that differ both in their size

T. Nickl-Jockschat (✉) · T. Wassink
Department of Psychiatry, Iowa Neuroscience Institute, Carver College of Medicine, University of Iowa, Iowa City, IA, USA
e-mail: thomas-nickl-jockschat@uiowa.edu; thomas-wassink@uiowa.edu

V. A. Diwadkar, S. B. Eickhoff (eds.), *Brain Network Dysfunction in Neuropsychiatric Illness*, https://doi.org/10.1007/978-3-030-59797-9_20

and in their frequency within the population. Smaller variations within the genetic code, especially single-nucleotide polymorphisms (SNPs) (variants, in which only one base pair is altered) are comparatively common within the population and usually exhibit only small odds ratios (usually roughly between 1.5 and 2.5) for the manifestation of an actual disorder (Cichon and Ripke 2016). In contrast to these, copy number variations (CNVs) are large deletions or duplications that can span over several thousand base pairs. They are found extremely rarely within the general population and also in psychiatric patients, but show high odds ratios (up to a 40-fold increase for the manifestation of a disorder) (Kirov 2015). It should be noted that this scheme is only a very coarse depiction with a lot of exceptions.

With regard to imaging, common variants and especially SNPs have been most extensively studied. The most widely used approach has been to study whether a given genetic variant is associated with differences in brain structure and/or function. Fewer studies, so far, have applied an inverse approach and conducted genome-wide association studies (GWAS) to link neuroimaging characteristics – for example, cortical or subcortical volumes – to specific genetic variants (Thompson et al. 2014), as these analyses usually require very large cohorts to achieve significant findings. Given their rarity, neuroimaging studies involving human carriers of CNVs are still scarce, although their high odds ratios have directed a lot of attention toward these variations recently. Finally, with the latest advances in the field of high-resolution animal magnetic resonance imaging (MRI) and the wide availability of genetic animal models, neuroimaging studies mainly on rodent models have emerged. This chapter aims to give an overview of these heterogeneous fields of imaging genetics and will critically discuss promises and pitfalls of these approaches.

Classical Imaging Genetics Approaches: Reverse Genetics

Given their frequency within patient samples, but also within the general population, it is little wonder that SNPs have been most extensively studied by researchers. The classical approach usually involves genotyping a given sample – for example, a patient cohort, healthy subjects, or a mix of both – and then selecting subsamples for (functional or structural) MR imaging and subsequent analyses according to their genotypes. Differences between the subsamples are then usually interpreted as effects of a given genotype. While the linear contrasts according to the number of risk alleles have been usually the most frequently applied in classical imaging genetics studies, it should be noted at this point that no authoritative consensus has been established about which contrasts to use, and so practices in the field vary accordingly.

It should also be noted at this point that the choice of the studied gene variants often reflects the state of knowledge in psychiatric genetics at the time the imaging genetics study was planned and conducted. The developments within the field of schizophrenia research (from which some of the most studied gene variants in imaging genetics stem) might serve as an illustrative example. In the late 1980s and

1990s, genetic research was driven by the idea that mono- or oligogenic forms of schizophrenia might exist and that they might be identified through studying families with multiple cases. This led to linkage studies in affected families using a few hundred genetic markers evenly spaced across the genome to locate disease genes by their chromosomal position alone. Another commonly used strategy was the application of candidate gene association studies that aimed to detect allelic variants that are more or less common in patient populations compared to healthy controls. As both techniques yielded unsatisfying results and the complete sequencing of the human genome in 2003 provided unforeseen opportunities for genetics, genome-wide association studies (GWAS) offered the opportunity to compare interindividual genetic variation between patients and a healthy comparison population across the entire genome. The enrollment of large cohorts led to the robust identification of more than 100 SNPs associated with the disorder (Schizophrenia Working Group of the Psychiatric Genomics Consortium 2014). Consequently, the evidence for an association between a genetic variant and a given disorder largely relies upon the type of study and the results of imaging genetics results should be interpreted in this light (Cichon and Ripke 2016, provide an overview over genetics in schizophrenia research).

Imaging genetics itself has been first introduced to neuroimaging by Tom Wassink and Nancy Andreasen in 1999 with a study on a potential association between brain-derived neurotrophic factor (BDNF) alleles and adult brain morphology. Variation in the BDNF gene had been repeatedly discussed as relevant to schizophrenia (Neves-Pereira et al. 2005), but GWAS findings did not corroborate that assumption (Schizophrenia Working Group of the Psychiatric Genomics Consortium 2014). Wassink and colleagues parcellated the brain into several large regions-of-interest (ROIs) including the major cerebral lobes, the ventricles, and the cerebellum. Volumes of these ROIs were extracted and their correlations with the absence or presence of a distinct BDNF allele were explored. The authors found indeed a correlation between BDNF genotype and parietal lobe volume, which they interpreted as evidence for an influence of this variant on brain morphology.

This new concept sparked a multitude of analogous studies. Daniel Weinberger and his lab were the first to take this new concept to functional MRI (fMRI) with a study on working memory-associated neural activations and a SNP in the catechol-o-methyltransferase (COMT) gene (Egan et al. 2001). This gene had been considered as a candidate gene for schizophrenia; no supportive evidence was found, however, in GWAS (Schizophrenia Working Group of the Psychiatric Genomics Consortium 2014). Studies on this variant – the rs4630 or more commonly Val158Met polymorphism – and neural correlates of working memory have become probably the most researched topic in imaging genetics (Mier et al. 2010). The COMT gene codes for an enzyme that mediates the degradation – and, therefore, inactivation – of monoamines, mainly dopamine, in the synaptic cleft and that is expressed primarily in the prefrontal cortex (Weinshilboum et al. 1999). A substitution of methionine by valine at the position 158 is correlated with increased thermostability and, thus, heightened activity of this enzyme (Lotta et al. 1995). This increased activity of the COMT enyzme leads to lower dopamine levels in the synaptic cleft. As dopaminergic neurotransmission is required for working memory, the Val158Met polymorphism has been suspected to modulate working memory performance. Egan and colleagues used a working memory paradigm – the n-back task – to assess working memory-related neural activations and their relation to COMT genotype. Consistent with their initial hypothesis, the authors found Met alleles associated with a reduced activation of the dorsolateral prefrontal cortex (DLPFC). They interpreted this as evidence for compensatory increase of DLPFC activation in carriers of the Val allele (which were expected to have lower prefrontal dopamine concentrations).

Substantially fewer studies exist on potential associations between SNPs and alterations in neural networks. An influential study investigated changes in functional connectivity due to ZNF804a genotype in healthy subjects (Esslinger et al. 2009). In contrast to the above-mentioned SNPs, this variant – rs1344706 – has been shown to be associated with schizophrenia by GWAS (O'Donovan et al. 2008). As little has been known about the actual functions of ZNF804a and the consequences of the rs1344706 variant, the researchers chose imaging genetics to further characterize this SNP. Motivated by the hypothesis that cerebral dysconnectivity might be the core pathophysiology in schizophrenia (Stephan et al. 2006), the aim of the study was to explore whether this ZNF804a SNP modulated functional connectivity of the DLPFC and the hippocampal formation during a working memory task. While regional brain activation itself remained remarkably unchanged due to genotype, the studied variant was found to change the connectivity of both seeds. Risk allele carriers showed increased connectivity both within the ipsilateral and contralateral DLPFC, and an uncoupling between the hippocampal formation and the DLPFC. The authors interpreted these findings as dysconnectivity as a core neurogenetic mechanism, with reduced DLPFC connectivity contributing to executive deficits in schizophrenia (Esslinger et al. 2009).

While in psychiatric research the study of schizophrenia-associated gene variants has been much in the focus of imaging genetics studies in general, and those related to the studies of particular networks in specific, polymorphisms associated with other disorders have also been studied. Variation in the serotonin transporter-linked polymorphic region (5-HTTLPR) within the SLC6A4 gene that codes for the serotonin transporter has been linked to anxiety (Lesch et al. 1996; Murphy Jr et al. 2004) and has also been repeatedly studied in the context of major depression (Schinka et al. 2004; Sen et al. 2004; Lotrich et al. 2003). It should be noted that the genome-wide significance for the association with depression is still lacking. Compared to the long allele (l), the short allele (s) in the 5-HTTLPR region is associated with both decreased SLC6A4 transcription and the level of serotonin transporter function (Lesch et al. 1996). The s-allele carriers have been reported to show increased vulnerability for emotion dysregulation, anxiety, and depression (Caspi et al. 2003; Canli and Lesch 2007). Given its suspected importance in emotion regulation and depression, one of the first imaging genetics studies focused on amygdala reactivity to fearful stimuli, namely, fearful faces, in healthy volunteers. Carriers of the s-allele were reported to indeed show greater activation of the right amygdala. The authors interpreted this as evidence for genetically driven variability in the response of the amygdala toward emotional stimuli (Hariri et al. 2002). More recent studies focused on modulations of cerebral networks associated with variation in the 5-HTTLPR region. One study described increased cerebral blood flow and reduced blood flow in the ventromedial prefrontal cortex in healthy homozygous s-allele carriers when compared to homozygous l-allele carriers (Rao et al. 2007). Remarkably, a study using the same emotional face processing task as Hariri and colleagues failed to find any significant changes in the amygdala due to 5-HTTLPR genotype (Cao et al. 2018). However, they found the number of s alleles to be correlated with functional connectivity of the visual-limbic network. The authors concluded that their data provided evidence for a functional network-based, rather than a brain region-based phenotype associated with this variant.

Besides functional connectivity, structural connectivity has also been a focus of classical imaging genetics studies. McIntosh and colleagues were one of the first to use diffusion tensor imaging (DTI) to study effects associated with a SNP located within the large Neuregulin-1 gene. This SNP, rs6994992, while implicated in the pathogenesis of psychosis by candidate studies (Li et al. 2006), nonetheless failed to reach genome-wide significance (Schizophrenia Working Group of the Psychiatric Genomics Consortium 2014). White matter integrity and density were reported to be reduced in the anterior limb of the internal capsule in association with the SNP (McIntosh et al. 2008).

It should be noted that besides the distinction between candidate genes derived from early studies versus variants with genome-wide significance, the field of DTI-based imaging genetics studies tends to be methodologically rather heterogeneous. This is usually due to the multiple options to analyze various DTI parameters. As tract-based spatial statistics (Smith et al. 2006, 2007) has become the gold standard for the alignment of diffusion-weighed sequences, it is little wonder that this method has been frequently applied in imaging genetics studies focusing on white matter (Nickl-Jockschat et al. 2012, 2015a; Darki et al. 2017; Squarcina et al. 2017). However, approaches using, for example, ROIs (Lippard et al. 2017) or probabilistic tractography (Berman et al. 2015) have also been introduced to the field. While this methodological heterogeneity certainly helps to exploit the wealth of information obtained by diffusion-weighed imaging and can provide valuable insights into white matter microstructure, it also poses a significant challenge that is pertinent across the field, but probably most pronounced when it comes to DTI-based studies. This certainly influences the comparability between the findings of the studies, which appears more and more as a critical pitfall of the field.

Pitfalls of Classical Imaging Genetics Studies

As depicted above, common variants, such as SNPs associated with psychiatric disorders, usually produce low odds ratios for the actual manifestation of a given disorder. This has led to the need for large cohorts in genome-wide discovery studies in the six-digit range for schizophrenia (Schizophrenia Working Group of the

Psychiatric Genomics Consortium 2014) or depression (Hyde et al. 2016). Classical imaging genetics studies usually rely upon sample sizes that are several magnitudes of order smaller than genetic epidemiological studies. This raises the question of whether the sample sizes are sufficient to reliably detect brain imaging changes due to genotype effects. Researchers have usually argued that genotype effects are more pronounced on a neuroimaging than on a behavioral level, as the former is closer in the causal chain to the genotype than the latter (Meyer-Lindenberg and Weinberger 2006), and therefore, much smaller sample sizes are sufficient to yield robust results. This necessarily requires much stronger effects of the genotype on the neuroimaging phenotype than on the behavioral level. If this assumption does not hold true, differences between the experimental groups might not be driven by effects of the studied genotype, but, for example, by sampling effects.

As genotype effects should be reproducible across cohorts and different research sites, findings should converge over the same brain regions. This hypothesis of spatially convergent findings can be tested objectively by coordinate-based meta-analyses (see Chap. XX). As of this date, only a few such studies have been published in the field of imaging genetics, but their results urge caution in the interpretation of single original studies.

As depicted above, a modulation of DLPFC activation associated with COMT genotype became one of the most popular hypotheses in imaging genetics, and it was claimed to have been repeatedly replicated by various subsequent studies. A recent coordinate-based meta-analysis capitalized on the comparative wealth of the literature on this subject and tested whether the reported results of COMT-related activation changes indeed spatially converged over a distinct brain region, namely, the DLPFC. However, neither the main contrast pooling over all studies nor multiple sub-analyses with regard to diagnostic status, age, and paradigm yielded significant or even suggestive results (Nickl-Jockschat et al. 2015b).

Another study from the same lab investigated another frequent claim from the imaging genetics literature that effects from common genetic variation are associated with a schizophrenia-related structural endophenotype. As genetic contributions to a structural endophenotype may be easier to delineate, when referring to biological rather than diagnostic criteria, this study used different options to group the studies for sub-meta-analyses, among others gene ontology (GO)-based groupings. Only one of the applied contrasts – enrolling studies on genotype X diagnosis interactions – yielded two significant clusters, one in the left lateral temporal lobe and one in the medial orbitofrontal gyrus (Janouschek et al. 2018). These regions certainly seem to be related to the neuroanatomy of schizophrenia.

While the latter study suggests that the results of imaging genetics studies can well converge above chance, once the fitting selection criteria are chosen, the predominant lack of significant results for both meta-analyses urges caution in several regards. Individual SNPs usually only exhibit low odds ratios with regard to an actual disorder, for example, schizophrenia (Schizophrenia Working Group of the Psychiatric Genomics Consortium 2014) or bipolar disorder (Hou et al.

2016), with an even more complicated situation for unipolar depression (Hyde et al. 2016). Consequently, sample sizes might not be large enough to capture comparatively subtle genotype effects. This is closely related to another central assumption in imaging genetics, the hypothesis that genotype effects might be easier to detect on a neuroimaging endophenotype than on a behavioral level (Mier et al. 2010). However, it should be emphasized at this point that the sparsity of meta-analytic corroboration should not be easily generalized. Though, this lack of findings should lead to a cautious interpretation of the results of the respective studies.

GWAS Studies: Forward Genetics

"Classical" imaging genetics studies follow a "reverse approach": a gene-of-interest gets characterized with regard to its impact on brain structure and/or function. They are, however, insufficient to systematically delineate genetic variants that contribute to structural or functional variation in the brain. As numerous studies with twin- or family-based designs have shown, various neuroimaging measures are highly heritable (Thompson et al. 2001; Hulshoff Pol et al. 2006; Kochunov et al. 2010). Despite this high heritability, little is still known about the specific genetic variants influencing these parameters.

In the first decade of the twenty-first century, GWAS approaches have revolutionized epidemiological psychiatric genetics. Thus, this powerful concept has also found its way to the field of imaging genetics. But how can GWAS be applied to neuroimaging data?

In general, GWAS relies upon the determination of individual variation within the human genome, particularly bi-allelic SNPs, which are the most abundant type of variation. With the deciphering of the genome, it became apparent that the majority of bases within the DNA do not differ between individuals, while there are approximately nine million SNPs that distinguish between subjects (The 1000 Genomes Project Consortium 2012). As some of these SNPs are part of the so-called haplotypes – that is, they covary with each other – it has been sufficient to genotype approximately one million SNPs to capture the entire variation across an individual genome. GWAS approaches nowadays usually rely upon chip-based arrays that allow the simultaneous genotyping of hundreds of thousands of SNPs in an individual at affordable costs (for a more detailed depiction of GWAS, please refer to Cichon and Ripke, in: Abel & Nickl-Jockschat 2016).

Once sequenced, genotypes, for example, of a patient group, can be compared to those of healthy controls and SNPs that are overrepresented in one group or the other can be determined. While in classical epidemiological approaches, GWAS have been conducted in binary settings (a subject is either assigned to a patient group or not), continuous variables, such as the volume of a given brain region, can also be included in these analyses. For the latter, an observer-independent search across the genome is conducted to determine which SNPs covary with the observed trait.

It should be noted that due to the high number of the determined SNPs, a rigorous correction for multiple comparisons is necessary: findings typically need to surpass a significance level of $p < 10^{-8}$. This drives a major development in the field since its introduction to imaging genetics: the need to gather large samples. Much of the development in the field, consequently, is related to the forming of large-scale collaborative consortia.

The very first studies, thus, were conducted enrolling comparatively small samples. One of the pioneering studies on genetic control over the resting brain enrolled 333 Mexican American subjects from the San Antonio metropolitan area who had undergone MR imaging. One great advantage of this population was that the researchers were able to recruit members from large family pedigrees that allowed an estimate of heritability for individual brain measures (indeed, the 333 subjects stemmed from only 29 extended pedigrees). This was a decisive advantage of this sample, as the identification of genetic factors to brain structural or functional variance requires a certain degree of heritability for the measured parameters. Within this sample, the researchers delineated the default mode network (DMN) and correlated covariation patterns for functional connectivity and also gray matter volumes between the components of the network. The authors found that distinct genetic effects drove covariation for structural and functional covariance within the DMN (Glahn et al. 2010).

The formation of the ENIGMA (Enhancing NeuroImaging Genetics through Meta-Analysis) marked the next important step in the gathering of large samples (Thompson et al. 2014). This large-scale collaboration gathered from multiple sites datasets providing neuroimaging and genotype information. This effort created a network of 30 sites worldwide that provides a platform not only for neuroimaging-driven GWAS but also for several other disease-working groups (http://enigma.ini.usc.edu). Given its multisite character, the need for unified image processing pipelines became evident. Having all sites adhere to the same image processing steps is a necessary prerequisite to the creation of datasets that can be jointly analyzed. This need has been bridged by elaborate protocols that largely rest upon FreeSurfer-based brain parcellation. It should be noted that the analyses themselves are usually based on meta-analytic approaches (i.e., a meta-analysis of results from each of the contributing sites).

A recent study by the ENIGMA consortium, for example, pooled over 11,840 individuals to determine genetic influences on subcortical brain volumes. Although the SNPs associated with subcortical brain volumes failed to reach significance after correction for multiple comparisons, the authors were able to identify candidate genes and compare them to risk SNPs for the manifestation of schizophrenia. As they did not observe an overlap, the authors concluded that brain structural changes might be an epiphenomenon and not directly related to the pathogenesis of schizophrenia (Franke et al. 2016).

Although this study enrolled a sample that was more than 35 times larger than the ones in the very first studies by David Glahn and Peter Fox (Glahn et al. 2010; see above), epidemiological GWAS studies still surpass this number by more than the factor 10 (cf. Schizophrenia Working Group of the Psychiatric Genomics Consortium 2014 and Wray et al. 2018 – the former study pooled over more than 130,000 subjects, the latter over more than 450,000). Thus, the formation of even larger consortia seems to be the next logical step in the development of neuroimaging-associated GWAS. One interesting effort in this regard is the initiation of the UK BioBank (http://www.ukbiobank.ac.uk). This ambitious project aims to collect various data, including genotypes and neuroimaging, on a total of 500,000 healthy volunteers (Collins 2012). It aims for a longitudinal characterization of these subjects, with the eventual promise of valuable insights. However, it should be noted that at the current time, the database still is being built up.

The field of forward genetics, in specific, GWAS neuroimaging, remains young and thus it is difficult to infer on required sample sizes for the robust detection of gene X endophenotype associations or to delineate typical pitfalls. As brain structure appears to be quite heritable at least in some brain regions (see above), identifying genetic variants that underlie this neuroanatomical variation might provide not only valuable insight into the genetic architecture of the brain, but also help to identify new therapeutic targets for neuropsychiatric disorders. Sample size appears as probably the most crucial factor for this endeavor. Future studies enrolling even larger numbers than today's approaches will have the ability to reveal how large a sample must be to yield robust results.

Beyond SNPs: Studying Copy Number Variations (CNVs)

Despite the heterogeneity in scientific questions, samples, methodologies, etc., all of the studies described above deal with a single type of genetic variation: single-nucleotide polymorphisms. Most SNPs are comparatively common within the general population, with those associated with disease producing low odds ratios for the manifestation of a given disorder (see above). This requires large sample sizes to yield robust results. Furthermore, polygenic effects – most likely a decisive pathomechanism in neuropsychiatric disorders – are challenging to model. Copy number variations (CNVs), in turn, occur rarely within the general population or even patient groups, but are highly associated with disease. In schizophrenia cases, for example, large deletions or duplications of chromosomal segments account only for about 1 per cent of disease cases, but exhibit an enormous robustness of their associations (Kirov 2015), and are usually subsumed under the term "rare variants."

With regard to imaging genetics studies, these CNVs have two important advantages when compared to SNPs:

1. Their high odds ratios
2. Their ability to model polygenic effects (CNVs often span over 20–40 genes)

Conversely, their primary disadvantage is obvious: their extreme rarity, which often requires consortia even for the recruitment of samples of double-digit numbers. This limitation might explain why only few neuroimaging studies on CNV carriers have been published so far. Given the attention that these approaches have received, some examples will be depicted here in brief.

Variation in the 22q11.2 region has long been known to be associated with neuropsychiatric disorders. 22q11.2 hemideletion is not only associated with velocardiofacial syndrome but also with increased manifestation of psychotic disorders (Schneider et al. 2014) and – less frequently – with neurodevelopmental disorders (Girirajan et al., 2011). With an occurrence of about 1 in 2000 live births, it is also one of the most commonly found CNVs (Grati et al. 2015). Interestingly, the (rarer) duplication of this chromosomal regions is also associated with neuropsychiatric disorders. In contrast to the hemideletion, however, the duplication seems to be protective for schizophrenia (Wenger et al. 2016) while increasing vulnerability for autism spectrum disorder (ASD). As the hemideletion would be presumed to be associated with a 50 per cent reduction in the expression of the affected genes, while the duplication would produce a 33 per cent increase, variation in this chromosomal deletion can be used to study the consequences of gene dosage effects on the brain. A recent study, therefore, compared brain morphology in carriers of 22q11.2 hemideletion and duplication to an age- and sex-matched control sample (Lin et al. 2017). The authors described variation in hippocampal regions and the basal ganglia that co-varied with gene dosage. Similar approaches studying gene dosage effects have been published, for example, for the 16p11.2 region (Qureshi et al. 2014; Maillard et al. 2015).

Publications on functional imaging studies involving CNV carriers are even more scarce. This is in part explained by the difficulties that emerge in multicenter fMRI studies, as it is hard to recruit enough patients at one site. As is the case for structural imaging, most studies available in this field are on 22q11.2 hemideletion, for example, on socio-emotional processing (Dubourg et al. 2018) or auditory sensory processing (Larsen et al. 2018). An interesting approach studied neural correlates of social cognition in carriers of 7q11.23 hemideletion (Meyer-Lindenberg et al. 2005). 7q11.23 hemideletion is associated with Williams Beuren syndrome that presents not only with distinctive facial features and cardiovascular problems but also with hypersocial behavior. In a pioneering study in this field, Meyer-Lindenberg and colleagues exposed the carriers to threatening scenes and faces in the fMRI scanner. Compared to healthy controls, 7q11.23 hemideletion carriers exhibited greater amygdala activation due to threatening scenes, but lower amygdala activation due to threatening faces. Also, the interaction of the orbitofrontal cortex with the amygdala was abnormal in CNV carriers. The authors concluded that this might be evidence for a genetically controlled neural circuit of human social behavior.

CNVs appear as a promising and up to now understudied subject of imaging genetics. Gathering sufficient sample sizes will be an important challenge for the future – a common theme that all imaging genetics approaches seem to share.

Animal Models

As of this date, neuroimaging of animals is still seen as a rather exotic though growing field within imaging genetics. The huge and obvious differences between, for example, rodent and human brains are certainly an important limitation for animal imaging studies. A number of important advantages, however, have led to the recent increase in publications:

1. The availability of valid genetic animal models due to the recent advances in psychiatric genetics.
2. The comparatively homogeneous genetic and environmental background that animal models provide compared to human subjects that will make it easier to rule out sampling effects as cause for the observed changes.
3. The direct accessibility of brain tissue, for example, for molecular follow-up investigations.
4. The recent advances in animal imaging.

Most animal imaging genetics studies rely upon mouse models, which are the most widely available model organisms. Given the size of the mouse brain, several challenges for the researchers emerge, as magnetic field strengths of 7 T or above and special sequences are required. In addition, animal housing facilities and trained technicians usually need to be present at a research site. These requirements often limit the option to implement animal imaging genetics approaches, as many neuroimaging labs either have no access to specialized animal scanners or do not have the expertise and/or interest to establish mouse colonies. Despite these constraints, a PubMed search query with the search terms "mouse+mri + psychiatry" shows a steep increase in publications, with a more than ten-fold increase in published manuscripts between 2005 and 2017 (retrieved on May 8th, 2018).

Structural imaging studies are still the most frequently published findings in this field. The choice between in vivo or ex vivo scans is largely driven by the research question and the desired resolution. High-resolution scans (i.e., Voxel-sizes of 40 μm^3 and below) usually are conducted ex vivo, as scanning times of several hours are required and breathing artifacts severely reduce image quality. In vivo scans are certainly necessary for longitudinal studies, although this comes at the cost of a coarser resolution.

Mass screenings of genetic mouse models with subsequent clustering approach-based neuroimaging parameters are one viable path in structural animal imaging genetics. A recent publication reported on 26 different genetic mouse models of ASD (Ellegood et al. 2015). Based on 62 brain regions, neuroanatomical variation was classified and the mouse models were clustered according to these findings. This analysis retrieved three clusters, which differed in both directions of the volume differences and the localization of the neuroanatomical phenotype. This clustering hinted at neuroanatomical similarities between several ASD-related risk genes, such as, for example, between X monosomy and MECP2. Such an approach, thus, may help to find previously unknown connections between risk genes based on neuroanatomical similarity.

Another possible direction for animal imaging genetics is the option to follow-up on molecular changes in structurally altered brain regions. A recent study on a mouse model of 16p11.2 hemideletion characterized brain structural changes and applied a novel analysis method, using gene expression maps from the Allen Mouse Brain Atlas to determine which expression patterns from the genes involved in the deletion spatially overlapped with the observed brain structural changes (Kumar et al. 2018). This analysis implicated 3 genes related to the MAPK pathway, which was indeed shown to be dysregulated in this mouse model (Grissom et al. 2018). Structural animal imaging genetics, therefore, has the potential to develop observer-independent hypotheses on molecular changes associated with neuroanatomical variation that can be subsequently tested by molecular approaches.

Advances in functional animal imaging have made it possible to conduct resting-state studies in genetic mouse models. It should be noted that these studies are usually conducted with the animals under anesthesia/sedation, as the noise within the scanner and the environment tends to cause a massive startle reaction. It stands to reason that anesthetic agents will have an impact on neural connectivity (if that wasn't the case, anesthesia wouldn't be effective). A frequent claim within the field, however, is that neural networks at resting state are rather stable, regardless of the anesthetic agent used. This claim, however, has been discussed critically, as the coupling between hemodynamic signals and actual neural activity might be rather weakly correlated (Winder et al. 2017).

In practice, rs-fMRI has been used, for example, to delineate differential properties of resting-state networks in genetic mouse models of schizophrenia (Gass et al. 2016). Another intriguing option is the use of optogenetics to manipulate distinct neural circuits and explore the effects on whole-brain resting-state connectivity. This approach has recently been demonstrated for the reward system (Brocka et al. 2018) and has the potential to yield further insight into the relationship between the activity of distinct neuronal circuits and changes in whole-brain connectivity as measured by the BOLD signal.

In summary, animal imaging genetics is the youngest, but also a very vital field of imaging genetics. It has the potential to yield some decisive insight, although important answers, such as its translational relevance, need to be yet answered.

Summary

Imaging genetics has become a highly popular approach within psychiatric research. The acquisition of sufficient sample sizes has crystalized as a critical factor to yield robust results. Animal imaging studies might help to corroborate human findings and to provide information beyond pure imaging modalities.

References

Abel T, Nickl-Jockschat T (Eds.). The Neurobiology of Schizophrenia. Academic Press. 2016.

Berman JI, Chudnovskaya D, Blaskey L, Kuschner E, Mukherjee P, Buckner R, Nagarajan S, Chung WK, Spiro JE, Sherr EH, Roberts TP. Abnormal auditory and language pathways in children with 16p11.2 deletion. Neuroimage Clin. 2015 Aug 1;9:50-7. doi: https://doi.org/10.1016/j.nicl.2015.07.006.

Brocka M, Helbing C, Vincenz D, et al. Contributions of dopaminergic and non-dopaminergic neurons to VTA-stimulation induced neurovascular responses in brain reward circuits. Neuroimage. 2018;177:88-97. doi:https://doi.org/10.1016/j.neuroimage.2018.04.059

Canli T, Lesch KP. Long story short: the serotonin transporter in emotion regulation and social cognition. Nat Neurosci. 2007 Sep;10(9):1103-9. doi: https://doi.org/10.1038/nn1964.

Cao H, Harneit A, Walter H, Erk S, Braun U, Moessnang C, Geiger LS, Zang Z, Mohnke S, Heinz A, Romanczuk-Seiferth N, Mühleisen T, Mattheisen M, Witt SH, Cichon S, Nöthen MM, Rietschel M, Meyer-Lindenberg A, Tost H. The 5-HTTLPR Polymorphism Affects Network-Based Functional Connectivity in the Visual-Limbic System in Healthy Adults. Neuropsychopharmacology. 2018 Jan;43(2):406-414. doi: https://doi.org/10.1038/npp.2017.121.

Caspi A, Sugden K, Moffitt TE, Taylor A, Craig IW, Harrington H, McClay J, Mill J, Martin J, Braithwaite A, Poulton R. Influence of life stress on depression: moderation by a polymorphism in the 5-HTT gene. Science. 2003 Jul 18;301(5631):386-9. doi: https://doi.org/10.1126/science.1083968

Cichon S, Ripke S. Insights From Genome-Wide Association Studies (GWAS). In: Abel T, Nickl-Jockschat T. The Neurobiology of Schizophrenia. Academic Press. 2016.

Collins R. What makes UK Biobank special? Lancet. 2012 Mar 31;379(9822):1173-4. doi: https://doi.org/10.1016/S0140-6736(12)60404-8.

Darki F, Massinen S, Salmela E, Matsson H, Peyrard-Janvid M, Klingberg T, Kere J. Human ROBO1 regulates white matter structure in corpus callosum. Brain Struct Funct. 2017 Mar;222(2):707-716. doi: https://doi.org/10.1007/s00429-016-1240-y.

Dubourg L, Vrticka P, Debbané M, Chambaz L, Eliez S, Schneider M. Neural correlates of socio-emotional perception in 22q11.2 deletion syndrome. J Neurodev Disord. 2018 Apr 10;10(1):13. doi: https://doi.org/10.1186/s11689-018-9232-2.

Egan MF, Goldberg TE, Kolachana BS, Callicott JH, Mazzanti CM, Straub RE, Goldman D, Weinberger DR. Effect of COMT Val108/158 Met genotype on frontal lobe function and risk for schizophrenia. Proc Natl Acad Sci U S A. 2001 Jun 5;98(12):6917-22. doi: https://doi.org/10.1073/pnas.111134598.

Ellegood J, Anagnostou E, Babineau BA, Crawley JN, Lin L, Genestine M, DiCicco-Bloom E, Lai JK, Foster JA, Peñagarikano O, Geschwind DH, Pacey LK, Hampson DR, Laliberté CL, Mills AA, Tam E, Osborne LR, Kouser M, Espinosa-Becerra F, Xuan Z, Powell CM, Raznahan A, Robins DM, Nakai N, Nakatani J, Takumi T, van Eede MC, Kerr TM, Muller C, Blakely RD, Veenstra-VanderWeele J, Henkelman RM, Lerch JP. Clustering autism: using neuroanatomical differences in 26 mouse models to gain insight into the heterogeneity. Mol Psychiatry. 2015 Feb;20(1):118-25. doi: https://doi.org/10.1038/mp.2014.98.

Esslinger C, Walter H, Kirsch P, Erk S, Schnell K, Arnold C, Haddad L, Mier D, Opitz von Boberfeld C, Raab K, Witt SH, Rietschel M, Cichon S, Meyer-Lindenberg A. Neural mechanisms of a genome-wide supported psychosis variant. Science. 2009 May 1;324(5927):605. doi: https://doi.org/10.1126/science.1167768

Franke B, Stein JL, Ripke S, Anttila V, Hibar DP, van Hulzen KJE, Arias-Vasquez A, Smoller JW, Nichols TE, Neale MC, McIntosh AM, Lee P, McMahon FJ, Meyer-Lindenberg A, Mattheisen M, Andreassen OA, Gruber O, Sachdev PS, Roiz-Santiañez R, Saykin AJ, Ehrlich S, Mather KA, Turner JA, Schwarz E, Thalamuthu A, Shugart YY, Ho YY, Martin NG, Wright MJ; Schizophrenia Working Group of the Psychiatric Genomics Consortium; ENIGMA Consortium, O'Donovan MC, Thompson PM, Neale BM, Medland SE, Sullivan PF. Genetic influences on schizophrenia and subcortical brain volumes: large-scale proof of concept. Nat Neurosci. 2016 Mar;19(3):420-431. doi: https://doi.org/10.1038/nn.4228.

Gass N, Weber-Fahr W, Sartorius A, Becker R, Didriksen M, Stensbøl TB, Bastlund JF, Meyer-Lindenberg A, Schwarz AJ. An acetylcholine alpha7 positive allosteric modulator rescues a schizophrenia-associated brain endophenotype in the 15q13.3 microdeletion, encompassing CHRNA7. Eur Neuropsychopharmacol. 2016 Jul;26(7):1150-60. doi: https://doi.org/10.1016/j.euroneuro.2016.03.013.

Glahn DC, Winkler AM, Kochunov P, Almasy L, Duggirala R, Carless MA, Curran JC, Olvera RL, Laird AR, Smith SM, Beckmann CF, Fox PT, Blangero J. Genetic control over the resting brain. Proc Natl Acad Sci U S A. 2010 Jan 19;107(3):1223-8. doi: https://doi.org/10.1073/pnas.0909969107.

Grati FR, Molina Gomes D, Ferreira JC, Dupont C, Alesi V, Gouas L, Horelli-Kuitunen N, Choy KW, García-Herrero S, de la Vega AG, Piotrowski K, Genesio R, Queipo G, Malvestiti B, Hervé B, Benzacken B, Novelli A, Vago P, Piippo K, Leung TY, Maggi F, Quibel T, Tabet AC, Simoni G, Vialard F. Prevalence of recurrent pathogenic microdeletions and microduplications in over 9500 pregnancies. Prenat Diagn. 2015 Aug;35(8):801-9. doi: https://doi.org/10.1002/pd.4613.

Grissom NM, McKee SE, Schoch H, Bowman N, Havekes R, O'Brien WT, Mahrt E, Siegel S, Commons K, Portfors C, Nickl-Jockschat T, Reyes TM, Abel T. Male-specific deficits in natural reward learning in a mouse model of neurodevelopmental disorders. Mol Psychiatry. 2018 Mar;23(3):544-555. doi: https://doi.org/10.1038/mp.2017.184.

Girirajan S, Brkanac Z, Coe BP, Baker C, Vives L, Vu TH, Shafer N, Bernier R, Ferrero GB, Silengo M, Warren ST, Moreno CS, Fichera M, Romano C, Raskind WH, Eichler EE. Relative burden of large CNVs on a range of neurodevelopmental phenotypes. PLoS Genet. 2011 Nov;7(11):e1002334.

Hariri AR, Mattay VS, Tessitore A, Kolachana B, Fera F, Goldman D, Egan MF, Weinberger DR. Serotonin transporter genetic variation and the response of the human amygdala. Science. 2002 Jul 19;297(5580):400-3. doi: https://doi.org/10.1126/science.1071829.

Hou L, Bergen SE, Akula N, Song J, Hultman CM, Landén M, Adli M, Alda M, Ardau R, Arias B, Aubry JM, Backlund L, Badner JA, Barrett TB, Bauer M, Baune BT, Bellivier F, Benabarre A, Bengesser S, Berrettini WH, Bhattacharjee AK, Biernacka JM, Birner A, Bloss CS, Brichant-Petitjean C, Bui ET, Byerley W, Cervantes P, Chillotti C, Cichon S, Colom F, Coryell W, Craig DW, Cruceanu C, Czerski PM, Davis T, Dayer A, Degenhardt F, Del Zompo M, DePaulo JR, Edenberg HJ, Étain B, Falkai P, Foroud T, Forstner AJ, Frisén L, Frye MA, Fullerton JM, Gard S, Garnham JS, Gershon ES, Goes FS, Greenwood TA, Grigoroiu-Serbanescu M, Hauser J, Heilbronner U, Heilmann-Heimbach S, Herms S, Hipolito M, Hitturlingappa S, Hoffmann P, Hofmann A, Jamain S, Jiménez E, Kahn JP, Kassem L, Kelsoe JR, Kittel-Schneider S, Kliwicki S, Koller DL, König B, Lackner N, Laje G, Lang M, Lavebratt C, Lawson WB, Leboyer M, Leckband SG, Liu C, Maaser A, Mahon PB, Maier W, Maj M, Manchia M, Martinsson L, McCarthy MJ, McElroy SL, McInnis MG, McKinney R, Mitchell PB, Mitjans M, Mondimore FM, Monteleone P, Mühleisen TW, Nievergelt CM, Nöthen MM, Novák T, Nurnberger JI Jr, Nwulia EA, Ösby U, Pfennig A, Potash JB, Propping P, Reif A, Reininghaus E, Rice J, Rietschel M, Rouleau GA, Rybakowski JK, Schalling M, Scheftner WA, Schofield PR, Schork NJ, Schulze TG, Schumacher J, Schweizer BW, Severino G, Shekhtman T, Shilling PD, Simhandl C, Slaney CM, Smith EN, Squassina A, Stamm T, Stopkova P, Streit F, Strohmaier J, Szelinger S, Tighe SK, Tortorella A, Turecki G, Vieta E, Volkert J, Witt SH, Wright A, Zandi PP, Zhang P, Zollner S, McMahon FJ. Genome-wide association study of 40,000 individuals identifies two novel loci associated with bipolar disorder. Hum Mol Genet. 2016 Aug 1;25(15):3383-3394. doi: https://doi.org/10.1093/hmg/ddw181.

Hulshoff Pol HE, Schnack HG, Posthuma D, Mandl RC, Baaré WF, van Oel C, van Haren NE, Collins DL, Evans AC, Amunts K, Bürgel U, Zilles K, de Geus E, Boomsma DI, Kahn RS. Genetic contributions to human brain morphology and intelligence. J Neurosci. 2006 Oct 4;26(40):10235-42. doi: https://doi.org/10.1523/JNEUROSCI.1312-06.2006.

Hyde CL, Nagle MW, Tian C, Chen X, Paciga SA, Wendland JR, Tung JY, Hinds DA, Perlis RH, Winslow AR. Identification of 15 genetic loci associated with risk of major depression in individuals of European descent. Nat Genet. 2016 Sep;48(9):1031-6. doi: https://doi.org/10.1038/ng.3623.

Janouschek H, Eickhoff CR, Mühleisen TW, Eickhoff SB, Nickl-Jockschat T. Using coordinate-based meta-analyses to explore structural imaging genetics. rain Struct Funct. 2018 Sep;223(7):3045-3061. doi: https://doi.org/10.1007/s00429-018-1670-9.

Kieseppä T, Partonen T, Haukka J, Kaprio J, Lönnqvist J. High concordance of bipolar I disorder in a nationwide sample of twins. Am J Psychiatry. 2004 Oct;161(10):1814-21. doi: https://doi.org/10.1176/ajp.161.10.1814.

Kirov G. CNVs in neuropsychiatric disorders. Hum Mol Genet. 2015 Oct 15;24(R1):R45–9. doi: https://doi.org/10.1093/hmg/ddv253.

Kochunov P, Glahn DC, Lancaster JL, Winkler AM, Smith S, Thompson PM, Almasy L, Duggirala R, Fox PT, Blangero J. Genetics of microstructure of cerebral white matter using diffusion tensor imaging. Neuroimage. 2010 Nov 15;53(3):1109-16. doi: https://doi.org/10.1016/j.neuroimage.2010.01.078.

Kumar VJ, Grissom NM, McKee SE, Schoch H, Bowman N, Havekes R, Kumar M, Pickup S, Poptani H, Reyes TM, Hawrylycz M, Abel T, Nickl-Jockschat T. Linking spatial gene expression patterns to sex-specific brain structural changes on a mouse model of 16p11.2 hemideletion. Transl Psychiatry. 2018 May 29;8(1):109. doi: https://doi.org/10.1038/s41398-018-0157-z.

Larsen KM, Mørup M, Birknow MR, Fischer E, Hulme O, Vangkilde A, Schmock H, Baaré WFC, Didriksen M, Olsen L, Werge T, Siebner HR, Garrido MI. Altered auditory processing and effective connectivity in 22q11.2 deletion syndrome. Schizophr Res. 2018 Jul;197:328-336. doi: https://doi.org/10.1016/j.schres.2018.01.026.

Lesch KP, Bengel D, Heils A, Sabol SZ, Greenberg BD, Petri S, Benjamin J, Müller CR, Hamer DH, Murphy DL. Association of anxiety-related traits with a polymorphism in the serotonin transporter gene regulatory region. Science. 1996 Nov 29;274(5292):1527-31. doi: https://doi.org/10.1126/science.274.5292.1527.

Li D, Collier DA, He L. Meta-analysis shows strong positive association of the neuregulin 1 (NRG1) gene with schizophrenia. Hum Mol Genet. 2006 Jun 15;15(12):1995-2002. doi: https://doi.org/10.1093/hmg/ddl122.

Lin A, Ching CRK, Vajdi A, Sun D, Jonas RK, Jalbrzikowski M, Kushan-Wells L, Pacheco Hansen L, Krikorian E, Gutman B, Dokoru D, Helleman G, Thompson PM, Bearden CE. Mapping 22q11.2 Gene Dosage Effects on Brain Morphometry. J Neurosci. 2017 Jun 28;37(26):6183-6199. doi: https://doi.org/10.1523/JNEUROSCI.3759-16.2017.

Lippard ETC, Jensen KP, Wang F, Johnston JAY, Spencer L, Pittman B, Gelernter J, Blumberg HP. Genetic variation of ANK3 is associated with lower white matter structural integrity in bipolar disorder. Mol Psychiatry. 2017 Sep;22(9):1225. doi: https://doi.org/10.1038/mp.2017.169.

Lotta T, Vidgren J, Tilgmann C, Ulmanen I, Melén K, Julkunen I, Taskinen J. Kinetics of human soluble and membrane-bound catechol O-methyltransferase: a revised mechanism and description of the thermolabile variant of the enzyme. Biochemistry. 1995 Apr 4;34(13):4202-10. doi: https://doi.org/10.1021/bi00013a008.

Lotrich FE, Pollock BG, Ferrell RE. Serotonin transporter promoter polymorphism in African Americans : allele frequencies and implications for treatment. Am J Pharmacogenomics. 2003;3(2):145-7. doi: https://doi.org/10.2165/00129785-200303020-00007.

Maillard AM, Ruef A, Pizzagalli F, Migliavacca E, Hippolyte L, Adaszewski S, Dukart J, Ferrari C, Conus P, Männik K, Zazhytska M, Siffredi V, Maeder P, Kutalik Z, Kherif F, Hadjikhani N, Beckmann JS, Reymond A, Draganski B, Jacquemont S; 16p11.2 European Consortium. The 16p11.2 locus modulates brain structures common to autism, schizophrenia and obesity. Mol Psychiatry. 2015 Feb;20(1):140-7. doi: https://doi.org/10.1038/mp.2014.145.

McGuffin P, Rijsdijk F, Andrew M, Sham P, Katz R, Cardno A. The heritability of bipolar affective disorder and the genetic relationship to unipolar depression. Arch Gen Psychiatry. 2003 May;60(5):497-502. doi: https://doi.org/10.1001/archpsyc.60.5.497.

McIntosh AM, Moorhead TW, Job D, Lymer GK, Muñoz Maniega S, McKirdy J, Sussmann JE, Baig BJ, Bastin ME, Porteous D, Evans KL, Johnstone EC, Lawrie SM, Hall J. The effects of a neuregulin 1 variant on white matter density and integrity. Mol Psychiatry. 2008 Nov;13(11):1054-9. doi: https://doi.org/10.1038/sj.mp.4002103.

Meyer-Lindenberg A, Hariri AR, Munoz KE, Mervis CB, Mattay VS, Morris CA, Berman KF. Neural correlates of genetically abnormal social cognition in Williams syndrome. Nat Neurosci. 2005 Aug;8(8):991-3. doi: https://doi.org/10.1038/nn1494.

Meyer-Lindenberg A, Weinberger DR. Intermediate phenotypes and genetic mechanisms of psychiatric disorders. Nat Rev Neurosci. 2006 Oct;7(10):818-27. doi: https://doi.org/10.1038/nrn1993.

Mier D, Kirsch P, Meyer-Lindenberg A. Neural substrates of pleiotropic action of genetic variation in COMT: a meta-analysis. Mol Psychiatry. 2010 Sep;15(9):918-27. doi: https://doi.org/10.1038/mp.2009.36.

Murphy GM Jr, Hollander SB, Rodrigues HE, Kremer C, Schatzberg AF. Effects of the serotonin transporter gene promoter polymorphism on mirtazapine and paroxetine efficacy and adverse events in geriatric major depression. Arch Gen Psychiatry. 2004 Nov;61(11):1163-9. doi: https://doi.org/10.1001/archpsyc.61.11.1163.

Neves-Pereira M, Cheung JK, Pasdar A, Zhang F, Breen G, Yates P, Sinclair M, Crombie C, Walker N, St Clair DM. BDNF gene is a risk factor for schizophrenia in a Scottish population. Mol Psychiatry. 2005 Feb;10(2):208-12.

Nickl-Jockschat T, Stöcker T, Markov V, Krug A, Huang R, Schneider F, Habel U, Zerres K, Nöthen MM, Treutlein J, Rietschel M, Shah NJ, Kircher T. The impact of a Dysbindin schizophrenia susceptibility variant on fiber tract integrity in healthy individuals: a TBSS-based diffusion tensor imaging study. Neuroimage. 2012 Apr 2;60(2):847-53. doi: https://doi.org/10.1016/j.neuroimage.2011.10.012.

Nickl-Jockschat T, Stöcker T, Krug A, Markov V, Maximov II, Huang R, Schneider F, Habel U, Eickhoff SB, Zerres K, Nöthen MM, Rietschel M, Jon Shah N, Treutlein J, Kircher T. Genetic variation in the G72 gene is associated with increased frontotemporal fiber tract integrity. Eur Arch Psychiatry Clin Neurosci. 2015a Jun;265(4):291-301. doi: https://doi.org/10.1007/s00406-014-0516-6.

Nickl-Jockschat T, Janouschek H, Eickhoff SB, Eickhoff CR. Lack of meta-analytic evidence for an impact of COMT Val158Met genotype on brain activation during working memory tasks. Biol Psychiatry. 2015b Dec 1;78(11):e43-6. doi: https://doi.org/10.1016/j.biopsych.2015.02.030.

O'Donovan MC, Craddock N, Norton N, Williams H, Peirce T, Moskvina V, Nikolov I, Hamshere M, Carroll L, Georgieva L, Dwyer S, Holmans P, Marchini JL, Spencer CC, Howie B, Leung HT, Hartmann AM, Möller HJ, Morris DW, Shi Y, Feng G, Hoffmann P, Propping P, Vasilescu C, Maier W, Rietschel M, Zammit S, Schumacher J, Quinn EM, Schulze TG, Williams NM, Giegling I, Iwata N, Ikeda M, Darvasi A, Shifman S, He L, Duan J, Sanders AR, Levinson DF, Gejman PV, Cichon S, Nöthen MM, Gill M, Corvin A, Rujescu D, Kirov G, Owen MJ, Buccola NG, Mowry BJ, Freedman R, Amin F, Black DW, Silverman JM, Byerley WF, Cloninger CR; Molecular Genetics of Schizophrenia Collaboration. Identification of loci associated with schizophrenia by genome-wide association and follow-up. Nat Genet. 2008 Sep;40(9):1053-5. doi: https://doi.org/10.1038/ng.201.

Otte C, Gold SM, Penninx BW, Pariante CM, Etkin A, Fava M, Mohr DC, Schatzberg AF. Major depressive disorder. Nat Rev Dis Primers. 2016 Sep 15;2:16065. doi: https://doi.org/10.1038/nrdp.2016.65.

Qureshi AY, Mueller S, Snyder AZ, Mukherjee P, Berman JI, Roberts TP, Nagarajan SS, Spiro JE, Chung WK, Sherr EH, Buckner RL; Simons VIP Consortium. Opposing brain differences in 16p11.2 deletion and duplication carriers. J Neurosci. 2014 Aug 20;34(34):11199-211. doi: https://doi.org/10.1523/JNEUROSCI.1366-14.2014.

Rao H, Gillihan SJ, Wang J, Korczykowski M, Sankoorikal GM, Kaercher KA, Brodkin ES, Detre JA, Farah MJ. Genetic variation in serotonin transporter alters resting brain function in healthy individuals. Biol Psychiatry. 2007 Sep 15;62(6):600-6. doi: https://doi.org/10.1016/j.biopsych.2006.11.028.

Schinka JA, Busch RM, Robichaux-Keene N. A meta-analysis of the association between the serotonin transporter gene polymorphism (5-HTTLPR) and trait anxiety. Mol Psychiatry 2004 Feb;9(2):197-202. doi: https://doi.org/10.1038/sj.mp.4001405.

Schizophrenia Working Group of the Psychiatric Genomics Consortium. Biological insights from 108 schizophrenia-associated genetic loci. Nature. 2014 Jul 24;511(7510):421-7. doi: https://doi.org/10.1038/nature13595.

Schneider M, Debbané M, Bassett AS, Chow EW, Fung WL, van den Bree M, Owen M, Murphy KC, Niarchou M, Kates WR, Antshel KM, Fremont W, McDonald-McGinn DM, Gur RE, Zackai EH, Vorstman J, Duijff SN, Klaassen PW, Swillen A, Gothelf D, Green T, Weizman A, Van Amelsvoort T, Evers L, Boot E, Shashi V, Hooper SR, Bearden CE, Jalbrzikowski M, Armando M, Vicari S, Murphy DG, Ousley O, Campbell LE, Simon TJ, Eliez S; International Consortium on Brain and Behavior in 22q11.2 Deletion Syndrome. Psychiatric disorders from childhood to adulthood in 22q11.2 deletion syndrome: results from the International Consortium on Brain and Behavior in 22q11.2 Deletion Syndrome. Am J Psychiatry. 2014 Jun;171(6):627-39. doi: https://doi.org/10.1176/appi.ajp.2013.13070864.

Sen S, Villafuerte S, Nesse R, Stoltenberg SF, Hopcian J, Gleiberman L, Weder A, Burmeister M. Serotonin transporter and GABAA alpha 6 receptor variants are associated with neuroticism. Biol Psychiatry. 2004 Feb 1;55(3):244-9. doi: https://doi.org/10.1016/j.biopsych.2003.08.006.

Smith SM, Jenkinson M, Johansen-Berg H, Rueckert D, Nichols TE, Mackay CE, Watkins KE, Ciccarelli O, Cader MZ, Matthews PM, Behrens TE. Tract-based spatial statistics: voxelwise analysis of multi-subject diffusion data. Neuroimage. 2006 Jul 15;31(4):1487-505. doi: https://doi.org/10.1016/j.neuroimage.2006.02.024

Smith SM, Johansen-Berg H, Jenkinson M, Rueckert D, Nichols TE, Miller KL, Robson MD, Jones DK, Klein JC, Bartsch AJ, Behrens TE. Acquisition and voxelwise analysis of multi-subject diffusion data with tract-based spatial statistics. Nat Protoc. 2007;2(3):499-503. doi: https://doi.org/10.1038/nprot.2007.45.

Squarcina L, Houenou J, Altamura AC, Soares J, Brambilla P. Association of increased genotypes risk for bipolar disorder with brain white matter integrity investigated with tract-based spatial statistics. J Affect Disord. 2017 Oct 15;221:312-317. doi: https://doi.org/10.1016/j.jad.2017.06.031.

Stephan KE, Baldeweg T, Friston KJ. Synaptic plasticity and dysconnection in schizophrenia. Biol Psychiatry. 2006 May 15;59(10):929-39. doi: https://doi.org/10.1016/j.biopsych.2005.10.005.

Sullivan PF, Kendler KS, Neale MC. Schizophrenia as a complex trait: evidence from a meta-analysis of twin studies. Arch Gen Psychiatry. 2003 Dec;60(12):1187-92. doi: https://doi.org/10.1001/archpsyc.60.12.1187.

The 1000 Genomes Project Consortium. An integrated map of genetic variation from 1,092 human genomes. Nature. 2012; 49:56–65

Thompson PM, Cannon TD, Narr KL, van Erp T, Poutanen VP, Huttunen M, Lönnqvist J, Standertskjöld-Nordenstam CG, Kaprio J, Khaledy M, Dail R, Zoumalan CI, Toga AW. Genetic influences on brain structure. Nat Neurosci. 2001 Dec;4(12):1253-8. doi: https://doi.org/10.1038/nn758.

Thompson PM, Stein JL, Medland SE, Hibar DP, Vasquez AA, Renteria ME, Toro R, Jahanshad N, Schumann G, Franke B, Wright MJ, Martin NG, Agartz I, Alda M, Alhusaini S, Almasy L, Almeida J, Alpert K, Andreasen NC, Andreassen OA, Apostolova LG, Appel K, Armstrong NJ, Aribisala B, Bastin ME, Bauer M, Bearden CE, Bergmann O, Binder EB, Blangero J, Bockholt HJ, Bøen E, Bois C, Boomsma DI, Booth T, Bowman IJ, Bralten J, Brouwer RM, Brunner HG, Brohawn DG, Buckner RL, Buitelaar J, Bulayeva K, Bustillo JR, Calhoun VD, Cannon DM, Cantor RM, Carless MA, Caseras X, Cavalleri GL, Chakravarty MM, Chang KD, Ching CR, Christoforou A, Cichon S, Clark VP, Conrod P, Coppola G, Crespo-Facorro B, Curran JE, Czisch M, Deary IJ, de Geus EJ, den Braber A, Delvecchio G, Depondt C, de Haan L, de Zubicaray GI, Dima D, Dimitrova R, Djurovic S, Dong H, Donohoe G, Duggirala R, Dyer TD,

Ehrlich S, Ekman CJ, Elvsåshagen T, Emsell L, Erk S, Espeseth T, Fagerness J, Fears S, Fedko I, Fernández G, Fisher SE, Foroud T, Fox PT, Francks C, Frangou S, Frey EM, Frodl T, Frouin V, Garavan H, Giddaluru S, Glahn DC, Godlewska B, Goldstein RZ, Gollub RL, Grabe HJ, Grimm O, Gruber O, Guadalupe T, Gur RE, Gur RC, Göring HH, Hagenaars S, Hajek T, Hall GB, Hall J, Hardy J, Hartman CA, Hass J, Hatton SN, Haukvik UK, Hegenscheid K, Heinz A, Hickie IB, Ho BC, Hoehn D, Hoekstra PJ, Hollinshead M, Holmes AJ, Homuth G, Hoogman M, Hong LE, Hosten N, Hottenga JJ, Hulshoff Pol HE, Hwang KS, Jack CR Jr, Jenkinson M, Johnston C, Jönsson EG, Kahn RS, Kasperaviciute D, Kelly S, Kim S, Kochunov P, Koenders L, Krämer B, Kwok JB, Lagopoulos J, Laje G, Landen M, Landman BA, Lauriello J, Lawrie SM, Lee PH, Le Hellard S, Lemaître H, Leonardo CD, Li CS, Liberg B, Liewald DC, Liu X, Lopez LM, Loth E, Lourdusamy A, Luciano M, Macciardi F, Machielsen MW, Macqueen GM, Malt UF, Mandl R, Manoach DS, Martinot JL, Matarin M, Mather KA, Mattheisen M, Mattingsdal M, Meyer-Lindenberg A, McDonald C, McIntosh AM, McMahon FJ, McMahon KL, Meisenzahl E, Melle I, Milaneschi Y, Mohnke S, Montgomery GW, Morris DW, Moses EK, Mueller BA, Muñoz Maniega S, Mühleisen TW, Müller-Myhsok B, Mwangi B, Nauck M, Nho K, Nichols TE, Nilsson LG, Nugent AC, Nyberg L, Olvera RL, Oosterlaan J, Ophoff RA, Pandolfo M, Papalampropoulou-Tsiridou M, Papmeyer M, Paus T, Pausova Z, Pearlson GD, Penninx BW, Peterson CP, Pfennig A, Phillips M, Pike GB, Poline JB, Potkin SG, Pütz B, Ramasamy A, Rasmussen J, Rietschel M, Rijpkema M, Risacher SL, Roffman JL, Roiz-Santiañez R, Romanczuk-Seiferth N, Rose EJ, Royle NA, Rujescu D, Ryten M, Sachdev PS, Salami A, Satterthwaite TD, Savitz J, Saykin AJ, Scanlon C, Schmaal L, Schnack HG, Schork AJ, Schulz SC, Schür R, Seidman L, Shen L, Shoemaker JM, Simmons A, Sisodiya SM, Smith C, Smoller JW, Soares JC, Sponheim SR, Sprooten E, Starr JM, Steen VM, Strakowski S, Strike L, Sussmann J, Sämann PG, Teumer A, Toga AW, Tordesillas-Gutierrez D, Trabzuni D, Trost S, Turner J, Van den Heuvel M, van der Wee NJ, van Eijk K, van Erp TG, van Haren NE, van 't Ent D, van Tol MJ, Valdés Hernández MC, Veltman DJ, Versace A, Völzke H, Walker R, Walter H, Wang L, Wardlaw JM, Weale ME, Weiner MW, Wen W, Westlye LT, Whalley HC, Whelan CD, White T, Winkler AM, Wittfeld K, Woldehawariat G, Wolf C, Zilles D, Zwiers MP, Thalamuthu A, Schofield PR, Freimer NB, Lawrence NS, Drevets W; Alzheimer's Disease Neuroimaging Initiative, EPIGEN Consortium, IMAGEN Consortium, Saguenay Youth Study (SYS) Group. The ENIGMA Consortium: large-scale collaborative analyses of neuroimaging and genetic data. Brain Imaging Behav. 2014 Jun;8(2):153-82. doi: https://doi.org/10.1007/s11682-013-9269-5.

Wassink TH, Nelson JJ, Crowe RR, Andreasen NC. Heritability of BDNF alleles and their effect on brain morphology in schizophrenia. Am J Med Genet. 1999 Dec 15;88(6):724-8. doi: https://doi.org/10.1002/(sici)1096-8628(19991215)88:6<724::aid-ajmg25>3.0.co;2-7.

Weinshilboum RM, Otterness DM, Szumlanski CL. Methylation pharmacogenetics: catechol O-methyltransferase, thiopurine methyltransferase, and histamine N-methyltransferase. Annu Rev Pharmacol Toxicol. 1999;39:19-52. doi: https://doi.org/10.1146/annurev.pharmtox.39.1.19.

Wenger TL, Miller JS, DePolo LM, de Marchena AB, Clements CC, Emanuel BS, Zackai EH, McDonald-McGinn DM, Schultz RT. 22q11.2 duplication syndrome: elevated rate of autism spectrum disorder and need for medical screening. Mol Autism. 2016 May 6;7:27. doi: https://doi.org/10.1186/s13229-016-0090-z.

Winder AT, Echagarruga C, Zhang Q, Drew PJ. Weak correlations between hemodynamic signals and ongoing neural activity during the resting state. Nat Neurosci. 2017 Dec;20(12):1761-1769. doi: https://doi.org/10.1038/s41593-017-0007-y.

Wray NR, Ripke S, Mattheisen M, Trzaskowski M, Byrne EM, Abdellaoui A, Adams MJ, Agerbo E, Air TM, Andlauer TMF, Bacanu SA, Bækvad-Hansen M, Beekman AFT, Bigdeli TB, Binder EB, Blackwood DRH, Bryois J, Buttenschøn HN, Bybjerg-Grauholm J, Cai N, Castelao E, Christensen JH, Clarke TK, Coleman JIR, Colodro-Conde L, Couvy-Duchesne B, Craddock N, Crawford GE, Crowley CA, Dashti HS, Davies G, Deary IJ, Degenhardt F, Derks EM, Direk N, Dolan CV, Dunn EC, Eley TC, Eriksson N, Escott-Price V, Kiadeh FHF, Finucane HK, Forstner AJ, Frank J, Gaspar HA, Gill M, Giusti-Rodríguez P, Goes FS, Gordon SD, Grove J, Hall LS,

Hannon E, Hansen CS, Hansen TF, Herms S, Hickie IB, Hoffmann P, Homuth G, Horn C, Hottenga JJ, Hougaard DM, Hu M, Hyde CL, Ising M, Jansen R, Jin F, Jorgenson E, Knowles JA, Kohane IS, Kraft J, Kretzschmar WW, Krogh J, Kutalik Z, Lane JM, Li Y, Li Y, Lind PA, Liu X, Lu L, MacIntyre DJ, MacKinnon DF, Maier RM, Maier W, Marchini J, Mbarek H, McGrath P, McGuffin P, Medland SE, Mehta D, Middeldorp CM, Mihailov E, Milaneschi Y, Milani L, Mill J, Mondimore FM, Montgomery GW, Mostafavi S, Mullins N, Nauck M, Ng B, Nivard MG, Nyholt DR, O'Reilly PF, Oskarsson H, Owen MJ, Painter JN, Pedersen CB, Pedersen MG, Peterson RE, Pettersson E, Peyrot WJ, Pistis G, Posthuma D, Purcell SM, Quiroz JA, Qvist P, Rice JP, Riley BP, Rivera M, Saeed Mirza S, Saxena R, Schoevers R, Schulte EC, Shen L, Shi J, Shyn SI, Sigurdsson E, Sinnamon GBC, Smit JH, Smith DJ, Stefansson H, Steinberg S, Stockmeier CA, Streit F, Strohmaier J, Tansey KE, Teismann H, Teumer A, Thompson W, Thomson PA, Thorgeirsson TE, Tian C, Traylor M, Treutlein J, Trubetskoy V, Uitterlinden AG, Umbricht D, Van der Auwera S, van Hemert AM, Viktorin A, Visscher PM, Wang Y, Webb BT, Weinsheimer SM, Wellmann J, Willemsen G, Witt SH, Wu Y, Xi HS, Yang J, Zhang F; eQTLGen; 23andMe, Arolt V, Baune BT, Berger K, Boomsma DI, Cichon S, Dannlowski U, de Geus ECJ, DePaulo JR, Domenici E, Domschke K, Esko T, Grabe HJ, Hamilton SP, Hayward C, Heath AC, Hinds DA, Kendler KS, Kloiber S, Lewis G, Li QS, Lucae S, Madden PFA, Magnusson PK, Martin NG, McIntosh AM, Metspalu A, Mors O, Mortensen PB, Müller-Myhsok B, Nordentoft M, Nöthen MM, O'Donovan MC, Paciga SA, Pedersen NL, Penninx BWJH, Perlis RH, Porteous DJ, Potash JB, Preisig M, Rietschel M, Schaefer C, Schulze TG, Smoller JW, Stefansson K, Tiemeier H, Uher R, Völzke H, Weissman MM, Werge T, Winslow AR, Lewis CM, Levinson DF, Breen G, Børglum AD, Sullivan PF; Major Depressive Disorder Working Group of the Psychiatric Genomics Consortium. Genome-wide association analyses identify 44 risk variants and refine the genetic architecture of major depression. Nat Genet. 2018 May;50(5):668-681. doi: https://doi.org/10.1038/s41588-018-0090-3.

Brain Networks and the Emergence of the Self: A Neurophenomenal Perspective

Georg Northoff

Introduction

Psychiatric disorders like depression and schizophrenia and bipolar disorders show major alterations in the sense of self, which have been characterized as 'basic disturbance', thus underlying all other psychopathological symptoms (Northoff 2014). The neural mechanisms of the 'basic disturbance of self' remain unclear though. At the same time, recent investigations show alterations in the various neural networks' resting state activity in psychiatric disorders. How are the neural networks' alterations related to the 'basic disturbance of self'? For that, we need to understand how the different neural networks mediate our sense of self in the healthy brain – that shall be the central focus in this chapter.

Investigations show that the default-mode network (DMN) in general and their cortical midline structures (CMS) in particular take on a special role in processing and mediating our sense of self, including our experience of self-specificity (see Northoff 2015; Northoff et al. 2006). The previous chapters in this book demonstrated major changes in DMN and CMS in the various psychiatric disorders which, psychopathologically, can also be featured by alterations in the sense of self (Northoff 2014).

The aim of the present chapter is to understand the neural correlates and mechanisms of our sense or experience of self, for example, the self as subject ('I'), and how it can be manifest in the various cognitive, motivational, affective and sensorimotor functions, for example, self as object ('Me') (see D'Argembeau 2013;

G. Northoff (✉)
University of Ottawa Institute of Mental Health Research, Ottawa, ON, Canada

Center for Cognition and Brain Disorders, Hangzhou Normal University, Hangzhou, China

Center for Brain and Consciousness, Taipeh Medical University (TMU), Taipeh, Taiwan

College for Humanities and Medicine, Taipeh Medical University (TMU), Taipeh, Taiwan

Seventh Hospital for Mental Health, Zhejiang University, Hangzhou, China
e-mail: georg.northoff@theroyal.ca; https://www.georgnorthoff.com

V. A. Diwadkar, S. B. Eickhoff (eds.), *Brain Network Dysfunction in Neuropsychiatric Illness*, https://doi.org/10.1007/978-3-030-59797-9_21

Klein 2012; Northoff 2013, 2014b; James 1890a, b). The focus of this chapter is on the 'healthy self' whose understanding of its underlying neuronal mechanisms carries major implications for its alterations in psychiatric disorders as it shall be indicated at the end of the chapter.

I first demonstrate that the self, for example, self-specificity, is manifest and processed throughout different neural networks (see part I) and how that is related to the CMS as core regions/network underlying our sense of self (see part II). As pointed out in the conclusion, we need to combine objective-neuronal, and subjective-experiential or phenomenal approach to investigate the subjective nature of our self, for example, sense or experience of self, and how it is mediated by CMS. That requires what I describe as 'neurophenomenal approach' (Northoff 2014, 2014a), which is central for understanding the 'basic disturbance' of the self in psychiatric disorders like depression and schizophrenia.

Part I./Neural Networks and Self – How Does Self-Specificity Modulate Other Functions?

As mentioned above, the self-specificity effect was first established in memory with self-related items being better memorized than non-self-related ones (see Cunningham et al. 2008, 2011). Such memory enhancement by the self is mediated by the medial prefrontal cortex (see Kim and Johnson 2010). This would still be well compatible with self-specificity being a higher order cognitive function. There are, however, findings showing that self-specificity modulates even more basic functions like perception (Sui et al. 2012a, 2013a, b, c; Tagliazucchi et al. 2013), action (Frings and Wentura 2014), emotions (Northoff et al. 2009; Phan et al. 2004) and reward (de Greck et al. 2008; Sui and Humphreys 2014). That argues against self-specificity being a higher-order cognitive function. How is such association or expansion of self-specificity mediated neurally? I will briefly review these studies in the following with a specific focus on their underlying neural correlates.

Sensorimotor Network and Self – Perception, Action and Self-Specificity

Sui et al. (2012a, b, 2013a, b, c; Tagliazucchi et al. 2013) investigated how the processing of geometrical shapes like circle, triangle and square can be modulated by associating them with a label signifying the self, a friend or a stranger. Theoretically, the assumption is here that the geometrical shapes reflect external stimuli that by themselves remain neutral with regard to the distinction between self and non-self. While the label signifies the self or self-specificity and may interact or associate with the geometrical shapes and thereby impact their subsequent processing. The task is thus about self-association or self-expansion in the domain of perception.

In their experiments, Sui and colleagues typically let subjects form associations between the three geometrical shapes and three labels. Afterwards, participants are presented pairs of shapes and labels (like circle-self, and so forth) that either matched or not matched (non-matching) with the learned associations. Reaction times were fastest and accuracy highest for those pairings where the self was involved with performance decreasing from self over friend to stranger. The data clearly demonstrate behavioural effect of the self-label on (self-) neutral perceptual stimuli like geometric shapes in both reaction time and accuracy.

Self-associations also appear to directly affect action. Frings and Wentura (2014) paired four different movements (right, left, up, down) with four different labels (self, mother, stranger and a neutral one). Subjects had to form (or learn) associations between movements and labels and afterwards had to perform the cue-based movements in conjunction with the brief presentation of a matching or non-matching label and subsequent judgment (whether they matched or not). Trials implicating the self (when compared to mother and stranger and neutral conditions) showed the fastest reaction times and were most accurate. Hence, as in the case of 'perception-self links', there are clear behavioural effects of self-specificity on action suggesting 'action-self links' (as Frings and Wentura 2014 say). Self-association effects can thus be observed behaviourally in both domains, perception and action.

Taken together, these results, albeit behavioural and lesion-based, suggest the involvement of the sensorimotor network in processing self-specificity. Self-specificity is processing in the sensorimotor network and, more specifically, it is associated with or assigned to sensory and motor stimuli. Sui et al. (2013a, b, c; Tagliazucchi et al. 2013) therefore speak of self-association meaning that self-specificity is associated with sensory and motor (and other) stimuli. We will come back to such self-association in the second part of this chapter. Before we will see that analogous self-association seems to occur with respect to other stimuli like emotional and rewarding stimuli as mediated by their respective networks, for example, affective and salience network.

Affective Network and Self – Emotions and Self-Specificity

Both positive and negative emotions have been shown to recruit the anterior cortical midline structures including the sub- and pregenual anterior cingulate cortex, as well as ventro- and dorsomedial prefrontal cortex (Phan et al. 2002; Roy et al. 2012). Since the very same regions are also recruited in the various studies on self, one can suggest overlap and interaction between self and emotions. Consider the study of Phan et al. (2004) who let subjects view emotional pictures while being scanned using fMRI. Afterwards, subjects were asked to evaluate the degree of self-relatedness or personal relevance of the same pictures in order to correlate them with the fMRI signal. Signal changes in the VMPFC during emotional picture presentation predicted the degree of personal relevance (or self-relatedness) subjects attributed to the same pictures post scanning.

Further support for direct neural interaction between self and emotion comes from a study by Moran and colleagues (2006). One early study by Moran et al. (2006) applied a two-by-two factorial design in which they presented words corresponding to personality characteristics varying emotional valence (positive or negative) and self-specificity (high and low) (which had to be judged by subjects as personally favourable or unfavourable). They showed an interaction between high degrees of self-specificity and (positive) emotional valence in specifically the pregenual and subgenual anterior cingulate cortex (Moran et al. 2006).

The central involvement of these regions was further supported by a study of ours (Northoff et al. 2009). To exclude task-related effects like the cognitive efforts related to judgments (and to focus on stimulus-related effects related to perception itself independent of subsequent judgment), we let subjects passively view emotional pictures in the scanner. Afterwards, outside the scanner, subjects had to judge the emotional arousal (i.e. excited/aroused or not) and valence (i.e. positive or negative), as well as the degree of self-specificity of the same pictures.

The results from the subjective ratings were then parametrically correlated with the neural activity changes during the perception of the same pictures. The correlation yielded a significant relationship between the neural activity in several subcortical/cortical regions and the pictures' degree of self-specificity. Subcortical regions included the tectum, colliculi, amygdala, ventral tegmental area (VTA), mediodorsal thalamus and ventral striatum (VS), while cortical regions included the perigenual anterior cingulate cortex (PACC), the ventromedial prefrontal cortex (VMPFC), the dorsomedial prefrontal cortex (DMPFC) and the posterior cingulate cortex (PCC; see Northoff et al. 2009). The higher the subjects' neural activity in these regions during the viewing of the emotional pictures, the higher the degree of self-specificity subjects assigned to the stimuli.

Is the parametric dependence of the neural activity in these midline regions really due to the stimuli's degree of self-specificity or, rather, to their affective-emotional components, that is, emotional arousal and valence? The regions whose neural activity was parametrically dependent on the degree of self-specificity also showed parametric dependence on the emotion dimensions like valence and intensity (arousal). However, the direction of their correlation was different in cortical and subcortical regions. Cortical regions like the PACC, VMPFC and DMPFC showed opposite directions in their dependence on self-specificity and arousal/valence. Increased degrees of self-specificity of the stimuli were associated with increased degrees of neural activity in the cortical regions, whereas increased degrees of arousal/valence of the same stimuli led to lower degrees of neural activity in the same regions.

These data contrasted with the activity patterns observed in the subcortical regions. Subcortical regions like the tectum, the periaqueductal grey, and amygdala did not show such opposite directionality of self-specificity and arousal/valence – both self-specificity and arousal/valence positively correlated with neural activity.

Taken together, these findings demonstrate a close relationship between emotion and self-specificity in neural activity of the affective network, which includes both anterior midline and subcortical midline regions including the amygdala (Kaiser et al. 2015). For that reason, Northoff and Panksepp (2008) proposed a subcortical-cortical midline system as being central for self-representation. Panksepp (1998a, b, 2011) proposed that emotional feelings are associated with activation in subcortical regions. Since emotional feelings (rather than mere emotions) are intrinsically subjective, they, so Panksepp, must presuppose some kind of basic primitive self, a primarily affective rather than cognitive self, which is closely related to the body and its homeostatic functions.

Salience Network and Self – Reward, Salience and Self-Specificity

Reward is considered a basic motivational function of the brain. If coupled to a particular reward (like money), certain stimuli are valued differently in that they are preferred over others that are not coupled to such reward even if the stimuli themselves and their respective contents (like a suitcase or bag) are identical. The neural correlates of reward can be found in subcortical regions like the ventral tegmental area (VTA) and the ventral striatum (VS), as well as in the ventromedial prefrontal cortex (VMPFC).

How about the relation between reward and self? Ersner-Hershfield et al. (2009) conducted an fMRI study in which subjects had to judge personality traits with regard to themselves and other people both in the present and the future (i.e. 10 years from now). After 1 week, subjects came back and had to perform a behavioural temporal discounting task in which they had to make choices between immediate monetary rewards and delayed but larger monetary gain. The VMPFC showed greater activity for self versus other trials, as well as for self-present versus self-future trials. Most relevant here the contrast between self-present and self-future activity in the VMPFC predicted the degree of temporal discounting in the reward task (i.e. the ratio between immediate and delayed rewards chosen): the greater the VMPFC activity for the present self (when compared to the future self), the greater the tendency to choose immediate (rather than delayed) monetary rewards (see analogous results by Mitchell et al. 2011).

These data suggest a close neural relationship between reward and self. The assumption of self-reward neural overlap is further supported by studies from our group. De Greck and colleagues (2008) presented different pictures (alcohol, gambling and food) which subjects had to evaluate either with regard to both their rewarding and self-related features. The focus was on the neural overlap of self-relatedness with reward (whereas the reverse focus, overlap of reward with self-relatedness was not tested). The reward system typically involved VS, VTA and VMPFC. Interestingly, stimuli evaluated as highly self-related by the subjects elicited high signal changes in the same regions while those assessed as low self-related did not show any signal changes in these regions. These findings suggest neural overlap of self-relatedness with reward-related tasks in the reward system.

Are self and reward identical? Is self-relatedness nothing but reward and apparent self-related activity changes in the VMPFC really due to the effects of reward (see also Northoff and Hayes 2011)? This has been argued by some authors such as D'Argembeau (2013) who suggests in his 'valuation hypothesis' that the VMPFC encodes personal value and assigns it to both internal and external stimuli (D'Argembeau 2013).

We have to be careful, however. Reward-self overlap in VMPFC does not entail reward-self-identity. Both may for instance dissociate from each other. Behaviourally, such dissociation has been demonstrated in a recent study by Sui and Humphreys (2014) where an increase in the size of the stimuli did not affect the reward bias but did increase the self-bias effect. There is thus behavioural dissociation between self and reward while both, reward bias and self-bias, were found to correlate with each other. Neurally, deGreck et al. (2009, 2010) demonstrated that reward-related activity signals in the three reward regions they observed were normal in recovered alcoholic and gambling patients while their self-related activity was virtually absent, showing no distinction between high and low self-related stimuli in VS, VTA and VMPFC.

Such dissociation of self-relatedness from reward in reward regions suggests that the same regions, VTA, VS and VMPFC, may be utilized in different ways by reward-related and self-related processing. This tentative conclusion requires further experimental support. In addition to any common regions for self and reward, there may also be different regions implicated. Enzi et al. (2009) took the same data set (as de Greck et al. 2008) and re-analysed the results with regard to self-specific regions by conducting exclusive masking (with masking of reward-related regions during self-relatedness condition, e.g. high versus low self). This revealed the premotor cortex and the insula (see also Modinos et al. 2009) as regions that were specifically related to the distinction between high and low self-related stimuli rather than to their rewarding features. That is even more interesting given that the insula is a core region of the salience network.

In sum, the data suggest regional overlap between self and reward in the regions of the reward system including VTA, VS and VMPFC. Though their exact relationship remains to be clarified, there is nevertheless some evidence that self-relatedness seems to recruit core regions (like VTA, VS and VMPFC) and processes (like valuation) of a basic motivational process like reward. There is at least some partial regional reward-self overlap but future work needs to explore whether such reward-self overlap translates and extends into self-expansion or self-association of rewarding stimuli.

More generally, the results on reward and self-specificity illustrate the central relevance of the salience network for self-specificity. Beyond reward and motivation, the salience network is implicated in linking intero- and exteroceptive stimulus processing and thus, more generally, body and environment. That makes it clear that, on a neuronal level, the self, metaphorically speaking, transcends the boundaries between brain, body and environment by linking and integrating their respective inputs and stimuli. Is the self therefore nothing but a salience detector? Yes, it allows for detecting salience in neuronal and intero- and exteroceptive stimuli coming from brain, body and environment by (seemingly) linking and integrating them. At the same time though, the self itself constructs salience for neuronal and intero- and exteroceptive stimuli in that it serves as baseline, reference or standard for measuring or assessing their degree of salience. How is that possible? For that we will now investigate the relationship between self-specificity and the default-mode network.

Central Executive Network and Self – Internally vs Externally Guided Decision-Making

Recent studies on decision-making by Nakao et al. (2012, 2013) focused on internally, for example, self, versus externally guided decision-making. They first compared all fMRI studies on decision-making that relied on external criteria (e.g. the amount of money) with those where subjects had to base their decision on internal criteria (e.g. subjective preferences as in moral judgment or occupation choice). The data showed that externally guided decision-making involved the lateral frontal and parietal cortical regions while internally guided decision-making implicated the CMS and especially VMPFC, PACC and PCC/precuneus. Since the latter regions are linked to self-processing, one would assume that internally guided decision-making implicates the self as manifest in subjective preference. This distinguished internally guided decision-making from externally guided DM that involves external, for example, non-self-specific rather than internal, for example, self-specific criteria for making decisions.

Does this imply that the resting state can account for subjective preference in internally guided decision-making? Nakao et al. (2012, 2013) recorded EEG data during rest and measured the N200[1] during both internally and externally guided decision-making (judgements of subjective preference and word frequency respectively). Although both conditions, for example, internally and externally guided decisions, showed similar amplitudes for the N200, only the N200 amplitude during the internally guided condition was predicted by the power (in delta, theta and beta activity) in the resting state. In contrast, the N200 for the externally guided condition did not correlate at all with resting state power. These data go beyond mere rest-self overlap and suggest that the resting state must contain some information about self-specificity in order to predict the N200 during internally guided decisions. The results are compatible with my proposal for rest-self containment (as distinguished from mere rest-self overlap).

Together, these findings show that internally and externally guided decision-making can be distinguished with regard to their regions/networks, for example, DMN versus central executive network/CEN and involvement of the resting state. It is well known that the DMN and CEN stand in a negative relation meaning that they anti-correlate with each other thus showing reciprocal modulation: if one is strong the other is weak and vice versa (see Northoff et al. 2004; Northoff and Sibille 2014 for details). How is this reciprocal modulation between DMN and CEN linked to

[1] A stimulus-evoked component associated with attention and stimulus saliency.

self-specificity? Sui et al. (2013a, b, c) show in their study on perception and self-association (see above) not only recruitment of VMPFC and LpSTS but also the bilateral DLPFC – part of the CEN. However, the DLPFC showed the opposite activity pattern: activity was strongest for stimuli associated with a stranger and weakest for stimuli associated to the self. The VMPFC showed the opposite pattern. This suggests that the reciprocal modulation with negative correlation between VMPFC and DLPFC that has been shown for emotion and cognition (Goel and Dolan 2003a, b; Northoff et al. 2004) applies also to self- and non-self-relatedness.

Part II./Default-Mode Network and the Self – How Are Self-Specificity and Its Phenomenal Features Constituted by Cortical Midline Structures?

'Rest-Self Overlap' – Resting State and Self-Specificity in Cortical Midline Structures

The cortical midline regions are the core part of the so-called default-mode network (DMN) (Buckner et al. 2008; Raichle et al. 2001). Originally, this network was defined in a metabolic way by high degrees of metabolism and oxygen extraction when compared to other regions (Raichle et al. 2001). Moreover, the DMN shows high degrees of intrinsic functional connectivity especially between the anterior and posterior cingulate cortex (Buckner et al. 2008), high degrees of variability (within and across individuals) (Lee et al. 2014), high degrees of dynamic functional connectivity to other regions and neural networks in the brain (de Pasquale et al. 2012) and a reciprocal relationship to other networks including the executive control network (CEN) in lateral prefrontal and parietal cortex (see Northoff et al. 2004; Northoff and Sibille 2014).

Psychologically, the DMN has been associated with a variety of different functions such as spontaneous internal mentation (Andrews-Hanna 2012; Andrews-Hanna et al. 2010a, b; Buckner et al. 2008; McKiernan et al. 2006); random thoughts (Andreasen et al. 1995), mind wandering (Christoff et al. 2009, 2011; Mason et al. 2007; Stawarczyk et al. 2011), episodic memory recall (Buckner et al. 2008) and social processing (Schilbach et al. 2012). In addition to these functions, the processing of self-related or personally relevant stimuli or tasks has also been suggested yet another function of the resting state activity in DMN and especially in CMS as it is supported by what I describe as 'rest-self overlap'.

D'Argembeau and colleagues (2005) conducted an H20 positron emission tomography (PET) investigation involving four conditions: thinking/reflection about one's own personality traits, thinking/reflection about another person's personality traits, thinking/reflection on social issues and a 'pure rest' condition where subjects could relax. A direct comparison of the self and rest conditions indicated strong overlap in the VMPFC for both rest and self-specificity, with similar degrees of rCBF increases in each case. In contrast, thinking about another person and about social issues condition induced rCBF decreases in the same region. A correlation of the post-scanning measures of self-referential thinking with the rCBF changes demonstrated a positive relationship in the VMPFC. The higher the rCBF in the VMPFC, the higher the degree of self-referentiality in the thoughts subjects reported.

The strong neural overlap between resting state and CMS regions like PACC and PCC recruited during self-specificity was further confirmed subsequent studies by Schneider et al. (2008), Whitfield-Gabriel et al. (2011) and Davey et al. (2016). Based on these findings, we conducted a meta-analysis of imaging studies on self and resting state (Qin and Northoff 2011). The regional activities during the processing of self-specific stimuli and those during the resting state overlapped especially in the PACC extending to the VMPFC, while no such regional overlap with the resting state was observed in the non-self-specific conditions (see also Davey et al. 2016 for analogous results). These results suggest neural overlap between self-specificity and resting-state activity in anterior midline regions like PACC/VMPFC and PCC. I, therefore, introduced the concept of 'rest-self overlap' (Northoff 2015) that describes the regional convergence between resting state activity and self-related activity in anterior and posterior cortical midline structures.

'Rest-Self Containment' – Encoding and Experience-Dependence of Self-Specificity in Cortical Midline Structures

The supposition of rest-self overlap is further supported and extended by recent studies showing that the resting state and pre-stimulus activity predict the degree of self-specificity. Bai et al. (2015) demonstrated in EEG that pre-stimulus alpha power (600–400 ms prior to stimulus onset) predicted the degree of self-specificity subjects assigned to subsequently presented emotional stimuli. Most interestingly, the degree of pre-stimulus alpha power predicting self-specificity was correlated with the level of glutamate in PACC (See Figs. 1a and 1b).

Such pre-stimulus activity preceding specifically self-specific stimuli was also observed in a magneto-enceophalo-graphic (MEG) study by Lou et al. (2010). They observed increase in midline functional connectivity (as strongest in the gamma range of 30–45 Hz) 900 ms before onset of self-specific stimuli while it was not observed during non-self-specific stimuli (see also Qin et al. 2016 for pre-stimulus increases prior to self-specific conditions). Yet another study demonstrated that the resting state itself, for example, the spontaneous activity's temporal structure as measured by scale-free activity with the power law index, predicted the degree of self-consciousness: the stronger especially the power in infraslow frequencies, the higher the degree of specifically private self-consciousness (as measured with the self-consciousness scale) (Huang et al. 2016; see also Wang et al. 2013).

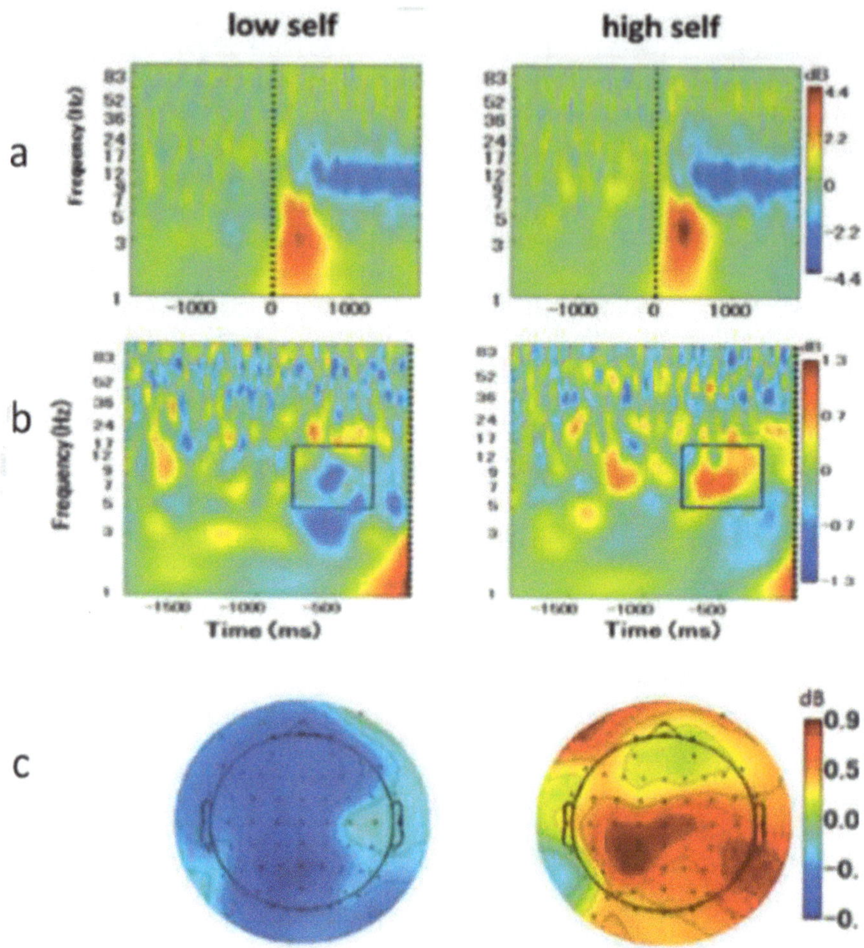

Fig. 1a Event-related spectral perturbations (ERSP) in post-stimulus (**a**) and pre-stimulus (**b**) periods during low (left) and high (right) self-specific stimuli (see Bai et al. 2015 for details)

Together, these data suggest that the resting state or spontaneous activity does not only overlap with self-related activity but, even stronger, contains or encodes some self-specific information by itself. I therefore speak of 'rest-self containment' (Northoff 2015) that seems to hold specifically for the CMS as core part of the DMN. Further support for rest-self containment comes from the cellular level in subgenual anterior cingulate cortex (see in Lipsman et al. 2014; Northoff 2014b). We have found that pre-stimulus local field power in the gamma range can predict whether subjects evaluated the subsequent stimulus as high- or low-self-related (higher pre-stimulus gamma power is associated with high-relatedness judgment). This further indicates that, even on the level of local field potentials, the pre-stimulus resting state already contains some information about self-specificity.

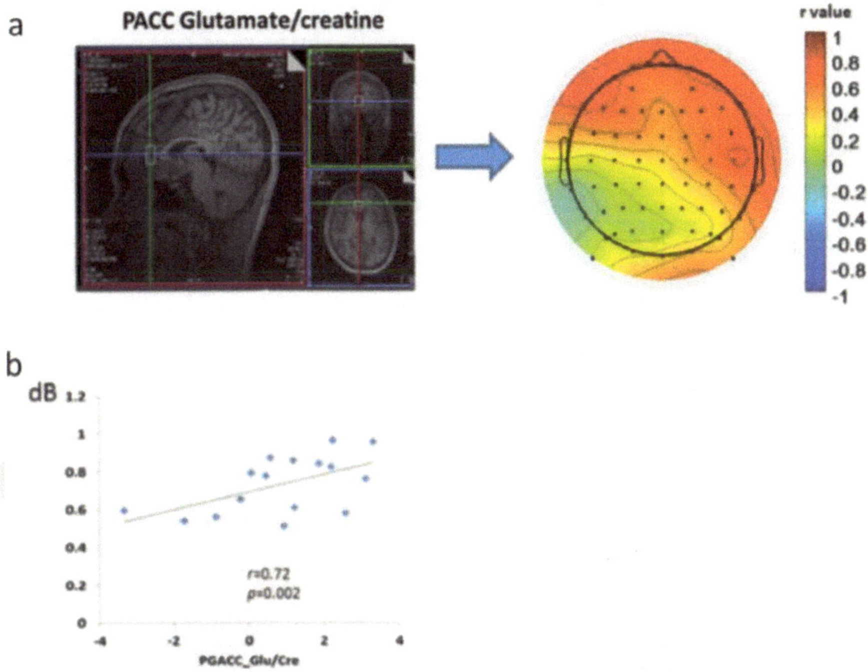

Fig. 1b Correlation (panel b lower left) of pre-stimulus ERSP (upper right in panel a) with glutamate level in perigenual anterior cingulate cortex (PACC) (upper left in panel a)

Finally, if the supposition of rest-self containment is correct, we would expect that highly self-specific stimuli like one's own name (or trait adjectives) do not induce any activity change in these regions. Data on this have recently been reported in an intracranial recording study by Lipsman et al. (2014): single neurons in the subgenual anterior cingulate cortex did not change their firing rates when the participant's own name was presented, when compared to the baseline spontaneous firing rates. Interestingly, this distinguished the subgenual cingulate cortical firing rates from the activity in a subcortical region, the nucleus subthalamicus, where the participant's own name induced significantly higher firing rates when compared to the baseline. I postulate that the null effect of own-name stimuli here occurs because spontaneous activity in the brain already contains some information about self-specificity and hence activity does not change when a highly self-specific stimulus is processed (See Fig. 2a, b).

Accordingly, processing self-specific stimuli in CMS is like seeing a very good old friend: both know each other very well for a long time so the friend does not need to change her/his behaviour at all and can thus be her/his own self. That is analogous to the resting state that does not change its activity level when 'seeing' or processing an old friend, a self-specific stimulus. Where and how though got the two friends, for example, self-specific stimuli and resting state got to know each other? As in a long-term friendship, this developed over time and is strongly depen-

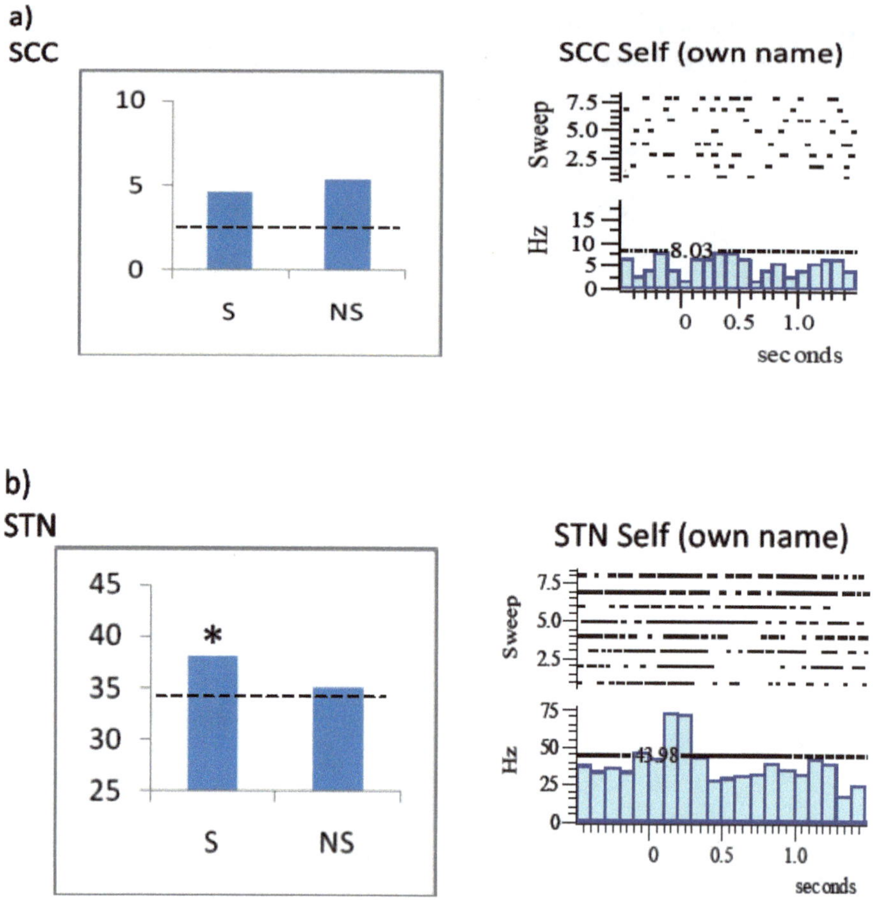

Fig. 2 Intracranial firing rates during self- and non-self-specific stimuli in subgenual cingulate cortex/SCC (**a**) and subthalamic nucleus/STN (**b**) (see Lipsmann et al. 2014 for details)

dent on each other's experience. In the same way the familiarity between two friends is encoded in their relationship, self-specificity is encoded in the spontaneous activity's spatiotemporal patterns entailing 'rest-self containment'.

Specifically, recent data by Duncan et al. (2015) and others (Sadaghiani and Kleinschmidt 2013) suggest that neural activity in the resting state's spatiotemporal patterns is strongly dependent upon the experience over time. For instance, conducting fMRI and MRS (Duncan et al. 2015) showed that the degree of entropy, for example, the degree of disorder or dissimilarity in the resting state's spatiotemporal activity patterns during adulthood, predicts the degree of early childhood trauma (as measured with the early childhood trauma questionnaire): the higher the degree of early childhood trauma, the higher the degree of entropy in PACC resting state activity during adulthood. Moreover, the degree of early childhood trauma was also related to the level of glutamate in PACC: lower adult levels of PACC glutamate were related with higher levels of early childhood trauma. This clearly indicates that neural activity in CMS seems to be strongly experience-dependence which, as I suggest, may provide the basis for encoding self-specificity into their spontaneous activity's spatiotemporal patterns.

'Default-Mode Functionality' – Cortical Midline Regions and Self-Expansion

What is the role of cortical midline structures (CMS) as part of the DMN for the rest of the brain? Raichle (2015; Raichle et al. 2001) supposed that they served as default-mode functionality for the brain and its neural activity. The concept of default-mode functionality means that the CMS serves as standard, reference or baseline for the other regions' and networks' activities: the latter's neural activity is set, matched, referenced or compared against the former. This means that the CMS exerts an impact on any neural activity in the other regions and networks of the brain. This is indeed well documented in that the CMS/DMN stand in reciprocal relationship to both CEN (Chen et al. 2013) and sensorimotor network (see Martino et al. 2016).

This is also reflected in the fact that, compared to the rest of the brain, the CMS show the highest degree of structural and functional connectivity to other regions and networks in the brain (See de Pasquale et al. 2010, 2012; Hagmann et al. 2008). Put metaphorically, the CMSs' role in the brain can be compared to the person standing in the middle of the crowd: in the same way the person standing in the middle of the crowd can communicate with everybody else, the CMSs, due to their position in the middle of the brain, are well suited to communicate and modulate neural activity in all other parts of the brain.

Why and how is such characterization of CMS important for processing self-specificity? Due to its extensive connectivity and communication, the CMS and its self-related processing can modulate stimulus-processing and neural activity in other with other regions and networks. This means that the latter's neural activity and their stimuli are modulated by the CMS' degree of self-relatedness: the latter such expands to or is associated with the former entailing what has been described as 'self-expansion' or 'self-association' (Sui and Humphreys 2014; Sui et al. 2013c). Based on the data demonstrated above, I hypothesize that such self-expansion or self-association is mediated by the relationship of the CMS to the respective region/network of the functional domain in which the stimuli are processed. More specifically, the CMS as part of the DMN may modulate neural processing of stimuli in the affective network, salience network, sensorimotor network and central executive network. That modulation, in turn, may allow to expand or associate the CMS' degree of self-relatedness with affective, perceptual, motivational and cognitive stimuli – the latter are thus associated or tagged as self-specific (See Fig. 3a).

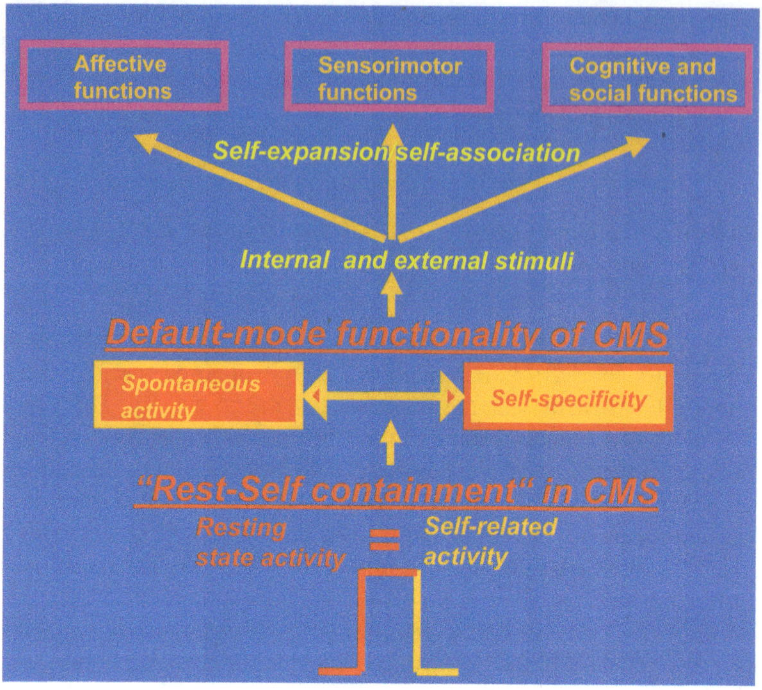

Fig. 3a 'Rest-self containment' and default-mode functionality of cortical midline structure (CMS)

Let us cite one example of empirical support (see also Qin et al. 2016). Sui et al. (2013a, b, c; Tagliazucchi et al. 2013) investigated the same paradigm in fMRI as described above. They observed two main regions, the ventromedial prefrontal cortex (VMPFC) and left posterior superior temporal sulcus (LpSTS) to be implicated in the self-association effects. More specifically, activity changes in the VMPFC were mainly related to the label of self by itself (during both matching and matching pairs) as distinguished from the other two, friend and stranger.

In contrast, the same region was not implicated in the external stimuli themselves regardless whether they were labelled by self or not. VMPFC activity was thus label-specific or self-specific. This was different for the LpSTS. This region responded to both self-label itself and those stimuli (the geometrical shapes) that were labelled by the self ('Self-tags'). To underline the behavioural relevance of both regions, their activity changes correlated with the respective reaction times with higher activity changes predicting faster reaction times. Functional connectivity between VMPFC and LpSTS was also increased during specifically matching pairs that labelled shapes by the self. This also correlated with behavioural measures showing that increasing VMPFC-LpSTS functional connectivity predicts increased accuracy and reaction times.

Based on their data the authors assume differential roles for VMPFC and LpSTS. The VMPFC may be implicated in internal self-representation, which then is linked to the processing of external stimuli as in LpSTS that is implicated in attributing salience or attention to stimuli. By coupling VMPFC and LpSTS, the internal self-representation can impact the external stimulus processing as in the self-labelled shapes. Considering their functional connectivity data, they assume that the VMPFC receives the visual input (from the external stimuli, the shapes) even earlier than visual associative areas (which though due to the sluggishness of the BOLD response in fMRI would need to be demonstrated in EEG). That in turn makes it possible for the VMPFC (and its self-specificity) to impact their subsequent processing in posterior regions like the LpSTS and to attribute higher degrees of salience or attention to them, which psychologically can be described as self-association.

Conclusion – Neurophenomenal Approach to CMS and Self in Psychiatric Disorders

The data show involvement of various regions and networks throughout the whole brain in self-specific processing. The CMS as core of the DMN seem to take on a central role in constituting self-specificity as such which, due to CMS' connections with other regions and networks, can be associated with and expand to other functional domains, for example, affective, cognitive, motivational and sensorimotor. Such self-expansion or self-association reflects the assignment of self-specificity to otherwise self-neutral stimuli (see above). The default-mode functionality that Raichle attributed to the CMS/DMN in neuronal terms can then be expanded on the psychological level with regard to self-specificity. In the same way the CMS/DMN neural activity serves as baseline or standard for the neural activity in the rest of the brain, self-specificity (as encoded in CMS) serves as baseline or standard for processing and tagging stimuli in cognitive, motivational, affective and sensorimotor domains. One thus want to speak of 'default mode functionality' of the self.

Self-expansion or self-association makes possible to link what conceptually has been described as 'self as subject' and 'self as object' (D'Argembeau 2013; Klein 2012; Northoff 2013, 2014b), which can be traced back to William James' distinction between 'I' (self as subject) and 'Me' (self as object) (James 1890a, b). The subjective and phenomenal features of the self as subject may be based upon the CMS. The CMS's role as default-mode functionality for the other regions and networks of the brain may allow for expansion of the self as subject – the latter becomes consecutively manifest in the objective affective, cognitive, sensorimotor and motivational functions as self as object. Disturbances in the former, for example, the self as subject, will then lead to alterations in the latter, the self as object (See Fig. 3b).

We can observe abnormal changes in the self as subject in psychiatric disorders like major depressive disorder (Northoff 2015, 2016; see Northoff and Sibille 2014 for details, as well as Vanhaudenhuyse et al. 2011), bipolar disorder (Martino et al. 2016) and schizophrenia (Northoff and Duncan 2016; see also Carhart-Harris et al. 2013; Carhart-Harris et al. 2014). As documented in the various chapters in this

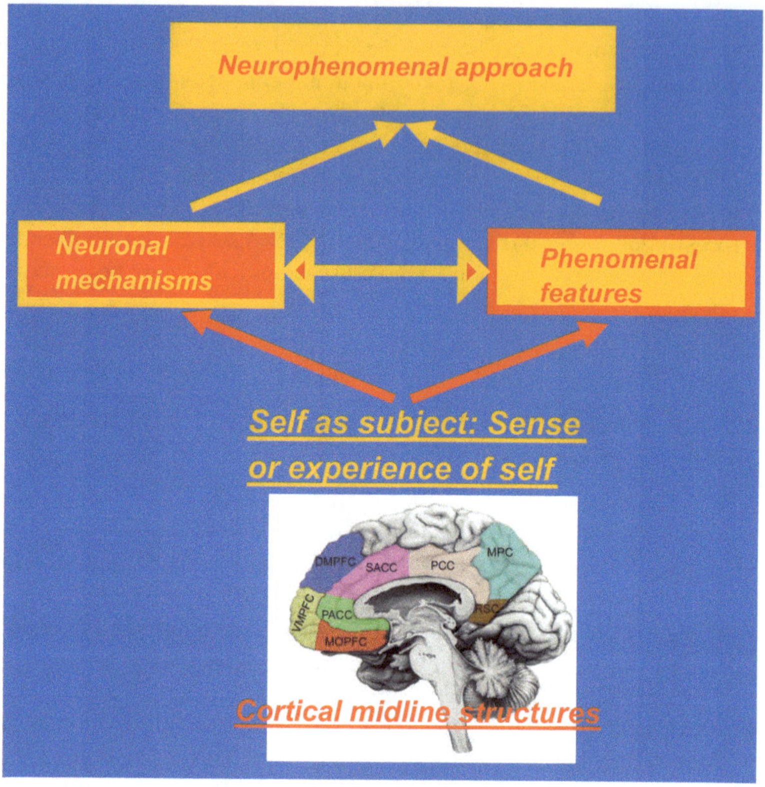

Fig. 3b Neurophenomenal approach

book, they all show changes in the CMS and their self as subject (see also Northoff 2014), which then expands to the self as object in the various domains, for example, affective, motivational, cognitive and sensorimotor.

Alterations of the self in psychiatric disorders may then be considered a 'basic disturbance' of the self as subject upon which the various cognitive, affective, sensorimotor and motivational symptoms as changes in the self as object are based and build. In order to fully understand such basic disturbance of the self as subject, one needs to combine its neuronal investigation with a phenomenal approach: such 'neurophenomenal approach' (see Northoff 2014, 2014a; Northoff and Stanghellini 2016) allows for linking the subjective experience of the own self and its relation to body, environment, time and space, for example, the self as subject, to corresponding spatiotemporal features in the neuronal activity of the brain and, more specifically, the CMS and its relation to other regions and networks.

References

Andreasen, N. C., O'Leary, D. S., Cizadlo, T., Arndt, S., Rezai, K., Watkins, G. L., … Hichwa, R. D. (1995). Remembering the past: two facets of episodic memory explored with positron emission tomography. *Am J Psychiatry*.

Andrews-Hanna, J. R. (2012). The brain's default network and its adaptive role in internal mentation. *Neuroscientist, 18*(3), 251–270. doi: https://doi.org/10.1177/1073858411403316

Andrews-Hanna, J. R., Reidler, J. S., Huang, C., & Buckner, R. L. (2010a). Evidence for the default network's role in spontaneous cognition. *J Neurophysiol, 104*(1), 322–335.

Andrews-Hanna, J. R., Reidler, J. S., Sepulcre, J., Poulin, R., & Buckner, R. L. (2010b). Functional-anatomic fractionation of the brain's default network. *Neuron, 65*(4), 550–562.

Bai, Y., Nakao, T., Xu, J., Qin, P., Chaves, P., Heinzel, A., … Tsai, S.-Y. (2015). Resting state glutamate predicts elevated pre-stimulus alpha during self-relatedness: A combined EEG-MRS study on "rest-self overlap". *Social neuroscience*, 1–15.

Buckner, R. L., Andrews-Hanna, J. R., & Schacter, D. L. (2008). The brain's default network: anatomy, function, and relevance to disease. *Ann.N.Y.Acad.Sci., 1124*, 1–38.

Carhart-Harris, R. L., Leech, R., Erritzoe, D., Williams, T. M., Stone, J. M., Evans, J., … Nutt, D. J. (2013). Functional connectivity measures after psilocybin inform a novel hypothesis of early psychosis. *Schizophr Bull, 39*(6), 1343–1351.

Carhart-Harris, R. L., Leech, R., Hellyer, P. J., Shanahan, M., Feilding, A., Tagliazucchi, E., … Nutt, D. (2014). The entropic brain: a theory of conscious states informed by neuroimaging research with psychedelic drugs. *Front Hum Neurosci, 8*.

Chen, A. C., Oathes, D. J., Chang, C., Bradley, T., Zhou, Z.-W., Williams, L. M., … Etkin, A. (2013). Causal interactions between fronto-parietal central executive and default-mode networks in humans. *Proceedings of the National Academy of Sciences, 110*(49), 19944–19949.

Christoff, K., Cosmelli, D., Legrand, D., & Thompson, E. (2011). Specifying the self for cognitive neuroscience. *Trends Cogn Sci., 15*(3), 104–112.

Christoff, K., Gordon, A. M., Smallwood, J., Smith, R., & Schooler, J. W. (2009). Experience sampling during fMRI reveals default network and executive system contributions to mind wandering. *Proc Natl Acad Sci U S A, 106*(21), 8719–8724. doi: https://doi.org/10.1073/pnas.0900234106

Cunningham, S. J., Brady-Van den Bos, M., & Turk, D. J. (2011). Exploring the effects of ownership and choice on self-memory biases. *Memory, 19*(5), 449–461.

Cunningham, S. J., Turk, D. J., Macdonald, L. M., & Macrae, C. N. (2008). Yours or mine? Ownership and memory. *Consciousness and Cognition, 17*(1), 312–318.

D'Argembeau, A., Collette, F., Van der, L. M., Laureys, S., Del Fiore, G., Degueldre, C., … Salmon, E. (2005). Self-referential reflective activity and its relationship with rest: a PET study. *Neuroimage., 25*(2), 616–624.

D'Argembeau, A. (2013). On the role of the ventromedial prefrontal cortex in self-processing: the valuation hypothesis. *Front Hum Neurosci, 7*.

Davey, C. G., Pujol, J., & Harrison, B. J. (2016). Mapping the self in the brain's default mode network. *Neuroimage, 132*, 390–397.

de Greck, M., Enzi, B., Prosch, U., Gantman, A., Tempelmann, C., & Northoff, G. (2010). Decreased neuronal activity in reward circuitry of pathological gamblers during processing of personal relevant stimuli. *Hum.Brain Mapp., 31*(11), 1802–1812.

de Greck, M., Rotte, M., Paus, R., Moritz, D., Thiemann, R., Proesch, U., … Northoff, G. (2008). Is our self based on reward? Self-relatedness recruits neural activity in the reward system. *Neuroimage, 39*(4), 2066–2075. doi: https://doi.org/10.1016/j.neuroimage.2007.11.006

de Greck, M., Supady, A., Thiemann, R., Tempelmann, C., Bogerts, B., Forschner, L., … Northoff, G. (2009). Decreased neural activity in reward circuitry during personal reference in abstinent alcoholics--a fMRI study. *Hum.Brain Mapp., 30*(5), 1691–1704.

de Pasquale, F., Della Penna, S., Snyder, A. Z., Lewis, C., Mantini, D., Marzetti, L., … Romani, G. L. (2010). Temporal dynamics of spontaneous MEG activity in brain networks. *Proceedings of the National Academy of Sciences, 107*(13), 6040–6045.

de Pasquale, F., Della Penna, S., Snyder, A. Z., Marzetti, L., Pizzella, V., Romani, G. L., & Corbetta, M. (2012). A cortical core for dynamic integration of functional networks in the resting human brain. *Neuron, 74*(4), 753–764.

Duncan, N. W., Hayes, D. J., Wiebking, C., Brice, T., Pietruska, K., Chen, D., … Northoff, G. (2015). Negative childhood experiences alter a prefrontal-insular-motor cortical network in healthy adults: A multimodal rsfMRI-fMRI-MRS-dMRI study. *Hum Brain Mapp, in press.*

Enzi, B., de Greck, M., Prosch, U., Tempelmann, C., & Northoff, G. (2009). Is our self nothing but reward? Neuronal overlap and distinction between reward and personal relevance and its relation to human personality. *PLoS One, 4*(12), e8429. doi: https://doi.org/10.1371/journal.pone.0008429

Ersner-Hershfield, H., Wimmer, G. E., & Knutson, B. (2009). Saving for the future self: neural measures of future self-continuity predict temporal discounting. *Soc Cogn Affect Neurosci, 4*(1), 85–92. doi: https://doi.org/10.1093/scan/nsn042

Frings, C., & Wentura, D. (2014). Self-priorization processes in action and perception.

Goel, V., & Dolan, R. J. (2003a). Explaining modulation of reasoning by belief. *Cognition, 87*(1), B11–22.

Goel, V., & Dolan, R. J. (2003b). Reciprocal neural response within lateral and ventral medial prefrontal cortex during hot and cold reasoning. *Neuroimage, 20*(4), 2314–2321.

Hagmann, P., Cammoun, L., Gigandet, X., Meuli, R., Honey, C. J., Wedeen, V. J., & Sporns, O. (2008). Mapping the structural core of human cerebral cortex. *PLoS Biol, 6*(7), e159. doi: https://doi.org/10.1371/journal.pbio.0060159

Huang, Z., Obara, N., Davis IV, H. H., Pokorny, J., & Northoff, G. (2016). The temporal structure of resting-state brain activity in the medial prefrontal cortex predicts self-consciousness. *Neuropsychologia.*

James, W. (1890a). The principles of psychology (Vol. 1). *New York: Holt.*

James, W. (1890b). The principles of psychology (Vol. 2). *New York: Holt.*

Kaiser, R. H., Andrews-Hanna, J. R., Wager, T. D., & Pizzagalli, D. A. (2015). Large-Scale Network Dysfunction in Major Depressive Disorder: A Meta-analysis of Resting-State Functional Connectivity. *JAMA psychiatry.*

Kim, K., & Johnson, M. K. (2010). Extended self: medial prefrontal activity during transient association of self and objects. *Soc Cogn Affect Neurosci*, nsq096.

Klein, S. B. (2012). Self, memory, and the self-reference effect: an examination of conceptual and methodological issues. *Pers Soc Psychol Rev, 16*(3), 283–300. doi: https://doi.org/10.1177/1088868311434214

Lee, T. W., Northoff, G., & Wu, Y. T. (2014). Resting network is composed of more than one neural pattern: An fMRI study. *Neuroscience, 274*(0), 198–208. doi: doi:https://doi.org/10.1016/j.neuroscience.2014.05.035

Lipsman, N., Nakao, T., Kanayama, N., Krauss, J. K., Anderson, A., Giacobbe, P., … Northoff, D. G. (2014). Neural overlap between resting state and self-relevant activity in human subcallosal cingulate cortex – Single unit recording in an intracranial study. *Cortex- in press*(0). doi: doi:https://doi.org/10.1016/j.cortex.2014.09.008

Lou, H. C., Gross, J., Biermann-Ruben, K., Kjaer, T. W., & Schnitzler, A. (2010). Coherence in consciousness: paralimbic gamma synchrony of self-reference links conscious experiences. *Hum.Brain Mapp., 31*(2), 185–192.

Martino, M., Magioncalda, P., Huang, Z., Conio, B., Piaggio, N., Duncan, N. W., … Wolff, A. (2016). Contrasting variability patterns in the default mode and sensorimotor networks balance in bipolar depression and mania. *Proceedings of the National Academy of Sciences, 113*(17), 4824–4829.

Mason, M. F., Norton, M. I., Van Horn, J. D., Wegner, D. M., Grafton, S. T., & Macrae, C. N. (2007). Wandering minds: the default network and stimulus-independent thought. *Science, 315*(5810), 393–395. doi: https://doi.org/10.1126/science.1131295

McKiernan, K. A., D'Angelo, B. R., Kaufman, J. N., & Binder, J. R. (2006). Interrupting the "stream of consciousness": an fMRI investigation. *Neuroimage, 29*(4), 1185–1191. doi: https://doi.org/10.1016/j.neuroimage.2005.09.030

Mitchell, J. P., Schirmer, J., Ames, D. L., & Gilbert, D. T. (2011). Medial prefrontal cortex predicts intertemporal choice. *J Cogn Neurosci, 23*(4), 857–866.

Modinos, G., Ormel, J., & Aleman, A. (2009). Activation of anterior insula during self-reflection. *PLoS One, 4*(2), e4618.

Moran, J. M., Macrae, C. N., Heatherton, T. F., Wyland, C. L., & Kelley, W. M. (2006). Neuroanatomical evidence for distinct cognitive and affective components of self. *J Cogn Neurosci, 18*(9), 1586–1594. doi: https://doi.org/10.1162/jocn.2006.18.9.1586

Nakao, T., Bai, Y., Nashiwa, H., & Northoff, G. (2013). Resting-state EEG power predicts conflict-related brain activity in internally guided but not in externally guided decision-making. *Neuroimage, 66*, 9-21. doi: https://doi.org/10.1016/j.neuroimage.2012.10.034

Nakao, T., Ohira, H., & Northoff, G. (2012). Distinction between Externally vs. Internally Guided Decision-Making: Operational Differences, Meta-Analytical Comparisons and Their Theoretical Implications. *Front Neurosci, 6*, 31. doi: https://doi.org/10.3389/fnins.2012.00031

Northoff, G. (2013). Brain and self-A neurophilosophical account. *Child Adolesc Psychiatry Ment Health, 7*, 28.

Northoff, G. (2014). How Is Our Self Altered in Psychiatric Disorders? A Neurophenomenal Approach to Psychopathological Symptoms. *Psychopathology*. doi: https://doi.org/10.1159/000363351

Northoff, G. (2014a). *Unlocking the Brain: Volume 1: Coding* (Vol. 1): Oxford University Press.

Northoff, G. (2014b). *Unlocking the Brain. Volume II: Consciousness*: Oxford: Oxford University Press.

Northoff, G. (2015). Is the self a higher-order of fundamental funciton of the brain? The "basis model of self-specifity" and its encoding by the brain's spontaneous activity. *Cognitive Neuroscience, Journal of.*

Northoff, G. (2016). Spatiotemporal psychopathology I: No rest for the brain's resting state activity in depression? Spatiotemporal psychopathology of depressive symptoms. *J Affect Disord, 190*, 854–866.

Northoff, G., & Duncan, N. W. (2016). How do abnormalities in the brain's spontaneous activity translate into symptoms in schizophrenia? From an overview of resting state activity findings to a proposed spatiotemporal psychopathology. *Prog Neurobiol*. doi: https://doi.org/10.1016/j.pneurobio.2016.08.003

Northoff, G., & Hayes, D. J. (2011). Is our self nothing but reward? *Biol Psychiatry, 69*(11), 1019–1025. doi: https://doi.org/10.1016/j.biopsych.2010.12.014

Northoff, G., Heinzel, A., Bermpohl, F., Niese, R., Pfennig, A., Pascual-Leone, A., & Schlaug, G. (2004). Reciprocal modulation and attenuation in the prefrontal cortex: an fMRI study on emotional-cognitive interaction. *Hum Brain Mapp, 21*(3), 202–212. doi: https://doi.org/10.1002/hbm.20002

Northoff, G., Heinzel, A., de Greck, M., Bermpohl, F., Dobrowolny, H., & Panksepp, J. (2006). Self-referential processing in our brain--a meta-analysis of imaging studies on the self. *Neuroimage., 31*(1), 440–457.

Northoff, G., & Panksepp, J. (2008). The trans-species concept of self and the subcortical-cortical midline system. *Trends Cogn Sci., 12*(7), 259–264.

Northoff, G., Schneider, F., Rotte, M., Matthiae, C., Tempelmann, C., Wiebking, C., ... Panksepp, J. (2009). Differential parametric modulation of self-relatedness and emotions in different brain regions. *Hum.Brain Mapp., 30*(2), 369–382.

Northoff, G., & Sibille, E. (2014). Why are cortical GABA neurons relevant to internal focus in depression? A cross-level model linking cellular, biochemical and neural network findings. *Mol Psychiatry, 19*(9), 966–977. doi: https://doi.org/10.1038/mp.2014.68

Northoff, G., & Stanghellini, G. (2016). How to link brain and experience? Spatiotemporal Psychopathology of the Lived Body. *Front Hum Neurosci, 10*, 172.

Panksepp, J. (1998a). *Affective neuroscience: The foundations of human and animal emotions*: Oxford university press.

Panksepp, J. (1998b). The preconscious substrates of consciousness: Affective states and the evolutionary origins of the SELF. *Journal of Consciousness Studies, 5*, 566–582.

Panksepp, J. (2011). Cross-species affective neuroscience decoding of the primal affective experiences of humans and related animals. *PLoS One, 6*(9), e21236.

Phan, K. L., Taylor, S. F., Welsh, R. C., Ho, S.-H., Britton, J. C., & Liberzon, I. (2004). Neural correlates of individual ratings of emotional salience: a trial-related fMRI study. *Neuroimage, 21*(2), 768-780.

Phan, K. L., Wager, T., Taylor, S. F., & Liberzon, I. (2002). Functional neuroanatomy of emotion: a meta-analysis of emotion activation studies in PET and fMRI. *Neuroimage, 16*(2), 331–348. doi: https://doi.org/10.1006/nimg.2002.1087

Qin, P., Grimm, S., Duncan, N. W., Fan, Y., Huang, Z., Lane, T., ... Northoff, G. (2016). Spontaneous activity in default-mode network predicts ascription of self-relatedness to stimuli. *Soc Cogn Affect Neurosci*, nsw008.

Qin, P., & Northoff, G. (2011). How is our self related to midline regions and the default-mode network? *Neuroimage., 57*(3), 1221–1233.

Raichle, M. E. (2015). The restless brain: how intrinsic activity organizes brain function. *Philosophical Transactions of the Royal Society of London B: Biological Sciences, 370*(1668), 20140172.

Raichle, M. E., MacLeod, A. M., Snyder, A. Z., Powers, W. J., Gusnard, D. A., & Shulman, G. L. (2001). A default mode of brain function. *Proc.Natl.Acad.Sci.U.S.A, 98*(2), 676–682.

Roy, M., Shohamy, D., & Wager, T. D. (2012). Ventromedial prefrontal-subcortical systems and the generation of affective meaning. *Trends Cogn Sci, 16*(3), 147–156.

Sadaghiani, S., & Kleinschmidt, A. (2013). Functional interactions between intrinsic brain activity and behavior. *Neuroimage, 80*(0), 379–386. doi: doi:https://doi.org/10.1016/j.neuroimage.2013.04.100

Schilbach, L., Bzdok, D., Timmermans, B., Fox, P. T., Laird, A. R., Vogeley, K., & Eickhoff, S. B. (2012). Introspective minds: using ALE meta-analyses to study commonalities in the neural correlates of emotional processing, social & unconstrained cognition. *PLoS One, 7*(2), e30920. doi: https://doi.org/10.1371/journal.pone.0030920

Schneider, F., Bermpohl, F., Heinzel, A., Rotte, M., Walter, M., Tempelmann, C., ... Northoff, G. (2008). The resting brain and our self: self-relatedness modulates resting state neural activity in cortical midline structures. *Neuroscience, 157*(1), 120–131.

Stawarczyk, D., Majerus, S., Maquet, P., & D'Argembeau, A. (2011). Neural correlates of ongoing conscious experience: both task-unrelatedness and stimulus-independence are related to default network activity. *PLoS One, 6*(2), e16997.

Sui, J., Chechlacz, M., & Humphreys, G. W. (2012b). Dividing the self: Distinct neural substrates of task-based and automatic self-prioritization after brain damage. *Cognition, 122*(2), 150–162.

Sui, J., Chechlacz, M., Rotshtein, P., & Humphreys, G. W. (2013b). Lesion-Symptom Mapping of Self-Prioritization in Explicit Face Categorization: Distinguishing Hypo-and Hyper-Self-Biases. *Cereb Cortex*, bht233.

Sui, J., He, X., & Humphreys, G. W. (2012a). Perceptual effects of social salience: evidence from self-prioritization effects on perceptual matching. *Journal of Experimental Psychology: Human Perception and Performance, 38*(5), 1105.

Sui, J., & Humphreys, G. W. (2014). Super-size me: self biases increase to larger stimuli. *Psychonomic bulletin & review*, 1–9.

Sui, J., Liu, M., Mevorach, C., & Humphreys, G. W. (2013a). The Salient Self: The Left Intraparietal Sulcus Responds to Social as Well as Perceptual-Salience After Self-Association. *Cereb Cortex*, bht302.

Sui, J., Rotshtein, P., & Humphreys, G. W. (2013c). Coupling social attention to the self forms a network for personal significance. *Proceedings of the National Academy of Sciences, 110*(19), 7607–7612.

Tagliazucchi, E., von Wegner, F., Morzelewski, A., Brodbeck, V., Jahnke, K., & Laufs, H. (2013). Breakdown of long-range temporal dependence in default mode and attention networks during deep sleep. *Proceedings of the National Academy of Sciences, 110*(38), 15419–15424.

Vanhaudenhuyse, A., Demertzi, A., Schabus, M., Noirhomme, Q., Bredart, S., Boly, M., ... Laureys, S. (2011). Two distinct neuronal networks mediate the awareness of environment and of self. *J Cogn Neurosci, 23*(3), 570–578. doi: https://doi.org/10.1162/jocn.2010.21488

Wang, C., Oyserman, D., Liu, Q., Li, H., & Han, S. (2013). Accessible cultural mind-set modulates default mode activity: Evidence for the culturally situated brain. *Social neuroscience, 8*(3), 203–216.

Whitfield-Gabrieli, S., Moran, J. M., Nieto-Castañón, A., Triantafyllou, C., Saxe, R., & Gabrieli, J. D. (2011). Associations and dissociations between default and self-reference networks in the human brain. *Neuroimage, 55*(1), 225–232.

Research Domains and Brain Network Dysfunction: Towards a New Taxonomy of Neuropsychiatric Illness

Sophia Frangou

Psychiatric disorders are conceptualized as syndromes defined by self-reported and observed abnormalities in perception, affective and non-affective mental processes and behaviour rather than by their biological substrates. However, advances in neuroimaging have the potential to transform the debate about the nature and validity of psychiatric nosology. This chapter begins with an overview of the imaging methods available to study mental states and cognitive-behavioural attributes and then outlines the current conceptual formulations of the psychiatric nosology, which provides the backdrop for understanding the contribution of neuroimaging findings. In presenting the neuroimaging literature, emphasis is placed on selected psychiatric disorders that have been studied most extensively, including schizophrenia, major depressive disorder, bipolar disorder, obsessive compulsive disorder and post-traumatic stress disorder, and on structural and functional magnetic resonance imaging findings in adults because these are better established. These findings are discussed in terms of their contribution to psychiatric nosology, the challenges they present and the future directions in this field.

Neuroimaging: A Window into the Brain and Mind

Neuroimaging has allowed the direct investigation of the structure and function of the brain in living human beings. Previously information about the in-vivo properties of the brain could only be inferred from post-mortem and lesion studies in humans and from experimentation in animals. Also prior to neuroimaging, biological validators of psychiatric diagnoses were sought exclusively in peripheral tissues but the relationship of these peripheral measures to events taking place within the

S. Frangou (✉)
Department of Psychiatry, Icahn School of Medicine at Mount Sinai, New York, NY, USA
e-mail: Sophia.Frangou@mssm.edu; https://www.frangougroup.org

V. A. Diwadkar, S. B. Eickhoff (eds.), *Brain Network Dysfunction in Neuropsychiatric Illness*, https://doi.org/10.1007/978-3-030-59797-9_22

brain ranges from uncertain to unlikely. In the last four decades, neuroimaging has become the main biological research tool in psychiatry mainly because of the rapid and extensive development of magnetic resonance imaging (MRI). The excellent safety profile and high patient acceptability of MRI have been major contributors to its widespread use in psychiatric research. In parallel, there has been an exponential increase in the availability of MR scanners, in the range of MRI modalities, in the strength of the magnetic fields used and in the sophistication of the analytical approaches to MRI-derived data.

T1-weighted structural MRI is used to examine brain morphometry, including regional volumes, cortical area and thickness, and gyrification. Diffusion-weighted imaging (DWI) is primarily used to examine the integrity of the white matter, either at the level of voxels or tracts. Extrapolating from the movement of water molecules within the cellular and extracellular brain components, DWI techniques offer information on the integrity of white matter tracks and on the density and orientation of neurites. Functional MRI (fMRI) is based on the paramagnetic properties of deoxyhaemoglobin to derive the blood oxygen level–dependent (BOLD) signal which can then be used to inform about brain activity and connectivity during spontaneous and task-related brain function. MR spectroscopy (MRS) is used to study brain metabolism and chemistry including the concentration of neurotransmitters such as g-aminobutyric acid (GABA) and glutamine or glutamate. Each of the MR methods described above provides distinct and complementary information about brain organization and function. Neuroimaging phenotypes in healthy individuals have been reliably associated with general intellectual function (e.g. Wilke et al. 2003; Frangou et al. 2004; Narr et al. 2007), perception (e.g. Kanwisher et al. 1997; Downing et al. 2001; Dick et al. 2017), affective and non-affective cognitive processing (e.g. Niendam et al. 2012; Barrett and Satpute 2013; Gratton et al. 2018) and internal mental states (e.g. Haynes and Rees 2006; Rubin et al. 2017). Whether these relationships are causative or associative remains arguable, and this debate potentially detracts from the value of neuroimaging in detecting these relationships which are crucial for shaping our understanding of human mental function.

Psychiatric Nosology: A Tortuous Journey

Diagnostic nosology in medicine is predicated on the presence of anatomical, physiological or biochemical abnormalities and increasingly incorporates information about aetiology. By contrast, psychiatric nosology is widely criticized for comprising diagnoses that are neither biologically validated nor etiologically determined. Psychiatric diagnoses currently derived from abnormalities in behaviour and in self-reported internal states and are defined using criteria enshrined in the International Statistical Classification of Diseases and Related Health Problems (ICD), developed by the World Health Organization (WHO), and in the Diagnostic and Statistical Manual of Mental Disorders (DSM), produced by the American Psychiatric Association (APA). From the 1980s onwards, successive versions of these classification systems have focused primarily on improving the reliability of psychiatric diagnoses. Their priority was to facilitate and maximize agreement among clinicians and scientists in the service of three specific aims: to provide a rationale for treatment planning in clinical care, to support the design and conduct of replicable clinical trials for the regulatory approval of pharmacologic treatments and to enable a principled examination of the biological underpinnings of mental illness in comparable clinical samples (Jablensky 2016).

The introduction of DSM-5 in 2013 (American Psychiatric Association 2013) and the 11th edition of the ICD (World Health Organization 2017) developed by the WHO have brought forth ongoing arguments about the validity of psychiatric diagnoses. Theoretical arguments for establishing the validity of psychiatric diagnoses begun with the pioneering work of Robins and Guze (1970) who listed five criteria: (a) clinical description, (b) laboratory studies, (c) delimitation from other disorders, (d) follow-up studies and (e) family studies. Kendler (1980) later proposed that validators used to populate these criteria should be considered in terms of antecedent (e.g. family history), concurrent (e.g. psychological tests) and predictive (e.g. longitudinal consistency). Following advances in biological technologies, and particularly in genetics and neuroimaging, Andreasen (1995) emphasized the need to incorporate biological validators based on genetic, cellular, molecular, neurophysiological and brain imaging findings.

Implicit in the conceptual arguments outlined above is the assumption that psychiatric disorders could be classified into discrete entities delimitated by 'zones of rarity' (Kendell and Jablensky 2003). An alternative model proposed that mental illness can be better understood in terms of psychopathology 'clusters' (Andrews et al. 2009a). This view is primarily based on robust epidemiological evidence of pervasive comorbidity and symptomatic overlap among psychiatric disorders (e.g. Kessler et al. 1994; Teesson et al. 2009; Boschloo et al. 2015). Early in the revision of DSM-5, a subgroup of the task force proposed a 'meta-structure' in which existing psychiatric diagnoses would be clustered on the basis of eleven validators; six were selected to reflect overlapping aetiology (shared genetic risk factors; familiality; shared specific environmental risk factors; shared neural substrates; shared biomarkers; shared temperamental antecedents) and five were selected on the basis of clinical commonalities (shared abnormalities of cognitive or emotional processing; symptom similarity; high rates of co-morbidity; course of illness; treatment response) (Andrews et al. 2009a). Five clusters were proposed in line with the distribution of these validators in the studies available: (1) the neurocognitive (delirium, dementias, amnesia and other cognitive disorders) (Sachdev et al. 2009), (2) the neurodevelopmental (mental retardation, learning, motor skills and communication disorders, and pervasive developmental disorders) (Andrews et al. 2009b), (3) psychosis (schizophrenia, bipolar disorder, schizotypal personality disorder) (Carpenter Jr. et al. 2009), (4) emotional (unipolar depression, anxiety, post-traumatic stress disorders, avoidant personality disorder) (Goldberg et al. 2009) and (5) externalizing (substance-related disorders, antisocial personality disorder and impulse–control disorders) (Krueger and South 2009). Many disorders however (e.g. eating, sleep and sexual disorders) were left outside this scheme and placed in a 'not assigned yet' cluster. The proposal was heavily criticized mostly on the basis of inadequate empirical support (Jablensky 2009a, b; Wittchen et al. 2009) and was eventually abandoned (Kupfer and Regier 2011). The final version of DSM-5 retained most of the diagnostic criteria that were in place in DSM-IV particularly for the major psychiatric diagnoses. Importantly, during the development of the DSM-5 field trials were undertaken to quantify the cross-sectional and longitudinal reliability of DSM-5 defined diagnostic categories. A total of 2246 patients with various diagnoses were interviewed (86% twice) by 279 clinicians in various disciplines. The task force provided reliability data on 15 adult and 8 child diagnoses using the kappa statistic, which is a measure of agreement between two independent raters; for example, a kappa value of 0.46 indicates inter-rater agreement of 85% of the time. Reliability for each diagnosis can be considered as very good (kappa range 0.60–0.79), good (kappa range 0.40–0.59), questionable (kappa range 0.20–0.39) and unacceptable (kappa < 0.20). Figure 1 shows the kappa values for the major psychiatric disorders which are encouraging with the exception of major depressive disorder (MDD), which has been a challenging phenotype in all classifications (details in Freedman et al. 2013).

In parallel to these clinician-led initiatives, the National Institute of Mental Health (NIMH) in the United States proposed a different classification framework, the Research Domain Criteria (RDoC) Project (Insel et al. 2010) predicated on the notion that biological, behavioural and psychological constructs should be used to 'sign' (i.e. characterize) mental disorders (National Institutes of Mental Health 2008, 2015). Proponents of this framework argued that clinical phenotypes in the APA and WHO classifications were poorly aligned with these 'signatures' of mental

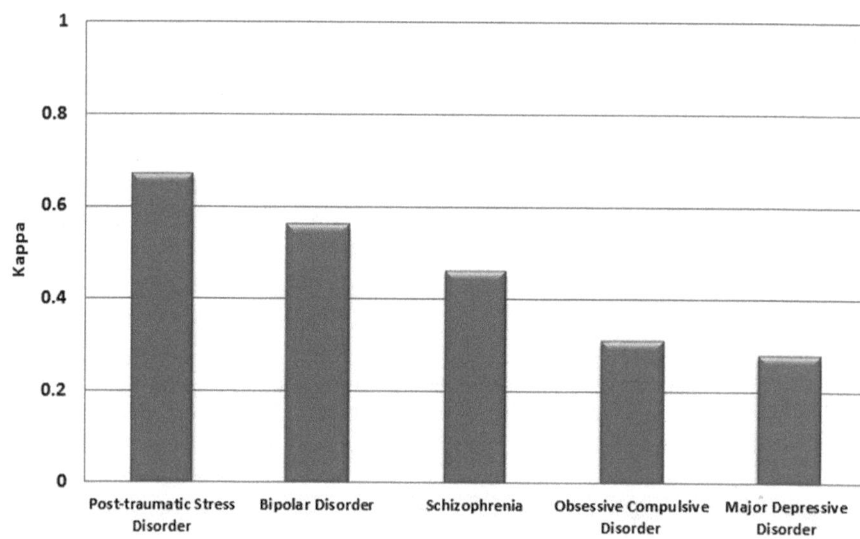

Fig. 1 Interrater reliability of DSM-5 diagnoses from the Initial Field Trials in adults. (Adapted from Freedman et al. 2013)

illness and that this lack of assignment was the major obstacle to progress (Cuthbert and Insel 2010; Kapur et al. 2012; Casey et al. 2013). In the RDoC framework, the key clinical phenotypes are transdiagnostic symptom dimensions and not diagnostic categories. These clinical phenotypes are the outputs of brain circuits, responsible for core mental processes, whose function may vary at different stages across the lifespan and is shaped by genetic and environmental factors. In the RDoC, core mental processes are represented by five 'domains': (1) Negative Valence Systems; (2) Positive Valence Systems; (3) Cognitive Systems; (4) Systems for Social Processes and (5) Arousal Systems (Cuthbert and Insel 2013; Cuthbert 2014). These domains can be investigated in terms of 'units of analysis' that comprise genes, molecules, cells, circuits, physiology, behaviour and self-reports (Cuthbert and Insel 2013). However, the RDoC framework has not been uniformly endorsed as an improvement on DSM/ICD classification. The RDoC may be helpful in capturing associations between its specified units of analyses and psychopathological dimensions, but the framework itself lacks any contextual criteria that would enable differentiation between a 'disorder' and normal 'variation' (Lilienfeld and Treadway 2016; Berenbaum 2013). Wakefield (2014) highlighted this problem by contrasting grief to a depressive episode; both may implicate changes in circuits relevant to negative affect, but the formal is normal and the latter is a disorder (Wakefield 2014). A further criticism concerns the discriminant validity of the domains at the heart of RDoC and their corresponding circuits. The relative independence of these domains and circuits is implicit in the structure of the RDoC framework, but this assumption is untested (Lilienfeld and Treadway 2016). Moreover, the RDoC framework is based on a 'modular' model of brain function, which considers brain areas as independent processors of specific mental operations (e.g. Kanwisher et al. 1997; Downing et al. 2001; Dick et al. 2017). Advances in neuroimaging and neuroscience have challenged this view (Fuster 2000) based on accumulating evidence which suggests that the human brain is intrinsically organized into domain-general, distributed networks; mental events, either spontaneous or task-related, arise from interactions within and between these domain-general networks (e.g. Niendam et al. 2012; Barrett and Satpute 2013; Gratton et al. 2018; Haynes and Rees 2006; Rubin et al. 2017; Bressler and Menon 2010). Therefore, many have voiced concerns that the main danger of the RDoC initiative is that its domains may be unjustifiably reified, as has been the case with the DSM categories. In that sense, the danger for RDoC is that it might become as restrictive as the 'epistemic prison' of the DSM (Hyman 2010) that it purports to replace.

Psychiatric Neuroimaging: A Journey of Discovery

We can trace the origins of modern psychiatric neuroimaging to a British paper by Johnstone, Crow, Frith and colleagues published in the Lancet in 1976, which demonstrated cerebral ventricular enlargement in people with chronic schizophrenia (Johnstone et al. 1976). Sir Dennis Hill, an eminent British professor at the Maudsley

Hospital in London, articulated the main objection of psychiatrists of the time in a letter to the Lancet (Hill 1976). He argued that since 'true schizophrenia' was a functional psychosis, the patients scanned were likely to have 'organic psychoses' secondary to neurological disorders that had been somehow overlooked by the authors. However, he conceded that the paper 'will prove a great stimulus to research in schizophrenia'.

Although multiple MRI technologies are available, structural MRI and fMRI studies, using case-control designs, dominate the psychiatric literature and provide the most appropriate material for the purpose of discussing the relationship between neuroimaging and psychiatric nosology. Both participants' caseness and their assignment to a specific diagnosis are mostly based on the diagnostic criteria of the fourth edition of the DSM (American Psychiatric Association 1994). These criteria are largely aligned with the tenth edition of the ICD10 (World Health Organization 1993), at least for the major diagnostic categories (Andrews et al. 1999). As already discussed, the validity of the DSM/ICD diagnoses has always been theoretically predicated on the degree to which they could be associated with explanatory variables relating to aetiology and/or pathophysiology. An important question therefore is whether neuroimaging has identified reliable brain correlates of DSM diagnoses.

Neuroimaging has been applied to all psychiatric disorders to variable degree. The focus here is on findings that can be considered robust because they have been reported in multiple primary studies. In this respect, the most notable research effort to identify the brain structural correlates of DSM disorders is the Enhancing Neuro Imaging Genetics through Meta-Analysis (ENIGMA) consortium (http://enigma.ini.usc.edu/). This is a consortium of researchers from around the world aimed to elucidate disorder-related brain alternations from large samples using harmonized analyses protocols (Thompson et al. 2014, 2017). Through the work of its disorder-specific working groups, the ENIGMA consortium has defined the most robust case–control differences (expressed in terms of Cohen's d effect size) in subcortical regional volumes in schizophrenia (SCZ; patients = 2028 and healthy individuals = 2540) (van Erp et al. 2016), bipolar disorder (BD; patients = 1710 and healthy individuals = 2594) (Hibar et al. 2016), MDD (patients = 1728 and healthy individuals = 7199) (Schmaal et al. 2016), obsessive-compulsive disorder (OCD; patients = 1495 and healthy individuals = 1472) (Boedhoe et al. 2017) and post-traumatic stress disorder (PTSD; patients = 794 and healthy individuals = 1074) (Logue et al. 2018) (Fig. 2). In addition, some of the ENIGMA working groups have reported case–control differences in cortical thickness in SCZ (patient = 4474 and healthy individuals = 5089) (van Erp et al. 2017), BD (patients = 1837 and healthy individuals = 2582) (Hibar et al. 2018), MDD (patients = 2148 and healthy individuals = 7957) (Schmaal et al. 2017) and OCD (patients = 1498 and healthy individuals = 1436) (PSW et al. 2017) (Fig. 3). A striking finding across all studies is that the effect sizes of case–control differences are uniformly small to medium, regardless of diagnosis, or brain structural phenotype. Interestingly, thalamic volume reduction was primarily noted in SCZ and BD, two disorders that are associated with increased propensity for psychotic symptoms. By contrast, hippocampal volume reduction was noted in all the diagnostic categories examined, which

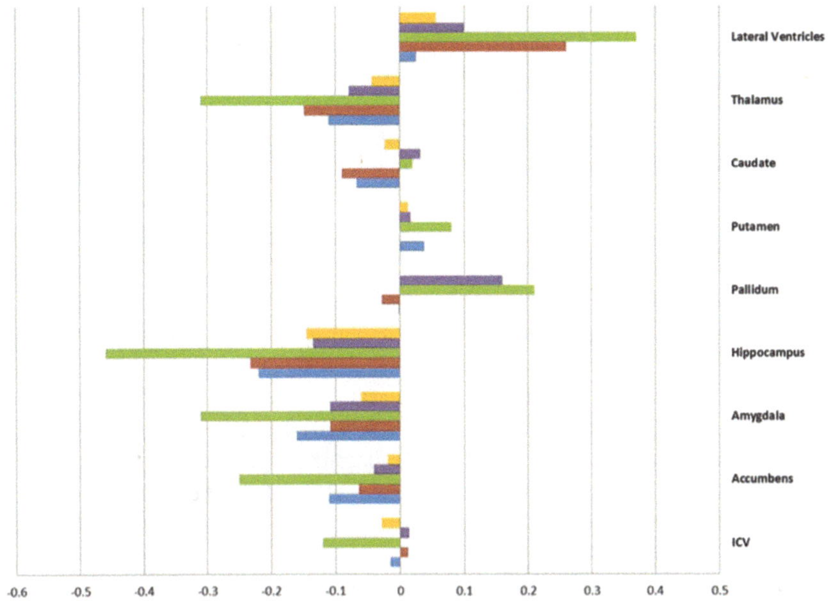

	ICV	Accumbens	Amygdala	Hippocampus	Pallidum	Putamen	Caudate	Thalamus	Lateral Ventricles
MDD	-0.029	-0.019	-0.06	-0.144	-0.001	0.012	-0.023	-0.044	0.056
OCD	0.014	-0.041	-0.108	-0.135	0.16	0.017	0.032	-0.079	0.1
SCZ	-0.12	-0.25	-0.31	-0.46	0.21	0.08	0.02	-0.31	0.37
BD	0.012	-0.064	-0.108	-0.232	-0.028	-0.001	-0.09	-0.148	0.26
PTSD	-0.015	-0.11	-0.16	-0.22	-0.002	0.038	-0.067	-0.11	0.026

Fig. 2 Summary of the results from the ENIGMA working groups on major depressive disorder (MDD), obsessive compulsive disorder (OCD), schizophrenia (SCZ), bipolar disorder (BD) and post-traumatic stress disorder (PTSD). The numerical values of the effect sizes are presented and also illustrated in the bar diagram

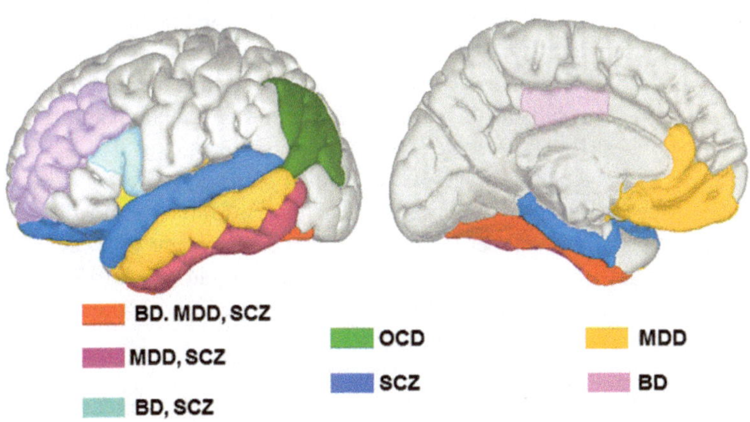

Fig. 3 Summary of the results of the ENIGMA working groups on major depressive disorder (MDD), obsessive compulsive disorder (OCD), schizophrenia (SCZ) and bipolar disorder (BD). The different colours indicate unique and shared regions of cortical thinning. Findings are based on the results of case–control comparisons undertaken separately for each disorder

indicates that this brain region is generally vulnerable to the presence of psychopathology. A similar assumption can be made with respect to cortical thinning, which emerged as another brain structural phenotype associated with psychopathology regardless of the DSM diagnosis. The ENIGMA studies provide evidence for diagnosis-related differences to the degree of cortical thinning; reductions were more widespread and pronounced in SCZ where case–control differences were mostly centred in the inferior frontal gyrus (pars opercularis), the fusiform gyrus, the lateral orbitofrontal cortex, the superior, medial and inferior temporal cortex, the posterior cingulate cortex, the parahippocampal gyrus and the insula. Cortical thinning frontal gyrus (pars opercularis), the fusiform gyrus was also detected in patients with BD but also involved the rostral middle frontal cortex. OCD was associated mostly with thinner parietal cortices while MDD showed the least pronounced changes mostly centred in the orbitofrontal cortex, the anterior and posterior cingulate and the insula. The different ENIGMA working groups have examined the effect of medication, as relevant to each diagnostic group, and have shown that generally antipsychotics are associated with increased striatal volume, particularly in SCZ (van Erp et al. 2016), while patients with BD that were prescribed lithium at the time of scanning showed less pronounced cortical thinning (Hibar et al. 2016).

The results of the ENIGMA consortium studies examine DSM diagnoses separately and any similarities or differences in their brain structural correlates can only be inferred from their deviance from healthy controls. Another approach is to synthesize data available through the published literature by applying meta-analytic techniques to the co-ordinates of case–control differences reported in structural imaging studies. Goodkind and colleagues (Goodkind et al. 2015) employed this approach on the co-ordinates of case–control differences extracted from 193 structural MRI studies, published between 2012 and 2015, on SCZ, BD, MDD, OCD, addiction and anxiety. They then used activation likelihood estimation (ALE) (Eickhoff et al. 2009, 2012; Turkeltaub et al. 2012) to determine the convergence of the coordinates reported in the different studies. The ALE first models these brain foci as probability distributions whose width is based on empirical estimates of the spatial uncertainty due to the between-subject and between-template variability of neuroimaging data and then assesses the results against a null-distribution of a random spatial association between studies (Eickhoff et al. 2009, 2012; Turkeltaub et al. 2012). The study found that all the psychiatric diagnoses examined were generally associated with grey matter reductions in cortical and subcortical regions (Table 1) with the exception of the striatum where grey matter increases were seen in patients, particularly those likely to be treated with antipsychotic medication (Goodkind et al. 2015).

In contrast to structural studies, fMRI designs are more amenable to the interrogation of the psychiatric nosology using the RDoC framework because task-fMRI is conventionally used to map mental processes to neural circuits. Case–control differences in task-related brain activity could then be used to make inferences about disease-related abnormalities in circuits corresponding to different domains. The Cognitive System domain of the RDoC framework focuses of specific processes that are considered under the construct of cognitive control and refer to mental operations relating to goal-selection, updating, representation and maintenance of goal-selection, response selection, inhibition and suppression and performance monitoring

Table 1 Regions of shared grey matter alternations in schizophrenia, bipolar disorder, major depressive disorder, substance abuse, obsessive compulsive disorder, anxiety disorders compared to healthy individuals

Results of the overall analysis	Results from targeted sub-analyses
Reduced GM in patients	
Cortical regions	
Dorsomedial prefrontal cortex	Reduction greater for schizophrenia compared to all other disorders
Ventromedial prefrontal cortex	Reduction greater for schizophrenia compared to all other disorders
Superior temporal gyrus	
Parietal operculum	
Dorsal anterior cingulate gyrus	Similarly reduced in all disorders
Insula	Similarly reduced in all disorders
Subcortical regions	
Thalamus	Reduction greater for schizophrenia compared to all other disorders
Amygdala	Reduction greater in schizophrenia; reduction greater in depression and anxiety disorders compared to bipolar disorder and addiction
Hippocampus	Reduction greater in schizophrenia; reduction greater in depression and anxiety disorders compared to bipolar disorder and addiction
Increased GM in patients	
Striatum	Mainly driven by studies on schizophrenia; may reflect antipsychotic treatment effect

From Goodkind et al. (2015)

(https://www.nimh.nih.gov/research-priorities/rdoc/constructs/cognitive-control.shtml). Several lines of evidence from studies in healthy individuals point to the existence of a ‘domain-general’ network, anchored in the lateral prefrontal, anterior cingulate and parietal cortices, that supports these broad and diverse cognitive control functions (Niendam et al. 2012; Gratton et al. 2018). Cognitive control is of relevance to psychiatric nosology for three main reasons. First, multiple behavioural studies have documented dysfunction in one or more of the processes involved across the majority of psychiatric disorders (e.g. Stefanopoulou et al. 2009; Polak et al. 2012; Abramovitch et al. 2013; Bora et al. 2013; Bourne et al. 2013; Gowin et al. 2013; Fatouros-Bergman et al. 2014; Bora 2018). Second, multiple fMRI studies have reported abnormal engagement of the cognitive control network in patients with SCZ (Minzenberg et al. 2009), BD (Chen et al. 2011), MDD (Miller et al. 2015; Hamilton et al. 2012), OCD (Rasgon et al. 2017) and anxiety disorders (Etkin and Wager 2007). Third, the structural MRI data already discussed (Figs. 2 and 3) suggest subtle but reliable volumetric deficits in regions that overlap with the cognitive control network.

Extending this line of reasoning, a recent meta-analysis of 251 studies sought to examine whether case–control differences in the functional engagement of the cognitive control network may overlap spatially across psychiatric disorders (McTeague et al. 2017). The primary studies collectively included 5728 healthy individuals and

5493 patients with DSM-IV diagnoses. SCZ, schizoaffective disorder and delusional disorder were considered under the category of psychotic disorders, while the category of non-psychotic disorders comprised BD, MDD, dysthymia, anxiety disorders (which included OCD and PTSD in contravention to the DSM-5) and substance-use disorders (mixed substance abuse and/or dependence). The tasks used in the primary studies were categorized according to the main cognitive control process they purport to assess into tasks of conflict monitoring ($n = 31$), performance monitoring ($n = 11$), response inhibition ($n = 42$), response selection ($n = 12$), set shifting ($n = 15$), verbal fluency ($n = 17$), recognition memory ($n = 37$), working memory ($n = 100$). Of note, the RDoC framework considers verbal fluency, episodic and working memory as constructs that are separate from cognitive control (https://www.nimh.nih.gov/research-priorities/rdoc/constructs/cognitive-systems.shtml). The coordinates of the case–control differences in task-related brain activation were meta-analysed using ALE. When data were restricted to adult samples (that comprised the majority of the data set), the overall analysis revealed a pattern of hypo-activation in the right ventrolateral prefrontal cortex/insula, the right parietal cortex, the left dorsal prefrontal cortex and the dorsal anterior cingulate cortex. There was also a cluster of hyper-activation in patients located in the anterior mid-cingulate/supplementary motor cortex that was associated with medication status (i.e. greater activation in medicated patients) and task performance (i.e. greater in studies in which patients' performance matched that of controls). The likelihood of aberrant activation in the left prefrontal cluster and in the mid-cingulate/pre-supplementary motor area was higher in psychotic disorders, whereas aberrant activation in right intraparietal sulcus and the right ventrolateral prefrontal cortex/cluster was more likely for non-psychotic disorders. Overall, this study offers no support for disorder-specific abnormalities within the cognitive control network but suggests that different DSM disorders mainly vary with respect to the degree of aberrant engagement of this network.

Another approach to interrogate the task-related fMRI literature is to test for disorder-specific abnormalities in task-related activation within multiple RDoC-defined circuits. This is an important question for psychiatric nosology because task-related abnormalities observed in fMRI studies comparing single DSM disorders to healthy individuals are often ascribed causal interpretations in an attempt to link disease expression and brain function. This issue has been debated in terms of forward and reverse inference (Aguirre 2003; Poldrack 2006). During a task-fMRI experiment, it is reasonable to infer that the patterns of brain activation observed are associated or even engendered by the mental operations involved in the task. However, inferring backwards and ascribing a specific mental process to a pattern of activation is problematic (Aguirre 2003; Poldrack 2006). This issue is of primary concern for the task-fMRI literature where there is often a tendency 'to reason backward from patterns of activation to infer the engagement of specific mental processes' (Aguirre 2003; Poldrack 2006) and by extension to infer abnormalities of specific mental processes in relation to different disorders (Paulus 2015). In order to test whether there is diagnostic specificity to the dysfunction RDoC-specified circuits, a recent study mapped coordinates of case–control differences derived from 537 task-fMRI studies in SCZ, BD, MDD, anxiety disorders and OCD comprising

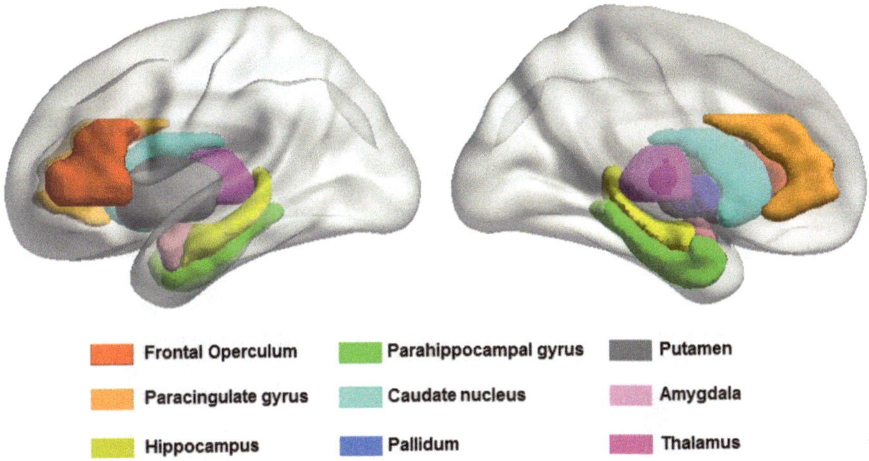

Fig. 4 These results were obtained following ranking all regions by frequency of reported case–control difference, adjusted for region size

observations derived from 21,427 participants to the same atlas (Sprooten et al. 2017). The fMRI tasks were further classified according to RDoC matrix. The inclusion of studies reporting results based on region-of-interest analyses or small-volume correction resulted in a significant effect of diagnosis relating to aberrant amygdala and caudate nucleus activation and an effect of RDoC domains and constructs for aberrant activity in the amygdala, hippocampus, putamen and nucleus accumbens. By contrast, when the analyses were restricted to study reported results from whole-brain analyses, there was no significant effect of diagnosis nor interaction between RDoC domains and diagnosis. Among whole-brain studies, the top 10 regions across all disorders (ranked by frequency of case–control difference adjusted for region size) were the nucleus accumbens, anterior insula, posterior parahippocampal gyrus, globus pallidus, amygdala, hippocampus, caudate, thalamus, paracingulate gyrus and putamen (Fig. 4).The general lack of diagnostic specificity was also supported by Spearmen rank correlation analyses (Fig. 5), as for each pair of diagnoses, the correlation coefficients across regions ranged between 0.42 and 0.82 and were highly significant.

How Has Neuroimaging Informed the Nosology of Mental Illness?

Neuroimaging has provided the most robust and direct evidence to date that psychiatric disorders are associated with alterations in the structure and function of the brain. In this respect, the neuroimaging findings reported strongly argue against the notion that psychiatric diagnoses are ‘labels’ devoid of biological substrate that

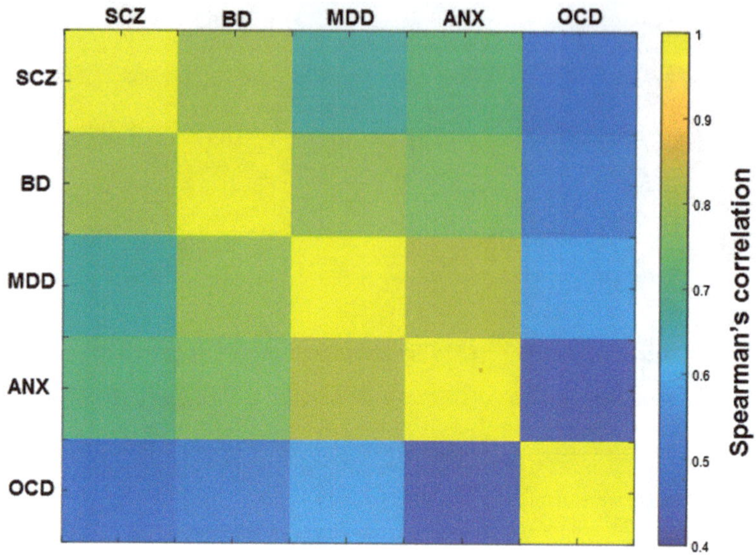

Fig. 5 Pair-wise Spearman's rank correlations between diagnoses of reported case–control differences per region, adjusted for region volume. *ANX* Anxiety Disorders, includes studies on patients with Generalized Anxiety Disorders, Panic Disorder, Post-Traumatic Stress Disorder, Specific Phobias and Social Anxiety Disorder, *BD* Bipolar Disorder, *MDD* Major Depressive Disorder, *OCD* Obsessive Compulsive Disorder, *SCZ* Schizophrenia. (From Sprooten et al. 2017)

simply represent social constructs. However, neither the structural nor the task-related fMRI literature have identified any 'zone of rarity between DSM-defined psychiatric syndromes. The neuroimaging data add to the evidence of transdiagnostic overlap across multiple scales of enquiry including genetic factors (Cross-Disorder Group of the Psychiatric Genomics Consortium 2013). However, the relationship between brain-based abnormalities and symptoms or syndromes is complex and unclear. One interpretation is that the brain structural and functional alterations reflect disorder-general conditions that facilitate the emergence and persistence of symptoms but are neither sufficient nor necessary for explaining psychiatric symptoms. Alternatively, the problem may be that of resolution. MR imaging cannot identify computations at the neuronal level; it can only detect changes in the structure and function of relatively large neuronal groupings. It is possible that the psychiatric symptoms are intimately linked to the exact nature of neuronal computations about which MRI is uninformative. Although improving this aspect of in vivo human imaging may be achievable in the long-term, the field is moving towards enhancing spatial resolution through the use of ultra-high field MR imaging (Kemper et al. 2018) and towards more detailed parcellation and segmentation atlases (Glasser et al. 2016). Current statistical modelling tends to ignore and in fact are tailored to minimize substantial inter-individual variation in anatomy and function. In healthy individuals, there is emerging evidence that inter-individual variation in brain organization may drive variation at the cognitive and even perceptual level. For instance, the surface area of early visual areas varies widely across individuals (Dougherty et al. 2003), and this variation is directly linked to inter-individual differences in visual ability (Schwarzkopf et al. 2011). To the extent, that psychopathology can be linked to more detailed, personalized and even microstructural changes in the brain, these approaches may yield novel and mechanistic insights. It is also important to note that the results of neuroimaging studies are critically depend on the choice of analysis models. The task-fMRI findings presented here challenge the notion that the RDoC-circuit-based approach is likely to revolutionize the fields of human neuroscience and psychiatric nosology. Task-fMRI assumes that, in response to task demands, some brain regions become 'active' and others do not. As previously mentioned, this modular model of the brain is being challenged both through review of the traditional literature and from emerging evidence that this spatial 'sparsity' of activation may be an artefact of MR data acquisition and analysis protocols. For example, Gonzalez-Castillo et al. (2012) demonstrated that if a large number of data points are obtained through long and repeated scanning, then between 70% and 90% of all voxels appear to be 'active' even during a simple visual discrimination task. This result raises the alarming possibility that the task-related circuits may have arisen because of underpowered study designs and that models allowing whole-brain involvement may be required. Although there are many problems to be solved, we appear to be on the cusp of ushering a new era for MR research.

References

Wilke M, Sohn JH, Byars AW, Holland SK. Bright spots: correlations of gray matter volume with IQ in a normal pediatric population. Neuroimage 2003;20(1):202–215.

Frangou S, Chitins X, Williams SC. Mapping IQ and gray matter density in healthy young people. Neuroimage 2004;23(3):800–805.

Narr KL, Woods RP, Thompson PM, Szeszko P, Robinson D, Dimtcheva T, Gurbani M, Toga AW, Bilder RM. Relationships between IQ and regional cortical gray matter thickness in healthy adults. Cereb Cortex 2007;17(9):2163–2171.

Kanwisher N, McDermott J, Chun MM. The fusiform face area: a module in human extrastriate cortex specialized for face perception. J. Neurosci 1997; 17(11): 4302–4311.

Downing PE, Jiang Y, Shuman M, Kanwisher N. A cortical area selective for visual processing of the human body. Science 2001;293(5539):2470–2473.

Dick FK, Lehet MI, Callaghan MF, Keller TA, Sereno MI, Holt LL. Extensive Tonotopic Mapping across Auditory Cortex Is Recapitulated by Spectrally Directed Attention and Systematically Related to Cortical Myeloarchitecture. J Neurosci 2017;37(50):12187–12201.

Niendam TA, Laird AR, Ray KL, Dean YM, Glahn DC, Carter CS. Meta-analytic evidence for a superordinate cognitive control network subserving diverse executive functions. Cogn Affect Behav Neurosci 2012;12(2):241–68.

Barrett LF, Satpute AB. Large-scale brain networks in affective and social neuroscience: towards an integrative functional architecture of the brain. Curr Opin Neurobiol 2013;23(3):361–372.

Gratton G, Cooper P, Fabiani M, Carter CS, Karayanidis F. Dynamics of cognitive control: Theoretical bases, paradigms, and a view for the future. Psychophysiology 2018;55(3) doi: https://doi.org/10.1111/psyp.13016

Haynes JD, Rees G. Decoding mental states from brain activity in humans. Nat Rev Neurosci 2006;7(7):523–534.

Rubin TN, Koyejo O, Gorgolewski KJ, Jones MN, Poldrack RA, Yarkoni T. Decoding brain activity using a large-scale probabilistic functional-anatomical atlas of human cognition. PLoS Comput Biol 2017;13(10):e1005649

Jablensky A. Psychiatric classifications: validity and utility. World Psychiatry 2016;15(1):26–31.

American Psychiatric Association. Diagnostic and statistical manual of mental disorders (5th ed.); 2013; Arlington, VA: American Psychiatric Publishing.

World Health Organization. International Statistical Classification of Diseases and Related Health Problems –draft of 11th edition; 2017; http://www.who.int/classifications/icd/revision/en/ and https://icd.who.int/dev11/l-m/en

Robins E, Guze SB. Establishment of diagnostic validity in psychiatric illness: its application to schizophrenia. Am J Psychiatry 1970; 126 (2): 107–111.

Kendler KS. The nosologic validity of paranoia (simple delusional disorder). Arch Gen Psychiatry 1980;37(6): 699–706.

Andreasen NC. The validation of psychiatric diagnosis: new models and approaches. Am J Psychiatry 1995; 152(2): 161–162.

Kendell RE, Jablensky A. Distinguishing between the validity and utility of psychiatric diagnoses. Am J Psychiatry 2003; 160(1): 4–12.

Andrews G, Goldberg DP, Krueger RF, Carpenter WT, Hyman SE, Sachdev P, Pine DS. Exploring the feasibility of a meta-structure for DSM-V and ICD-11: could it improve utility and validity? Psychol Med 2009a;39(12):1993–2000.

Kessler RC, McGonagle KA, Zhao S, Nelson CB, Hughes M, Eshleman S, Wittchen H-U, Kendler KS. Lifetime and 12-month prevalence of DSM-III-R psychiatric disorders in the United States: Results from the national comorbidity survey. Arch Gen Psychiatry 1994; 51(1):8–19.

Teesson M, Slade T, Mills K. Comorbidity in Australia: findings of the 2007 National Survey of Mental Health and Wellbeing. Aust N Z J Psychiatry 2009;43:606–614

Boschloo L, van Borkulo CD, Rhemtulla M, Keyes KM, Borsboom D, Schoevers RA. The Network Structure of Symptoms of the Diagnostic and Statistical Manual of Mental Disorders. PLoS One 2015;10(9): e0137621.

Sachdev P, Andrews G, Hobbs MJ, Sunderland M, Anderson TM. Neurocognitive disorders: Cluster 1 of the proposed meta-structure for DSM-V and ICD-11. Psychol Med 2009;39(12):2001–2012.

Andrews G, Pine DS, Hobbs MJ, Anderson TM, Sunderland M. Neurodevelopmental disorders: Cluster 2 of the proposed meta-structure for DSM-V and ICD-11. Psychol Med 2009b;39(12):2013–2023.

Carpenter Jr. WT, Bustillo JR, Thaker GK, van Os J, Krueger RF, Green MJ. Psychoses: Cluster 3 of the proposed meta-structure for DSM-V and ICD-11. Psychol Med 2009;39(12):2025–2042.

Goldberg DP, Krueger RF, Andrews G, Hobbs MJ. Emotional disorders: cluster 4 of the proposed meta-structure for DSM-V and ICD-11. Psychol Med 2009;39(12):2043–2059.

Krueger RF, South SC. Externalizing disorders: cluster 5 of the proposed meta-structure for DSM-V and ICD-11. Psychol Med 2009;39(12):2061–2070.

Jablensky M. Reorganizing the diagnostic groupings in DSM-V and ICD-11: a cost/benefit analysis. Psychol Med 2009a;39(12):2091–2097.

Jablensky A. A meta-commentary on the proposal for a meta-structure for DSM-V and ICD-11. Psychol Med 2009b;39(12): 2099–2103.

Wittchen H, Beesdo K and Gloster A. A new meta-structure of mental disorders: a helpful step into the future or a harmful step back to the past? Psychol Med 2009;39(12): 2083–2089.

Kupfer DJ, Regier DA. Neuroscience, clinical evidence, and the future of psychiatric classification in DSM-5. Am J Psychiatry 2011; 168(7):672–674.

Freedman R, Lewis DA, Michels R, Pine DS, Schultz SK, Tamminga CA, Gabbard GO, Gau SS, Javitt DC, Oquendo MA, Shrout PE, Vieta E, Yager J. The initial field trials of DSM-5: new blooms and old thorns. Am J Psychiatry 2013;170(1):1–5.

Insel T, Cuthbert B, Garvey M, Heinssen R, Pine DS, Quinn K, Sanislow C, Wang P. Research domain criteria (RDoC): toward a new classification framework for research on mental disorders. Am J Psychiatry 2010; 167(7):748–751.

National Institutes of Mental Health. Strategic plan for research. NIMH 2008; https://www.nimh.nih.gov/news/science-news/2008/new-nimh-strategic-plan-accelerates-mental-health-research.shtml.

National Institutes of Mental Health. Strategic plan for research. NIMH 2015; https://www.nimh.nih.gov/about/strategic-planning-reports/nimh_strategicplanforresearch_508compliant_corrected_final_149979.pdf

Cuthbert BN, Insel TR. Toward new approaches to psychotic disorders: the NIMH Research Domain Criteria project. Schizophr Bull 2010; 36(6):1061–1062.

Kapur S, Phillips AG, Insel TR. Why has it taken so long for biological psychiatry to develop clinical tests and what to do about it? Mol Psychiatry 2012;17(12):1174–1179.

Casey BJ, Craddock N, Cuthbert BN, Hyman SE, Lee FS, Ressler KJ. DSM-5 and RDoC: progress in psychiatry research? Nat Rev Neurosci 2013; 14(11):810–814.

Cuthbert BN, Insel TR. Toward the future of psychiatric diagnosis: the seven pillars of RDoC. BMC Med 2013;11:126.

Cuthbert BN. Research domain criteria: toward future psychiatric nosology. Asian J Psychiatry 2014;7(1):4–5.

Lilienfeld SO, Treadway MT. Clashing Diagnostic Approaches: DSM-ICD Versus RDoC. Annu Rev Clin Psychol 2016;12:435-463.

Berenbaum H. Classification and psychopathology research. J Abnorm Psychol 2013; 122:894–901.

Wakefield JC. Wittgenstein's nightmare: why the RDoC grid needs a conceptual dimension World Psychiatry 2014;13(1):38–40.

Fuster JM. The module: crisis of a paradigm. Neuron 2000; 26:51–53.

Bressler SL, Menon V. Large-scale brain networks in cognition: emerging methods and principles. Trends Cogn Sci 2010;14(6):277–290.

Hyman SE. The diagnosis of mental disorders: the problem of reification. Annual Review of Clinical Psychology 2010; 6:155–179.

Johnstone EC, Crow TJ, Frith CD, Husband J, Kreel L. Cerebral ventricular size and cognitive impairment in chronic schizophrenia. Lancet 1976;2:924–926.

Hill D. Cerebral atrophy and cognitive impairment in chronic schizophrenia. Lancet 1976;2:1132

American Psychiatric Association. Diagnostic and Statistical Manual of Mental Disorders (4th edn) (DSM-IV);Washington, DC; 1994; American Psychiatric Publishing

World Health Organization. The ICD-10 Classification of Mental and Behavioral Disorders: Diagnostic Criteria for Research. Geneva, WHO; 1993

Andrews G, Slade T, Peters L. Classification in psychiatry: ICD-10 versus DSM-IV. Br J Psychiatry 1999;174:3–5.

Thompson PM, Stein JL, Medland SE, Hibar DP, Vasquez AA, Renteria ME, Toro R, Jahanshad N, Schumann G, Franke B, Wright MJ, Martin NG, Agartz I, Alda M, Alhusaini S, Almasy L, Almeida J, Alpert K, Andreasen NC, Andreassen OA, Apostolova LG, Appel K, Armstrong NJ, Aribisala B, Bastin ME, Bauer M, Bearden CE, Bergmann O, Binder EB, Blangero J, Bockholt HJ, Bøen E, Bois C, Boomsma DI, Booth T, Bowman IJ, Bralten J, Brouwer RM, Brunner HG, Brohawn DG, Buckner RL, Buitelaar J, Bulayeva K, Bustillo JR, Calhoun VD, Cannon DM, Cantor RM, Carless MA, Caseras X, Cavalleri GL, Chakravarty MM, Chang KD, Ching CR, Christoforou A, Cichon S, Clark VP, Conrod P, Coppola G, Crespo-Facorro B, Curran JE, Czisch M, Deary IJ, de Geus EJ, den Braber A, Delvecchio G, Depondt C, de Haan L, de Zubicaray GI, Dima D, Dimitrova R, Djurovic S, Dong H, Donohoe G, Duggirala R, Dyer TD, Ehrlich S, Ekman CJ, Elvsåshagen T, Emsell L, Erk S, Espeseth T, Fagerness J, Fears S, Fedko I, Fernández G, Fisher SE, Foroud T, Fox PT, Francks C, Frangou S, Frey EM, Frodl T, Frouin V, Garavan H, Giddaluru S, Glahn DC, Godlewska B, Goldstein RZ, Gollub RL, Grabe HJ, Grimm O, Gruber O, Guadalupe T, Gur RE, Gur RC, Göring HH, Hagenaars S, Hajek T, Hall GB, Hall J, Hardy J, Hartman CA, Hass J, Hatton SN, Haukvik UK, Hegenscheid K, Heinz A,

Hickie IB, Ho BC, Hoehn D, Hoekstra PJ, Hollinshead M, Holmes AJ, Homuth G, Hoogman M, Hong LE, Hosten N, Hottenga JJ, Hulshoff Pol HE, Hwang KS, Jack CR Jr, Jenkinson M, Johnston C, Jönsson EG, Kahn RS, Kasperaviciute D, Kelly S, Kim S, Kochunov P, Koenders L, Krämer B, Kwok JB, Lagopoulos J, Laje G, Landen M, Landman BA, Lauriello J, Lawrie SM, Lee PH, Le Hellard S, Lemaître H, Leonardo CD, Li CS, Liberg B, Liewald DC, Liu X, Lopez LM, Loth E, Lourdusamy A, Luciano M, Macciardi F, Machielsen MW, Macqueen GM, Malt UF, Mandl R, Manoach DS, Martinot JL, Matarin M, Mather KA, Mattheisen M, Mattingsdal M, Meyer-Lindenberg A, McDonald C, McIntosh AM, McMahon FJ, McMahon KL, Meisenzahl E, Melle I, Milaneschi Y, Mohnke S, Montgomery GW, Morris DW, Moses EK, Mueller BA, Muñoz Maniega S, Mühleisen TW, Müller-Myhsok B, Mwangi B, Nauck M, Nho K, Nichols TE, Nilsson LG, Nugent AC, Nyberg L, Olvera RL, Oosterlaan J, Ophoff RA, Pandolfo M, Papalampropoulou-Tsiridou M, Papmeyer M, Paus T, Pausova Z, Pearlson GD, Penninx BW, Peterson CP, Pfennig A, Phillips M, Pike GB, Poline JB, Potkin SG, Pütz B, Ramasamy A, Rasmussen J, Rietschel M, Rijpkema M, Risacher SL, Roffman JL, Roiz-Santiañez R, Romanczuk-Seiferth N, Rose EJ, Royle NA, Rujescu D, Ryten M, Sachdev PS, Salami A, Satterthwaite TD, Savitz J, Saykin AJ, Scanlon C, Schmaal L, Schnack HG, Schork AJ, Schulz SC, Schür R, Seidman L, Shen L, Shoemaker JM, Simmons A, Sisodiya SM, Smith C, Smoller JW, Soares JC, Sponheim SR, Sprooten E, Starr JM, Steen VM, Strakowski S, Strike L, Sussmann J, Sämann PG, Teumer A, Toga AW, Tordesillas-Gutierrez D, Trabzuni D, Trost S, Turner J, Van den Heuvel M, van der Wee NJ, van Eijk K, van Erp TG, van Haren NE, van 't Ent D, van Tol MJ, Valdés Hernández MC, Veltman DJ, Versace A, Völzke H, Walker R, Walter H, Wang L, Wardlaw JM, Weale ME, Weiner MW, Wen W, Westlye LT, Whalley HC, Whelan CD, White T, Winkler AM, Wittfeld K, Woldehawariat G, Wolf C, Zilles D, Zwiers MP, Thalamuthu A, Schofield PR, Freimer NB, Lawrence NS, Drevets W; Alzheimer's Disease Neuroimaging Initiative, EPIGEN Consortium, IMAGEN Consortium, Saguenay Youth Study (SYS) Group. The ENIGMA Consortium: large-scale collaborative analyses of neuroimaging and genetic data. Brain Imaging Behav 2014;8(2):153–182.

Thompson PM, Andreassen OA, Arias-Vasquez A, Bearden CE, Boedhoe PS, Brouwer RM, Buckner RL, Buitelaar JK, Bulayeva KB, Cannon DM, Cohen RA, Conrod PJ, Dale AM, Deary IJ, Dennis EL, de Reus MA, Desrivieres S, Dima D, Donohoe G, Fisher SE, Fouche JP, Francks C, Frangou S, Franke B, Ganjgahi H, Garavan H, Glahn DC, Grabe HJ, Guadalupe T, Gutman BA, Hashimoto R, Hibar DP, Holland D, Hoogman M, Pol HEH, Hosten N, Jahanshad N, Kelly S, Kochunov P, Kremen WS, Lee PH, Mackey S, Martin NG, Mazoyer B, McDonald C, Medland SE, Morey RA, Nichols TE, Paus T, Pausova Z, Schmaal L, Schumann G, Shen L, Sisodiya SM, Smit DJA, Smoller JW, Stein DJ, Stein JL, Toro R, Turner JA, van den Heuvel MP, van den Heuvel OL, van Erp TGM, van Rooij D, Veltman DJ, Walter H, Wang Y, Wardlaw JM, Whelan CD, Wright MJ, Ye J; ENIGMA Consortium. ENIGMA and the individual: Predicting factors that affect the brain in 35 countries worldwide. Neuroimage 2017;145(Pt B):389–408.

van Erp TG, Hibar DP, Rasmussen JM, Glahn DC, Pearlson GD, Andreassen OA, Agartz I, Westlye LT, Haukvik UK, Dale AM, Melle I, Hartberg CB, Gruber O, Kraemer B, Zilles D, Donohoe G, Kelly S, McDonald C, Morris DW, Cannon DM, Corvin A, Machielsen MW, Koenders L, de Haan L, Veltman DJ, Satterthwaite TD, Wolf DH, Gur RC, Gur RE, Potkin SG, Mathalon DH, Mueller BA, Preda A, Macciardi F, Ehrlich S, Walton E, Hass J, Calhoun VD, Bockholt HJ, Sponheim SR, Shoemaker JM, van Haren NE, Pol HE, Ophoff RA, Kahn RS, Roiz-Santiañez R, Crespo-Facorro B, Wang L, Alpert KI, Jönsson EG, Dimitrova R, Bois C, Whalley HC, McIntosh AM, Lawrie SM, Hashimoto R, Thompson PM, Turner JA. Subcortical brain volume abnormalities in 2028 individuals with schizophrenia and 2540 healthy controls via the ENIGMA consortium. Mol Psychiatry 2016;21(4):547–552.

Hibar DP, Westlye LT, van Erp TG, Rasmussen J, Leonardo CD, Faskowitz J, Haukvik UK, Hartberg CB, Doan NT, Agartz I, Dale AM, Gruber O, Krämer B, Trost S, Liberg B, Abé C, Ekman CJ, Ingvar M, Landén M, Fears SC, Freimer NB, Bearden CE; Costa Rica/Colombia Consortium for Genetic Investigation of Bipolar Endophenotypes, Sprooten E, Glahn DC, Pearlson GD, Emsell L, Kenney J, Scanlon C, McDonald C, Cannon DM, Almeida J, Versace

A, Caseras X, Lawrence NS, Phillips ML, Dima D, Delvecchio G, Frangou S, Satterthwaite TD, Wolf D, Houenou J, Henry C, Malt UF, Bøen E, Elvsåshagen T, Young AH, Lloyd AJ, Goodwin GM, Mackay CE, Bourne C, Bilderbeck A, Abramovic L, Boks MP, van Haren NE, Ophoff RA, Kahn RS, Bauer M, Pfennig A, Alda M, Hajek T, Mwangi B, Soares JC, Nickson T, Dimitrova R, Sussmann JE, Hagenaars S, Whalley HC, McIntosh AM, Thompson PM, Andreassen OA. Subcortical volumetric abnormalities in bipolar disorder. Mol Psychiatry 2016;21(12):1710–1716.

Schmaal L, Veltman DJ, van Erp TG, Sämann PG, Frodl T, Jahanshad N, Loehrer E, Tiemeier H, Hofman A, Niessen WJ, Vernooij MW, Ikram MA, Wittfeld K, Grabe HJ, Block A, Hegenscheid K, Völzke H, Hoehn D, Czisch M, Lagopoulos J, Hatton SN, Hickie IB, Goya-Maldonado R, Krämer B, Gruber O, Couvy-Duchesne B, Rentería ME, Strike LT, Mills NT, de Zubicaray GI, McMahon KL, Medland SE, Martin NG, Gillespie NA, Wright MJ, Hall GB, MacQueen GM, Frey EM, Carballedo A, van Velzen LS, van Tol MJ, van der Wee NJ, Veer IM, Walter H, Schnell K, Schramm E, Normann C, Schoepf D, Konrad C, Zurowski B, Nickson T, McIntosh AM, Papmeyer M, Whalley HC, Sussmann JE, Godlewska BR, Cowen PJ, Fischer FH, Rose M, Penninx BW, Thompson PM, Hibar DP. Subcortical brain alterations in major depressive disorder: findings from the ENIGMA Major Depressive Disorder working group. Mol Psychiatry 2016;21(6):806–812.

Boedhoe PS, Schmaal L, Abe Y, Ameis SH, Arnold PD, Batistuzzo MC, Benedetti F, Beucke JC, Bollettini I, Bose A, Brem S, Calvo A, Cheng Y, Cho KI, Dallaspezia S, Denys D, Fitzgerald KD, Fouche JP, Giménez M, Gruner P, Hanna GL, Hibar DP, Hoexter MQ, Hu H, Huyser C, Ikari K, Jahanshad N, Kathmann N, Kaufmann C, Koch K, Kwon JS, Lazaro L, Liu Y, Lochner C, Marsh R, Martínez-Zalacaín I, Mataix-Cols D, Menchón JM, Minuzzi L, Nakamae T, Nakao T, Narayanaswamy JC, Piras F, Piras F, Pittenger C, Reddy YC, Sato JR, Simpson HB, Soreni N, Soriano-Mas C, Spalletta G, Stevens MC, Szeszko PR, Tolin DF, Venkatasubramanian G, Walitza S, Wang Z, van Wingen GA, Xu J, Xu X, Yun JY, Zhao Q; ENIGMA OCD Working Group, Thompson PM, Stein DJ, van den Heuvel OA. Distinct Subcortical Volume Alterations in Pediatric and Adult OCD: A Worldwide Meta- and Mega-Analysis. Am J Psychiatry 2017;174(1):60–69.

Logue MW, van Rooij SJH, Dennis EL, Davis SL, Hayes JP, Stevens JS, Densmore M, Haswell CC, Ipser J, Koch SBJ, Korgaonkar M, Lebois LAM, Peverill M, Baker JT, Boedhoe PSW, Frijling JL, Gruber SA, Harpaz-Rotem I, Jahanshad N, Koopowitz S, Levy I, Nawijn L, O'Connor L, Olff M, Salat DH, Sheridan MA, Spielberg JM, van Zuiden M, Winternitz SR, Wolff JD, Wolf EJ, Wang X, Wrocklage K, Abdallah CG, Bryant RA, Geuze E, Jovanovic T, Kaufman ML, King AP, Krystal JH, Lagopoulos J, Bennett M, Lanius R, Liberzon I, McGlinchey RE, McLaughlin KA, Milberg WP, Miller MW, Ressler KJ, Veltman DJ, Stein DJ, Thomaes K, Thompson PM, Morey RA. Smaller Hippocampal Volume in Posttraumatic Stress Disorder: A Multisite ENIGMA-PGC Study: Subcortical Volumetry Results From Posttraumatic Stress Disorder Consortia. Biol Psychiatry 2018;83(3):244–253.

van Erp T, Hibar D, Walton E, Jahanshad N, Schmaal L, Jiang WH, Thompson PM, Turner J, ENIGMA Schizophrenia Working Group. Cortical Abnormalities in Schizophrenia: An ENIGMA Schizophrenia Working Group Meta-Analysis. Proceedings of Human Brain Mapping Annual meeting 2017; https://ww5.aievolution.com/hbm1701/index.cfm?do=abs.viewAbs&abs=3712

Hibar DP, Westlye LT, Doan NT, Jahanshad N, Cheung JW, Ching CRK, Versace A, Bilderbeck AC, Uhlmann A, Mwangi B, Krämer B, Overs B, Hartberg CB, Abé C, Dima D, Grotegerd D, Sprooten E, Bøen E, Jimenez E, Howells FM, Delvecchio G, Temmingh H, Starke J, Almeida JRC, Goikolea JM, Houenou J, Beard LM, Rauer L, Abramovic L, Bonnin M, Ponteduro MF, Keil M, Rive MM, Yao N, Yalin N, Najt P, Rosa PG, Redlich R, Trost S, Hagenaars S, Fears SC, Alonso-Lana S, van Erp TGM, Nickson T, Chaim-Avancini TM, Meier TB, Elvsåshagen T, Haukvik UK, Lee WH, Schene AH, Lloyd AJ, Young AH, Nugent A, Dale AM, Pfennig A, McIntosh AM, Lafer B, Baune BT, Ekman CJ, Zarate CA, Bearden CE, Henry C, Simhandl C, McDonald C, Bourne C, Stein DJ, Wolf DH, Cannon DM, Glahn DC, Veltman DJ,

Pomarol-Clotet E, Vieta E, Canales-Rodriguez EJ, Nery FG, Duran FLS, Busatto GF, Roberts G, Pearlson GD, Goodwin GM, Kugel H, Whalley HC, Ruhe HG, Soares JC, Fullerton JM, Rybakowski JK, Savitz J, Chaim KT, Fatjó-Vilas M, Soeiro-de-Souza MG, Boks MP, Zanetti MV, Otaduy MCG, Schaufelberger MS, Alda M, Ingvar M, Phillips ML, Kempton MJ, Bauer M, Landén M, Lawrence NS, van Haren NEM, Horn NR, Freimer NB, Gruber O, Schofield PR, Mitchell PB, Kahn RS, Lenroot R, Machado-Vieira R, Ophoff RA, Sarró S, Frangou S, Satterthwaite TD, Hajek T, Dannlowski U, Malt UF, Arolt V, Gattaz WF, Drevets WC, Caseras X, Agartz I, Thompson PM, Andreassen OA. Cortical abnormalities in bipolar disorder: an MRI analysis of 6503 individuals from the ENIGMA Bipolar Disorder Working Group. Mol Psychiatry 2018;23(4):932–942.

Schmaal L, Hibar DP, Sämann PG, Hall GB, Baune BT, Jahanshad N, Cheung JW, van Erp TGM, Bos D, Ikram MA, Vernooij MW, Niessen WJ, Tiemeier H, Hofman A, Wittfeld K, Grabe HJ, Janowitz D, Bülow R, Selonke M, Völzke H, Grotegerd D, Dannlowski U, Arolt V, Opel N, Heindel W, Kugel H, Hoehn D, Czisch M, Couvy-Duchesne B, Rentería ME, Strike LT, Wright MJ, Mills NT, de Zubicaray GI, McMahon KL, Medland SE, Martin NG, Gillespie NA, Goya-Maldonado R, Gruber O, Krämer B, Hatton SN, Lagopoulos J, Hickie IB, Frodl T, Carballedo A, Frey EM, van Velzen LS, Penninx BWJH, van Tol MJ, van der Wee NJ, Davey CG, Harrison BJ, Mwangi B, Cao B, Soares JC, Veer IM, Walter H, Schoepf D, Zurowski B, Konrad C, Schramm E, Normann C, Schnell K, Sacchet MD, Gotlib IH, MacQueen GM, Godlewska BR, Nickson T, McIntosh AM, Papmeyer M, Whalley HC, Hall J, Sussmann JE, Li M, Walter M, Aftanas L, Brack I, Bokhan NA, Thompson PM, Veltman DJ. Cortical abnormalities in adults and adolescents with major depression based on brain scans from 20 cohorts worldwide in the ENIGMA Major Depressive Disorder Working Group. Mol Psychiatry 2017;22(6):900–909.

Boedhoe PSW, Schmaal L, Abe Y, Alonso P, Ameis SH, Anticevic A, Arnold PD, Batistuzzo MC, Benedetti F, Beucke JC, Bollettini I, Bose A, Brem S, Calvo A, Calvo R, Cheng Y, Cho KIK, Ciullo V, Dallaspezia S, Denys D, Feusner JD, Fitzgerald KD, Fouche JP, Fridgeirsson EA, Gruner P, Hanna GL, Hibar DP, Hoexter MQ, Hu H, Huyser C, Jahanshad N, James A, Kathmann N, Kaufmann C, Koch K, Kwon JS, Lazaro L, Lochner C, Marsh R, Martínez-Zalacaín I, Mataix-Cols D, Menchón JM, Minuzzi L, Morer A, Nakamae T, Nakao T, Narayanaswamy JC, Nishida S, Nurmi E, O'Neill J, Piacentini J, Piras F, Piras F, Reddy YCJ, Reess TJ, Sakai Y, Sato JR, Simpson HB, Soreni N, Soriano-Mas C, Spalletta G, Stevens MC, Szeszko PR, Tolin DF, van Wingen GA, Venkatasubramanian G, Walitza S, Wang Z, Yun JY; ENIGMA-OCD Working Group, Thompson PM, Stein DJ, van den Heuvel OA; ENIGMA OCD Working Group. Cortical Abnormalities Associated With Pediatric and Adult Obsessive-Compulsive Disorder: Findings From the ENIGMA Obsessive-Compulsive Disorder Working Group. Am J Psychiatry 2017; doi: https://doi.org/10.1176/appi.ajp.2017.17050485 and https://ww5.aievolution.com/hbm1701/index.cfm?do=abs.viewAbs&abs=2799

Goodkind M, Eickhoff SB, Oathes DJ, Jiang Y, Chang A, Jones-Hagata LB, Ortega BN, Zaiko YV, Roach EL, Korgaonkar MS, Grieve SM, Galatzer-Levy I, Fox PT, Etkin A. Identification of a common neurobiological substrate for mental illness. JAMA Psychiatry 2015;72(4):305–315.

Eickhoff SB, Laird AR, Grefkes C, Wang LE, Zilles K, Fox PT. Coordinate-based activation likelihood estimation meta-analysis of neuroimaging data: a random-effects approach based on empirical estimates of spatial uncertainty. Hum Brain Mapp 2009; 30(9):2907–2926.

Eickhoff SB, Bzdok D, Laird AR, Kurth F, Fox PT. Activation likelihood estimation meta-analysis revisited. Neuroimage 2012; 59(3):2349–2361.

Turkeltaub PE, Eickhoff SB, Laird AR, Fox M, Wiener M, Fox P. Minimizing within-experiment and within-group effects in activation likelihood estimation meta-analyses. Hum Brain Mapp 2012; 33(1):1–13.

Stefanopoulou E, Manoharan A, Landau S, Geddes JR, Goodwin G, Frangou S. Cognitive functioning in patients with affective disorders and schizophrenia: a meta-analysis. Int Rev Psychiatry 2009;21(4):336–356.

Polak AR, Witteveen AB, Reitsma JB, Olff M. The role of executive function in posttraumatic stress disorder: a systematic review. J Affect Disord 2012;141(1):11–21.

Abramovitch A, Abramowitz JS, Mittelman A. The neuropsychology of adult obsessive-compulsive disorder: a meta-analysis. Clin Psychol Rev 2013;33(8):1163–1171.

Bora E, Harrison BJ, Yücel M, Pantelis C. Cognitive impairment in euthymic major depressive disorder: a meta-analysis. Psychol Med 2013;43(10):2017–2026.

Bourne C, Aydemir Ö, Balanzá-Martínez V, Bora E, Brissos S, Cavanagh JT, Clark L, Cubukcuoglu Z, Dias VV, Dittmann S, Ferrier IN, Fleck DE, Frangou S, Gallagher P, Jones L, Kieseppä T, Martínez-Aran A, Melle I, Moore PB, Mur M, Pfennig A, Raust A, Senturk V, Simonsen C, Smith DJ, Bio DS, Soeiro-de-Souza MG, Stoddart SD, Sundet K, Szöke A, Thompson JM, Torrent C, Zalla T, Craddock N, Andreassen OA, Leboyer M, Vieta E, Bauer M, Worhunsky PD, Tzagarakis C, Rogers RD, Geddes JR, Goodwin GM. Neuropsychological testing of cognitive impairment in euthymic bipolar disorder: an individual patient data meta-analysis. Acta Psychiatr Scand 2013;128(3):149–162.

Gowin JL, Mackey S, Paulus MP. Altered risk-related processing in substance users: imbalance of pain and gain. Drug Alcohol Depend 2013;132(1-2):13–21.

Fatouros-Bergman H, Cervenka S, Flyckt K, Edman G, Farde L. Meta-analysis of cognitive performance in drug-naïve patients with schizophrenia. Schizophr Res 2014;158(1-3):156–162.

Bora E. Neurocognitive features in clinical subgroups of bipolar disorder: A meta-analysis. J Affect Disord 2018;229:125–134.

Minzenberg MJ, Laird AR, Thelen S, Carter CS, Glahn DC. Meta-analysis of 41 functional neuroimaging studies of executive function in schizophrenia. Arch Gen Psychiatry 2009;66(8):811–822.

Chen CH, Suckling J, Lennox BR, Ooi C, Bullmore ET. A quantitative meta-analysis of fMRI studies in bipolar disorder. Bipolar Disord 2011;13(1):1–15.

Miller CH, Hamilton JP, Sacchet MD, Gotlib IH. Meta-analysis of Functional Neuroimaging of Major Depressive Disorder in Youth. JAMA Psychiatry 2015;72(10):1045–1053.

Hamilton JP, Etkin A, Furman DJ, Lemus MG, Johnson RF, Gotlib IH. Functional neuroimaging of major depressive disorder: a meta-analysis and new integration of base line activation and neural response data. Am J Psychiatry 2012;169(7):693–703.

Rasgon A, Lee WH, Leibu E, Laird A, Glahn D, Goodman W, Frangou S. Neural correlates of affective and non-affective cognition in obsessive compulsive disorder: A meta-analysis of functional imaging studies. Eur Psychiatry 2017;46:25–32.

Etkin A, Wager TD. Functional neuroimaging of anxiety: a meta-analysis of emotional processing in PTSD, social anxiety disorder, and specific phobia. Am J Psychiatry 2007;164(10):1476–1488.

McTeague LM, Huemer J, Carreon DM, Jiang Y, Eickhoff SB, Etkin A. Identification of Common Neural Circuit Disruptions in Cognitive Control Across Psychiatric Disorders. Am J Psychiatry 2017;174(7):676–685.

Henson R. What can functional neuroimaging tell the experimental psychologist? Q J Exp Psychol A. 2005; 58(2):193–233.

Aguirre GK. Functional imaging in behavioral neurology and cognitive neuropsychology. In: Feinberg TE, Farah MJ, editors. Behavioral Neurology and Cognitive Neuropsychology. McGraw-Hill; 2003.

Poldrack RA. Can cognitive processes be inferred from neuroimaging data? Trends Cogn Sci 2006; 10(2):59–63.

Paulus M. Pragmatism instead of mechanism: A call for impactful biological psychiatry. JAMA Psychiatry 2015; 72(7):631–632.

Sprooten E, Rasgon A, Goodman M, Carlin A, Leibu E, Lee WH, Frangou S. Addressing reverse inference in psychiatric neuroimaging: Meta-analyses of task-related brain activation in common mental disorders. Hum Brain Mapp 2017;38(4):1846–1864.

Cross-Disorder Group of the Psychiatric Genomics Consortium, Identification of risk loci with shared effects on five major psychiatric disorders: A genome-wide analysis. Lancet 2013;381 (9875):1371–1379.

Kemper VG, De Martino F, Emmerling TC, Yacoub E, Goebel R. High resolution data analysis strategies for mesoscale human functional MRI at 7 and 9.4T. Neuroimage 2018;164:48–58

Glasser MF, Coalson TS, Robinson EC, Hacker CD, Harwell J, Yacoub E, Ugurbil K, Andersson J, Beckmann CF, Jenkinson M, Smith SM, Van Essen DC. A multi-modal parcellation of human cerebral cortex. Nature 2016;536(7615):171–178.

Dougherty RF, Koch VM, Brewer AA, Fischer B, Modersitzki J, Wandell BA. Visual field representations and locations of visual areas V1/2/3 in human visual cortex. J Vis 2003;3(10):586–598.

Schwarzkopf DS, Song C, Rees G. The surface area of human V1 predicts the subjective experience of object size. Nat Neurosci 2011;14(1):28–30.

Gonzalez-Castillo J, Saad ZS, Handwerker DA, Inati SJ, Brenowitz N, Bandettini PA. Whole-brain, time-locked activation with simple tasks revealed using massive averaging and model-free analysis Proc Natl Acad Sci U S A. 2012;109(14):5487–992.

Index

V. A. Diwadkar, S. B. Eickhoff (eds.), *Brain Network Dysfunction in Neuropsychiatric Illness*, https://doi.org/10.1007/978-3-030-59797-9

C

D

N

O

P

Q

R

FSC
www.fsc.org
MIX
Papier | Fördert gute Waldnutzung
FSC® C083411

Kolibri 360 GmbH
Ferdinand-Jühlke-Straße 7
99095 Erfurt, Deutschland
produktsicherheit@kolibri360.de